Maxwell's Demon 2
Entropy, Classical and Quantum Information, Computing

# Maxwell's Demon 2

## Entropy, Classical and Quantum Information, Computing

Edited by

**Harvey S Leff**

*Physics Department,*
*California State Polytechnic University*

and

**Andrew F Rex**

*Physics Department,*
*University of Puget Sound*

INSTITUTE OF PHYSICS PUBLISHING
BRISTOL AND PHILADELPHIA

*British Library Cataloguing-in-Publication Data*

A catalogue record for this book is available from the British Library.

ISBN 0 7503 0759 5

*Library of Congress Cataloging-in-Publication Data are available*

First edition published 1990 as *Maxwell's Demon: Entropy, Information, Computing*

Commissioning Editor: James Revill
Production Editor: Simon Laurenson
Production Control: Sarah Plenty
Cover Design: Frédérique Swist
Marketing: Nicola Newey and Verity Cooke

Published by Institute of Physics Publishing, wholly owned by The Institute of Physics, London

Institute of Physics Publishing, Dirac House, Temple Back, Bristol BS1 6BE, UK

US Office: Institute of Physics Publishing, The Public Ledger Building, Suite 929, 150 South Independence Mall West, Philadelphia, PA 19106, USA

Typeset in $\LaTeX\,2_\varepsilon$ by Text 2 Text, Torquay, Devon
Printed in the UK by MPG Books Ltd, Bodmin, Cornwall

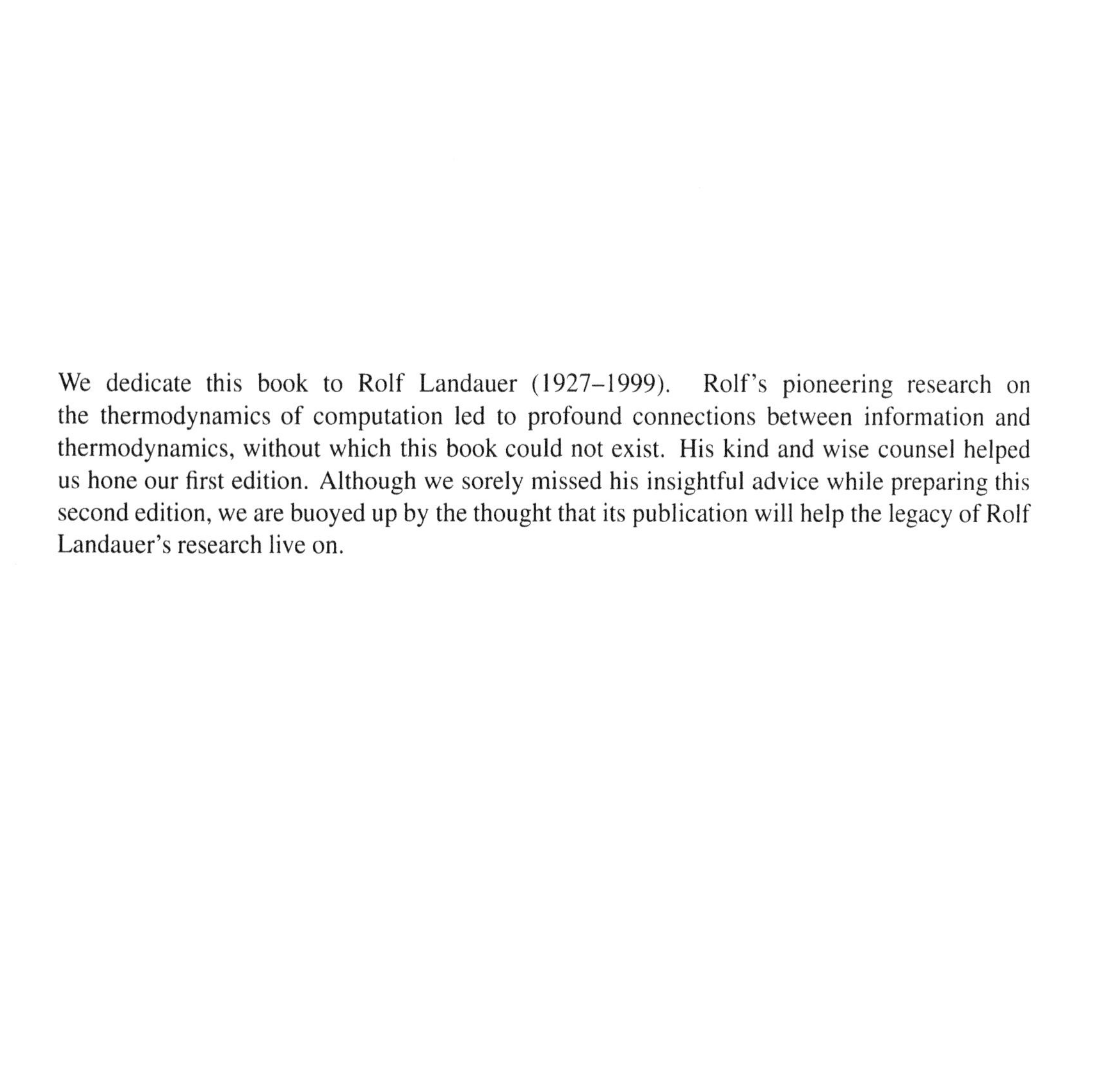

We dedicate this book to Rolf Landauer (1927–1999). Rolf's pioneering research on the thermodynamics of computation led to profound connections between information and thermodynamics, without which this book could not exist. His kind and wise counsel helped us hone our first edition. Although we sorely missed his insightful advice while preparing this second edition, we are buoyed up by the thought that its publication will help the legacy of Rolf Landauer's research live on.

# Contents

# Preface to the First Edition

*Note: The chapter organization alluded to here differs in part from that in the second edition*

Most of us have had the experience (probably while waiting for an overdue flight) of wondering whether a particular scientific conference was worth a week of valuable research time. Yet scientists continue to attend these meetings, because every once in a while something very worthwhile will come out of one. The two of us met at the AAPT Summer meeting at Columbus in 1986. We both gave papers at a session on thermodynamics, and both papers included some mention of Maxwell's demon. That inspired a fruitful correspondence in which we explored our mutual interests in Maxwell's demon, exchanged bibliographies, and began the collaboration that resulted in this book.

We quickly discovered our mutual frustrations in trying to learn more about the demon. It[1] has appeared time and again in a wide range of scientific and technical publications since its invention by James Clerk Maxwell in 1867. Our demon is cat-like; several times it has jumped up and taken on a new character after being left for dead. This made our literature searches quite difficult. We found that there was no single source that explained the many faces of Maxwell's demon in any detail. The subject is treated poorly (if at all) in the standard thermodynamics and statistical mechanics texts. As college teachers we were simply looking for a good example to illustrate the limitations and statistical character of the second law of thermodynamics. What we found was that Maxwell's demon does that and much, much more. But it is by no means simple.

Our motivation for putting together this collection was to spare other college teachers the difficult literature search we have done, so that they may easily communicate some understanding of the demon to their students. This is a worthy goal for a number of reasons. First and foremost it still serves Maxwell's original intention of demonstrating the statistical nature of the second law. In our experience students who have failed to appreciate the probabilistic arguments used to develop the laws of thermodynamics often show greater understanding of the role of probability and statistics after learning of Maxwell's demon. Those students who like to think for themselves are then forced to wonder whether Maxwell was correct and whether a demon could defeat the laws of thermodynamics to create a perpetual motion machine. It is then necessary to bring quantum theory into the picture to obtain at least partial relief from the conundrum. We are convinced that there is no teaching tool that shows the intimate connection between classical thermodynamics and quantum theory as well as Maxwell's demon. Although some of the more subtle arguments will be best understood by graduate students, many of these ideas would not be wasted on advanced undergraduates. Finally, students with some experience with computers can be shown the results of Landauer and Bennett, in which Maxwell's demon is related to the idea of dissipationless computing. The demon really has something for everyone, and that helps make it a very powerful teaching tool.

It is so powerful, in fact, that we should not limit its use to college and graduate students and their teachers. Because of the connection with computing just mentioned, researchers in computer science

[1] We assign the demon neuter characteristics, lest those of either gender feel slighted.

and information theory will benefit from the background this reprint book provides. We also believe that this book will provide useful background and references for scientists and engineers who work in thermodynamics. Biologists and psychologists will be able to appreciate many of the articles, particularly those relating the second law to life and the human mind. Last but certainly not least, there will be a wealth of information for those who study the history and philosophy of science. A subject that has occupied the minds of Maxwell, Thomson, Szilard, and other luminaries can scarcely be ignored by those trying to piece together the history of our discipline.

We have assembled this book in such a way that it should be useful to all of the interested parties just mentioned. Chapter 1 is an overview of the Maxwell's demon phenomenon. It is designed to familiarize the reader with the most important aspects of the demon. Chapter 1 is also intended to serve as a guide to Chapters 2–4, which contain the reprinted articles. In Chapter 2 the connection between the demon and the laws of thermodynamics is explored thoroughly. This leads unavoidably to a consideration of the role of quantum phenomena in thermodynamics, information gathering, and, ultimately, life processes. The papers reprinted in Chapter 3 focus more narrowly on the crucial question of how the demon gathers and stores information. These problems in information gathering and (more important, as it turns out) information erasure provide the basis for understanding the role of Maxwell's demon in the theory of computing (Chapter 4). We have selected these sentinel papers with the idea of exposing the reader to a wide scope of ideas and problems concerning the demon in its various guises. There are sufficient references in Chapters 1–4 to allow those with particular subject interests to proceed with more in depth studies. The chronological bibliography should be of use to the historians and to anyone else interested in better understanding the sequence of ideas as they have developed over the last century and a quarter. Those interested in seeing a fairly complete Maxwell's demon subject bibliography are referred to our American Journal of Physics resource letter[2].

Although there is necessarily some variation in the level of difficulty of the reprinted articles, most of the material should be accessible to anyone who has had a good upper division course in statistical mechanics and thermodynamics. A few articles (particularly some in Chapter 2) are appropriate for anyone who has had at least an introductory physics course. Clearly this book is not a textbook, but it could be used as the primary resource for an advanced undergraduate, graduate, or faculty seminar.

Where then does Maxwell's demon stand today? What has it accomplished since it was conceived by Maxwell? It has certainly not achieved the demonic goal of circumventing the second law. If anything the second law is stronger for the challenge to it. The words of Eddington[3] serve as a sharp reminder to those who would imagine otherwise:

> The law that entropy always increases—the second law of thermodynamics—holds, I think, the supreme position among the laws of Nature. If someone points out to you that your pet theory of the universe is in disagreement with Maxwell's equations—then so much the worse for Maxwell's equations. If it is found to be contradicted by observation, well, these experimentalists do bungle things sometimes. But if your theory is found to be against the second law of thermodynamics I can give you no hope; there is nothing for it but to collapse in deepest humiliation.

In spite of that failure the demon has brought us a host of new and exciting ideas unimagined by Maxwell. In that way this is a wonderful example of the scientific enterprise working as we believe it should, not toward any particular result but wherever the truth may lead.

We are indebted to James Revill of Adam Hilger and Edward Tenner of Princeton University Press for their strong support of this project. Thoughtful reviews of Chapter 1 by Rolf Landauer, Peter

[2] Leff H S and Rex A F 1990 Resource letter MD-1: Maxwell's demon *Am. J. Phys.* **58** 201–9.

[3] Eddington A S 1948 *The Nature of the Physical World* (New York: Macmillan) p 74.

Landsberg, and Oliver Penrose led to significant improvements. We thank Arthur Fine for providing us with bibliographic information and are grateful for the receipt of camera-ready copies of articles from Charles Bennett, Edward Daub, Martin Klein, Rolf Landauer, Elihu Lubkin, Stefan Machlup, Jerome Rothstein, Myron Tribus, Alvin Weinberg, and Wojciech Zurek. One of us (HSL) thanks Paul Zilsel for sparking his curiosity in Maxwell's demon about 20 years ago and Alvin Weinberg for revitalizing that interest in 1982. The other author (AR) thanks Greg Shubert for several inspiring discussions in 1985. Finally, we acknowledge worthwhile communications on Maxwell's demon with many others too numerous to mention here. They have all helped shape this book.

**Harvey Leff** Pomona, California
**Andrew Rex** Tacoma, Washington
November, 1989

# Preface to the Second Edition

Since publication of the first edition of *Maxwell's Demon: Entropy, Information, Computing*, approximately 13 years ago, more new relevant bibliographic references were generated than in the first 120 years of Maxwell's demon's life! The total number of references now exceeds 550. This explosion of scholarship provides the impetus for this second edition.

The existence of many articles suitable for reprinting, the enormous number of new bibliographic entries, and our desire to keep the size and cost of the second edition manageable led us to some hard decisions. In order to add 16 new reprints, we have reluctantly omitted 10 of the 25 first edition reprints. However, we have included the full, expanded bibliography, not only in the traditional alphabetical format, but also in chronological order, with annotations, to assist readers who wish to follow the historical flow of ideas. The omitted articles, which are listed in our bibliographies as 'Reprinted in MD1,' are:

P M Heimann (1970) Molecular Forces, Statistical Representation and Maxwell's Demon

K Denbigh (1981) How Subjective is Entropy?

A M Weinberg (1982) On the Relation Between Information and Energy Systems: A Family of Maxwell's Demons

R C Raymond (1951) The Well-informed Heat Engine

C Finfgeld and S Machlup (1960) Well-informed Heat Engine: Efficiency and Maximum Power

P Rodd (1964) Some Comments on Entropy and Information

D Gabor (1951) Light and Information (Section 5 and Appendix IV)

C H Bennett (1973) Logical Reversibility of Computation

R Laing (1973) Maxwell's Demon and Computation

A F Rex (1987) The Operation of Maxwell's Demon in a Low Entropy System

The general structure of Chapter 1—Overview has been retained. Most of the major additions are in the last 3 sections namely: 1.6 Foundations of Landauer's Principle, 1.7 Quantum Mechanics and Maxwell's Demon, and 1.8 Other Aspects of Maxwell's Demon.

Since publication of the first edition, we have developed a greater appreciation of Oliver Penrose's pioneering work on Maxwell's demon. In his 1970 book *Foundations of Statistical Mechanics*, he independently discovered the importance of erasing the demon's memory in order to assess thermodynamic changes solely in terms of the environment. In essence he discovered Landauer's principle by 1970. Charles Bennett's work relating Maxwell's demon to Landauer's principle was published in

1982. In private communication Penrose confirmed that he had been unaware of Landauer's 1961 paper, and Landauer and Bennett were both unaware of Penrose's work until 1989, when the first edition of this book was in preparation. It seems fitting to now refer to the Landauer–Penrose–Bennett (LPB) solution of the Maxwell's demon puzzle.

The term 'Landauer's principle' is not used in the first edition but it has become so common in the literature that it pervades this second edition. Rolf Landauer's work lies at the very core of modern day assessments of Maxwell's demon. In addition to the two articles (1961, 1987a) by Rolf reprinted in edition one, we have added two more here (1996a, 1996b). We are honored to have known him and to have shared many physics ideas with him until his untimely death in 1999.

In July 2002, shortly before completing this edition, we had the good fortune to participate in the First International Conference on Quantum Limits to the second law at the University of San Diego. That experience brought us face to face with many researchers engaged in studies of the limits of the second law, informed us of the latest findings, and positively influenced the final form of this edition. We are grateful to the organizers, Vladislav Čápek, Theo Nieuwenhuizen, Alexey Nikulov, and Daniel Sheehan for inviting us to speak at that exciting conference.

We owe a debt of gratitude to our editor, James Revill, who strongly encouraged the production of this second edition. We also thank the following authors of the newly reprinted papers, who kindly provided us with camera-ready copy: Charles Bennett, Carlton Caves, Neil Gershenfeld, Seth Lloyd, Barbara Piechocinska, Rüdiger Schack, Kousuke Shizume, Panayotis Skordos, Vlatko Vedral, and Wojciech Zurek. We are particularly grateful to Charles Bennett, who provided us with camera-ready copy of the two newly reprinted articles by Rolf Landauer.

We also thank Charles Bennett, Jeffrey Bub, Theo Nieuwenhuizen, Martin Plenio, and Daniel Sheehan for illuminating correspondence and conversations.

**Harvey Leff** Pomona, California
**Andrew Rex** Tacoma, Washington
September, 2002

# Acknowledgments and Copyright Information

We are grateful to the copyright holders listed below for granting permission to reprint materials that are the core of this book. Section numbers refer to the contents list.

2.1 'The kinetic theory of the dissipation of energy' by William Thomson. Reprinted by permission from *Nature* **IX**, 441–444 (1874). Read before the Royal Society of Edinburgh in 1874.
2.2 'Maxwell's demon,' by E E Daub. Reprinted by permission from *Stud. Hist. Phil. Sci.* **1**, 213–227 (1970).
2.3 'Maxwell, his demon, and the second law of thermodynamics,' by M J Klein. Reprinted by permission from *Am. Sci.* **58**, 84–97 (1970).
2.4 'Life, thermodynamics, and cybernetics,' by L Brillouin. Reprinted by permission from *Am. Sci.* **37**, 554–568 (1949).
2.5 'Information, measurement, and quantum mechanics,' by J Rothstein. Reprinted by permission from *Science* **114**, 171–175 (1951)
3.1 'Maxwell's demon, rectifiers, and the second law: Computer simulation of Smoluchowski's trapdoor,' by P A Skordos and W H Zurek. Reprinted by permission from *Am. J. Phys.* **60**, 876–882 (1992).
3.2 'Entropy and information for an automated Maxwell's demon,' by A F Rex and R Larsen. Reprinted by permission from *Proceedings of the Workshop on Physics and Computation (PhysComp '92)*, (IEEE Press, Los Alamitos, 1992), pp 93–101.
3.3 'On the decrease of entropy in a thermodynamic system by the intervention of intelligent beings,' by L. Szilard. English translation by A Rapoport and M Knoller, Reprinted by permission from *Behavioral Science* **9**, 301–310 (1964).
3.4 'Maxwell's demon cannot operate: Information and entropy. I,' by L Brillouin. Reprinted by permission from *J. Appl. Phys.* **22**, 334–337 (1951).
3.5 'Entropy, information and Szilard's paradox,' by J M Jauch and J G Báron. Reprinted by permission from *Helv. Phys. Acta* **45**, 220–232 (1972).
3.6 'Information theory and thermodynamics,' by O Costa de Beauregard and M Tribus. Reprinted by permission from *Helv. Phys. Acta* **47**, 238–247 (1974).
4.1 'Irreversibility and heat generation in the computing process,' by R Landauer. Reprinted by permission from *IBM J. Res. Dev.* **5**, 183–191 (1961).
4.2 'Entropy of Measurement and Erasure: Szilard's Membrane Model Revisited,' by H S Leff and A F Rex, Reprinted by permission from *Am. J. Phys.* **63**, 994–1000 (1994).
4.3 'Heat generation required by erasure,' by K Shizume. Reprinted by permission from *Phys. Rev. E* **52**, 3495–3499 (1995).
4.4 'Information erasure,' by B Piechocinska. Reprinted by permission from *Phys. Rev. A* **61**, 62314, 1–9 (2000).
5.1 'Maxwell's demon, Szilard's engine and quantum measurements,' by W H Zurek. Reprinted by

permission from *Frontiers of Nonequilibrium Statistical Physics*, ed G T Moore and M O Scully (Plenum Press, New York, 1984), pp 151–161.

5.2 ‘Keeping the entropy of measurement: Szilard revisited,’ by E Lubkin. Reprinted by permission from *Int. J. Theor. Phys.* **26**, 523–535 (1987).

5.3 ‘Use of mutual information to decrease entropy: Implications for the second law of thermodynamics,’ by S Lloyd. Reprinted by permission from *Phys. Rev. A* **39**, 5378–5386 (1989).

5.4 ‘Quantum-mechanical Maxwell’s demon,’ by S Lloyd. Reprinted by permission from *Phys. Rev. A* **56**, 3374–3382 (1997).

5.5 ‘Landauer’s erasure, error correction and entanglement,’ by V Vedral. Reprinted by permission from *Proc. R. Soc. Lond. A* **456**, 969–984 (2000).

6.1 ‘Information and entropy,’ by C M Caves. Reprinted by permission from *Phys. Rev. E* **47**, 4010–4017 (1993).

6.2 ‘Algorithmic information and simplicity in statistical physics,’ by R Schack. Reprinted by permission from *Int. J. Theoret. Phys.* **36**, (1) 209–226 (1997).

6.3 ‘Algorithmic randomness, physical entropy, measurements, and the demon of choice,’ by W H Zurek. Reprinted by permission from *Feynman and Computation: Exploring the Limits of Computers*, ed J G Hey, (Perseus, Reading, 1999), pp 393–410.

7.1 ‘The thermodynamics of computation—a review,’ by C H Bennett. Reprinted by permission from *Int. J. Theor. Phys.* **21**, 905–940 (1982).

7.2 ‘Computation: A fundamental physical view,’ by R Landauer. Reprinted by permission from *Phys. Scr.* **35**, 88–95 (1987).

7.3 ‘Notes on the history of reversible computation,’ by C H Bennett. Reprinted by permission from *IBM J. Res. Dev.* **32**, 16–23 (1988).

7.4 ‘The physical nature of information,’ by R Landauer. Reprinted by permission from *Phys. Lett. A* **217**, 188–193 (1996).

7.5 ‘Minimal energy requirements in communication,’ by R Landauer. Reprinted by permission from *Science* **272**, 1914–1918 (1996).

7.6 ‘Information physics in cartoons,’ by C H Bennett. Reprinted by permission from *Superlattices and Microstructures* **23**, 367–372 (1998).

7.7 ‘Signal entropy and the thermodynamics of computation,’ by N Gershenfeld. Reprinted by permission from *IBM Sys. J.* **35**, 577–586 (1996).

7.8 ‘Ultimate physical limits to computation,’ by S Lloyd. Reprinted by permission from *Nature* **406**, (6799) 1047–1054 (2000).

The figures appearing in Chapter 1 have been reproduced from the following sources and we gratefully acknowledge the copyright holders.

Figure 1 *Cybernetics A to Z*, Pekelis V, (Mir Publishers, Moscow, 1974), p 106. Copyright 1974, Mir Publishers.

Figure 2 *Symbols, Signals and Noise*, Pierce J R, (Harper and Brothers, New York, 1961), p 199.

Figures 3 and 4 From *Order and Chaos* by Stanley W Angrist and Loren G Hepler, ©1967 by Basic Books, Inc., Publishers, New York.

Figure 5 *Fundamentals of Cybernetics*, Lerner A Y, (Plenum Pub. Corp., New York, 1975), p 257.

Figure 6 *On Maxwell’s Demon*, Darling L and Hulbert E O, *Am. J. Phys.* **23**, 470 (1955).

Figure 7 Illustration from *Entropy for Biologists: An Introduction to Thermodynamics* by Harold J Morowitz, copyright ©1979 by Harcourt Brace Jovanovich, Inc. Reprinted by permission of the publisher.

Figure 8 *Entropy and Energy Levels*, Gasser R P H and Richards W G (Clarendon Press, Oxford, 1974), pp 117–118.

Figure 10 Adapted from Feyerabend P K, 'On the possibility of a perpetuum mobile of the second kind.' In *Mind, Matter, and Method: Essays in Philosophy and Science in Honor of Herbert Feigl*, P K Feyerabend and G Maxwell (University of Minnesota Press, Minneapolis, 1966), p 411.

Figure 11 *Entropy, Information and Szilard's Paradox*, Jauch J M and Báron J G, *Helv. Phys. Acta* **45**, 231 (1972).

Figure 12 'Generalized entropy, boundary conditions, and biology,' Rothstein J. In *The Maximum entropy Formalism*, Levine R J D and Tribus M. (The MIT Press, Cambridge, MA, 1979), p 467.

# Chapter 1

# Overview

**Maxwell's demon** ... *[after J. C. Maxwell, its hypothecator]: 'A hypothetical being of intelligence but molecular order of size imagined to illustrate limitations of the second law of thermodynamics.' Webster's Third New International Dictionary*

## 1.1 Introduction

Maxwell's demon lives on. After more than 130 years of uncertain life and at least two pronouncements of death, this fanciful character seems more vibrant than ever. As the dictionary entry above shows, Maxwell's demon is no more than a simple idea. Yet it has challenged some of the best scientific minds, and its extensive literature spans thermodynamics, statistical physics, quantum mechanics, information theory, cybernetics, the limits of computing, biological sciences, and the history and philosophy of science.

Despite this remarkable scope and the demon's longevity, coverage in standard physics, chemistry, and biology textbooks typically ranges from cursory to nil. Because its primary literature is scattered throughout research journals, semipopular books, monographs on information theory, and a variety of specialty books, Maxwell's demon is somewhat familiar to many but well known only to relatively few. Two Scientific American articles on the demon (Ehrenberg, 1967; Bennett, 1987) have been helpful, but they only scratch the surface of the existing literature. A recent popularization of the subject (von Baeyer, 1998) helps but, being directed at a lay audience, is limited in both depth and breadth.

The main purpose of this reprint collection is to place in one volume: (1) important original papers covering Maxwell's demon, (2) an overview of the demon's life and current status, and (3) an annotated bibliography that provides perspective on the demon plus a rich trail of citations for further study.

The life of Maxwell's demon can be viewed usefully in terms of three major phases. The first phase covers its 'birth' in approximately 1867 through the first 62 years of its relatively quiet existence. The flavor of the early history is reflected in Thomson's classic paper on the dissipation of energy (1874, Article 2.1). A modern perspective of automatic (non-intelligent) mechanical demons along the line of those addressed by Smoluchowski (1912, 1914) is given by Skordos and Zurek (1992, Article 3.1) and Rex and Larsen (1992, Article 3.2).

The second phase began with an important paper by Leo Szilard (1929, Article 3.3). The entry on Szilard in Scribner's Dictionary of Scientific Biography cites his '...famous paper of 1929, which established the connection between entropy and information, and foreshadowed modern cybernetic theory.' Notably, Szilard discovered the idea of a 'bit' of information, now central in computer science. His discovery seems to have been independent of earlier identifications of logarithmic forms for information by Nyquist (1924) and Hartley (1928). The term 'bit' (= binary digit) was suggested approximately 15 years after Szilard's work by John Tukey. Subsequently Rothstein formulated fundamental information-theoretic interpretations of thermodynamics, measurement, and quantum theory (1951, Article 2.5). The history of the demon during the first two phases is described by Daub (1970, Article 2.2), Klein (1970, Article 2.3), and Heimann (1970, reprinted in MD1).

After a hiatus of about 20 years, Leon Brillouin (1949, Article 2.4) became involved in the Maxwell's demon puzzle through his interest in finding a scientific framework to explain intelligent life. At roughly the same time Dennis Gabor (1964, reprinted in MD1), based on his 1951 lecture, and Brillouin (1951, Article 3.4) focused on the demon's acquisition of information. Both Brillouin and Gabor assumed the use of light signals in the demon's attempt to defeat the second law of thermodynamics. The result was a proclaimed 'exorcism' of Maxwell's demon, based upon the edict that information acquisition is dissipative, making it impossible for a demon to violate the second law.

Independently of Brillouin, Raymond (1951, reprinted in MD1) published an account of a clever variant of Maxwell's demon that did not explicitly entail light signals—a 'well-informed heat engine' using density fluctuations in a gas. Raymond found that 'an outside observer creates in the system a negative information entropy equal to the negative entropy change involved in the operation of the engine'. His work, though less influential than Brillouin's, also made the important connection between information and entropy. Finfgeld and Machlup (1960, reprinted in MD1) analyzed Raymond's model further, assuming that the necessary demon uses light signals, and also obtained an estimate of its power output.

The impact of Brillouin's and Szilard's work was far-reaching, inspiring numerous investigations of Maxwell's demon, including those by Rodd (1964) and Rex (1987), both reprinted in MD1. Weinberg (1982, reprinted in MD1) broadened the demon's realm to include 'macroscopic' and 'social' demons. Despite some critical assessments of the connections between information and entropy by Denbigh, (1981, reprinted in MD1), Jauch and Báron (1972, Article 3.5), and Costa de Beauregard and Tribus (1974, Article 3.6), those linkages and applications to the Maxwell's demon puzzle became firmly entrenched in the scientific literature and culture.

The third phase of the demon's life began at age 94 when Rolf Landauer made the important discovery (1961, Article 4.1) that memory erasure in computers feeds entropy to the environment. This is now called 'Landauer's principle'. Landauer referred to Brillouin's argument that measurement requires a dissipation of order $kT$, and observed: 'The computing process ... is closely akin to a measurement'. He also noted that: '...the arguments dealing with the measurement process do not define measurement very well, and avoid the very essential question: When is a system $A$ coupled to a system $B$ performing a measurement? The mere fact that two physical systems are coupled does not in itself require dissipation'.

Landauer's work inspired Charles Bennett to investigate logically reversible computation, which led to Bennett's important demonstration (Bennett, 1973, reprinted in MD1) that reversible computation, which avoids erasure of information, is possible in principle. The direct link between Landauer's and Bennett's work on computation and Maxwell's demon came with Bennett's observation (1982, Article 7.1) that a demon 'remembers' the information it obtains, much as a computer records data in its memory. Bennett argued that erasure of a demon's memory is the fundamental act that saves the second law because of Landauer's principle. This was a surprising, remarkable event in the history of Maxwell's demon. Subsequent analyses of memory erasure for a quantum mechanical Szilard's model by Zurek (1984, Article 5.1) and Lubkin (1987, Article 5.2) support Bennett's finding.

A key point in Bennett's work is that, in general, the use of light signals for information acquisition can be avoided. That is, although such dissipative information gathering is sufficient to save the second law of thermodynamics, it is not necessary. Bennett's argument nullifies Brillouin's 'exorcism', which was so ardently believed by a generation of scientists. The association of the Maxwell's demon puzzle with computation greatly expanded the audience for the demon, and articles by Bennett (1988, Article 7.3) and Landauer (1961, 1987a, 1996a, 1996b, Articles 4.1, 7.2, 7.4, 7.5, respectively) illustrating that association are reprinted here.

These three phases of Maxwell's demon life are described in further detail in Sections 1.2–1.5. Section 1.6 provides an examination of the foundations of Landauer's principle and Section 1.7 contains a discussion of the role of quantum mechanics in the Maxwell's demon puzzle. Section 1.8 deals with aspects of the demon not treated in the earlier sections. This includes the concept of *process* demons that do not collect information, but nevertheless threaten the second law of thermodynamics. Chapters 2–7 contain reprinted articles covering, respectively: historical and philosophical considerations; information acquisition; and information erasure and computing. This is followed by a chronological bibliography, with selected annotations and quotations that provide a colorful perspective on the substantial impacts of Maxwell's demon. An alphabetical bibliography and an extensive index are also included.

## 1.2 The Demon and Its Properties

### 1.2.1 Birth of the Demon

The demon was introduced to a public audience by James Clerk Maxwell in his 1871 book, *Theory of Heat*. It came near the book's end in a section called 'Limitation of the Second Law of Thermodynamics'. In one of the most heavily quoted passages in physics, Maxwell wrote:

> Before I conclude, I wish to direct attention to an aspect of the molecular theory which deserves consideration.
>
> One of the best established facts in thermodynamics is that it is impossible in a system enclosed in an envelope which permits neither change of volume nor passage of heat, and in which both the temperature and the pressure are everywhere the same, to produce any inequality of temperature or of pressure without the expenditure of work. This is the second law of thermodynamics, and it is undoubtedly true as long as we can deal with bodies only in mass, and have no power of perceiving or handling the separate molecules of which they are made up. But if we conceive a being whose faculties are so sharpened that he can follow every molecule in its course, such a being, whose attributes are still as essentially finite as our own, would be able to do what is at present impossible to us. For we have seen that the molecules in a vessel full of air at uniform temperature are moving with velocities by no means uniform, though the mean velocity of any great number of them, arbitrarily selected, is almost exactly uniform. Now let us suppose that such a vessel is divided into two portions, A and B, by a division in which there is a small hole, and that a being, who can see the individual molecules, opens and closes this hole, so as to allow only the swifter molecules to pass from A to B, and only the slower ones to pass from B to A. He will thus, without expenditure of work, raise the temperature of B and lower that of A, in contradiction to the second law of thermodynamics.
>
> This is only one of the instances in which conclusions which we have drawn from our experience of bodies consisting of an immense number of molecules may be found not to be applicable to the more delicate observations and experiments which we may suppose made by one who can perceive and handle the individual molecules which we deal with only in large masses.
>
> In dealing with masses of matter, while we do not perceive the individual molecules, we are compelled to adopt what I have described as the statistical method of calculation, and to abandon the strict dynamical method, in which we follow every motion by the calculus.

Maxwell's thought experiment dramatized the fact that the second law is a statistical principle that holds almost all the time for a system composed of many molecules. That is, there is a nonzero probability that anisotropic molecular transfers, similar to those accomplished by the demon, will occur if the hole is simply left open for a while. Maxwell had introduced this idea in a 1867 letter to Peter Guthrie Tait (Knott, 1911) '...to pick a hole' in the second law. There he specified more detail about the sorting strategy intended for the demon:

> Let him first observe the molecules in *A* and when he sees one coming the square of whose velocity is less than the mean sq. vel. of the molecules in *B* let him open the hole and let go into *B*. Next let him watch for a molecule of *B*, the square of whose velocity is greater than the mean sq. vel. in *A*, and when it comes to the hole let him draw the slide and let it go into *A*, keeping the slide shut for all other molecules.

This allows a molecule to pass from *A* to *B* if its kinetic energy is less than the average molecular kinetic energy in *B*. Passage from *B* to *A* is allowed only for molecules whose kinetic energies exceed the average kinetic energy per molecule in *A*. In the same letter Maxwell emphasized the quality of 'intelligence' possessed by the demon:

> Then the number of molecules in *A* and *B* are the same as at first, but the energy in *A* is increased and that in *B* diminished, that is, the hot system has got hotter and the cold colder and yet no work has been done, only the intelligence of a very observant and neat-fingered being has been employed.

William Thomson (1874, Article 2.1) subsequently nicknamed Maxwell's imaginary being 'Maxwell's intelligent demon.' He apparently did not envisage the creature as malicious: 'The definition of a demon, according to the use of this word by Maxwell, is an intelligent being endowed with free-will and fine enough tactile and perceptive organization to give him the faculty of observing and influencing individual molecules of matter.' He expounded further on his view of 'the sorting demon of Maxwell' (Thomson, 1879):

> The word 'demon', which originally in Greek meant a supernatural being, has never been properly used to signify a real or ideal personification of malignity.
>
> Clerk Maxwell's 'demon' is a creature of imagination having certain perfectly well defined powers of action, purely mechanical in their character, invented to help us to understand the 'Dissipation of Energy' in nature.
>
> He is a being with no preternatural qualities and differs from real living animals only in extreme smallness and agility. ... He cannot create or annul energy; but just as a living animal does, he can store up limited quantities of energy, and reproduce them at will. By operating selectively on individual atoms he can reverse the natural dissipation of energy, can cause one-half of a closed jar of air, or of a bar of iron, to become glowingly hot and the other ice cold; can direct the energy of the moving molecules of a basin of water to throw the water up to a height and leave it there proportionately cooled ...; can 'sort' the molecules in a solution of salt or in a mixture of two gases, so as to reverse the natural process of diffusion, and produce concentration of the solution in one portion of the water, leaving pure water in the remainder of the space occupied; or, in the other case separate the gases into different parts of the containing vessel.
>
> 'Dissipation of Energy' follows in nature from the fortuitous concourse of atoms. The lost motivity is essentially not restorable otherwise than by an agency dealing with individual atoms; and the mode of dealing with the atoms to restore motivity is essentially a process of assortment, sending this way all of one kind or class, that way all of another kind or class.

Following Thomson's introduction of the term 'demon,' Maxwell clarified his view of the demon (quoted in Knott, 1911) in an undated letter to Tait:

> Concerning Demons.
>
> 1. Who gave them this name? Thomson.
> 2. What were they by nature? Very small BUT lively beings incapable of doing work but able to open and shut valves which move without friction or inertia.
> 3. What was their chief end? To show that the 2nd Law of Thermodynamics has only a statistical certainty.
> 4. Is the production of an inequality of temperature their only occupation? No, for less intelligent demons can produce a difference in pressure as well as temperature by merely allowing all particles going in one direction while stopping all those going the other way. This reduces the demon to a valve. As such value him. Call him no more a demon but a valve like that of the hydraulic ram, suppose.

In light of Maxwell's intentions, it is interesting to examine the accuracy of dictionary definitions. The *Webster's Third New International Dictionary* definition quoted at the beginning of this chapter, though brief, properly cites Maxwell's intention to 'illustrate limitations of the second law of thermodynamics.' In contrast, *The Random House Dictionary of the English Language* (Second Edition, 1988) contains the definition:

> A hypothetical agent or device of arbitrarily small mass that is considered to admit or block selectively the passage of individual molecules from one compartment to another according to their speed, constituting a violation of the second law of thermodynamics.

And the Second Edition (1989) of The Oxford English Dictionary describes it in the entry for James Clerk Maxwell:

> ...a being imagined by Maxwell as allowing only fast-moving molecules to pass through a hole in one direction and only slow-moving ones in the other direction, so that if the hole is in a partition dividing a gas-filled vessel into two parts one side becomes warmer and the other cooler, in contradiction to the second law of thermodynamics.

Despite the emphasis on violating rather than illustrating limitations on the second law in these two definitions, there is no indication that Maxwell intended his hypothetical character to be a serious challenge to that law. Nevertheless, the latter two definitions reflect the interpretation by many subsequent researchers that Maxwell's demon was a puzzle that must be solved: If such a demon cannot defeat the second law, then why not? And if it can defeat the second law, then how does that affect that law's status?

Maxwell did not relate his mental construction to entropy. In fact, he evidently misunderstood the Clausius definition of entropy and went out of his way to adopt a different definition in early editions of his *Theory of Heat*. He wrote: 'Clausius has called the remainder of the energy, which cannot be converted into work, the Entropy of the system. We shall find it more convenient to adopt the suggestion of professor Tait, and give the name of Entropy to the part which can be converted into mechanical work.' He then argued that entropy *decreases* during spontaneous processes. Later Maxwell recanted: 'In former editions of this book the meaning of the term Entropy, as introduced by Clausius, was erroneously stated to be that part of the energy which cannot be converted into work. The book then proceeded to use the term as equivalent to the available energy; thus introducing great confusion into the language of thermodynamics.'

Maxwell's discomfort and confusion with entropy is ironic, for his demon has had a profound effect on the way entropy is viewed. In particular, Maxwell's demon led to an important linkage between entropy and information. Unfortunately, Maxwell did not live long enough to see this outgrowth of his thought experiment. It is also noteworthy that his originally adopted definition of entropy gave rise to a function that decreases during spontaneous processes. Many years later, Brillouin found it useful for interpretive purposes to define a function, negentropy ($=$ −entropy), with this property (see Section 1.4 for more on negentropy).

### 1.2.2 Temperature and Pressure Demons

Maxwell's specification of the demon was brief enough to leave considerable room for interpretation. As envisioned, his creature was a temperature-demon that acts within a thermally isolated system of which it is an integral part. Its task was to generate a temperature difference without performing work on the gas. In effect this is the equivalent of producing heat flow from a lower to a higher temperature with no other effect, in conflict with the Clausius form of the second law.

In his later clarification (recall 'Concerning demons' in Section 1.2.1), Maxwell recognized that 'less intelligent' demons could generate differences in pressure. The first detailed investigation of a pressure-demon was by Leo Szilard (1929, Article 3.1). Szilard's work is discussed further in Section 1.3. A pressure demon operates in a system linked to a constant-temperature reservoir, with the sole net effect of converting energy transferred as heat from that reservoir to work on an external object, in conflict with the Kelvin-Planck form of the Second Law. The 'Maxwell's demon puzzle' is to show why neither a temperature nor pressure demon can operate outside the limits imposed by the second law of thermodynamics.

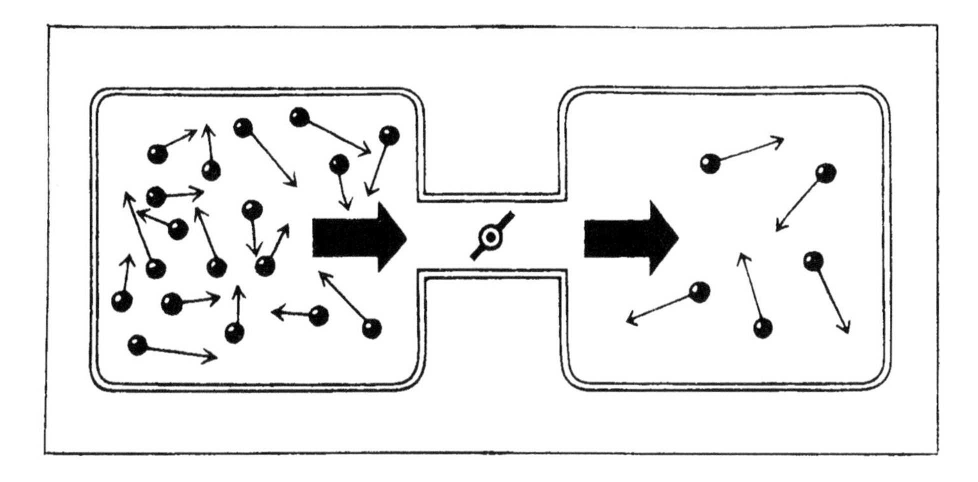

**Figure 1.**

### 1.2.3 Depictions of the Demon

Maxwell described his proposed being as 'small'. The dictionary definitions above suggest 'molecular' or 'arbitrarily small' size. Various authors have included cartoon depictions of Maxwell's demon with their writings. Figures 1–9 illustrate some of the ways the demon has been portrayed. Figure 1 (Pekelis, 1974) is in accord with Maxwell's view that the demon is nothing more than a valve, but does not show any control mechanism. Figures 2 (Pierce, 1961) and 3 (Angrist and Hepler, 1967) show characters operating trap doors manually from within one of the chambers, but without any obvious means of detecting molecules. Figure 4 (Angrist and Hepler, 1967) shows the demon wearing a helmet with a built in light source. Figure 5 (Lerner, 1975) shows a satanic character with a flashlight, operating a shutter from inside one of the chambers.

Figure 6 (Darling and Hulburt, 1955) shows the demon *outside* the two chambers. Figure 7 (Morowitz, 1970) depicts a pressure demon controlling a shutter between two chambers that are in contact with a constant-temperature heat bath. Figure 8 (Gasser and Richards, 1974) shows yet another view of an external demon, here operating a valve, allowing one species of a two component gas (hot and cold) through a partition separating a gas from an initially evacuated chamber. Only fast molecules are allowed through, resulting in a cold gas in one chamber and a hot gas in the other. More cartoons can be found in 'Information physics in cartoons,' honoring Rolf Landauer (Bennett, 1998, Article 7.6).

Typically, cartoon depictions show the demon as being relatively large compared to the shutter, sometimes with a light source to detect molecules, and sometimes located outside the system. Placing a temperature-demon outside the gas is questionable because of the need for thermal isolation. Depictions with a light source are not surprising in view of Maxwell's specification of a 'being who can see the individual molecules ....' Because his intent was to dramatize the statistical nature of the second law rather than to exorcise the demon, Maxwell had no reason to address the question of whether a demon could detect molecules by any means other than vision.

### 1.2.4 Means of Detection

Leon Brillouin (1951, Article 3.4), closely following the work of Pierre Demers (1944, 1945), took Maxwell's specification of 'seeing' molecules seriously and assumed the use of light signals. Dennis Gabor did the same, apparently independently. Others have considered detecting molecules via their

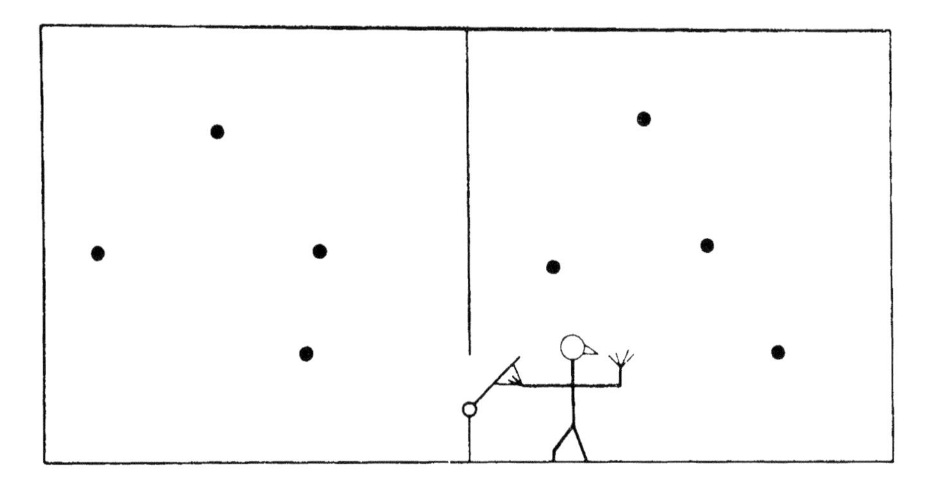

**Figure 2.**

**Figure 3.**

magnetic moments (Bennett, 1982, Article 7.1), Doppler-shifted radiation (Denur, 1981; Chardin, 1984; Motz, 1983), and even via purely mechanical means (Bennett, 1987). The prevailing modern view is that one must not prejudice the demon's operation by assuming the use of light signals, for that is too restrictive. The fundamental question is whether measurement in general is necessarily irreversible.

The clever mechanical detector (Bennett, 1987) proposed in the context of Szilard's 1929 model suggests that, in principle, the presence of a molecule can be detected with arbitrarily little work and dissipation. Bennett's scheme is compelling, but is limited to a one-molecule gas. The general question of whether measurement in a many-particle gas must be irreversible lacks a correspondingly compelling

**Figure 4.**

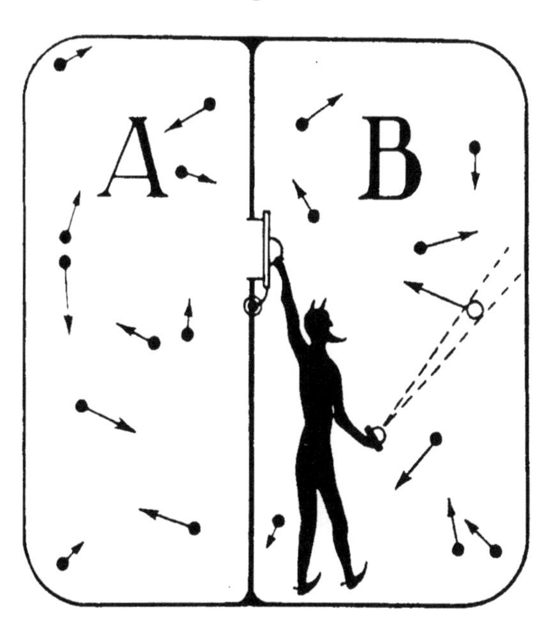


**Figure 5.**

answer. Maxwell's original temperature-demon must distinguish between molecular velocities among numerous molecules, a more complex task than detecting the presence of a single molecule. To our knowledge no specific device that can operate with arbitrarily little work and dissipation has been proposed for such velocity measurements. Given this void, the possibility of measurement without entropy generation in a macroscopic system is not universally accepted. See for example Rothstein (1988), Porod, *et al* (1984a) and responses thereto (1984b). Yet it is also true that there is no reason to believe that work and entropy thresholds exist for such measurements.

**Figure 6.**

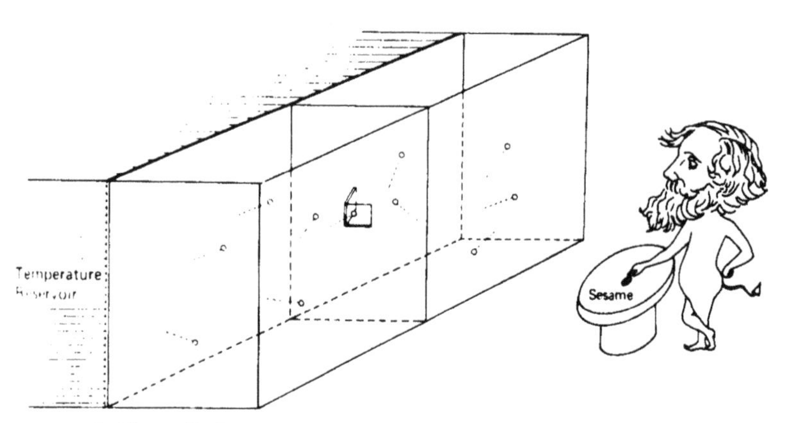


A Maxwell demon controlling a door between two chambers each initially at temperature $T_1$ and pressure $P_1$

**Figure 7.**

### 1.2.5 Thermal Equilibrium and Fluctuations

The demon must be in thermal equilibrium with the gas in which it resides, or an irreversible energy transfer between gas and demon would occur, clouding the basic puzzle. As a temperature-demon generates a temperature gradient within a gas, its own temperature presumably changes with its host gas. The heat capacity of a 'small' demon is presumably much less than that of the gas, and its temperature can vary with that of the gas via negligibly small energy exchanges. Except for the receipt of light signals, energy exchanges with the demon are usually neglected.

A demon that continually receives energy input via light signals (or other means) must eventually experience a temperature rise unless it transfers energy (via heat) to its surroundings. Additionally, if a lamp is used to generate light signals, photons that miss the demon will add energy to the gas and or container walls. Such phenomena threaten the assumption of constant-temperature operation, and most treatments of Maxwell's temperature-demon ignore these details. Of course such photons could heat one chamber directly, with no need for a demon—a phenomenon for which there is obviously no challenge to the second law.

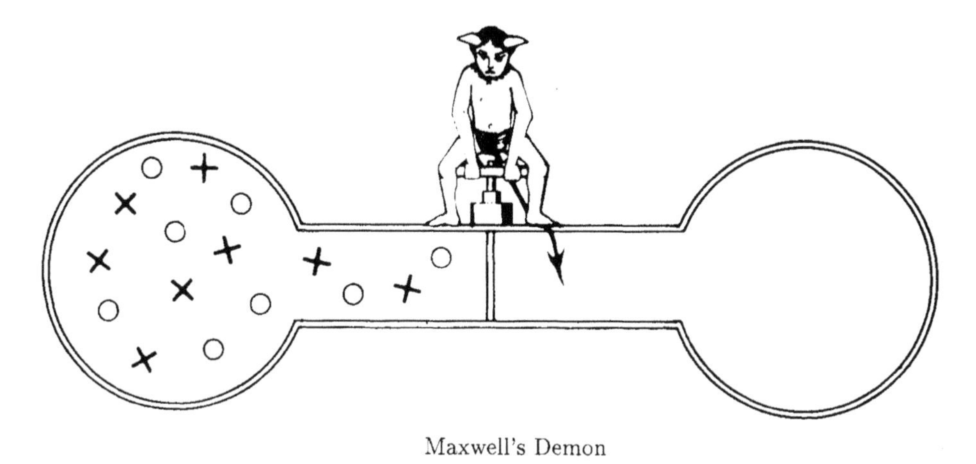

Maxwell's Demon

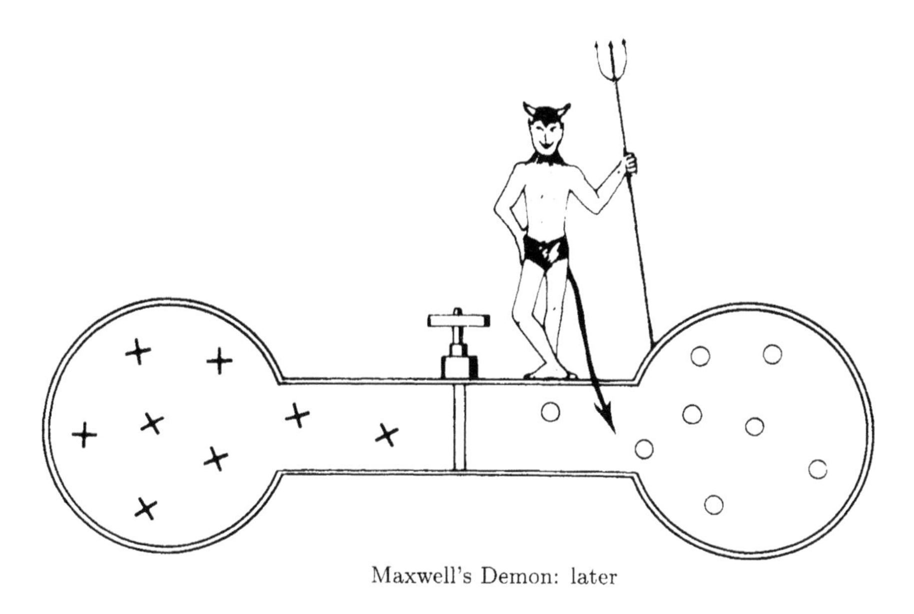

Maxwell's Demon: later

**Figure 8.**

Located within a gas, a Maxwell's demon is continually bombarded by gas molecules and by photons from the blackbody radiation field within the container. It can be jostled around by this bombardment, impeding the accuracy of its measuring activities. Long ago it was pointed out (Smoluchowski, 1912, 1914) that thermal fluctuations would prevent an automatic device from operating successfully as a Maxwell's demon.

A modern discussion of Smoluchowski's ideas was given by Richard Feynman (1963), who compared Maxwell's demon with a ratchet and pawl and an electrical rectifier, neither of which can systematically transform internal energy from a single reservoir to work. He wrote: 'If we assume that the specific heat of the demon is not infinite, it must heat up. It has but a finite number of internal gears and wheels, so it cannot get rid of the extra heat that it gets from observing the molecules. Soon it is shaking from Brownian motion so much that it cannot tell whether it is coming or going, much less whether the molecules are coming or going, so it does not work.'

If a demon absorbs energy, periodic dumping of energy to an external reservoir is needed to keep its temperature approximately equal to the temperature of the gas in which it resides. For a temperature-demon this violates assumed thermal isolation, and the 'system' must be expanded to include the demon, gas, and external reservoir. Of course feeding entropy to the reservoir helps to keep the second law intact. Modern day computer simulations dramatically reveal the fluctuation phenomena envisaged by Smoluchowski and Feynman (Skordos and Zurek, 1992, Article 3.1; Rex and Larsen, 1992, Article 3.2).

Smoluchowski's observation regarding thermal fluctuations suggested that Maxwell's demon ought to be buried and forgotten. But that did not happen, apparently because Smoluchowski left open the possibility that somehow, a perpetual motion machine operated by an intelligent being might be achievable. It was the fascinating idea of using intelligence that captured Leo Szilard's interest.

### 1.2.6 Intelligence

The demon must have sufficient 'intelligence' to discern fast from slow molecules, right-moving from left-moving molecules, or (in Szilard's model) simply the presence or non-presence of a molecule. In normal parlance intelligence is considered to include, among other things, ability to learn, reason, and understand relationships. But none of these seems to be required by Maxwell's demon. One feature associated with intelligence that is needed by a demon is memory: it must 'remember' what it measures, even if only briefly. Indeed without somehow recording a result, one can argue that a measurement has not been completed.

Despite the title 'The decrease of entropy by intelligent beings' of his classic 1929 paper, Leo Szilard wrote that physics is not capable of properly accounting for the biological phenomena associated with human intervention. Szilard asserted, 'As long as we allow intelligent beings to perform the intervention, a direct test (of the second law) is not possible. But we can try to describe simple nonliving devices that effect such coupling, and see if indeed entropy is generated and in what quantity.' In 1929, prior to the development of solid state electronics, that was a fanciful thought.

If the demon were an automaton, it would perform preprogrammed functions upon receipt of certain well-defined signals. Evidently a Maxwell's demon need not be any more intelligent than an electronic computing machine connected to some type of transducer that detects molecular phenomena and puts out electrical signals signifying detection. Certainly it need not possess human intelligence. The concept of Maxwell's demon as a computer automaton was explored by Laing (1974; reprinted in MD1) who, unfortunately, was unaware of Landauer's important finding (1961, Article 4.1) that memory erasure in computers feeds entropy to the environment.

In recent years some researchers have investigated the feasibility of quantum mechanical computers that operate via changes in the states of individual atoms. Feynman (1986) wrote '... we are going to be even more ridiculous later and consider bits written on one atom instead of the present $10^{11}$ atoms. Such nonsense is very entertaining to professors like me. I hope you will find it interesting and entertaining also ... it seems that the laws of physics present no barrier to reducing the size of computers until bits are the size of atoms, and quantum behavior holds dominant sway.' This suggests the possibility of Maxwell's demon being a quantum automaton of microscopic size, if such a microscopic demon could avoid devastation from fluctuations.

### 1.2.7 Interplay Between the First and Second Laws

Before continuing with the demon's history, it is helpful to examine implications of the first and second laws of thermodynamics on its actions. Consider first a temperature-demon that sorts molecules, lowering the entropy of a gas without altering its energy. The term 'demon' here includes any peripheral equipment used to effect sorting. What do the first and second laws of thermodynamics imply? Because the demon-gas system is energetically isolated, the second law requires the demon's entropy to increase at least as much as the gas entropy decreases (which is assumed here) during sorting. The first law implies that a temperature-demon's energy is unchanged by sorting because the gas and gas-demon system energies are both fixed. Thus, the demon's entropy must increase at fixed energy.

Can the demon be returned to its initial state without disturbing the gas? Such 'resetting' of the demon is desirable for two reasons. First, if the demon is to operate repeatedly, its entropy cannot be allowed to increase indefinitely or it will ultimately become too 'disordered' and unable to operate (see Section 1.2.5). Second, resetting simplifies the thermodynamic analysis, which can focus on the gas and its environment, without regard for the demon's details. Resetting the demon requires an exchange of energy with other objects. For example, the demon's excess entropy might be dumped as heat to a constant-temperature reservoir, with an external work source subsequently increasing the demon's energy at constant entropy, returning it to its initial state.

Evidently the first and second laws of thermodynamics assure that: (1) a temperature-demon cannot sort molecules without increasing its entropy, (2) the demon cannot return to its initial state without external energy exchanges, and (3) the combination of sorting and resetting generates an energy transfer from an energy source to a reservoir.

Next consider a pressure-demon, operating a cyclic process in a constant temperature ideal gas. Contact with a thermal reservoir assures that the temperature will be constant. Initially the gas pressures and densities are equal on each side of a central partition. The cyclic process is defined as follows:

(a) The demon reduces the gas entropy at fixed temperature and energy by letting molecules through the partition in one direction only. This sorting process generates pressure and density differences across the partition.
(b) The gas returns to its initial state by doing isothermal, reversible work on an external load. Specifically, the partition becomes a frictionless piston coupled to a load, moving slowly to a position of mechanical equilibrium (away from the container's center) with zero pressure and density gradients across the piston. The piston is then withdrawn and reinserted at the container's center.
(c) The demon is returned to its initial state.

What do the laws of thermodynamics imply? The process sequence (a)–(c) results in a load with increased energy. The first law of thermodynamics requires that this energy come from some well-defined source. It cannot be supplied by the reservoir or the entropy of the universe would decrease in the cyclic process, in violation of the second law. Apparently, resetting the demon in (c) requires use of a work source which, in effect, supplies the energy to the load. It is helpful to look at the thermodynamic details of steps (a)–(c).

The second law implies that the demon's entropy increases in (a) to 'pay' for the entropy decrease of the gas. That is, sorting must increase the pressure-demon's entropy. In (b) work $W$ is done by the gas on the load, inducing heat $Q = W$ from reservoir to gas. The load's energy increases with its entropy unchanged, and the gas is returned to its initial state. Withdrawing and replacing the piston has zero thermodynamic effect. In step (b) the work on the load is compensated by the diminished energy (and entropy) of the reservoir. The demon's entropy increase offsets the reservoir's entropy decrease to maintain the second law's integrity. Now suppose that in (c) the demon is reset, returning to its initial state by energy exchanges with the reservoir and a reversible work source, with work $E$ done on the demon.

The demon's entropy decrease here must be compensated by an entropy increase in the reservoir. We conclude that resetting the demon results in heat to the reservoir.

Overall, in (a)–(c) the entropy change of the universe equals that of the reservoir. The second law guarantees this is nonnegative; i.e., the reservoir cannot lose energy. The cyclic process results in an increased load energy and a reservoir internal energy that is no lower than its initial value. The first law implies that the work source loses sufficient internal energy to generate the above gains; in particular, the source does positive work in (c). The relevant energy transfers during the cycle are: work $W > 0$ by gas on load, work $E > 0$ by work source on demon, and energy $E - W \geq 0$ added to the reservoir. The entropy change of the universe is $(E - W)/T \geq 0$, where $T$ is the reservoir temperature. Maxwell apparently envisioned a being who could run on arbitrarily little energy, an assumption that is implicit in most treatments of Maxwell's demon. Thomson assumed demons could store limited quantities of energy for later use, implying a need for refueling. Our analysis here illustrates that if the first and second laws of thermodynamics are satisfied, the refueling (i.e. resetting) energy to a Maxwell's pressure-demon is transferred to the load as the gas and demon traverse their cycles. This suggests that resetting a demon is of fundamental importance, a view that is strengthened considerably in Section 1.5.

## 1.3 Szilard's Model: Entropy and Information Acquisition

Sixty-two years after Maxwell's demon was conceived, Leo Szilard introduced his famous model in which an 'intelligent' being operates a heat engine with a one-molecule working fluid (1929, Article 3.3). We briefly outline that model. Initially the entire volume $V$ of a cylinder is available to the fluid, as shown in Figure 9a. Step 1 consists of placing a partition into the cylinder, dividing it into two equal chambers. In Step 2 a Maxwell's demon determines which side of a partition the one-molecule fluid is on (for the sake of illustration, Figure 9b shows the molecule captured on the right side), and records this result. In Step 3 the partition is replaced by a piston, and the recorded result is used to couple the piston to a load upon which work $W$ is then done (Figures 9d and 9c). Strictly speaking, the load should be varied continuously to match the average force on the piston by the fluid, enabling a quasistatic, reversible work process. The gas pressure moves the piston to one end of the container, returning the gas volume to its initial value, $V$ (Figure 9a). In the process the one-molecule gas has energy $Q = W$ delivered to it via heat from a constant-temperature heat bath.

After Step 3 the gas has the same volume and temperature it had initially. The heat bath, which has transferred energy to the gas, has a lower entropy than it had initially. It appears that without some other effect, the second law of thermodynamics has been violated during the cyclic process. Szilard observed: 'One may reasonably assume that a measurement procedure is fundamentally associated with a certain definite average entropy production, and that this restores concordance with the second law. The amount of entropy generated by the measurement may, of course, always be greater than this fundamental amount, but not smaller.' He further identified the 'fundamental amount' to be $k \ln 2$. His observation was the beginning of information theory.

The ingenuity of Szilard's engine is striking. His tractable model allows thermodynamic analysis and interpretation, but at the same time entails a binary decision process. Thus, long before the existence of modern information theory and the computer age, Szilard had the foresight to focus attention on the 'information' associated with a binary process. In doing so he discovered what is now called the binary digit—or 'bit'—of information. Szilard's observation that an inanimate device could effect the required tasks—obviating the need to analyze the thermodynamics of complex biological systems—was a precursor to cybernetics.

Szilard examined two other models involving memory in his 1929 paper. Unfortunately, his arguments are sometimes difficult to follow, and it is unclear whether the thermodynamic cost is from

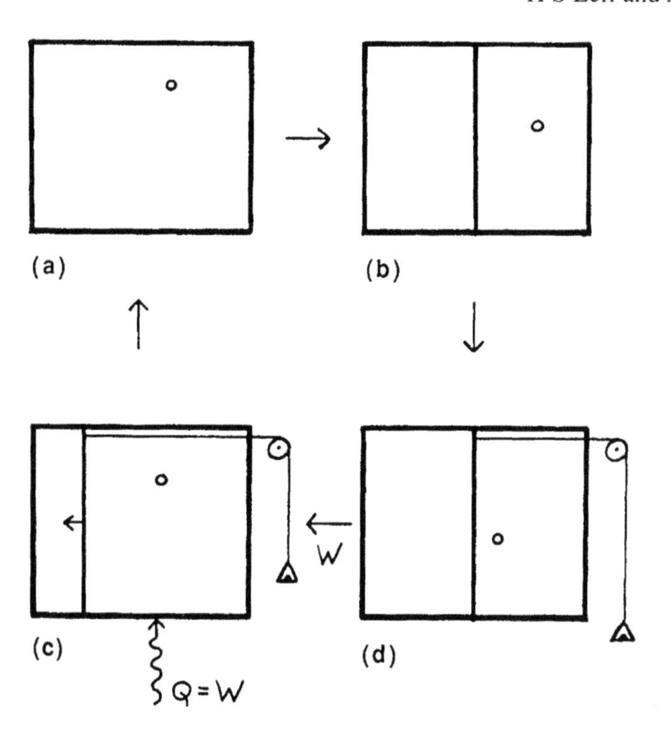


**Figure 9.**

measurement, remembering, or forgetting. In the course of his analyses, Szilard observed: 'Having already recognized that the only important factor (of intervention) is a certain characteristic type of coupling, a 'measurement', we need not construct any complicated models which imitate the intervention of living beings in detail. We can be satisfied with the construction of this particular type of coupling which is accompanied by memory.' His concluding sentence is, 'We have examined the 'biological phenomena' of a nonliving device and have seen that it generates exactly that quantity of entropy which is required by thermodynamics.'

Thus, Szilard regarded memory as an important feature in a demon's operation, but he did not identify its specific role in saving the second law. His writing implies the production of entropy during measurement, along with an undefined, but important, effect of the memory process. While he did not fully solve the puzzle, the tremendous import of Szilard's 1929 paper is clear: *He identified the three central issues related to information-gathering Maxwell's demons as we understand them today—measurement, information, and memory—and he established the underpinnings of information theory and its connections with physics.*

Szilard's work met with mixed response. Some researchers felt that it put the final nail in the coffin of Maxwell's demon. For example, Jordan (1949) wrote, 'This ... stands rather isolated apart from the flow of modern physical ideas; but I am inclined to regard it as one of the greatest achievements of modern theoretical physics, and believe that we are still very far from being able to evaluate all its consequences.' Much later Peter Landsberg (1982) wrote, 'Maxwell's demon died at the age of 62 (when a paper by Leo Szilard appeared), but it continues to haunt the castles of physics as a restless and lovable poltergeist.' Although it is unclear how much he was influenced by the work of Szilard, Brillouin broke new ground by developing an extensive mathematical theory connecting measurement and information. Demers (1944, 1945), Brillouin (1951a,b, 1956), and Gabor (1964) were led to the conclusion that the second law is linked to the quantum nature of light.

On the negative side, there were criticisms of Szilard's efforts to link entropy and information. Popper (1974) described Szilard's suggestion that knowledge and entropy are related as 'spurious'. Similar criticisms may be found elsewhere (see Popper, 1957; Feyerabend, 1966; Chambadal, 1971; Jauch and Báron, 1972, Article 3.5). Some objections emanate from the view that thermodynamical entropy is a measurable quantity (within an additive constant) that is independent of an observer's knowledge, and any other definition of entropy that is observer-dependent is unacceptable.

Rothstein (1957) clarified this point as follows: 'Demons do not lower entropy; the information they act on defines a lower entropy state of the system than one not subject to the restrictions in which the information consists.' Later, Rothstein (1971) elaborated further on this point: 'Physical information and its associated entropy reduction, localized in the system to which the information refers, can be expressed via specifications or constraints taken as part of the description of a system, or can be obtained from measurement. Measuring a system and thus finding it to be in some state is formally equivalent ... to preparing it to be in that state, specifying it to be in that state, or constraining it in a manner so that it can be in no other state (the state in question can, of course, be mixed).' The intimate connections between entropy, a system property, and information, a property of the observer, are discussed also by Morowitz (1970).

Along similar lines, E T Jaynes (1979) wrote, 'The entropy of a thermodynamic system is a measure of the degree of ignorance of a person whose sole knowledge about its microstate consists of the values of the macroscopic quantities $X_i$ which define its thermodynamic state. This is a completely "objective" quantity, in the sense that it is a function only of the $X_i$, and does not depend on anybody's personality. There is then no reason why it cannot be measured in the laboratory.'

Jaynes (1965) also observed that a given physical system corresponds to many different thermodynamic systems. Entropy is not a property simply of the system, but of the experiments chosen for it. One normally controls a set of variables, and measures entropy for that set. A solid with $N$ atoms has approximately $6N$ degrees of freedom, of which only a few (e.g., temperature, pressure, magnetic field) are usually specified to get the entropy. By expanding that set (say, to include components of the strain tensor), we could get a sequence of entropy values, each of which corresponds to a different set of constraints. Extension of this process ultimately gets one outside the normal domain of thermodynamics, for which the number of degrees of freedom greatly exceeds the number of thermodynamic variables.

A one-molecule system never satisfies the latter description because experiments do not even exist. Thus the use of the entropy concept—or any other thermodynamic concept—must be clarified. One possible clarification envisages an ensemble of similar systems, the average behavior of which is related to a 'typical' single system. In ordinary statistical mechanics of macroscopic systems, the system of interest is typically in contact with a constant-temperature reservoir. Energy exchanges between system and reservoir go on continually, and observations over a long time period can in principle detect fluctuations about a well-defined, time-averaged energy. The ensemble description replaces the single system, viewed over an extended time by a collection of many similar systems, all viewed at a chosen time. Use of the ensemble approach assumes the equality of time and ensemble averages.

In the present context one may consider taking a large number of one-molecule gases through Szilard's cycle. Using statistical mechanics, entropy and average pressure may be defined as meaningful thermodynamic properties of the ensemble. One must choose 'appropriate' variables of the system over which to take statistical averages. In the Szilard cycle, left ($L$) and right ($R$) side indexes are appropriate. In a sense these are measurable 'macroscopic' variables. Being outside the normal domain of thermodynamics, the Szilard model can be criticized as having no thermodynamic significance. An alternative viewpoint, which we take here, is that it gives an opportunity to extend thermodynamic concepts into interesting new territory, with some help from information theory. Jordan (1949) described this well: '...the tendency of Szilard's views is to acknowledge also a microphysical applicability of thermodynamics.' We discuss this further in Section 1.5.

In summary, Szilard's contributions have influenced the way we think about entropy. Through Szilard's ideas, Maxwell's demon led to the concept of a 'bit' of information and to key concepts in information theory, cybernetics, and computing. In a remarkable and fitting reciprocation, modern day theories of computing have led to a new understanding of the Maxwell's demon puzzle. A new, fundamentally different resolution of that conundrum involves erasure of the demon's memory, a point that Szilard just narrowly missed in 1929. We return to this in Section 1.5.

## 1.4 Information Acquisition via Light Signals: A Temporary Resolution

As mentioned earlier, Leon Brillouin (1951, Article 3.4) and (independently) Dennis Gabor (1951; published in 1964, reprinted in MD1) followed up on the measurement aspect of Maxwell's demon about 20 years later using the quantum nature of light. Because quantum theory was still not invented during his lifetime, Maxwell could not have foreseen that his demon would provide a path to the quantum domain. But a small demon exists in a sea of gas molecules and photons. The photons, quanta of the blackbody electromagnetic radiation within the vessel, have a well-defined energy distribution dictated by quantum theory. In the mid 1940s, Pierre Demers recognized that because of this, a high temperature lamp is needed to provide signals that are distinguishable from the existing blackbody radiation. Brillouin, who was influenced by Demers' studies, adopted this assumption.

Consider the nonuniform wavelength distribution of blackbody radiation. For a gas temperature $T$, Wien's law gives the wavelength of maximum spectral density, $\lambda_m(T) \approx 2900/T$ [$\mu$m]. Assuming an ambient temperature $T = 290$ K, the wavelength region in the vicinity of $\lambda_m(290) \approx 10\ \mu$m can be avoided by having the lamp emit substantial radiation power with $\lambda \ll \lambda_m$. Using a constant-emissivity approximation, a lamp with radiating temperature 1500 K has $\lambda_m(1500) \approx 2\ \mu$m, and an incandescent light bulb with filament temperature 3,000 K has $\lambda_m(3,000) \approx 0.1\mu$m. Whether a lamp's radiation is distinguishable from ambient blackbody radiation depends on the power incident to the demon's eyes in the low wavelength region with and without the lamp. The radiating area of the lamp and geometrical considerations can be important, but the details of this complicated problem do not appear to have been pursued in the literature. It is clear that a lamp giving distinguishable signals can be chosen: humans regularly use high-temperature incandescent lamps.

Could a low temperature radiator, say, with $T = 100$ K and $\lambda_m = 29\ \mu$m be used? This is less satisfactory for two reasons. First, the total power radiated is proportional to $A_s T^4$, where $A_s$ is the radiating surface area—and a low-temperature lamp must have a much larger radiating surface area to emit the same total power as a high temperature source. The radiating surface of a 100 K radiator must be 810,000 times larger than that for a 3,000 K lamp with the same total power output. Second, higher wavelength radiation is accompanied by more pronounced diffraction effects than low wavelength light, decreasing the demon's ability to resolve signals.

Brillouin assumed a high temperature lamp and melded the developing field of information theory with the Maxwell's demon puzzle. The first assumption, together with judicious use of the quantum nature of radiation, enabled an explicit demonstration that information gathering via light signals is accompanied by an entropy increase. This increase is sufficient to save the second law.

The mathematical theory of information had been solidified by Claude Shannon (Shannon and Weaver, 1949) in connection with communication processes. Shannon introduced a mathematical function, which he called information entropy, to analyze the information carrying capacity of communication channels. Although Shannon's function bears a striking mathematical resemblance to the canonical ensemble entropy of statistical mechanics, Shannon's stimulus, method of attack, and interpretation were very different. Brillouin boldly postulated a direct connection between information entropy and thermodynamic entropy.

Suppose a physical system can be in any of $P_0$ states with equal likelihood, and we do not know which state is actually occupied. Brillouin assigned information $I_0 = 0$ to signify total ignorance. If by measurement we eliminate some of the states as possibilities, reducing the number to $P_1 < P_0$, the information so gathered is defined as $I_1 \equiv K' \ln(P_0/P_1) > 0$. $K'$ is an undesignated positive constant. Had the number of states increased, $I_1$ would be negative; i.e., we would have lost information. These ideas are described in more detail in Article 3.4.

Five years after his path-breaking article, Brillouin published *Science and Information Theory*, which solidified his ideas on the subject. There he distinguished between two kinds of information, 'free' and 'bound', in order to handle information that did not have thermodynamic significance. Free information ($I_f$) was regarded as abstract and without physical significance. Bound information ($I_b$) was defined in terms of the possible states of a physical system. Brillouin gave as an example of free information the knowledge possessed by an individual. That knowledge is transformed into bound information when it is transmitted from one individual to another via physical signals.

According to Brillouin it is the physical character of signals that makes the information they carry 'bound.' In the communication process, the information might get distorted or partially lost; i.e., $I_b$ can decrease. When the resulting bound information is received by another individual, it is again considered to be free information. Brillouin linked changes in bound information to changes in entropy of a physical system via the hypothesis: $I_{b1} - I_{b0} \equiv k(\ln P_0 - \ln P_1) = S_0 - S_1 > 0$, where the initially arbitrary constant $K'$ has been chosen to be Boltzmann's constant, $k$; and $S_0$ and $S_1$ are the initial and final entropy values for the physical system. Choosing $K' = k$ makes information entropy and physical entropy comparable in the sense that they have the same units.

Brillouin's hypothesis implies that gaining bound information about a physical system decreases its physical entropy. He then made two further important steps. First, he defined 'negentropy' $\equiv N \equiv -(entropy)$ and thus negentropy change $\Delta N = -$ (entropy change) $= -\Delta S$. Second, he applied his negentropy principle of information to an isolated physical system. Suppose this system's entropy is $S_1 = S_0 - I_{b1}$, as above. The second law of thermodynamics is then written:

$$\Delta S_1 = \Delta(S_0 - I_{b1}) = \Delta S_0 - \Delta I_{b1} = -\Delta N_0 - \Delta I_{b1} \geq 0,$$

or simply,

$$\Delta(N_0 + I_{b1}) \leq 0.$$

With this last result Brillouin gave a new interpretation of the second law of thermodynamics: The quantity (negentropy + information) can never increase, and in a reversible transformation, the sum remains fixed. He applied these ideas to 'exorcise' Maxwell's demon.

As might have been anticipated, Brillouin's proposal to generalize and reinterpret the second law got considerable attention, splitting the scientific community into groups of believers and nonbelievers. If the subsequent literature accurately reflects level of belief, the believers are more numerous, for Brillouin's method is widely quoted (see for example: Barrow, 1986; Bell, 1968; Ehrenberg, 1967; Dugdale, 1966; Rex, 1987; Waldram, 1985; Zemansky, 1981; Yu, 1976; Rodd, 1964). Unqualified acceptance is evident in a paragraph labeled, 'Obituary: Maxwell's Demon (1871-c.1949)' (Bent, 1965), reprinted in the chronological bibliography. Acceptance is evident also in all 20 articles that comprise Chapters 4–7 in this book.

Though smaller in numbers, nonbelievers leveled thoughtful criticisms of the subjectivity implied by Brillouin's theory. (In contrast, recall arguments illustrating and supporting objectivity of entropy within the informational approach in Section 1.3.) Among the most vociferous critics of Brillouin's theory is Kenneth Denbigh, (1981, reprinted in MD1) who rejects the view that entropy is subjective. He emphasizes that Brillouin's exorcism of Maxwell's demon can be accomplished solely using thermodynamic principles, without need for information theory or negentropy. Denbigh's dismay with subjectivism led to a book on the subject (Denbigh and Denbigh, 1985). Karl Popper has leveled harsh criticisms at attempts to link information and thermodynamics (Popper, 1957, 1974, 1982). Much of this is focused on Szilard's 1929 paper, which began the process of associating information and entropy (see Section 1.3).

Rudolph Carnap (1977) wrote, 'Although the general identification of entropy (as a physical concept) with the negative amount of information cannot be maintained, there are certainly important relations between these two concepts.' He praised Szilard's work analyzing the paradox of Maxwell's demon as showing an important connection between entropy and information. He summarized Brillouin's ideas, which '...are certainly interesting and clarify the situation with respect to Maxwell's paradox in the direction first suggested by Szilard.' Despite this commendation Carnap also took issue with Brillouin's identification of negentropy with information: 'However, when Brillouin proceeds to identify negentropy with an amount of information, I cannot follow him any longer ... He does not seem to be aware that the definition of $S$ which he uses (and which he ascribes to Boltzmann and Planck) makes $S$ a logical rather than a physical concept.' We return to connections between logical and physical concepts in the next section. More recent critiques of the information-entropy connection are discussed in Section 1.8.1.

In work complementary to Brillouin's, Gabor (1964, reprinted in MD1) analyzed the use of light signals to operate the Szilard engine. Although that work was not published until 1964, it was actually reported in lectures Gabor presented the same month that Brillouin's paper was published. The point of Gabor's treatment was to illustrate that a Maxwell's demon could in principle violate the second law if the light used satisfies *classical* laws. Employing a cleverly designed system with an incandescent lamp, mirrors, and photodetector, Gabor found that if the light intensity can be made arbitrarily large relative to the background blackbody radiation, then the second law is vulnerable. He argued, however, that this is prohibited by quantum theory because 'Very weak beams of light cannot be concentrated.'

Attempted resolution of the Maxwell's demon puzzle by focusing on information acquisition was an important phase of the demon's life. It is interesting that the focus on information acquisition seemed to eliminate all interest in the memory aspects that Szilard emphasized. This is nowhere more clear than in Brillouin's decision to define two types of information, one of which ('free' information, $I_f$) was designed explicitly to deal with 'knowledge,' and the other ('bound' information, $I_b$) was linked to entropy changes. In effect this inhibited considerations of the *physical* aspects of memory. Ironically, it is these physical effects of memory that subsequently led to an overthrow of the resolutions proposed by Brillouin and Gabor!

## 1.5 Computers and Erasure of Information: A New Resolution

### 1.5.1 Memory Erasure and Logical Irreversibility

Recall that after Step 3 in the Szilard model discussed in Section 1.3, the demon retains the memory of its finding, plus any other effects of the measurement process. We assume the demon has experienced zero temperature change and negligible, if any, 'other' effects of the measurement. In order to make the process within the gas-demon system cyclic, thereby restricting all thermodynamic changes to the environment, the memory evidently must be erased. The thermodynamic consequences of this process become of fundamental interest.

Landauer (1961, Article 4.1) introduced the concept of 'logical irreversibility' in connection with information-discarding processes in computers. Memory erasure, which takes a computer memory from an (arbitrary) existing state $A$, to a unique, standard reference state $L$ discards information in a logically irreversible way. Logical irreversibility means that the prescription 'Map the existing state $A$ to the state $L$' has no unique inverse because state $A$ can be any of many possible states in the computer's memory.

Put differently, starting from state $L$, one cannot get to the state $A$ without using further information - e.g., the computer program and the initial data that led to state $A$ in the first place.

Landauer argued that to each logical state there must correspond a physical state. Logical irreversibility carries the implication of a reduction of physical degrees of freedom, resulting in 'dissipation'. This is a subtle concept. We show shortly that logical irreversibility does not necessarily imply physical irreversibility in the thermodynamic sense. Rather, it can manifest itself in terms of a thermodynamically reversible conversion of work to heat. That is, the work of erasure $W$ can result in heat $Q = W$ going to the environment. Landauer also argued that computation steps that do not discard information, e.g., writing and reading, can be thermodynamically reversible in principle.

Charles Bennett (1973, reprinted in MD1) extended Landauer's work, arguing that a computing automaton can be made logically reversible at every step. This allows an in-principle thermodynamically reversible computer that saves all intermediate results, avoiding irreversible erasure, prints out the desired output, and reversibly disposes of all undesired intermediate results by retracing the program's steps in reverse order, restoring the machine to its original condition.

Subsequently Bennett (1982, Article 7.1) argued that a demon's memory may be viewed as a two-state system that is set in a standard state prior to measurement. The measurement process increases the available phase space of the memory from one state to two (e.g., in an ensemble of systems, measurement can lead to either state). Memory erasure returns it to the standard state $L$, thereby compressing a two-state phase space to a single state. This is a logically irreversible act that is accompanied by an entropy transfer to the reservoir.

Bennett showed that if all steps in the Szilard model are carried out slowly, the resulting entropy increase of the reservoir compensates exactly for the entropy decrease of the demon's memory and saves the second law. Strictly speaking this cyclic process is thermodynamically reversible: the gas, demon, and reservoir are all returned to their initial states. Bennett uses a phase space that contains both the states of the particle (left or right) and demon memory (also left and right). Fahn (1994, 1996) expanded this idea, and argued that the expansion step, which decorrelates the gas from the memory is the source of the dissipation that saves the second law.

In his 1970 book *Foundations of Statistical Mechanics*, Oliver Penrose independently recognized the importance of 'setting' operations that bring all members of an ensemble to the same observational state. Applied to Szilard's heat engine, this is nothing more than memory erasure. Penrose was unaware of Bennett's work, and it is interesting that he arrived at similar conclusions but by rather different arguments. Penrose wrote:

> The large number of distinct observational states that the Maxwell demon must have in order to make significant entropy reductions possible may be thought of as a large memory capacity in which the demon stores the information about the system which he acquires as he works reducing its entropy. As soon as the demon's memory is completely filled, however, ... he can achieve no further reduction of the Boltzmann entropy. He gains nothing for example, by deliberately forgetting or erasing some of his stored information in order to make more memory capacity available; for the erasure being a setting process, itself increases the entropy by an amount at least as great as the entropy decrease made possible by the newly available memory capacity.

Penrose did not go as far as Bennett, who argued that measurement can be done with arbitrarily little dissipation and that erasure is the fundamental act that saves Maxwell's demon. Published within a rather abstract, advanced treatment of statistical mechanics, Penrose's modest but important treatment of memory erasure went largely unnoticed among Maxwell's demon enthusiasts. We discuss his approach further in Section 1.6.1.

### 1.5.2 Logical versus Thermodynamic Irreversibility

Because the concept of memory erasure has generated considerable debate, further clarification is appropriate. Motivated by the Szilard model, suppose we choose our memory device to be a box of volume $V$, partitioned down its middle, and containing a single molecule. The molecule is either in the left ($L$) side or the right ($R$) side, and the container walls are maintained at temperature $T$. In effect the molecule is in a double potential well whose middle barrier potential is infinite. As before, let the standard reference state in this example be $L$, and consider an ensemble (see Section 1.3) of demon memories in which some of the ensemble members can occupy state $L$ and others occupy state $R$.

Erasure and resetting takes each memory from its existing state and bring it to the standard state $L$. A crucial observation is this: It is not possible to use a specific erasure process for an $L$ state and a different one for the $R$ state. Why? Because that would necessitate first determining the state of each memory. After erasure, the knowledge from that determination would remain; i.e., erasure would not really have been accomplished. (In Section 1.8.2 an attempt to circumvent this restriction is addressed.)

An acceptable erasure and resetting process must work equally well for either initial memory state ($L$ or $R$). For example, this can be accomplished by the following two-step algorithm applied to each ensemble member:

(i) To effect erasure, remove the central partition from each ensemble member.
(ii) To effect resetting, slowly compress each gas isothermally to the left half of the box.

The diffusion process in the erasure step (i), eradicates the initial memory state. Despite the fact that this process is logically irreversible, it is thermodynamically reversible for the special case where the ensemble has half its members in state $L$ and half in state $R$. This is evident from the fact that partition replacement leads to the initial thermodynamic state (assuming fluctuations are negligibly small). Isothermal compression in (ii) means that the walls of the box are maintained at temperature $T$. Each gas molecule's energy, on average, is determined by the wall temperature, and the work of compression on each memory results in a transfer of energy to the constant-temperature reservoir. For the ensemble, the average work $W$ must equal the average heat $Q$ to the reservoir. Thermodynamically, work has been 'converted' to heat, and entropy $\Delta S_{res} = Q/T = W/T = k \ln 2$ has been delivered to the reservoir. This example illustrates how the act of blurring the distinction between $L$ and $R$ can be linked to the delivery of entropy to the reservoir.

How has the ensemble entropy of the memory changed during the erasure process? Under our assumptions, the initial ensemble entropy per memory associated with the equally likely left and right states is $S_{LR}(\text{initial}) = k \ln 2$. After erasure and resetting, each ensemble member is in state $L$, and $S_{LR}(\text{final}) = 0$. Therefore, $\Delta S_{LR} = -k \ln 2 = -\Delta S_{res}$. In this sense the process is *thermodynamically reversible*; i.e., the entropy change of the universe is zero. This counterintuitive result is a direct consequence of the assumed uniform initial distribution of ensemble members among $L$ and $R$ states. During erasure, work from an external source has been used to effect energy transfer to the reservoir, but without altering the entropy of the universe.

Further understanding of the erasure-resetting procedure's thermodynamically reversible character for a uniform initial distribution of $L$ and $R$ states can be gained by reversing the steps of that procedure. Starting with all ensemble memories in state $L$ ($S_{LR} = 0$), let each gas in the ensemble slowly expand isothermally to the full volume $V$. The performance of average work $W = kT \ln 2$ by each gas on its work source (now a work *recipient*) induces energy transfer $Q = W$ from the reservoir to the gas. The average gas entropy increases by $\Delta S_{LR} = k \ln 2 = -\Delta S_{res}$. Subsequent placement of the partition has zero entropic effect, because (approximately) half the ensemble members are likely to end up in each of the two states.

The fact that some specific systems that were initially $L$ become $R$, and vice versa, illustrates that the process is *logically* irreversible. However, it is *thermodynamically* reversible in the sense that carrying out the steps in reversed order: (a) re-establishes the initial distribution of $L$ and $R$ states among ensemble members; (b) returns energy $Q$, transferred from gas to reservoir during resetting, back to the gas; (c) returns energy $W = Q$ used to effect erasure-resetting back to the external work source; and (d) leaves the entropy of the universe unaltered.

We emphasize that memory erasure and resetting is always *logically* irreversible. It is *thermodynamically* reversible only when the initial memory ensemble is distributed uniformly among $L$ and $R$ states. To see how erasure can be thermodynamically irreversible, consider the case where all ensemble memories are initially in state $L$. In the above two-step erasure-resetting procedure, partition removal in step (i) is thermodynamically irreversible, with the entropy change of the universe equaling $\Delta S_{LR} = k \ln 2$. During the subsequent compression of each ensemble member in step (ii), external work $W$ results in heat $Q = W$ to the reservoir. The initial and final ensemble entropy values of the gas are both zero, and the average entropy change of the universe equals that of the reservoir, namely, $k \ln 2$, which is attributable to irreversible partition removal. Similar reasoning shows that the erasure-setting combination is both thermodynamically *and* logically irreversible whenever the initial ensemble of memories is not distributed equally among $L$ and $R$ states.

One might argue that prior to an erasure procedure, the memory of a single memory device (rather than an ensemble of memory devices) is in a fixed state and its entropy $S_{LR}$ must be zero. (Note, however, that it is shown in Section 1.8.1 that *algorithmic* entropy is *nonzero* for a given memory state.) With this view, erasure brings the memory to another single state with zero entropy, and the entropy change of the memory is zero. The only entropy change is the positive one in the reservoir, and the process must be viewed as *thermodynamically* irreversible. Whether this or the previous interpretation is used, the crucial point is that memory erasure saves the second law, and discarding information results in energy dissipation as heat to the environment.

The foregoing analysis suggests that the entropy of a collection of Szilard gases does not change when partitions are installed or removed. Without partitions installed, and without the use of special measurements, we expect half the boxes in our ensemble to have their molecules on the left and half on the right at any chosen time, giving an ensemble entropy $S_{LR} = k \ln 2$. This is unchanged by placement of a partition in each box and is unchanged again upon partition removal. Thus, for both replacement and removal of partitions, the change in the ensemble entropy of the gas is zero. John von Neumann (1955; originally published in German, 1932) recognized this in his *Mathematical Foundations of Quantum Mechanics*, writing:

> ...if the molecule is in the volume $V$, but it is known whether it is in the right side or left side ... then it suffices to insert a partition in the middle and allow this to be pushed ... to the left or right end of the container ... In this case, the mechanical work $kT \ln 2$ is performed, i.e., this energy is taken from the heat reservoir. Consequently, at the end of the process, the molecule is again in the volume $V$, but we no longer know whether it is on the left or right ... Hence there is a compensating entropy decrease of $k \ln 2$ (in the reservoir). That is, we have exchanged our knowledge for the entropy decrease $k \ln 2$. Or, the entropy is the same in the volume $V$ as in the volume $V/2$, provided that we know in the first mentioned case, in which half of the container the molecule is to be found. Therefore, if we knew all the properties of the molecule before diffusion (position and momentum), we could calculate for each moment after the diffusion whether it is on the right or left side, i.e., the entropy has not decreased. If, however, the only information at our disposal was the macroscopic one that the volume was initially $V/2$, then the entropy does increase upon diffusion.

It is notable that von Neumann associated entropy decrease with the demon's knowledge. Had he addressed the process of *discarding* information, which is needed to bring the demon back to its initial state, he might have discovered the Landauer–Penrose–Bennett (LPB) resolution of the puzzle decades earlier.

The idea that neither partition placement nor removal changes the one-molecule gas entropy is supported and clarified by a quantum mechanical analysis of entropy changes for the gas (and memory) given by Zurek (1984, Article 5.1). His work was evidently inspired by a criticism of the Szilard model by Jauch and Báron (1972; Article 3.5), who argued that the Szilard model is outside the realm of statistical physics, and should be dismissed altogether! That opinion was subsequently rebuked by Costa de Beauregard and Tribus (1974; Article 3.6). Zurek viewed partition insertion in terms of the introduction of a thin potential barrier of increasing strength $V_0$. When $V_0 = 0$ there is no barrier, and when $V_0$ is made sufficiently large, the barrier is effectively impenetrable. As $V_0$ is increased, the wave function of the molecule distorts, and Zurek shows the entropy to be unchanged by partition insertion. Zurek's work is discussed further in Section 1.7.2.

### 1.5.3 Role of Measurement in Szilard's Model

Some authors have argued that in Szilard's engine no measurement is needed prior to coupling the piston to the external load. They argue that clever design of the engine would enable the proper coupling to be made automatically. For example, Chambadal (1971) wrote as follows:

> As far as the location of the molecule is concerned, that is determined after its first collision with the piston, since the latter experiences a very small displacement in one direction or the other. We may suppose that the work supplied by the molecule can be absorbed by two gears situated in the two parts of the cylinder. After the piston has experienced the first impact we connect it, according to the direction of its motion, to one or another of these gears which will thereafter absorb the work supplied by the movement of the molecule. This coupling of the piston to the parts which it drives can also be achieved automatically.
>
> But, in fact, it is not even necessary to solve the problem of the location of the molecule. Indeed we can, without altering the principle of the apparatus at all, visualize it in the following way. When the piston is placed in the cylinder, we fix two shafts on its axis, one on either side. These shafts make contact with the piston, but are not connected to it. Consequently, whatever the position of the molecule, the piston, moving in either direction, pushes one of the two shafts and so engages the gears which make use of the work produced.

Chambadal concluded that neither entropy nor information is involved in this model.

Popper (1974) and Feyerabend (1966) proposed similarly modified Szilard engines that couple the piston via pulleys to equal weights on either side of it. The weights can be lifted by the pulley system but are constrained such that they cannot be lowered (see Figure 10). If the engine's molecule is in the left chamber, the piston moves to the right, raising the left weight, leaving the right weight unmoved. If the molecule is in the right chamber, the reverse happens; i.e., the right weight gets raised and the left weight stays fixed. Feyerabend wrote 'The process can be repeated indefinitely ... We have here a "perpetual source of income" of the kind von Smoluchowski did not think to be possible.'

Jauch and Báron (1972, Article 3.5) imagined a similar situation (see Figure 11), writing: 'Near the mid-plane of the cylinder and on both its sides are electrical contacts in its walls. When activated by the piston's motion along them, they operate mechanisms which attach a weight to the piston in whichever direction it moves. Thus a weight is lifted and the engine performs work, without interference by a conscious observer.'

An ingenious coupling was illustrated by Rothstein (1979). His intent was not to argue against Szilard's work but rather to rebut Popper who had attempted to do so. Rothstein couples the piston to two racks that alternately engage a pinion gear as it moves left or right (see Figure 12). When it moves left, one

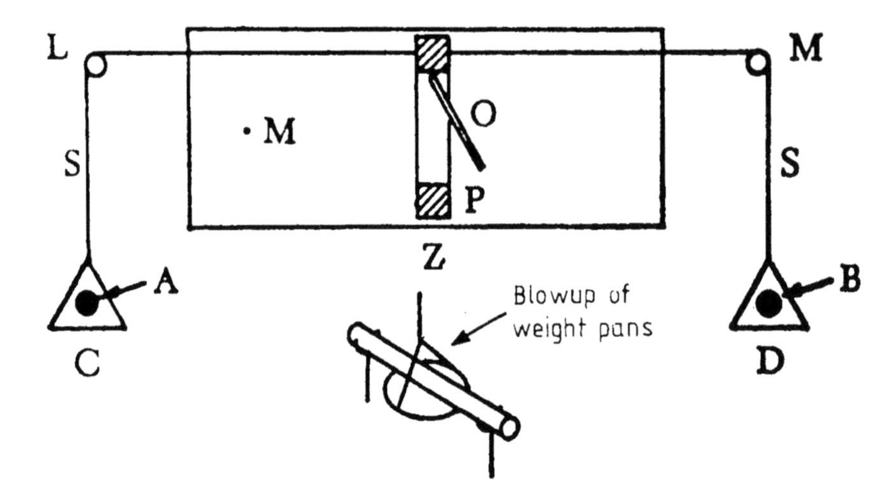


Figure 10.

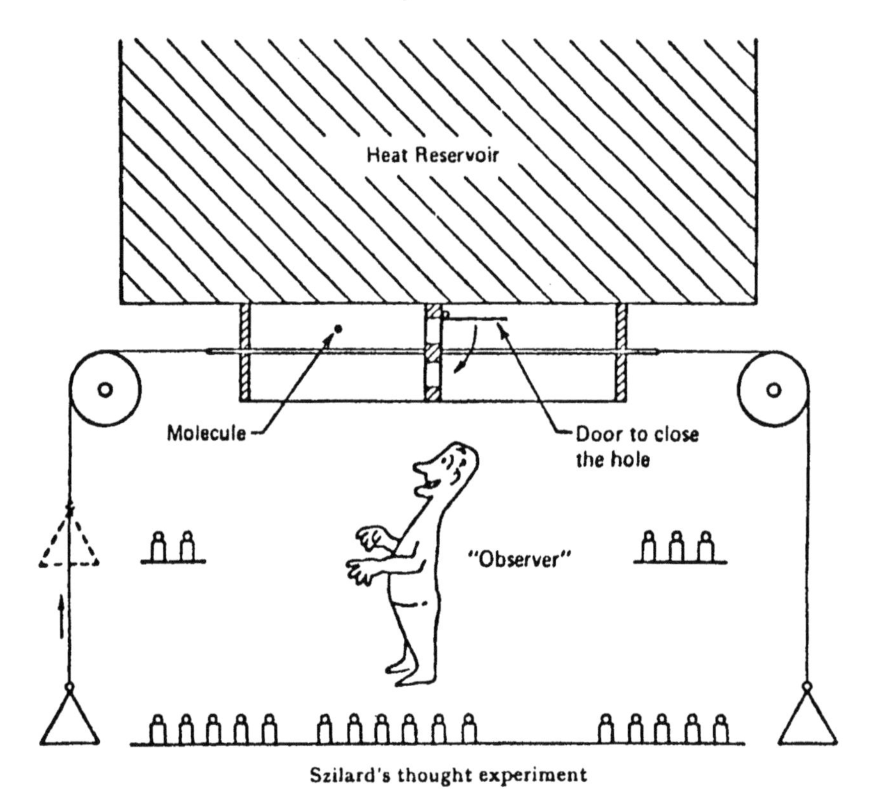


Figure 11.

rack rotates the pinion gear counterclockwise while the other rack is disengaged. When the piston moves right, the second rack (diametrically opposed to the first) rotates the pinion gear, again counterclockwise. Thus, regardless of whether the molecule is in the left or right chamber, the design of the racks assures

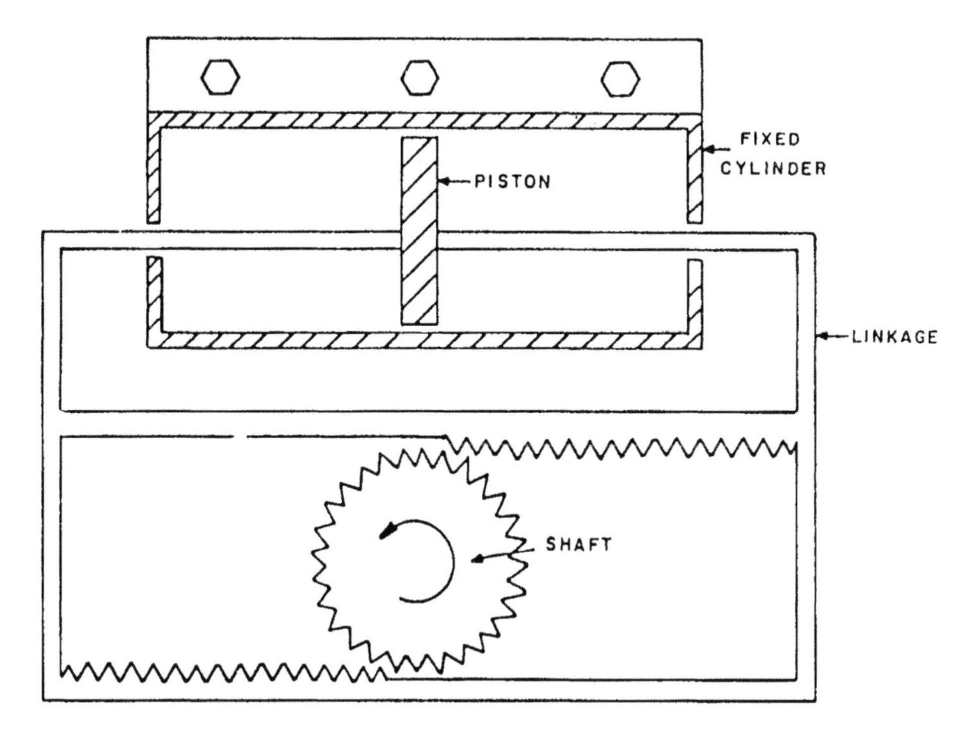


Thought experiment illustrating Popper's refutation of Szilard's assignment of an entropy equivalent to physical information.

**Figure 12.**

counterclockwise motion, suggesting an automatic machine for converting heat from a single reservoir to work.

The above examples are clever and do indeed *seem* to challenge the second law. However, there is more to the story. In Figure 10, after the work has been done, one of the weight hangers is raised. The pulley string at the other side is relaxed and limp. In essence, this configuration stores information about the molecule's previous location: it serves as a memory. Put differently, the process is not truly cyclic. The memory must be reset enabling commencement of the next cycle. Because resetting of the pulley system was not accounted for, the arguments by Popper and Feyerabend—while rather convincing at first look—must be considered incomplete.

Along the same lines, the Jauch-Báron idea leads to an asymmetric situation (Figure 11), with a weight attached on one side only. This is a physical embodiment of a memory that must be reset in order to make the process cyclic. Chambadal's analysis similarly overlooks the need for resetting the apparatus. In Rothstein's example, Figure 12, each cycle moves the rack assembly either left or right, where it stays until it is reset. Again this constitutes a memory register that must be zeroed periodically.

In summary, Szilard's engine requires a binary decision process in order to couple the piston and load. This requires information acquisition, memory, and subsequent information erasure. Although examples based upon macroscopic analogues involving gears and pulleys suggest that resetting can be done at arbitrarily little cost, that is misleading. Maxwell's demon actually entails a memory in which the relevant energy modes are *microscopic*. Erasure must act upon hidden degrees of freedom, without knowledge of their existing states. One cannot simply examine a register and zero it in the least energy-consuming or entropy-producing way. That examination would transfer information to another memory that would still have to be erased subsequently.

Our algorithm must be such that erasure occurs independently of the existing state of the memory. As the example above suggests, this can entail first randomizing the memory's hidden energy modes and then using a work process to bring the memory to the standard state. In a perceptive discussion of information and thermodynamics, Rothstein (1952a) observed: 'From an information viewpoint quantity of heat is thus energy transferred in a manner which has eluded mechanical description, about which information is lacking in terms of mechanical categories.' Given this observation, it is interesting that memory erasure-resetting via rearrangement of hidden degrees of freedom entails energy transfer (to the environment) via heat. In Section 1.8.2 we present a somewhat different view of memory erasure and Landauer's principle in connection with other criticisms of the LPB approach to the Maxwell's demon puzzle.

### 1.5.4 Entropy of Measurement Revisited

What about the entropy of measurement? As discussed earlier, Landauer showed that, in contrast with memory erasure, most computer operations could in principle be performed with arbitrarily little energy dissipation per bit. Bennett argued that a demon can do its measurements with arbitrarily little dissipation, in analogy with reading instructions in a computer. The act of 'reading' can in fact be viewed as a measurement process. Bennett proposed idealized magnetic and mechanical detection devices to buttress his argument that a Maxwell's demon can accomplish dissipationless measurement.

Were dissipationless means of detection not possible, we could simply append Bennett's erasure-based resolution to Brillouin's measurement-based resolution. But if detection can in fact be done with arbitrarily little dissipation, the Landauer–Penrose–Bennett (LPB) viewpoint implies that the exorcism accepted by a generation of researchers and teachers must now be rejected. In retrospect it is clear that assuming the use of light signals is not sufficient to rule out *all* demonic operations.

Remarkably, this lack of generality was not recognized by most researchers prior to Bennett's work (for an exception, see Penrose, 1970, p 236). Light signals became widely accepted as *the* way a Maxwell's demon collects information. In *Science and Information Theory*, after showing that detection via light signals saves the second law, Brillouin extrapolated his result: 'We have ... discovered a very important physical law ... every physical measurement requires a corresponding entropy increase, and there is a lower limit below which the measurement becomes impossible.'

Why generalization based upon a special case achieved such wide acceptance is puzzling. Landauer (1989b) wrote in this regard, 'Brillouin ... and others found dissipative ways of transferring information, and without further justification, assumed that they had discovered a minimally dissipative process. It is one of the great puzzles in the sociology of science why this obviously inadequate argument met with wide and uncritical acceptance. Only in recent years have clearer discussions emerged, and these are not yet widely appreciated.'

Had the use of light signals been questioned earlier, Brillouin's method of attack might have achieved far less credibility. Yet, despite its overly narrow view, Brillouin's work brought Maxwell's demon considerable attention in 1951 and subsequent years. The demon's popularity seemed to grow as its perceived challenge to the second law diminished. Bennett's 1982 overthrow of Brillouin's exorcism enhanced the popularity of Maxwell's demon, and more reason to retain it as a tool for understanding.

As mentioned already there is not universal agreement on the thesis that measurement can in principle be accomplished with arbitrarily little dissipation. Rothstein (1952) argued that 'the accuracy of any measurement is limited by how much entropy can be usefully expended in order to perform the measurement.' More recently (Rothstein,1988) he wrote:

> Despite several ingenious attempts to achieve reversible computation, including conceptual designs for quantum mechanical computers, we remain convinced that an entropy price for unavoidable selection, measurement, or preparation acts must be paid for every such act in physical communication or computation. ... We are willing to grant that for limited kinds of computation physical systems can be set up in principle whose dynamical equations will generate a succession of states isomorphic to the computation, and, as an idealization, such systems can be reversible. We deny that possibility for a true general purpose computer. Information must be generated and stored until it needs to be consulted. The storage is writing, i.e., a preparation of some subsystem. The consultation is reading, i.e., a measurement on some subsystem. Both kinds of operation are selective and thus demand their entropy costs.

To close this section we point out that the distinction between entropy of data acquisition and entropy of data erasure is not sharp. In Szilard's model, when a demon determines the side ($L$ or $R$) in which the molecule resides, its own memory state changes from a unique, known reference state to either of two possible states. This generates an entropy increase, in an ensemble sense, that in essence 'pays' entropically for the diminished entropy of the gas. This 'entropy of measurement' is stored by the demon, and ultimately becomes entropy of erasure—which is passed on to the environment when the demon's memory is reset. In this sense, the entropy of erasure feeds to the environment entropy gained by the demon via information acquisition.

## 1.6 Foundations of Landauer's Principle

### 1.6.1 Landauer's Pioneering Work and Penrose's Independent Development

In his seminal article, Landauer (1961, Article 4.1) introduced what is now commonly known as Landauer's principle, namely, that erasure of one bit of information increases the entropy of the environment by at least $k \ln 2$. Landauer views storage of a bit of information in terms of a particle in a bistable potential well whose potential barrier is high relative to $kT$, to assure that a ONE (ZERO) will remain a ONE (ZERO) over time with high probability. Erasure entails modification of the bistable potential so as to bring all ensemble members to the state ONE. This compresses the available phase space and lowers the entropy of the memory. From the Clausius inequality (assuming the second law of thermodynamics holds), for any isothermal entropy change $\Delta S$ of the memory, $\Delta S \geq Q/T$, where $Q$ is the heat *to* the memory. If $\Delta S < 0$, then $Q_{res} \equiv -Q \geq T|\Delta S|$. That is the heat ($Q_{res}$) *to* the reservoir is at least $T|\Delta S|$. In the case where the entropy of the demon's memory has been reduced by one bit, namely, $|\Delta S| = k \ln 2$, this gives Landauer's principle: $Q_{res} \geq k \ln 2$. Landauer's arguments in Article 4.1 are based upon these ideas, aided by simple models, under the assumption that the second law of thermodynamics is valid.

As mentioned already, independently of Landauer's work, and prior to Bennett's publications on Maxwell's demon, Oliver Penrose (1970) argued in his book, *Foundations of Statistical Mechanics*, that a demon must store information in a memory, and erasure of that memory sends entropy of at least $k \ln 2$ to the environment. This solution to the Maxwell's demon puzzle emanated from Penrose's examination of the possibility that a *perpetual motion machine* could be operated using statistical fluctuations such as Brownian motion. Penrose considers a composite system $R' = R + M$, where $R$ is a composite system, consisting of the system of interest, plus the constant-temperature reservoir, plus the load to be lifted. $M$ is a 'machine' that monitors statistical fluctuations and uses them to do work on the load. $R'$ is an isolated system—i.e., a mini-universe.

Penrose defines the entropy of $R'$ at time $t$ to be $\tilde{S}_t = \langle S_B \rangle_t + S_d(t)$, where $\langle S_B \rangle_t$ is the Boltzmann entropy, averaged over observational states at time $t$, and $S_d(t)$ is the statistical entropy of the distribution for observational states at time $t$. Using the underlying postulates of his unique development, he shows that $\tilde{S}_{t'} \geq \tilde{S}_t$ for $t' \geq t$, consistent with the second law. In essence the second law is implied by Penrose's postulates.

The main point is that an entropy reduction in $\langle S_B \rangle_t$ requires a compensating increase $\Delta S_d(t) \geq -\Delta\langle S_B \rangle_t$ in the entropy of the distribution. This increase in $S_d(t)$ is interpreted as the number of *observational* states the machine $M$ needs to reduce $\langle S_B \rangle_t$. Penrose calls $\Delta S_d(t)$ a 'latent contribution' to the entropy, and argues that completion of the cyclic process requires 'resetting' the machine $M$ to its original state. The *latent contribution* to the entropy then becomes manifest as entropy of the environment. It is clear that Penrose discovered Landauer's principle, albeit in a somewhat different context than that used by Landauer.

Typically, the number of such observational states required to reduce the entropy of a macroscopic system *measurably* is enormous. An example is the cooling of 0.001 kg of water by 1 K, which requires at least $exp(10^{21})$ observational states! Penrose applies the above logic to a traditional Maxwell's demon who operates a shutter, letting faster molecules go one way and slower ones the other. He relates the increased number of observational states to the demon's memory, which stores the observed information, and calls for its erasure. This erasure saves the second law of thermodynamics.

### 1.6.2 Recent Proofs of Landauer's Principle

In recent years, a very different proof of Landauer's principle was given by Shizume (1995, Article 4.3), with no explicit appeal to any form of the second law of thermodynamics. He considers a *classical* particle in a bistable potential well. The particle is subjected to three forces: a random thermal force, a damping force, and the force from the bistable potential well. Shizume uses the Langevin equation of motion and the concomitant Fokker–Planck equation to prove that erasure of one bit of information leads to energy $Q_{res} \geq kT \ln 2$ sent to the constant-temperature reservoir. This is Landauer's principle. The details are in Article 4.3.

Piechocinska (2000, Article 4.4) provides proofs of Landauer's principle in the domains of both classical and quantum mechanics. In the quantum mechanical proof, she makes the following assumptions: (1) the memory consists of a particle in a bistable potential well that is in contact with a constant-temperature reservoir. (2) The canonical ensemble formalism of equilibrium statistical mechanics holds. (3) The erasure process is governed by a unitary time evolution operator. (4) The reservoir decoheres in the sense that its off diagonal density matrix elements vanish. With these assumptions, Piechocinska exploits the convexity of the exponential function $exp(-x)$ to obtain Landauer's principle. The details are in Article 4.4.

These proofs of Landauer's principle suggest that its validity is *universal*. However, in Section 1.7.4 we find that this is not necessarily so under extreme quantum conditions.

### 1.6.3 Zero-Work Erasure Using a Reservoir

Lubkin (1987, Article 5.2) examines erasure by putting the system in contact with a reservoir at temperature $T$, and others (Vedral, 2000, Article 5.5; Plenio and Vitelli, 2001) have done similarly. We assume here that the reservoir temperature is *low* relative to room temperature. The result of this zero-work type of erasure is a bit different from that for Landauer's principle. Before erasure, the memory state is mixed, with density operator $\rho$. Each pure component has zero entropy. During erasure each component undergoes entropy change $\Delta S_{sys} = -k_B tr(\omega \ln \omega)$ where $\omega = exp(-H/k_B T)/Z$. The concomitant entropy change of the reservoir is $\Delta S_{res} = k_B[tr(\omega \ln \omega) - tr(\rho \ln \omega)] \geq k_B[tr(\omega \ln \omega) - tr(\rho \ln \rho)]$. The total entropy change for erasure is $\Delta S = -k_B tr(\rho \ln \omega) \geq -k_B tr(\rho \ln \rho)$. This result extends

Landauer's principle to zero-work erasure with a low-temperature reservoir, showing that erasure yields a *total* entropy increase at least as large as the memory's entropy before erasure.

## 1.7 Quantum Mechanics and Maxwell's Demon

### 1.7.1 Quantum Conundrums: Szilard and Einstein

Albert Einstein devised a Gedankenexperiment that bears resemblance to Szilard's 1929 model heat engine. Fine (1986) discusses the idea as outlined in a letter from Einstein to Schrödinger in 1935. A ball is located in one of two closed boxes, but it is not known which. We might expect that the probability is $1/2$ for either possibility. Einstein was concerned with two concepts, incompleteness and separation. By definition, a complete theory would have unit probability for the ball being in one of the boxes. Otherwise the theory is incomplete. Separation means that if two objects are separated spatially by a sufficient amount, they must become independent of one another. If knowledge about one of the boxes provides information regarding the other, then separation does not hold.

Suppose the ball cannot be destroyed or duplicated, and the two boxes are separated. If observation of box 1 gives a complete answer (YES or NO) as to whether it contains the ball, it is known instantly whether the distant box 2 contains the ball. Thus completeness implies that separation is impossible. On the other hand, if separation holds, then a measurement of one box cannot give a certain answer, for that would lead to an inconsistency. That is, separation implies that completeness is impossible. The conclusion is that separation and completeness are incompatible, a perplexing finding if one believes that both must be valid. The act of observation seems inconsistent with expected physical properties.

A familiar problem of the same ilk arises when a measurement of which box contains the ball is interpreted in terms of 'collapse of the wave function.' That is, if the particle is found in box 1, the wave function collapses to zero in box 2, and it appears that the system's state is intimately connected with the observer's knowledge. Yet we expect the state of the system to be independent of the observer's knowledge.

In the Szilard problem, a possible interpretation is that the entropy decreases by $k \ln 2$ upon determination of which chamber the molecule is in. Once again the state of the observer seems to determine the state of the system. With this approach, the measurement induces an increase in the demon's accessible phase space, and thus the entropy of its memory. This increase equals the entropy decrease of the gas (assuming a nondissipative measurement). Is this a subjectivist view that 'works' but should be avoided? Not necessarily, because *any* demon making a measurement of the molecule's location would reach the same conclusion (see Section 1.3). The gas entropy is 'known' to be less to any observer who measures the left-right state.

Another interpretation is suggested for an observer *outside* the gas–demon system. To this observer, the coupling of the gas to the demon produces a correlation between the two objects but no entropy change overall, consistent with a nondissipative measurement. This outsider cannot discuss the demon and gas independently because they are linked. The coupling between gas and demon is broken when the gas does work against the piston, expanding to fill the volume. Then to the outside observer, the gas has attained the same entropy it had initially. The demon's entropy has increased, compensating for the entropy decrease of the reservoir during the isothermal expansion.

Such conundrums are not easy to resolve, or even to accept. They seem to be an integral part of microscopic, probabilistic physics, and often lead to mental discomfort. They extend from quantum mechanics to thermal physics through Maxwell's playful, imaginary demon.

### 1.7.2 Quantum Mechanical Treatments of Szilard's Engine

As mentioned already, Zurek (1984, Article 5.1) examined a quantum mechanical version of the Szilard one-particle gas. He replaced the partition with a slowly generated, narrow potential barrier of height $V_0$, centered about the container's center. The probability density $|\psi(x)|^2$ is symmetric about the container's center, as expected. Zurek assumes high temperature ($kT \gg \epsilon$), where $\epsilon$ is the ground state energy; i.e., the particle exists in the *classical* domain.

The quantum mechanical features of Zurek's treatment lie in his representation of the particle's position ket vectors $|L_n\rangle$ and $|R_n\rangle$ when the particle is in quantum state $n$ and on the right or left sides, respectively. The corresponding demon ket vectors are $|D_L\rangle$ and $|D_R\rangle$. Zurek argues that the demon's density operator goes from $\rho_D = |D_0\rangle\langle D_0|$ initially (i.e., a pure standard state) to $\rho' = (|D_L\rangle\langle D_L| + |D_R\rangle\langle D_R|)/2$ after it measures, with a concomitant entropy increase of $k \ln 2$. If the measurement is reversible, then the total entropy change of the gas plus demon cannot change (the reservoir is not involved here), and Zurek argues that there is a compensating increase in *mutual* information. Interested readers are referred to article 5.1 for the details, and to the work of Lloyd (1989, Article 5.3) for a detailed treatment of the mutual information concept. It is fair to say that the quantum elements of Zurek's treatment in Article 5.1 are primarily useful as *interpretive* rather than *calculational* tools.

A very different quantum model has been published by Lloyd (Lloyd, 1997, Article 5.4). This is a magnetic resonance model of a quantum Maxwell's demon, which focuses on particles with spin-1/2 and magnetic moment $\mu$ in a constant magnetic field $B$ pointing in the negative-$z$ direction. The magnetic interaction energy is $(-\mu B, +\mu B)$ when the magnetic moment and field are (parallel, antiparallel). Experimental magnetic resonance techniques can induce spin flips from $|\uparrow\rangle$ to $|\downarrow\rangle$ or *vice versa* using a so-called $\pi$ pulse with Larmor precession angular frequency $\omega = 2\mu B/\hbar$. If a spin is flipped from an antiparallel to parallel configuration, its spin state goes from $|\uparrow\rangle$ to $|\downarrow\rangle$, and the interaction energy with the field decreases by $2\mu B$. A photon of energy $\hbar\omega$ is emitted, and the spin does positive work, adding energy to the electromagnetic radiation field.

Using established conditional spin-flipping techniques, one can effect measurement of a particle's spin using a *reference* spin, and can subsequently use a $\pi$ pulse to induce spin flips that result in work done on the electromagnetic radiation field. Measurement generates an 'entangled' state (see Section 1.7.3), and decoherence occurs when the reference spin is erased. Lloyd examines the complex details, using two electromagnetic-field reservoirs, in which case the demon's information gathering facilitates operation of a heat engine, and he proposes ways to achieve such a Maxwell's demon in a laboratory.

Evidently all quantum mechanical treatments of the Szilard one-particle engine to date have treated the constant-temperature reservoir in classical terms—as a nearly independent entity that interacts weakly with the system. In contrast, one might imagine a situation where the system and reservoir are inextricably linked in an 'entangled' quantum state. This is the topic of Section 1.7.4. But first we provide a brief introduction to quantum information theory.

### 1.7.3 Quantum Information, Maxwell's Demon, and Landauer's Principle

Landauer strongly and repeatedly made the point that 'information is physical'. That is, information is stored in material systems in accordance with the laws of physics. (Landauer, 1996a,b, Articles 7.4, 7.5) Information is also transferred from place to place by physical means—e.g., electromagnetic radiation. Information in *not* a mystical, non-physical entity.

At a fundamental level, the world we live in is quantum mechanical. Like quantum systems, quantum systems can store information—in this case, *quantum* information (Nielsen and Chuang, 2000). Quantum information is described using vectors in Hilbert space and must reduce to classical information in the classical limit. Thus the Hilbert space *contains* the classical information and, generally, also contains *much more* information than that.

The measure of classical information is the bit, an amount of information associated with binary choices between, say, 0 and 1. In contrast, the measure of quantum information is the *qubit* (quantum bit), the amount of information associated with the direction of a vector, whose tip is confined to a sphere of radius 1. Sometimes the qubit state is written $|\psi\rangle = \alpha|0\rangle + \beta|1\rangle$, where the complex numbers $\alpha$ and $\beta$ satisfy $|\alpha|^2 + |\beta|^2 = 1$. An example is an atomic electron that is in a superposition of its ground state $|0\rangle$ and its first excited state $|1\rangle$.

Unlike classical information, which can be copied perfectly, quantum information *cannot* generally be copied with perfect fidelity. Noise from the environment can cause continual bit flips within qubits. Furthermore the noise problem is exacerbated by the fact that *observing* a quantum state to see if it needs correction generally destroys that state! Nevertheless, error correcting codes that operate on *probabilistic* information can diminish such unwanted effects.

Interestingly, error correction can be viewed as a *friendly* Maxwell's demon that observes and then corrects errors. Suppose the system starts in state $i$ and is driven, via noise, to state $j$, with entropy $S_j > S_i$. The demon (i.e., error-correcting code) measures the state and finds state $m$ with probability $p_m$. This demon then operates on the system, which means it applies a unitary operator that brings the system to the final state $f$. The demon is successful only if $f = i$, in which case the system has undergone a cyclic process. The demon's memory contains information from the measurement process and must be erased to make the process cyclic for the system + demon. It can be shown (Nielsen, 2000) that, on average, the entropy sent to the environment by erasure is at least as great as the entropy reduction achieved by the demon's error-correcting operations. Thus the second law of thermodynamics is satisfied amidst the benevolent actions of the friendly demon. This is yet another example of the utility of Landauer's principle.

The domain of application of Landauer's principle is larger yet. Vedral (2000, Article 5.5) has argued that measurement can involve quantum entanglement (defined later in this section) between a system and the measuring apparatus. In such cases one might expect the degree of entanglement to be related to the amount of information that gets stored in the memory of the apparatus. Once again Landauer's principle links the latter information to the entropy sent to the environment upon erasure. Vedral argues that such erasure entropy limits the possible increase in entanglement. He goes on to link the second law of thermodynamics to the principle that 'entanglement cannot increase locally.'

In an article entitled, 'The physics of forgetting: Landauer's erasure principle and information theory,' Plenio and Vitelli (2001) argue that Landauer's principle is important throughout the new field called *the physics of information*. They argue that the amount of classical information that can be encoded in a *mixed* quantum state, described by the density operator, $\hat{\rho} = \sum_j q_j |e_j\rangle\langle e_j|$, is given by the von Neumann entropy $-tr[\hat{\rho}\log_2 \hat{\rho}]$. The argument is similar to that in Section 1.6.3. They then use Landauer's principle to obtain a form of the so-called Holevo bound of information theory (see Nielsen and Chuang, 2000). They also use Landauer's principle to argue that 'the efficiency of quantum data compression is limited by the von Neumann entropy ... just as classical data compression is limited by the Shannon entropy.'

In 1935, Erwin Schrödinger wrote, 'I would not call [entanglement] *one* but rather *the* characteristic trait of quantum mechanics, the one that enforces its entire departure from classical lines of thought.' Two particles are entangled quantum mechanically if their state vector cannot be written as a tensor product of the state vector for one times the state vector for the other. An example is the pure state $|\psi_{ent}\rangle = \frac{1}{\sqrt{2}}(|0\rangle_A|0\rangle_B + |1\rangle_A|1\rangle_B)$, where subscripts $A$ and $B$ connote particles $A$ and $B$.

Entangled states exhibit nonlocal correlations that cannot be understood classically. The density operator for $|\psi_{ent}\rangle$ is $\rho_{AB} = |\psi_{ent}\rangle\langle\psi_{ent}|$. If we 'trace out' particle $B$, to get the density operator $\rho_A$, we find the remarkable result, $\rho_A = \frac{1}{2}(|0\rangle_A\langle 0|_A + |1\rangle_A\langle 1|_A)$. The quantity in parentheses is a sum of two projection operators that span the $A$-space; i.e., it is the identity operator. Thus measurements of particle $A$ yield each of the two possible $A$-states half the time and provide the observer with *no* information about the state $|\psi_{ent}\rangle$. Indeed an entangled state is a weird thing.

If one could arrange conditions such that a system of interest becomes entangled with its environment, we might then not be surprised to find extraordinary, nonclassical behavior. This is precisely the case in theoretical work by Allahverdyan and Nieuwenhuizen, described in the next section.

### 1.7.4 Quantum Entanglement and the Second Law: Can the Demon Win?

Is the second law of thermodynamics sacrosanct? Recent theoretical work by Allahverdyan and Nieuwenhuizen (2000a, 2000b, 2001, 2002, 2002) suggests that the answer is *no*. They find that the second law of thermodynamics is violated for some models in the extreme quantum domain. The source of the violation is *quantum entanglement* between the system and the constant-temperature reservoir with which it interacts. As described in Section 1.7.3 when such entanglement exists, the system and reservoir cannot be usefully examined separately; the entangled systems act as a single entity.

For example, energy of (system + reservoir) $\neq$ energy of (system) + energy of (reservoir), as is usually assumed in thermodynamics. The last sentence holds also if we replace energy by entropy. Allahverdyan and Nieuwenhuizen show in particular that the Clausius inequality is violated and, of most interest here, Landauer's principle does not hold. Thus in contrast to the results of Section 1.6, which suggest the *universal* validity of Landauer's principle, the results of Allahverdyan and Nieuwenhuizen indicate that there are exceptions.

Allahverdyan and Nieuwenhuizen (2001, 2002) discuss a Brownian particle in a harmonic potential, in thermal interaction with a large reservoir. Their method of attack is via *microscopic* physics, namely quantum mechanics, with relevant thermodynamics *emerging* from the analysis rather than being assumed at the outset. Traditional statistical mechanics, for example in the form of the canonical ensemble, cannot be used because it is based on three assumptions that do not apply here: (i) the entropy for the system and bath is additive, (ii) the system plus reservoir achieve thermal equilibrium quickly, with $dQ_{res} = TdS_{res}$, and (iii) the interaction energy is sufficiently 'small.'

Allahverdyan and Nieuwenhuizen use the Wigner phase space distribution to show that at sufficiently low temperatures, the system can have $dQ > 0$ while $dS < 0$. This implies that, as already stated, Clausius's inequality and Landauer's principle are both violated. This violation is referred to by Allahverdyan and Nieuwenhuizen as being due to an 'entanglement demon.' A Maxwell's demon of this ilk does not gather information, but rather functions because of the idiosyncrasies of quantum mechanics. Other such 'process' or 'disembodied' demons are discussed further in Section 1.8.3.

## 1.8 Other Aspects of Maxwell's Demon and Computation

### 1.8.1 Algorithmic Entropy

Charles Bennett (1982, Article 7.1) observed that under certain circumstances algorithmic entropy, defined within the framework of algorithmic information theory, is a microscopic analogue of ordinary statistical entropy. Zurek (1989a, 1989b) extended the definition of algorithmic entropy to physical systems. He argued that physical entropy must consist of two distinct contributions: (i) a term $I(s)$ that represents the randomness of known aspects of the system in microstate $s$; and (ii) a term $H$, representing the remaining ignorance of the observer about the system's actual state. In his words, 'This recognition of the dual nature of physical entropy allows one to consider 'engines' operated by a modern day 'Maxwell demon'—a universal Turing machine capable of measuring and processing information—without endangering the validity of the second law.'

To a Maxwell's demon, after a measurement, a reliable memory is in a *specific* memory state and no ensemble is necessary to describe it. The quantity $I(s)$ in the previous paragraph is the so-called algorithmic entropy (also called algorithmic information), defined as the length, in bits, of the shortest computer program that runs on a *universal* computer and fully specifies the state $s$. For a memory, state $s$ is a string of ONES and ZEROES. Algorithmic entropy is maximal for a *random* string $s$, in which case $I(s) = \log($*length of the string in bits*$)$.

Despite the fact that algorithmic entropy is defined for a system in a *specific* microstate, it is useful nevertheless to consider an ensemble of $M$ systems. If state $s_i$ occurs with probability $p_i$ in the ensemble, the *average* algorithmic entropy is defined as $\langle I \rangle \equiv \sum p_i I(s_i)$, where $i = 1, 2, \ldots M$. The reason this is useful is the following important result from algorithmic information theory: For sufficiently large $M$, $\langle I \rangle \approx -\sum p_i \log_2 p_i = H$, the Shannon entropy in bits. These ideas, and those that follow, are elaborated in work by Caves (1993, Article 6.1) and Zurek (1999, Article 6.3). Mathematical details concerning the goodness of the approximation $\langle I \rangle \approx H$ are given by Schack (1997, Article 6.2).

The Shannon entropy $H$ is proportional to the Gibbs entropy $S_G$. Specifically, $S_G = k_B(\ln 2)H$ and thus, $S_G = -k_B \sum p_i \ln p_i = k_B(\ln 2)H \approx k_B(\ln 2)\langle I \rangle$; i.e., $\langle I \rangle \approx H = S_G/(k_B \ln 2)$. This result shows that the *average* algorithmic entropy $\langle I \rangle$ for an ensemble of similar systems can be well approximated using the Gibbs entropy obtained via statistical mechanics. This is fortuitous because there is no known way to systematically *calculate* $I$. If $I$ for a system is well approximated by $\langle I \rangle$ for the ensemble, then $I$ can be determined from a knowledge of $S_G$. In such cases, algorithmic entropy provides a definition of entropy for a microstate, just as an energy eigenvalue is defined for each microstate.

The above definition of physical entropy, in bits, written as $S_{phys} = H + I$, is useful to discuss Maxwell's demon. If the demon obtains information about a gas reversibly, then we might expect the 'physical entropy' $S_{phys}$ to be unchanged by the measurement. Specifically we expect that measurement increases $I$ (the information known about the system) and therefore must diminish $H$ (the missing information) *in the view of the demon*. This expectation is borne out by the following mathematical result: For an ensemble of similar gases and demons, *reversible* information gathering induces $\Delta S_{phys} = \Delta H + \langle \Delta I \rangle \approx 0$.

Thermodynamically, the isothermal work $W$ made possible by a Shannon entropy reduction $-\Delta H$ is limited to $W \leq -k_BT(\ln 2)\Delta H$. Given the average erasure cost of $W_{erase} \geq k_BT(\ln 2)\langle \Delta I \rangle$ from Landauer's principle, the *net* work possible is $W_{net} = W - W_{erase} \leq -k_BT(\ln 2)\Delta S_{phys} \approx 0$, or $W \leq W_{erase}$. Thus the erasure cost is at least as great as the work made possible by the demon's observations, and the demon's actions *cannot* turn heat from the reservoir into an equal quantity of work with no other effect. This is consistent with the second law of thermodynamics. Zurek (1999, Article 6.3) applies these ideas to a *demon of choice* that places the partition in the Szilard gas well off center, trying to obtain work only when unlikely states (with the particle in the smaller volume) are observed. Zurek shows that this strategy does not succeed, and that the demon cannot defeat the second law with such shenanigans.

### 1.8.2 Challenges to the LPB View

Although the Landauer–Penrose–Bennett (LPB) solution of the Maxwell's demon puzzle has been accepted widely, some researchers have argued against its validity. One such challenge has come from an in-depth critical analysis by Earman and Norton (1998, 1999). They allege a number of deficiencies in the LPB solution, writing:

> ...exorcisms of the demon provide largely illusory benefits ... they either return a presupposition that can be had without information theoretic consideration or they postulate a broader connection between information and entropy than can be sustained.

> In so far as the demon is a thermodynamic system already governed by the second law, no further supposition about information and entropy is needed to save the second law. In so far as the demon fails to be such a system, no supposition about the entropy cost of information acquisition and processing can save the second law from the demon.

A major point of criticism by Earman and Norton is that Landauer's principle emerges from the demand that the second law of thermodynamics must be satisfied; i.e., it is not *independent* of the second law. Thus, it appears that the second law is being used to save itself! They were evidently unaware of the proof of Landauer's principle by Shizume (1995, Article 4.3), described in Section 1.6.2. That proof does *not* assume the second law. Earman and Norton's work also preceded the proofs (classical and quantum mechanical) by Piechocinska (2000, Article 4.4), described in Section 1.6.2. Those proofs also did not entail an explicit assumption of the second law. Earman and Norton also seem to have been unaware of the work of Penrose (1970), whose postulational development of statistical mechanics leads to a proof of the second law in the form of the principle of entropy increase, and then to Landauer's principle. The existence of these proofs weakens Earman and Norton's criticisms of the LPB approach.

Another thought-provoking criticism by Earman and Norton concerns a modified Szilard engine, which requires information gathering by a Maxwell's demon, but which (they argue) does *not* require logically irreversible erasure. It is claimed that this model violates the second law of thermodynamics and is a counterexample to the LPB framework. Specifically, the model is defined as follows.

The standard memory state is taken to be left ($L$). The demon places the partition in the container's center and then determines the side where the particle resides. If the result is $L$, the memory remains in the standard state, and the demon fits the engine with pulleys that enable it to do isothermal work $W = kT \ln 2$ on an external load. Memory erasure is unnecessary because the memory state is already $L$, the standard memory state. If the particle is on the right ($R$), the demon fits the engine with pulleys appropriately, enabling it to do work $W$. In this case, Earman and Norton argue, the memory is *known* to be in state $R$ because the $R$-algorithm (subroutine) is being run. It is possible to design the $R$-algorithm to transform $R$ to $L$ with arbitrarily little entropy production and outside work.

Earman and Norton propose including such memory-specific algorithms within the $L$ and $R$ subroutines so that the memory ends up in state $L$ regardless of the particle's observed position—and with arbitrarily little entropy sent to the reservoir. Landauer's principle does not apply here, and they argue that work $W$ is done by the gas *without* sending entropy $k \ln 2$ to the environment. Their main point is that the net result is transfer of energy $Q$ from the reservoir and performance of work $W = Q$, with no other effect. If this argument is correct, then the second law is violated.

The Earman and Norton argument has been rebutted by Bub (2001) and Bennett (2002). In the operation of a computer, various types of information-bearing degrees of freedom exist. One type, discussed already, relates to data bits stored in memory. Another relates to *control variables*, for example the position of the magnetic head in a Turing machine. In this kind of computer a program is written in a sequence of squares on a linear magnetic tape (Bennett, 1988, Article 7.3), and the magnetic head advances sequentially through the program until it is directed to a subroutine. Then the head must find, and execute, that subroutine, which is located in some region of the tape. In this case, the position of the magnetic head deviates from its sequential path, because it must move to the location of the subroutine, and when the subroutine is completed, it must return to the main program.

If the phase volume of *any* information-bearing degree of freedom is compressed, there must be a corresponding increase in the phase volume of another part of the universe. Such compression occurs whenever two possible predecessor paths connect to one successor path, effecting a compression of phase space volume (Bennett, 2002). This is precisely what happens as the magnetic head of a Turing machine moves from the last step of a subroutine, back to the main program. The head could have come from *either* of the two subroutines and the step that returns it to the main program is logically irreversible. Effectively this step constitutes an erasure of information.

Applying Landauer's principle, this compression of the head position's phase space must be accompanied by an entropy increase 'elsewhere.' In the Szilard model, the only possible 'elsewhere' is the reservoir. When the entropy increase, $k \ln 2$, of the reservoir is taken into account, the *apparent* violation of the second law disappears. What seemed initially to be a second law violation is transformed into an instructive example that illustrates the importance of accounting for *all* information-bearing degrees of freedom. Bennett's main point is that if one does a full accounting along the lines described here, the second law of thermodynamics remains intact.

Earman and Norton leveled other thoughtful criticisms at the LPB approach to the Maxwell's demon puzzle. Interested readers are encouraged to read their articles (Earman and Norton, 1998, 1999).

Gyftopoulos leveled two different criticisms against the LPB solution of the Maxwell's demon puzzle. One criticism (Gyftopoulos, 2002a, 1993) is purely thermodynamic while the other (Gyftopoulos, 2002b) is attributed to quantum mechanics. In the first, Gyftopoulos appeals to his own development of thermodynamics (Gyftopoulos and Beretta, 1991). In particular, he examines an energy *versus* entropy diagram for a container of air, with stable equilibrium states lying on a convex curve with positive first and second derivatives. The slope of this curve is the absolute temperature.

The action of a Maxwell's demon would move the air's thermodynamic state from a point on the latter curve, leftward to a non-equilibrium state with the same energy, but lower entropy. Gyftopoulos assumes the demon does its sorting 'without any contribution on his behalf,' interpreting Maxwell's specification of the demon to mean that it must operate with neither energy expenditure nor anything else (including entropy). Disallowing any entropic effects for the demon itself, he concludes correctly that the process is impossible. Alternatively, he argues that the demon can be considered to undergo a cyclic process, and comes to the same conclusion because at fixed energy, the entropy cannot decrease. Anticipating claims that his argument is circular because he uses the laws of thermodynamics to show that they cannot be violated, Gyftopoulos enunciates six points of rebuttal and an example to ward off such claims.

The second criticism (Gyftopoulos, 2002b) is of a very different nature. He adopts a *unified* theory, in which quantum theoretic and thermodynamic laws do *not* apply to density operators for mixed states. With this choice, he is restricted to use *homogeneous* ensembles for which the ensemble average of an observable whose hermitian operator is $A$ is $\langle A\rangle = Tr[\rho A] = \sum a_j/N$ where $a_j$ is the measured result for ensemble member $j$, with $j = 1, 2 \ldots M$, and $M \to \infty$.

With these assumptions, it is shown that the average value of any momentum component of a molecule is identically zero. Gyftopoulos argues that there are in fact no swift and slow molecules for the demon to sort and that, in fact, each molecule is 'at a standstill.' Gyftopoulos again anticipates criticism by readers that the vanishing of the expectation value does not imply that *each* ensemble member will have zero momentum, and he provides a rebuttal in advance. To accept this argument, one must be willing to reject the existence of mixed states.

The critiques by Earman and Norton and by Gyftopoulos are well researched, articulately written, and well worth reading. Many thoughtful and thought-provoking issues are raised. A similar statement holds for pointed criticisms addressed by Shenker (1999b, 2000). Interested readers are encouraged to study the articles by these authors to develop an appreciation of the points of contention. Indeed the Maxwell's demon literature shows many foibles, and there are grounds for criticism and skepticism on various issues. Critical assessments can give rise to healthy debate and sometimes, deeper understanding of fundamental issues. To date the criticisms alluded to have not convinced the bulk of the scientific community that the LPB framework should be rejected.

In their book, *The Refrigerator and the Universe*, Goldstein and Goldstein (1993) argue that the distinction made between non-intelligent and intelligent demons is artificial. They cite the spring-loaded trapdoor as an example for which (they claim) measurement occurs when a particle pushes the door open; information storage is related to the energy given to the door; and erasure is effected by the subsequent sharing of this energy with many other particles. However, this so-called information storage is uncontrollable and ephemeral, and therefore fundamentally different from *reliable* information storage using a potential well (Leff, 1995b). Thus Goldstein and Goldstein's conclusion, 'It is ironic that von Smoluchowski fully resolved the paradox of the demon in 1912 and then undermined the force of his reasoning by suggesting that it might not apply to intelligent beings,' is difficult to accept by advocates of the LPB approach.

Biedenharn and Solem (1995) examine a quantum mechanical Szilard engine, namely, a one-dimensional variant of the model examined by Zurek (1984, Article 5.1). They assume observation is adiabatic but, in contrast with Jauch and Báron (1972, Article 3.5) and Popper (1957), they argue that measurement *is* necessary to extract work. The wave function collapses when the particle is observed in one side of the container, and all energy eigenvalues become 4 times their original value. Biedenharn and Solem write that the quantum state does not change (quantum mechanical adiabaticity) and thus the energy and temperature both increase by a factor of 4. They assume the subsequent gas expansion is also adiabatic, so that the complete cycle proceeds with zero entropy change. Distinguishing between *physical* and *information* entropy, they argue that the $k$ in the information entropy $k \ln 2$ cannot be Boltzmann's constant. They write that 'there is simply no way to compare the [gas and information] entropies.' Biedenharn and Solem do not mention, and evidently do not accept, Landauer's principle; therefore they do not account for the changed state of the demon's memory.

### 1.8.3 Process (Disembodied) Maxwell's Demons

Maxwell's original description of his *demon* as 'a being who can see the individual molecules' suggests to some that it is an entity that gathers information. Certainly the demons considered in connection with the Szilard one-particle engine have been of this type and, indeed, this book focuses on information-gathering demons. In contrast Allahverdyan and Nieuwenhuizen have argued that Maxwell intended his demon to be something other than a miniature physical system. They posit that Maxwell was motivated by theological considerations, and intended his *demon* to be a psyche lacking a body. Such an ethereal entity could operate without input of energy and would not require erasure of information because it has no physical memory. With this interpretation, Maxwell's *original* demon can be a *physical process* rather than a small physical system. The notion of this type of demon goes back at least to Whiting (1885).

A substantial class of such demons has surfaced in the literature. These *process* or *disembodied* demons threaten the second law of thermodynamics, not by gathering information and using it cleverly, but rather by taking advantage of specific conditions, including, but not limited to, special geometries, force fields, or the oddities of quantum physics—such as quantum entanglement.

In recent years, Sheehan and co-workers have proposed a variety of process demon challenges to the second law. These proposals are notable in that they are aimed at ultimate experimental testing. In each case, *naturally-occurring* steady state potential gradients (electrostatic, chemical, gravitational) induce matter flows from which macroscopic work can be generated. For each it appears possible that a cyclic process can convert heat to an equivalent amount of work, with no other effect, in violation of the second law of thermodynamics. In some cases, preliminary experiments have been performed. The experiments are difficult and (with the exception of the model in the next paragraph) require uncommon, sometimes extreme, physical conditions. Primarily because the conditions are so different from ambient conditions, entropy generation by the apparatus has exceeded the entropy decrease that is theoretically possible.

The most recent proposal (Sheehan, *et al* 2002) deals with a 'solid state Maxwell's demon.' The main apparatus is a solid-state electrostatic motor that utilizes the electric field energy in an open-gap p-n junction. The model has undergone detailed numerical studies using a commercial semiconductor device-simulator code, yielding results that agree with those for a soluble one-dimensional analytic model. An experimental test is thought to be feasible within the next 5 years. Other second law violation proposals by Sheehan's group deal with plasmas (Sheehan, 1995, 1996a,b, 2002a; Sheehan and Means, 1998; Čápek and Sheehan, 2002); chemical physics (Sheehan, 1998a, 2001, 2002a); and systems under the influence of gravity (Sheehan, Glick, and Means, 2000; Sheehan, *et al* 2002, Sheehan, 2002a).

Various second law violations, based upon theoretical models, have been reported. An example is a violation of Thomson's formulation of the second law, discovered by Allahverdyan, Balian, and Nieuwenhuizen (2002). They consider a thermodynamic system interacting with a macroscopic reservoir at temperature $T$, and able to exchange energy with a work source. They find that when the work source is *mesoscopic*, a violation of Thomson's principle is possible. They call this a 'mesoscopicity demon'.

Čápek and co-workers have investigated a number of abstract models of open quantum systems, defined by specific second quantized Hamiltonians. These systems show second law violations, in that potential gradients are self-generated or the principle of detailed balance is violated under isothermal conditions (Čápek, 1997a,b, 1998, 2001, 2002; Čápek and Bok, 1998, 1999, 2001, 2002; Čápek and Tributsch, 1999; Čápek and Mancal, 1999, 2002; Čápek and Frege, 2000, 2002).

Zhang and Zhang (1992) proposed a second law-defying mechanical model with velocity-dependent forces, whose classical phase space volume is *not* invariant under time evolution. Liboff (1997) found behavior inconsistent with the second law for a two dimensional classical mechanical model consisting of three disks that can move in any of three narrow (one-dimensional) channels. This weird geometry causes the three disks to collect in one of the channels, and the motion is *not* reversible.

Gordon (1981) considers a molecular machine, patterned after the Szilard engine, that generates a chemical potential gradient and seems to function as a perpetual motion machine. More recently, Gordon (2002) examines a molecular rotor model that generates a temperature gradient from Brownian motion in an equilibrated, isolated system, in violation of the second law. Goychuk and Hanggi (2000) investigate fluctuation-induced currents in a non-dissipative system with initially localized particles.

The examples above are illustrative and surely not exhaustive. Other examples exist in the bibliography and the general literature. Research on possible violations of the second law of thermodynamics has become a burgeoning field.

We close this section with mention of a related line of theoretical research. Scully and coworkers (2001, 2002a, 2002) have studied a quantum heat engine whose working fluid is a single-mode radiation field, and work is extracted from a single reservoir. However, in this case there is no violation of the second law because entropy is generated as the heat engine operates.

### 1.8.4 Maxwell's Demon, Efficiency, Power, and Time

Although the bulk of the work on Maxwell's demon has centered on its exorcism, one can ask how effective a demon can be whether or not it can defeat the second law. The following questions have been posed (Leff, 1987a, 1990) in this regard: What rate of energy transfer is attainable by a Maxwell's demon who sorts gas molecules serially, and how much time does it take it to achieve a designated temperature difference, $T$, across a partition? The assumption of serial processing enables an estimate of minimal effectiveness. By the use of two or more demons operating in parallel, improved performance is possible.

Numerical estimates have been made using the energy-time form of Heisenberg's uncertainty principle and also using classical kinetic theory. For a dilute gas at 300 K, the uncertainty principle implies that power $< 1.5 \times 10^{-6}$ W. If the gas volume is the size of a large room, and $\Delta T = 2$ K, then the demon's processing time $> 10^3$ years. With similar assumptions classical kinetic theory implies much tighter bounds, namely, power $< 10^{-9}$ watt and processing time $> 4 \times 10^6$ years. The latter power level, which is comparable to the average dissipation per neuron in a human brain, illustrates the impotence of a lone Maxwell's demon using serial processing.

Once a temperature difference exists between two portions of a gas, it is possible in principle to run a heat engine using this difference. The available energy and efficiency for delivery of this energy as work has been determined (Leff, 1987a). The maximum efficiency for operating a reversible heat engine between two identical chambers at initial temperatures $T_+$ and $T_- < T_+$, and equal final temperatures, has the simple form $\eta = 1 - (T_-/T_+)^{1/2}$. As expected, this maximum efficiency is lower than that from a Carnot cycle operating between infinite reservoirs at fixed temperatures $T_-$ and $T_+$. It is noteworthy that the same efficiency expression arises in other contexts, including the reversible Otto and Joule cycles (Leff, 1987b), other reversible cycles (Landsberg and Leff, 1989), and the irreversible Curzon–Ahlborn cycle (Curzon and Ahlborn, 1975) at maximum power.

### 1.8.5 Limits of Computation

Landauer's principle places a lower bound on the energy dissipated to the environment for each information bit that is erased. Real computers dissipate much more energy per erased bit. Gershenfeld (1996, Article 7.7) points out that laptop computers consume about 10 W, desktop computers consume $\sim$100 W, and supercomputers consume as much as $\sim$100 kW of electric power intermittently, pushing the limits of energy transfer via heat to avoid excessive temperatures.

As computer speeds increase and chip sizes diminish, it is natural to ask: What are the fundamental *thermodynamic* limits to computation? Physicists are typically interested in the limits of single logic gates, while engineers tend to focus on incremental design improvements. Gershenfeld begins a study of fundamental *thermodynamic* limits, sketching a theory of computation that can handle both *information-bearing* and *thermal* degrees of freedom. In order to approach the physical limits of computation, Gershenfeld observes that a computer should not erase its internal states, its processing speed should be no faster than required, and reliability should be no greater than necessary. He provides some interesting back of the envelope calculations to help make his case.

Lloyd (2000, Article 7.8) investigates *ultimate* physical limits on computation in terms of the speed of light $c$, Planck's (reduced) constant $\hbar \equiv h/(2\pi)$, the universal gravitational constant $G$, and Boltzmann's constant $k$. Lloyd observes that if Moore's law is extrapolated into the future, 'then it will take only 250 years to make up the 40 orders of magnitude in performance between current computers that perform $10^{10}$ operations per second on $10^{10}$ bits and our 1-kg ultimate laptop that performs $10^{51}$ operations per second on $10^{31}$ bits.'

### 1.8.6 Physics Outlaw or Physics Teacher?

As we have seen, Maxwell's demon was invented to illustrate the statistical nature of the second law of thermodynamics. It ultimately became viewed as a potential physics outlaw that had to be defeated. Now, more than 130 years later, the widely accepted Landauer–Penrose–Bennett solution has shown that information gathering demons do not threaten the second law, except in the extreme quantum domain, where there are indications that standard thermodynamics—including the second law—seem to break down. On the other hand, process (i.e., disembodied) Maxwell's demons are alive and well, and continue to pose potential threats to the second law. It is impossible to know how this will play out, but it will surely be intriguing to follow.

In the closing paragraph of his Scientific American review article, Ehrenberg (1967) captured the spirit of what has kept Maxwell's demon alive: 'Let us stop here and be grateful to the good old Maxwellian demon, even if he does not assist in providing power for a submarine. Perhaps he did something much more useful in helping us to understand the ways of nature and our ways of looking at it.'

Regardless of whether Maxwell's demon is a physics outlaw, it has surely been a potent physics teacher! Though merely a simple idea, it has been a vehicle for relating measurement and information to thermodynamics, quantum mechanics, and biology. Modern electronic computing seems at first thought to be totally unrelated to Maxwell's demon. Yet important connections exist, with the demon illuminating the binary decision process and the computer amplifying the importance of information erasure. Remarkably, Maxwell's microscopic demon has even played a role in the development of black hole thermodynamics (Bekenstein, 1972, 1980).

Will Maxwell's demon play a role in future progress? Considering its rich history and present research trends, a continuing active life is likely. Whether or not it is proven 'guilty' of being an outlaw, we expect Maxwell's demon to remain a potent teacher for many years to come!

# Chapter 2

# Early History and Philosophical Considerations

## KINETIC THEORY OF THE DISSIPATION OF ENERGY

IN abstract dynamics an instantaneous reversal of the motion of every moving particle of a system causes the system to move backwards, each particle of it along its old path, and at the same speed as before when again in the same position—that is to say, in mathematical language, any solution remains a solution when $t$ is changed into $-t$. In physical dynamics, this simple and perfect reversibility fails on account of forces depending on friction of solids; imperfect fluidity of fluids; imperfect elasticity of solids; inequalities of temperature and consequent conduction of heat produced by stresses in solids and fluids; imperfect magnetic retentiveness; residual electric polarisation of dielectrics; generation of heat by electric currents induced by motion; diffusion of fluids, solution of solids in fluids, and other chemical changes; and absorption of radiant heat and light. Consideration of these agencies in connection with the all-pervading law of the conservation of energy proved for them by Joule, led me twenty-three years ago to the theory of the dissipation of energy, which I communicated first to the Royal Society of Edinburgh in 1852, in a paper entitled "On a Universal Tendency in Nature to the Dissipation of Mechanical Energy.

The essence of Joule's discovery is the subjection of physical phenomena to dynamical law. If, then, the motion of every particle of matter in the universe were precisely reversed at any instant, the course of nature would be simply reversed for ever after. The bursting bubble of foam at the foot of a waterfall would reunite and descend into the water: the thermal motions would reconcentrate their energy and throw the mass up the fall in drops reforming into a close column of ascending water. Heat which had been generated by the friction of solids and dissipated by conduction, and radiation with absorption, would come again to the place of contact and throw the moving body back against the force to which it had previously yielded. Boulders would recover from the mud the materials required to rebuild them into their previous jagged forms, and would become reunited to the mountain peak from which they had formerly broken away. And if also the materialistic hypothesis of life were true, living creatures would grow backwards, with conscious knowledge of the future, but no memory of the past, and would become again unborn. But the real phenomena of life infinitely transcend human science, and speculation regarding consequences of their imagined reversal is utterly unprofitable. Far otherwise, however, is it in respect to the reversal of the motions of matter uninfluenced by life, a very elementary consideration of which leads to the full explanation of the theory of dissipation of energy.

To take one of the simplest cases of the dissipation of energy, the conduction of heat through a solid—consider a bar of metal warmer at one end than the other and left to itself. To avoid all needless complication, of taking loss or gain of heat into account, imagine the bar to be varnished with a substance impermeable to heat. For the sake of definiteness, imagine the bar to be first given with one half of it at one uniform temperature, and the other half of it at another uniform temperature. Instantly a diffusing of heat commences, and the distribution of temperature becomes continuously less and less unequal, tending to perfect uniformity, but never in any finite time attaining perfectly to this ultimate condition. This process of diffusion could be perfectly prevented by an army of Maxwell's "intelligent demons"* stationed at the surface, or interface as we may call it with Prof. James Thomson, separating the hot from the cold part of the bar. To see precisely how this is to be done, consider rather a gas than a solid, because we have much knowledge regarding the molecular motions of a gas, and little or no knowledge of the molecular motions of a solid. Take a jar with the lower half occupied by cold air or gas, and the upper half occupied with air or gas of the same kind, but at a higher temperature, and let the mouth of the jar be closed by an air-tight lid. If the containing vessel were perfectly impermeable to heat, the diffusion of heat would follow the same law in the gas as in the solid, though in the gas the diffusion of heat takes place chiefly by the diffusion of molecules, each taking its energy with it, and only to a small proportion of its whole amount by the interchange of energy between molecule and molecule; whereas in the solid there is little or no diffusion of substance, and the diffusion of heat takes place entirely, or almost entirely, through the communication of energy from one molecule to another. Fourier's exquisite mathematical analysis expresses perfectly the statistics of the process of diffusion in each case, whether it be "conduction of heat," as Fourier and his followers have called it, or the diffusion of substance in fluid masses (gaseous or liquid) which Fick showed to be subject to Fourier's formulæ. Now, suppose the weapon of the ideal army to be a club, or, as it were, a molecular cricket-bat; and suppose for convenience the mass of each demon with his weapon to be several times greater than that of a molecule. Every time he strikes a molecule he is to send it away with the same energy as it had immediately before. Each demon is to keep as nearly as possible to a certain station, making only such excursions from it as the execution of his orders requires. He is to experience no forces except such as result from collisions with molecules, and mutual forces between parts of his own mass, including his weapon: thus his voluntary movements cannot influence the position of his centre of gravity, otherwise than by producing collision with molecules.

The whole interface between hot and cold is to be divided into small areas, each allotted to a single demon. The duty of each demon is to guard his allotment, turning molecules back or allowing them to pass through from either side, according to certain definite orders. First, let the orders be to allow no molecules to pass from either side. The effect will be the same as if the interface were stopped by a barrier impermeable to matter and to heat. The pressure of the gas being, by hypothesis, equal in the hot and cold parts, the resultant momentum taken by each demon from any considerable number of molecules will be zero; and therefore he may so time his strokes that he shall never move to any considerable distance from his station. Now, instead of stopping and turning all the molecules from crossing his allotted area, let each demon permit a hundred molecules chosen arbitrarily to cross it from the hot side; and the same number of molecules, chosen so as to have the same entire amount of energy and the same resultant momentum, to cross the other way from the cold side. Let this be done over and over again within certain small equal consecutive intervals of time, with care that if the specified balance of energy and momentum is not exactly fulfilled in respect to each successive hundred molecules crossing each way, the error will be carried forward, and as nearly as may be corrected, in respect to the next hundred. Thus, a certain perfectly regular diffusion of the gas both ways across the interface goes on, while the original different temperatures on the two sides of the interface are maintained without change.

Suppose, now, that in the original condition the temperature and pressure of the gas are each equal throughout the vessel, and let it be required to disequalise the temperature but to leave the pressure the same in any two portions $A$ and $B$ of the whole space. Station the army on the interface as previously described. Let the orders now be that each demon is to stop all molecules from crossing his area in either direction except 100 coming from $A$, arbitrarily chosen to be let pass into $B$, and a greater number, having among them less energy but equal momentum, to cross from $B$ to $A$. Let this be repeated over and over again. The temperature in $A$ will be continually diminished and the number of molecules in it continually increased, until there are not in $B$ enough of molecules with small enough velocities to fulfil the condition with reference to permission to pass from $B$ to $A$. If after that no molecule be allowed to pass the

* The definition of a "demon," according to the use of this word by Maxwell, is an intelligent being endowed with free will, and fine enough tactile and perceptive organisation to give him the faculty of observing and influencing individual molecules of matter

interface in either direction, the final condition will be very great condensation and very low temperature in $A$; rarefaction and very high temperature in $B$; and equal temperature in $A$ and $B$. The process of disequalisation of temperature and density might be stopped at any time by changing the orders to those previously specified (2), and so permitting a certain degree of diffusion each way across the interface while maintaining a certain uniform difference of temperatures with equality of pressure on the two sides.

If no selective influence, such as that of the ideal "demon," guides individual molecules, the average result of their free motions and collisions must be to equalise the distribution of energy among them in the gross; and after a sufficiently long time from the supposed initial arrangement the difference of energy in any two equal volumes, each containing a very great number of molecules, must bear a very small proportion to the whole amount in either; or, more strictly speaking, the probability of the difference of energy exceeding any stated finite proportion of the whole energy in either is very small. Suppose now the temperature to have become thus very approximately equalised at a certain time from the beginning, and let the motion of every particle become instantaneously reversed. Each molecule will retrace its former path, and at the end of a second interval of time, equal to the former, every molecule will be in the same position, and moving with the same velocity, as at the beginning; so that the given initial unequal distribution of temperature will again be found, with only the difference that each particle is moving in the direction reverse to that of its initial motion. This difference will not prevent an instantaneous subsequent commencement of equalisation, which, with entirely different paths for the individual molecules, will go on in the average according to the same law as that which took place immediately after the system was first left to itself.

By merely looking on crowds of molecules, and reckoning their energy in the gross, we could not discover that in the very special case we have just considered the progress was towards a succession of states in which the distribution of energy deviates more and more from uniformity up to a certain time. The number of molecules being finite, it is clear that small finite deviations from absolute precision in the reversal we have supposed would not obviate the resulting disequalisation of the distribution of energy. But the greater the number of molecules, the shorter will be the time during which the disequalising will continue; and it is only when we regard the number of molecules as practically infinite that we can regard spontaneous disequalisation as practically impossible. And, in point of fact, if any finite number of perfectly elastic molecules, however great, be given in motion in the interior of a perfectly rigid vessel, and be left for a sufficiently long time undisturbed except by mutual impacts and collisions against the sides of the containing vessel, it must happen over and over again that (for example) something more than nine-tenths of the whole energy shall be in one half of the vessel, and less than one-tenth of the whole energy in the other half. But if the number of molecules be very great, this will happen enormously less frequently than that something more than 6-10ths shall be in one half, and something less than 4-10ths in the other. Taking as unit of time the average interval of free motion between consecutive collisions, it is easily seen that the probability of there being something more than any stated percentage of excess above the half of the energy in one half of the vessel during the unit of time, from a stated instant, is smaller the greater the dimensions of the vessel and the greater the stated percentage. It is a strange but nevertheless a true conception of the old well-known law of the conduction of heat to say that it is very improbable that in the course of 1,000 years one half the bar of iron shall of itself become warmer by a degree than the other half; and that the probability of this happening before 1,000,000 years pass is 1,000 times as great as that it will happen in the course of 1,000 years, and that it certainly will happen in the course of some very long time. But let it be remembered that we have supposed the bar to be covered with an impermeable varnish. Do away with this impossible ideal, and believe the number of molecules in the universe to be infinite; then we may say one half of the bar will never become warmer than the other, except by the agency of external sources of heat or cold. This one instance suffices to explain the philosophy of the foundation on which the theory of the dissipation of energy rests.

Take however another case in which the probability may be readily calculated. Let a hermetically-sealed glass jar of air contain 2,000,000,000,000 molecules of oxygen, and 8,000,000,000,000 molecules of nitrogen. If examined any time in the infinitely distant future, what is the number of chances against one that all the molecules of oxygen and none of nitrogen shall be found in one stated part of the vessel equal in volume to 1-5th of the whole? The number expressing the answer in the Arabic notation has about 2,173,220,000,000 of places of whole numbers. On the other hand the chance against there being exactly 2-10ths of the whole number of particles of nitrogen, and at the same time exactly 2-10ths of the whole number of particles of oxygen in the first specified part of the vessel is only $4021 \times 10^9$ to 1.

---

[*Appendix.—Calculation of Probability respecting Diffusion of Gases.*]

For simplicity I suppose the sphere of action of each molecule to be infinitely small in comparison with its average distance from its nearest neighbour: thus, the sum of the volumes of the spheres of action of all the molecules will be infinitely small in proportion to the whole volume of the containing vessel. For brevity, space external to the sphere of action of every molecule will be called free space: and a molecule will be said to be in free space at any time when its sphere of action is wholly in free space; that is to say, when its sphere of action does not overlap the sphere of action of any other molecule. Let $A$, $B$ denote any two particular portions of the whole containing vessel, and let $a$, $b$ be the volumes of those portions. The chance that at any instant one individual molecule of whichever gas shall be in $A$ is $\frac{a}{a+b}$, however many or few other molecules there may be in $A$ at the same time; because its chances of being in any specified portions of free space are proportional to their volumes; and, according to our supposition, even if all the other molecules were in $A$, the volume of free space in it would not be sensibly diminished by their presence. The chance that of $n$ molecules in the whole space there shall be $i$ stated individuals in $A$, and that the other $n-i$ molecules shall be at the same time in $B$, is

$$\left(\frac{a}{a+b}\right)^i \left(\frac{b}{a+b}\right)^{n-i}, \text{ or } \frac{a^i b^{n-i}}{(a+b)^n}$$

Hence the probability of the number of molecules in $A$ being exactly $i$, and in $B$ exactly $n-i$, irrespectively of individuals, is a fraction having for denominator $(a+b)^n$, and for numerator the term involving $a^i b^{n-i}$ in the expansion of this binomial; that is to say it is—

$$\frac{n(n-1) \; . \; . \; . \; . \; (n-i+1)}{1.2 \; . \; . \; . \; . \; i} \left(\frac{a}{a+b}\right)^i \left(\frac{b}{a+b}\right)^{n-i}$$

If we call this $T_i$ we have

$$T_{i+1} = \frac{n-i}{i+1}\frac{a}{b}\, T_{i+1}$$

Hence $T_i$ is the greatest term if $i$ is the smallest integer which makes

$$\frac{n-i}{i+1} < \frac{b}{a}$$

this is to say, if $i$ is the smallest integer which exceeds

$$n\frac{a}{a+b} - \frac{b}{a+b}$$

Hence if $a$ and $b$ are commensurable the greatest term is that for which

$$i = n\frac{a}{a+b}$$

To apply these results to the cases considered in the preceding article, put in the first place

$$n = 2 \times 10^{12}$$

this being the number of particles of oxygen; and let $i = n$. Thus, for the probability that all the particles of oxygen shall be in $A$, we find

$$\left(\frac{a}{a+b}\right)^{8 \times 10^{12}}$$

Similarly, for the probability that all the particles of nitrogen are in the space $B$, we find

$$\left(\frac{b}{a+b}\right)^{2 \times 10^{12}}$$

Hence the probability that all the oxygen is in $A$ and all the nitrogen in $B$ is

$$\left(\frac{a}{a+b}\right)^{2 \times 10^{12}} \times \left(\frac{b}{a \times b}\right)^{8 \times 10^{12}}$$

Now by hypothesis

$$\frac{a}{a+b} = \frac{2}{10}$$

and therefore

$$\frac{b}{a+b} = \frac{8}{10}$$

hence the required probability is

$$\frac{2^{26 \times 10^{12}}}{10^{10^{13}}}$$

Call this $\frac{1}{N}$, and let log denote common logarithm. We have $\log N = 10^{13} - 26 \times 10^{12} \times \log. 2 = (10 - 26 \log. 2) \times 10^{12} = 2173220 \times 10^{6}$. This is equivalent to the result stated in the text above. The logarithm of so great a number, unless given to more than thirteen significant places, cannot indicate more than the number of places of whole numbers in the answer to the proposed question, expressed according to the Arabic notation.

The calculation of $T_i$ when $i$ and $n - i$ are very large numbers is practicable by Stirling's Theorem, according to which we have approximately

$$1.2 \ldots . i = i^{i + \frac{1}{2}} \epsilon^{-i} \sqrt{2\pi}$$

and therefore

$$\frac{n(n-1) \ldots . (n-i+1)}{1.2 \quad \ldots . \quad i} = \frac{n^{n+\frac{1}{2}}}{\sqrt{2\pi}\, i^{(i+\frac{1}{2})}(n-i)^{n}}$$

Hence for the case

$$i = n \frac{a}{a+b}$$

which, according to the preceding formulæ, gives $T_i$ its greatest value, we have

$$T_i = \frac{1}{\sqrt{2\pi n e f}}$$

where

$$e = \frac{a}{a+b} \text{ and } f = \frac{b}{a+b}$$

Thus, for example, let $n = 2 \times 10^{12}$;

$$e = {\cdot}2,\ f = {\cdot}8$$

we have

$$T_i = \frac{1}{800000\sqrt{\pi}} = \frac{1}{1418000}$$

This expresses the chance of there being $4 \times 10^{11}$ molecules of oxygen in $A$, and $16 \times 10^{11}$ in $B$. Just half this fraction expresses the probability that the molecules of nitrogen are distributed in exactly the same proportion between $A$ and $B$, because the number of molecules of nitrogen is four times greater than of oxygen.

If $n$ denote the molecules of one gas, and $n'$ that of the molecules of another, the probability that each shall be distributed between $A$ and $B$ in the exact proportion of the volume is

$$\frac{1}{2\pi e f \sqrt{n\, n'}}$$

The value for the supposed case of oxygen and nitrogen is

$$\frac{1}{2\pi \times {\cdot}16 \times 4 \times 10^{12}} = \frac{1}{4021 \times 10^{9}}$$

which is the result stated at the conclusion of the text above.

WILLIAM THOMSON

*EDWARD E. DAUB*

# MAXWELL'S DEMON

IN HIS presentation of the 'two cultures' issue, C. P. Snow relates that he occasionally became so provoked at literary colleagues who scorned the restricted reading habits of scientists that he would challenge them to explain the second law of thermodynamics. The response was invariably a cold negative silence.[1] The test was too hard. Even a scientist would be hard-pressed to explain Carnot engines and refrigerators, reversibility and irreversibility, energy dissipation and entropy increase, Gibbs free energy and the Gibbs rule of phase, all in the span of a cocktail party conversation. How much more difficult, then, for a non-scientist. Even Henry Adams, who sought to find an analogy for his theory of history in the second law of thermodynamics, had great difficulty in understanding the rule of phase.

When Adams sought help with his manuscript 'The Rule of Phase Applied to History', he, too, encountered a cold silence. After months of search he complained to his brother Brooks that he had yet to discover a physicist 'who can be trusted to tell me whether my technical terms are all wrong'.[2] James F. Jameson, editor of the *American Historical Review*, responding to Henry's plea to find him 'a critic . . . a scientific, physico-chemical proof-reader', also met several rebuffs before he found the right man, Professor Henry A. Bumstead of Yale, a former student of Gibbs.[3] Bumstead's twenty-seven pages of detailed commentary must have satisfied Adams's hunger for 'annihilation by a competent hand',[4] as his revised version appeared only posthumously.[5] In it the chastened historian wrote that 'Willard Gibbs helped to change the face of science, but his Phase was not the Phase of History'.[6] Attracted to Gibbs's terminology because of the purely verbal agreement between physical phases and the epochs of Comtean history,[7] Adams erroneously adopted the phase rule as a scientific analogy for the progressive mutations of history.[8] If Maxwell had read Adams's misinterpretation of Gibbs's thought, he might have repeated his quip that the value of metaphysics is inversely proportional to the author's 'confidence in reasoning from the names of things',[9] but he would doubtless have been amused at the antics Adams attributed to his demon in history.

Adams once wrote to his brother Brooks that 'an atom is a man' and that 'Clerk Maxwell's demon who runs the second law of Thermodynamics ought to be made President'.[10] On another occasion he found Maxwell's demon a useful illustration for the behaviour of the German nation. 'Do you know the kinetic theory of gases?' he asked a British friend. 'Of course you do, since Clerk Maxwell was an Oxford man, I suppose. Anyway, Germany is and always has been a remarkably apt illustration of Maxwell's conception of "sorting demons". By bumping against all its neighbours, and being bumped in turn, it gets and gives at last a common motion.'[11] But such an aggressive mobile demon as the German nation was very different from the one Maxwell had conceived, a being who did not jostle atoms but arranged to separate them, not for the sake of generating some common motion but rather to illustrate Maxwell's contention that the second law of thermodynamics was statistical in character.

**Maxwell's Tiny Intelligence and the Statistical Second Law**

The fundamental basis for the second law of thermodynamics was Clausius's axiom that it is impossible for heat to pass from a colder to a warmer body unless some other change accompanies the process. To show that this law was only statistically true, Maxwell proposed a thought experiment in which a gas at uniform temperature and pressure was separated by a partition, equipped with a frictionless sliding door and operated by a tiny intelligence who could follow the movements of individual molecules. Although the temperature of the gas was uniform, the velocities of the gas molecules need not be, since temperature is the average kinetic energy of the molecules. The velocities should in fact vary, because the molecules would inevitably be exchanging energy in collisions. The demon might therefore circumvent the axiom regarding the behaviour of heat merely by separating the faster molecules from the slower. By permitting only fast molecules to enter one half and only slow molecules to leave it, Maxwell's tiny intelligence could create a temperature difference and a flow of heat from lower to higher temperatures. Maxwell concluded that the second law 'is undoubtedly true as long as we can deal with bodies only in mass, and have no power of perceiving or handling the separate molecules of which they are made up'. In the absence of such knowledge, we are limited to the statistical behaviour of molecules.[12]

Maxwell did not reach this insight immediately upon conceiving the idea of a tiny intelligence regulating the motions of molecules. His thought progressed (as did his characterizations) in the letters to Tait, Thomson and Rayleigh, where he first discussed this quaint creature. Upon introducing him to Tait in 1867 as a 'very observant and neat-fingered being', Maxwell prefaced his description by suggesting to Tait, who was deeply engrossed writing his *Sketch of Thermodynamics* at the time, that in his book Tait might 'pick a hole—say in the second law of $\theta\Delta^{cs}$., that if two things are in contact the hotter cannot take heat from the colder without external agency'. If only we were clever enough, Maxwell suggested, we too might mimic the neat-fingered one.[13] A month later, discussing his newly designated 'pointsman for flying molecules', he teased Thomson with the provocative thought, 'Hence energy need not be always dizzy-pated as in the present wasteful world'.[14]

Not, however, until three years later, when the 'intelligence' had grown to a mature 'doorkeeper . . . exceedingly quick', did Maxwell reach what became his enduring verdict: 'Moral. The 2nd law of thermodynamics has the same degree of truth as the statement that if you throw a tumblerful of water into the sea, you cannot get the same tumblerful of water out again.'[15] Thus was born Maxwell's prophetic insight that the second law of thermodynamics could never be given a mechanical interpretation based on the laws of pure dynamics which follow the motion of every particle. The second law was true only for matter *en masse*, and that truth was only statistical, not universal.

If Maxwell's demon was thus limited to operating a door for the sole purpose of demonstrating the statistical nature of the second law, where did Henry Adams get the idea of a Germanic bouncing demon who could generate a common motion? In his only public reference to the demon, Adams introduced the idea while criticizing the mechanistic view for omitting mind from the universe. Mind would be the only possible source for direction in an otherwise chaotic universe, Adams maintained, noting that 'The sum of motion without direction is zero, as in the motion of a kinetic gas where only Clerk Maxwell's demon of thought could create a value'.[16] This image of a demon who operates amidst molecules to create value from chaos stemmed from Adams's reading of William Thomson's ideas in 'The Sorting Demon of Maxwell'.[17]

## Thomson's Demon and the Dissipation of Energy

It was Thomson who baptized and popularized the creature of Maxwell's imagination in an essay in 1874. Maxwell had introduced his brainchild in the span of a few pages in his *Theory of Heat*, describing him simply as a 'being whose faculties are so sharpened that he can follow every molecule in its course',[18] and Thomson went on to christen him the 'intelligent demon'.[19] Whereas Maxwell had stationed his lonely being at the single minute hole in a partitioning wall, there to separate the fleet from the slow, Thomson recruited a whole army of demons to wage war with cricket bats and drive back an onrushing hoard of diffusing molecules.[20] In his second essay, Thomson's description became even more anthropomorphic:

He is a being with no preternatural qualities, and differs from real living animals only in extreme smallness and agility. He can at pleasure stop, or strike, or push, or pull away any atom of matter, and so moderate its natural course of motion. Endowed equally with arms and hands—two hands and ten fingers suffice—he can do as much for atoms as a pianoforte player can do for the keys of the piano—just a little more, he can push or pull each atom in any direction.[21]

Thomson's amazing creature could even subdue the forces of chemical affinity by absorbing kinetic energy from moving molecules and then applying that energy to sever molecular bonds. 'Let him take in a small store of energy by resisting the mutual approach of two compound molecules, letting them press as it were on his two hands and store up energy as in a bent spring; then let him apply the two hands between the oxygen and double hydrogen constituents of a compound molecule of vapour of water, and tear them asunder.'[22] The exploits of Thomson's demon give the impression that his main role was to restore dissipated energy. Is motion lost in viscous friction? Simply sort out the molecules moving in one direction and motion reappears. Is heat lost by conduction? Simply separate the faster and slower moving molecules to restore the temperature gradient. Is chemical energy dissipated as heat? Simply use the kinetic energy of the molecules to tear asunder the chemical bonds.

Such an interpretation of the demon's activity appealed to Thomson, for he had been the first to suggest the rather dire image of a universe ruled by the inexorable second law of thermodynamics. In his 1852 paper on the universal dissipation of mechanical energy, Thomson had observed that mechanical energy is continually being dissipated into heat by friction and heat is continually being dissipated by conduction. In this state of affairs, the Earth and its life are caught in a vicious cycle of energy dissipation and decline unless there is action by some non-mechanical agency.[23] Maxwell's thought-experiment showed, however, that energy dissipation need not be irrevocable from the point of view of an acute and designing mind. ' "Dissipation of Energy" ', Thomson wrote, 'follows in nature from the fortuitous concourse of atoms. The lost motivity is not restorable otherwise than by an agency dealing with atoms; and the mode of dealing with the atoms is essentially a process of assortment.'[24]

Such an understanding of the nature of dissipation was not original with Thomson. Maxwell had drawn the same analogy in far clearer terms, though without reference to any demonic activity. In discussing the difference between dissipated and available energy, Maxwell showed that these concepts were relative to the extent of our knowledge:

It follows . . . that the idea of dissipation of energy depends on the extent of our knowledge. Available energy is energy which we can direct into any desired channel. Dissipated energy is energy which we cannot lay hold of and direct at pleasure, such as the energy of the confused agitation of molecules which we call heat. Now, confusion, like the correlative term order, is not a property of material things in themselves, but only in relation to the mind which perceives them. A memorandum-book does not, provided it is neatly written, appear confused to an illiterate person, or to the owner who understands it thoroughly, but to any other person able to read it appears to be inextricably confused. Similarly the notion of dissipated energy would not occur to a being who could not turn any of the energies of nature to his own account, or to one who could trace the motion of every molecule and seize it at the right moment. It is only to a being in the intermediate stage, who can lay hold of some forms of energy while others elude his grasp, that energy appears to be passing inevitably from the available to the dissipated state.[25]

It was thus Maxwell, not Thomson, who assigned the demon the role of illuminating the nature of dissipated energy and of showing that all energy remains available for a mind able to 'trace the motion of every molecule and seize it at the right moment'. No doubt Maxwell first conceived his tiny intelligence because he was concerned about energy dissipation.

Why else would Maxwell have suggested to Tait in 1867 that he should pick a hole in the second law of thermodynamics, namely, 'that if two things are in contact the hotter cannot take heat from the colder without external agency'? Why should Maxwell choose this problem and why did he suggest that if man were clever enough he might mimic Maxwell's thought-child? Why did Maxwell tease Thomson with the suggestion that energy need not always be 'dizzypated as in the present wasteful world'? Since Maxwell did not finally conclude that the second law is statistical until three years later, he must have had some other reason for beginning this chain of thought. The reason is to be found in Maxwell's concern about energy dissipation. It is significant in this connection that Josef

Loschmidt, the scientist on the Continent who most abhorred the image of a decaying universe, also invented a 'demon' to thwart dissipation, even before Maxwell.

## Loschmidt's Non-Demon and the Dynamical Interpretation of the Second Law

Ludwig Boltzmann, Loschmidt's colleague in Vienna, gives the following report of Loschmidt's invention in his account of Loschmidt's varied efforts to obviate the dire theoretical consequences of the second law:

> On another occasion he imagined a tiny intelligent being who would be able to see the individual gas molecules and, by some sort of contrivance, to separate the slow ones from the fast and thereby, even if all activity [*Geschehen*] in the universe had ceased, to create new temperature differences. As we all know, this idea, which Loschmidt only hinted in a few lines of an article, was later proposed in Maxwell's *Theory of Heat* and was widely discussed.[26]

Boltzmann's memory, however, failed him on two counts. Loschmidt had devoted more than a few lines to the topic, but he had conceived no tiny intelligent creature. Boltzmann was recalling, not Loschmidt's original idea, but the later arguments he had had with Loschmidt concerning Maxwell's creation. In one of these discussions Boltzmann told Loschmidt that no intelligence could exist in a confined room at uniform temperature, at which point Josef Stefan, who had been listening quietly to the dispute, remarked to Loschmidt, 'Now I understand why your experiments in the basement with glass cylinders are such miserable failures'.[27] Loschmidt had been trying to observe gravitational concentration gradients in salt solutions as evidence for refuting the second law and its prediction of final uniformity.

Loschmidt's non-demon appeared in 1869.[28] It represented an attempt to do exactly the kind of thing which Maxwell had suggested to Tait, namely, 'to pick a hole' in the second law of thermodynamics. Loschmidt was also aiming at Clausius's statement that 'It is impossible for heat to pass from a colder to a warmer body without an equivalent compensation'.[29] Although the axiom was admittedly supported by ordinary experience, Loschmidt proposed to show that it was not true for all conceivable cases. Imagine, he said, a large space $V$ with molecules moving about at various velocities, some above and some below the mean velocity $c$, plus a small adjoining space $v$ that is initially empty. Consider now, he continued, a small surface element of the wall separating the two compartments and the succession of molecules striking it. Given the initial conditions of all the molecules, the order of their collisions with that surface element should be so fixed and determined that the element could be instructed to open and close in a pattern that would admit only the faster molecules into the empty space *v*.

> Thus we can obviously conceive of these exchanges as so ordered that only those molecules whose velocities lie above the average value *c* may trespass into *v*, and it would further be possible to allow their number so to increase, that the density of the gas in *v* may become greater than that in *V*. It is therefore not theoretically impossible, without the expenditure of work or other compensation, to bring a gas from a lower to a higher temperature or even to increase its density.[30]

Thus, Loschmidt's conception was both earlier and far less anthropomorphic than Maxwell's doorkeeper.

In some of his later writings, Maxwell moved in Loschmidt's direction. When Maxwell wrote to Rayleigh in 1870, just before the only public appearance of his idea in the *Theory of Heat*, he noted, 'I do not see why even intelligence might not be dispensed with and the thing made self-acting'.[31] In a final undated summary statement entitled 'Concerning Demons', he reduced his creature to a valve:

> Is the production of an inequality their only occupation? No, for less intelligent demons can produce a difference of pressure as well as temperature by merely allowing all particles going in one direction while stopping all those going the other way. This reduces the demon to a valve. As such value him. Call him no more a demon but a valve. . . .[32]

But although Maxwell had revised his thinking and moved from a tiny intelligence to a more mechanical device, he still claimed the same important and distinctive role for his thought experiment. What, he asked, is the chief end of my creature? 'To show that the 2nd Law of Thermodynamics has only a statistical certainty.'[33]

Loschmidt drew a very different conclusion from the ability of his non-demon to create temperature differences. Since the absolute validity of Clausius's axiom had been rendered doubtful, he argued, the second law of thermodynamics must be established on other grounds, namely, on those very dynamical foundations[34] which Maxwell's demon had led Maxwell to reject. Thus, despite their common origin in the desire to pick a hole in the second law of thermodynamics, Maxwell's demon and Loschmidt's non-demon performed strikingly different roles. In Maxwell's view, the demon did not undermine Clausius's axiom as a basis for the second law of thermodynamics. Since the axiom was statistically true, the predictions based upon it, namely, the irreversible increase in entropy and the increasing dissipation of energy, were valid conclusions. Since these truths were only statistical, however, Maxwell scorned supposed proofs of the second law based upon dynamical studies that traced the motions of individual atoms.

In his *Theory of Heat* Maxwell had raised the following question: 'It would be interesting to enquire how far those ideas . . . derived from the dynamical method . . . are applicable to our actual knowledge of concrete things, which . . . is of an essentially statistical character'.[35] Although that query stood unchanged throughout the four editions of Maxwell's book, his own conviction became clear in candid comments to Tait with regard to various attempts, notably by Clausius and Boltzmann, to explain the second law of thermodynamics by means of Hamilton's equations in dynamics. In 1873 he jested in lyrical fashion:

> But it is rare sport to see those learned Germans contending for the priority in the discovery that the second law of $\theta\Delta^{cs}$ is the Hamiltonsche Princip. . . . The Hamiltonsche Princip the while soars along in a region unvexed by statistical considerations while the German Icari flap their waxen wings in nephelococcygin, amid those cloudy forms which the ignorance and finitude of human science have invested with the incommunicable attributes of the invisible Queen of Heaven.[36]

In 1876, he was pungently clear. No pure dynamical statement, he said, 'would submit to such an indignity'.[37] There in essence lay the true scientific role of Maxwell's doorkeeper, to show the folly of seeking to prove a statistical law as though it expressed the ordered behaviour of traditional dynamic models.

In Loschmidt's thought, a similar devotion to mechanics led to a very different orientation. Since his non-demon disproved the unconditional validity of Clausius's axiom, the appearances, he felt, must be deceptive. The true basis for the second law must lie in traditional dynamical principles. Loschmidt stated his fundamental position most clearly in 1876 when he said: 'Since the second law of the mechanical theory of heat, just like the first, should be a principle of analytical mechanics, then it must be valid, not only under the conditions which occur in nature, but also with complete generality, for systems with any molecular form and under any assumed forces, both intermolecular and external'.[38] He then conceived a variety of models in which columns of atoms in thermal equilibrium would exhibit a temperature gradient and thus contradict the usual thermodynamic axiom proposed by Clausius. Loschmidt concluded that

Clausius's axiom was an inadequate basis for the second law, since these molecular models, which fulfilled all the requisite conditions for proving the second law from Hamilton's principle, did not entail the truth of that axiom.[39]

Loschmidt rejoiced that the threatening implications of the second law for the fate of the universe were finally disproved:

> Thereby the terrifying *nimbus* of the second law, by which it was made to appear as a principle annihilating the total life of the universe, would also be destroyed; and mankind could take comfort in the disclosure that humanity was not solely dependent upon coal or the Sun in order to transform heat into work, but would have an inexhaustible supply of transformable heat at hand in all ages.[40]

When Loschmidt pressed his case against Clausius's axiom even further, however, by raising the so-called reversibility paradox, he forced Boltzmann to conclude that no dynamical interpretation of the second law is possible.

## Loschmidt's Reversibility Paradox and the Statistical Second Law

The dynamical interpretations of the second law which Loschmidt favoured were restricted to those cases where entropy is conserved in the universe. Such processes are generally called reversible because they are so ideally contrived that the original conditions may always be completely recovered. No entropy increase or dissipation of energy occurs in reversible cycles. Interpretations of the second law based on Clausius's axiom, however, consider another type of process, the irreversible case in which entropy irrevocably increases and energy dissipates. According to Clausius, there were three possible cases corresponding to negative, zero and positive changes in entropy. The negative entropy change was impossible because that would be equivalent to a flow of heat from a cold to a hot body, contrary to Clausius's axiom. Entropy was a quantity, therefore, which could only remain constant or increase, depending on whether the process was reversible or irreversible.[41] Loschmidt's dynamical interpretation would only countenance the reversible case. Boltzmann, however, sought a dynamical interpretation for the irreversible increase in entropy as well, and it was to refute that possibility that Loschmidt created his reversibility paradox.

In order to have a dynamical interpretation of the irreversible case, there would have to be a mathematical function which showed a unilateral change to a maximum, after which it would remain constant, thereby reflecting the irreversible increase of entropy to a maximum at equilibrium. Boltzmann had derived such a function from an analysis of the collisions between molecules. To refute such an idea, only a single counter-example is necessary. One need only demonstrate that there exists at least one distribution of molecular velocities and positions from which the opposite behaviour would proceed. Loschmidt provided just such a thought-experiment. Imagine, he said, a system of particles where all are at rest at the bottom of the container except for one which is at rest some distance above. Let that particle fall and collide with the others, thus initiating motion among them. The ensuing process would lead to increasingly disordered motion until the system reached an apparently static equilibrium, just as the proponents of irreversible entropy change to an equilibrium would predict. But now imagine the instantaneous reversal of every single velocity and the very opposite process becomes inevitable. At first, little change would be evident, but the system would gradually move back towards the initially ordered situation in which all particles were at rest on the bottom and only one rested at a height above them.[42] Boltzmann labelled this Loschmidt's paradox, and a real paradox it became, since Boltzmann managed to reverse all of Loschmidt's conclusions.

The paradox revealed to Boltzmann that his attempt to find a dynamical function of molecular motion which would mirror the behaviour of entropy could only lead to a dead end, for whatever the mathematical function might be, the mere reversal of velocities would also reverse the supposedly unidirectional behaviour of that function. Boltzmann concluded that no purely dynamical proof of the second law would ever be possible and that the irreversible increase of entropy must reflect, not a mechanical law, but states of differing probabilities. Systems move towards equilibrium simply because the number of molecular states which correspond to equilibrium is vastly greater than the number of more ordered states of low entropy. Boltzmann offered an analogy to the Quinterns used in Lotto. Each Quintern has an equal probability of appearing, but a Quintern with a disordered arrangement of numbers is far more likely to appear than one with an ordered arrangement such as 12345. Boltzmann therefore provided the key for quantifying the statistical interpretation of the second law in terms of the relative numbers of molecular states that correspond to equilibrium and non-equilibrium.[43]

Thus, the chief end of Maxwell's creature, 'to show that the 2nd Law of Thermodynamics has only a statistical certainty', became established as a cardinal principle of classical physics by way of Loschmidt's non-demon and his reversibility paradox. Although Maxwell's demon has served to popularize the statistical interpretation of the second law through generations of thermodynamics textbooks, a new thought-experiment involving information theory has now challenged the demon's traditional role. The arguments made there suggest that Maxwell's brainchild must, alas, be laid to rest.

## The Price of Information

Consider the case, often discussed in information theory and originally introduced by Szilard in 1929,[44] where but a single molecule is involved. If the second law is merely statistical, then it certainly should fail to meet the test of this simplest of all arrangements. Pierce has described a modified version of Szilard's innovation as follows. Consider a piston equipped with a large trapdoor and with arrangements to allow the piston to lift weights by moving either left or right. Initially the piston is not connected to any weights; it is moved to the centre of the cylinder while the trap door is kept open, thus assuring that no collisions with the lone molecule in the cylinder can occur and that, therefore, no work will be required. Then the trap door is closed and the piston clamped into its central position. The molecule must be entrapped on one of the two sides of the piston, and Maxwell's demon informs us whether it is on the left or the right. With that information in hand, the piston is released and the molecule is made to do work by driving the weightless piston and a suitably suspended weight towards the empty side. The maximum possible work may be readily calculated; it would amount to $W=0.693\ kT$.[45] In Maxwell's day, the demon would have chalked up another victory, but not now.

'Did we get this mechanical work free?' Pierce succinctly asks. 'Not quite!'

> In order to know which pan to put the weight on, we need one bit of information, specifying which side the molecule is on.... What is the very least energy needed to transmit one bit of information at the temperature $T$?... exactly 0.693 $kT$ joule, just equal to the most energy the machine can generate.... Thus, we use up all the output of the machine in transmitting enough information to make the machine run![46]

Thus, the reign of Maxwell's brainchild in physics, designed to demonstrate that the second law of thermodynamics has only statistical validity, has come to an end.

The mood of the scientific community had changed. The second law was no longer subject to reproach for the indignities Maxwell and Loschmidt supposed it would inflict on pure dynamics. It was natural, therefore, to extend the province of the law and challenge all imagined contradictions. Szilard was the first to stress that any manipulator of molecules would have to rely on measurement and memory. If one assumed that the demon could perform such operations without causing any changes in the system, one would by that very assumption deny the second law, which requires equivalent compensations for all decreases in entropy.[47] Szilard therefore proposed that whatever negative entropy Maxwell's demon might be able to create should be considered as compensated by an equal entropy increase due to the measurements the demon had to make. In essence, Szilard made Maxwell's doorkeeper mortal—no longer granting this tiny intelligence the ability to 'see' molecules without actually seeing them, *i.e.*, without the sensory exchanges of energy that all other existences require. Szilard took this step for the sake of a grander vision, the dream that the adoption of his principle would lead to the discovery of a more general law of entropy in which there would be a completely universal relation for all measurements.[48] Information theory has brought that vision to reality.

One puzzling question, however, remains. Why did Maxwell not realize that his creature required energy in order to detect molecules? Brillouin has suggested that Maxwell did not have an adequate theory of radiation at his disposal. 'It is not surprising', he said, 'that Maxwell did not think of including radiation in the system in equilibrium at temperature *T*. Black body radiation was hardly known in 1871, and it was thirty years before the thermodynamics of radiation was clearly understood.'[49] It is certainly true that a quantitative expression for radiant energy and the entropy of information would require an adequate theory of black body radiation, but the absence of such a detailed theory does not explain why Maxwell failed to realize that some energy exchanges were required.

If we were able to ask Maxwell, 'Why did you not require your tiny intelligence to use energy in gathering his information?', Maxwell would no doubt reply, 'Of course! Why didn't I think of that?'[50] Why didn't Maxwell think of that? Because his demon was the creature of his theology.

### The Demon and Theology

Maxwell's demon is the very image of the Newtonian God who has ultimate dominion over the world and senses the world in divine immediacy. Newton wrote in his General Scholium:

It is allowed by all that the Supreme God exists necessarily, and by the same necessity he exists *always* and *everywhere*. Whence also he is all similar, all eye, all ear, all brain, all arm, all power to perceive, to understand, and to act; but in a manner not at all human, in a manner not at all corporeal, in a manner utterly unknown to us. As a blind man has no idea of colour, so we have no idea of the manner by which the all-wise God perceives and understands all things.[51]

How natural it was for Maxwell, faced with the idea of a universe destined towards dissipation, to conceive a being on the model of God, for whom the universe always remains ordered and under his rule. A memorandum-book, Maxwell said, does not appear confused to its owner though it does to any other reader. Nor would the notion of dissipated energy occur to 'a being who could trace the motion of every molecule and seize it at the right moment'. Maxwell's demon was not mortal because he was made in the image of God. And like God, he could see without seeing and hear without hearing. In short, he could acquire information without any expenditure of energy.

Upon being asked by a clergyman for a viable scientific idea to explain how, in the *Genesis* account, light could be created on the first day although the Sun did not appear until the third day, Maxwell replied that he did not favour reinterpreting the text in terms of prevailing scientific theory. To tie a religious idea to a changeable scientific view, Maxwell said, would only serve to keep that scientific idea in vogue long after it deserved to be dead and buried.[52] Thus Maxwell would certainly sever the demon's ties to theology if he were faced with Szilard's requirements for the cost of information. Witnessing the demise of his creature, he would not long mourn at the grave but rather be grateful for the exciting years his thought-child had enjoyed. Steeped in the knowledge and love of biblical imagery, Maxwell would doubtless take pleasure in the thought that by becoming mortal his doorkeeper had prepared the way for new life in science.[53]

*University of Kansas*

## NOTES

[1] C. P. Snow, *The Two Cultures and the Scientific Revolution* (New York, 1961), 15–16.

[2] H. D. Cater, ed., *Henry Adams and his Friends* (Boston, 1947), 640.

[3] *Ibid.*, 646–7, 650n.

[4] *Ibid.*, 647.

[5] E. Samuels, *Henry Adams: The Major Phase* (Cambridge, 1964), 450: 'Adams minutely revised the essay during the next year or two to meet Bumstead's more specific criticisms. . . . The firm outlines of the script would suggest that all of these changes were made before his stroke in 1912.

No indication remains that he submitted the revised essay to the *North American Review*. Not until 1919, a year after his death, did it appear in Brooks Adams's edition of Henry's "philosophical writings", *The Degradation of the Democratic Dogma*. . . .'.

[6] H. Adams, *The Degradation of the Democratic Dogma*, ed. B. Adams (New York, 1919), 267.

[7] W. H. Jordy, *Henry Adams: Scientific Historian* (New Haven, 1952), 166.

[8] *Ibid.*, 169, 170.

[9] Maxwell–Tait Correspondence, Cambridge University Library; letter to Tait, 23 December 1867: 'I have read some metaphysics of various kinds and find it more or less ignorant discussion of mathematical and physical principles, jumbled with a little physiology of the senses. The value of metaphysics is equal to the mathematical and physical knowledge of the author divided by his confidence in reasoning from the names of things.'

[10] Letter to Brooks Adams, 2 May 1903, *op. cit.*, note 2, 545.

[11] H. Adams, *Letters of Henry Adams (1892–1918)*, ed. N. C. Ford (Boston, 1938). Letter to Cecil Spring Rice, 11 November 1897, 135–6.

[12] J. C. Maxwell, *The Theory of Heat*, 2nd edition (London, 1872), 308–9.

[13] Letter from Maxwell to Tait, 11 December 1867, quoted in C. G. Knott, *Life and Scientific Work of Peter Guthrie Tait* (Cambridge, 1911), 213–14.

[14] Letter from Maxwell to William Thomson, 16 January 1868, Edinburgh University Library.

[15] Letter from Maxwell to Strutt, 6 December 1870, quoted in R. J. Strutt, *John William Strutt* (London, 1924), 47.

[16] Adams, *op. cit.*, note 6, 279.

[17] W. Thomson, 'The Sorting Demon of Maxwell', reprinted in *Popular Lectures and Addresses*, Vol. 1 (London, 1889), 137–41.

[18] Maxwell, *op. cit.*, note 12, 308.

[19] W. Thomson, 'Kinetic Theory of the Dissipation of Energy', *Nature*, **9** (1874), 442.

[20] *Ibid.*

[21] Thomson, *op. cit.*, note 17, 137–8.

[22] *Ibid.*, 140.

[23] W. Thomson, 'On a Universal Tendency in Nature to the Dissipation of Mechanical Energy', *Philosophical Magazine*, **4** (1852), 256–60.

[24] Thomson, *op. cit.*, note 17, 139.

[25] J. C. Maxwell, 'Diffusion', *Encyclopedia Britannica*, 9th edition (New York, 1878), vol. 7, 220.

[26] L. Boltzmann, 'Zur Errinerung an Josef Loschmidt', in *Populäre Schriften* (Leipzig, 1905), 231. (I am indebted to R. Dugas, *La théorie physique au sens Boltzmann* (Neuchatel, 1959), 171, note 2, for this reference.)

[27] *Ibid.*

[28] J. Loschmidt, 'Der zweite Satz der mechanischen Wärmetheorie', *Akademie der Wissenschaften, Wien. Mathematisch-Naturwissenschaftliche Klasse, Sitzungsberichte*, **59**, Abth. 2 (1869), 395–418.

[29] *Ibid.*, 399. The editor must have slipped, however, since the text reads 'Die Wärme geht niemals aus einem heisseren in einen kälteren über ohne eine äquivalente Compensation'.

[30] *Ibid.*, 401.

[31] Quoted in Strutt, *op. cit.*, note 15, 47.

[32] Quoted in Knott, *op. cit.*, note 13, 215. Knott seems to have erred in suggesting that this undated letter was penned at about the same time that Maxwell originally proposed his idea to Tait in December 1867.

[33] *Ibid.*

[34] Loschmidt, *op. cit.*, note 28, 401–6.

[35] Maxwell, *op. cit.*, note 12, 309.

[36] Letter from Maxwell to Tait, Tripos, December 1873, Maxwell-Tait Correspondence, Cambridge University Library.

[37] Letter from Maxwell to Tait, 13 October 1876, *ibid.* Martin Klein's recent article, 'Maxwell, his Demon, and the Second Law of Thermodynamics', *American Scientist*, **58** (1970), 84–97, treats these attempts to reduce the second law to dynamics in considerable detail.

[38] J. Loschmidt, 'Ueber den Zustand des Wärmegleichgewichtes eines System von Körpern', *Sitzungsberichte* (see note 28), **73**, Abth. 2 (1876), 128.

[39] *Ibid.*, 128–35.

[40] *Ibid.*, 135.

[41] R. Clausius, 'Ueber eine veränderte Form des zweiten Hauptsatzes der mechanischen Wärmetheorie', *Annalen der Physik*, **93** (1854), 481–506. Clausius did not use the word 'entropy' until 1865. In the original paper he spoke of the sum of transformation values $\int dQ/T = N$ for a cycle, where $N < 0$ was impossible, $N = 0$ was the reversible case, and $N > 0$ was the irreversible cycle.

[42] Loschmidt, *op. cit.*, note 38, 137–9.

[43] L. Boltzmann, 'Bemerkungen über einige Probleme der mechanischen Wärmetheorie' (1877), reprinted in *Wissenschaftliche Abhandlungen von Ludwig Boltzmann*, ed. F. Hasenohrl (Leipzig, 1909), Vol. II, 120.

[44] L. Szilard, 'Ueber die Entropieverminderung in einem thermodynamischen System bei Eingriffen intelligenter Wesen', *Zeitschrift für Physik*, **53** (1929), 840–56.

[45] J. R. Pierce, *Symbols, Signals, and Noise* (New York, 1961), 198–201.

[46] *Ibid.*, 201.

[47] Szilard, *op. cit.*, note 44, 842.

[48] *Ibid.*, 843.

[49] L. Brillouin, *Science and Information Theory* (New York, 1956), 164.

[50] I am indebted to my colleague Richard Cole of the Kansas University Philosophy Department for this thought-experiment.

[51] I. Newton, *Newton's Philosophy of Nature*, ed. H. S. Thayer (New York, 1953), 44.

[52] L. Campbell and W. Garnett, *The Life of James Clerk Maxwell* (London, 1882), 394.

[53] The support of the National Science Foundation is gratefully acknowledged.

Martin J. Klein

# Maxwell, His Demon, and the Second Law of Thermodynamics

*Maxwell saw the second law as statistical, illustrated this with his demon, but never developed its theory*

By the mid-sixties of the last century the science of thermodynamics had reached a certain level of maturity. One could already describe it, in the words James Clerk Maxwell would use a few years later, as "a science with secure foundations, clear definitions, and distinct boundaries" (*1*). As one sign of the maturity of the subject, Rudolf Clausius, one of its creators, reissued his principal papers on thermodynamics as a book in 1864 (*2*). He also set aside his protracted efforts to find the simplest and most general form of the second law of thermodynamics in order to meet the need for convenient working forms of the thermodynamic equations, forms suitable for dealing with the variety of experimental situations to which the theory was being applied (*3*).

Another indication of the maturity of the subject was the appearance in 1868 of Peter Guthrie Tait's book, *Sketch of Thermodynamics* (*4*). This consisted of a revision of two articles Tait had already published on the history of the recent developments in the theory of heat, supplemented by a brief treatment of the principles of thermodynamics. Tait did not claim to have written a comprehensive treatise, but his book did make the basic concepts and methods available to students.

One of his reasons for writing the book was, in fact, his feeling of "the want of a short and elementary textbook" for use in his own classes (*4*, p. iii). Another reason was Tait's desire to set the historical record straight, which for him meant urging the claims of his compatriots, James Prescott Joule and William Thomson (later Lord Kelvin), against those of the Germans who had contributed along similar lines—Julius Robert Mayer, Hermann von Helmholtz, and Clausius. Since Tait admitted in his preface that he might have taken "a somewhat too British point of view" (*4*, p. v), it is not surprising that his book became the center of a stormy controversy, but that controversy is not our concern here (*5*).

Before sending his manuscript off to the publisher, Tait wrote to Maxwell, asking him to apply his critical powers to it. Tait was already expecting trouble over his assignment of priorities and credit, since both Clausius and Helmholtz had been sent parts of the manuscript and had reacted negatively (*5a*, pp. 216–17; *6*). Maxwell was an old friend; the two men had been together at school, at the University of Edinburgh and at Cambridge. They shared a variety of scientific interests and carried on a particularly lively and vigorous correspondence (*7*). It was not at all unusual for Maxwell to read the manuscripts or proofs of his friends' books and to "enrich them by notes, always valuable and often of the quaintest character," as Tait himself wrote (*8*). This time Maxwell provided his enrichment even before he saw Tait's book.

Maxwell wrote Tait that he would be glad to see his manuscript, although he did not "know in a controversial manner the history of thermodynamics" and so was not prepared to join his friend in waging the priority wars. "Any contributions I could make to that study," he went on, "are in the way of altering the point of view here and there for clearness or variety, and picking holes here and there to ensure strength and stability" (*9*). Maxwell proceeded to

*Martin J. Klein, Professor of the History of Physics at Yale, teaches both physics and the history of science at the University. A graduate of Columbia, he holds a Ph.D. from M.I.T. (1948). He has worked on various problems in thermodynamics and statistical mechanics, and most recently has been studying the development of physics in the nineteenth and twentieth centuries. His papers on Planck, Einstein, Gibbs, and others have appeared in a number of journals. His book,* Paul Ehrenfest, *Volume 1,* The Making of a Theoretical Physicist *(North-Holland Publishing Co., Amsterdam), will appear this winter. Address: Department of the History of Science and Medicine, Yale University, 56 Hillhouse Avenue, New Haven, Conn. 06520.*

pick such a hole—in the second law of thermodynamics itself.

Since its original appearance in Sadi Carnot's memoir of 1824, the principle that would eventually become the second law was always formulated as a completely general statement, as free of exceptions as Newton's laws of motion or his law of universal gravitation. What Maxwell challenged was just this universal, invariable validity of the second law. His challenge took the strange form of what we now call Maxwell's demon, an argument of the "quaintest character" indeed. This was Maxwell's way of expressing his insight into the peculiar nature of the second law of thermodynamics: it was not a law that could be reduced ultimately to dynamics, but it expressed instead the statistical regularity of systems composed of unimaginably large numbers of molecules.

Maxwell's views on the nature of the second law, expressed in brief passages and passing remarks, were never developed into a systematic statistical mechanics, but they show how clearly he saw to the heart of the matter. He was insisting on the statistical character of the second law at a time when Rudolf Clausius and Ludwig Boltzmann were trying to show that it was a strictly mechanical theorem. His writings on this subject show, in their fragmentary character as well as in their penetration, that quality which Tait summed up in a sentence: "It is thoroughly characteristic of the man that his mind could never bear to pass by any phenomenon without satisfying itself of at least its general nature and causes" (*8*, pp. 319–20).

## The demon and molecular statistics

Carnot formulated his general result in these words: "The motive power of heat is independent of the agents employed to realize it; its quantity is fixed solely by the temperatures of the bodies between which is effected, finally, the transfer of the caloric" (*10*). As the last word indicates, Carnot was using the caloric theory of heat, so that for him heat was a conserved fluid: as much heat was rejected at the low temperature as had been absorbed at the high temperature, when work was performed by a cyclic process. Carnot's proof made use of the impossibility of perpetual motion, the impossibility, that is, of "an unlimited creation of motive power without consumption either of caloric or of any other agent whatever" (*10*, p. 12).

When the possibility of transforming work into heat, or heat into work, in a fixed proportion was conclusively demonstrated in the 1840s, the basis for Carnot's principle seemed to be lost (*11*). Caloric was not conserved and Carnot's proof no longer held. It was Clausius who saw that the equivalence of work and heat could be made compatible with Carnot's principle, if the latter were modified only slightly (*12*). Whereas Carnot had said that heat, $Q$, must be transferred from a hot body to a cold one when work, $W$, is done in a cyclic process, one could now say instead that when heat $Q$ is absorbed from the hot body, only the difference, $Q - W$, is rejected as heat to the cold body. This revised form of Carnot's assumption allowed for the equivalence of heat and work, and, according to Clausius, it could still serve as the basis for a proof of Carnot's theorem. But something more than the usual impossibility of perpetual motion had to be invoked as a postulate to carry out the proof.

Clausius repeated Carnot's indirect reasoning, proving his result by showing that assuming the converse led one to an evidently intolerable conclusion. For Carnot this had been the appearance of perpetual motion; for Clausius it was something different. "By repeating these two processes alternately it would be possible, without any expenditure of force or any other change, to transfer as much heat as we please from a *cold* to a *hot* body, and this is not in accord with the other relations of heat, since it always shows a tendency to equalize temperature differences and therefore to pass from *hotter* to *colder* bodies" (*13*). This was the new assumption Clausius needed for the creation of a thermodynamics based on both the equivalence of heat and work (the first law) and Carnot's principle. He phrased it more compactly a few years later: "Heat can never pass from a colder to a warmer body without some other change, connected therewith, occurring at the same time" (*14*). Modern textbook formulations of this second law of thermodynamics are careful to specify that cyclic processes are being considered and that the "other change" in question is the performance of external work on the system, but these specifications were implicit in Clausius' statement. The point to be stressed here is the universality of the statement, the presence of the word "never," for example, in the second quotation from Clausius.

It was here that Maxwell chose to "pick a hole" when he wrote to Tait in December 1867. He suggested a conceivable way in which, "if two things are in contact, the hotter" *could* "take heat from the colder without external agency." Maxwell considered a gas in a vessel divided into two sections, A and B, by a fixed diaphragm. The gas in A was assumed to be hotter than the gas in B, and Maxwell looked at the implications of this assumption from the molecular point of view. A higher temperature meant a higher average value of the kinetic energy of the gas molecules in A compared to those in B. But, as Maxwell had shown some years earlier, each sample of gas would necessarily contain molecules having velocities of all possible magnitudes. "Now," Maxwell wrote, "conceive a finite being who knows the paths and velocities of all the molecules by simple inspection but who can do no work except open and close a hole in the diaphragm by means of a slide without mass." This being would be assigned to open the hole for an approaching molecule in A only when that molecule had a velocity less than the root mean square velocity of the molecules in B. He would allow a molecule from B to pass through the hole into A only when its velocity exceeded the root mean square velocity of the molecules in A. These two procedures were to be carried out alternately, so that the numbers of molecules in A and B would not change. As a result of this process, however, "the energy in A is increased and that in B diminished; that is, the hot system has got hotter and the cold colder and yet no work has been done, only the intelligence of a very observant and neat-fingered being has been employed." If one could only deal with the molecules directly and individually in the manner of this supposed being, one could violate the second law. "Only we can't," added Maxwell, "not being clever enough" (*9*).

This is the first time that Maxwell's talented little being appeared on paper. Thomson immediately gave him the name "demon," by which he has been known ever since (*5a*, p. 214; *15*). Thomson used this name only in its original meaning, a supernatural being, and did not want to suggest any evil intentions on the part of this being who could reverse the common tendency of nature. In his letter to Tait, Maxwell probably did not make sufficiently clear his reason for introducing this fanciful construction. He described "the chief end" of his demon soon afterwards in these words: "to show that the 2nd law of thermodynamics has only a statistical certainty" (*5a*, p. 215). He meant that while it would require the action of the demon to produce an observable flow of heat from a cold body to a hotter one, this process is occurring spontaneously all the time, on a submicroscopic scale. The possibility of the demon's existence is less significant than the actuality of the statistical distribution of molecular velocities in a gas at equilibrium. This statistical character of a system composed of an enormous number of molecules, which would form the basis for the demon's actions, necessarily leads to spontaneous fluctuations, including fluctuations that take heat from a cold body to a hotter one.

Maxwell came a little closer to saying these things explicitly when he repeated his argument a few years later in a letter to John William Strutt (later Lord Rayleigh). This time he referred to the possibility of a complete reversal of the motion of all particles in the world, thereby reversing the sequence in which events normally occur. This possibility was quite consistent with the assumption that "this world is a purely dynamical system." In this time-reversed state the trend would always be away from equilibrium; such behavior would be in flagrant violation of the second law. Maxwell knew, of course, that "the possibility of executing this experiment is doubtful," but he did not "think it requires such a feat to upset the 2nd law of thermodynamics." It could be done more easily with the help of the demon, now described more graphically as "a doorkeeper very intelligent and exceedingly quick, with microscopic eyes," who would be "a mere guiding agent, like a pointsman on a railway with perfectly acting switches who should send the express along one line and the goods along another." The demon argument was introduced, however, by this sentence: "For if there is any truth in the dynamical theory of gases, the different molecules in a gas of uniform temperature are moving with very different velocities." That was the essential thing; the demon only served to make its implications transparently clear. Maxwell even drew an explicit "moral" from his discussion: "The 2nd law of thermodynamics has the same degree of truth as the statement that if you throw a tumblerful of water into the sea, you cannot get the same tumblerful of water out again" (*16*).

It is, after all, not surprising that Maxwell should have been ready to take the consequences of the velocity distribution so seriously: he was the one who had introduced this concept into physics. Maxwell became interested in the molecular theory of gases in 1859 when he read Clausius' papers on the subject in the *Philosophical Magazine* (*17*). Clausius had made use of the idea of the average distance traveled by a molecule between its collisions with the other molecules of the gas. It was Maxwell, however, who showed how the theory could be subjected to experimental test, as he derived the relationships between Clausius' mean free path and the measurable coefficients of viscosity, thermal conductivity, and diffusion. Again, although Clausius had remarked several times that "actually the greatest possible variety exists among the velocities of the several molecules" in a gas, he made no attempt to analyze the nature of this "variety" but simply used the average molecular velocity. Maxwell saw at once that an analysis of how molecular velocities were distributed would be fundamental to the development of the theory.

As early as May 1859, Maxwell wrote to George Gabriel Stokes about his work on the theory of gases, commenting, "Of course my particles have not all the same velocity, but the velocities are distributed according to the same formula as the errors are distributed in the theory of least squares" (*18*). He presented a derivation of this velocity distribution law in his first paper on gases in 1860, pointing out that molecular collisions would continually alter the individual velocities, and that what he found was "the average number of particles whose velocities lie between given limits after a great number of collisions among a great number of equal particles" (*19*). This relationship became even clearer in a new derivation of the distribution law that Maxwell published in 1866 (*20*). This time he calculated the effect of collisions on the distribution law and found the equilibrium distribution as that distribution which would be unchanged by collisions—the stationary law. This velocity distribution formed the basis of all the calculations of the properties of gases that Maxwell carried out, and it is quite natural that it should also have been the basis for his criticism of the second law of thermodynamics. Although this statistical approach seemed natural to Maxwell, it was very different from contemporary efforts to probe into the meaning of the second law.

## Attempts at mechanical explanation

In 1866 Ludwig Boltzmann wrote a paper, "On the Mechanical Meaning of the Second Law of Thermodynamics" (*21*). It was his first scientific work of any significance and the first serious attempt to give a general mechanical interpretation of the second law. Boltzmann began by referring to the "peculiarly exceptional position" of the second law and to the "roundabout and uncertain methods" by which physicists had tried to establish it, contrasting this with the first law, whose identity with the energy principle had been known "for a long time already." This "long time" was only about fifteen years, but Boltzmann was only twenty-two himself. He announced his goal with all the self-assurance of his age: "It is the purpose of this article to give a purely analytical, completely general proof of the second law of thermodynamics, as well as to discover the theorem in mechanics that corresponds to it."

Boltzmann was pursuing the classic program of physics, "tracing the phenomena of nature back to the simple laws of mechanics" (*22*). Since the first law could be clearly understood by introducing the energy of molecular motions, why not the second, too? Boltzmann apparently persuaded himself that he had achieved his goal, but for all his sweeping claims his success was very limited. To evaluate it we must state more precisely what he was trying to explain.

The statement that "heat can never pass from a colder to a warmer body without some other change occurring" had been an adequate basis for Clausius' first thermodynamic studies, but he felt that it was "incomplete, because we cannot recognize therein, with sufficient clearness, the real nature of the theorem" (*14*, p. 81). He was more satisfied with another form which he deduced from this a few years later and then expressed very compactly in 1865, in a paper which might well have stimulated Boltzmann's work (*3*). Clausius' results were these: For every thermodynamic system there exists a function, $S$, which is a function of the thermodynamic state of the system. $S$, which he named the entropy, is defined by the differential relationship,

$$dS = \frac{đQ}{T} \tag{1}$$

where $đQ$ is the heat supplied to the system in a *reversible* infinitesimal process and $T$ is the absolute temperature of the system. (The notation $đQ$, not used by Clausius, emphasizes that $đQ$ is not an exact differential.) For processes that are *irreversible*, one can only state the inequality,

$$dS > \frac{đQ}{T}. \tag{2}$$

In order to give the second law a mechanical explanation, Boltzmann would have had to find mechanical expressions for all of the quantities appearing in (1) and (2) and would have had to show that these mechanical quantities, functions of the molecular coordinates and momenta, do indeed satisfy the relationships satisfied by their thermodynamic counterparts. All of this should be done for a mechanical system of very general character. Boltzmann had certainly tried to do just these things. He was least successful in dealing with irreversible processes; his brief discussion of one particular example threw no light on the molecular basis for irreversibility.

Boltzmann did give a detailed justification for the idea, already expressed by others, that temperature was a measure of the average kinetic energy of a molecule. He also proved a very interesting mechanical theorem, a generalization of the principle of least action, on the strength of which he claimed to have given a mechanical explanation of the second law. This theorem will be discussed later; it is enough for the present to point out that Boltzmann could prove it only under a very restrictive assumption. He had to limit himself to periodic systems. For a system whose entire molecular configuration repeats itself after a time, $\tau$, he found a mechanical counterpart to the entropy, $S$. Boltzmann's equation for the entropy of such a system had the form,

$$S = \Sigma \ln(T\tau)^2 + \text{constant}, \qquad (3)$$

where the sum is over all molecules. The absolute temperature, $T$, is the kinetic energy of one molecule averaged over the period,

$$T = \frac{1}{\tau}\int_0^\tau \left(\frac{1}{2}mv^2\right) dt. \qquad (4)$$

Since the period, $\tau$, appears explicitly in the entropy equation one is inclined to guess that the periodicity of the system is an essential aspect of the theorem. Boltzmann tried to extend his proof to nonperiodic systems, where the particle orbits are not closed, but his argument was not cogent. It concluded rather lamely with the remark that "if the orbits are not closed in a finite time, one may still regard them as closed in an infinite time" (*23*). (It may be worth noting, however, that one can evaluate the entropy, with the help of (3), for a very simple model of a gas—a collection of particles bouncing back and forth between the walls of a container of volume $V$, all particles moving with the same speed, $v$. The result does depend linearly on the quantity ln $(VT^{3/2})$, as it should for an ideal gas of point particles.)

Boltzmann was not the only physicist who tried to reduce the second law to a theorem in mechanics. Rudolf Clausius had been thinking about the second law since 1850; "the second law of thermodynamics is much harder for the mind to grasp than the first," he wrote once (*3*, p. 353). This is not the place to follow the evolution of his ideas, but it must be pointed out that Clausius arrived at a very particular way of looking at the second law (*24*). He introduced a concept that he named the disgregation, intended as a measure of "the degree in which the molecules of a body are dispersed." It was to be a thermodynamic state function, denoted by $Z$, related to the total work, internal as well as external, done by a system in a reversible process. This relationship was fixed by the equation,

$$dL = TdZ \qquad (5)$$

where $dL$ represents the total work, an inexact differential. (Clausius considered the internal energy, $U$, of a system to be the sum of two terms, both state functions—the heat in the body, $H$, and the internal work, $I$. The total work $đL$ was the sum of $dI$ and the usual external work, $đW$, or $PdV$.) It was (5) that Clausius saw as the ultimate thermodynamic formulation of the second law: "the effective force of heat is proportional to the absolute temperature" (*25*). And it was (5) for which Clausius tried to find a mechanical interpretation.

Clausius' first attempt, in 1870, did not succeed, but it could hardly be called a failure since it led him to the virial theorem (*25*). The following year he was convinced that he had solved the problem, as he entitled his paper, "On the Reduction of the Second Law of Thermodynamics to General Mechanical Principles" (*26*). This title sounds very much like that of Boltzmann's paper of 1866, and with good reason. Clausius had independently discovered the theorem published by Boltzmann five years earlier. Although the results were the same, there were some differences in approach. Boltzmann had been looking for a mechanical interpretation of entropy and its properties, while Clausius gave disgregation the central position in the theory. The two functions are simply related, however, and the entropy Clausius derived from his mechanical expression for disgregation was identical with the one Boltzmann had found directly (see (3) above).

Boltzmann wasted no time in pointing out his considerable priority in having established this mechanical analogue of the second law (*27*). He showed that all of Clausius' basic equations were identical with his own, apart from notation, reinforcing the point by reproducing verbatim the relevant pages of his earlier article. "I think I have established my priority," Boltzmann concluded, adding, "I can only express my pleasure that an authority with Mr. Clausius' reputation is helping to spread the knowledge of my work on thermodynamics" (*27*, p. 236).

Clausius had, of course, overlooked Boltzmann's original memoir. He had moved from Zürich to Würzburg in 1867 and again to Bonn in 1869, and the resulting "extraordinary demands" on his time and energy had prevented him from keeping up with the literature properly. He obviously granted Boltzmann's claim to priority for all the results common to both papers, but Clausius was not convinced that Boltzmann's arguments were as general or as sound as his own, and he proceeded to discuss some of these matters in detail (*28*).

## Limitations of the second law

Maxwell observed this and later disputes over the mechanical interpretation of the second law with detachment—and no little amusement. "It is rare sport to see those learned Germans contending for the priority of the discovery that the 2nd law of $\theta\Delta$cs [thermodynamics] is the Hamiltonsche Princip," he wrote to Tait, "when all the time they *assume* that the temperature of a body is but another name for the vis viva of one of its molecules, a thing which was suggested by the labours of Gay Lussac, Dulong, etc., but first deduced from dynamical statistical considerations by *dp/dt* [i.e. Maxwell: see Appendix]. The Hamiltonsche Princip, the while, soars along in a region unvexed by statistical considerations, while the German Icari flap their waxen wings in nephelococcygia amid those cloudy forms which the ignorance and finitude of human science have invested with the incommunicable attributes of the invisible Queen of Heaven" (*29*).

The prize for which "those learned Germans" were contending was an illusion: the second law was not a dynamical theorem at all, but an essentially statistical result.

Maxwell expressed some of his own ideas on the meaning of the second law in his *Theory of Heat*, published in 1871 (*30*). This book appeared as one of a series, described by the publisher as "text-books of science adapted for the use of artisans and of students in public and science schools." They were meant to be "within the comprehension of working men, and suited to their wants," with "every theory . . . reduced to the stage of direct and useful application" (*31*). Maxwell's book did not quite fit this description. Tait thought that some of it was probably "more difficult to follow than any other of his writings" (!) and that as a whole it was "*not* an elementary book." His explanation of this fact was interesting, and probably accurate. "One of the few knowable things which Clerk-Maxwell did not know," he wrote, "was the distinction which most men readily perceive between what is easy and what is hard. What *he called* hard, others would be inclined to call altogether unintelligible." As a consequence Maxwell's book contained "matter enough to fill two or three large volumes without undue dilution (perhaps we should rather say, *with the necessary dilution*) of its varied contents" (*8*, p. 320). Maxwell did not hesitate to include discussions of the latest work in thermodynamics in the successive editions of his book; one wonders what his intended audience of "artisans and students" made of something like Willard Gibbs's thermodynamic surface, which Maxwell treated at some length, less than two years after the publication of Gibbs's paper (*32*).

In discussing the second law, Maxwell emphasized that Carnot's principle does not follow from the first law, but must instead be deduced from an independent assumption, hence a second law. He quoted Thomson's and Clausius' versions of this law, both of which asserted, in effect, the impossibility of a cyclic process converting the heat of a body into work without allowing heat to pass from that body to a colder one. He advised the student to compare the various statements of the law so as "to make himself master of the fact which they embody, an acquisition which will be of much greater importance to him than any form of words on which a demonstration may be more or less compactly constructed." And then in his next two paragraphs Maxwell went to the crux of the question.

> Suppose that a body contains energy in the form of heat, what are the conditions under which this energy or any part of it may be removed from the body? If heat in a body consists in a motion of its parts, and if we were able to distinguish these parts, and to guide and control their motions by any kind of mechanism, then by arranging our apparatus so as to lay hold of every moving part of the body, we could, by a suitable train of mechanism, transfer the energy of the moving parts of the heated body to any other body in the form of ordinary motion. The heated body would thus be rendered perfectly cold, and all its thermal energy would be converted into the visible motion of some other body.
>
> Now this supposition involves a direct contradiction to the second law of thermodynamics, but is consistent with the first law. The second law is therefore equivalent to a denial of our power to perform the operation just described, either by a train of mechanism, or by any other method yet discovered. Hence, if the heat consists in the motion of its parts, the separate parts which move must be so small or so impalpable that we cannot in any way lay hold of them to stop them [*30*, pp. 153–54].

This argument, so deceptively simple in appearance, brought out the point that had escaped Boltzmann and Clausius, among others. Those processes that are declared to be impossible by the second law are perfectly consistent with the laws of mechanics. The second law must, therefore, express some essentially nonmechanical aspect of nature. If it is to receive an explanation at the molecular level, that explanation must refer to the smallness of molecules, or, equivalently, to their enormous numbers. This is really the same point of view that Maxwell had already expressed privately in his demon argument. He made the connection himself in the last chapter of his book, which dealt briefly with the molecular theory of matter. In a section entitled "Limitation of the Second Law of Thermodynamics," Maxwell remarked that this law "is undoubtedly true as long as we can deal with bodies only in mass, and have no power of perceiving or handling the separate molecules of which they are made up." The status of the second law would be very different if we had the powers of "a being whose faculties are so sharpened that he can follow every molecule in its course." Maxwell described how the second law could be flouted by such a demon, just as he had described it in his letters to Tait and Strutt. As a consequence, one had to be careful about extending conclusions "drawn from our experience of bodies consisting of an immense number of molecules" to the "more delicate observations and experiments" that might be made if individual molecules could be perceived and handled. Unless that should become possible we were "compelled to adopt ... the statistical method of calculation, and to abandon the strict dynamical method" (*30*, pp. 328–29).

## The inadequacy of mechanical explanation

Maxwell returned to these considerations a few years later, when he was asked to review the second edition of Tait's *Sketch of Thermodynamics*. Written only a year or so before Maxwell's fatal illness, this review contains his final reflections on the subject of thermodynamics, presented in his most brilliant style. (As Tait himself remarked in another connection, "No living man has shown a greater power of condensing the whole marrow of a question into a few clear and compact sentences than Maxwell" [*8*, p. 321]). Much of Maxwell's review was devoted to the second law, since he thought that the manner in which an author handled this thorny subject formed "the touchstone of a treatise on thermodynamics" (*1*, p. 667).

The review offered a new variation on Maxwell's old theme of the irreducibility of the second law to pure dynamics. The whole point of the second law depends on the distinction between two modes of communicating energy from one body to another, heat and work. "According to the molecular theory," Maxwell observed, "the only difference between these two kinds of communication of energy is that the motions and displacements which are concerned in the communication of heat are those of molecules, and are so numerous, so small individually, and so irregular in their distribution, that they quite escape all our methods of observation; whereas when the motions and displacements are those of visible bodies consisting of great numbers of molecules moving altogether, the communication of energy is called work" (*1*, p. 669). If one supposed that individual molecular motions could be followed, as one would have to do in a dynamical molecular theory, then "the distinction between work and heat would vanish." At the molecular level there is no mechanical distinction between work and heat, and so no attempt to deduce the second law from purely dynamical principles, however satisfactory it might appear, could be a "sufficient explanation" of this law (*1*, p. 671).

This conclusion has a certain irony about it, and Maxwell was probably quite aware of that. The second law denies the possibility of perpetual motion machines of the second kind: no matter how ingenious one of these devices may seem to be, one can be certain that in some way or other its existence would violate the laws of nature. Maxwell had cast his conclusion in precisely the same form: no matter how convincing a purely dynamical proof of the second law might seem to be, it could not adequately explain the limited convertibility of heat into work.

Maxwell might have observed—but did not—that one of the subjects for dispute between "those learned Germans" was precisely the proper way of describing work and heat at the molecular level. Boltzmann (*21*, pp. 24–28) imagined a process in which each molecule was supplied with the same amount of energy, $\epsilon$. The sum of $\epsilon$ for all molecules was his way of representing the heat added to the system, but he gave no indication of *how* this $\epsilon$ was to be supplied to each molecule. Boltzmann did have to require that this energy went either into work done against external forces or into an increase of the kinetic energy of that particular molecule, explicitly excluding its partial transfer to other molecules. This might be reasonable enough on the average, but the nature of this average was never analyzed. Boltzmann also made no attempt to clarify how the work done by the individual molecules, represented only very schematically, was to be related to the work done by the gas in displacing a piston under pressure. The coordinate of the piston, that is to say, the volume of the gas, not being a molecular coordinate, never appeared in his equations.

Clausius (*28*) criticized Boltzmann's treatment of

some of these points, and this was the reason he did not grant Boltzmann's absolute priority for the mechanical derivation of the second law. His own way of treating the work involved making the potential energy function depend on a parameter. This parameter, in turn, which varied from one orbit to another, was assumed to change uniformly with time as the particle went through a complete orbit. It must be remembered that Clausius, like Boltzmann, could make even reasonably precise arguments only when the particles moved in closed orbits, that is, for a periodic system. Boltzmann was not happy about Clausius' way of "allowing the laws of nature to vary with time," and never adopted this procedure in his later discussions of the problem (*33*).

Perhaps the best way of seeing the difficulty of separating work and heat in a mechanical theory is to look at the particularly clear exposition of the Boltzmann-Clausius theorem given by Paul Ehrenfest (*34*). One considers a system whose Lagrangian depends on a set of generalized coordinates, $q$, the corresponding generalized velocities, $\dot{q}$, and a parameter, $a$, such as the volume in the case of a gas. The system is assumed to be periodic, and to remain periodic (but with a changed period) in a variation from the original orbit to a neighboring orbit. This variation is of the kind considered in the principle of least action, i.e. nonsimultaneous, but the two orbits are also assumed to differ because the parameter has the value $a + \Delta a$ rather than $a$ on the new orbit. By the usual variational procedures Ehrenfest derived the equation,

$$\Delta \int_0^\tau 2T\, dt = \tau(\Delta E + \bar{A}\,\Delta\, a). \qquad (6)$$

The $\Delta$ represents the change in the generalized variation just described. $T$ is the kinetic energy, $E$ is the total energy, and $\tau$ is the period of the system. The quantity $\bar{A}$ is defined by the equations,

$$\bar{A} = \frac{1}{\tau}\int_0^\tau A\,dt; \qquad A = \frac{\partial L}{\partial a}, \qquad (7)$$

so that $(-A)$ is the generalized force that must be exerted on the system to hold the parameter, $a$, constant, and $\bar{A}$ is its time average. In the standard treatment of the principle of least action there is no variation of a parameter, $a$, and the variation of the path is carried out at constant total energy, $E$. As a result the variation of the action, which is just the time integral of the kinetic energy, vanishes.

When the parameter, $a$, is varied slowly, so that the variation occurs during a time long compared to the period of the motion, the external work done by the system is just $\bar{A}\,\Delta a$. In this case the change in energy, $\Delta E$, is just the negative of the work done, and the complete variation of the action, (6), still vanishes. This slow variation is the Ehrenfest adiabatic process, so important in the old quantum theory (*35*). When the variation is not carried out at this very slow pace the right-hand side of (6) need not vanish. *If* one still calls $\bar{A}\Delta a$ the work, $\Delta W$, done by the system, *then* it is proper to identify $\Delta E + \Delta W$ with $\Delta Q$, the heat supplied to the system. In that case (6) leads directly to the equation,

$$2\Delta(\tau\bar{T}) = \tau\Delta Q. \qquad (8)$$

This is easily rewritten in the form,

$$\frac{\Delta Q}{\bar{T}} = \Delta \ln(\bar{T}\tau)^2, \qquad (9)$$

which is equivalent to (3), the Boltzmann-Clausius result, when the average kinetic energy is set equal to the temperature.

The argument hinges on the identification of $\bar{A}\Delta a$ with the work done by the system, but there is no compelling reason for this identification. The work is the integral, $\int_a^{a+\Delta a} A da$, and this is not equivalent to $\bar{A}\,\Delta a$ when the time of the variation is of the order of, or less, than, the period. There is no obvious way of distinguishing two separate modes of energy transfer as heat and work in such a process. This is an illustration of Maxwell's general statement that mechanical explanations of the second law can never be adequate, no matter how impressive their outward appearance may be.

## Boltzmann and statistical mechanics

Maxwell was not the one to develop the statistical mechanics that does explain the second law. He was apparently satisfied to have recognized that "the truth of the second law [is] of the nature of a strong probability. . . not an absolute certainty" (*1*, p. 671), that it depended in an essential way on the unimaginably large number of molecules that constitute macroscopic bodies. It was Boltzmann who undertook the task of creating a statistical theory that would make clear just how a mechanical system of so many particles could show the behavior required by the laws of thermodynamics. But Boltzmann found his starting point in Maxwell's work, for which he had a profound admiration.

At the time when Boltzmann tried to account for the

second law in purely mechanical terms, he was apparently unfamiliar with what Maxwell had written on the theory of gases. When he quoted the relationship between the pressure of a gas and the velocities of its molecules, he referred to Krönig, Rankine, and Clausius, but did not mention Maxwell or his law for the distribution of molecular velocities (*36*). We know that Boltzmann was learning to read English at this time—in order to read Maxwell's papers on electromagnetism (*37*). He returned to the theory of gases in 1868, however, with fresh ideas prompted by his study of Maxwell's new analysis of the kinetic theory (*38*). Boltzmann had embarked on a series of lengthy memoirs in which the statistical description of the gas was absolutely central to his treatment.

By the time Clausius was criticizing the fine points of his mechanical interpretation of the second law, Boltzmann was already committed to the statistical approach. In that same year, 1871, he offered a new proof of the law, but this time it was a statistical mechanical, rather than a mechanical, proof (*39*). This new proof dealt only with the equilibrium aspect of the second law—that is, with the existence of the entropy as a thermodynamic state function related to heat and temperature in the proper way, through equation (1). Work and heat were now distinguished with the help of the distribution function: roughly speaking, a change in the total energy produced by a change in the molecular potential energy with a fixed distribution was work, whereas heat was identified with a change in the energy due only to a change in the distribution. The exponential energy distribution at equilibrium played an important part in the analysis. The essential features of Boltzmann's argument have survived the transition from classical to quantum physics, and it is still to be found in textbooks on statistical mechanics (*40*).

A year later Boltzmann was able to provide a molecular basis for the other aspect of the second law, the increase of the entropy of an isolated system whenever an irreversible process occurs (*41*). This came as a by-product of an investigation into the uniqueness of the Maxwell-Boltzmann equilibrium distribution. Boltzmann's result, the famous $H$-theorem, was based on his analysis of how the distribution evolved in time as a result of binary molecular collisions. Although much more limited in its scope than his discussion of equilibrium, the $H$-theorem offered the first insight into the nature of irreversibility.

It must be emphasized, however, that while Boltzmann made constant use of the statistical distribution of molecular velocities and energies, the result he asserted was supposed to be a certainty and not just a probability. The $H$-theorem of 1872 said that entropy would *always* increase (*41*, p. 345). Boltzmann did not revise this statement until 1877, when Josef Loschmidt pointed out that it was untenable in its original form (*42*). Since the equations of motion of the molecules do not distinguish between the two senses in which time might flow (that is, they are invariant to the replacement of $t$ by $-t$), the time reversal of any actual motion is an equally legitimate solution of the equations. As a consequence, for any motion of the molecules in which the entropy of the gas increased with time, one could construct another motion in which the entropy decreased with time, in apparent contradiction to the second law. It was only in his reply to Loschmidt's criticism that Boltzmann recognized "how intimately the second law is connected to the theory of probability," and that the increase in entropy can only be described as very highly probable, but not as certain. He elaborated this view in a major article later that same year, in which he formulated the relationship between the entropy of a thermodynamic state and the number of molecular configurations compatible with that state (*43*). Entropy became simply a measure of this "thermodynamic probability."

Boltzmann's work can be viewed legitimately as a development of the insight first expressed by Maxwell—an extended, detailed, and quantitative development of a brief qualitative insight, to be sure, but a natural development, all the same. Nevertheless Maxwell seems to have taken little or no interest in these papers by Boltzmann. He did follow some particular aspects of Boltzmann's work in statistical mechanics; Maxwell's last scientific work was a new analysis of the equipartition theorem, taking Boltzmann's 1868 memoir as its starting point and presenting an alternate way of handling some of the difficult assumptions (*44*). Maxwell was certainly aware of much of Boltzmann's other work. Henry William Watson's little book on the kinetic theory, which appeared in 1876 (*45*), followed Boltzmann's methods and included a modified version of Boltzmann's first statistical derivation of the existence of an entropy function. Watson acknowledged "much kind assistance" from Maxwell in his Preface, and Maxwell also reviewed the book for *Nature* (*46*). But nowhere did Maxwell discuss Boltzmann's statistical mechanical theory of the second law of thermodynamics. It is not even mentioned in his review of Tait, where it would have been completely appropriate.

There is no obvious explanation for the peculiar lack of intellectual contact between Maxwell and Boltzmann at this one point. It is true that their styles were very different. Boltzmann found Maxwell hard to understand because of his terseness (*38*, p. 49); Maxwell found the great length of Boltzmann's discussions to be "an equal stumbling block" and wanted to reduce one of his papers to "about six lines" (*47*). It may well be that Maxwell simply never read Boltzmann's memoirs of 1877, and so missed the pleasure of seeing the relationship between entropy and probability.

## Fluctuations and the demon

Maxwell conjured up his demon as a way of giving life to the difficult idea that the second law of thermodynamics is a statistical truth and not a dynamical certainty. This idea caught on quickly in Maxwell's circle. William Thomson incorporated it into his own thinking and elaborated it in a paper on the kinetic theory of energy dissipation. He went so far as to imagine an army of Maxwell's demons stationed at the interface of a hot and a cold gas, each member of this "ideal army" being armed with "a club, or, as it were, a molecular cricket bat," and described the variety of ways in which such an army could control molecular processes (*48*). But even though Thomson was playful enough to endow the demon "with arms and hands and fingers—two hands and ten fingers suffice" (*49*), he knew very well why they had been called into being. Thomson's paper of 1874 raised the reversibility paradox with which Loschmidt would confront Boltzmann a few years later, and also showed that it was no real paradox, in much the way Boltzmann would. Thomson went further. He calculated the probability of the fluctuations that are an inescapable aspect of the statistical theory, showing that while it was not impossible that one would find all the oxygen molecules in a vessel of air in the left-most fifth of the container, the probability of this event was staggeringly small.

Tait also absorbed the statistical molecular approach to the second law. He pointed out in his lectures that molecular processes taking energy from a colder to a warmer body are "constantly going on, but on a very limited scale, in every mass of a gas." This fact did not interfere with the validity of Carnot's theorem and could be safely ignored for "the enormous number of particles in a cubic inch of even the most rarefied gas" (*15*, pp. 119–20). But Tait also found the existence of such fluctuations to be a convenient stick with which to beat his old antagonist, Clausius, commenting that the demon argument was "absolutely fatal to Clausius' reasoning."

This was, of course, an exaggerated version of what Maxwell had originally said. He had looked for possible limitations of the second law, but he had certainly never intended to destroy it. Clausius, used to Tait's onslaughts and practiced in countering them with equal vigor, had a proper answer to this new attack: "I believe I can summarize my reply to Mr. Tait's objection in the brief remark that my law is concerned, not with what heat can do with the help of demons, but rather with what it can do by itself" (*50*). He saw no reason why the existence of fluctuations should be considered as contradicting the average behavior described by the second law.

Maxwell's own view was not essentially different. He saw no conflict between the two very different ways of establishing thermodynamics. "The second law," he wrote, "must either be founded on our actual experience in dealing with real bodies of sensible magnitude, or else deduced from the molecular theory of these bodies, on the hypothesis that the behaviour of bodies consisting of millions of molecules may be deduced from the theory of the encounters of pairs of molecules, by supposing the relative frequency of different kinds of encounters to be distributed according to the laws of probability (*1*, pp. 669–70).

Real experimental support for the statistical molecular theory did not come until the early years of this century, but then it came in strength. A whole series of fluctuation phenomena was studied quantitatively—the Brownian motion, critical opalescence, density fluctuations in colloidal suspensions—and the statistical theory accounted for all of them. By 1912, Maryan von Smoluchowski could give an invited lecture with the "revolutionary sounding" title "Experimentally Verifiable Molecular Phenomena that Contradict Ordinary Thermodynamics" (*51*). A decade earlier "it would have been foolhardy to have spoken so disrespectfully of the traditional formulation of thermodynamics," at least on the continent of Europe. There was now good reason for taking Maxwell's statistical approach completely seriously. The hole Maxwell had picked in the second law had become plainly visible (*52*).

Smoluchowski's brilliant analysis of the fluctuation phenomena showed that one could observe violations of almost all the usual statements of the second law by dealing with sufficiently small systems. The trend toward equilibrium, the increase of entropy, and so on, could not be taken as certainties. The one statement that could be upheld, even in the presence of fluctuations, was the impossibility of a perpetual motion of the second kind. No device could ever be made that would use the existing fluctuations to convert heat completely into work on a macroscopic scale. For any such device would have to be constituted of molecules and would therefore itself be subject to the same chance fluctuations. To *use* the molecular fluctuations one would have to construct a device whose action would not be limited by them. "Only we can't," as Maxwell had said, "not being clever enough."

## Appendix. On Maxwell's Signature

James Clerk Maxwell often signed letters to such friends as Thomson and Tait with the signature $dp/dt$. He also used this derivative as a nom de plume for the witty verses on scientific subjects that he published in *Nature* (*53*). It is not hard to guess that he was making private reference to an equation of the form

$$\frac{dp}{dt} = JCM \tag{A1}$$

so that $dp/dt$ simply represented his initials. But what was the equation, and where did it come from? The late D. K. C. MacDonald recently described it as one of Maxwell's four thermodynamic relations (*54*), derived by him in the *Theory of Heat* from the geometry of isothermal and adiabatic curves (*30*, pp. 165–69), and now commonly derived from the equality of the mixed second derivatives of the various thermodynamic potentials. MacDonald's statement is technically correct but historically misleading. To Maxwell the relationship in (A1) expressed the second law of thermodynamics itself.

Carnot's statement of his fundamental theorem said that the motive power of heat depends only on the temperatures between which heat is transferred in the working of an engine. This was translated into mathematical terms by Émile Clapeyron in 1834 (*55*). A slightly modified version of the argument runs in the following way.

Consider an infinitesimal Carnot cycle consisting of an isothermal expansion at temperature $(t + dt)$ an adiabatic expansion which cools the system down to temperature $t$, an isothermal compression at temperature $t$, and a final adiabatic compression back to the initial state, all processes being carried out reversibly. Since the cycle is infinitesimal it can be represented as a parallelogram in the pressure-volume diagram. The area of this parallelogram is the work, $\Delta W$, done per cycle and it has the value

$$\Delta W = (dp)\,(dV) \tag{A2}$$

where $dp$ and $dV$ are, respectively, the difference in pressure between the two isotherms and the difference in volume in either isothermal step. The efficiency of the cycle is the ratio of $\Delta W$ to $\Delta Q$, where $\Delta Q$ is the heat absorbed in the isothermal expansion. Carnot's theorem states that this efficiency, "the motive power of heat," depends only on the temperatures $t$ and $t + dt$, which means that it can be expressed as $C(t)dt$ since $dt$ is infinitesimal. Here $C(t)$ is a universal function (the Carnot function) of the temperature $t$, this temperature, $t$, being measured on any arbitrary scale. If one introduces the mechanical equivalent of heat, $J$, to allow for the fact that heat and work were expressed in different units, the equation expressing Carnot's theorem can be written in the form

$$\frac{\Delta W}{J\Delta Q} = C(t)\ dt. \tag{A3}$$

The quantity of heat absorbed is, however, proportional to the extent of the expansion, $dV$, so that one has the equation

$$\Delta Q = M\ dV \tag{A4}$$

where $M$ is the coefficient of proportionality, the heat absorbed per unit volume change in an isothermal expansion. With the help of (A2) and (A4), the efficiency equation now becomes

$$\frac{(dp)\quad (dV)}{J\ M\ (dV)} = C\ dt \tag{A5}$$

or, with a slight rearrangement,

$$\frac{dp}{dt} = JCM. \tag{A1}$$

This is the basis of the assertion made by Maxwell's and Tait's biographers that (A1) represents the second law of thermodynamics (*56*).

There is, however, a little more to the story than that. It is obvious that Maxwell's use of $dp/dt$ depends in an essential way on the notation used to express this result, but not all the notation in (A1) was standard. Pressure and temperature were commonly denoted by $p$ and $t$. The quantities on the other side of the equation had no standard designation. In his first papers on thermodynamics, William Thomson had used $J$ for the mechanical equivalent, probably in honor of Joule, and $M$ for the quantity then usually denoted by $dQ/dV$ (*57*). But it is $C$ that is really interesting.

Clapeyron first made explicit use of the universal function of temperature whose existence is called for by Carnot's theorem. He did call it $C(t)$, but Clapeyron's $C(t)$ is essentially the *reciprocal* of the function defined by (A3). The central importance of this universal function was recognized by Clapeyron, who evaluated it numerically from several different kinds of data to establish its universality. Clausius used $C$ as Clapeyron had, and also confirmed that $C$ was equal to the ideal gas temperature (*58*). In modern notation $C(t)$, as used by Clapeyron and Clausius, is proportional to the absolute temperature.

Thomson did not use this notation. He wrote the equation we have called (A1) in the form

$$\frac{dp}{dt} = \mu M, \tag{A6}$$

where $\mu$ was the universal function of temperature, which Thomson called the Carnot function. The function $\mu$ incorporates the mechanical equivalent and is the integrating factor for the inexact differential $dQ$.

Where, then, did Maxwell get the particular form of the Carnot theorem which allowed him to use $dp/dt$ as his signature? The first use of $C(t)$ in the form of (A3) and the first appearance of (A1) occur together in that same *Sketch of Thermodynamics* which called forth the Maxwell demon. It was Tait who renamed the Carnot function, making his $C(t)$ the reciprocal of the absolute temperature, and who thereby put $JCM$ into the second law of thermodynamics (*4*, p. 91). It is not very likely that this was fortuitous. The Tait–Maxwell correspondence is full of verbal joking of various sorts. Both men wrote well, and often playfully; both were sensitive to questions of terminology and notation.

Tait actually commented on his old friend's thermodynamic signature in his notice of Maxwell's scientific work: "This nom de plume was suggested to him by me from the occurrence of his initials in the well-known expression of the second Law of Thermodynamics (for whose establishment on thoroughly valid grounds he did so much) $dp/dt = JCM$" (*8*, p. 321).

The first appearance of $dp/dt$ in the Maxwell–Tait correspondence is on a postcard from Tait to Maxwell early in 1871 (*59*). Tait addressed Maxwell as $dp/dt$, and added an explanatory parenthesis: "[$= JCM$ (T′'s Thermodynamics § 162]." This is a reference to the paragraph of his *Sketch* in which (A1) is stated. T′ is the shorthand designation for Tait, just as T was used to designate Thomson, their famous book being referred to as T and T′ by the inner circle.

There is, however, an even earlier use of $dp/dt$ by Maxwell. It occurs in a letter to Thomson in April 1870 (*60*), almost a year before Tait's card pointing out that $dp/dt$ is equivalent to $JCM$, but well over a year after the publication of Tait's book. This suggests that Maxwell had already noticed the equation that Tait probably redesigned for his benefit and did not need to have it pointed out explicitly to him. He was certainly clever enough.

## Acknowledgments

This work was supported in part by a grant from the National Science Foundation. I thank Dr. Elizabeth Garber for many informative discussions on Maxwell, and my colleague Professor Derek J. de Solla Price for the use of his copy of the Cambridge University collection of Maxwell correspondence.

## *Notes*

1. J. C. Maxwell, "Tait's *Thermodynamics*," *Nature 17* (1878), 257. Reprinted in *The Scientific Papers of James Clerk Maxwell*, ed. W. D. Niven (Cambridge, 1890), Vol. 2, 660–71. Quotation from p. 662.
2. R. Clausius, *Abhandlungen über die mechanische Wärmetheorie* (Braunschweig, 1864).
3. R. Clausius, "Über verschiedene für die Anwendung bequeme Formen der Hauptgleichungen der mechanischen Wärmetheorie," *Pogg. Ann. 125* (1865), 353.
4. P. G. Tait, *Sketch of Thermodynamics* (Edinburgh, 1868).
5. For further discussion of this controversy see: (a) C. G. Knott, *Life and Scientific Work of Peter Guthrie Tait* (Cambridge, 1911), 208–26. Referred to below as Knott, *Tait*. (b) M. J. Klein, "Gibbs on Clausius," *Historical Studies in the Physical Sciences 1* (1969), 127–49.
6. R. Clausius, *Die Mechanische Wärmetheorie*, 2d ed. Vol. 2 (Braunschweig, 1879), 324–30.
7. Knott, *Tait*, contains extensive excerpts from this correspondence as well as a description of the relationship of Tait and Maxwell. Also see L. Campbell and W. Garnett, *The Life of James Clerk Maxwell* (London, 1882).
8. P. G. Tait, "Clerk-Maxwell's Scientific Work," *Nature 21* (1880), 321.
9. J. C. Maxwell to P. G. Tait, 11 December 1867. Reprinted in Knott, *Tait*, 213–14.
10. S. Carnot, *Reflections on the Motive Power of Fire*, trans. R. H. Thurston, ed. E. Mendoza (New York, 1960), 20. This volume also contains papers by É. Clapeyron and R. Clausius. It is referred to below as Mendoza reprint.
11. See, e.g., E. Mach, *Die Principien der Wärmelehre*, 3d ed. (Leipzig, 1919), 269–71. Also see M. J. Klein, n. 5, for further references.
12. R. Clausius, "Über die bewegende Kraft der Wärme, und die Gesetze, welche sich daraus für die Wärmelehre selbst ableiten lassen," *Pogg. Ann. 79* (1850), 368,500. Trans. by W. F. Magie in Mendoza reprint. Also see J. W. Gibbs, "Rudolf Julius Emanuel Clausius," *Proc. Amer. Acad. 16* (1889), 458. Reprinted in *The Scientific Papers of J. Willard Gibbs* (New York, 1906), Vol. 2, 261.
13. R. Clausius in Mendoza reprint, 134.
14. R. Clausius, "On a Modified Form of the Second Fundamental Theorem in the Mechanical Theory of Heat," *Phil. Mag. 12* (1856), 86.
15. Also see P. G. Tait, *Lectures on Some Recent Advances in Physical Science*, 2d ed. (London, 1876), 119, and Sir William Thomson, Baron Kelvin, *Mathematical and Physical Papers*, Vol. 5 (Cambridge, 1911), 12, 19.
16. J. C. Maxwell to J. W. Strutt, 6 December 1870. Reprinted in R. J. Strutt, *Life of John William Strutt, Third Baron Rayleigh* (Madison, 1968), 47.
17. R. Clausius, "The Nature of the Motion Which We Call Heat," *Phil. Mag. 14* (1857), 108; "On the Mean Lengths of the Paths Described by the Separate Molecules of Gaseous Bodies," *Phil. Mag. 17* (1859), 81. Both are reprinted in S. G. Brush, *Kinetic Theory*, Vol. 1 (Oxford, 1965).
18. J. C. Maxwell to G. G. Stokes, 30 May 1859. Quoted by Brush (n. 17), 26–27.
19. J. C. Maxwell, "Illustrations of the Dynamical Theory of Gases," *Phil. Mag. 19* (1860), 19; *20* (1860), 21. *Scientific Papers*, Vol. 1, 377–409. Quotation from p. 380.
20. J. C. Maxwell, "On the Dynamical Theory of Gases," *Phil. Mag. 32* (1866), 390; *35* (1868) 129, 185. *Scientific Papers*, Vol. 2, 26–78.
21. L. Boltzmann, "Über die mechanische Bedeutung des zweiten Hauptsatzes der Wärmetheorie," *Wiener Berichte 53* (1866), 195. Reprinted in L. Boltzmann, *Wissenschaftliche*

*Abhandlungen*, ed. F. Hasenöhrl (Leipzig, 1909), Vol. 1, 9–33. This collection will be referred to as Boltzmann, *Wiss. Abh.*, and page references to Boltzmann's work are made to this reprint.

22. H. Hertz, *The Principles of Mechanics Presented in a New Form*, trans. D. E. Jones and J. T. Walley (Reprinted New York, 1956), Author's Preface.
23. Boltzmann, *Wiss. Abh.*, Vol. 1, 30. Some years later Boltzmann proved that nonperiodic systems (open orbits) could not be handled by these methods. See L. Boltzmann, "Bemerkungen über einige Probleme der mechanischen Wärmetheorie," *Wiener Berichte 75* (1877), 62, or *Wiss. Abh.*, Vol. 2, 122–48.
24. R. Clausius, "On the Application of the Theorem of the Equivalence of Transformations to the Internal Work of a Mass of Matter," *Phil. Mag. 24* (1862) 81, 201. Also see M. J. Klein (n. 5) and E. E. Daub, "Atomism and Thermodynamics," *Isis 58* (1967), 293.
25. R. Clausius, "On a Mechanical Theorem Applicable to Heat," *Phil. Mag. 40* (1870), 122. Reprinted in Brush, *Kinetic Theory*, 172–78.
26. R. Clausius, "Über die Zurückführung des zweiten Hauptsatzes der mechanischen Wärmetheorie auf allgemeine mechanische Principien," *Pogg. Ann. 142* (1871), 433.
27. L. Boltzmann, "Zur Priorität der Auffindung der Beziehung zwischen dem zweiten Hauptsatze der mechanischen Wärmetheorie und dem Prinzip der kleinsten Wirkung," *Pogg. Ann. 143* (1871), 211; *Wiss. Abh.*, Vol. 1, 228–36.
28. R. Clausius, "Bemerkungen zu der Prioritätsreclamation des Hrn. Boltzmann," *Pogg. Ann. 144* (1871), 265.
29. J. C. Maxwell to P. G. Tait, 1 December 1873. Reprinted in Knott, *Tait*, 115–16.
30. J. C. Maxwell, *Theory of Heat* (London, 1871). My quotations are from the 4th London edition of 1875.
31. *Ibid.* Advertisement of the publisher (Longmans, Green and Co.) for the series, printed at the end of the text.
32. Maxwell, *Heat*, 4th ed., 195–208. For a detailed discussion of Gibbs's influence on Maxwell and Maxwell's appreciation of Gibbs see E. Garber, "James Clerk Maxwell and Thermodynamics," *Amer. Jour. Phys. 37* (1969) 146.
33. L. Boltzmann, *Vorlesungen über die Prinzipe der Mechanik* (Leipzig, 1904), Vol. 2, 162–64.
34. P. Ehrenfest, "On Adiabatic Changes of a System in Connection with the Quantum Theory," *Proc. Acad. Amst. 19* (1916), 591. Reprinted in P. Ehrenfest, *Collected Scientific Papers*, ed. M. J. Klein (Amsterdam, 1959), 393–97.
35. M. J. Klein, *Paul Ehrenfest. The Making of a Theoretical Physicist* (Amsterdam, 1970), chapters 10–11.
36. L. Boltzmann, n. 21, p. 21.
37. L. Boltzmann, *Populäre Schriften* (Leipzig, 1905), 96.
38. L. Boltzmann, "Studien über das Gleichgewicht der lebendigen Kraft zwischen bewegten materiellen Punkten," *Wiener Berichte 58* (1868), 517. *Wiss. Abh.*, Vol. 1, 49–96.
39. L. Boltzmann, "Analytischer Beweis des zweiten Hauptsatzes der mechanischen Wärmetheorie aus den Sätzen über das Gleichgewicht der lebendigen Kraft," *Wiener Berichte 63* (1871), 712. *Wiss. Abh.*, Vol. 1, 288–308.
40. See, e.g., E. Schrödinger, *Statistical Thermodynamics* 2d ed. (Cambridge, 1952), 10–14.
41. L. Boltzmann, "Weitere Studien über das Wärmegleichgewicht unter Gasmolekülen," *Wiener Berichte 66* (1872), 275. *Wiss. Abh.*, Vol. 1, 316–402.
42. L. Boltzmann, "Bemerkungen über einige Probleme der mechanischen Wärmetheorie," *Wiener Berichte 75* (1877), 62. *Wiss. Abh.*, Vol. 2, 116–22.
43. L. Boltzmann, Über die Beziehung zwischen dem zweiten Hauptsatze der mechanischen Wärmetheorie und der Wahrscheinlichkeitsrechnung respektive den Sätzen über das Wärmegleichgewicht," *Wiener Berichte 76* (1877), 373. *Wiss. Abh.*, Vol. 2, 164–223.
44. J. C. Maxwell, "On Boltzmann's Theorem on the Average Distribution of Energy in a System of Material Points," *Trans. Cambr. Phil. Soc. 12* (1879), 547. *Scientific Papers*, Vol. 2 713–41.
45. H. W. Watson, *A Treatise on the Kinetic Theory of Gases* (Oxford, 1876).
46. J. C. Maxwell, "The Kinetic Theory of Gases," *Nature 16* (1877), 242. This is not reprinted in the *Scientific Papers*.
47. J. C. Maxwell to P. G. Tait, August 1873. Quoted in Knott, *Tait*, 114.
48. W. Thomson, "The Kinetic Theory of the Dissipation of Energy," *Nature 9* (1874), 441. *Papers*, Vol. 5, 13.
49. W. Thomson, "The Sorting Demon of Maxwell," *Papers*, Vol. 5, 21.
50. R. Clausius, n. 6, p. 316. Clausius' comments on Maxwell's demon are discussed by L. Rosenfeld, "On the Foundations of Statistical Thermodynamics," *Acta Physica Polonica 14* (1955), 37–38.
51. M. v. Smoluchowski, "Experimentell nachweisbare, der üblichen Thermodynamik widersprechende Molekularphänomene," *Phys. Zeits. 13* (1912), 1069.
52. A recent analysis of the relationship between Maxwell's demon and information theory can be found in L. Brillouin, *Science and Information Theory* 2d ed. (New York, 1962), 162–82.
53. These verses are reprinted in the Campbell and Garnett biography of Maxwell (n. 7).
54. D. K. C. MacDonald, *Faraday, Maxwell, and Kelvin* (New York, 1964), 62–63, 98–99. The derivative $dp/dt$ would now be written more carefully as $(\partial p/\partial t)_V$.
55. É. Clapeyron, "Memoir on the Motive Power of Heat," in Mendoza reprint, pp. 73–105.
56. Knott, *Tait*, p. 101; Campbell and Garnett, *Maxwell*, pp. ix–x.
57. W. Thomson, "On the Dynamical Theory of Heat," *Trans. Roy. Soc. Edinburgh 20* (1851), 261. Reprinted in W. Thomson, *Mathematical and Physical Papers*, Vol. 1 (Cambridge, 1882), 174–316. See especially p. 187.
58. R. Clausius, n. 12.
59. P. G. Tait to J. C. Maxwell, 1 February 1871.
60. J. C. Maxwell to W. Thomson, 14 April 1870. (This letter and the preceding one are in the Cambridge University collection of Maxwell correspondence.)

# LIFE, THERMODYNAMICS, AND CYBERNETICS

By L. BRILLOUIN
Cruft Laboratory, Harvard University*

HOW is it possible to understand life, when the whole world is ruled by such a law as the second principle of thermodynamics, which points toward death and annihilation? This question has been asked by many scientists, and, in particular, by the Swiss physicist, C. E. Guye, in a very interesting book.[1] The problem was discussed at the Collège de France in 1938, when physicists, chemists, and biologists met together and had difficulty in adjusting their different points of view. We could not reach complete agreement, and at the close of the discussions there were three well defined groups of opinion:

(*A*) Our present knowledge of physics and chemistry is practically complete, and these physical and chemical laws will soon enable us to explain life, without the intervention of any special "life principle."

(*B*) We know a great deal about physics and chemistry, but it is presumptuous to pretend that we know all about them. We hope that, among the things yet to be discovered, some new laws and principles will be found that will give us an interpretation of life. We admit that life obeys all the laws of physics and chemistry at present known to us, but we definitely feel that something more is needed before we can understand life. Whether it be called a "life principle" or otherwise is immaterial.

(*C*) Life cannot be understood without reference to a "life principle." The behavior of living organisms is completely different from that of inert matter. Our principles of thermodynamics, and especially the second one, apply only to dead and inert objects; life is an exception to the second principle, and the new principle of life will have to explain conditions contrary to the second law of thermodynamics.

Another discussion of the same problems, held at Harvard in 1946, led to similar conclusions and revealed the same differences of opinion.

In summarizing these three points of view, I have of course introduced some oversimplifications. Recalling the discussions, I am certain that opinions *A* and *B* were very clearly expressed. As for opinion *C*, possibly no one dared to state it as clearly as I have here, but it was surely in the minds of a few scientists, and some of the points introduced in the discussion lead logically to this opinion. For instance, consider a living organism; it has special properties which enable it to resist destruction, to heal its wounds, and to cure occasional sickness. This is very strange behavior, and nothing similar can be observed about inert matter. Is such behavior an exception to the second principle? It appears so, at least superficially, and we must be prepared to

* *Now Director of Electronic Education, International Business Machines Corporation, New York, N. Y.*

[1] *L'évolution physico-chimique* (Paris, E. Chiron, 1922).

accept a "life principle" that would allow for some exceptions to the second principle. When life ceases and death occurs, the "life principle" stops working, and the second principle regains its full power, implying demolition of the living structure. There is no more healing, no more resistance to sickness; the destruction of the former organism goes on unchecked and is completed in a very short time. Thus the conclusion, or question: What about life and the second principle? Is there not, in living organisms, some power that prevents the action of the second principle?

### *The Attitude of the Scientist*

The three groups as defined in the preceding section may be seen to correspond to general attitudes of scientists towards research: (*A*) strictly conservative, biased against any change, and interested only in new development and application of well established methods or principles; (*B*) progressive, open-minded, ready to accept new ideas and discoveries; (*C*) revolutionary, or rather, metaphysical, with a tendency to wishful thinking, or indulging in theories lacking solid experimental basis.

In the discussion just reviewed, most non-specialists rallied into group *B*. This is easy to understand. Physicists of the present century had to acquire a certain feeling for the unknown, and always to be very cautious against over-confidence. Prominent scientists of the previous generation, about 1900, would all be classed in group *A*. Common opinion about that time was that everything was known and that coming generations of scientists could only improve on the accuracy of experiments and measure one or two more decimals on the physical constants. Then some new laws were discovered: quanta, relativity, and radioactivity. To cite more specific examples, the Swiss physicist Ritz was bold enough to write, at the end of the nineteenth century, that the laws of mechanics could not explain optical spectra. Thirty years passed before the first quantum mechanical explanation was achieved. Then, about 1922, after the first brilliant successes of quantum mechanics, things came to a standstill while experimental material was accumulating. Some scientists (Class *A*) still believed that it was just a question of solving certain very complicated mathematical problems, and that the explanation would be obtained from principles already known. On the contrary, however, we had to discover wave mechanics, spinning electrons, and the whole structure of the present physical theories. Now, to speak frankly, we seem to have reached another dead-end. Present methods of quantum mechanics appear not to be able to explain the properties of fundamental particles, and attempts at such explanations look decidedly artificial. Many scientists again believe that a new idea is needed, a new type of mathematical correlation, before we can go one step further.

All this serves to prove that every physicist must be prepared for many new discoveries in his own domain. Class *A* corresponds to cautiousness. Before abandoning the safe ground of well established ideas, says the cautious scientist, it must be proved that these ideas do not check with experiments. Such was the case with the Michelson-Morley experiment. Nevertheless, the same group of people were extremely reluctant to adopt relativity.

Attitude *B* seems to be more constructive, and corresponds to the trend of scientific research through past centuries; attitude *C*, despite its exaggeration, is far from being untenable. We have watched many cases of new discoveries leading to limitations of certain previous "laws." After all, a scientific law is not a "decree" from some supernatural power; it simply represents a systematization of a large number of experimental results. As a consequence, the scientific law has only a limited validity. It extends over the whole domain of experimentation, and maybe slightly beyond. But we must be prepared for some strange modifications when our knowledge is expanded much farther than this. Many historical examples could be introduced to support this opinion. Classical mechanics, for instance, was one of the best-established theories, yet it had to be modified to account for the behavior of very fast particles (relativity), atomic structure, or cosmogony.

Far from being foolish, attitude *C* is essentially an exaggeration of *B;* and any scholar taking attitude *B* must be prepared to accept some aspects of group *C*, if he feels it necessary and if these views rest upon a sound foundation.

To return to the specific problem of life and thermodynamics, we find it discussed along a very personal and original line in a small book published by the famous physicist E. Schrödinger.[2] His discussion is very interesting and there are many points worth quoting. Some of them will be examined later on. In our previous classification, Schrödinger without hesitation joins group *B*:

> We cannot expect [he states] that the "laws of physics" derived from it [from the second principle and its statistical interpretation] suffice straightaway to explain the behavior of living matter. . . . We must be prepared to find a new type of physical law prevailing in it. Or are we to term it a non-physical, not to say a super-physical law?[3]

The reasons for such an attitude are very convincingly explained by Schrödinger, and no attempt will be made to summarize them here. Those who undertake to read the book will find plenty of material for reflection and discussion. Let us simply state at this point that there is a problem about "life and the second principle." The answer is not obvious, and we shall now attempt to discuss that problem systematically.

### *The Second Principle of Thermodynamics, Its Successes and Its Shortcomings*

Nobody can doubt the validity of the second principle, no more than he can the validity of the fundamental laws of mechanics. However, the question is to specify its domain of applicability and the

[2] E. Schrödinger, *What is Life?* (London, Cambridge University Press, and New York, The Macmillan Company, 1945).

[3] *Ibid.*, p. 80.

chapters of science or the type of problems for which it works safely. We shall put special emphasis on all cases where the second principle remains silent and gives no answer. It is a typical feature of this principle that it has to be stated as an inequality. Some quantity, called "entropy," cannot decrease (under certain conditions to be specified later); but we can never state whether "entropy" simply stays constant, or increases, or how fast it will increase. Hence, the answer obtained from the second principle is very often evasive, and keeps a sibyllic character. We do not know of any experiment telling *against* the second principle, but we can easily find many cases where it is useless and remains dumb. Let us therefore try to specify these limitations and shortcomings, since it is on this boundary that life plays.

*Both principles of thermodynamics apply only to an isolated system,* which is contained in an enclosure through which no heat can be transferred, no work can be done, and no matter nor radiation can be exchanged.[4] The first principle states that the total energy of the system remains constant. The second principle refers to another quantity called "entropy," $S$, that may only increase, or at least remain constant, but can never decrease. Another way to explain the situation is to say that the total amount of energy is conserved, but not its "quality." Energy may be found in a high-grade quality, which can be transformed into mechanical or electrical work (think of the energy of compressed air in a tank, or of a charged electric battery); but there are also low-grade energies, like heat. The second principle is often referred to as a principle of energy degradation. The increase in entropy means a decrease in quality for the total energy stored in the isolated system.

Consider a certain chemical system (a battery, for instance) and measure its entropy, then seal it and leave it for some time. When you break the seal, you may again measure the entropy, and you will find it increased. If your battery were charged to capacity before sealing, it will have lost some of its charge and will not be able to do the same amount of work after having been stored away for some time. The change may be small, or there may be no change; but certainly the battery cannot increase its charge during storage, unless some additional chemical reaction takes place inside and makes up for the energy and entropy balance. On the other hand, life feeds upon high-grade energy or "negative entropy."[5] A decrease in high-grade energy is tantamount to a loss of food for living organisms. Or we can also say that living organisms automatically destroy first-

[4] The fundamental definition must always start with an isolated system, whose energy, total mass, and volume remain constant. Then, step by step, other problems may be discussed. A body at "constant temperature" is nothing but a body enclosed in a large thermostat, that is, in a big, closed, and isolated tank, whose energy content is so large that any heat developed in the body under experience cannot possibly change the average temperature of the tank. A similar experimental device, with a closed tank containing a large amount of an ideal gas, leads to the idea of a body maintained at constant pressure and constant temperature. These are secondary concepts derived from the original one.

[5] Schrödinger, p. 72.

quality energy, and thus contribute to the different mechanisms of the second principle. If there are some living cells in the enclosure, they will be able for some time to feed upon the reserves available, but sooner or later this will come to an end and death then becomes inevitable.

The second principle means *death by confinement,* and it will be necessary to discuss these terms. Life is constantly menaced by this sentence to death. The only way to avoid it is to prevent confinement. Confinement implies the existence of perfect walls, which are necessary in order to build an ideal enclosure. But there are some very important questions about the problem of the existence of perfect walls. Do we really know any way to build a wall that could not let any radiation in or out? This is theoretically almost impossible; practically, however, it can be done and is easily accomplished in physical or chemical laboratories. There is, it is true, a limitation to the possible application of the second principle, when it comes to highly penetrating radiation, such as ultra-hard rays or cosmic rays; but this does not seem to have any direct connection with the problem of life and need not be discussed here.

*Time and the second principle.* The second principle is a death sentence, but it contains no time limit, and this is one of the very strange points about it. The principle states that in a closed system, $S$ will increase and high-grade energy must decrease; but it does not say how fast. We have even had to include the possibility that nothing at all might happen, and that $S$ would simply remain constant. The second principle is an arrow pointing to a direction along a one-way road, with no upper or lower speed limit. A chemical reaction may flash in a split second or lag on for thousands of centuries.

Although time is not a factor in the second principle, there is, however, a very definite connection between that principle and the definition of time. One of the most important features about time is its irreversibility. Time flows on, never comes back. When the physicist is confronted with this fact he is greatly disturbed. All the laws of physics, in their elementary form, are reversible; that is, they contain the time but not its sign, and positive or negative times have the same function. All these elementary physical laws might just as well work backward. It is only when phenomena related to the second principle (friction, diffusion, energy transferred) are considered that the irreversibility of time comes in. The second principle, as we have noted above, postulates that time flows always in the same direction and cannot turn back. Turn the time back, and your isolated system, where entropy ($S$) previously was increasing, would now show a decrease of entropy. This is impossible: evolution follows a one-way street, where travel in the reverse direction is strictly forbidden. This fundamental observation was made by the founders of thermodynamics, as, for instance, by Lord Kelvin in the following paragraphs:

If, then, the motion of every particle of matter in the universe were precisely reversed at any instant, the course of nature would be simply reversed forever after. The bursting bubble of foam at the foot of a waterfall would reunite and descend into the water; the thermal motions would reconcentrate their energy and throw the mass up the fall in drops re-forming into a close column of ascending water. Heat which had been generated by the friction of solids and dissipated by conduction, and radiation with absorption, would come again to the place of contact and throw the moving body back against the force to which it had previously yielded. Boulders would recover from the mud the materials required to rebuild them into their previous jagged forms, and would become reunited to the mountain peak from which they had formerly broken away. And if, also, the materialistic hypothesis of life were true, living creatures would grow backward, with conscious knowledge of the future but with no memory of the past, and would become again, unborn.

But the real phenomena of life infinitely transcend human science, and speculation regarding consequences of their imagined reversal is utterly unprofitable. Far otherwise, however, is it in respect to the reversal of the motions of matter uninfluenced by life, a very elementary consideration of which leads to the full explanation of the theory of dissipation of energy.

This brilliant statement indicates definitely that Lord Kelvin would also be classed in our group *B*, on the basis of his belief that there is in life something that transcends our present knowledge. Various illustrations have been given of the vivid description presented by Lord Kelvin. Movie-goers have had many opportunities to watch a waterfall climbing up the hill or a diver jumping back on the springboard; but, as a rule, cameramen have been afraid of showing life going backward, and such reels would certainly not be authorized by censors!

In any event, it is a very strange coincidence that life and the second principle should represent the two most important examples of the impossibility of time's running backward. This reveals the intimate relation between both problems, a question that will be discussed in a later section.

*Statistical interpretation of the second principle.* The natural tendency of entropy to increase is interpreted now as corresponding to the evolution from improbable toward most probable structures. The brilliant theory developed by L. Boltzmann, F. W. Gibbs, and J. C. Maxwell explains entropy as a physical substitute for "probability" and throws a great deal of light upon all thermodynamical processes. This side of the question has been very clearly discussed and explained in Schrödinger's book[2] and will not be repeated here.

Let us, however, stress a point of special interest. With the statistical theory, entropy acquires a precise mathematical definition as the logarithm of probability. It can be computed theoretically, when a physical model is given, and the theoretical value compared with experiment. When this has been found to work correctly, the same physical model can be used to investigate problems outside the reach of classical thermodynamics and especially problems involving time. The questions raised

[6] William Thompson (Lord Kelvin) in the *Proceedings of the Royal Society of Edinburgh 8:* 325-331, 1874. Quoted in *The Autobiography of Science,* F. R. Moulton and J. J. Shifferes, Editors (New York, 1945), p. 468.

in the preceding section can now be answered; the rate of diffusion for gas mixtures, the thermal conductivity of gases, the velocity of chemical reactions can be computed.

In this respect, great progress has been made, and in a number of cases it can be determined *how fast* entropy will actually increase. It is expected that convenient models will eventually be found for all of the most important problems; but this is not yet the case, and we must distinguish between those physical or chemical experiments for which a detailed application of statistical thermodynamics has been worked out, and other problems for which a model has not yet been found and for which we therefore have to rely on classical thermodynamics without the help of statistics. In the first group, a detailed model enables one to answer the most incautious questions; in the second group, questions involving time cannot be discussed.

*Distinction between two classes of experiments where entropy remains constant.* The entropy of a closed system, as noted, must increase, or at least remain constant. When entropy increases, the system is undergoing an *irreversible* transformation; when the system undergoes a *reversible* transformation, its total entropy remains constant. Such is the case for reversible cycles discussed in textbooks on thermodynamics, for reversible chemical reactions, etc. However, there is *another case* where no entropy change is observed, a case that is usually ignored, about which we do not find a word of discussion in textbooks, simply because scientists are at a loss to explain it properly. This is the case of systems in *unstable equilibrium*. A few examples may serve to clarify the problem much better than any definition.

In a private kitchen, there is a leak in the gas range. A mixture of air and gas develops (unstable equilibrium), but nothing happens until a naughty little boy comes in, strikes a match, and blows up the roof. Instead of gas, you may substitute coal, oil, or any sort of fuel; all our fuel reserves are in a state of unstable equilibrium. A stone hangs along the slope of a mountain and stays there for years, until rains and brooklets carry the soil away, and the rock finally rolls downhill. Substitute waterfalls, water reservoirs, and you have all our reserves of “white fuel.” Uranium remained stable and quiet for thousands of centuries; then came some scientists, who built a pile and a bomb and, like the naughty boy in the kitchen, blew up a whole city. Such things would not be permitted if the second principle were an active principle and not a passive one. Such events could not take place in a world where this principle was strictly enforced.

All this makes one thing clear. All our so-called power reserves are due to systems in unstable equilibrium. They are really reserves of negative entropy—structures wherein, by some sort of miracle, the normal and legitimate increase of entropy does not take place, until man, acting like a catalytic agent, comes and starts the reaction.

Very little is known about these systems of unstable equilibrium. No explanation is given. The scientist simply mumbles a few embarrassed words about "obstacles" hindering the reaction, or "potential energy walls" separating systems that should react but do not. There is a hint in these vague attempts at explanation, and, when properly developed, they should constitute a practical theory. Some very interesting attempts at an interpretation of *catalysis,* on the basis of quantum mechanics, have aroused great interest in scientific circles. But the core of the problem remains. How is it possible for such tremendous negative entropy reserves to stay untouched? What is the mechanism of *negative catalysis,* which maintains and preserves these stores of energy?

That such problems have a stupendous importance for mankind, it is hardly necessary to emphasize. In a world where oil simply waits for prospectors to come, we already watch a wild struggle for fuel. How would it be if oil burned away by itself, unattended, and did not wait passively for the drillers?

### *Life and Its Relations with the Second Principle*

We have raised some definite questions about the significance of the second principle, and in the last section have noted certain aspects of particular importance. Let us now discuss these, point by point, in connection with the problem of life maintenance and the mechanism of life.

*Closed systems.* Many textbooks, even the best of them, are none too cautious when they describe the increase of entropy. It is customary to find statements like this one: "The entropy of the universe is constantly increasing." This, in my opinion, is very much beyond the limits of human knowledge. Is the universe bounded or infinite? What are the properties of the boundary? Do we know whether it is tight, or may it be leaking? Do entropy and energy leak out or in? Needless to say, none of these questions can be answered. We know that the universe is expanding although we understand very little of how and why. Expansion means a moving boundary (if any), and a moving boundary is a leaking boundary; neither energy nor entropy can remain constant within. Hence, it is better not to speak about the "entropy of the universe." In the last section we emphasized the limitations of physical laws, and the fact that they can be safely applied only within certain limits and for certain orders of magnitude. The whole universe is too big for thermodynamics and certainly exceeds considerably the reasonable order of magnitude for which its principles may apply. This is also proved by the fact that the Theory of Relativity and all the cosmological theories that followed always involve a broad revision and drastic modification of the laws of thermodynamics, before an attempt can be made to apply them to the universe as a whole. The only thing that we can reasonably discuss is the entropy of a conceivable closed structure. Instead of the very mysterious universe, let us speak of our home, the earth. Here we stand on familiar ground. The earth is not a closed system. It is constantly receiving energy and negative entropy from outside—radiant heat from the sun, gravitational energy from sun and moon (provoking sea tides), cosmic radiation from unknown origin, and so on. There is also a certain amount of outward leak, since the earth itself radiates energy and entropy. How does the balance stand? Is it positive or negative? It is very doubtful whether any scientist can answer this question, much less such a question relative to the universe as a whole.

The earth is not a closed system, and life feeds upon energy and negative entropy leaking into the earth system. Sun heat and rain make crops (remember April showers and May flowers), crops provide food, and the cycle reads: first, creation of unstable equilibriums (fuels, food, waterfalls, etc.); then, use of these reserves by all living creatures.

Life acts as a catalytic agent to help destroy unstable equilibrium, but it is a very peculiar kind of catalytic agent, since it profits by the operation. When black platinum provokes a chemical reaction, it does not seem to care, and does not profit by it. Living creatures care about food, and by using it they maintain their own unstable equilibrium. This is a point that will be considered later.

The conclusion of the present section is this: that the sentence to "death by confinement" is avoided by living in a world that is not a confined and closed system.

*The role of time.* We have already emphasized the silence of the second principle. The direction of any reaction is given, but the velocity of the reaction remains unknown. It may be zero (unstable equilibrium), it may remain small, or it may become very great. Catalytic agents usually increase the velocity of chemical reactions; however, some cases of "anticatalysis" or "negative catalysis" have been discovered, and these involve a slowing down of some important reactions (e.g., oxidation).

Life and living organisms represent a most important type of catalysis. It is suggested that a systematic study of positive and negative catalysts might prove very useful, and would in fact be absolutely necessary before any real understanding of life could be attained.

The *statistical interpretation* of entropy and *quantum mechanics* are undoubtedly the tools with which a theory of catalysis should be built. Some pioneer work has already been done and has proved extremely valuable, but most of it is restricted, for the moment, to the most elementary types of chemical reactions. The work on theoretical chemistry should be pushed ahead with great energy.

Such an investigation will, sooner or later, lead us to a better understanding of the mechanisms of "unstable equilibrium." New negative catalysts may even make it possible to stabilize some systems that otherwise would undergo spontaneous disintegration, and to preserve new types of energies and negative entropies, just as we now know how to preserve food.

We have already emphasized the role of living organisms as catalytic agents, a feature that has long been recognized. Every biochemist now thinks of ferments and yeasts as peculiar living catalysts, which help release some obstacle and start a reaction, in a system in unstable equilibrium. Just as catalysts are working within the limits of the second principle, so living organisms are too. It should be noted, however, that catalytic action in itself is something which is not under the jurisdiction of the second principle. Catalysis involves the velocity of chemical reactions, a feature upon which the second principle remains silent. Hence, in this first respect, life is found to operate along the border of the second principle.

However, there is a second point about life that seems to be much more important. Disregard the very difficult problem of birth and reproduction. Consider an adult specimen, be it a plant or an animal or man. This adult individual is a most extraordinary example of a chemical system in unstable equilibrium. The system is unstable, undoubtedly, since it represents a very elaborate organization, a most improbable structure (hence a system with very low entropy, according to the statistical interpretation of entropy). This instability is further shown when death occurs. Then, suddenly, the whole structure is left to itself, deprived of the mysterious power that held it together; within a very short time the organism falls to pieces, rots, and goes (we have the wording of the scriptures) back to the dust whence it came.

Accordingly, a living organism is a chemical system in unstable equilibrium maintained by some strange "power of life," which manifests itself as a sort of *negative catalyst.* So long as life goes on, the organism maintains its unstable structure and escapes disintegration. It slows down to a considerable extent (exactly, for a lifetime) the normal and usual procedure of decomposition. Hence, a new aspect of life. Biochemists usually look at living beings as possible catalysts. But this same living creature is himself an unstable system, held together by some sort of internal anticatalyst! After all, a poison is nothing but an active catalyst, and a good drug represents an anticatalyst for the final inevitable reaction: death.

N. Wiener, in his *Cybernetics,* takes a similar view when he compares enzymes or living animals to Maxwell demons, and writes: "It may well be that enzymes are metastable Maxwell demons, decreasing entropy. . . . We may well regard living organisms, such as Man himself, in this light. Certainly the enzyme and the living organism are alike metastable: the stable state of an enzyme is to be deconditioned, and the stable state of a living organism is to be dead. All catalysts are ultimately poisoned: they change rates of reaction, but not true equilibrium. Nevertheless, catalysts and Man alike have sufficiently definite states of metastability to deserve the recognition of these states as relatively permanent conditions."[7]

### *Living Organisms and Dead Structures*

In a discussion at Harvard (1946), P. W. Bridgman stated a funda-

[7] Norbert Wiener, *Cybernetics, or Control and Communication in the Animal and the Machine* (New York, John Wiley and Sons, 1948), p. 72.

mental difficulty regarding the possibility of applying the laws of thermodynamics to any system containing living organisms. How can we compute or even evaluate the entropy of a living being? In order to compute the entropy of a system, it is necessary to be able to create or to destroy it in a reversible way. We can think of no reversible process by which a living organism can be created or killed: both birth and death are irreversible processes. There is absolutely no way to define the change of entropy that takes place in an organism at the moment of its death. We might think of some procedure by which to measure the entropy of a dead organism, albeit it may be very much beyond our present experimental skill, but this does not tell us anything about the entropy the organism had just before it died.

This difficulty is fundamental; it does not make sense to speak of a quantity for which there is no operational scheme that could be used for its measurement. The entropy content of a living organism is a completely meaningless notion. In the discussion of all experiments involving living organisms, biologists always avoid the difficulty by assuming that the entropy of the living objects remains practically constant during the operation. This assumption is supported by experimental results, but it is a bold hypothesis and impossible to verify.

To a certain extent, a living cell can be compared to a flame: here is matter going in and out, and being burned. The entropy of a flame cannot be defined, since it is not a system in equilibrium. In the case of a living cell, we may know the entropy of its food and measure the entropy of its wastes. If the cell is apparently maintained in good health and not showing any visible change, it may be assumed that its entropy remains practically constant. All experimental measures show that the entropy of the refuse is larger than that of the food. The transformation operated by the living system corresponds to an increase of entropy, and this is presented as a verification of the second principle of thermodynamics. But we may have some day to reckon with the underlying assumption of constant entropy for the living organism.

There are many strange features in the behavior of living organisms, as compared with dead structures. The evolution of species, as well as the evolution of individuals, is an irreversible process. The fact that evolution has been progressing from the simplest to the most complex structures is very difficult to understand, and appears almost as a contradiction to the law of degradation represented by the second principle. The answer is, of course, that degradation applies only to the whole of an isolated system, and not to one isolated constituent of the system. Nevertheless, it is hard to reconcile these two opposite directions of evolution. Many other facts remain very mysterious: reproduction, maintenance of the living individual and of the species, free will, etc.

A most instructive comparison is presented by Schrödinger[8] when he points to similarities and differences between a living organism, such as a cell, and one of the most elaborate structures of inanimate matter,

[8] Schrödinger, pp. 3 and 78.

a crystal. Both examples represent highly organized structures concontaining a very large number of atoms. But the crystal contains only a few types of atoms, whereas the cell may contain a much greater variety of chemical constituents. The crystal is always more stable at very low temperatures, and especially at absolute zero. The cellular organization is stable only within a given range of temperatures. From the point of view of thermodynamics this involves a very different type of organization.

When distorted by some stress, the crystal may to a certain extent repair its own structure and move its atoms to new positions of equilibrium, but this property of self-repair is extremely limited. A similar property, but exalted to stupendous proportions, characterizes living organisms. The living organism heals its own wounds, cures its sicknesses, and may rebuild large portions of its structure when they have been destroyed by some accident. This is the most striking and unexpected behavior. Think of your own car, the day you had a flat tire, and imagine having simply to wait and smoke your cigar while the hole patched itself and the tire pumped itself to the proper pressure, and you could go on. This sounds incredible.[9] It is, however, the way nature works when you "chip off" while shaving in the morning. There is no inert matter possessing a similar property of repair. That is why so many scientists (class *B*) think that our present laws of physics and chemistry do not suffice to explain such strange phenomena, and that something more is needed, some very important law of nature that has escaped our investigations up to now, but may soon be discovered. Schrödinger, after asking whether the new law required to explain the behavior of living matter (see page 556) might not be of a superphysical nature, adds: "No, I do not think that. For the new principle that is involved is a genuinely physical one. It is, in my opinion, nothing else than the principle of quantum theory over again."[10] This is a possibility, but it is far from certain, and Schrödinger's explanations are too clever to be completely convincing.

There are other remarkable properties characterizing the ways of living creatures. For instance, let us recall the paradox of Maxwell's demon, that submicroscopical being, standing by a trapdoor and opening it only for fast molecules, thereby selecting molecules with highest energy and temperature. Such an action is unthinkable, on the submicroscopical scale, as contrary to the second principle.[11] How does it

[9] The property of self-repairing has been achieved in some special devices. A self-sealing tank with a regulated pressure control is an example. Such a property, however, is not realized in most physical structures and requires a special control device, which is a product of human ingenuity, not of nature.

[10] Schrödinger, p. 81.

[11] Wiener discusses very carefully the problem of the Maxwell demon (*Cybernetics*, pp. 71-72). One remark should be added. In order to choose the fast molecules, the demon should be able to see them: but he is in an enclosure in equilibrium at constant temperature, where the radiation must be that of the black body, and it is impossible to see anything in the interior of a black body. The demon simply does not see the particles, unless we equip him with a torchlight, and a torchlight is obviously a source of radiation not at equilibrium. It pours negative entropy into the system.

(*Continued on following page*)

become feasible on a large scale? Man opens the window when the weather is hot and closes it on cold days! Of course, the answer is that the earth's atmosphere is not in equilibrium and not at constant temperature. Here again we come back to the unstable conditions created by sunshine and other similar causes, and the fact that the earth is not a closed isolated system.

The very strange fact remains that conditions forbidden on a small scale are permitted on a large one, that large systems can maintain unstable equilibrium for large time-intervals, and that life is playing upon all these exceptional conditions on the fringe of the second principle.

### *Entropy and Intelligence*

One of the most interesting parts in Wiener's *Cybernetics* is the discussion on "Time series, information, and communication," in which he specifies that a certain "amount of information is the negative of the quantity usually defined as entropy in similar situations."[12]

This is a very remarkable point of view, and it opens the way for some important generalizations of the notion of entropy. Wiener introduces a precise mathematical definition of this new negative entropy for a certain number of problems of communication, and discusses the question of time prediction: when we possess a certain number of data about the behavior of a system in the past, how much can we predict of the behavior of that system in the future?

In addition to these brilliant considerations, Wiener definitely indicates the need for an extension of the notion of entropy. "Information represents negative entropy"; but if we adopt this point of view, how can we avoid its extension to all types of intelligence? We certainly must be prepared to discuss the extension of entropy to scientific knowledge, technical know-how, and all forms of intelligent thinking. Some examples may illustrate this new problem.

Take an issue of the New York *Times,* the book on Cybernetics, and an equal weight of scrap paper. Do they have the same entropy? According to the usual physical definition, the answer is "yes." But for an intelligent reader, the amount of information contained in these three bunches of paper is very different. If "information means negative entropy," as suggested by Wiener, how are we going to measure this new contribution to entropy? Wiener suggests some practical and numerical definitions that may apply to the simplest possible problems of this kind. This represents an entirely new field for investigation and a most revolutionary idea.

Under these circumstances, the demon can certainly extract some fraction of this negative entropy by using his gate at convenient times. Once we equip the demon with a torchlight, we may also add some photoelectric cells and design an automatic system to do the work, as suggested by Wiener. The demon need not be a living organism, and intelligence is not necessary either. The preceding remarks seem to have been generally ignored, although Wiener says: the demon can only act on information received and this information represents a negative entropy.

[12] Wiener, Chap. III, p. 76.

Many similar examples can be found. Compare a rocky hill, a pyramid, and a dam with its hydroelectric power station. The amount of "know-how" is completely different, and should also correspond to a difference in "generalized entropy," although the physical entropy of these three structures may be about the same. Take a modern large-scale computing machinery, and compare its entropy with that of its constituents before the assembling. Can it reasonably be assumed that they are equal? Instead of the "mechanical brain," think now of the living human brain. Do you imagine its (generalized) entropy to be the same as that for the sum of its chemical constituents?

It seems that a careful investigation of these problems, along the directions initiated by Wiener, may lead to some important contributions to the study of life itself. Intelligence is a product of life, and a better understanding of the power of thinking may result in a new point of discussion concerning this highly significant problem.

Let us try to answer some of the questions stated above, and compare the "value" of equal weights of paper: scrap paper, New York *Times*, *Cybernetics*. To an illiterate person they have the same value. An average English-reading individual will probably prefer the New York *Times*, and a mathematician will certainly value the book on Cybernetics much above anything else. "Value" means "generalized negative entropy," if our present point of view be accepted. The preceding discussion might discourage the reader and lead to the conclusion that such definitions are impossible to obtain. This hasty conclusion, however, does not seem actually to be correct. An example may explain the difficulty and show what is really needed.

Let us try to compare two beams of light, of different colors. The human eye, or an ultraviolet photo-cell, or an infrared receiving cell will give completely different answers. Nevertheless, the entropy of each beam of light can be exactly defined, correctly computed, and measured experimentally. The corresponding definitions took a long time to discover, and retained the attention of most distinguished physicists (e.g., Boltzman, Planck). But this difficult problem was finally settled, and a careful distinction was drawn between the intrinsic properties of radiation and the behavior of the specific receiving set used for experimental measurements. Each receiver is defined by its "absorption spectrum," which characterizes the way it reacts to incident radiations. Similarly, it does not seem impossible to discover some criterion by which a definition of generalized entropy could be applied to "information," and to distinguish it from the special sensitivity of the observer. The problem is certainly harder than in the case of light. Light depends only upon one parameter (wave length), whereas a certain number of independent variables may be required for the definition of the "information value," but the distinction between an absolute intrinsic value of information and the absorption spectrum of the receiver is indispensable. Scientific information represents certainly a sort of negative entropy for Wiener, who knows how to use it for prediction, and may be of no value whatsoever to a non-scientist. Their respective absorption spectra are completely different.

Similar extensions of the notion of entropy are needed in the field of biology, with new definitions of entropy and of some sort of absorption spectrum. Many important investigations have been conducted by biologists during recent years, and they can be summarized as "new classifications of energies." For inert matter, it suffices to know energy and entropy. For living organisms, we have to introduce the "food value" of products. Calories contained in coal and calories in wheat and meat do not have the same function. Food value must itself be considered separately for different categories of living organisms. Cellulose is a food for some animals, but others cannot use it. When it comes to vitamins or hormones, new properties of chemical compounds are observed, which cannot be reduced to energy or entropy. All these data remain rather vague, but they all seem to point toward the need for a new leading idea (call it principle or law) in addition to current thermodynamics, before these new classifications can be understood and typical properties of living organisms can be logically connected together. Biology is still in the empirical stage and waits for a master idea, before it can enter the constructive stage with a few fundamental laws and a beginning of logical structure.

In addition to the old and classical concept of physical entropy, some bold new extensions and broad generalizations are needed before we can reliably apply similar notions to the fundamental problems of life and of intelligence. Such a discussion should lead to a reasonable answer to the definition of entropy of living organisms and solve the paradox of Bridgman (pages 563-564).

A recent example from the physical sciences may explain the situation. During the nineteenth century, physicists were desperately attempting to discover some mechanical models to explain the laws of electromagnetism and the properties of light. Maxwell reversed the discussion and offered an electromagnetic theory of light, which was soon followed by an electromagnetic interpretation of the mechanical properties of matter. We have been looking, up to now, for a physico-chemical interpretation of life. It may well happen that the discovery of new laws and of some new principles in biology could result in a broad redefinition of our present laws of physics and chemistry, and produce a complete change in point of view.

In any event, two problems seem to be of major importance for the moment: a better understanding of catalysis, since life certainly rests upon a certain number of mechanisms of negative catalysis; and a broad extension of the notion of entropy, as suggested by Wiener, until it can apply to living organisms and answer the fundamental question of P. W. Bridgman.

# Information, Measurement, and Quantum Mechanics[1]

Jerome Rothstein

*Solid State Devices Section, Thermionics Branch, Evans Signal Laboratory, Belmar, New Jersey*

RECENT DEVELOPMENTS IN COMMUNICATION THEORY use a function called the entropy, which gives a measure of the quantity of information carried by a message. This name was chosen because of the similarity in mathematical form between informational entropy and the entropy of statistical mechanics. Increasing attention is being devoted to the connection between information and physical entropy (*1–9*), Maxwell's demon providing a typical opportunity for the concepts to interact.

It is the purpose of this paper to present a short history of the concepts of entropy and information, to discuss information in physics and the connection between physical and informational entropies, and to demonstrate the logical identity of the problem of measurement and the problem of communication. Various implications for statistical mechanics, thermodynamics, and quantum mechanics, as well as the possible relevance of generalized entropy for biology, will be briefly considered. Paradoxes and questions of interpretation in quantum mechanics, as well as reality, causality, and the completeness of quantum mechanics, will also be briefly examined from an informational viewpoint.

## Information and Entropy

Boltzmann's discovery of a statistical explanation for entropy will always rank as one of the great achievements in theoretical physics. By its use, he was able to show how classical mechanics, applied to billiard-ball molecules or to more complicated mechanical systems, could explain the laws of thermodynamics. After much controversy (arising from the reversibility of mechanics as opposed to the irreversibility of thermodynamics), during which the logical basis of the theory was recast, the main results were firmly based on the abstract theory of measurable sets. The function Boltzmann introduced depends on how molecular positions and momenta range over their possible values (for fixed total energy), becoming larger with increasing randomness or spread in these molecular parameters, and decreasing to zero for a perfectly sharp distribution. Entropy often came to be described in later years as a measure of disorder, randomness, or chaos. Boltzmann himself saw later that statistical entropy could be interpreted as a measure of missing information.

A number of different definitions of entropy have been given, the differences residing chiefly in the employment of different approximations or in choosing a classical or quantal approach. Boltzmann's classical and Planck's quantal definitions, for example, are, respectively,

$$S = -k\int f \log f \, d\tau,$$
$$\text{and} \quad S = k \log P.$$

Here $k$ is Boltzmann's constant, $f$ the molecular distribution function over coordinates and momenta, $d\tau$ an element of phase space, and $P$ the number of independent wave functions consistent with the known energy of the system and other general information, like requirements of symmetry or accessibility.

Even before maturation of the entropy concept, Maxwell pointed out that a little demon who could "see" individual molecules would be able to let fast ones through a trap door and keep slow ones out. A specimen of gas at uniform temperature could thereby be divided into low and high temperature portions, separated by a partition. A heat engine working between them would then constitute a *perpetuum mobile* of the second kind. Szilard (*1*), in considering this problem, showed that the second law of thermodynamics could be saved only if the demon paid for the information on which he acted with entropy increase elsewhere. If, like physicists, the demon gets his information by means of measuring apparatus, then the price is paid in full. He was led to ascribe a thermodynamical equivalent to an item of information. If one knew in which of two equal volumes a molecule was to be found, he showed that the entropy could be reduced by $k \log 2$.

Hartley (*2*), considering the problem of transmitting information by telegraph, concluded that an appropriate measure of the information in a message is the logarithm of the number of equivalent messages that might have been sent. For example, if a message consists of a sequence of $n$ choices from $k$ symbols, then the number of equivalent messages is $k^n$, and transmission of any one conveys an amount of information $n \log k$. In the hands of Wiener (*3,4*), Shannon (*5*), and others, Hartley's heuristic beginnings become a general, rigorous, elegant, and powerful theory related to statistical mechanics and promising to revolutionize communication theory. The ensemble of possible messages is characterized by a quantity

[1] Presented in shorter form before the American Physical Society Feb. 1, 1951, and before a physics seminar at Purdue University Feb. 21, 1951.

completely analogous to entropy and called by that name, which measures the information conveyed by selection of one of the messages. In general, if a subensemble is selected from a given ensemble, an amount of information equal to the difference of the entropies of the two ensembles is produced. A communication system is a means for transmitting information from a source to a destination and must be capable of transmitting any member of the ensemble from which the message is selected. Noise introduces uncertainty at the destination regarding the message actually sent. The difference between the a priori entropy of the ensemble of messages that might have been selected and the a posteriori entropy of the ensemble of messages that might have given rise to the received signal is reduced by noise so that less information is conveyed by the message.

It is clear that Hartley's definition of quantity of information agrees with Planck's definition of entropy if one correlates equivalent messages with independent wave functions. Wiener and Shannon generalize Hartley's definition to expressions of the same form as Boltzmann's definition (with the constant $k$ suppressed) and call it entropy. It may seem confusing that a term connoting lack of information in physics is used as a measure of amount of information in communication, but the situation is easily clarified. If the message to be transmitted is known in advance to the recipient, no information is conveyed to him by it. There is no initial uncertainty or doubt to be resolved; the ensemble of a priori possibilities shrinks to a single case and hence has zero entropy. The greater the initial uncertainty, the greater the amount of information conveyed when a definite choice is made. In the physical case the message is not sent, so to speak, so that physical entropy measures how much physical information is missing. Planck's entropy measures how uncertain we are about what the actual wave function of the system is. Were we to determine it exactly, the system would have zero entropy (pure case), and our knowledge of the system would be maximal. The more information we lose, the greater the entropy, with statistical equilibrium corresponding to minimal information consistent with known energy and physical make-up of the system. We can thus equate physical information and negative entropy (or negentropy, a term proposed by Brillouin [*9*]). Szilard's result can be considered as giving thermodynamical support to the foregoing.

## Measurement and Communication

Let us now try to be more precise about what is meant by information in physics. Observation (measurement, experiment) is the only admissible means for obtaining valid information about the world. Measurement is a more quantitative variety of observation; e.g., we observe that a book is near the right side of a table, but we measure its position and orientation relative to two adjacent table edges. When we make a measurement, we use some kind of procedure and apparatus providing an ensemble of possible results. For measurement of length, for example, this ensemble of a priori possible results might consist of: ($a$) too small to measure, ($b$) an integer multiple of a smallest perceptible interval, ($c$) too large to measure. It is usually assumed that cases ($a$) and ($c$) have been excluded by selection of instruments having a suitable range (on the basis of preliminary observation or prior knowledge). One can define an entropy for this a priori ensemble, expressing how uncertain we are initially about what the outcome of the measurement will be. The measurement is made, but because of experimental errors there is a whole ensemble of values, each of which could have given rise to the one observed. An entropy can also be defined for this a posteriori ensemble, expressing how much uncertainty is still left unresolved after the measurement. We can define the quantity of physical information obtained from the measurement as the difference between initial (a priori) and final (a posteriori) entropies. We can speak of position entropy, angular entropy, etc., and note that we now have a quantitative measure of the information yield of an experiment. A given measuring procedure provides a set of alternatives. Interaction between the object of interest and the measuring apparatus results in selection of a subset thereof. When the results of this process of selection become known to the observer, the measurement has been completed.

It is now easy to see that there is an analogy between communication and measurement which actually amounts to an identity in logical structure. Fig. 1

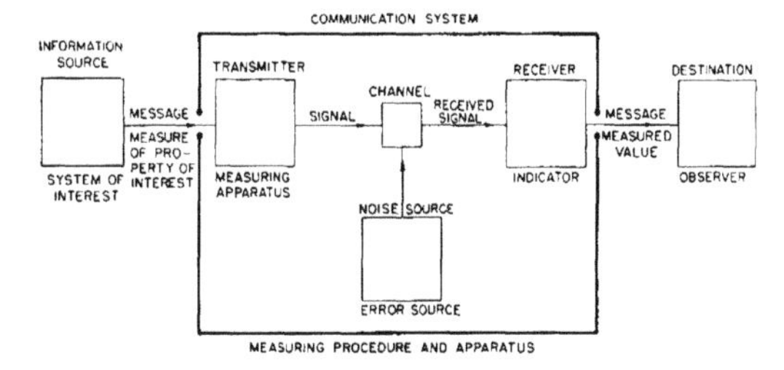


Fig. 1.

shows this less abstractly. The blocks and upper captions follow Shannon's characterization of a communication system; the lower captions give analogous terms for a measuring apparatus. The system of interest corresponds to the information source, the observer to the destination for which the message is intended. The message corresponds to a measure of the property of interest, which is often encoded by the transmitter or measuring apparatus into a signal consisting of information-bearing variations of some physical quantity, often quite distinct from the one of direct interest. The signal, corrupted by noise or errors, is decoded by the receiver or indicator and presented as a message or measured value at the output of the system. Calibration in measurement is, in part, the analog of distortion correction in communication. In practice a communication or measuring system often consists of a number of subsystems in series, intermediate ones serving as destinations for their predecessors and as sources for their successors. The sensory and nervous apparatus of the observer can be considered the ultimate system, which, together with instruments, operations, and apparatus, constitutes the means whereby the mind of the scientist communicates with, or acquires information about, the universe.

## Physical Consequences of the Information Viewpoint

Some implications of the informational viewpoint must be considered. First of all, the entropy of information theory is, except for a constant depending on choice of units, a straightforward generalization of the entropy concept of statistical mechanics. Information theory is abstract mathematics dealing with measurable sets, with choices from alternatives of an unspecified nature. Statistical mechanics deals with sets of alternatives provided by physics, be they wave functions, as Planck's quantal definition, or the complexions in phase space of classical quantum statistics. Distinguishing between identical particles (which leads to Gibbs' paradox and nonadditivity of entropy) is equivalent to claiming information that is not at hand, for there is no measurement yielding it. When this nonexistent information is discarded, the paradox vanishes. Symmetry numbers, accessibility conditions, and parity are additional items of (positive or negative) information entering into quantal entropy calculations.

Second, we can formulate the statistical expression of the second law of thermodynamics rather simply in terms of information: Our information about an isolated system can never increase (only by measurement can new information be obtained). Reversible processes conserve, irreversible ones lose information.

Third, all physical laws become relationships between types of information, or information functions collected or constructed according to various procedures. The difference between classical or quantum mechanics, on one hand, and classical or quantum statistics, on the other, is that the former is concerned with theoretically maximal information, the latter with less than the maximal. From the present viewpoint, therefore, classical and quantum mechanics are limiting cases of the corresponding statistics, rather than separate disciplines. The opposite limiting cases —namely, minimum information or maximum entropy —relate to the equilibrium distributions treated in texts on statistical mechanics. The vast, almost virgin field of nonequilibrium physics lies between these two extremes.

It is tempting to speculate that living matter is distinguished, at least in part, by having a large amount of information coded in its structure. This information would be in the form of "instructions" (constraints) restricting the manifold of possibilities for its physicochemical behavior. Perhaps instructions for developing an organism are "programmed" in the genes, just as the operation of a giant calculating machine, consisting of millions of parallel or consecutive operations, is programmed in a control unit. Schroedinger, in a fascinating little book, *What Is Life?* views living matter as characterized by its "disentropic" behavior, as maintaining its organization by feeding on "negative entropy," the thermodynamic price being a compensating increase in entropy of its waste products. Gene stability is viewed as a quantum effect, like the stability of atoms. In view of previous discussion, the reader should have no trouble fitting this into the informational picture above.

Returning to more prosaic things, we note, fourth, that progress either in theory of measurement or in theory of communication will help the other. Their logical equivalence permits immediate translation of results in one field to the other. Theory of errors and of noise, of resolving power and minimum detectable signal, of best channel utilization in communication and optimal experimental design are three examples of pairs where mutual cross-fertilization can be confidently expected.

Fifth, absolutely exact values of measured quantities are unattainable in general. For example, an infinite amount of information is required to specify a quantity capable of assuming a continuum of values. Only an ensemble of possibly "true" or "real" values is determined by measurement. In classical mechanics, where the state of a system is specified by giving simultaneous positions and momenta of all particles in the system, two assumptions are made at this point—namely, that the entropy of individual measurements can be made to approach zero, and furthermore that this can be done simultaneously for all quantities needed to determine the state of the system. In other words, the ensembles can be made arbitrarily sharp in principle, and these sharp values can be taken as "true" values. In current quantum mechanics the first assumption is retained, but the second is dropped. The ensembles of position and momenta values cannot be made sharp simultaneously by any measuring procedure. We are left with irreducible ensembles of possible "true" values of momentum, consistent with the position information on hand from previous measurements. It thus seems natural, if not unavoidable, to conclude that quantum mechanics describes the ensemble of systems consistent with the information specifying a state rather than a single system. The wave function is a kind of generating function for all the information deducible from operational specification of the mode of preparation of the system, and from it the probabilities of obtaining possible values of measurable quantities can be calculated. In communication terminology, the stochastic nature of the message source—i.e., the ensemble of possible messages and their probabilities—is specified, but not the individual message. The entropy of a given state for messages in $x$-language, $p$-language, or any other language, can be calculated in accordance with the usual rules. It vanishes in the language of a given observable if, and only if, the state is an eigenstate of that observable. For an eigenstate of an operator commuting with the Hamiltonian, all entropies are constant in time, analogous to equilibrium distributions in statistical mechanics. This results from the fact that change with time is expressed by a unitary transformation, leaving inner products in Hilbert space invariant. The corresponding classical case is one with maximal information where the entropy is zero and remains so. For a wavepacket representing the result of a position measurement, on the other hand, the distribution smears out more and more as time goes on, and its entropy of position increases. We conjecture, but have not proved, that this is a special case of a new kind of quantal $H$-theorem.

Sixth, the informational interpretation seems to resolve some well-known paradoxes (*10*). For example, if a system is in an eigenstate of some observable, and a measurement is made on an incompatible observable, the wave function changes instantaneously from the original eigenfunction to one of the second observable. Yet Schroedinger's equation demands that the wave function change continuously with time. In fact, the system of interest and the measuring equipment can be considered a single system that is unperturbed and thus varying continuously. This causal anomaly and action-at-a-distance paradox vanishes in the information picture. Continuous variation occurs so long as no new information is obtained incompatible with the old. New information results from measurement and requires a new representative ensemble. The system of interest could "really" change continuously even though our information about it did not. It does no harm to believe, as Einstein does, in a "real" state of an individual system, so long as one remembers that quantum mechanics does not permit an operational definition thereof. The Einstein-Podolsky-Rosen paradox (*11*), together with Schroedinger's (*12*) sharpening of it, seems to be similarly resolved. Here two systems interact for a short time and are then completely separated. Measurement of one system determines the state of the other. But the kind of measurement is under the control of the experimenter, who can, for example, choose either one of a pair of complementary observables. He obtains one of a pair of incompatible wave functions under conditions where an objective or "real" state of the system cannot be affected. If the wave function describes an individual system, one must renounce all belief in its objective or "real" state. If the wave function only bears information and describes ensembles consistent therewith, there is no paradox, for an individual system can be compatible with both of two inequivalent ensembles, as long as they have a nonempty intersection. The kind of information one gets simply varies with the kind of measurement one chooses to make.

## Reality, Causality, and the Completeness of Quantum Mechanics

We close with some general observations.

First, it is possible to believe in a "real" objective state of a quantum-mechanical system without contradiction. As Bohr and Heisenberg have shown, states of simultaneous definite position and definite momentum in quantum mechanics are incompatible because they refer to simultaneous results of two mutually exclusive procedures. But, if a variable is not measured, its corresponding operation has not been performed, and so unmeasured variables need not correspond to operators. Thus there need be no conflict with the quantum conditions. If one denies simultaneous reality to position and momentum then EPR forces the conclusion that one or the other assumes reality only when measured. In accepting this viewpoint, should one not assume, for consistency, that electrons in an atom have no reality, because any attempt to locate one by a photon will ionize the atom? The electron then becomes real (i.e., is "manufactured") only as a result of an attempt to measure its position. Similarly, one should also relinquish the continuum of space and time, for one can measure only a countable infinity of locations or instants, and even this is an idealization, whereas a continuum is uncountable. If one admits as simultaneously real all positions or times that *might* be measured, then for consistency simultaneous reality of position and momentum must be admitted, for either one might be measured.

Second, it is possible to believe in a strictly causal universe without contradiction. Quantum indeterminacy can be interpreted as reflecting the impossibility of getting enough information (by measurement) to permit prediction of unique values of all observables. A demon who can get physical information in other ways than by making measurements might then see a causal universe. Von Neumann's proof of the impossibility of making quantum mechanics causal by the introduction of hidden parameters assumes that these parameters are values that internal variables can take on, the variables themselves satisfying quantum conditions. Causality and reality (i.e., objectivity) have thus been rejected on similar grounds. Arguments for their rejection need not be considered conclusive for an individual system if quantum mechanics be viewed as a Gibbsian statistical mechanics of ensembles.

The third point is closely connected with this—namely, that quantum mechanics is both incomplete in Einstein's sense and complete in Bohr's sense (*13*). The former demands a place in the theory for the "real" or objective state of an individual system; the latter demands only that the theory correctly describe what will result from a specified operational procedure—i.e., an ensemble according to the present viewpoint. We believe there is no reason to exclude the possibility that a theory may exist which is complete in Einstein's sense and which would yield quantum mechanics in the form of logical inferences. In the communication analogy, Bohr's operational viewpoint corresponds to demanding that the ensemble of possible messages be correctly described by theory when the procedure determining the message source is given with maximum detail. This corresponds to the attitude of the telephone engineer who is concerned with transmitting the human voice but who is indifferent to the meaning of the messages. Einstein's attitude implies that the messages may have meaning, the particular meaning to be conveyed determining what message is selected. Just as no amount of telephonic circuitry will engender semantics, so does "reality" seem beyond experiment as we know it. It seems arbitrary, however, to conclude that the problem of reality is meaningless or forever irrelevant to science. It is conceivable, for example, that a long sequence of alternating measurements on two noncommuting variables carried out on a single system might suggest new kinds of regularity. These would, of course, have to yield the expectation values of quantum mechanics.

**References**


1. SZILARD, L. *Z. Physik*, **53**, 840 (1929).
2. HARTLEY, R. V. L. *Bell System Tech. J.*, **7**, 535 (1928).
3. WIENER, N. *Cybernetics*. New York: Wiley (1950).
4. ———. *The Extrapolation, Interpolation and Smoothing of Time Series*. New York: Wiley (1950).
5. SHANNON, C. E. *Bell System Tech. J.*, **27**, 279, 623 (1948); *Proc. I. R. E.*, **37**, 10 (1949).
6. TULLER, W. G. *Proc. I. R. E.*, **37**, 468 (1949).
7. GABOR, D. *J. Inst. Elec. Engrs. (London)*, Pt. III, **93**, 429 (1946); *Phil. Mag.*, **41**, 1161 (1950).
8. MACKAY, D. M. *Phil. Mag.*, **41**, 289 (1950).
9. BRILLOUIN, L. *J. Applied Phys.*, **22**, 334, 338 (1951).
10. REICHENBACH, H. *Philosophical Foundations of Quantum Mechanics*. Berkeley: Univ. Calif. Press (1944).
11. EINSTEIN, A., PODOLSKY, B., and ROSEN, N. *Phys. Rev.*, **47**, 777 (1935).
12. SCHROEDINGER, E. *Proc. Cambridge Phil. Soc.*, **31**, 555 (1935); **32**, 466 (1936).
13. BOHR, N. *Phys. Rev.*, **48**, 696 (1935).

# Chapter 3

# Smoluchowski Trap Doors and Information Acquisition

# Maxwell's demon, rectifiers, and the second law: Computer simulation of Smoluchowski's trapdoor

P. A. Skordos
*Santa Fe Institute, Santa Fe, New Mexico and Massachusetts Institute of Technology, 545 Technology Square, Cambridge, Massachusetts 02139*

W. H. Zurek
*Theoretical Division, T-6, MS B288, Los Alamos National Laboratory, Los Alamos, New Mexico 87545 and Santa Fe Institute, Santa Fe, New Mexico*

(Received 5 September 1991; accepted 13 March 1992)

An automated version of Maxwell's demon inspired by Smoluchowski's ideas of 1912 is simulated numerically. Two gas chambers of equal volume are connected via an opening that is covered by a trapdoor. The trapdoor can open to the left but not to the right, and is intended to rectify naturally occurring fluctuations in density between the two chambers. The simulation results confirm that though the trapdoor behaves as a rectifier when large density differences are imposed by external means, it cannot extract useful work from the thermal motion of the molecules when left on its own.

## I. INTRODUCTION

The second law of thermodynamics has been a subject of debate ever since it was formulated. It says that the entropy of a closed system can only increase with time, and thus natural phenomena are irreversible. In other words, a system left on its own can only evolve in one direction, toward equilibrium. This is in contrast to time-reversible dynamics and raises the question of how reversible dynamics can lead to macroscopic irreversibility. An answer can be furnished using probabilistic arguments in statistical mechanics, but the arguments are difficult to translate into a rigorous proof without postulating a new axiom about nature, the *stosszahlansatz*, also called the assumption of molecular chaos,[1,2] which is at odds with dynamical reversibility. As a result, the origin of the second law of thermo-

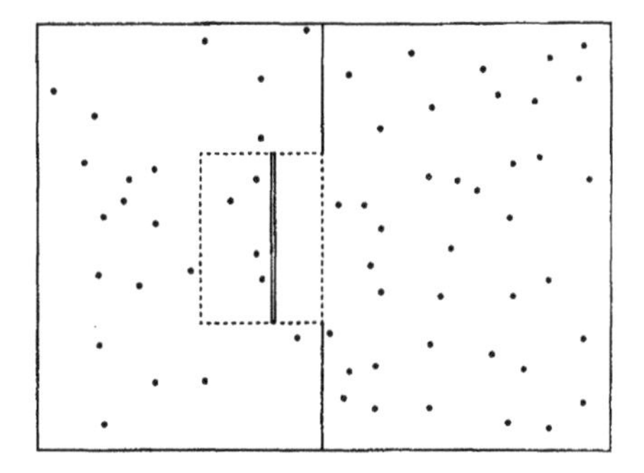

Fig. 1. The automated Maxwell's demon we simulated numerically was inspired by Smoluchowski's trapdoor. The dashed lines show the region where the trapdoor can move.

dynamics remains elusive and provides a source of interesting discussions on the foundations of physics.

A popular way of challenging the second law is the idea of "perpetual motion of the second kind," that is the extraction of useful work in a closed cycle from the perpetual thermal motion of gas molecules. This is prohibited by the second law because if it were possible to convert thermal energy into useful work in a closed equilibrated system, then the entropy of an isolated system could be made to decrease. However, it is puzzling why a microscopic device cannot be constructed to take advantage of spontaneous variations in density between microdomains of gas, to bring a system from a state of maximum disorder (equilibrium) into an ordered state, and eventually to convert thermal energy into useful work.

The history of microengines that convert thermal energy into useful work began when J. C. Maxwell proposed a "demon"-operated microengine at the end of his book *Theory of Heat.*[3,4] Since then, the term "demon" has become standard. Maxwell's demon works by opening and closing a tiny door between two gas chambers, based on the information that the demon has about individual molecules. The method used to obtain information is not specified. The demon allows only fast molecules to pass from left to right, and only slow ones to pass from right to left. This leads to a temperature difference between the two gas chambers, which can be used to convert thermal energy into useful work. It is currently believed, however, that Maxwell's demon cannot violate the second law[5–8] because the information needed to operate the demon's door does not come without a price. As Bennett explains in Ref. 5 following an idea that goes back to Refs. 6 and 9, the demon must dissipate energy into a heat bath in order to erase the information that it obtains by examining molecules. The heat bath may be the molecules that the demon examines or it may be another system that is colder than the demon and the gas molecules. The energy lost to the heat bath is always greater than or equal to the energy that can be extracted after the demon has finished its operations. This limitation applies even to "intelligent" demons—that is, capable of universal computing ability—as was shown by Zurek in Refs. 10 and 11. Thus Maxwell's demon cannot violate the second law.

There is an another class of microengines, however, that attempt to violate the second law, which is not plagued by issues of measurement and information erasure as is the original Maxwell's demon. This class of microengines consists of purely mechanical devices, also called automated Maxwell's demons, which are described completely within a theoretical framework such as Newtonian physics. In particular, there is no measurement mechanism outside of the mechanical model. An example of such an engine is the trapdoor mechanism discussed by Smoluchowski in Refs. 12 and 13. Another example, similar in the spirit to Smoluchowski's trapdoor, is the ratchet and pawl mechanism discussed by Feynman in Ref. 14.

Our paper describes the computer simulation of a trapdoor mechanism inspired by Smoluchowski's ideas. Our results confirm Smoluchowski's insight that though the trapdoor acts as a rectifier when large density differences are imposed by external means, it cannot extract useful work from the thermal motion of the molecules when left on its own. The next section describes our trapdoor mechanism and the simulation program. Following this, we discuss how the trapdoor succeeds at rectifying large density differences which are imposed by external means. Then, we discuss how our trapdoor system fails to work successfully when left to operate on its own. Finally, we discuss how the trapdoor can be modified to work successfully as a pump that creates a density difference in a system that starts initially with an equal number of molecules in the two chambers. However, the modification requires dissipation of the trapdoor's random motions, and it can be accomplished only by "opening" the system; that is, for example, by keeping the trapdoor at a lower temperature than the molecules. Of course, such a pump is no more a threat to the second law than is a refrigerator.

## II. DESCRIPTION OF THE MODEL

The system we have simulated is shown in Fig. 1. It consists of two gas chambers of equal area, connected via an opening that is covered by a trapdoor. The simulation is two dimensional and the gas molecules are billiard balls moving on the plane and colliding with each other. All the collisions conserve energy and momentum, except for particle–wall collisions that reflect a particle's momentum just as light rays reflect off mirrors. The collision forces are derived from infinite hard core potentials, and all forces act radially through the center of the colliding balls. Angular momentum (spin of the billiard balls) is not included in the model.

The trapdoor is constrained to move between two *door stops*, one of which is located on the middle wall, and the other is located inside the left chamber. The location of the door stops allows the trapdoor to open to the left but not to the right. This endows the trapdoor with rectifying properties, as we shall see in the next section. Geometrically, the trapdoor is a line segment of zero width that is impenetrable by the molecules and is oriented vertically at all times. The trapdoor *moves horizontally* at constant speed, and reverses its direction when it comes in contact with the door stops. Collisions between the trapdoor and the molecules conserve energy and momentum except for collisions at the edges of the trapdoor which do not conserve the $y$ component of momentum. The reason is that the trapdoor does not move in the $y$ direction and does not rotate at all—it can be thought of as sliding along ideal rails that are attached to the infinitely massive box containing the gas.

To evolve the system of molecules and the trapdoor we use the following algorithm: Given the positions and velocities at time $t_0$, we find all the collisions that are about to occur. We select the shortest collision time $\Delta t$, and move all the particles and the door freely during $\Delta t$. At time $t_0+\Delta t$ we perform the collision that occurs at this time, and then repeat the cycle looking for the next shortest collision time. The algorithm works because collisions in our system are instantaneous. The types of collisions that can occur are of four types: particle against particle, particle against wall, particle against door, and door against door stop. The collision equations are discussed in detail below.

The evolution algorithm can be implemented efficiently on a computer if we are careful to avoid unnecessary computations. For example, we do not need to examine all pairs of particles at every time step. If we see that two particles are far from each other, then we need not examine them again until a number of time steps have elapsed. Only then will the two particles have another chance of being near each other and being able to collide. Also, if we compute the collision time for a pair of particles that are near each other, and another pair of particles collides before them, we need not discard the first collision time. We simply decrement the first collision time by the time interval by which the whole system is evolved. A word of caution, however, is necessary. The process of decrementing collision times should not be repeated more than a few times because the roundoff error in subtracting small intervals of time becomes significant very quickly. Also, a collision involving some particle must invalidate all precomputed collision times involving that particle.

Two kinds of elementary formulas are used in the evolution algorithm: *collision equations* giving the new velocities in terms of the old velocities, and *timing equations* giving the time interval until an upcoming collision. The timing equations are the simpler of the two. They are derived from geometrical constraints and the fact that the particles and the door move at constant velocity between collisions. For example, to compute the collision time between two particles, we "draw a straight line" from the current position of the particles to the point where the particles are tangential to each other. The geometric constraint of tangency allows two possibilities, and we have to choose the one that occurs first and is the physical one. Algebraically we have to solve a quadratic equation, and to pick the smallest positive solution. The timing equations for the other types of collisions in our system are derived in a similar manner.

The collision equations are a little more complicated than the timing equations. The collision equations are derived from conservation of kinetic energy, conservation of linear momentum, and the condition that forces act radially. The last condition means that the force vector must pass through the center of the particle disk that is colliding, and hence momentum is exchanged along this direction. For nearly all collisions in our system the radial force condition is satisfied automatically in setting up the geometry of the problem. However, there is one type of collision that requires explicit use of the radial condition. This occurs when a particle disk collides with the edges of the moving trapdoor. Since it is not discussed in most textbooks, we review briefly the equations.

The radial force condition requires that the change in $y$ momentum divided by the change in $x$ momentum equals the tangent of the angle $\theta$ formed by the center of the colliding disk, the point of contact, and the $x$ axis. The point of contact is the edge of the moving trapdoor. If $v_x$, $v_y$ are the old velocities and $v_x'$, $v_y'$ are the new velocities of the colliding disk, we have the equation,

$$(v_x-v_x')=(\cos\theta/\sin\theta)(v_y-v_y'). \tag{1}$$

To find the velocities following a collision in terms of the velocities before the collision, we use Eq. (1) together with kinetic energy and $x$-momentum conservation. The $y$ momentum is not conserved because the trapdoor moves on ideal rails, and its $y$ velocity is always zero. After some algebra, we get the following equations for the new velocities.

$$v_x'=\frac{-2csv_y+2c^2V_x+(s^2-\delta c^2)v_x}{(\gamma c^2+s^2)}, \tag{2}$$

$$v_y'=\frac{-2csv_x+2csV_x+(\gamma c^2-s^2)v_y}{(\gamma c^2+s^2)}, \tag{3}$$

$$V_x'=V_x+\frac{m}{M}(v_x-v_x'), \tag{4}$$

where

$$c=\cos\theta,$$

$$s=\sin\theta,$$

$$\gamma=(1+m/M),$$

$$\delta=(1-m/M),$$

and where $M$, $V_x$ are the mass and $x$ velocity of the trapdoor; $m$, $v_x$, $v_y$ are the mass and velocities of the particle; and $\theta$ is the angle formed by the center of the particle, the colliding edge of the trapdoor, and the $x$ axis. The collision equations for all other types of collisions in our system can be found in textbooks.[15]

The numbers we used in our simulations were chosen to correspond to a standard gas like nitrogen. We experimented with different values for the size of the gas chambers, molecular speeds, and other quantities, and the qualitative behavior of the gas was the same for all choices. We looked at systems containing a number of molecules ranging from 20 to 500. We chose the radius of the molecules to be $3\times10^{-8}$ cm, mass $4.7\times10^{-23}$ gm, and velocities of the order $10^4$ cm/s. We chose the size of the gas chambers to give a mean free path between collisions of the order of the size of the chambers. Specifically in the case of 500 molecules, the width of each chamber was $13.5\times10^{-6}$ cm and the height was $18\times10^{-6}$ cm. The mean free path at equilibrium in each chamber can be estimated by the ratio,[16]

$$\lambda=\text{Area}/(n\times2R),$$

which gives $\lambda=(13.5\times18\times10^{-12})/(250\times2\times3\times10^{-8})=16.2\times10^{-6}$ cm.

We experimented with different masses for the trapdoor, and in the results reported below the mass of the trapdoor is of the order of three to four times the mass of one particle, unless otherwise indicated. Using a trapdoor mass that is comparable to the molecular mass leads to an average speed for the trapdoor that is comparable to that of the molecules, and facilitates numerical simulation. A very

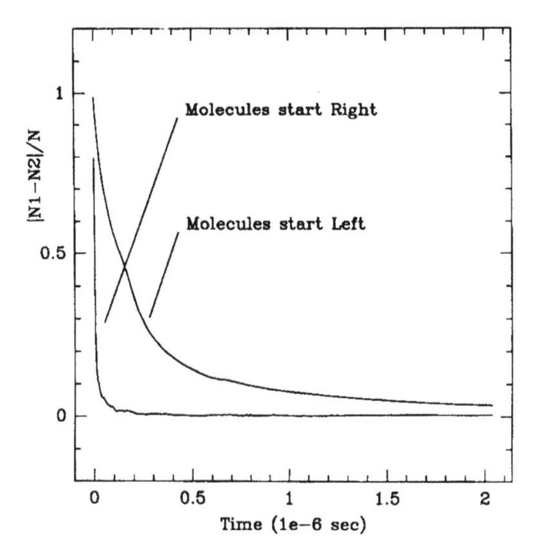


Fig. 2. The absolute value of the fractional difference in the number of particles between the two chambers is plotted against time, as the system approaches equilibrium. Two curves are shown, one for the case when the particles start in the left chamber, and one for the case when the particles start in the right chamber.

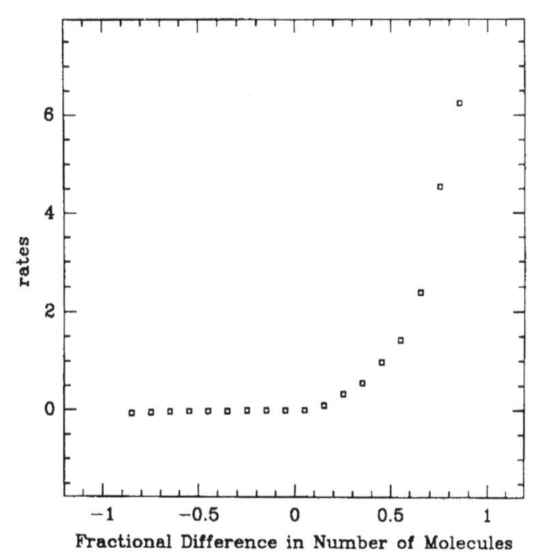


Fig. 3. The rate of particles passing from one chamber to the other is plotted against the fractional difference in the number of particles between the two chambers. The resulting curve resembles qualitatively the voltage-current characteristic of an electrical diode. The $y$ axis is in units of $25\times10^9$ particles/s. The "conducting region" (positive $x$ axis) corresponds to higher densities in the right chamber, which is expected because the trapdoor opens to the left.

light trapdoor moves too fast and increases the numerical roundoff error; while an excessively heavy trapdoor (orders of magnitude heavier than a molecule) can delay the approach to equilibrium and requires longer time averaging. It should be emphasized that our qualitative conclusions are expected to be independent of the mass of the trapdoor, and we certainly do not have any indications in our simulations that would lead us to question this expectation. Thus the choice of the mass of the door was dictated primarily by the above considerations of numerical convenience. Our simulations took typically several days using standard Unix workstations, and the relative error in the total energy of the system was kept less than $10^{-10}$ using double precision arithmetic. The running times were dictated primarily by the desire to gather good statistics. In contrast, the rectifying behavior of the equilibration experiments described in the next section can be seen after a running time of a few minutes to 1 h; depending of course on the computer that is used, the number of particles that are simulated, and the geometry of the trapdoor.

## III. THE TRAPDOOR AS A RECTIFIER

This section discusses the behavior of the trapdoor when large density differences between the two chambers are imposed by external means. It is found that under these circumstances the trapdoor acts as a diode, and prolongs the duration of states of higher density in the left chamber. There are a number of ways to exhibit this rectifying behavior, and we shall describe three of them.

The first way measures the equilibration time to an initial density difference, for example when all molecules start in the left chamber. To be precise, we place all molecules along the outermost wall of one chamber with the trapdoor set motionless in the closed position. We release the molecules with random initial velocities, and we monitor the number of molecules in each chamber as a function of time until the populations in the two chambers equalize. It should be pointed out that the motionless trapdoor immediately becomes "thermalized"—it acquires kinetic energy and starts moving after the first collision with a molecule. For example in the case of molecules starting in the left chamber, when the first molecule strikes the trapdoor, it transfers momentum to the trapdoor in the positive $x$ direction. At the next instant, the trapdoor's momentum is reversed in the negative $x$ direction as the trapdoor comes in contact with the middle wall door stop. A multiple collision between particle, trapdoor, and door stop is avoided by positioning the trapdoor initially at an infinitesimal distance away from the door stop. Thus the initially motionless trapdoor picks up energy and starts moving. Subsequently, the action of the molecules in the left chamber keeps the trapdoor near the middle wall most of the time, bouncing between the middle wall door stop and the large number of molecules in the left chamber. By contrast, when the molecules start in the right chamber, the trapdoor is pushed immediately all the way inside the left chamber and is kept mostly near the left door stop, providing a large opening for the molecules to enter from the right chamber into the left chamber.

Figure 2 plots the absolute value of the difference in the number of molecules between the two chambers against time. The difference in the number of molecules is normalized by the total number of molecules, which is 500 in this experiment. Two curves are shown, one for the case when all molecules start in the left chamber, and one for the case

Table I. The flow of molecules through the middle wall opening in forward and reverse bias conditions, for systems of 100 and 500 molecules. The molecules crossing from left to right are counted positive, and those crossing from right to left are counted negative.

| | $N=500$ | $N=100$ |
|---|---|---|
| Reverse bias | $4.26\times10^9$ | $1.05\times10^9$ |
| Forward bias | $-9.31\times10^{10}$ | $-1.81\times10^{10}$ |
| ratio | 1:22 | 1:17 |

when all molecules start in the right chamber. We see that in the latter case the populations equalize "immediately." In other words, the density difference between the chambers vanishes much more quickly when the molecules start in the right chamber than when the molecules start in the left chamber.

The second way of observing the rectifying behavior of the trapdoor is shown in Fig. 3. The data come from the same kind of equilibration experiment as Fig. 2, where all the particles are positioned initially along the outermost wall of one chamber. Now we look at the time interval it takes 25 molecules to pass from one chamber to the other as a function of the density difference. If $T$ is the time interval, then the ratio $25/T$ measures the current of particles—the rate—at which they pass through the middle wall opening in response to the density difference during that time. Figure 3 plots the particle current against the density difference between the two chambers for a system of 500 particles. The resulting curve resembles qualitatively the voltage-current characteristic of an electrical diode, and indicates that the trapdoor acts as a rectifier when large density differences are imposed by external means.[4,17]

The precise quantities plotted in Fig. 3 are the rates, inverse time intervals $1/(T_{i+1}-T_i)$, as a function of the mean fractional difference in the number of particles between the two chambers $(\Delta N_{i+1}+\Delta N_i)/(2N)$ during the time interval $(T_{i+1}-T_i)$. Here, $N$ is the total number of particles, $\Delta N_i$ is the difference in the number of particles at the starting time $T_i$, and $\Delta N_{i+1}$ is the difference in the number of particles at the finishing time $T_{i+1}$ when 25 molecules have moved from the source chamber (high density) to the sink chamber (low density). Positive values of $\Delta N_i$ correspond to higher densities in the right chamber. Thus the positive $x$ axis of Fig. 3, the "conducting region," corresponds to higher densities in the right chamber, which is expected because the trapdoor opens to the left. The $y$ axis is in units of $25\times10^9$ particles/s. The $x$ axis is in units of the fractional difference in the number of particles, so that an interval of size 0.1 corresponds to 25 particles moving from one chamber to the other $[0.1\times500\times(1/2)=25]$. The intervals $(-1,-0.9)$ and $(0.9,1.0)$ are not included in the plot because the times immediately after the release of the system from our initial conditions do not correspond to smooth flow.

The third and last method of exhibiting the rectifying behavior of the trapdoor focuses on steady state behavior. In contrast to the equilibration experiments above, this method measures the time averaged flow of molecules through the middle wall opening when a large density difference is maintained artificially. In particular, we continuously "reverse bias" the system by removing the molecules that hit the rightmost wall of the right chamber and reinserting them in the left side of the left chamber. This results in a density difference that pushes the trapdoor towards the closed position. In an opposite experiment, we forward bias the system by reinserting molecules from the leftmost wall into the right chamber.

Table I lists the flow of molecules (number of particles per second) passing through the middle wall opening under reverse and forward bias conditions. Molecules passing left to right are counted positive and molecules passing right to left are counted negative. We list the results for two different systems, a system of 500 particles and a system of 100 particles. The time averaging is done over $10^{-5}$ s for the 100 particle system and $10^{-6}$ s for the 500 particle system, which are both large enough to guarantee convergence; namely, the averages will not change over longer time intervals. We have checked this by plotting the time averages against time, and seeing that the curves approach a horizontal slope and a constant value. The values in Table I show that the flow allowed by the trapdoor in the forward bias condition is 22 times larger than the flow allowed in the reverse bias condition for 500 particles, and 17 times larger in the case of 100 particles. In other words, the trapdoor acts as a rectifier.

It is worth pointing out that the rectifying behavior of the trapdoor depends greatly on the geometry of the system. Experimentally, we have found that our trapdoor becomes a better rectifier the longer the trapdoor is, and the more molecules there are near the trapdoor. When many collisions take place exclusively on one side of the trapdoor during the time interval it takes the trapdoor to move significantly, the trapdoor is pushed and kept near one door stop. The probability of moving significantly away from that door stop is very small. For example, if many collisions take place exclusively on the left side of the trapdoor, the trapdoor will be kept near the middle wall opening, bouncing between the middle wall door stop and the large number of particles on its left side. The trapdoor performance can be improved further by placing one door stop slightly inside the right chamber. This centers the jittering of the door exactly on the middle wall and decreases the chance of a molecule leaking from the high-density left chamber into the low-density right chamber. For similar reasons we expect that making the trapdoor have finite width, that is using a two-dimensional trapdoor in the shape of a rectangle will result in even better rectifying behavior for large density differences.

## IV. VERIFICATION OF THE SECOND LAW

Having seen that the trapdoor acts as a rectifier under external bias, we now consider what happens when the trapdoor and molecules are left to evolve on their own in an isolated container. The asymmetry of the trapdoor's location, opening to the left but not to the right, intends to hinder the passage of molecules from left to right while providing an easy access from right to left. In this way, the trapdoor attempts to exploit the naturally occurring fluctuations in density between the two chambers and to make states of high density in the left chamber last longer than corresponding states of high density in the right chamber. The ultimate goal of the trapdoor is to maintain on the average a higher number of molecules in the left chamber than in the right chamber.

Our simulations show that the trapdoor does not succeed. When the trapdoor and molecules are left to evolve on their own, the time-averaged number of molecules in the left chamber is actually smaller than the time-averaged number of molecules in the right chamber. However, this imbalance is not a true density difference between the two chambers and does not violate the second law. The reason for the unequal number of molecules between the two chambers is that the presence of the trapdoor in the left chamber occupies space, which makes the available area in the left chamber slightly smaller than the available area in the right chamber.

The effect of the excluded area by the trapdoor can be estimated theoretically and also measured experimentally. In our simulations we performed an experiment with 20 particles where each chamber measured $13.5\times10^{-7}$ cm horizontally and $18\times10^{-7}$ cm vertically. The particle radius in this experiment was $6\times10^{-6}$ cm giving a mean free path in the order of $20\times10^{-7}$ cm. The length of the trapdoor (vertical direction) was $10\times10^{-7}$ cm. Based on these numbers we can estimate theoretically the average number of particles in the left chamber by assuming that the density is uniform between the two chambers (equilibrium) in a time average sense. If $N_L$ is the time averaged number of particles in the left chamber and $A_L$ the available area in the left chamber, we have

$$\frac{N_L}{A_L}=\frac{N_R}{A_R}=\frac{1-N_L}{A_R},$$

which gives $N_L=A_L/(A_L+A_R)$. To estimate the available area in each chamber, $A_L$ and $A_R$, we consider the area where the *centers* of the particle disks can travel. Thus $A_R=[(13.5\times18)-(13.5+13.5+18)\times0.06]\times10^{-14}$ cm$^2$ for the right chamber, and for the left chamber we subtract from the above $A_R$ the area excluded by the trapdoor $[(10\times0.06)+\pi(0.06)^2]\times10^{-14}$ cm$^2$. Putting these numbers together we find $N_L=0.484$. In our simulations, we found the time-averaged number of particles in the left chamber to be 0.486 in excellent agreement with our theoretical estimate.[18]

It is worth mentioning that there is also an alternative way of checking that the difference in the number of particles between the two chambers is not a true density difference, and it cannot lead to a violation of the second law. The idea is to open a second hole in the middle wall, in addition to the opening covered by the trapdoor. If the trapdoor could lead to a true density difference, pumping molecules from one chamber to the other through the opening covered by the trapdoor, then the second free opening should exhibit the *return flux* of molecules and lead to perpetual flow between the two chambers. Our simulations of this situation did not show any *return flux*.

Our simulations show that the operation of the trapdoor is consistent with the second law of thermodynamics in the sense that the particles are distributed uniformly in the available area on the average, and the entropy of the system is maximized. Our simulations have also shown that the time-averaged temperature is the same in each chamber and equal to the temperature of the trapdoor (average kinetic energy divided by the number of degrees of freedom, two for the particles, and one for the trapdoor). Finally, we have confirmed that the time averaged velocities in each chamber are distributed as Gaussian distributions in $v_x$ and $v_y$. The two Gaussian distributions are identical to each other and identical between the two chambers, which is consistent with the Maxwell–Boltzmann distribution law and equipartition of energy.

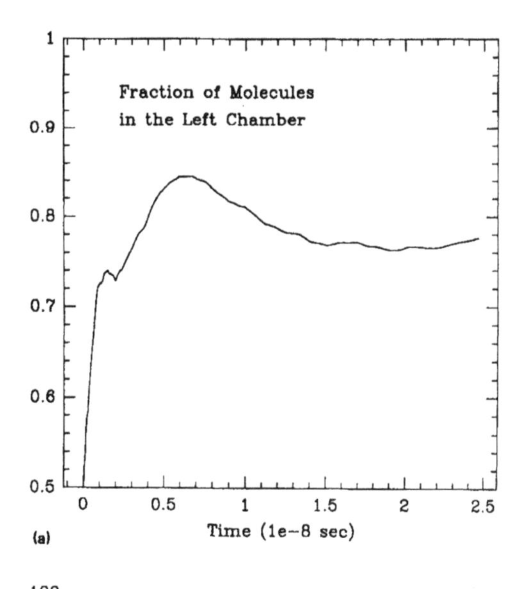


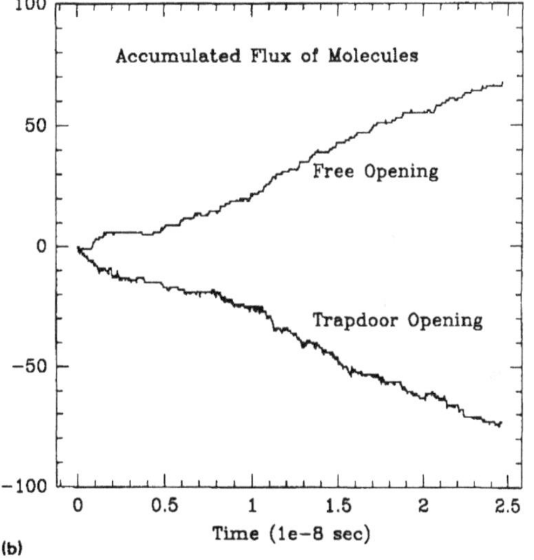


Fig. 4. A trapdoor with a cooling mechanism acts as a pump. The graph on the top (a) shows how the fraction of the total number of molecules in the left chamber builds up after the system is released from an initial state of equal number of molecules in each chamber. The graph on the bottom (b) shows the accumulated flux of molecules through the trapdoor opening (negative slope) and the accumulated flux of molecules through an additional opening that allows free passage (positive slope).

An intuitive explanation of why our trapdoor system fails to work when operating in an isolated container of gas molecules is that the trapdoor becomes thermalized—its temperature equals the temperature of the particles—and the trapdoor's thermal motion prevents the rectifying behavior.[12–14] By contrast, a macroscopic trapdoor works successfully as a rectifier because it can get rid of excess energy through dissipation. Following this analogy further, we may expect that our trapdoor would work successfully if a reservoir of lower temperature than the particles were available to dissipate its motion. The trapdoor would then act as a pump, letting the molecules pass from one chamber to the other more easily in one direction than the other. We have tested this idea in simulations, and we report our results in the next section.

## V. THE TRAPDOOR AS A PUMP

To convert our trapdoor system into an effective pump, we modify the evolution algorithm to cool the trapdoor by removing energy in small increments. In particular, we scale the trapdoor's velocity by 0.5 every $\Delta t$ time interval with $\Delta t$ sufficiently small. The lost energy is reinserted in equal amounts to all the particles by scaling their velocities, conserving the total energy of the system. Moreover, the cooling of the trapdoor is performed only when the trapdoor is near the closed position, which makes the trapdoor tend to remain closed. The goal of the cooled trapdoor is to pump molecules from the right chamber into the left chamber.

Our simulations show that this design works successfully. Further, our simulations show that the mass of the trapdoor in relation to the mass of each particle is crucial for efficient operation. If the trapdoor mass is much smaller than the mass of one particle, then the action of a single particle coming from the right chamber is enough to open the trapdoor and let the particle through, even though some energy is lost by interacting with the trapdoor. Similarly, the action of a single particle coming from the left chamber immediately closes the trapdoor and traps the molecule in the left chamber. On the other hand, if the trapdoor mass is much larger than the mass of one particle, many particles must collide with the trapdoor in a short period of time in order to move the trapdoor—in particular to open the trapdoor from the right chamber. Clearly, such collision situations do not occur very frequently, and so a heavy trapdoor does not work very well. Our simulations show that a very light trapdoor with dissipation can act effectively as a one-way door, opening to particles from the right, and closing to particles from the left. A heavier trapdoor with dissipation works also, but not as well.

In Fig. 4, we report results for a trapdoor system with dissipation, where the mass of the trapdoor is $4.7\times10^{-24}$ gm, or one tenth of the mass of one particle. The time interval which controls the rate of energy dissipation is $2.5\times10^{-13}$ s, while the mean free path and mean collision time in the left chamber are of the order of $20\times10^{-7}$ cm and $5\times10^{-11}$ s. The system contains 20 particles, the length of the trapdoor is $6\times10^{-7}$ cm, and each chamber measures $13.5\times10^{-7}$ cm horizontally and $18\times10^{-7}$ cm vertically. In this experiment, we have also included a second hole in the middle wall, of size $1\times10^{-7}$ cm, in addition to the hole covered by the trapdoor. The purpose of the additional hole is to verify that the trapdoor indeed acts as a pump of molecules from right to left, by exhibiting the *return flux* of molecules left to right. Graph (a) of Fig. 4 shows how the fraction of molecules in the left chamber builds up as soon as the system is released from a state of equal number of molecules in each chamber. Simulations that were run much longer than Fig. 4 show that the time average of the number of molecules in the left chamber stabilizes around 0.76 (averaging time over $10^{-5}$ s). The time averaged temperature of the trapdoor is 11-deg Kelvin, compared to 270-deg Kelvin for the particles. Graph (b) of Fig. 4 shows the accumulated flux of particles through the trapdoor covered opening, and the accumulated flux of particles through the second opening that allows free passage—this is the *return flux*. The absolute slope of both flux curves (measured over $10^{-5}$ s) is approximately $2\times10^{9}$ particles/s. These results show that the trapdoor can operate successfully as a rectifier when a heat reservoir at lower temperature is available. However, as discussed previously the trapdoor cannot operate successfully when run at the same temperature as the gas molecules, in accordance with the second law of thermodynamics.


[1]P. C. W. Davies, *The Physics of Time Asymmetry* (Univ. of California, Berkeley and Los Angeles, 1977), Chap. 3.

[2]P. Ehrenfest and T. Ehrenfest, *The Conceptual Foundations of the Statistical Approach in Mechanics*, translated by M. Moravcsik (Cornell U.P., Ithaca, NY, 1959).

[3]J. C. Maxwell, *Theory of Heat* (Longmans, Green, London, 1871), pp. 308–309.

[4]W. Ehrenberg, "Maxwell's demon," Sci. Am. **217**, 103–110 (1967).

[5]C. H. Bennett, "Demons, engines, and the second law," Sci. Am. 108–116 (1987).

[6]L. Szilard, "On the decrease of entropy in a thermodynamic system by the intervention of intelligent beings," (translation) in *Quantum Theory and Measurement*, edited by J. A. Wheeler and W. H. Zurek (Princeton U.P., Princeton, NJ, 1983), pp. 539–548. [L. Szilard, "Uber die entropieverminderung in einem thermodynamischen system bei eingriffen intelligenter wesen," Z. Phys. **53**, 840–856 (1929).]

[7]L. Brillouin, *Science and Information Theory* (Academic, New York, 1962).

[8]H. Leff and A. Rex, "Resource letter MD-1: Maxwell's demon," Am. J. Phys. **58**, 201–209 (1990); *Maxwell's Demon: Entropy, Information*, edited by H. S. Leff and A. F. Rex (Hilger/Princeton U.P., Princeton, NJ, 1990).

[9]R. Landauer, "Irreversibility and heat generation in the computing process," IBM J. Res. Dev. **5**, 183–191 (1961).

[10]W. H. Zurek, "Thermodynamic cost of computation, algorithmic complexity and the information metric," Nature **341**, 119–124 (1989); W. H. Zurek, "Algorithmic randomness and physical entropy," Phys. Rev. A **40**, 4731–4751 (1989).

[11]C. M. Caves, W. G. Unruh, and W. H. Zurek, "Comment on quantitative limits on the ability of a Maxwell demon to extract work from heat," Phys. Rev. Lett. **65**, 1387 (1990).

[12]M. Smoluchowski, "Experimentell nachweisbare der ublichen thermodynamik widersprechende Molekularphanomene," Phys. Z. **13**, 1069–1080 (1912).

[13]M. Smoluchowski, "Vortgage uber die Kinetische Theorie der Materie und der Elektrizitat," edited by M. Planck (Teubner und Leipzig, Berlin, 1914), pp. 89–121.

[14]R. P. Feynman, R. B. Leighton, and M. Sands, *The Feynman Lectures on Physics—Vol. 1* (Addison-Wesley, Reading, MA, 1963, 1965), Chap. 1-46.

[15]L. B. Loeb, *The Kinetic Theory of Gases* (Dover, New York, 1961), pp. 64–66.


[16]A derivation of the mean free path formula in three dimensions can be found in C. Kittel and H. Kroemer, *Thermal Physics* (Freeman, New York, 1980), 2nd ed., pp. 395–396. To derive a corresponding formula in two dimensions, we replace the area $\pi(2R)^2$ swept by a molecule in three dimensions, in the sense of collisions with other molecules, with the length $2R$.


[17]L. Brillouin, "Can the rectifier become a thermodynamical demon?" Phys. Rev. **78**, 627–628 (1950).


[18]It should be noted that there is an additional small correction to the available area in each chamber because the presence of one particle in some location excludes other particles from that location. However, this correction is small in our system, less than $10\pi(0.06)^2\times10^{-14}$ $cm^2$, and does not change the first three decimal places of the theoretical estimate.

# Entropy and Information for an Automated Maxwell's Demon

Andy Rex
Ross Larsen
Physics Department
University of Puget Sound
Tacoma, WA 98416

**Abstract**

In an attempt to create a Maxwell's demon that does not depend on information, we have developed a computational simulation of a demon that uses an automated, refrigerated trapdoor. We keep track of and compare the entropy reduction by the demon and the entropy cost of running the refrigerator.

**Background and Rationale**

Skordos and Zurek[1] have developed an ingenious type of Maxwell's demon, one that operates without any information on molecular positions or speeds. Their work is based on a suggestion made by Smoluchowski[2] in 1912 that a mechanical, perhaps spring-loaded trapdoor separating two chambers of ideal gas molecules might act as a one-way valve for faster molecules. Smoluchowski concluded that the Second Law would lead us to believe that thermal fluctuations would in fact prohibit such a mechanical demon from creating on average a net entropy reduction, although his work by no means proved this fact conclusively.

The idea of a mechanical demon is enticing, because of the conclusive work by Landauer, Bennett, and others in the physics of computation. Their work has shown that there is not necessarily any entropy increase associated with the gathering of information, but that in a reversible computation cycle there is an entropy increase associated with information erasure or the resetting of a memory. Their work has in a sense rendered obsolete the ideas of Szilard and Brillouin, not to mention the computational studies one of us (Rex) did in the 1980s.

But with a mechanical demon there should be no information to consider in the first place. The device developed by Skordos and Zurek consists of two gas chambers of equal area, with a hole between those chambers and a trapdoor that can just cover the hole (Figure 1). The door is free to slide back and forth on rails between two stops, one of them at the chamber interface and the other in the interior of one chamber. In this way an asymmetry is created, with molecules on one

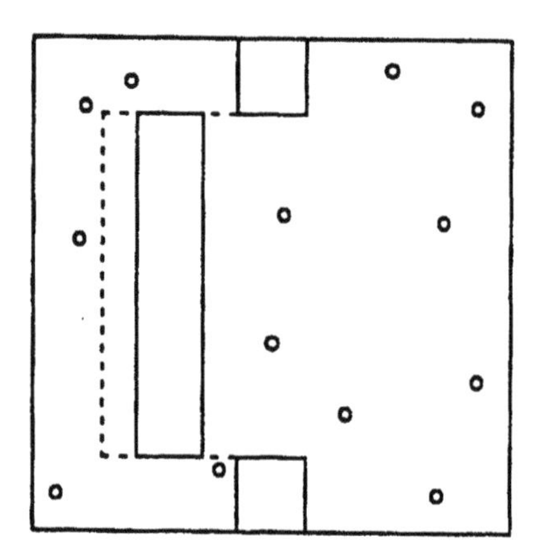

Figure 1. Schematic diagram of the mechanical demon system.

side tending to push the door open, but those on the other side tending to push it closed. A "run" consists of watching a computer-simulated evolution of the molecules as they collide elastically with the chamber walls, the trapdoor, and each other. Skordos and Zurek showed that their device could act as a "rectifier". That is, there is a natural tendency for molecules to flow one way through the trapdoor opening but a reluctance to flow in the other direction. Skordos and Zurek further showed that their device is not successful in defeating the Second Law, because it cannot be used to create a density difference between the two sides, due to the random thermal motion of the door itself, thus verifying what Smoluchowski had suspected 80 years ago. Following this logical line, they used a scheme for "refrigerating" their trapdoor, taking energy away from it when it approached the "closed" position. They discovered that this scheme could be used to create a density imbalance. Of course this does not in principle violate the Second Law, because there is some energy expense involved in slowing the door's motion.

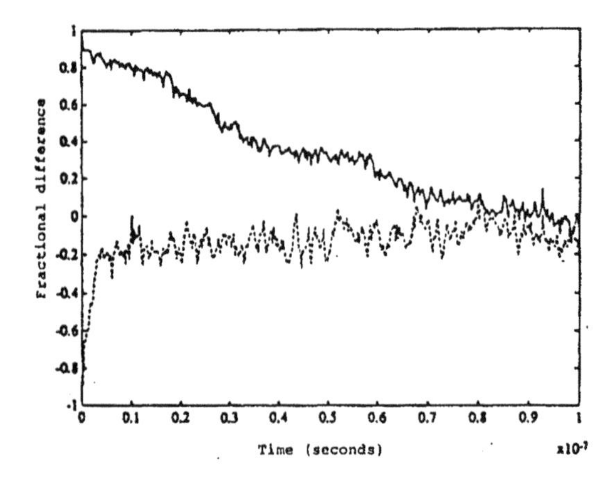


Figure 2. Rectifying behavior of the mechanical demon. When a run is started with all particles in the left chamber (top curve), the approach to equilibrium is slower then when the run is started with all on the right (bottom curve).

We have developed a simulation along the lines of the one made by Skordos and Zurek. We began by quickly verifying that our device could act as a rectifier (Figure 2). But instead of looking further at the qualitative behavior of the system, our work has focused on the measurement of entropy. If this refrigerated mechanical demon creates a density difference between the two sides of the container, it should be possible to compute the entropy reduction from fundamental relations. Specifically, Boltzmann's entropy equation S = k log w (where w is the number of microstates that will generate a particular macrostate) can be used to compute the entropy associated with various configurations of molecules. Actually the quantity w must be adjusted slightly by the fact that the door occupies a small fraction of the volume of one side of the container. The computation of the change in statistical entropy from one macrostate to another is then straightforward.

According to the Second Law, there must be an offsetting entropy increase, and in this case it is due to the energy dissipated in running the refrigerator. In the process of refrigeration we have extracted some energy from the door. In our model we refrigerate by letting the door (when it reaches the closed position) interact with a fairly low temperature heat bath. Then the entropy added to the heat bath is simply the energy lost by the door divided by the temperature of the bath, according to the well-known thermodynamic relation. We are then able to compare the thermodynamic entropy increase with the statistical entropy reduction mentioned above, or rather, compute their sum to find the net entropy change of the entire system.

**Implementation**

To implement our scheme, it is necessary to derive the collision and timing equation of the system. The collision equations give the new velocities of the colliding bodies in terms of their initial velocities. We assume perfectly rigid walls so that when a particle hits a wall, we simply reverse the component of its velocity that is perpendicular to the wall. This has the effect of treating the particles as light rays reflecting off of a mirror. In collisions between two particles, or between a particle and the trapdoor, the collision equations are derived using the conservation of energy, conservation of momentum, and the fact that the force of collision acts along the line connecting the particles' centers. This calculation is much easier if it is performed in the rest frame of one particle, so in our computer program we transform into the rest frame of one particle and calculate the new velocities before transforming back to the box's reference frame.

The timing equations tell us the amount of time until the next collision. For collisions involving a particle with a wall, another particle, or the trapdoor, these are straightforward to derive. Once the time to the next collision has been computed, the particles and trapdoor are moved along the straight line trajectories dictated by their velocities. The collision equations then change the velocities of the colliding bodies, and we calculate the time to the next collision.

Before we can run the program, it is necessary to initialize the positions and speeds of each particle. Since our program is supposed to model the real world, we choose to initialize the velocities so that the distribution of particle velocities closely approximates the two dimensional analog of the Maxwell-Boltzmann speed distribution. This is a straightforward procedure once the correct distribution is known. By modifying the arguments used in derivations of the Maxwell-Boltzmann distribution, it is straightforward to derive the two dimensional distribution:

$$F(v) = \frac{mv}{kT} \exp\left[-\frac{mv^2}{2kT}\right] \qquad (1)$$

This function, similar in shape to the well-known Maxwell-Boltzmann distribution, is known as the Rayleigh distribution.

The particles are assigned speeds according to the Rayleigh speed distribution F(v). Velocity is a function of speed and direction so once a particle has a speed, we assign it a direction using a random number generator. Having initialized the particle velocities, we now need to initialize their positions. We assign each particle to a random position on a predetermined side of the box. The number of particles to be placed in each side is varied from run to run. To initialize the door's motion, we give it a speed that we would expect it to have if it were at the same temperature, T, as the gas. We use the door's single degree of freedom and the equipartition theorem to compute the door's likeliest speed

$$v = \sqrt{kT/m} \qquad (2)$$

and give the door this speed initially.

### The Door as a Pressure Demon

To test whether the trapdoor can create large pressure differences between the two sides, we record the fractional difference between the number of particles in each half as a function of time. The fractional difference is defined as the number of particles in the left half minus the number of particles in the right half, divided by the total number of particles. The system rapidly approaches an equilibrium about which it fluctuates. The equilibrium value of the fractional difference is nonzero because the trapdoor's finite width keeps some area from being occupied by the particles. Our simulations show that the fractional difference approaches exactly the value expected at equilibrium. This means that the trapdoor does not act as a pressure demon, and we are prompted to ask why this is the case. We have found that the door's speed increases over time, leaving the door open so often that a pressure difference cannot be maintained. This is consistent with the behavior Smoluchowski predicted for automated demons.

The above result suggests that the trapdoor cannot decrease the entropy of the system when left to itself, because the door's high speed lets particles pass through the opening. To test whether or not the speed of the door really affects the ability of the trapdoor to operate as a demon, we changed our program to remove energy from the door periodically, effectively cooling it. The energy removal scheme pulls energy out by slowing the door at the instant it collides with the center partition. This has the effect of keeping the door near the closed position longer than would otherwise occur. Our findings, which resemble Skordos and Zurek's, show that the cooled door can create large pressure differences between the two sides (Figure 3). Since we can create a measurable pressure difference, we need to find a way to see if the entropy change associated with cooling the door offsets the decrease associated with the pressure difference. We believe that this is the first time that actual measurement of the entropy change caused by

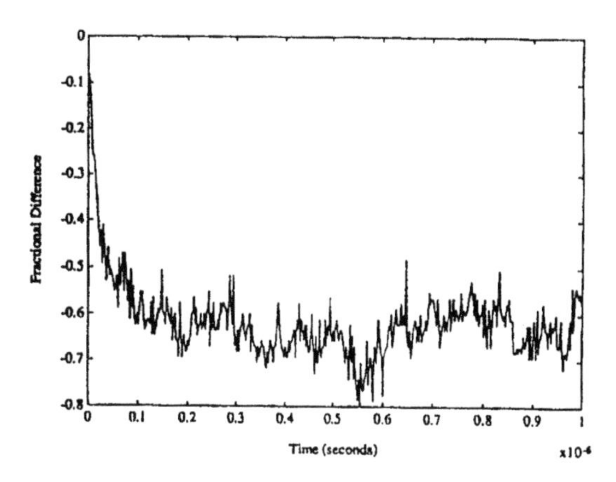


Figure 3. A pressure difference is created by the use of a refrigerated door.

the operation of an automated demon has been done, and we expect this to make our analysis of the demon's behavior less dependent on intuition than earlier analyses of automated demons. Below we outline a method of measuring the actual entropy change in our system, taking into account the energy removed from the system.

## Entropy in the Operation of the Heat Pump

To fully test the ability of the trapdoor to decrease the entropy of the box/heat bath system, we must first formulate a precise measure of the entropy of the system. It is straightforward to measure an entropy change due to a change in the number of particles on each side of the box. Less probable configurations correspond to a lower entropy than more probable configurations, a fact which is quantified using the Boltzmann distribution

$$S = k \log(w)$$

where w is the standard statistical weight used in statistical mechanics. The derivation of w is complicated by the fact that the two halves of the box are not of equal volume, due to volume excluded by the trapdoor. Let the left half of the box have volume $V_L$, the right volume $V_R$, and the box itself have volume V. The probability that L out of N particles will be on the left side in a random initial configuration is given by elementary probability theory as:

$$\frac{(V_L/V)^L (V_R/V)^{N-L} N!}{L! (N-L)!}$$

The number of macrostates is simply the total number of microstates times the probability of a single macrostate occurring. So, the number of microstates for the macrostate with L particles on the left is given by

$$w = \frac{2^N (V_L/V)^L (V_R/V)^{N-L} N!}{L! (N-L)!} \quad (3)$$

The above w lets us define the entropy to be

$$S = k \log \frac{2^N (V_L)^L (V_R/V)^{N-L} N!}{L! (N-L)!} \quad (4)$$

We shall henceforth call this value the statistical entropy. The change in statistical entropy is easily calculated since we compute $S_0$, the initial value of S, before we begin the time-evolution of our system.

We also need to measure the entropy change due to the energy removed from the system each time we slow the door. The change due to cooling the door will, for reasons that become clear below, be called the thermodynamic entropy. The magnitude of this entropy change depends on what method is used to cool the door. To keep the simulation physically reasonable, we assume that the trapdoor is momentarily put in contact with a heat bath at temperature $T_b$ each time the door reaches its closed position. If the door had the same temperature as the heat bath, we would expect its speed to be:

$$v = \sqrt{kT_b/m} \quad (5)$$

To cool the door, we try to bring its speed closer to the speed we expect the door to have if it is at the temperature of the bath. To do this we decrease the door's speed to an amount equal to the average of the door's initial speed and the expected speed every time the door has a collision with the center dividers; this removes energy from the door by an amount that increases as the temperature difference increases. The cooling process effectively takes a certain amount of energy, dQ, and dumps it into the heat bath. The well-known thermodynamic relation then gives us the entropy change associated with this process as:

$$dS = dQ/T_b \qquad (6)$$

Now that we have determined the form of both the thermodynamic and statistical entropy changes in our system, we can compute the total entropy change associated with each run. The change in the statistical entropy will often be negative, but we expect this to be offset by the entropy increase from the very cooling process that lets the door create the pressure difference. A bonus is that we make a direct connection between the entropy concept as it is defined in statistical mechanics and the entropy commonly used in thermodynamics. It is important to note that our statistical entropy is only defined to within an additive constant, as is the thermodynamic entropy. Since we are only interested in the change in entropy, there is no need to invoke the Third Law of Thermodynamics to determine the additive constant.

**Entropy and Information in the Trapdoor System**

Our simulations use 170 particles to keep computing time to a minimum. Larger particle numbers are not expected to change our results, but it is important to keep in mind that the trapdoor operates in a high vacuum. Studying the result of a representative simulation, that was started with approximately half of the particles on each side of the box, we find that there is a decrease of the net entropy (thermodynamic + statistical) of the system equal to $-2.4 \times 10^{-22}$ J/K (Figure 4). This data suggests that the trapdoor system does operate as a demon, easily violating the Second Law of Thermodynamics in virtually every run.

What will save the Second Law? It is the connection with information, the quantity we sought to avoid by using a mechanical demon. By starting a run in a maximum entropy configuration, we have in fact supposed that we have knowledge of the positions of all the particles. The maximum entropy state is indeed the most probable macrostate, but it is by no means the only likely one. Since we have restricted ourselves to a relatively small number of particles, the probabilities of other macrostates are not negligible. Therefore, the observed entropy decrease occurs because we start the system near its maximum entropy configuration. We need to store some information about the system to give the box this initial configuration, and this must be erased to complete a cycle. This erasure has a large entropy increase associated with it, according to Landauer's principle, that more than makes up for the decrease we see within the system.

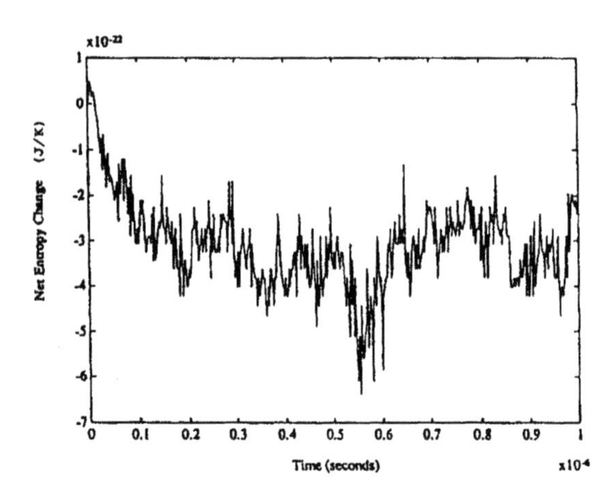


Figure 4. The demon creates a net entropy reduction when the run begins with equal numbers of particles on each side of the partition.

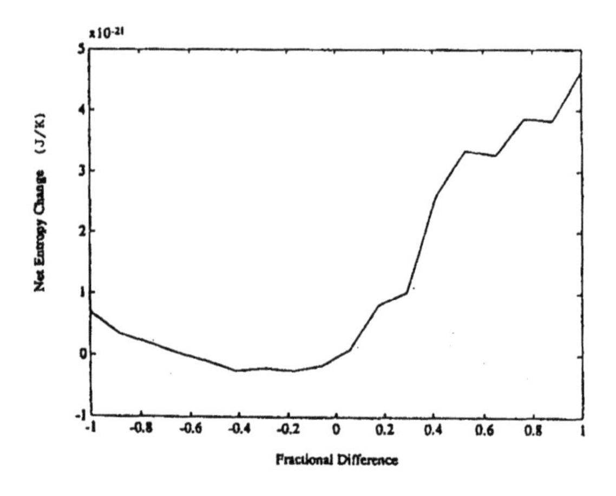


Figure 5. The net entropy change as a function of the initial fractional difference. Notice that there is a net entropy decrease for only a few special initial configurations, and for other initial configurations the entropy increase can be large.

**Operation of the Demon in the Absence of Information**

For the system to be truly automated, and to have a fully mechanical demon, we have to operate the demon without using any information about its initial state. To do this, the simulation must be run with initial conditions that range from all particles on the left to all particles on the right. Our procedure involves starting all of the particles on the left, then running the simulation. Then we place ten particles on the right and the rest on the left and run the simulation. We continue this process until we have run simulations with initial conditions ranging from all particles on the left to all particles on the right. Figure 5 shows the net change of entropy as a function of the initial fractional difference. Since each initial configuration is not equally likely, we compute the weighted average by multiplying the net change for a given fractional difference by the probability of this state occurring naturally. In our initialization procedure the particles are placed in equal volumes on each side of the box. Therefore, the probability of the various initial configurations follows the binomial distribution:

$$P(L) = \frac{N!}{2^N \; L! \; (N - L)!} \tag{7}$$

It must be noted that our initialization scheme carries no information cost, since we can simply mask the forbidden regions before we toss the particles into the box. The weighted entropy curve (Figure 6) is simply the data from Figure 5 where each point has been multiplied by the appropriate weighting factor. By summing over this curve, we find the average entropy associated with each blind run of the door to be about 2.75 x $10^{-24}$ J/K. In the case shown, the entropy change associated with the operation of our totally mechanical demon is (barely) positive. We must note that the error bars on this data point are large enough to include the point $\Delta S = 0$. This data has been obtained after an exhaustive search for an optimal door configuration, which we believe we have reached, given the result $\Delta S \approx 0$.

## Conclusion

We believe that the computational scheme and results presented in this paper provide some insight into the production of entropy by a Maxwell's demon. Further, we have presented a unique view of the connection between statistical entropy, thermodynamic entropy, and information.

## Acknowledgement

We gratefully acknowledge the support given to this project by the Murdock Charitable Trust.

## References

1. P.A. Skordos and W.H.Zurek, Am. J. Phys. 60, 876-882 (1992).

2. M. v. Smoluchowski, Physik. Z. 13, 1069-1080 (1912).

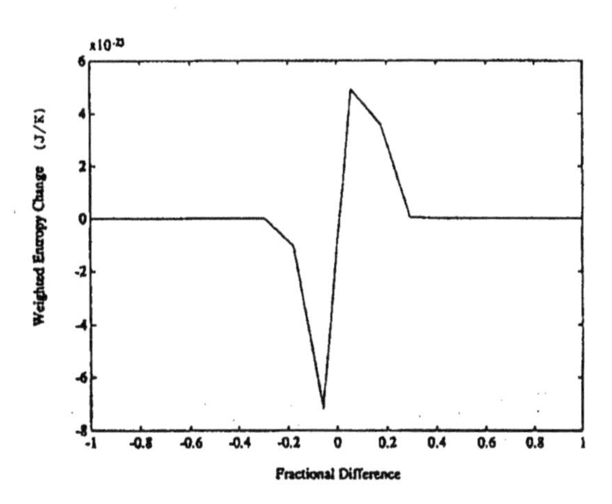


Figure 6. The net entropy change as a function of the initial configuration, weighted by the probability of the corresponding initial configuration occurring naturally. The sum of the weighted entropy changes over all possible configurations (basically the integral of this curve) is approximately zero.

# ON THE DECREASE OF ENTROPY IN A THERMODYNAMIC SYSTEM BY THE INTERVENTION OF INTELLIGENT BEINGS

LEO SZILARD

*Translated by Anatol Rapoport and Mechthilde Knoller from the original article "Über die Entropieverminderung in einem thermodynamischen System bei Eingriffen intelligenter Wesen." Zeitschrift für Physik, 1929, 53, 840–856.*

The objective of the investigation is to find the conditions which apparently allow the construction of a perpetual-motion machine of the second kind, if one permits an intelligent being to intervene in a thermodynamic system. When such beings make measurements, they make the system behave in a manner distinctly different from the way a mechanical system behaves when left to itself. We show that it is a sort of a memory faculty, manifested by a system where measurements occur, that might cause a permanent decrease of entropy and thus a violation of the Second Law of Thermodynamics, were it not for the fact that the measurements themselves are necessarily accompanied by a production of entropy. At first we calculate this production of entropy quite generally from the postulate that full compensation is made in the sense of the Second Law (Equation [1]). Second, by using an inanimate device able to make measurements—however under continual entropy production—we shall calculate the resulting quantity of entropy. We find that it is exactly as great as is necessary for full compensation. The actual production of entropy in connection with the measurement, therefore, need not be greater than Equation (1) requires.

THERE is an objection, already historical, against the universal validity of the Second Law of Thermodynamics, which indeed looks rather ominous. The objection is embodied in the notion of Maxwell's demon, who in a different form appears even nowadays again and again; perhaps not unreasonably, inasmuch as behind the precisely formulated question quantitative connections seem to be hidden which to date have not been clarified. The objection in its original formulation concerns a demon who catches the fast molecules and lets the slow ones pass. To be sure, the objection can be met with the reply that man cannot in principle foresee the value of a thermally fluctuating parameter. However, one cannot deny that we can very well measure the value of such a fluctuating parameter and therefore could certainly gain energy at the expense of heat by arranging our intervention according to the results of the measurements. Presently, of course, we do not know whether we commit an error by not including the intervening man into the system and by disregarding his biological phenomena.

Apart from this unresolved matter, it is known today that in a system left to itself no "perpetuum mobile" (perpetual motion machine) of the second kind (more exactly, no "automatic machine of continual finite work-yield which uses heat at the lowest temperature") can operate in spite of the fluctuation phenomena. A perpetuum mobile would have to be a machine which in the long run could lift a weight at the expense of the heat content of a reservoir. In other words, if we want to use the fluctuation phenomena in order to gain energy at the expense of heat, we are in the same position as playing a game of chance, in which we may win certain amounts now and then, although the expectation value of the winnings is zero or negative. The same applies to a system where the intervention from outside is performed strictly periodically, say by periodically moving machines. We consider this as established (Szilard, 1925) and intend here only to consider the difficulties that occur when intelligent beings intervene in a system. We shall try to discover the quantitative relations having to do with this intervention.

Smoluchowski (1914, p. 89) writes: "As far as we know today, there is no automatic, permanently effective perpetual motion machine, in spite of the molecular fluctuations, but such a device might, perhaps, function regularly if it were appropriately operated by intelligent beings...."

A perpetual motion machine therefore is possible if—according to the general method of physics—we view the experimenting man as a sort of *deus ex machina*, one who is continuously and exactly informed of the existing state of nature and who is able to start or interrupt the macroscopic course of nature at any moment without expenditure of work. Therefore he would definitely not have to possess the ability to catch single molecules like Maxwell's demon, although he would definitely be different from real living beings in possessing the above abilities. In eliciting any physical effect by action of the sensory as well as the motor nervous systems a degradation of energy is always involved, quite apart from the fact that the very existence of a nervous system is dependent on continual dissipation of energy.

Whether—considering these circumstances—real living beings could continually or at least regularly produce energy at the expense of heat of the lowest temperature appears very doubtful, even though our ignorance of the biological phenomena does not allow a definite answer. However, the latter questions lead beyond the scope of physics in the strict sense.

It appears that the ignorance of the biological phenomena need not prevent us from understanding that which seems to us to be the essential thing. We may be sure that intelligent living beings—insofar as we are dealing with their intervention in a thermodynamic system—can be replaced by nonliving devices whose "biological phenomena" one could follow and determine whether in fact a compensation of the entropy decrease takes place as a result of the intervention by such a device in a system.

In the first place, we wish to learn what circumstance conditions the decrease of entropy which takes place when intelligent living beings intervene in a thermodynamic system. We shall see that this depends on a certain type of coupling between different parameters of the system. We shall consider an unusually simple type of these ominous couplings.[1] For brevity we shall talk about a "measurement," if we succeed in coupling the value of a parameter $y$ (for instance the position co-ordinate of a pointer of a measuring instrument) at one moment with the simultaneous value of a fluctuating parameter $x$ of the system, in such a way that, from the value $y$, we can draw conclusions about the value that $x$ had at the moment of the "measurement." Then let $x$ and $y$ be uncoupled after the measurement, so that $x$ can change, while $y$ retains its value for some time. Such measurements are not harmless interventions. A system in which such measurements occur shows a sort of memory

[1] The author evidently uses the word "ominous" in the sense that the possibility of realizing the proposed arrangement threatens the validity of the Second Law.—*Translator*

faculty, in the sense that one can recognize by the state parameter $y$ what value another state parameter $x$ had at an earlier moment, and we shall see that simply because of such a memory the Second Law would be violated, if the measurement could take place without compensation. We shall realize that the Second Law is not threatened as much by this entropy decrease as one would think, as soon as we see that the entropy decrease resulting from the intervention would be compensated completely in any event if the execution of such a measurement were, for instance, always accompanied by production of $k \log 2$ units of entropy. In that case it will be possible to find a more general entropy law, which applies universally to all measurements. Finally we shall consider a very simple (of course, not living) device, that is able to make measurements continually and whose "biological phenomena" we can easily follow. By direct calculation, one finds in fact a continual entropy production of the magnitude required by the above-mentioned more general entropy law derived from the validity of the Second Law.

The first example, which we are going to consider more closely as a typical one, is the following. A standing hollow cylinder, closed at both ends, can be separated into two possibly unequal sections of volumes $V_1$ and $V_2$ respectively by inserting a partition from the side at an arbitrarily fixed height. This partition forms a piston that can be moved up and down in the cylinder. An infinitely large heat reservoir of a given temperature $T$ insures that any gas present in the cylinder undergoes isothermal expansion as the piston moves. This gas shall consist of a single molecule which, as long as the piston is not inserted into the cylinder, tumbles about in the whole cylinder by virtue of its thermal motion.

Imagine, specifically, a man who at a given time inserts the piston into the cylinder and somehow notes whether the molecule is caught in the upper or lower part of the cylinder, that is, in volume $V_1$ or $V_2$. If he should find that the former is the case, then he would move the piston slowly downward until it reaches the bottom of the cylinder. During this slow movement of the piston the molecule stays, of course, above the piston. However, it is no longer constrained to the upper part of the cylinder but bounces many times against the piston which is already moving in the lower part of the cylinder. In this way the molecule does a certain amount of work on the piston. This is the work that corresponds to the isothermal expansion of an ideal gas—consisting of one single molecule—from volume $V_1$ to the volume $V_1 + V_2$. After some time, when the piston has reached the bottom of the container, the molecule has again the full volume $V_1 + V_2$ to move about in, and the piston is then removed. The procedure can be repeated as many times as desired. The man moves the piston up or down depending on whether the molecule is trapped in the upper or lower half of the piston. In more detail, this motion may be caused by a weight, that is to be raised, through a mechanism that transmits the force from the piston to the weight, in such a way that the latter is always displaced upwards. In this way the potential energy of the weight certainly increases constantly. (The transmission of force to the weight is best arranged so that the force exerted by the weight on the piston at any position of the latter equals the average pressure of the gas.) It is clear that in this manner energy is constantly gained at the expense of heat, insofar as the biological phenomena of the intervening man are ignored in the calculation.

In order to understand the essence of the man's effect on the system, one best imagines that the movement of the piston is performed mechanically and that the man's activity consists only in determining the altitude of the molecule and in pushing a lever (which steers the piston) to the right or left, depending on whether the molecule's height requires a down- or upward movement. This means that the intervention of the human being consists only in the coupling of two position co-ordinates, namely a co-ordinate $x$, which determines the altitude of the molecule, with another co-ordinate $y$, which determines the position of the lever and therefore also whether an upward or downward motion is imparted to the piston. It is best to imagine the mass of the piston as large and its speed sufficiently great, so that the thermal agitation of the piston at the temperature in question can be neglected.

In the typical example presented here, we wish to distinguish two periods, namely:

1. The period of *measurement* when the piston has just been inserted in the middle of the cylinder and the molecule is trapped either in the upper or lower part; so that if we choose the origin of co-ordinates appropriately, the $x$-co-ordinate of the molecule is restricted to either the interval $x > 0$ or $x < 0$;

2. The period of *utilization of the measurement*, "the period of decrease of entropy," during which the piston is moving up or down. During this period the $x$-co-ordinate of the molecule is certainly not restricted to the original interval $x > 0$ or $x < 0$. Rather, if the molecule was in the upper half of the cylinder during the period of measurement, i.e., when $x > 0$, the molecule must bounce on the downward-moving piston in the lower part of the cylinder, if it is to transmit energy to the piston; that is, the co-ordinate $x$ has to enter the interval $x < 0$. The lever, on the contrary, retains during the whole period its position toward the right, corresponding to downward motion. If the position of the lever toward the right is designated by $y = 1$ (and correspondingly the position toward the left by $y = -1$) we see that during the period of measurement, the position $x > 0$ corresponds to $y = 1$; but afterwards $y = 1$ stays on, even though $x$ passes into the other interval $x < 0$. We see that in the utilization of the measurement the coupling of the two parameters $x$ and $y$ disappears.

We shall say, quite generally, that a parameter $y$ "measures" a parameter $x$ (which varies according to a probability law), if the value of $y$ is directed by the value of parameter $x$ at a given moment. A measurement procedure underlies the entropy decrease effected by the intervention of intelligent beings.

One may reasonably assume that a measurement procedure is fundamentally associated with a certain definite average entropy production, and that this restores concordance with the Second Law. The amount of entropy generated by the measurement may, of course, always be greater than this fundamental amount, but not smaller. To put it precisely: we have to distinguish here between two entropy values. One of them, $\bar{S}_1$, is produced when during the measurement $y$ assumes the value 1, and the other, $\bar{S}_2$, when $y$ assumes the value $-1$. We cannot expect to get general information about $\bar{S}_1$ or $\bar{S}_2$ separately, but we shall see that *if* the amount of entropy produced by the "measurement" is to compensate the entropy decrease affected by utilization, the relation must always hold good.

$$e^{-\bar{S}_1/k} + e^{-\bar{S}_2/k} \leqq 1 \tag{1}$$

One sees from this formula that one can make one of the values, for instance $\bar{S}_1$, as small as one wishes, but then the other value $\bar{S}_2$ becomes correspondingly greater. Furthermore, one can notice that the magnitude of the interval under consideration is of no consequence. One can also easily understand that it cannot be otherwise.

Conversely, as long as the entropies $\bar{S}_1$ and $\bar{S}_2$, produced by the measurements, satisfy the inequality (1), we can be sure that the expected decrease of entropy caused by the later utilization of the measurement will be fully compensated.

Before we proceed with the proof of inequality (1), let us see in the light of the above mechanical example, how all this fits together. For the entropies $\bar{S}_1$ and $\bar{S}_2$ produced by the measurements, we make the following Ansatz:

$$\bar{S}_1 = \bar{S}_2 = k \log 2 \tag{2}$$

This ansatz satisfies inequality (1) and the mean value of the quantity of entropy produced by a measurement is (of course in this special case independent of the frequencies $w_1$, $w_2$ of the two events):

$$\bar{S} = k \log 2 \tag{3}$$

In this example one achieves a decrease of entropy by the isothermal expansion:[2]

$$\begin{aligned} -\bar{s}_1 &= -k \log \frac{V_1}{V_1 + V_2}; \\ -\bar{s}_2 &= -k \log \frac{V_2}{V_1 + V_2}, \end{aligned} \tag{4}$$

[2] The entropy generated is denoted by $\bar{s}_1$, $\bar{s}_2$.

depending on whether the molecule was found in volume $V_1$ or $V_2$ when the piston was inserted. (The decrease of entropy equals the ratio of the quantity of heat taken from the heat reservoir during the isothermal expansion, to the temperature of the heat reservoir in question). Since in the above case the frequencies $w_1$, $w_2$ are in the ratio of the volumes $V_1$, $V_2$, the mean value of the entropy generated is (a negative number):

$$\bar{s} = w_1 \cdot (+\bar{s}_1) + w_2 \cdot (+\bar{s}_2) =$$

$$\frac{V_1}{V_1 + V_2} k \log \frac{V_1}{V_1 + V_2} + \frac{V_2}{V_1 + V_2} k \log \frac{V_1}{V_1 + V_2} \tag{5}$$

As one can see, we have, indeed

$$\frac{V_1}{V_1 + V_2} k \log \frac{V_1}{V_1 + V_2} + \frac{V_2}{V_1 + V_2} \cdot k \log \frac{V_2}{V_1 + V_2} + k \log 2 \geqq 0 \tag{6}$$

and therefore:

$$\bar{S} + \bar{s} \geqq 0. \tag{7}$$

In the special case considered, we would actually have a full compensation for the decrease of entropy achieved by the utilization of the measurement.

We shall not examine more special cases, but instead try to clarify the matter by a general argument, and to derive formula (1). We shall therefore imagine the whole system—in which the co-ordinate $x$, exposed to some kind of thermal fluctuations, can be measured by the parameter $y$ in the way just explained—as a multitude of particles, all enclosed in one box. Every one of these particles can move freely, so that they may be considered as the molecules of an ideal gas, which, because of thermal agitation, wander about in the common box independently of each other and exert a certain pressure on the walls of the box—the pressure being determined by the temperature. We shall now consider two of these molecules as chemically different and, in principle, separable by semipermeable walls, if the co-ordinate $x$ for one molecule is in a preassigned interval while the corresponding co-ordinate of the other molecule falls outside that interval. We also shall look upon them as chemically different, if they differ only in that the $y$ co-ordinate is $+1$ for one and $-1$ for the other.

We should like to give the box in which the "molecules" are stored the form of a hollow cylinder containing four pistons. Pistons $A$ and $A'$ are fixed while the other two are movable, so that the distance $BB'$ always equals the distance $AA'$, as is indicated in Figure 1 by the two brackets. $A'$, the bottom, and $B$, the cover of the container, are impermeable for all "molecules," while $A$ and $B'$ are semipermeable; namely, $A$ is permeable only for those "molecules" for which the parameter $x$ is in the preassigned interval, i.e., $(x_1, x_2)$, $B'$ is only permeable for the rest.

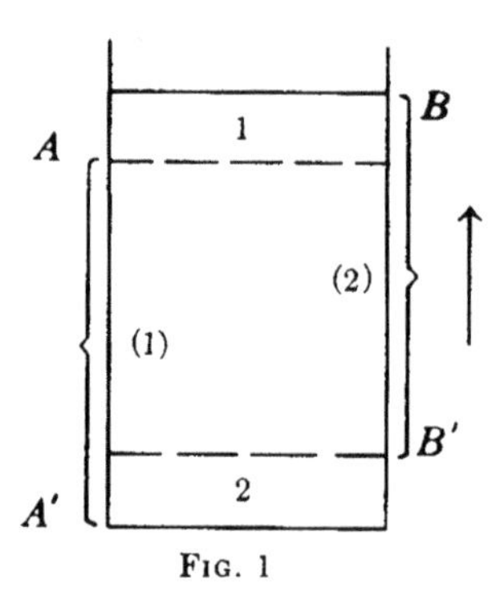


FIG. 1

In the beginning the piston $B$ is at $A$ and therefore $B'$ at $A'$, and all "molecules" are in the space between. A certain fraction of the molecules have their co-ordinate $x$ in the preassigned interval. We shall designate by $w_1$ the probability that this is the case for a randomly selected molecule and by $w_2$ the probability that $x$ is outside the interval. Then $w_1 + w_2 = 1$.

Let the distribution of the parameter $y$ be over the values $+1$ and $-1$ in any proportion but in any event independent of the $x$-values. We imagine an intervention by an intelligent being, who imparts to $y$ the value 1 for all "molecules" whose $x$ at that moment is in the selected interval. Otherwise the value $-1$ is assigned. If then, because of thermal fluctuation, for any "molecule," the parameter $x$ should come out of the preassigned interval or, as we also may put it, if the "molecule" suffers a monomolecular chemical reaction with regard to $x$ (by which it is transformed from a species that can pass the semipermeable piston $A$ into a species for which the piston is impermeable), then the parameter $y$ retains its value 1 for the time being, so that the "molecule," because of the value of the parameter $y$, "remembers" during the whole following process that $x$ originally was in the preassigned interval. We shall see immediately what part this memory may play. After the intervention just discussed, we move the piston, so that we separate the two kinds of molecules without doing work. This results in two containers, of which the first contains only the one modification and the second only the other. Each modification now occupies the same volume as the mixture did previously. In one of these containers, if considered by itself, there is now no equilibrium with regard to the two "modifications in $x$." Of course the ratio of the two modifications has remained $w_1 : w_2$. If we allow this equilibrium to be achieved in both containers independently and at constant volume and temperature, then the entropy of the system certainly has increased. For the total heat release is 0, since the ratio of the two "modifications in $x$" $w_1 : w_2$ does not change. If we accomplish the equilibrium distribution in both containers in a reversible fashion then the entropy of the rest of the world will decrease by the same amount. Therefore the entropy increases by a negative value, and, the value of the entropy increase per molecule is exactly:

$$\bar{s} = k(w_1 \log w_1 + w_2 \log w_2). \qquad (9)$$

(The entropy constants that we must assign to the two "modifications in $x$" do not occur here explicitly, as the process leaves the total number of molecules belonging to the one or the other species unchanged.)

Now of course we cannot bring the two gases back to the original volume without expenditure of work by simply moving the piston back, as there are now in the container—which is bounded by the pistons $BB'$—also molecules whose $x$-co-ordinate lies outside of the preassigned interval and for which the piston $A$ is not permeable any longer. Thus one can see that the calculated decrease of entropy (Equation [9]) does not mean a contradiction of the Second Law. *As long as we do not use the fact that the molecules in the container $BB'$, by virtue of their co-ordinate $y$, "remember" that the $x$-co-ordinate for the molecules of this container originally was in the preassigned interval, full compensation exists for the calculated decrease of entropy,* by virtue of the fact that the partial pressures in the two containers are smaller than in the original mixture.

*But now we can use the fact that all molecules in the container $BB'$ have the $y$-co-ordinate 1, and in the other accordingly $-1$, to bring all molecules back again to the original volume.* To accomplish this we only need to replace the semipermeable wall $A$ by a wall $A^*$, which is semipermeable not with regard to $x$ but with regard to $y$, namely so that it is permeable for the molecules with the $y$-co-ordinate 1 and impermeable for the others. Correspondingly we replace $B'$ by a piston $B'^*$, which is impermeable for the molecules with $y = -1$ and permeable for the others. Then both containers can be put into each other again without expenditure of energy. The distribution of the $y$-co-ordinate with regard to 1 and $-1$ now has become statistically independent of the $x$-values and besides we are able to re-establish the original distribution over 1 and $-1$. Thus we would have gone through a complete cycle. The only change that we have to register is the resulting decrease of entropy given by (9):

$$\bar{s} = k(w_1 \log w_1 + w_2 \log w_2). \qquad (10)$$

If we do not wish to admit that the Second Law has been violated, we must conclude *that the intervention which establishes the coupling between $y$ and $x$, the measurement of $x$ by $y$, must be accompanied by a production of entropy.* If a definite way of achieving this coupling is adopted and if the quantity of entropy that is inevitably produced is designated by $S_1$ and $S_2$, where $S_1$ stands for the mean increase in entropy that occurs when $y$ acquires the value 1, and accordingly $S_2$ for the increase that occurs when $y$ acquires the value $-1$, we arrive at the equation:

$$w_1 S_1 + w_2 S_2 = \bar{S} \qquad (11)$$

In order for the Second Law to remain in force, this quantity of entropy must be greater than the decrease of entropy $\bar{s}$, which according to (9) is produced by the utilization of the measurement. Therefore the following unequality must be valid:

$$\bar{S} + \bar{s} \geqq 0$$

$$w_1 S_1 + w_2 S_2 \qquad (12)$$

$$+ k(w_1 \log w_1 + w_2 \log w_2) \geqq 0$$

This equation must be valid for any values of $w_1$ and $w_2$,[3] and of course the constraint $w_2 + w_2 = 1$ cannot be violated. We ask, in particular, for which $w_1$ and $w_2$ and given $S$-values the expression becomes a minimum. For the two minimizing values $w_1$ and $w_2$ the inequality (12) must still be valid. Under the above constraint, the minimum occurs when the following equation holds:

$$\frac{S_1}{k} + \log w_1 = \frac{S_2}{k} + \log w_2 \qquad (13)$$

But then:

$$e^{-S_1/k} + e^{-S_2/k} \leqq 1. \qquad (14)$$

This is easily seen if one introduces the notation

$$\frac{S_1}{k} + \log w_1 = \frac{S_2}{k} + \log w_2 = \lambda; \quad (15)$$

then:

$$w_1 = e^{\lambda} \cdot e^{-S_1/k}; \quad w_2 = e^{\lambda} \cdot e^{-S_2/k}. \quad (16)$$

If one substitutes these values into the inequality (12) one gets:

$$\lambda e^{\lambda}(e^{-S_1/k} + e^{-S_2/k}) \geqq 0. \qquad (17)$$

Therefore the following also holds:

$$\lambda \geqq 0. \qquad (18)$$

If one puts the values $w_1$ and $w_2$ from (16) into the equation $w_1 + w_2 = 1$, one gets

$$e^{-S_1/k} + e^{-S_2/k} = e^{-\lambda}. \qquad (19)$$

And because $\lambda \geqq 0$, the following holds:

$$e^{-S_1/k} + e^{-S_2/k} \leqq 1. \qquad (20)$$

This equation must be universally valid, if thermodynamics is not to be violated.

As long as we allow intelligent beings to perform the intervention, a direct test is not possible. But we can try to describe simple nonliving devices that effect such coupling, and see if indeed entropy is generated and in what quantity. Having already recognized that the only important factor is a certain characteristic type of coupling, a "measurement," we need not construct any complicated models which imitate the intervention of living beings in detail. We can be satisfied with the construction of this particular type of coupling which is accompanied by memory.

In our next example, the position co-ordinate of an oscillating pointer is "measured" by the energy content of a body $K$. The pointer is supposed to connect, in a purely mechanical way, the body $K$—by whose energy content the position of the pointer is to be measured—by heat conduction with one of two intermediate pieces, $A$ or $B$. The body is connected with $A$ as long as the co-ordinate—which determines the position of the pointer—falls into a certain preassigned, but otherwise arbitrarily large or small interval $a$, and otherwise if the co-ordinate is in the interval $b$, with $B$. Up to a certain moment, namely the moment of the "measurement," both intermediate pieces will be thermally connected with a heat reservoir at temperature $T_0$. At this moment the insertion $A$ will be cooled reversibly to the temperature $T_A$, e.g., by a periodically functioning mechanical device. That is, after successive contacts with heat reservoirs of intermediate temperatures, $A$ will be brought into contact with a heat reservoir of the temperature $T_A$. At the same time the insertion $B$ will be heated in the same way to temperature $T_B$. Then the intermediate pieces will again be isolated from the corresponding heat reservoirs.

We assume that the position of the pointer changes so slowly that all the operations that we have sketched take place while the position of the pointer remains unchanged. If the position co-ordinate of the pointer fell in the preassigned interval, then the body was connected with the insertion $A$ during the above-mentioned operation, and consequently is now cooled to temperature $T_A$.

In the opposite case, the body is now heated to temperature $T_B$. Its energy content becomes—according to the position of

[3] The increase in entropy can depend only on the types of measurement and their results but not on how many systems of one or the other type were present.

the pointer at the time of "measurement"—small at temperature $T_A$ or great at temperature $T_B$ and will retain its value, even if the pointer eventually leaves the preassigned interval or enters into it. After some time, while the pointer is still oscillating, one can no longer draw any definite conclusion from the energy content of the body $K$ with regard to the momentary position of the pointer but one can draw a definite conclusion with regard to the position of the pointer at the time of the measurement. Then the measurement is completed.

After the measurement has been accomplished, the above-mentioned periodically functioning mechanical device should connect the thermally isolated insertions $A$ and $B$ with the heat reservoir $T_0$. This has the purpose of bringing the body $K$—which is now also connected with one of the two intermediate pieces—back into its original state. The direct connection of the intermediate pieces and hence of the body $K$—which has been either cooled to $T_A$ or heated to $T_B$—to the reservoir $T_0$ consequently causes an increase of entropy. This cannot possibly be avoided, because it would make no sense to heat the insertion $A$ reversibly to the temperature $T_0$ by successive contacts with the reservoirs of intermediate temperatures and to cool $B$ in the same manner. After the measurement we do not know with which of the two insertions the body $K$ is in contact at that moment; nor do we know whether it had been in connection with $T_A$ or $T_B$ in the end. Therefore neither do we know whether we should use intermediate temperatures between $T_A$ and $T_0$ or between $T_0$ and $T_B$.

The mean value of the quantity of entropy $S_1$ and $S_2$, per measurement, can be calculated, if the heat capacity as a function of the temperature $\bar{u}(T)$ is known for the body $K$, since the entropy can be calculated from the heat capacity. We have, of course, neglected the heat capacities of the intermediate pieces. If the position co-ordinate of the pointer was in the preassigned interval at the time of the "measurement," and accordingly the body in connection with insertion $A$, then the entropy conveyed to the heat reservoirs during successive cooling was

$$\int_{T_A}^{T_0} \frac{1}{T} \frac{d\bar{u}}{dT}. \qquad (21)$$

However, following this, the entropy withdrawn from the reservoir $T_0$ by direct contact with it was

$$\frac{\bar{u}(T_0) - \bar{u}(T_A)}{T_0}. \qquad (22)$$

All in all the entropy was increased by the amount

$$S_A = \frac{\bar{u}(T_A) - \bar{u}(T_0)}{T_0} + \int_{T_A}^{T_0} \frac{1}{T} \frac{d\bar{u}}{dT} dT. \qquad (23)$$

Analogously, the entropy will increase by the following amount, if the body was in contact with the intermediate piece $B$ at the time of the "measurement":

$$S_B = \frac{\bar{u}(T_B) - \bar{u}(T_0)}{T_0} + \int_{T_B}^{T_0} \frac{1}{T} \frac{d\bar{u}}{dT} dT. \qquad (24)$$

We shall now evaluate these expressions for the very simple case, where the body which we use has only two energy states, a lower and a higher state. If such a body is in thermal contact with a heat reservoir at any temperature $T$, the probability that it is in the lower or upper state is given by respectively:

$$\left.\begin{aligned} p(T) &= \frac{1}{1 + ge^{-u/kT}} \\ q(T) &= \frac{ge^{-u/kT}}{1 + ge^{-u/kT}} \end{aligned}\right\} \qquad (25)$$

Here $u$ stands for the difference of energy of the two states and $g$ for the statistical weight. We can set the energy of the lower state equal to zero without loss of generality. Therefore:[4]

$$\left.\begin{aligned} S_A &= q(T_A)\, k \log \frac{q(T_A)\, p(T_0)}{q(T_0)\, p(T_A)} + k \log \frac{p(T_A)}{p(T_0)} \\ S_B &= p(T_B)\, k \log \frac{q(T_0)\, p(T_B)}{q(T_B)\, p(T_0)} + k \log \frac{q(T_B)}{q(T_0)} \end{aligned}\right\}. \qquad (26)$$

Here $q$ and $p$ are the functions of $T$ given

[4] See the Appendix.

by equation (25), which are here to be taken for the arguments $T_0$, $T_A$, or $T_B$.

If (as is necessitated by the above concept of a "measurement") we wish to draw a dependable conclusion from the energy content of the body $K$ as to the position co-ordinate of the pointer, we have to see to it that the body surely gets into the lower energy state when it gets into contact with $T_B$. In other words:

$$p(T_A) = 1, q(T_A) = 0; \quad p(T_B) = 0, q(T_B) = 1. \tag{27}$$

This of course cannot be achieved, but may be arbitrarily approximated by allowing $T_A$ to approach absolute zero and the statistical weight $g$ to approach infinity. (In this limiting process, $T_0$ is also changed, in such a way that $p(T_0)$ and $q(T_0)$ remain constant.) The equation (26) then becomes:

$$S_A = -k \log p(T_0); \quad S_B = -k \log q(T_0) \tag{28}$$

and if we form the expression $e^{-S_A/k} + e^{-S_B/k}$, we find:

$$e^{-S_A/k} + e^{-S_B/k} = 1. \tag{29}$$

Our foregoing considerations have thus just realized the smallest permissible limiting care. The use of semipermeable walls according to Figure 1 allows a complete utilization of the measurement: inequality (1) certainly cannot be sharpened.

As we have seen in this example, a simple inanimate device can achieve the same essential result as would be achieved by the intervention of intelligent beings. We have examined the "biological phenomena" of a nonliving device and have seen that it generates exactly that quantity of entropy which is required by thermodynamics.

## APPENDIX

In the case considered, when the frequency of the two states depends on the temperature according to the equations:

$$p(T) = \frac{1}{1 + ge^{-u/kT}}; \quad q(T) = \frac{ge^{-u/kT}}{1 + ge^{-u/kT}} \tag{30}$$

and the mean energy of the body is given by:

$$\bar{u}(T) = uq(T) = \frac{uge^{-u/kT}}{1 + ge^{-u/kT}}, \tag{31}$$

the following identity is valid:

$$\frac{1}{T}\frac{d\bar{u}}{dT} = \frac{d}{dT}\left\{\frac{\bar{u}(T)}{T} + k \log\left(1 + e^{-u/kT}\right)\right\}. \tag{32}$$

Therefore we can also write the equation:

$$B_A = \frac{\bar{u}(T_A) - \bar{u}(T_0)}{T_0} + \int_{T_A}^{T_0} \frac{1}{T}\frac{d\bar{u}}{dT}\,dT \tag{33}$$

as

$$S_A = \frac{\bar{u}(T_A) - \bar{u}(T_0)}{T_0} + \left\{\frac{\bar{u}(T)}{T} + k \log(1 + ge^{-u/kT})\right\}_{T_A}^{T_0}, \tag{34}$$

and by substituting the limits we obtain:

$$S_A = \bar{u}(T_A)\left(\frac{1}{T_0} - \frac{1}{T_A}\right) + k \log \frac{1 + ge^{-u/kT_0}}{1 + ge^{-u/kT_A}}. \tag{35}$$

If we write the latter equation according to (25):

$$1 + ge^{-u/kT} = \frac{1}{p(T)} \tag{36}$$

for $T_A$ and $T_0$, then we obtain:

$$S_A = \bar{u}(T_A)\left(\frac{1}{T_0} - \frac{1}{T_A}\right) + k \log \frac{p(T_A)}{p(T_0)} \tag{37}$$

and if we then write according to (31):

$$\bar{u}(T_A) = uq(T_A) \tag{38}$$

we obtain:

$$S_A = q(T_A)\left(\frac{u}{T_0} - \frac{u}{T_A}\right) + k \log \frac{p(T_A)}{p(T_0)}. \tag{39}$$

If we finally write according to (25):

$$\frac{u}{T} = -k \log \frac{q(T)}{gp(T)} \tag{40}$$

for $T_A$ and $T_0$, then we obtain:

$$S_A = q(T_A)\, k \log \frac{p(T_0)\, q(T_A)}{q(T_0)\, p(T_A)} + k \log \frac{p(T_A)}{p(T_0)}. \tag{41}$$

We obtain the corresponding equation for $S_B$, if we replace the index $A$ with $B$. Then we obtain:

$$S_B = q(T_B)\,k\log\frac{p(T_0)}{q(T_0)}\frac{q((T_B)}{p((T_B)} + k\log\frac{p(T_B)}{p(T_0)}. \qquad (42)$$

Formula (41) is identical with (26), given, for $S_A$, in the text.

We can bring the formula for $S_B$ into a somewhat different form, if we write:

$$q(T_B) = 1 - p(T_B), \qquad (43)$$

expand and collect terms, then we get

$$S_B = p(T_B)\,k\log\frac{q(T_0)\,p(T_B)}{p(T_0)\,q(T_B)} + k\log\frac{q(T_B)}{q(T_0)}. \qquad (44)$$

This is the formula given in the text for $S_B$.

## REFERENCES

Smoluchowski, F. *Vorträge über die kinetische Theorie der Materie u. Elektrizitat*. Leipzig: 1914.

Szilard, L. Zeitschrift fur Physik, 1925, 32, 753.

# Maxwell's Demon Cannot Operate: Information and Entropy. I

L. Brillouin
*International Business Machines Corporation, Poughkeepsie and New York, New York*
(Received September 18, 1950)

In an enclosure at constant temperature, the radiation is that of a "blackbody," and the demon cannot see the molecules. Hence, he cannot operate the trap door and is unable to violate the second principle. If we introduce a source of light, the demon can see the molecules, but the over-all balance of entropy is positive. This leads to the consideration of a cycle

Negentropy→Information→Negentropy

for Maxwell's demon as well as for the scientist in his laboratory. Boltzmann's constant $k$ is shown to represent the smallest possible amount of negative entropy required in an observation.

## I. MAXWELL'S DEMON

The Sorting demon was born in 1871 and first appeared in Maxwell's *Theory of Heat* (p. 328), as "a being whose faculties are so sharpened that he can follow every molecule in his course, and would be able to do what is at present impossible to us···. Let us suppose that a vessel is divided into two portions $A$ and $B$ by a division in which there is a small hole, and that a being who *can see the individual molecules* opens and closes this hole, so as to allow only the swifter molecules to pass from $A$ to $B$, and only the slower ones to pass from $B$ to $A$. He will, thus, without expenditure of work raise the temperature of $B$ and lower that of $A$, in contradiction to the second law of thermodynamics."[1]

The paradox was considered by generations of physicists, without much progress in the discussion, until Szilard[2] pointed out that the demon actually transforms "information" into "negative entropy"—we intend to investigate this side of the problem in a moment.

Another contribution is found in a recent paper by the present author.[3]

In order to select the fast molecules, the demon should be able to see them (see Maxwell, passage reproduced in italics); but he is in an enclosure in equilibrium at constant temperature, where the radiation must be that of the blackbody, and it is impossible to see anything in the interior of a black body. It would not help to raise the temperature. At "red" temperature, the radiation has its maximum in the red and obtains exactly the same intensity, whether there are no molecules or millions of them in the enclosure. Not only is the intensity the same but also the fluctuations. The demon would perceive radiation and its fluctuations, he would never see the molecules.

No wonder Maxwell did not think of including radiation in the system in equilibrium at temperature $T$. Blackbody radiation was hardly known in 1871, and it took 30 more years before the thermodynamics of radiation was clearly understood and Planck's theory developed.

The demon cannot see the molecules, hence, he cannot operate the trap door and is unable to violate the second principle.

## II. INFORMATION MEANS NEGATIVE ENTROPY

Let us, however, investigate more carefully the possibilities of the demon. We may equip him with an electric torch and enable him to see the molecules. The torch is a source of radiation not in equilibrium. It pours negative entropy into the system. From this negative entropy the demon obtains "informations." With these informations he may operate the trap door and rebuild negative entropy, hence, completing a cycle:

$$\text{negentropy} \rightarrow \text{information} \rightarrow \text{negentropy}. \quad (1)$$

We coined the abbreviation "negentropy" to characterize entropy with the opposite sign. This quantity is very useful to consider and already has been introduced by some authors, especially Schrödinger.[4] Entropy must always increase, and negentropy always decreases. Negentropy corresponds to "grade" of energy in Kelvin's discussion of "degradation of energy."

[1] The full passage is quoted by J. H. Jeans, *Dynamical Theory Gases* (Cambridge University Press, London, 1921), third edition, p. 183.
[2] L. Szilard, Z. Physik **53**, 840–856 (1929).
[3] L. Brillouin, Am. Scientist **37**, 554–568 (1949), footnote to p. 565; **38**, 594 (1950).
[4] E. Schrödinger, *What is Life?* (Cambridge University Press, London, and The Macmillan Company, New York, 1945).

We shall discuss more carefully the new cycle (1) for the demon and show later on how it extends to man and scientific observation.

The first part of the cycle, where negentropy is needed to obtain information, seems to have been generally overlooked. The second transformation of information into negentropy was very clearly discussed by L. Szilard[2] who did the pioneer work on the question.

Our new cycle (1) compares with C. E. Shannon's[5] discussion of telecommunications, which can be stated this way:

information→telegram→negentropy on the cable→
telegram received→information received. (2)

Shannon, however, compares information with positive entropy, a procedure which seems difficult to justify since information is lost during the process of transmission, while entropy is increased. Norbert Wiener[6] recognized this particular feature and emphasized the similarity between information and negentropy.

Our new cycle (1) adds another example to the general theory of information. We shall now discuss this problem in some detail.

### III. ENTROPY BALANCE FOR MAXWELL'S DEMON

In order to discuss an entropy balance, the first question is to define an isolated system, to which the second principle can be safely applied. Our system is composed of the following elements:

1. A charged battery and an electric bulb, representing the electric torch.
2. A gas at constant temperature $T_0$, contained in Maxwell's enclosure, with a partition dividing the vessel into two portions and a hole in the partition.
3. The demon operating the trap door at the hole. The whole system is insulated and closed.

The battery heats the filament at a high temperature $T_1$.

$$T_1 \gg T_0. \tag{3}$$

This condition is required, in order to obtain visible light,

$$h\nu_1 \gg kT_0, \tag{4}$$

that can be distinguished from the background of blackbody radiation in the enclosure at temperature $T_0$. During the experiment, the battery yields a total energy $E$ and no entropy. The filament radiates $E$ and an entropy $S_f$,

$$S_f = E/T_1. \tag{5}$$

If the demon does not intervene, the energy $E$ is absorbed in the gas at temperature $T_0$, and we observe a global increase of entropy

$$S = E/T_0 > S_f > 0 \tag{6}$$

Now let us investigate the work of the demon. He can detect a molecule when at least one quantum of energy $h\nu_1$ is scattered by the molecule and absorbed in the eye of the demon (or in a photoelectric cell, if he uses such a device).[7] This represents a final increase of entropy

$$\Delta S_d = h\nu_1/T_0 = kb \quad h\nu_1/kT_0 = b \gg 1 \tag{7}$$

according to condition (4).

Once the information is obtained, it can be used to decrease the entropy of the system. The entropy of the system is

$$S_0 = k \ln P_0 \tag{8}$$

according to Boltzmann's formula, where $P_0$ represents the total number of microscopic configurations (Planck's "complexions") of the system. After the information has been obtained, the system is more completely specified. $P$ is decreased by an amount $p$ and

$$P_1 = P_0 - p \quad \Delta S_i = S - S_0 = k\Delta(\log P) = -k(p/P_0). \tag{9}$$

It is obvious that $p \ll P_0$ in all practical cases. The total balance of entropy is

$$\Delta S_d + \Delta S_i = k(b - p/P_0) > 0, \tag{10}$$

since $b \gg 1$ and $p/P_0 \ll 1$. The final result is still an increase of entropy in the isolated system, as required by the second principle. All the demon can do is to recuperate a small part of the entropy and use the information to decrease the degradation of energy.

In the first part of the process [Eq. (7)], we have an increase of entropy $\Delta S_d$, hence, a change $\Delta N_d$ in the negentropy:

$$\Delta N_d = -kb < 0, \quad \text{a decrease.} \tag{7a}$$

From this lost negentropy, a certain amount is changed into information, and in the last step of the process [Eq. (9)], this information is turned into negentropy again:

$$\Delta N_i = k(p/P_0) > 0, \quad \text{an increase.} \tag{9a}$$

This justifies the general scheme (1) of Sec. II.

Let us discuss more specifically the original problem of Maxwell. We may assume that, after a certain time, the demon has been able to obtain a difference of temperature $\Delta T$:

$$\begin{aligned} &T_B > T_A \qquad &&T_B = T + \tfrac{1}{2}\Delta T \\ &T_B - T_A = \Delta T \qquad &&T_A = T - \tfrac{1}{2}\Delta T. \end{aligned} \tag{11}$$

On the next step, the demon selects a fast molecule in $A$ [kinetic energy $\frac{3}{2}kT(1+\epsilon_1)$] and directs it into $B$. Then he selects a slow molecule in $B$, $\frac{3}{2}kT(1-\epsilon_2)$ and lets it enter into $A$. In order to see these two molecules,

[5] C. E. Shannon and Warren Weaver, *The Mathematical Theory of Communication* (University of Illinois Press, Urbana, Illinois, 1949).

[6] N. Wiener, *Cybernetics* (John Wiley and Sons, Inc., New York, 1948).

[7] We may replace the demon by an automatic device with a "magic eye" which opens the trap door at convenient instants of time. This is a mere problem of devising some ingenious gadget, and it does not modify the general conditions of the problem.

the demon had to use two light quanta, hence, an increase of entropy, similar to the one computed in Eq. (7):

$$\Delta S_d = 2kb, \quad b = h\nu/kT \gg 1. \tag{12}$$

The exchange of molecules results in an energy transfer

$$\Delta Q = \tfrac{3}{2}kT(\epsilon_1 + \epsilon_2) \tag{13}$$

from $A$ to $B$, which corresponds to a decrease of the total entropy, on account of (11):

$$\Delta S_i = \Delta Q\left(\frac{1}{T_B} - \frac{1}{T_A}\right) = -\Delta Q\frac{\Delta T}{T^2} = -\tfrac{3}{2}k(\epsilon_1 + \epsilon_2)\frac{\Delta T}{T}. \tag{14}$$

The quantities $\epsilon_1$ and $\epsilon_2$ will usually be small but may exceptionally reach a value of a few units. $\Delta T$ is much smaller than $T$, hence,

$$\Delta S_i = -\tfrac{3}{2}k\eta \quad \eta \ll 1$$

and

$$\Delta S_d + \Delta S_i = k(2b - \tfrac{3}{2}\eta) > 0. \tag{15}$$

Carnot's principle is actually satisfied.

Szilard, in a very interesting paper[2] published in 1929, discussed the second part of the process (1). He first considered simplified examples, where he could prove that additional information about the structure of the system could be used to obtain work and represented a potential decrease of entropy of the system. From these general remarks he inferred that the process of physical measurement, by which the information could be obtained, must involve an increase of entropy so that the whole process would satisfy Carnot's principle. Szilard also invented a very curious (and rather artificial) model on which he could compute the amount of entropy corresponding to the first part of the process, the physical measurement yielding the information.

In the examples selected by Szilard, the most favorable one was a case with just two complexions in the initial state,

$$P_0 = 2. \tag{16}$$

The information enabled the observer to know which one of the two possibilities was actually obtained.

$$P_1 = 1,$$

hence,

$$\Delta S_i = k(\ln P_0 - \ln P_1) = k \ln 2 < k, \tag{17}$$

since the natural logarithms of base $e$ are used in Boltzmann's formula. Even in such an oversimplified example, we still obtain an inequality similar to (10),

$$\Delta S_d + \Delta S_i = k(b - \ln 2) > 0, \tag{18}$$

and Carnot's principle is not violated.

Szilard's discussion was completed by J. von Neumann and by P. Jordan in connection with statistical problems in quantum mechanics.[8]

[8] J. von Neumann, ***Math. Grundlagen der Quantum Mechanik*** (Verlag. Julius Springer, Berlin, 1932). P. Jordan, Philosophy of Science **16**, 269–278 (1949).

### IV. ENTROPY AND OBSERVATION IN THE LABORATORY

The physicist in his laboratory is no better off than the demon. Every observation he makes is at the expense of the negentropy of his surroundings. He needs batteries, power supply, compressed gases, etc., all of which represent sources of negentropy. The physicist also needs light in his laboratory in order to be able to read ammeters or other instruments.

What is the smallest possible amount of negentropy required in an experiment? Let us assume that we want to read an indication on an ammeter. The needle exhibits a brownian motion of oscillation, with an average kinetic energy $(\frac{1}{2})kT$ and an average total energy

$$\bar{E}_t = kT. \tag{19}$$

G. Ising[9] considers that an additional energy

$$E_m \geqslant 4\bar{E}_t = 4kT \tag{20}$$

is required to obtain a correct reading. Let us be more optimistic and assume

$$\Delta E_m \geqslant kT. \tag{21}$$

This energy is dissipated in friction and viscous damping in the ammeter, after the physicist has performed the reading. This means an increase of entropy $\Delta S$ (or a decrease of negentropy)

$$-\Delta N = \Delta S \geqslant kT/T = k. \tag{22}$$

Boltzmann's constant $k$ thus appears as the lower limit of negentropy required in an experiment.

Instead of reading an ammeter, the physicist may be observing a radiation of frequency $\nu$. The smallest observable energy is $h\nu$, and it must be absorbed somewhere in the apparatus in order to be observed, hence,

$$-\Delta N = \Delta S \geqslant h\nu/T, \tag{23}$$

a quantity that must be greater than $k$, if it is to be distinguished from blackbody radiation at temperature $T$. We thus reach the conclusion that

$$-\Delta N = \Delta S \geqslant \begin{cases} k \\ h\nu/T \end{cases} \text{(whichever is greater)}, \tag{24}$$

represents the limit of possible observation.

A more compact formula can be obtained by the following assumptions: let us observe an oscillator of frequency $\nu$. At temperature $T$, it obtains an average number

$$n = 1/(e^{h\nu/kT} - 1) \tag{25}$$

of quanta of energy $h\nu$. In order to make an observation, this number $n$ must be increased by a quantity that is materially larger than $n$ and its fluctuations; let us assume

$$\Delta n \geqslant n + 1. \tag{26}$$

[9] G. Ising, Phil. Mag. **51**, 827–834 (1926); M. Courtines, *Les Fluctuations dans les Appareils de Mesures* (Congres Intern. d'Electricité, Paris, 1932), Vol. 2.

This means at least one quantum for high frequencies $\nu$, when $n$ is almost zero. For very low frequencies, formula (26) requires $\Delta n$ of the order of $n$, as in formula (21).

Observation means absorption of $\Delta n$ quanta $h\nu$, hence, a change in entropy

$$-\Delta N = \Delta S = \frac{\Delta n}{T} h\nu \geqslant \frac{h\nu}{T}\left[\frac{1}{e^{h\nu/kT}-1}+1\right] \tag{27}$$

or

$$-\Delta N = \Delta S \geqslant k\left[\frac{x}{e^{x}-1}+x\right] = k\frac{x}{1-e^{-x}} \quad x = h\nu/kT. \tag{28}$$

This formula satisfies the requirements stated in Eq. (24).

R. C. Raymond[10] recently published some very interesting remarks on similar subjects. He was especially concerned with the original problem of Shannon, Wiener, and Weaver, namely, the transmission of information through communication channels. He succeeded in applying thermodynamic definitions to the entropy of information and discussed the application of the second principle to such problems.

The limitations to the possibilities of measurements, as contained in our formulas, have nothing to do with the uncertainty relations. They are based on entropy, statistical thermodynamics, Boltzmann's constant, all quantities that play no role in the uncertainty principle.

The physicist making an observation performs the first part of process (1); he transforms negative entropy into information. We may now ask the question: can the scientist use the second part of the cycle and change information into negentropy of some sort? Without directly answering the question, we may risk a suggestion. Out of experimental facts, the scientist builds up scientific knowledge and derives scientific laws. With these laws he is able to design and build machines and equipment that nature never produced before; these machines represent most unprobable structures,[11] and low probability means negentropy? Let us end here with this question mark, without attempting to analyze any further the process of scientific thinking and its possible connection with some sort of generalized entropy.

[10] R. C. Raymond, Phys. Rev. **78**, 351 (1950); Am. Scientist **38**, 273–278 (1950).

## APPENDIX

A few words may be needed to justify the fundamental assumption underlying Eq. (26). We want to compute the order of magnitude of the fluctuations in a quantized resonator. Let $n_a$ be the actual number of quanta at a certain instant of time, and $\bar{n}$ the average number as in Eq. (26).

$$n_a = \bar{n} + m, \tag{29}$$

where $m$ is the fluctuation. It is easy to prove the following result

$$\langle m^2\rangle_{\text{Av}} = \bar{n}(\bar{n}+1) = \bar{n}^2 + \bar{n}. \tag{30}$$

For large quantum numbers, $\langle m^2\rangle_{\text{Av}}^{\frac{1}{2}}$ is of the order of magnitude of $\bar{n}$; hence, when we add $n$ more quanta $h\nu$, we obtain an increase of energy that is only of the order of magnitude of the average fluctuations. This certainly represents the lower limit for an observation.

Formula (29) is obtained in the following way: We first compute the partition function $Z$ (Planck's Zustandssumme) for the harmonic oscillator

$$Z = \Sigma e^{-nx} = \frac{1}{1-e^{-x}} \quad x = h\nu/kT. \tag{31}$$

Next we compute the average $\bar{n}$

$$\bar{n} = \frac{\Sigma n e^{-nx}}{z} = -\frac{1}{z}\frac{\partial z}{\partial x} = \frac{e^{-x}}{1-e^{-x}} = \frac{1}{e^{x}-1}, \tag{32}$$

which is our Eq. (25). Then we obtain

$$\langle n^2\rangle_{\text{Av}} = \frac{1}{z}\Sigma_1 n^2 e^{-nx} = \frac{1}{z}\frac{\partial^2 z}{\partial x^2} = \frac{e^{-x}+e^{-2x}}{(1-e^{-x})^2}, \tag{33}$$

and finally

$$\langle m^2\rangle_{\text{Av}} = \langle n^2\rangle_{\text{Av}} - (\bar{n})^2 = \frac{e^{-x}+e^{-2x}-e^{-2x}}{(1-e^{-x})^2} = \frac{1}{1-e^{-x}}\cdot\frac{1}{e^{x}-1} = \bar{n}^2 + \bar{n}, \tag{34}$$

which is our formula (30). Hence, our assumption (26) is completely justified.

[11] L. Brillouin, Am. Scientist **38**, 594 (1950).

# Entropy, Information and Szilard's Paradox

by **J. M. Jauch**

Dept. of Theoretical Physics, University of Geneva,
and Dept. of Mathematics, University of Denver

and **J. G. Báron**

Rye, New York

(15. XII. 71)

This essay is presented in homage to Professor Markus Fierz, whose long-standing interest in statistical physics is well known, on the occasion of his 60th birthday.

*Abstract.* Entropy is defined as a general mathematical concept which has many physical applications. It is found useful in classical thermodynamics as well as in information theory. The similarity of the formal expressions in the two cases has misled many authors to identify entropy of information (as measured by the formula of Shannon) with negative physical entropy. The origin of the confusion is traced to a seemingly paradoxical thought experiment of Szilard, which we analyze herein. The result is that this experiment cannot be considered a justification for such identification and that there is no paradox.

## 1. Introduction

There is a widespread belief that the physical entropy used in thermodynamics is more or less closely related to the concept of information as used in communication theory.

This thesis has been made precise and explicit, primarily by Brillouin [1], who is of the opinion that both concepts should be united by identifying information (suitably normalized) by establishing an equivalence relation with negative physical entropy (called 'negentropy' by him), which then together satisfy a generalized principle of Clausius.

This point of view, however, is not universally accepted by those physicists who have thought about the question. We quote here as an example an explicit denial of such identification, by ter Haar [2], who writes in his textbook on statistical mechanics:

> 'The relationship between entropy and lack of information has led many authors, notably Shannon, to introduce "entropy" as a measure for the information transmitted by cables and so on, and in this way entropy has figured largely in recent discussions on information theory. It must be stressed here that the entropy introduced in information theory is *not* a thermodynamic quantity and that the use of the same term is rather misleading. It was probably introduced because of a rather loose use of the term "information".'

We want to elaborate ter Haar's point of view and discuss the reasons why we believe that the two concepts should not be identified.

One can trace the origin of this identification to a paper by Szilard [3], published in 1929, which discusses a particular version of Maxwell's demon and an apparent violation of the second law of thermodynamics.

The emphasis in that paper is on the intelligence of the 'demon', who, by utilizing the 'information' gained by observation of the detailed properties of a thermodynamic system, could use this information for the manipulation of a macroscopic gadget which could extract mechanical energy from the fluctuations of a thermodynamic system and thus produce a *perpetuum mobile* of the second kind.

Szilard based his version on a remark by Smoluchowski which was published in the latter's lectures on the kinetic theory of matter [4]. Smoluchowski said, 'As far as our present knowledge is concerned there does not exist a permanently working *automatic perpetuum mobile* in spite of molecular fluctuations, but such a contraption could function *if it were operated by intelligent beings in a convenient manner* . . .' (italics ours).

This statement seems to imply that the second law of thermodynamics could somehow be violated in the presence of intelligent beings and that this possible violation would be associated with the acquisition and retention of knowledge by such beings. It is with this in mind that Szilard contructed his thought experiment.

Although in a subsequent passage, Smoluchowski expressed considerable doubt ('recht zweifelhaft') about this possibility, Szilard proposed to elucidate the conjectured role of the intelligent being in creating the uncompensated entropy decrease. He described an idealized heat engine that seemingly functioned with continuous decrease of entropy. In order to save the second law, Szilard conjectured that the intelligent being (we shall call him 'the observer') must perform measurements in order to operate the engine, and that this process is in principle connected with a compensating increase of entropy.

We shall discuss Szilard's thought experiment in section 4 of this paper. Here we merely point out that this experiment provoked much discussion and, in our opinion, misinterpretation. We mention in particular the discussion by von Neumann [5], who transferred considerations of this kind into the realm of quantum mechanics with reference to the measuring process.

Many aspects of the measuring process in quantum mechanics are still controversial. Szilard's conjecture mentioned above has led many commentators [6] to believe that the measuring process in quantum mechanics is connected in an essential manner with the presence of a conscious observer who registers in his mind an effect, and that this conscious awareness is responsible for the oft-discussed, paradoxical 'reduction of the wave packet'.

We expect to show that the presence of a conscious observer in Szilard's experiment is not necessary; he can be replaced by an automatic device with no consciousness at all. Interestingly, Szilard noted this himself; toward the end of his paper, he concluded:

> 'As we have seen with this example, a simple, inanimate device can do exactly the same, as far as the essentials are concerned, as the intervention of an intelligent being would accomplish.'

It is strange that Szilard seemed not to realize that an automatic version of the intelligent observer contradicts the conclusion of Smoluchowski, according to which such mechanisms are not possible. As a matter of fact, the solution of the paradox in the case of the living observer is the same as that which Smoluchowski indicated for the explanation of the mechanical demon: The demon is himself subject to fluctuations, just as the system which he tries to control. To use a medical analogy, the demon who wants to operate the molecular trap is like a patient with a severe case of Parkinson's disease trying to thread a fast-vibrating needle!

We shall not question the analysis of the problem given by Smoluchowski; we shall consider this aspect of the problem as solved. From this it follows that Szilard's conjecture is not proven by his experiment.

## 2. The Classical Notion of Entropy

Entropy as a basic notion of science was introduced by Clausius to summarize thermal behavior of systems in equilibrium or changing in reversible fashion in the second principle of thermodynamics.

Boltzmann [8] and Gibbs [9] defined entropy of non-equilibrium states and entropy changes of irreversible processes in purely mechanical terms. Their theory was more general; it also explained how the same thermodynamic process can be irreversible from the phenomenological point of view—and completely reversible from the purely mechanical point of view. This paradoxical situation was cleared up by statistical interpretation of thermodynamic entropy.

Increase in generality resulted in some ambiguity of the notion of entropy. The reason for this is that in any statistical consideration a more or less arbitrary model must be used. Expressed differently, the system may be described at different levels (see H. Grad [7]).

We shall return to the significance of these ambiguities later in this section. First we briefly review some special features of the statistical interpretation of thermodynamic entropy.

In thermodynamics one may specify a homogeneous thermal system by a certain number of extensive variables $x_1, \ldots x_n$ which usually have a simple physical interpretation (volume, surface, magnetic moment, etc.).

The generalized forces $y_1, \ldots y_n$ associated with these variables are homogeneous functions of them, such that the element of work $\delta A$ delivered by the system to the surrounding is related to the differentials $dx_r$ $(r = 1, \ldots n)$ by

$$\delta A = \sum_{r=1}^{n} y_r dx_r. \tag{1}$$

Mathematically, (1) is a differential form defined on an open region of $\mathbb{R}^n$, the Euclidean space of $n$ dimensions.

The first principle of thermodynamics which expresses conservation of energy for a conservative system states that for an adiabatic system (that is, a thermally isolated system), this differential form is total. That means that there exists a function $U(x_1, \ldots x_n)$ of the extensive variables $x_r$, which is itself extensive, such that

$$\delta A = -dU \tag{2}$$

Physically interpreted, this equation says that the work delivered to the outside by an adiabatic system is exactly compensated by the loss of internal energy.

If the system is not adiabatic, then equation (2) is no longer true and must be generalized to

$$\delta Q = dU + \delta A \tag{3}$$

where now $\delta Q$ is the differential of the amount of heat added to the system in a reversible manner.

We may consider equation (3) as a new differential form in $n+1$ variables where $U = x_0$ may be defined as the new variable. Each of the generalized forms $y_r$ is then a function of all the variables $x_0, x_1, \ldots, x_n$.

The differential form (3) is of a special kind which admits an integrating factor $T(x_0, x_1, \ldots x_n)$ such that the form

$$dS = \frac{\delta Q}{T} \tag{4}$$

is the total differential of an extensive function $S(x_0, x_1, \ldots x_n)$. This function is the entropy and the integrating factor (suitably normalized) is the absolute temperature of the system.[1])

The second principle of thermodynamics says that in a spontaneous evolution of a closed system not in equilibrium, the entropy always increases and attains its maximum value for the state of equilibrium.

Definition (4) determines the thermodynamic entropy only up to a constant of integration.

Boltzmann's statistical definition is given by the famous formula,

$$S = k \ln W \tag{5}$$

where $W$ represents a probability for the system specified by the thermodynamic variables based on some appropriate statistical mode. This formula was apparently never written down by Boltzmann; yet it appears on his tombstone and indeed is one of the most important advances in statistical physics. For the practical application of this formula, one usually goes through the following procedure:

a) One assumes (explicitly or implicitly) an a priori probability. In phase space of a classical system it is given by a convenient selection of a volume element.

b) One then imposes constraints in agreement with a certain number of external parameters characterizing the thermodynamic state of the system.

c) One then calculates the probability of such constrained systems on the basis of the a priori probability field assumed.

d) Finally, one calculates a maximum value of this probability under the assumed constraints to obtain an expression for $W$.

The $W$ thus calculated is in an arbitrary normalization. This arbitrariness corresponds to the constant of integration for the thermodynamic entropy $S$.

Boltzmann's theoretical interpretation of the entropy gives immediate insight into two important properties which are characteristic for the thermodynamic, and, as we shall see, for all other forms of entropy. They are:

a) *Extensity*

If there are two independent systems with their respective probabilities $W_1$ and $W_2$, then the joint system has a probability

$$W = W_1 W_2. \tag{6}$$

---

[1]) It seems not to be generally known that the existence of the integrating factor, hence the existence of the function entropy, is a consequence of the first principle of thermodynamics for conservative systems under reversible quasistatic variations. This was discovered by T. Ehrenfest [12]. A new proof of this statement will be given in a subsequent publication.

Hence

$$S = S_1 + S_2 = k \ln (W_1 W_2). \tag{7}$$

b) *Maximum property*

Any system outside the equilibrium state will have a probability $W < W_0$, the equilibrium probability.
Hence

$$S = k \ln W < k \ln W_0 = S_0, \tag{8}$$

since $\ln W$ is a monotonic function.

The arbitrariness in the definition of $W$, and thus the ambiguity of $S$, is brought out explicitly if we turn now to the definition for $W$ used by Boltzmann and others in the derivation of the so-called $H$-theorem.

Here one considers the phase space $\Gamma$ of a classical system endowed with a probability measure $\rho(P)$, $P \in \Gamma$. $\rho(P)$ is assumed to be a positive function, normalized by

$$\int_{\Gamma} \rho(P) d\Omega = 1 \tag{9}$$

and interpreted to represent the probability of finding the system at the point $P$ in phase space. We have written $d\Omega$ for the volume element in phase space.

One can define a quantity $\eta = \ln \rho$ and its average

$$\sigma = \int_{\Gamma} \rho \ln \rho \, d\Omega = \bar{\eta} \tag{10}$$

and one can then show that this quantity reaches its maximum value under the subsidiary condition

$$\int \epsilon \rho d\Omega = E = \text{constant} \tag{11}$$

provided it has the form of the canonical distribution

$$\rho = e^{(\psi - \epsilon)/\theta}. \tag{12}$$

However, the quantity $\sigma$ is a constant under the evolution in time. This is true for any $\sigma$ of the form (10) for any $\rho$.

One obtains a more suitable statistical definition for the entropy if one uses the process of 'coarse-graining'. Physically, this corresponds to the process which one would use for describing a system, the state of which is incompletely known. It is carried out in the following manner:

One divides the phase space into a certain number of cells with volume $\Omega_i$ and defines

$$P_i = \frac{1}{\Omega_i} \int_{\Omega_i} \rho \, d\Omega.$$

It follows then that

$$\sum_i P_i \Omega_i = 1.$$

The coarse-grained density is defined as

$$P(x) = P_i \quad \text{if } x \in \Omega_i$$

and

$$\sum \equiv \sum_i P_i \ln P_i \Omega_i = \int_\Gamma P \ln P \, d\Omega.$$

Hence

$$\sum = \overline{\ln P}.$$

One can then prove that $\sum$ is a decreasing function in time, reaching its minimum value for the canonical distribution provided that the mean energy is kept constant.

This suggests that $-k\sum$ could be identified with the thermodynamic entropy when the system is away from equilibrium.

This result shows quite clearly that there are several ways of defining and interpreting statistical entropy. The arbitrariness is connected with both the assumed probability field and with the nature of constraints used in the coarse-graining process. Both Boltzmann [8] and Gibbs [9] were aware of these ambiguities in the statistical interpretation of thermodynamic variables. Boltzmann says, for example:

> 'I do not believe that one is justified to consider this result as final, at least not as long as one has not defined very precisely what one means by the most probable state distribution.'

Gibbs is still more explicit:

> 'It is evident that there may be more than one quantity defined for finite values of the degrees of freedom, which approach the same limiting form for infinitely many degrees of freedom. There may be, therefore, and there are, other quantities which may be thought to have some claim to be regarded as temperature and entropy with respect to systems of a finite number of freedoms.'

As an example of two different ways of interpreting entropy, we mention the mixing problem discussed by Gibbs.

If a blue and a red liquid are mixed, there will result after a while a purplish mixture which cannot be unmixed by further agitation.

In this case the statistical entropy increases if the coarse-graining size is much larger than the average size of the volume element that has a definite (unmixed) color. On the other hand, the statistical entropy remains constant if the volume element having a definite color is much larger than the coarse-graining size. Only on the molecular level, e.g., in the diffusion process of two different gases, can this entropy be related to thermodynamic entropy; it increases in the diffusion process by

$$S = R \log 2,$$

provided that the volumes of the two kinds of gases are equal. The so-called paradox of Gibbs results from the confusion of two different kinds of entropy, based on different statistical models.

All that we have said in this section has been said before. Nevertheless, we felt reiteration useful to emphasize the polymorphic nature of statistical entropy. It is precisely this non-uniqueness which makes the concept so versatile.

### 3. The Definition of Entropy as a Mathematical Concept

In this section we will define the notion of entropy in an abstract setting without reference to its interpretations. It will be seen that a suitable definition will immediately reveal the range of its applications to a variety of situations, including thermodynamic entropy and the notion of 'information'.

Before proceeding with the formal definition, let us give a heuristic description of what we seek in order to achieve sufficient motivation for the mathematical definitions.

We wish to establish between two measures a relationship which represents a quantitative expression for the variability of one with respect to the other. As an example, let us consider the probability of the outcome of one of two alternatives, head or tail in flipping a coin. If the coin is 'true' then the a priori probability for either of the two events is $\frac{1}{2}$. However, if the experiment is made and the coin has been observed, then this probability is changed and is now 0 for one and 1 for the other event. The observation, or the specification, has restricted the variability and as a result a certain quantity which we wish to define—entropy—has decreased.

This quantity should have the property that it behaves additively for independent events, so that if the coin is flipped twice, then the entropy for the two observations should be twice that for one. As we shall see, these properties determine the quantity almost uniquely.

Proceeding now to the formal definition, what we need first of all is a measure space; by this we mean the triplet $(X, \mathscr{S}, \mu)$ where $X$ is a non-empty set, $\mathscr{S}$ a collection of subsets of $X$, and $\mu$ a normalized finite measure on $\mathscr{S}$. We assume that the subsets of $\mathscr{S}$ are closed under the formation of complements and countable unions (hence also countable intersections), and we call such a collection a *field* of subsets.

The measure $\mu$ is a positive, countably additive set function defined for the sets of $\mathscr{S}$ such that

$$\mu(X) = 1, \quad \mu(\varnothing) = 0 \quad (\varnothing = \text{null set}).$$

Let $S \in \mathscr{S}$ be such that $\mu(S) = 0$, then we say the set $S$ has $\mu$-measure zero. Two different measures $\mu$ and $\nu$ on $\mathscr{S}$ are said to be *equivalent* if they have exactly the same sets of measure zero. We write $\mu \sim \nu$ so that

$$\mu \sim \nu \Leftrightarrow \{\mu(S) = 0 \Leftrightarrow \nu(S) = 0\}.$$

A more general concept is *absolute continuity*. We say $\mu$ is absolutely continuous with respect to another measure $\nu$ if

$$\nu(R) = 0 \Rightarrow R \text{ is } \mu\text{-measurable and } \mu(R) = 0.$$

We then write

$$\mu \propto \nu.$$

The relation $\propto$ is a partial order relation between different measures since

1. $\mu \propto \mu$
2. $\mu \propto \nu \quad \text{and} \quad \nu \propto \rho \Rightarrow \mu \propto \rho.$

Two measures are equivalent if and only if $\mu \propto \nu$ and $\nu \propto \mu$.

A function $f$ on $X$ is called $L^1(\nu)$ if it is measurable with respect to $\nu$ and if

$$\int_X |f|\, d\nu < \infty.$$

It is convenient to identify functions which differ only on a set of measure zero with respect to a measure $\nu$. In this case we write for two such functions

$$f_1 = f_2 \text{ a.e. } [\nu].$$

where a.e. stands for 'almost everywhere'.

For any measure $\nu$ and $f \in L^1(\nu)$ ($f \geqslant 0$ a.e. $[\nu]$) we can define a measure $\mu_f$ by setting

$$\mu_f(R) = \int_R f \, d\nu \tag{13}$$

and it is easy to verify that $\mu_f < \nu$.

More interesting is the converse, which is the content of one of the Radon-Nikodym theorems [10].

If $\mu \propto \nu$ then there exists a uniquely defined $f \geqslant 0$ a.e. $[\nu]$ and $f \in L^1(\nu)$ such that

$$\mu = \mu_f$$

The function $f$ so defined is called the Radon-Nikodym derivative and is often written as

$$f = \frac{d\mu}{d\nu}. \tag{14}$$

If $X_1$ and $X_2$ are two measure spaces with measures $\mu_1$ and $\mu_2$, respectively, then we can define the product measure $\mu_{12}$ on $X_1 \times X_2$ as follows: On every set of the form $S_1 \times S_2$ with $S_1 \in \mathscr{S}_1$ and $S_2 \in \mathscr{S}_2$ we set

$$\mu_{12}(S_1 \times S_2) = \mu_1(S_1)\, \mu_2(S_2).$$

There is then a unique continuation of this measure by additivity to the field of all sets generated by the sets of the form $S_1 \times S_2$. This is called the *product measure.*

If $\mu_1 \propto \nu_1$ and $\mu_2 \propto \nu_2$ then one can prove that

$$\mu_{12} < \nu_{12} \quad \text{and} \quad f_{12} = \frac{d\mu_{12}}{d\nu_{12}} = \frac{d\mu_1}{d\nu_1}\frac{d\mu_2}{d\nu_2} = f_1 f_2.$$

We have now all the concepts needed for the definition of *entropy*. If $\mu \propto \nu$ and $f = d\mu/d\nu$ we define entropy of $\mu$ with respect to $\nu$ by

$$H(\mu, \nu) = \int f \ln f \, d\nu. \tag{15}$$

It has the following properties

1. $H(\mu, \mu) = 0$.
2. $H(\mu, \nu) \geqslant 0$ for $\mu \propto \nu$.
3. $H(\mu_{12}, \nu_{12}) = H(\mu_1, \nu_1) + H(\mu_2, \nu_2)$.

We verify them as follows:

1. is obvious, since $f = 1$ for $\mu = \nu$,
2. we prove as follows: Let $f = 1 + \phi$. Since

$$\int_X d\mu = \int_X \frac{d\mu}{d\nu}\, d\nu = 1 = \int_X d\nu$$

it follows that $\int \phi d\nu = 0$. Therefore

$$0 = \int_X f \ln (f - \phi) d\nu = \int_X f \ln f \left(1 - \frac{\phi}{f}\right) d\nu$$

$$= \int_X f \ln f + \int_X f \ln \left(1 - \frac{\phi}{f}\right) d\nu$$

Here we use the inequality

$$\ln \left(1 - \frac{\phi}{f}\right) \leqslant \frac{-\phi}{f}$$

so that the second term is

$$\int_X f \ln \left(1 - \frac{\phi}{f}\right) d\nu \leqslant - \int \phi \, d\nu = 0.$$

Thus we have verified

$$0 \leqslant \int f \ln f d\nu$$

3. follows from the definition $f_{12} = f_1 f_2$ and Fubini's theorem:

$$\int f_1 f_2 \ln (f_1 f_2) \, d\nu_{12} = \int f_1 f_2 (\ln f_1 + \ln f_2) \, d\nu_1 \, d\nu_2$$
$$= \int f_1 \ln f_1 \, d\nu_1 + \int f_2 \ln f_2 \, d\nu_2$$
$$= H(\mu_1, \nu_1) + H(\mu_2, \nu_2).$$

We illustrate the foregoing with a few examples:

As a first example we consider a finite sample space $X$ consisting of $n$ objects. For the a priori measure $\nu$ we choose the value $1/n$ for each element of $X$. The family of sets $\mathscr{S}$ is the class of all subsets of $X$. The measure of any subset $S$ with $k$ elements is then given by $\nu(S) = k/n$.

For the measure $\mu$ we choose the value 1 on one particular element of $X$ and zero for all others. Let $A$ denote this element and denote by $B$ any other element of $X$. We have then

$$\nu(B) = \frac{1}{n} \quad B \in X, \quad \mu(A) = 1$$
$$\mu(B) = 0 \text{ for } A \neq B.$$

If $S \in \mathscr{S}$ is any subset of $X$ and if $\nu(S) = 0$ then it follows that $S = \phi$. Therefore $\mu(S) = 0$ also, and $\mu$ is absolutely continuous with respect to $\nu$: $\mu \propto \nu$. We can easily calculate the values of the Radon-Nikodym derivative $f = d\mu/d\nu$. It has the values

$$f(A) = n, \quad f(B) = 0 \quad \text{for } A \neq B$$

where $A$ is that particular element for which $\nu(A) = 1$.

We obtain now for the entropy as we have defined it

$$H(\mu, \nu) = \int f \ln f \, d\nu = \sum_B f(B) \ln f(B) \cdot \frac{1}{n} = \ln n. \tag{16}$$

We observe that this is a very simple, special case of Boltzmann's formula, since $n$ can be interpreted as the number of 'states' which are contained in a set $X$ of $n$ elements.

One can generalize this example a little to bring it closer to the use of 'entropy in information theory'. Let us assume as before that the a priori probability on the elements $A_i$ of $X$ has the value $\nu(A_i) = 1/n$ $(i = 1, \ldots, n)$. For the measure $\mu$, however, we assume $\mu(A_i) = p_i$ where $p_i \geqslant 0$ and $\sum_i^n = 1, p_i = 1$. The positive number $p_i$ represents the probability with which the element $A_i$ appears for instance in a message with an alphabet of $n$ letters. If $p_i = 1/n$ the message is completely garbled; if it is $=1$ for one $A_i$ and $=0$ for the others, it is clear. We consider now the intermediate case.

If $\nu(S) = 0$ for some set $S \in \mathscr{S}$ then $S = \phi$; hence again as before $\mu(S) = 0$ so that $\mu$ is absolutely continuous with respect to $\nu : \mu \propto \nu$.

Furthermore

$$\frac{du}{d\nu}(A_i) \equiv f(A_i) = np_i.$$

This follows from the formula

$$\int_S \frac{d\mu}{d\nu} \, d\nu = \mu(S).$$

When we calculate the entropy in this case, we obtain

$$H(\mu, \nu) = \sum_{i=1}^{n} np_i \ln(np_i) \cdot \frac{1}{n} = \ln n + \sum_{i=1}^{n} p_i \ln p_i. \tag{17}$$

This expression reaches its minimum value 0 for $p_i = 1/n$ and its maximum value $\ln n$ for $p_k = 1$ for $k = i$, $p_k = 0$ for $k \neq i$.

In information theory [11] one uses the quantity $I = -\sum_{i=1}^{n} p_i \ln p_i$ as a measure of the information contained in the one-letter message and we may therefore write

$$H(\mu, \nu) = \ln n - I.$$

Our definition of entropy is therefore in this case, apart from a constant, $\ln n$ equal to the negative entropy of the measure $\mu$ with respect to $\nu$.

The abstract mathematical concept of entropy introduced here is at the basis of numerous different applications of this notion in physics as well as in other fields. This was recently emphasized by Grad [7]. Mathematicians use this concept as a versatile research tool in mathematical probability theory.

As has been emphasized, the use of the word entropy should not lead to confusion of the mathematical concept defined here with the physical concept of thermodynamic entropy. The concept introduced here relates two measures, one of which is absolutely continuous with respect to the other and has at this stage of abstraction nothing to do with any particular physical system. The misleading use of the same word for mathematical and for physical entropy is well-entrenched; it is now unavoidable. We use it in the general sense; its special meanings should be clear from the context.

## 4. Szilard's Paradox

Many authors who have tried to identify thermodynamic entropy with information (or rather, its negative) refer explicitly to the thought experiment of Szilard which seemed to lead to a violation of the second principle of thermodynamics unless the loss of entropy in the hypothetical experiment, so it is alleged, is compensated by a gain of information by some observer. In this section we describe Szilard's thought experiment and present an analysis to show that it cannot be considered a basis for the alleged identity and interchangeability of these two kinds of entropy.

In our description of Szilard's thought experiment, we follow closely von Neumann's version. We make only minor, physically irrelevant changes in the experimental set-up to make it more convenient for analysis and to avoid extreme idealizations.

In the experiment, a rigid, hollow, heat-permeable cylinder, closed at both ends, is used. It is fitted with a freely moveable piston with a hole large enough for the molecule to pass easily through it. The hole can be closed from outside. All motions are considered reversible and frictionless. The cylinder is in contact with a very large heat reservoir to keep the temperature of the entire machine constant.

Within the cylinder is a gas consisting of a single molecule. At the beginning of the experiment, the piston is in the middle of the cylinder and its hole is open so that the molecule can move (almost) freely from one side of the piston to the other. The hole is then closed, trapping the molecule in one half of the cylinder.

The observer now determines the location of the molecule by a process called by Szilard 'Messung', meaning measurement. If it is found to the left of the piston (see figure), the observer attaches a weight to the piston with a string over a pulley so that the pressure is almost counterbalanced. He then moves the piston very slowly to the right, thereby raising the weight. When the piston reaches the end of the cylinder, the hole in the piston is opened and the piston is moved back to the middle of the cylinder, reversibly and without effect on the gas. At the end of this process the starting position has been reached, except that a certain amount of heat energy $Q$ from the heat reservoir has been transformed into potential energy $A$ of the weight lifted.

Although Szilard does not mention this, obviously the same procedure can be used if the molecule happens to be trapped on the other side of the cylinder after the closing of the piston's hole.

By repeating the process a large number of times (say, $N$), an arbitrarily large quantity of heat energy $Q = NA$ from the reservoir is transformed into potential energy without any other change in the system. This violates the second principle of thermodynamics.

In order to 'save' the second law, Szilard assumes that the observation of the molecule, for determining in which half of the cylinder it is contained, is in principle connected with an exactly compensating increase of entropy of the observer.

Before analyzing the experiment, a few remarks are in order concerning admissibility of procedures in idealized experiments. In this we are helped by a statement of von Neumann (p. 359 of his book, Ref. [5]): 'In phenomenological thermodynamics each conceivable process constitutes valid evidence, provided that it does not conflict with the two fundamental laws of thermodynamics.' To this we add: 'If an idealization is in conflict with a law that is basic to the second law, then it cannot be used as evidence for violation of the second law.'

Now this is precisely what is happening in the case of Szilard's experiment. Obviously, frictionless motion, reversible expansion and heat transfer, and infinitely large reservoir are all admissible under these criteria. Even the single-molecule gas is admissible so long as it satisfies the gas laws. However, at the exact moment when the piston is in the middle of the cylinder and the opening is closed, the gas violates the law of Gay-Lussac because the gas is compressed to half its volume without expenditure of energy. We therefore conclude that the idealizations in Szilard's experiment are inadmissible in their actual context.

It is of further interest to analyze that phase of the experiment during which the hole in the piston is closed. This phase is almost a replica of the expansion phase of a

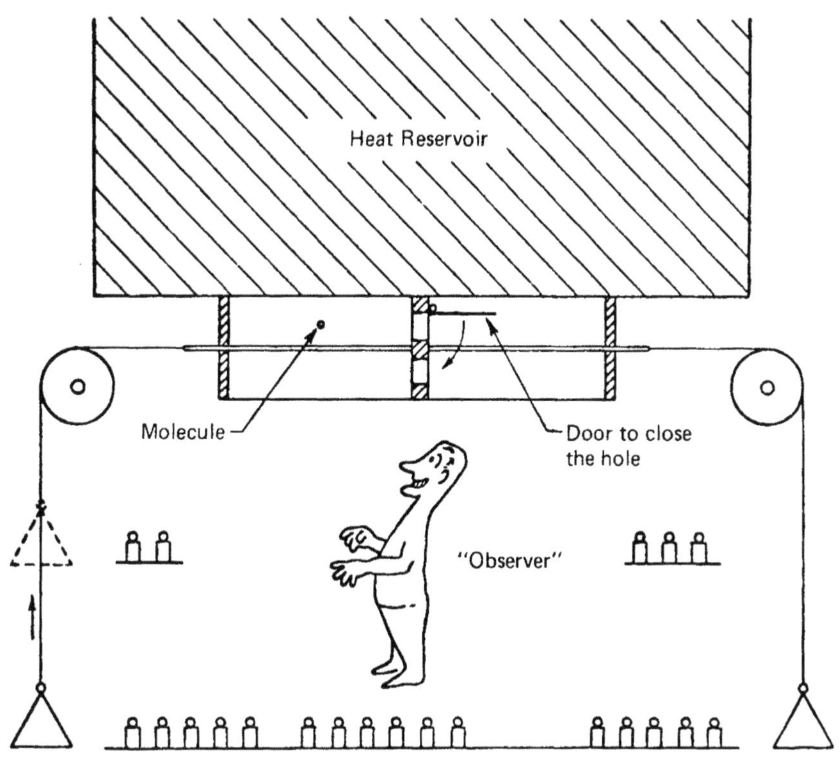


Szilard's thought experiment

Carnot cycle. Szilard believed that during this phase there is an uncompensated decrease in entropy, since during this interval the entropy of the heat reservoir decreases while at the same time the entropy of the gas increases by the same amount. The entropy of the entire system (piston, gas and reservoir) remains constant because the system is closed and the changes are reversible. Thus there is nothing to be compensated for by the alleged increase of entropy of the observer during observation.

Finally, and perhaps most importantly, there is an unacceptable assumption in Szilard's interpretation of his experiment. He believed that the observer must *know* on which side of the piston the molecule is located in order that he may start the piston moving in the right direction. This knowledge is unnecessary, as is the pushing of the piston by the observer. The piston starts moving—under the idealized conditions of the experiment—by the pressure of the gas.

The automatic device, referred to in section 1, which can completely replace the observer, could work as follows:

Near the mid-plane of the cylinder and on both its sides are electrical contacts in its walls. When activated by the piston's motion along them, they operate mechanisms which attach a weight to the piston in whichever direction it moves. Thus a weight is lifted and the engine performs work, without interference by a conscious observer.

## 5. Summary and Concluding Remarks

Entropy is a fundamental mathematical concept, which relates two measures on a measure space in a certain manner.

The concept has many different applications, including thermodynamics (where it was first discovered) and information theory. It is also applicable in quantal systems, for example, and in random variables of any kind. The fact that entropy can be applied to many fields is no excuse for confusing its different meanings when applied to physical systems or mathematical constructions.

In particular, the identification of entropy of information (as defined by Shannon) as equivalent with negative thermodynamic entropy is unfounded and a source of much confusion. We have traced the origin of this confusion to the paradox of Szilard. Analysis of Szilard's paradox has shown specifically that:

1. Szilard's experiment is based on an inadmissible idealization; therefore it cannot be used for examining the principles of thermodynamics.

2. The observer needs no information about the location of the molecule at the beginning of the experiment.

3. There is no uncompensated entropy change in the system during the expansion phase of the experiment.

4. Thus, Szilard's thought experiment does not work; it is no paradox; and has nothing to do with information.

REFERENCES

[1] L. Brillouin, *Science and Information Theory* (Academic Press, New York, N.Y. 1956).
[2] D. ter Haar, *Elements of Statistical Mechanics* (Rinehart and Co., New York, N.Y. 1954, p. 161.
[3] L. Szilard, Z. Phys. *53*, 840 (1929).
[4] B. von Smoluchowski, *Vorträge über die kinetische Theorie der Materie und Elektrizität* (Leipzig 1914), esp. p. 89.
[5] J. von Neumann, *Mathematische Grundlagen der Quantenmechanik*, p. 212 (1932).
[6] F. London and E. Bauer, *La Théorie de l'Observation en Mécanique Quantique*, Actual. scient. ind. *775* (Hermann, Paris 1939).
[7] H. Grad, *The Many Faces of Entropy*, Comm. Pure Appl. Math. *XIV*, 323 (1961).
[8] L. Boltzmann, Collected Papers, No. 42, p. 193.
[9] W. Gibbs, *Elementary Principles in Statistical Mechanics*, Dover Press, p. 169.
[10] S. K. Berberian, *Measure and Integration* (MacMillan & Co., New York, N.Y. 1962), esp. p. 160 seq.
[11] C. L. Shannon and W. Weaver, *Mathematical Theory of Communication*. (Univ. of Illinois Press, 1949).
[12] P. and T. Ehrenfest, *The Conceptual Foundations of the Statistical Approach in Mechanics*, transl. from the German by M. T. Moravcsik (Cornell Univ. Press, 1959), esp. Preface to the translation, by T. Ehrenfest-Afanassjewa. See also T. Ehrenfest, *Die Grundlagen der Thermodynamik* (Leiden 1956); A. Lande, *Axiomatische Begründung der Thermodynamik*, Handbuch der Physik *IX*; A. H. Wilson, *Thermodynamics and Statistical Mechanics* (Cambridge Univ. Press, 1960), esp. §2.5, p. 24 seq.

# Information Theory and Thermodynamics

by **Olivier Costa de Beauregard**

Laboratoire de Physique Theorique associé an CNRS, Institut Henri Poincaré, 11 rue Pierre et Marie Curie, 75005 Paris, France

and **Myron Tribus**

Xerox Corporation, Webster, New York, USA

(19. XI. 73)

*Abstract.* In answer to a recent article by Jauch and Baron bearing this same title, the information theory approach to thermodynamics is here upheld. After a brief historical survey and an outline of the derivation as formulated by one of us in previous publications, Jauch and Baron's critique of Szilard's argument pertaining to the 'well-informed heat engine' is discussed.

## I. Introduction

In a recent paper Jauch and Baron [1] argue against the identification of the thermodynamic entropy concept as defined by Clausius and the information theory entropy concept as defined by Shannon [2]. In support of their thesis they present a new discussion of Szilard's thought experiment of the 'well-informed heat engine' [4].

We definitely belong to the other school of thought and intend to say briefly why.

While admitting with others [5] that a general mathematical definition of the entropy concept can be produced and used in many fields, Jauch and Baron do not mention the fact that the entropy concept of statistical mechanics has actually been deduced from the information concept. The corresponding papers are not quoted in their article, where Brillouin (1957) [4] is the only author mentioned as asserting the identity of the Clausius and the Shannon entropy concepts.

Jaynes (1957) [6], much inspired [7] by an article of Cox (1946) [8], later expanded in a book [9], is the recognized author of the derivation of the fundamental equations of equilibrium statistical mechanics (or 'thermostatistics' in the terminology of one of us [10]) from the inductive reasoning probability concept as introduced by Bayes [11], Laplace [12] and Hume [13]. In his two pioneering articles [6, 7] Jaynes presents his deduction in terms of, respectively, classical and quantal statistical mechanics. In the latter case he uses of course von Neumann's [14, 15] density matrix.

What is perhaps less known is that the same line of reasoning had been lucidly presented and used, also in von Neumann's quantum statistical formalism, as early as 1937 by Elsasser [16], who quotes Fisher (1929) [17] as one of his inspirers.

One point of interest in the Elsasser–Jaynes quantum information formalism is that the density matrix, and corresponding negentropy or information attached to the system under study, are calculated from the results of a set of 'macroscopic' measurements simultaneously performed, these being interpreted as the (quantum mechanical) mean values $\langle F_i \rangle$ attached to *not necessarilly commuting* operators $F_i$.

For the sake of completeness we mention that the history of the inductive reasoning probability concept continues beyond Fisher, Cox and Shannon; Watanabe [18], Carnap [19], Kemeny [20], Jeffreys [21] and others can be added to the list.

In statistical physics Szilard is not the only forerunner of Elsasser, Brillouin and Jaynes. Lewis (1930) [22], in a paper devoted to time symmetry, has the sentence 'Gain in entropy always means loss of information, and nothing more. It is a subjective concept.' Van der Waals (1911) [23] derives the time asymmetry in the $H$-theorem from the classical time asymmetric use of Bayes' conditional probabilities, an idea also expressed by Gibbs [24] in an often quoted sentence. Jaynes has followed up his pioneering articles by more comprehensive publications [25], as has also one of us [26, 27]. Among other authors using the information theoretical approach in probability theory or in statistical mechanics we quote Kinchin [28], Yaglom and Yaglom [29], Katz [30], Hobson [31], and Baierlein [32].

In Section II we outline the information-theoretical derivation of the laws of equilibrium in statistical mechanics.

As for the more special topic of Maxwell's demon and Szilard's 'well-informed heat engine' we have brief comments in Section III with a reference to Brillouin.

## II. Outline of the Information Theory Basis for Physical Theory

Part of Cox's contribution may be summarized as follows: suppose we wish to inform someone else of our *incomplete* knowledge of a subject. Is there a unique code which enables us to say neither more or less than we really know? Cox saw that instead of trying to *find* such a code, it would be necessary to *design* it. To design implies generation of alternative designs and the selection among them according to criteria. But what criteria should be used for the code? Cox chose criteria equivalent to the following:

1. Consistency
2. Freedom from Ambiguity
3. Universality
4. Honesty

What is surprising is that these criteria are necessary and sufficient to develop unique functional equations (27, Chapter I). For example, one of Cox's functional equations is:

$$[AB|E] = F([A|BE], [B|E]) = F([B|AE], [A|E]) \tag{1}$$

$F$ is a function to be determined, [ ] is a measure. Cox's solutions are the ordinary equations of mathematical probability theory:

$$p(A|E) + p(\sim A|E) = 1 \quad 0 < p < 1 \tag{2a}$$

$$p(AB|E) = p(A|BE)\,p(B|E) \tag{2b}$$

[$A$ and $B$ are propositions, $\sim A$ is the denial to $A$, $E$ is the evidence and $p$ is 'a numerical encoding of what the evidence $E$ implies'.] Equations (2a) and (2b) can be used to develop all of probability calculus [27].

Cox's approach is unique in his deliberate attempt to *design a code* for the communication of partial information. The code is constrained to obey certain functional equations, which turn out to yield the equations of the calculus of probabilities. This result sheds new light on an old controversy; namely, the 'meaning' of the concept 'probability'. Because these equations are obtained by design, the interpretation of the function $p$ is clear.

In the notation $p(\bullet|E)$, $p$ is 'an encoding of knowledge about $\bullet$'. $E$ represents the knowledge to be encoded. Only a measure which obeys the above two equations can be used as an encoding to satisfy the desired criteria.

The definition of $p$, due to Cox, is free of two limitations. On the one hand, it is not defined by reference to physical objects such as balls in urns or frequency of numbers in the toss of dice. On the other hand, it is not developed as an 'element on a measure space', devoid of all reference to anything but mathematical context. In no sense do we wish to minimize the importance of being able to put mathematical probability properly in perspective with respect to the rest of mathematics. But if we are to say what we 'mean' by 'probability', we must go beyond merely stating the mathematical properties of the function $p(\,)$.

The interpretation '$p$ is a numerical encoding of what the evidence $E$ implies' is critical to all that follows.

If we take the Cox interpretation of $p$ as fundamental and general, the question naturally arises: What are the rules for translating a statement $E$ (normally made in a 'natural' language) into an assignment of a set of numbers represented by $p$? This question is the central task of statistics.

Jaynes' principle enters as a synthesis of Cox's result, just given, and Shannon's result in communication theory. If the knowledge $E$ has been encoded as a set of $p$'s, the measure $S$ indicates how much is yet left to be learned.

$$S = -k \sum P_i \ln p_i \tag{3}$$

Proofs of the uniqueness and generality of $S$ abound [2, 4, 28].

What interests us here is that Shannon's measure uniquely measures the *incompleteness* of the knowledge represented by $E$. If $E$ were *complete*, the knowledge would be *deterministic*; i.e., would leave no residue of uncertainty. When $E$ is deterministic the calculus reduces to sets of $p$'s which are either 0 or 1 and the logic becomes purely deductive rather than inductive. For cases in which $E$ is incomplete, Jaynes proposed, therefore, the principle of minimum prejudice as follows [6, 7]:

*The minimally prejudiced assignment of probabilities is that which maximizes the entropy*

$$S = -k \sum p_i \ln p_i \tag{4}$$

*subject to the given information*

The particularization of this principle for any field of inquiry depends upon the information to be encoded (i.e., upon $E$ and the set $A_i$). In common parlance we say the uncertainty measure has to be applied to a well-defined question (which $A_i$ is true?) and well-defined evidence ($E$). Statisticians say 'define the sample space'; thermodynamicists say 'define the system'. Both are saying the same thing, i.e. define the set $A_i$ and be explicit about the experiment ($E$).

The entropy of Clausius becomes a special case of Shannon's entropy if we ask the right question. The question is put in the following form: suppose an observer knows he is dealing with a system in a volume V which may be in a quantum state '$i$' characterized by $N_{ai}$ particles of type $a$, $N_{bi}$ particles of type $b$, etc., and an energy $\epsilon_i$. He knows his instruments are too crude to say precisely what $\epsilon_i$, $N_{ai}$, $N_{bi}$, etc., truly are. All he can usefully observe are the repeatable measurements on $\epsilon_i$, $N_{ai}$, $N_{bi}$, etc.

To apply Jaynes' principle requires the definition of the set of all possible answers. For the system postulated the question is therefore:

$Q$ = 'In what quantum state is the system? ($V$ is given)'.

The set of possible answers is:

$A_i$ = 'It is in the $i$th quantum state, for which the energy is $\epsilon_i$, the number of particles of type $a$ is $N_{ai}$, the number of particles type $b$ is $N_{bi}$, etc.'

For illustration in this paper we shall confine our attention to systems in which electricity, magnetism and gravity play no part. The generalization to these phenomena has been given [6, 7, 26].

According to the Cox–Shannon–Jaynes development, the observer should 'encode' his knowledge in a probability assignment. We identify the 'repeatable measurements' with a mathematical 'expectation' for within the theory no other quantity can be so identified. The knowledge, $E$, therefore, is encoded by maximizing

$$S = -k \sum p_i \ln p_i \tag{4}$$

subject to

$$\sum p_i = 1 \tag{5}$$

$$\sum p_i \epsilon_i = \langle \epsilon \rangle \tag{6}$$

$$\sum p_i N_{ci} = \langle N_c \rangle \quad c = a, b, \ldots \tag{7}$$

where $p_i$ is the probability that the system is in state $i$.

By the usual mathematical methods we find

$$p_i = \exp(-\Omega - \beta\epsilon_i - \alpha_a N_{ai} - \alpha_b N_{bi} - \ldots) \tag{8}$$

which is recognized as Gibbs Grand Canonical Distribution. The application of Jaynes' principle has served to introduce three new constructs, represented by $\Omega$, $\beta$ and the set $\{\alpha_c\}$.

The system of four equations above may be used to replace the set of probabilities and thereby exhibit a set of necessary and sufficient relations among the four constructs $S, \Omega, \beta, \{\alpha_c\}$.

$$S = \Omega + \beta\langle\epsilon\rangle + \sum_c \alpha_c \langle N_c \rangle \tag{9}$$

$$\Omega = \ln \sum_i \exp\left(-\beta\epsilon_i - \sum_c \alpha_c N_{ci}\right) \tag{10}$$

$$\partial\Omega/\partial\beta = -\langle\epsilon\rangle \tag{11}$$

$$\partial\Omega/\partial\alpha_c = -\langle N_c \rangle \tag{12}$$

Two things should be pointed out here. First, there has been no real use of physics. Thus far $\langle\epsilon\rangle$ and $\langle N_c \rangle$ have been defined only by identifying them with 'repeatable' measurements of energy and composition without saying what 'energy' and 'composition' are. The mathematical results should be familiar to all who have studied statistical mechanics and their extension to more general cases should be obvious. Since we have not introduced or made use of any of the properties of energy or the particles, the results are all due to the rules of statistical inference inherent in the Cox–Shannon–Jaynes formulation. There is no reference to ensembles, heat baths, 'thermodynamic systems of which ours is an example drawn at random', etc.

This feature of being able to separate clearly which results are due to statistics and which are due to physics is a particular advantage in the information theory approach. The statistical quantities $p_i$, $S$, $\langle\epsilon\rangle$ and $\langle N_c\rangle$ generated Lagrange multipliers $\Omega$, $\beta$, $\{\alpha_c\}$ *independent of the physical properties* associated with $\langle\epsilon\rangle$ and $\langle N_c\rangle$; the form of the four equations given comes only from the general procedure for inference laid down by Jaynes. That principle comes from logic; not reasoning about physical systems.

To describe the physical behavior of $S$, $\Omega$, $\beta$ and $\{\alpha_c\}$ we have to define the rules for changes in $\langle\epsilon\rangle$ and $\langle N_c\rangle$, i.e. put physics into the description. This has been done, for example, in Ref. [26].

The detailed derivation is given in the references. We quote only some general results to illustrate that the maximum entropy encoding of knowledge about the system of volume $V$ to which we attach expectations $\langle\epsilon\rangle$ and $\langle N_c\rangle$ leads to the concepts associated with classical thermodynamics.

In this derivation the zeroth, first, second and third laws become *consequences*, not premises. Such a conclusion is indeed far reaching and it is no wonder the idea has been resisted for the dozen years since it was first put forward [10].

The properties of the Grand Canonical Distribution were first given by Gibbs who referred to his distributions as 'analogues' [24]. Denbigh [33], in his famous textbook on thermodynamics, also makes it quite clear that his statistical descriptions are *analogies* to classical thermodynamics. All workers who have dealt with statistical mechanics without basing their work squarely upon Shannon's information theory, as used by Jaynes and Cox, have either been silent on the connection between Clausius' entropy and the statistically defined entropy or been careful to disclaim any *necessary* connection between the two.

The important clue to understanding why the results given are more than a mere analogy is the recognition that we are actually defining that elusive state called 'equilibrium'. In this treatment, the encoding and all deductions from it, are valid only for those systems for which 'repeatable' measurements on $\epsilon_i$ and $N_{ci}$ are possible and for which knowledge of $\langle\epsilon\rangle$ and $\langle N_c\rangle$ *are sufficient to define the macrostate of the system in volume* $V$. We can use the derivation to describe the mathematical properties of this state, which we call 'equilibrium', i.e. how changes occur on passage from one equilibrium state to another, how two or more systems interact, etc. These results are valid for systems which satisfy the premises of the theory, i.e. for which 'equilibrium' exists. It remains for experiment to decide if there exist states of physical systems for which the premises can be met. Such a situation is quite common in physics. For example, in mechanics it is postulated that in a 'Newtonian frame': force = mass × acceleration. The question of whether Newtonian frames exist or if a particular frame of reference is Newtonian is settled by seeing if all the deductions from 'force = mass × acceleration' are satisfied.

This circularity in all physical theories is usually glossed over in the education of physicists. It is the subject of very careful scrutiny in Norwood Hanson's illuminating inquiry into theory building [34]. It is in a similar way that the general rules of inference (the maximum entropy encoding) are used to develop a description of 'equilibrium'. Mathematical relations are shown to be a consequence of the encoding process; behavior consistent with these equations proves that 'equilibrium' exists.

It is not generally understood that the concept of equilibrium is ill-defined in the literature of classical thermodynamics just as 'Newtonian frame' was glossed over in pre-relativity days. Attempts to define equilibrium, when they are made at all, are usually based on the idea of waiting for a long time. This idea of waiting a long time is not useful. Geological samples from the earth's interior have been found to be in thermodynamic disequilibrium. If substances that have been around for times comparable to the life of the earth are not in equilibrium, surely 'waiting' doesn't guarantee equilibrium. The proof of disequilibrium is the failure of the material to satisfy the phase rule. And the phase rule is, of course, a consequence of thermodynamics.

There is no way out of the dilemma that equilibrium is defined via thermodynamic constructs which constructs were in turn defined for the equilibrium state. We have no way of telling if a system is 'at equilibrium' except by making experiments which rely on the constructs which are defined by the theory of equilibrium. It is this dilemma which has inspired the comment 'there is no such thing as an immaculate perception'. What we see does depend on what we think.

This is not a special weakness of the information theory approach. It is inherent in the work of Gibbs, who defined equilibrium as the condition of maximum entropy.

The definition of 'equilibrium' thus given does not depend on the physics associated with $\epsilon_i$ and $N_{ci}$. Indeed, these symbols could stand for *anything* and the results would satisfy the desiderata. In references [10] and [27] it is demonstrated that the needed physics is obtained by introducing the following ideas:

1. The $\epsilon_i$ are additive, conserved.
2. The $\epsilon_i$ depend on $i$ and the dimensions which determine $V$.

From these ideas and the previous equations we may derive all the known relations of classical thermodynamics.

Because the information theory's simplicity and freedom from such artificial constructs as ensembles, heat baths, etc., enables us to keep separate which results come from physics and which from statistics; important differences, often lost in non-information theory treatments are kept in the foreground. For example, irreversibility is traced to the difference between 'force' and 'expected (or equilibrium) force'. There is maintained a distinction between the principle of conservation of energy (a deterministic addition of energies of Newtonian systems) and the First Law of Thermodynamics (a statistical treatment of energy). The roles of $\beta(=1/RT)$ in diffusion of energy and of $\alpha(=-\mu/RT)$ in diffusion of particles are seen to be identical.

Heat is usually treated as a pre-existing idea to be 'explained' by physics. In the information theory treatment it is seen that someone who believes in the principle of conservation of energy (and who wishes to retain consistency in his ideas) *must invent* the concept of heat if he tries to make statistical descriptions. In this treatment entropy is taken as *primitive* and heat *derived* from the resulting statistical descriptions. Also, heat is introduced without the need to define it in terms of temperature or adiabatic walls (and not developed as if 'adiabatic' were understood before heat)!

The information theory treatment thus inverts the usual procedure, in which heat, temperature, and work are taken as primitive, and in which energy and entropy are derived. This break in tradition is hard to accept for those steeped in a tradition that treats the laws of thermodynamics as though they were 'discovered' by experimentalists who knew instinctively about heat, temperature, work, equilibrium and equations of state. But the laws of physics are not 'discovered', they are 'invented' or – better yet – 'designed' according to criteria such as:

1. Consistency
2. Freedom from Ambiguity
3. Universality
4. Honesty

and are, therefore, constrained to the same results as produced simply, elegantly and directly by the information theory approach.

The information theory approach tells us *why* our ancestors *had* to invent temperature, heat, and reversible processes. It elucidates the 'paradoxes' of Szilard, Gibbs and many others. But most of all, it unifies our understanding of many phenomena. And it does so by showing that the entropy of thermodynamics is but a special case of the entropy of information theory.

## III. Critique of Jauch and Baron's Discussion of Szilard's Thought Experiment

We definitely do not follow Jauch and Baron in their rebuttal of Szilard's argument. We understand the question as follows:

1. *A matter of semantics.* Jauch and Baron borrow from von Neumann the statement that 'in phenomenological thermodynamics each conceivable process constitutes valid evidence, provided that it does not conflict with the two fundamental laws of thermodynamics', and then add from their own that 'If an idealization is in conflict with a law that is basic to the second law, then it cannot be used as evidence for violation of the second law'.

It is certainly obvious that a self-consistent theory (and macroscopic, phenomenological thermodynamics is a consistent theory) cannot be criticized from within. But this does not forbid the production of thought experiments based on knowledge from *outside* the domain of the theory in order to criticize it. In his criticism of Aristotelian mechanical conceptions Galileo largely used thought experiments based on information that was at hand to everybody, *but* was outside the realm of Aristotelian physics.

The very concept of bouncing point molecules, and of pressure as integrated momentum exchange per sec cm$^2$, certainly is *outside* the domain of phenomenological, macroscopic, thermodynamics – not to speak of the consideration of one single molecule, and of learning in what half of the cylinder it is found.

Therefore, let us imitate Socrates in his criticism of Zeno, and proceed.

2. With Jauch and Baron, let us not get involved in quantum subtleties, but instead use the point particle concept of classical mechanics and of the classical kinetic theory of gases.

Let us first discard Jauch and Baron's large heat reservoir; we will bring it in later on.

There is no question that if Szilard's single molecule is (nineteenth-century language) or is known to be (post Szilard and Lewis language) in one definite half of the cylinder, the entropy of the single molecule gas is smaller than when the molecule is allowed to move all through the whole cylinder. If the transition from one state to the other is secured by opening the door in the piston, then the entropy of the gas inside the cylinder increases by $k\mathrm{Ln}2$.

Does this statement contradict the classical thermodynamical statement that (in Jauch and Baron's terms) 'The entropy of the...system piston, gas...remains constant because it is closed and the changes are reversible'? In fact it does not, because the mere (frictionless) opening of the door entails an 'irreversible' change. Though energetically isolated, our (limited) system is not informationally isolated. However, with this distinction, we are certainly stepping outside the domain of classical thermodynamics.

As a result of opening the door, the pressure inside the cylinder falls down to one-half of what it was; this is because the molecule, while retaining by hypothesis its kinetic energy, now spends only half of its time in each half of the cylinder. Whence the classical expression for the change in entropy. (Note that while the molecule is not 'aware' that the door has been opened, the observer is, hence the entropy increase is computed as of the opening of the door.)

3. Jauch and Baron write: "However, at the exact moment when the piston is in the middle of the cylinder and the opening is closed, the gas violates the law of Gay Lussac because it is compressed to half its volume without expenditure of energy. We therefore conclude that the idealizations in Szilard's experiment are inadmissible in their actual context."

Socrates' walking and walking certainly was inadmissible in the context of Zeno's arguments. Nevertheless he *could* walk. If, by definition, our single molecule is a Newtonian point particle, nothing on earth can prevent us from suddenly closing our frictionless, massless, impenetrable door, and learning afterwards in what half of the cylinder we have trapped our molecule. There is *absolutely nothing* self-contradictory in this, not even the idealized concept of the door, which can be approached at will without even contradicting *classical* thermodynamics.

Moreover, no *classical* thermodynamicist would object to the reversed procedure: opening the door. This is adiabatic expansion, the subject of an experiment by Joule.

Why then had we termed adiabatic expansion an 'irreversible' process, and are we now speaking of it as time symmetric to the process of trapping the single molecule by closing the door? Because macroscopic irreversibility is one thing, and microscopic reversibility another thing, which have to be reconciled. It would lead us too far astray to delve here in this problem; we simply refer the reader to a recent discussion of it in terms that are consonant to those we are using here [35].

4. Finally we bring in Jauch and Baron's pulleys, strings and scales, which will restore the macroscopic irreversibility of the whole process.

From now on, as soon as we suddenly close the door, we do *not* allow the (single molecule) gas to expand adiabatically: we harness it, and have it lift (reversibly) a weight. This will cause its temperature to drop down, so that the heat reservoir becomes extremely useful if we want to lift many weights. So we bring it in also.

And *now* the *whole* system, supposed to be isolated, undergoes a (macroscopically) *reversible* change, so that its entropy remains constant.

However, *potential energy has been gained at the expenditure of heat* so that, at first sight, there is something wrong in the entropy balance. What are we overlooking?

We are overlooking that the observer has to *know*, that is, to *learn* which way the piston begins moving, and *then* push a weight on the right scale. This is his *free decision*, in accord with the *general decision* he has made, to use the heat in the reservoir for lifting weights. However, we are not delving here in a discussion of the twin Aristotelian aspects of information: cognizance and will [36].

Suffice it to say that, in order to have the entropy balance right, we *must* include in it the *information* gained by the observer. But this was Szilard's statement.

5. Finally, together with Brillouin [4] and others, Jauch and Baron point out rightly that the preceding 'system' can be transformed into a 'robot', the functioning of which, however, will require drawing negentropy from an existing source, by an amount at least as large as the negentropy created by lifting the weights.

The point is that the robot does not just step out of nowhere – no more than does a refrigerator or a heat engine. This brings into the problem Brillouin's 'structural negentropy', also Brillouin's 'information contained in the expression of the physical laws' that the engineer has used, and, finally, the problem of all the thinking and decision-making on the engineer's part.

In other words, if you push information out of the door by appealing to the robot, then it will come back right through the adiabatic wall.

6. *Concluding* this section, we believe that, while it is possible, and very informative, to deduce Thermodynamics from a general theory of Information, the converse is not possible. Thermodynamics is too rooted in specifics of physics to produce a general theory of information.

But this is a mere instance of the way scientific progress goes on...

## REFERENCES

[1] J. M. JAUCH and J. G. BARON, Helv. Phys. Acta *45*, 220 (1972).
[2] C. E. SHANNON, Bell Syst. Techn. Journ. *27*, 379, 623 (1948); C. E. SHANNON and W. WEAVER, *The Mathematical Theory of Communication* (Univ. of Illinois Press, Urbana 1949).
[3] L. Szilard, Zeits. für Phys. *53*, 840 (1929).
[4] L. BRILLOUIN, *Science and Information Theory* (Academic Press, New York 1957).
[5] H. GRAD, Comm. Pure Appl. Math. *14*, 323 (1971).
[6] E. T. JAYNES, Phys. Rev. *106*, 620 (1957); *108*, 171 (1957).
[7] E. T. JAYNES, Amer. Journ. Phys. *31*, 66 (1963).
[8] R. T. COX, Amer. Journ. Phys. *14*, 1 (1946).
[9] R. T. COX, *The Algebra of Probable Inference* (John Hopkins Press, Baltimore 1961).
[10] M. TRIBUS, J. Appl. Mech. March 1961, pp. 1–8; M. TRIBUS and R. B. Evans, Appl. Mech. Rev. *16* (10), 765–769 (1963); M. TRIBUS, P. T. SHANNON and R. B. Evans, AEChI Jour. March 1966, pp. 244–248; M. TRIBUS, Amer. Sci. *54* (2) (1966).
[11] R. T. BAYES, Philos. Trans. *53*, 370 (1763).
[12] S. P. DE LAPLACE, *Essai Philosophique sur les Probabilitiés*, 1774. See also *A Philosophical Essay on Probabilities* (Dover, New York 1951).
[13] D. HUME, *An Enquiry Concerning Human Destiny*, 1758.
[14] J. VON NEUMANN, Nachr. Akad. Wiss. Goettingen, Math.-Phys. Kl., 245 and 273 (1927).
[15] J. VON NEUMANN, *Mathematische Grundlagen der Quantenmachanik* (Springer, Berlin 1972). See also *Mathematical Foundations of Quantum Mechanics* (Princeton Univ. Press, Princeton, N.J. 1955).
[16] W. M. ELSASSER, Phys. Rev. *52*, 987 (1937).
[17] R. A. FISHER, Proc. Camb. Phil. Soc. *26*, 528 (1929); *28*, 257 (1932).
[18] S. WATANABE, Zeits. für Phys. *113*, 482 (1939).
[19] R. CARNAP, *The Continuum of Inductive Methods* (Univ. of Chicago Press 1952).
[20] J. KEMENY, Journ. Symbolic Logic *20*, 263 (1955).
[21] H. JEFFREYS, *Theory of Probability* (Oxford, Clarendon Press 1939).
[22] G. N. LEWIS, Science *71*, 569 (1939).
[23] J. D. VAN DER WAALS, Phys. Zeits. *12*, 547 (1911).
[24] J. W. GIBBS, *Elementary Principles in Statistical Mechanics* (Yale Univ. Press, New Haven, Conn. 1914), p. 150.
[25] E. T. JAYNES, *Foundations of probability theory and statistical mechanics*, in *Studies in the Foundations of Methodology and Philosophy of Science*, Vol. 1, Delaware Seminar in the Foundations of Physics (Springer, New York 1967); *Information theory and statistical mechanics*, in *Brandeis Lectures* (Benjamin, Inc., New York 1963), p. 181.

[26] M. Tribus, *Thermostatics and Thermodynamics* (D. van Nostrand Co., Princeton, N.J. 1961). See also N. Tepmoctatnka, *Tepmoznhamnka* (Moscow 1970).
[27] M. Tribus, *Rational Descriptions, Decisions and Designs* (Pergamon, Oxford and New York 1970). See also *Decisions Rationnelles dans l'Incertain* (Masson, Paris 1972).
[28] A. I. Khinchin, *Mathematical Foundations of Information Theory* (Dover, New York 1957).
[29] A. M. Yaglom and I. M. Yaglom, *Probabilité et Information* (Dunod, Paris 1959).
[30] A. Katz, *Principles of Statistical Mechanics* (W. H. Freeman and Co., San Francisco and London 1967).
[31] A. Hobson, *Concepts in Statistical Mechanics* (Gordon and Breach, New York 1971).
[32] R. Baierlein, *Atoms and Information Theory* (W. H. Freeman and Co., San Francisco and London 1971).
[33] K. G. Denbigh, *The Principles of Chemical Equilbrium* (Cambridge University Press 1955).
[34] N. R. Hanson, *Patterns of Discovery* (Cambridge University Press 1958).
[35] O. Costa de Beauregard (reporter): *Discussion on temporal asymmetry in thermodynamics and cosmology*, in *Proceedings of the International Conference on Thermodynamics*, held in Cardiff, edited by P. T. Landsberg (Butterworths, London 1970).
[36] O. Costa de Beauregard, *Is there a paradox in the theory of time anisotropy*, in *A Critical Review of Thermodynamics*, edited by E. B. Stuart, B. Gal-Or and A. J. Brainard (Mono Book Corps., Baltimore 1970), pp. 463–472. See also Studium Generale *54*, 10 (1971).

# Chapter 4

# Information Erasure: Landauer's Principle

# Irreversibility and Heat Generation in the Computing Process

**Abstract: It is argued that computing machines inevitably involve devices which perform logical functions that do not have a single-valued inverse. This logical irreversibility is associated with physical irreversibility and requires a minimal heat generation, per machine cycle, typically of the order of $kT$ for each irreversible function. This dissipation serves the purpose of standardizing signals and making them independent of their exact logical history. Two simple, but representative, models of bistable devices are subjected to a more detailed analysis of switching kinetics to yield the relationship between speed and energy dissipation, and to estimate the effects of errors induced by thermal fluctuations.**

## 1. Introduction

The search for faster and more compact computing circuits leads directly to the question: What are the ultimate physical limitations on the progress in this direction? In practice the limitations are likely to be set by the need for access to each logical element. At this time, however, it is still hard to understand what physical requirements this puts on the degrees of freedom which bear information. The existence of a storage medium as compact as the genetic one indicates that one can go very far in the direction of compactness, at least if we are prepared to make sacrifices in the way of speed and random access.

Without considering the question of access, however, we can show, or at least very strongly suggest, that information processing is inevitably accompanied by a certain minimum amount of heat generation. In a general way this is not surprising. Computing, like all processes proceeding at a finite rate, must involve some dissipation. Our arguments, however, are more basic than this, and show that there is a minimum heat generation, independent of the rate of the process. Naturally the amount of heat generation involved is many orders of magnitude smaller than the heat dissipation in any practically conceivable device. The relevant point, however, is that the dissipation has a real function and is not just an unnecessary nuisance. The much larger amounts of dissipation in practical devices may be serving the same function.

Our conclusion about dissipation can be anticipated in several ways, and our major contribution will be a tightening of the concepts involved, in a fashion which will give some insight into the physical requirements for logical devices. The simplest way of anticipating our conclusion is to note that a binary device must have at least one degree of freedom associated with the information. Classically a degree of freedom is associated with $kT$ of thermal energy. Any switching signals passing between devices must therefore have this much energy to override the noise. This argument does not make it clear that the signal energy must actually be dissipated. An alternative way of anticipating our conclusions is to refer to the arguments by Brillouin and earlier authors, as summarized by Brillouin in his book, *Science and Information Theory*,[1] to the effect that the measurement process requires a dissipation of the order of $kT$. The computing process, where the setting of various elements depends upon the setting of other elements at previous times, is closely akin to a measurement. It is difficult, however, to argue out this connection in a more exact fashion. Furthermore, the arguments concerning the measurement process are based on the analysis of specific models (as will some of our arguments about computing), and the specific models involved in the measurement analysis are rather far from the kind of mechanisms involved in data processing. In fact the arguments dealing with the measurement process do not define *measurement* very well, and avoid the very essential question: When is a system $A$ coupled to a system $B$ performing a measurement? The mere fact that two physical systems are coupled does not in itself require dissipation.

Our main argument will be a refinement of the following line of thought. A simple binary device consists of a particle in a bistable potential well shown in Fig. 1. Let us arbitrarily label the particle in the left-hand well as the ZERO state. When the particle is in the right-hand well, the device is in the ONE state. Now consider the operation

RESTORE TO ONE, which leaves the particle in the ONE state, regardless of its initial location. If we are told that the particle is in the ONE state, then it is easy to leave it in the ONE state, without spending energy. If on the other hand we are told that the particle is in the ZERO state, we can apply a force to it, which will push it over the barrier, and then, when it has passed the maximum, we can apply a retarding force, so that when the particle arrives at ONE, it will have no excess kinetic energy, and we will not have expended any energy in the whole process, since we extracted energy from the particle in its downhill motion. Thus at first sight it seems possible to RESTORE TO ONE without any expenditure of energy. Note, however, that in order to avoid energy expenditure we have used two different routines, depending on the initial state of the device. This is not how a computer operates. In most instances a computer pushes information around in a manner that is independent of the exact data which are being handled, and is only a function of the physical circuit connections.

Can we then construct a single time-varying force, $F(t)$, which when applied to the conservative system of Fig. 1 will cause the particle to end up in the ONE state, if it was initially in either the ONE state or the ZERO state? Since the system is conservative, its whole history can be reversed in time, and we will still have a system satisfying the laws of motion. In the time-reversed system we then have the possibility that for a single initial condition (position in the ONE state, zero velocity) we can end up in at least two places: the ZERO state or the ONE state. This, however, is impossible. The laws of mechanics are completely deterministic and a trajectory is determined by an initial position and velocity. (An initially unstable position can, in a sense, constitute an exception. We can roll away from the unstable point in one of at least two directions. Our initial point ONE is, however, a point of stable equilibrium.) Reverting to the original direction of time development, we see then that it is not possible to invent a single $F(t)$ which causes the particle to arrive at ONE regardless of its initial state.

If, however, we permit the potential well to be lossy, this becomes easy. A very strong positive initial force applied slowly enough so that the damping prevents oscillations will push the particle to the right, past ONE, regardless of the particle's initial state. Then if the force is taken away slowly enough, so that the damping has a chance to prevent appreciable oscillations, the particle is bound to arrive at ONE. This example also illustrates a point argued elsewhere[2] in more detail: While a heavily overdamped system is obviously undesirable, since it is made sluggish, an extremely underdamped one is also not desirable for switching, since then the system may bounce back into the wrong state if the switching force is applied and removed too quickly.

## 2. Classification

Before proceeding to the more detailed arguments we will need to classify data processing equipment by the means used to hold information, when it is not interacting or being processed. The simplest class and the one to which all the arguments of subsequent sections will be addressed consists of devices which can hold information without dissipating energy. The system illustrated in Fig. 1 is in this class. Closely related to the mechanical example of Fig. 1 are ferrites, ferroelectrics and thin magnetic films. The latter, which can switch without domain wall motion, are particularly close to the one-dimensional device shown in Fig. 1. Cryotrons are also devices which show dissipation only when switching. They do differ, however, from the device of Fig. 1 because the ZERO and ONE states are not particularly favored energetically. A cryotron is somewhat like the mechanical device illustrated in Fig. 2, showing a particle in a box. Two particular positions in the box are chosen to represent ZERO and ONE, and the preservation of information depends on the fact that Brownian motion in the box is very slow. The reliance on the slowness of Brownian motion rather than on restoring forces is not only characteristic of cryotrons, but of most of the more familiar forms of information storage: Writing, punched cards, microgroove recording, etc. It is clear from the literature that all essential logical functions can be performed by

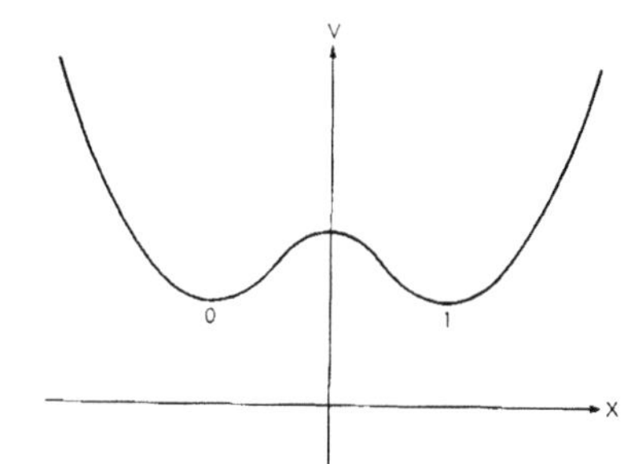


*Figure 1* **Bistable potential well.**
*x is a generalized coordinate representing quantity which is switched.*

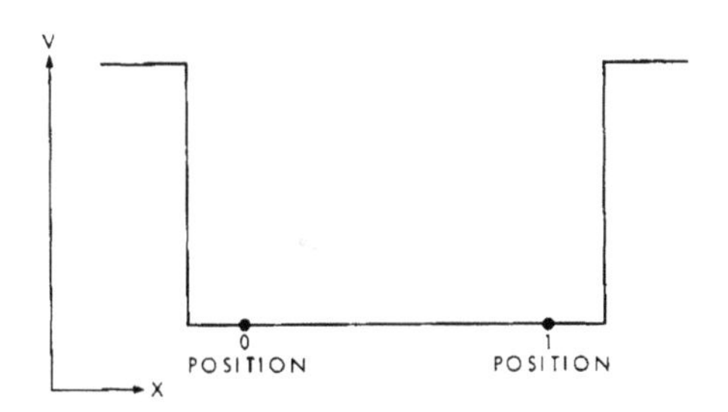


*Figure 2* **Potential well in which ZERO and ONE state are not separated by barrier.**
*Information is preserved because random motion is slow.*

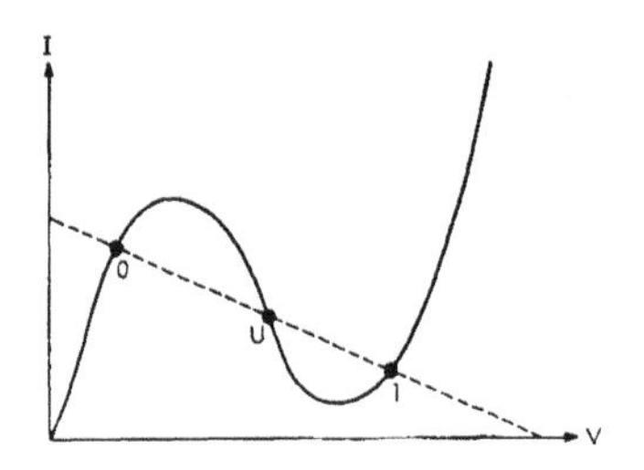


*Figure 3* **Negative resistance characteristic (solid line) with load line (dashed).**
*ZERO and ONE are stable states, U is unstable.*

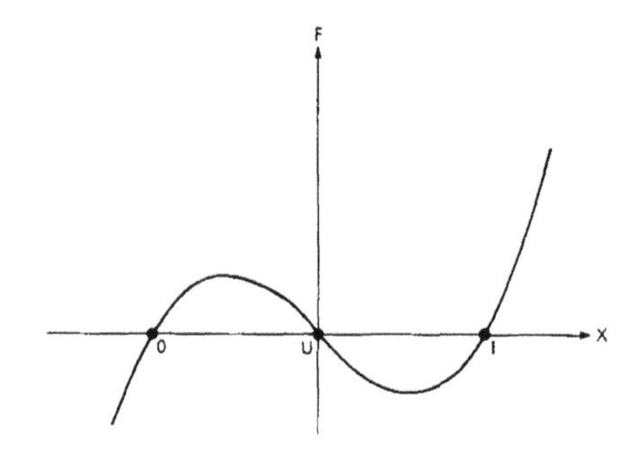


*Figure 4* **Force versus distance for the bistable well of Fig. 1.**
*ZERO and ONE are the stable states, U the unstable one.*

devices in this first class. Computers can be built that contain either only cryotrons, or only magnetic cores.[3,4]

The second class of devices consists of structures which are in a steady (time invariant) state, but in a dissipative one, while holding on to information. Electronic flip-flop circuits, relays, and tunnel diodes are in this class. The latter, whose characteristic with load line is shown in Fig. 3, typifies the behavior. Two stable points of operation are separated by an unstable position, just as for the device in Fig. 1. It is noteworthy that this class has no known representatives analogous to Fig. 2. All the active bistable devices (latches) have built-in means for restoration to the desired state. The similarity between Fig. 3 and the device of Fig. 1 becomes more conspicuous if we represent the bistable well of Fig. 1 by a diagram plotting force against distance. This is shown in Fig. 4. The line $F=0$ intersects the curve in three positions, much like the load line (or a line of constant current), in Fig. 3. This analogy leads us to expect that in the case of the dissipative device there will be transitions from the desired state, to the other stable state, resulting from thermal agitation or quantum mechanical tunneling, much like for the dissipationless case, and as has been discussed for the latter in detail by Swanson.[5] The dissipative device, such as the single tunnel diode, will in general be an analog, strictly speaking, to an unsymmetrical potential well, rather than the symmetrical well shown in Fig. 1. We can therefore expect that of the two possible states for the negative resistance device only one is really stable, the other is metastable. An assembly of bistable tunnel diodes left alone for a sufficiently long period would eventually almost all arrive at the same state of absolute stability.

In general when using such latching devices in computing circuits one tries hard to make the dissipation in the two allowed states small, by pushing these states as closely as possible to the voltage or current axis. If one were successful in eliminating this dissipation almost completely during the steady state, the device would become a member of our first class. Our intuitive expectation is, therefore, that in the steady state dissipative device the dissipation per switching event is at least as high as in the devices of the first class, and that this dissipation per switching event is supplemented by the steady state dissipation.

The third and remaining class is a "catch-all"; namely, those devices where time variation is essential to the recognition of information. This includes delay lines, and also carrier schemes, such as the phase-bistable system of von Neumann.[6] The latter affords us a very nice illustration of the need for dissipative effects; most other members of this third class seem too complex to permit discussion in simple physical terms.

In the von Neumann scheme, which we shall not attempt to describe here in complete detail, one uses a "pump" signal of frequency $\omega_0$, which when applied to a circuit tuned to $\omega_0/2$, containing a nonlinear reactance, will cause the spontaneous build-up of a signal at the lower frequency. The lower frequency signal has a choice of two possible phases (180° apart at the lower frequency) and this is the source of the bistability. In the von Neumann scheme the pump is turned off after the subharmonic has developed, and the subharmonic subsequently permitted to decay through circuit losses. This decay is an essential part of the scheme and controls the direction in which information is passed. Thus at first sight the circuit losses perform an essential function. It can be shown, however, that the signal reduction can be produced in a lossless nonlinear circuit, by a suitably phased pump signal. Hence it would seem adequate to use lossless nonlinear circuits, and instead of turning the pump off, change the pump phase so that it causes signal decay instead of signal growth. The directionality of information flow therefore does not really depend on the existence of losses. The losses do, however, perform another essential function.

The von Neumann system depends largely on a coupling scheme called *majority logic,* in which one couples to three subharmonic oscillators and uses the sum of their oscillations to synchronize a subharmonic oscillator whose pump will cause it to build up at a later time than the initial three. Each of the three signals which are added together can have one of two possible phases. At most two of the signals can cancel, one will always survive, and thus there will always be a phase determined for the build-up of the next oscillation. The synchronization signal can, therefore, have two possible magnitudes. If all three of the inputs agree we get a synchronization signal three times as big as in the case where only two inputs have a given phase. If the subharmonic circuit is lossless the subsequent build-up will then result in two different amplitudes, depending on the size of the initial synchronization signal. This, however, will interfere with the basic operation of the scheme at the next stage, where we will want to combine outputs of three oscillators again, and will want all three to be of equal amplitude. We thus see that the absence of the losses gives us an output amplitude from each oscillator which is too dependent on inputs at an earlier stage. While perhaps the deviation from the desired amplitudes might still be tolerable after one cycle, these deviations could build up, through a period of several machine cycles. The losses, therefore, are needed so that the unnecessary details of a signal's history will be obliterated. The losses are essential for the standardization of signals, a function which in past theoretical discussions has perhaps not received adequate recognition, but has been very explicitly described in a recent paper by A. W. Lo.[7]

### 3. Logical irreversibility

In the Introduction we analyzed Fig. 1 in connection with the command RESTORE TO ONE and argued that this required energy dissipation. We shall now attempt to generalize this train of thought. RESTORE TO ONE is an example of a logical truth function which we shall call *irreversible*. We shall call a device *logically irreversible* if the output of a device does not uniquely define the inputs. We believe that devices exhibiting logical irreversibility are essential to computing. Logical irreversibility, we believe, in turn implies physical irreversibility, and the latter is accompanied by dissipative effects.

We shall think of a computer as a distinctly finite array of $N$ binary elements which can hold information, without dissipation. We will take our machine to be synchronous, so that there is a well-defined machine cycle and at the end of each cycle the $N$ elements are a complicated function of their state at the beginning of the cycle.

Our arguments for logical irreversibility will proceed on three distinct levels. The first-level argument consists simply in the assertion that present machines do depend largely on logically irreversible steps, and that therefore any machine which copies the logical organization of present machines will exhibit logical irreversibility, and therefore by the argument of the next Section, also physical irreversibility.

The second level of our argument considers a particular class of computers, namely those using logical functions of only one or two variables. After a machine cycle each of our $N$ binary elements is a function of the state of at most two of the binary elements before the machine cycle. Now assume that the computer is logically reversible. Then the machine cycle maps the $2^N$ possible initial states of the machine onto the same space of $2^N$ states, rather than just a subspace thereof. In the $2^N$ possible states each bit has a ONE and a ZERO appearing with equal frequency. Hence the reversible computer can utilize only those truth functions whose truth table exhibits equal numbers of ONES and ZEROS. The admissible truth functions then are the identity and negation, the EXCLUSIVE OR and its negation. These, however, are not a complete set[8] and do not permit a synthesis of all other truth functions.

In the third level of our argument we permit more general devices. Consider, for example, a particular three-input, three-output device, i.e., a small special purpose computer with three bit positions. Let $p$, $q$, and $r$ be the variables before the machine cycle. The particular truth function under consideration is the one which replaces $r$ by $p \cdot q$ if $r=0$, and replaces $r$ by $\overline{p \cdot q}$ if $r=1$. The variables $p$ and $q$ are left unchanged during the machine cycle. We can consider $r$ as giving us a choice of program, and $p$, $q$ as the variables on which the selected program operates. This is a logically reversible device, its output always defines its input uniquely. Nevertheless it is capable of performing an operation such as AND which is not, in itself, reversible. The computer, however, saves enough of the input information so that it supplements the desired result to allow reversibility. It is interesting to note, however, that we did not "save" the program; we can only deduce what it was.

Now consider a more general purpose computer, which usually has to go through many machine cycles to carry out a program. At first sight it may seem that logical reversibility is simply obtained by saving the input in some corner of the machine. We shall, however, label a machine as being logically reversible, if and only if all its individual steps are logically reversible. This means that every single time a truth function of two variables is evaluated we must save some additional information about the quantities being operated on, whether we need it or not. Erasure, which is equivalent to RESTORE TO ONE, discussed in the Introduction, is not permitted. We will, therefore, in a long program clutter up our machine bit positions with unnecessary information about intermediate results. Furthermore if we wish to use the reversible function of three variables, which was just discussed, as an AND, then we must supply in the initial programming a separate ZERO for every AND operation which is subsequently required, since the "bias" which programs the device is not saved, when the AND is performed. The machine must therefore have a great deal of extra capacity to store both the extra "bias" bits and the extra outputs. Can it be given adequate capacity to make all intermediate steps reversible? If our machine is capable, as machines are generally understood to be, of a nonterminating program, then it is clear that the capacity for preserving all the information about all the intermediate steps cannot be there.

Let us, however, not take quite such an easy way out. Perhaps it is just possible to devise a machine, useful in the normal sense, but not capable of embarking on a nonterminating program. Let us take such a machine as it normally comes, involving logically irreversible truth functions. An irreversible truth function can be made into a reversible one, as we have illustrated, by "embedding" it in a truth function of a large number of variables. The larger truth function, however, requires extra inputs to bias it, and extra outputs to hold the information which provides the reversibility. What we now contend is that this larger machine, while it is reversible, is not a useful computing machine in the normally accepted sense of the word.

First of all, in order to provide space for the extra inputs and outputs, the embedding requires knowledge of the number of times each of the operations of the original (irreversible) machine will be required. The usefulness of a computer stems, however, from the fact that it is more than just a table look-up device; it can do many programs which were not anticipated in full detail by the designer. Our enlarged machine must have a number of bit positions, for every embedded device of the order of the number of program steps and requires a number of switching events during program loading comparable to the number that occur during the program itself. The setting of bias during program loading, which would typically consist of restoring a long row of bits to say ZERO, is just the type of nonreversible logical operation we are trying to avoid. Our unwieldy machine has therefore avoided the irreversible operations during the running of the program, only at the expense of added comparable irreversibility during the loading of the program.

#### 4. Logical irreversibility and entropy generation

The detailed connection between logical irreversibility and entropy changes remains to be made. Consider again, as an example, the operation RESTORE TO ONE. The generalization to more complicated logical operations will be trivial.

Imagine first a situation in which the RESTORE operation has already been carried out on each member of an assembly of such bits. This is somewhat equivalent to an assembly of spins, all aligned with the positive $z$-axis. In thermal equilibrium the bits (or spins) have two equally favored positions. Our specially prepared collections show much more order, and therefore a lower temperature and entropy than is characteristic of the equilibrium state. In the adiabatic demagnetization method we use such a prepared spin state, and as the spins become disoriented they take up entropy from the surroundings and thereby cool off the lattice in which the spins are embedded. An assembly of ordered bits would act similarly. As the assembly thermalizes and forgets its initial state the environment would be cooled off. Note that the important point here is not that all bits in the assembly initially agree with each other, but only that there is a single, well-defined initial state for the collection of bits. The well-defined initial state corresponds, by the usual statistical mechanical definition of entropy, $S = k \log_e W$, to zero entropy. The degrees of freedom associated with the information can, through thermal relaxation, go to any one of $2^N$ states (for $N$ bits in the assembly) and therefore the entropy can increase by $kN \log_e 2$ as the initial information becomes thermalized.

Note that our argument here does not necessarily depend upon connections, frequently made in other writings, between entropy and information. We simply think of each bit as being located in a physical system, with perhaps a great many degrees of freedom, in addition to the relevant one. However, for each possible physical state which will be interpreted as a ZERO, there is a very similar possible physical state in which the physical system represents a ONE. Hence a system which is in a ONE state has only half as many physical states available to it as a system which can be in a ONE or ZERO state. (We shall ignore in this Section and in the subsequent considerations the case in which the ONE and ZERO are represented by states with different entropy. This case requires arguments of considerably greater complexity but leads to similar physical conclusions.)

In carrying out the RESTORE TO ONE operation we are doing the opposite of the thermalization. We start with each bit in one of two states and end up with a well-defined state. Let us view this operation in some detail.

Consider a statistical ensemble of bits in thermal equilibrium. If these are all reset to ONE, the number of states covered in the ensemble has been cut in half. The entropy therefore has been reduced by $k \log_e 2 = 0.6931\, k$ per bit. The entropy of a closed system, e.g., a computer with its own batteries, cannot decrease; hence this entropy must appear elsewhere as a heating effect, supplying $0.6931\, kT$ per restored bit to the surroundings. This is, of course, a minimum heating effect, and our method of reasoning gives no guarantee that this minimum is in fact achievable.

Our reset operation, in the preceding discussion, was applied to a thermal equilibrium ensemble. In actuality we would like to know what happens in a particular computing circuit which will work on information which has not yet been thermalized, but at any one time consists of a well-defined ZERO or a well-defined ONE. Take first the case where, as time goes on, the reset operation is applied to a random chain of ONES and ZEROS. We can, in the usual fashion, take the statistical ensemble equivalent to a time average and therefore conclude that the dissipation per reset operation is the same for the timewise succession as for the thermalized ensemble.

A computer, however, is seldom likely to operate on random data. One of the two bit possibilities may occur more often than the other, or even if the frequencies are equal, there may be a correlation between successive bits. In other words the digits which are reset may not carry the maximum possible information. Consider the extreme case, where the inputs are all ONE, and there is no need to carry out any operation. Clearly then no entropy changes occur and no heat dissipation is involved. Alternatively if the initial states are all ZERO they also carry no information, and no entropy change is involved in resetting them all to ONE. Note, however, that the reset operation which sufficed when the inputs were all ONE (doing nothing) will not suffice when the inputs are all ZERO. When the initial states are ZERO, and we wish to go to ONE, this is analogous to a phase transformation between two phases in equilibrium, and can, presumably, be done reversibly and without an entropy increase in the universe, but only by a procedure specifically designed for that task. We thus see that when the initial states do not have their fullest possible diversity, the necessary entropy increase in the RESET operation can be reduced, but only by taking advantage of our knowledge about the inputs, and tailoring the reset operation accordingly.

The generalization to other logically irreversible operations is apparent, and will be illustrated by only one additional example. Consider a very small special-purpose computer, with three binary elements $p$, $q$, and $r$. A machine cycle replaces $p$ by $r$, replaces $q$ by $r$, and replaces $r$ by $p \cdot q$. There are eight possible initial states, and in thermal equilibrium they will occur with equal probability. How much entropy reduction will occur in a machine cycle? The initial and final machine states are shown in Fig. 5. States $\alpha$ and $\beta$ occur with a probability of 1/8 each: states $\gamma$ and $\delta$ have a probability of occurrence of 3/8 each. The initial entropy was

$$S_i = k \log_e W = -k\Sigma p \log_e p$$
$$= -k\Sigma \tfrac{1}{8} \log_e \tfrac{1}{8} = 3k \log_e 2 \,.$$

The final entropy is

$$S_f = -k\Sigma p \log_e p$$
$$= -k(\tfrac{1}{8} \log \tfrac{1}{8} + \tfrac{1}{8} \log \tfrac{1}{8} + \tfrac{3}{8} \log \tfrac{3}{8} + \tfrac{3}{8} \log \tfrac{3}{8}) \,.$$

The difference $S_i - S_f$ is 1.18 $k$. The minimum dissipation, if the initial state has no useful information, is therefore 1.18 $kT$.

The question arises whether the entropy is really reduced by the logically irreversible operation. If we really map the possible initial ZERO states and the possible initial ONE states into the same space, i.e., the space of ONE states, there can be no question involved. But, perhaps, after we have performed the operation there can be some small remaining difference between the systems which were originally in the ONE state already and those that had to be switched into it. There is no harm in such differences persisting for some time, but as we saw in the discussion of the dissipationless subharmonic oscillator, we cannot tolerate a cumulative process, in which differences between various possible ONE states become larger and larger according to their detailed past histories. Hence the physical "many into one" mapping, which is the source of the entropy change, need not happen in full detail during the machine cycle which performed the logical function. But it must eventually take place, and this is all that is relevant for the heat generation argument.

## 5. Detailed analysis of bistable well

To supplement our preceding general discussion we shall give a more detailed analysis of switching for a system representable by a bistable potential well, as illustrated, one-dimensionally, in Fig. 1, with a barrier large compared to $kT$. Let us, furthermore, assume that switching is accomplished by the addition of a force which raises the energy of one well with respect to the other, but still leaves a barrier which has to be surmounted by thermal activation. (A sufficiently large force will simply eliminate one of the minima completely. Our switching forces are presumed to be smaller.) Let us now consider a statistical ensemble of double well systems with a nonequilibrium distribution and ask how rapidly equilibrium will be approached. This question has been analyzed in detail in an earlier paper,[2] and we shall therefore be satisfied here with a very simple kinetic analysis which leads to the same answer. Let $n_A$ and $n_B$ be the number of ensemble members in Well $A$ and Well $B$ respectively. Let $U_A$ and $U_B$ be the energies at the bottom of each well and $U$ that of the barrier which has to be surmounted. Then the rate at which particles leave Well $A$ to go to Well $B$ will be of the form $\nu n_A \exp[-(U-U_A)/kT]$. The flow from $B$ to $A$ will be $\nu n_B \exp[-(U-U_B)/kT]$. The two frequency factors have been taken to be identical. Their differences are, at best, unimportant compared to the differences in exponents. This yields

| BEFORE CYCLE p | q | r | | AFTER CYCLE $p_1$ | $q_1$ | $r_1$ | FINAL STATE |
|---|---|---|---|---|---|---|---|
| 1 | 1 | 1 | → | 1 | 1 | 1 | $\alpha$ |
| 1 | 1 | 0 | → | 0 | 0 | 1 | $\beta$ |
| 1 | 0 | 1 | → | 1 | 1 | 0 | $\gamma$ |
| 1 | 0 | 0 | → | 0 | 0 | 0 | $\delta$ |
| 0 | 1 | 1 | → | 1 | 1 | 0 | $\gamma$ |
| 0 | 1 | 0 | → | 0 | 0 | 0 | $\delta$ |
| 0 | 0 | 1 | → | 1 | 1 | 0 | $\gamma$ |
| 0 | 0 | 0 | → | 0 | 0 | 0 | $\delta$ |

*Figure 5* **Three input - three output device which maps eight possible states onto only four different states.**

$$\frac{dn_A}{dt} = -n_A\nu \exp[-(U-U_A)/kT] + n_B\nu \exp[-(U-U_B)/kT] \,,$$

$$\frac{dn_B}{dt} = n_A\nu \exp[-(U-U_A)/kT] - n_B\nu \exp[-(U-U_B)/kT] \,. \tag{5.1}$$

We can view Eqs. (5.1) as representing a linear transformation on $(n_A, n_B)$, which yields $\left(\frac{dn_A}{dt}, \frac{dn_B}{dt}\right)$. What are the characteristic values of the transformation? They are:

$$\lambda_1 = 0, \quad \lambda_2 = -\nu \exp[(U-U_A)/kT] - \nu \exp[-(U-U_B)/kT] \,.$$

The eigenvalue $\lambda_1 = 0$ corresponds to a time-independent well population. This is the equilibrium distribution

$$n_A = n_B \exp \frac{1}{kT}[U_B - U_A].$$

The remaining negative eigenvalue must then be associated with deviations from equilibrium, and $\exp(-\lambda_2 t)$ gives the rate at which these deviations disappear. The relaxation time $\tau$ is therefore in terms of a quantity $U_0$, which is the average of $U_A$ and $U_B$

$$\frac{1}{\tau} = \lambda_2 = \nu \exp[-(U-U_0)/kT \cdot \{\exp[-(U_0-U_A)kT] + \exp[(U_0-U_B)kT]\}. \tag{5.2}$$

The quantity $U_0$ in Eq. (5.2) cancels out, therefore the validity of Eq. (5.2) does not depend on the definition of $U_0$. Letting $\Delta = \frac{1}{2}(U_A - U_B)$, Eq. (5.2) then becomes

$$\frac{1}{\tau} = 2\nu \exp[-(U-U_0)/kT] \cosh \Delta/kT. \tag{5.3}$$

To first order in the switching force which causes $U_A$ and $U_B$ to differ, $(U-U_0)$ will remain unaffected, and therefore Eq. (5.3) can be written

$$\frac{1}{\tau} = \frac{1}{\tau_0} \cosh \Delta/kT, \tag{5.4}$$

where $\tau_0$ is the relaxation time for the symmetrical potential well, when $\Delta = 0$. This equation demonstrates that the device is usable. The relaxation time $\tau_0$ is the length of time required by the bistable device to thermalize, and represents the maximum time over which the device is usable. $\tau$ on the other hand is the minimum switching time. Cosh $\Delta/kT$ therefore represents the maximum number of switching events in the lifetime of the information. Since this can be large, the device can be useful. Even if $\Delta$ is large enough so that the first-order approximation needed to keep $U-U_0$ constant breaks down, the exponential dependence of cosh $\Delta/kT$ on $\Delta$, in Eq. (5.3) will far outweigh the changes in $\exp[(U-U_0)kT]$, and $\tau_0/\tau$ will still be a rapidly increasing function of $\Delta$.

Note that $\Delta$ is one-half the energy which will be dissipated in the switching process. The thermal probability distribution within each well will be about the same before and after switching, the only difference is that the final well is $2\Delta$ lower than the initial well. This energy difference is dissipated and corresponds to the one-half hysteresis loop area energy loss generally associated with switching. Equation (5.4) therefore confirms the empirically well-known fact that increases in switching speed can only be accomplished at the expense of increased dissipation per switching event. Equation (5.4) is, however, true only for a special model and has no really general significance. To show this consider an alternative model. Let us assume that information is stored by the position of a particle along a line, and that $x = \pm a$ correspond to ZERO and ONE, respectively. No barrier is assumed to exist, but the random diffusive motion of the particle is taken to be slow enough, so that positions will be preserved for an appreciable length of time. (This model is probably closer to the behavior of ferrites and ferroelectrics, when the switching occurs by domain wall motion, than our preceding bistable well model. The energy differences between a completely switched and a partially switched ferrite are rather small and it is the existence of a low domain-wall mobility which keeps the particle near its initial state, in the absence of switching forces, and this initial state can almost equally well be a partially switched state, as a completely switched one. On the other hand if one examines the domain wall mobility on a sufficiently microscopic scale it is likely to be related again to activated motion past barriers.) In that case, particles will diffuse a typical distance $s$ in a time $\tau \sim s^2/2D$. $D$ is the diffusion constant. The distance which corresponds to information loss is $s \sim a$, the associated relaxation time is $\tau_0 \sim a^2/2D$. In the presence of a force $F$ the particle moves with a velocity $\mu F$, where the mobility $\mu$ is given by the Einstein relation as $D/kT$. To move a particle under a switching force $F$ through a distance $2a$ requires a time $\tau_s$ given by

$$\mu F \tau_s = 2a, \tag{5.5}$$

or

$$\tau_s = 2a/\mu F. \tag{5.6}$$

The energy dissipation $2\Delta$, is a 2aF. This gives us the equations

$$\tau_s = 2a^2/\mu\Delta. \tag{5.7}$$

$$\tau_s/\tau_0 = 4kT/\Delta, \tag{5.8}$$

which show the same direction of variation of $\tau_s$ with $\Delta$ as in the case with the barrier, but do not involve an exponential variation with $\Delta/kT$. If all other considerations are ignored it is clear that the energy bistable element of Eq. (5.4) is much to be preferred to the diffusion stabilized element of Eq. (5.8).

The above examples give us some insight into the need for energy dissipation, not directly provided by the arguments involving entropy consideration. In the RESTORE TO ONE operation we want the system to settle into the ONE state regardless of its initial state. We do this by lowering the energy of the ONE state relative to the ZERO state. The particle will then go to this lowest state, and on the way dissipate any excess energy it may have had in its initial state.

## 6. Three sources of error

We shall in this section attempt to survey the relative importance of several possible sources of error in the computing process, all intimately connected with our preceding considerations. First of all the actual time allowed for switching is finite and the relaxation to the desired state will not have taken place completely. If $T_s$ is the actual time during which the switching force is applied and $\tau_s$ is the relaxation time of Eq. (5.4) then $\exp(-T_s/\tau_s)$ is the probability that the switching will not have taken place. The second source of error is the one considered in detail in an earlier paper by J. A. Swanson,[5] and represents the fact that $\tau_0$ is finite and information will decay while it is supposed to be sitting quietly in its initial state. The relative importance of these two errors is a matter of design compromises. The time $T_s$, allowed for switching, can always be made longer, thus making the switching relaxation more complete. The total time available for a program is, however, less than $\tau_0$, the relaxation time for stored information, and therefore increasing the time allowed for switching decreases the number of steps in the maximum possible program.

A third source of error consists of the fact that even if the system is allowed to relax completely during switching there would still be a fraction of the ensemble of the order $\exp(-2\Delta/kT)$ left in the unfavored initial state. (Assuming $\Delta \gg kT$.) For the purpose of the subsequent discussion let us call this Boltzmann error. We shall show that no matter how the design compromise between the first two kinds of errors is made, Boltzmann error will never be dominant. We shall compare the errors in a rough fashion, without becoming involved in an enumeration of the various possible exact histories of information.

To carry out this analysis, we shall overestimate Boltzmann error by assuming that switching has occurred in every machine cycle in the history of every bit. It is this upper bound on the Boltzmann error which will be shown to be negligible, when compared to other errors. The Boltzmann error probability, per switching event is $\exp(-2\Delta/kT)$. During the same switching time bits which are not being switched are decaying away at the rate $\exp(-t/\tau_0)$. In the switching time $T_s$, therefore, unswitched bits have a probability $T_s/\tau_0$ of losing their information. If the Boltzmann error is to be dominant

$$T_s/\tau_0 < \exp(-2\Delta/kT) \,. \tag{6.1}$$

Let us specialize to the bistable well of Eq. (5.4). This latter equation takes (6.1) into the form

$$\frac{2T_s}{\tau_s}\exp(-\Delta/kT) < \exp(-2\Delta/kT) \,, \tag{6.2}$$

or equivalently

$$\frac{T_s}{\tau_s} < \tfrac{1}{2}\exp(-\Delta/kT) \,. \tag{6.3}$$

Now consider the relaxation to the switched state. The error incurred due to incomplete relaxation is $\exp(-T_s/\tau_s)$, which according to Eq. (6.3) satisfies

$$\exp(-T_s/\tau_s) > \exp[-\tfrac{1}{2}\exp(-\Delta/kT)] \,. \tag{6.4}$$

The right-hand side of this inequality has as its argument $\frac{1}{2}\exp(-\Delta/kT)$ which is less than $\frac{1}{2}$. Therefore the right-hand side is large compared to $\exp(-2\Delta/kT)$, the Boltzmann error, whose exponent is certainly larger than unity. We have thus shown that if the Boltzmann error dominates over the information decay, it must in turn be dominated by the incomplete relaxation during switching.

A somewhat alternate way of arguing the same point consists in showing that the accumulated Boltzmann error, due to the maximum number of switching events permitted by Eq. (5.4), is small compared to unity.

Consider now, instead, the diffusion stabilized element of Eq. (5.8). For it, we can find instead of Eq. (6.4) the relationship

$$\exp(-T_s/\tau_s) > \exp[(-\Delta/4kT)\exp(-2\Delta/kT)] \,, \tag{6.5}$$

and the right-hand side is again large compared to the Boltzmann error, $\exp(-2\Delta/kT)$. The alternative argument in terms of the accumulated Boltzmann error exists also in this case.

When we attempt to consider a more realistic machine model, in which switching forces are applied to coupled devices, as is done for example in diodeless magnetic core logic,[4] it becomes difficult to maintain analytically a clean-cut breakdown of error types, as we have done here. Nevertheless we believe that there is still a somewhat similar separation which is manifested.

## Summary

The information-bearing degrees of freedom of a computer interact with the thermal reservoir represented by the remaining degrees of freedom. This interaction plays two roles. First of all, it acts as a sink for the energy dissipation involved in the computation. This energy dissipation has an unavoidable minimum arising from the fact that the computer performs irreversible operations. Secondly, the interaction acts as a source of noise causing errors. In particular thermal fluctuations give a supposedly switched element a small probability of remaining in its initial state, even after the switching force has been applied for a long time. It is shown, in terms of two simple models, that this source of error is dominated by one of two other error sources:

1) Incomplete switching due to inadequate time allowed for switching.

2) Decay of stored information due to thermal fluctuations.

It is, of course, apparent that both the thermal noise and the requirements for energy dissipation are on a scale which is entirely negligible in present-day computer components. The dissipation as calculated, however, is an absolute minimum. Actual devices which are far from minimal in size and operate at high speeds will be likely to require a much larger energy dissipation to serve the purpose of erasing the unnecessary details of the computer's past history.

## Acknowledgments

Some of these questions were first posed by E. R. Piore a number of years ago. In its early stages[2-5] this project was carried forward primarily by the late John Swanson. Conversations with Gordon Lasher were essential to the development of the ideas presented in the paper.

**References**

1. L. Brillouin, *Science and Information Theory*, Academic Press Inc., New York, New York, 1956.

2. R. Landauer and J. A. Swanson, *Phys. Rev.*, **121**, 1668 (1961).

3. K. Mendelssohn, *Progress in Cyrogenics*, Vol. I, Academic Press Inc., New York, New York, 1959. Chapter I by D. R. Young, p. 1.

4. L. B. Russell, *IRE Convention Record*, p. 106 (1957).

5. J. A. Swanson, *IBM Journal*, **4**, 305 (1960).
We would like to take this opportunity to amplify two points in Swanson's paper which perhaps were not adequately stressed in the published version.

(1) The large number of particles ($\sim 100$) in the optimum element are a result of the small energies per particle (or cell) involved in the typical cooperative phenomenon used in computer storage. There is no question that information can be stored in the position of a single particle, at room temperature, if the activation energy for its motion is sufficiently large ($\sim$ several electron volts).

(2) Swanson's optimum volume is, generally, not very different from the common sense requirement on $U$, namely: $\nu t \exp(-U/kT) \ll 1$, which would be found without the use of information theory. This indicates that the use of redundancy and complicated coding methods does not permit much additional information to be stored. It is obviously preferable to eliminate these complications, since by making each element only slightly larger than the "optimum" value, the element becomes reliable enough to carry information without the use of redundancy.

6. R. L. Wigington, *Proceedings of the IRE*, **47**, 516 (1959).

7. A. W. Lo, Paper to appear in *IRE Transactions on Electronic Computers*.

8. D. Hilbert and W. Ackermann, *Principles of Mathematical Logic*, Chelsea Publishing Co., New York, 1950, p. 10.

*Received October 5, 1960*

# Entropy of measurement and erasure: Szilard's membrane model revisited

Harvey S. Leff[a)]
*Physics Department, California State Polytechnic University, Pomona, 3801 West Temple Avenue, Pomona, California 91768*

Andrew F. Rex[b)]
*Physics Department, University of Puget Sound, Tacoma, Washington 98416*

(Received 16 February 1994; accepted 18 May 1994)

It is widely believed that measurement is accompanied by irreversible entropy increase. This conventional wisdom is based in part on Szilard's 1929 study of entropy decrease in a thermodynamic system by intelligent intervention (i.e., a Maxwell's demon) and Brillouin's association of entropy with information. Bennett subsequently argued that information *acquisition* is not necessarily irreversible, but information *erasure* must be dissipative (Landauer's principle). Inspired by the ensuing debate, we revisit the membrane model introduced by Szilard and find that it can illustrate and clarify (1) reversible measurement, (2) information storage, (3) decoupling of the memory from the system being measured, and (4) entropy increase associated with memory erasure and resetting.

## I. INTRODUCTION

In his *Theory of Heat*, Maxwell proposed a thought experiment designed to illustrate that the second law of thermodynamics is statistical.[1] The chief character in that idea became known as Maxwell's demon.[2,3] For approximately 120 years the puzzle of whether a Maxwell's demon can or cannot violate the second law has been studied and debated.[4] This rich history inspired a resource letter on Maxwell's demon in this journal.[5] A resolution based largely on the work of Szilard[6] and Brillouin[7,8] gained wide acceptance: *A demon must gather information in order to operate, and information acquisition is sufficiently dissipative to "save" the second law.* Following the widespread acceptance of this reasoning, it appeared that Maxwell's demon was "dead."

Thirty-one years after Brillouin's work a very different solution to the Maxwell's demon puzzle was proposed by Bennett.[9] Asserting that information acquisition is not necessarily dissipative, Bennett found Brillouin's resolution to be incomplete. Drawing upon the seminal work of Landauer on energy limits in the computational process,[10] he proposed an alternative resolution of the puzzle using the notion that a demon requires a memory and must clear that memory periodically. According to Bennett the second law is saved by the fact that erasing and resetting a memory is a dissipative operation.

The importance of memory had been discussed in 1929 by Szilard.[6] He introduced the idea of measuring a system parameter $x$ and recording the measurement's result via a corresponding value $y$ in a memory register. He wrote, "Then let $x$ and $y$ be uncoupled after the measurement, so that $x$ can change, while $y$ retains its value for some time. Such measurements are not harmless interventions. A system in which such measurements occur shows a sort of memory faculty, in the sense that one can recognize by the state parameter $y$ what value another state parameter $x$ had at an earlier moment, and we shall see that simply because of such a memory the second law would be violated, if the measurement could take place without compensation."

Szilard identified both "intervention" and "memory" as key ingredients, providing seeds for Brillouin's and Bennett's subsequent work. One of Szilard's three examples, which involved a heat engine with a one-molecule working fluid, became quite popular and was examined further by a variety of researchers. Expounding on that model in his book *Foundations of Statistical Mechanics*, Penrose[11] noted that memory erasure is dissipative. He wrote, "The large number of distinct observational states that the Maxwell demon must have in order to make significant entropy reductions possible may be thought of as a large memory capacity in which the demon stores the information about the system which he acquires as he works reducing its entropy. As soon as the demon's memory is completely filled, however, ... he can achieve no further reduction of the Boltzmann entropy. He gains nothing, for example, by deliberately forgetting or erasing some of his stored information in order to make more memory capacity available; for the erasure, being a setting process, itself increases the entropy by an amount at least as great as the entropy decrease made possible by the newly available memory capacity."

Bennett's subsequent ideas regarding irreversible memory erasure and also the possibility of reversible information acquisition were developed independently, and have received enthusiastic support.[12–14] Unfortunately, a difficulty with Bennett's thesis is that relatively few known thermodynamical models clearly illustrate nondissipative information acquisition. Identification of unambiguous examples would help resolve criticisms[4] of the subtle points involved.

Our purpose here is to re-examine a simple, tractable membrane model introduced by Szilard in 1929 along with his one-molecule gas model. In contrast with the latter model, which has been discussed widely, the membrane model has received little attention. We show that a simple modification of his analysis can shed light on the issue of information acquisition and erasure. The modified membrane model is related to the Maxwell's demon puzzle in the sense that it takes a working fluid through a cyclic process during which: (a) physical measurements (which are *reversible* in this case) are made, (b) the results are *stored* in a memory register, (c) the memory is then *decoupled* from the working fluid, and (d) the memory register is erased and reset *irreversibly*.

Szilard's membrane model differs from the Maxwell's demon problem in that it does not involve conversion of heat to work. Although it is too specialized to allow *general* conclusions, it clearly illustrates the central points that link the efforts of Szilard, Landauer, and Bennett. In Secs. II and III the original form of the model is revisited, and two questionable parts of Szilard's analysis are scrutinized. In Sec. IV we introduce a simplified variant of Szilard's model that succinctly illustrates information acquisition, storage, and erasure. Section V contains a discussion of our main results.

## II. SZILARD'S MEMBRANE MODEL

The Szilard membrane model[6] consists of an N-molecule dilute gas contained in a volume $V$. This system is in thermal contact with an energy reservoir at temperature $T$. The gas molecules can exist in either of two forms, labeled "1" and "2," and transitions $1 \leftrightarrow 2$ can be induced over relatively long time intervals (relative to typical observation times) via interactions with neighboring gas molecules. A real physical example is provided by molecular hydrogen, which can exist as *ortho*-hydrogen, with total nuclear spin$=1$ (and $z$ components $-1$, 0, $+1$) or as *para*-hydrogen, with total nuclear spin$=0$. Equilibration times for hydrogen are on the order of years. This example was mentioned in Ehrenberg's 1967 Scientific American article on Maxwell's demon.[15]

Interactions between species 1 and 2 are assumed to be sufficiently weak that the fluid behaves as a mixture of two ideal gases. The internal energy of the gas can be written

$$U_{\text{gas}} = U_1(T) + U_2(T), \tag{1a}$$

where each term is of the form,

$$U_i(T) = N_i e_i(T) \quad \text{for } i = 1,2. \tag{1b}$$

$e_i(T)$ is the average energy per molecule for species $i$, $N_i$ is the molecule number for this species, and $N_1 + N_2 = N$. In thermodynamic equilibrium the average species fractions are $w_1(T)$ and $w_2(T)$, with $w_1(T) + w_2(T) = 1$, and $0 < w_i(T) < 1$ for $i = 1, 2$, i.e.,

$$N_i \equiv N_i(T) \equiv w_i(T) N \tag{2}$$

for $i = 1, 2$.

The gas entropy is

$$S_{\text{gas}} = S_1 + S_2, \tag{3}$$

where each single-component entropy has the ideal gas form[16]

$$S_i = N_i k \ln[f_i(T) V / N_i] \quad \text{for } i = 1,2. \tag{4}$$

$f_1$ and $f_2$ can differ from one another (as can $e_1$ and $e_2$) but they are functions only of $T$, and the argument of the logarithm is dimensionless. Szilard did not write down Eqs. (1), (2), and (4) explicitly, but they are very useful. In what follows we adopt the shorthand notation $w_1 = w_1(T)$ and $w_2 = w_2(T)$.

The container is assumed to have two telescoping sections, $AA'$ and $B'B$ as illustrated in Fig. 1. Initially, the sections are telescoped together [Fig. 1(a)] giving the (minimum possible) volume $V$, the gas is in thermal equilibrium, and $N_i = w_i N$ for $i = 1, 2$. End piece $A'$ of section $AA'$ is a semipermeable membrane that passes only species 1 molecules; and end piece $B'$ of $B'B$ is permeable only to species 2 molecules. Although *real* semipermeable membranes are not perfect sorters, Szilard assumed *ideal* membranes that are. We retain this assumption throughout this article. Szilard prescribed a four-step cyclic process that entails information acquisition, storage of information in a memory register, decoupling of the system and memory register, and erasure of the information. In Sec. III we describe his steps, provide some details omitted from his paper, and suggest two modifications to make his analysis more useful.

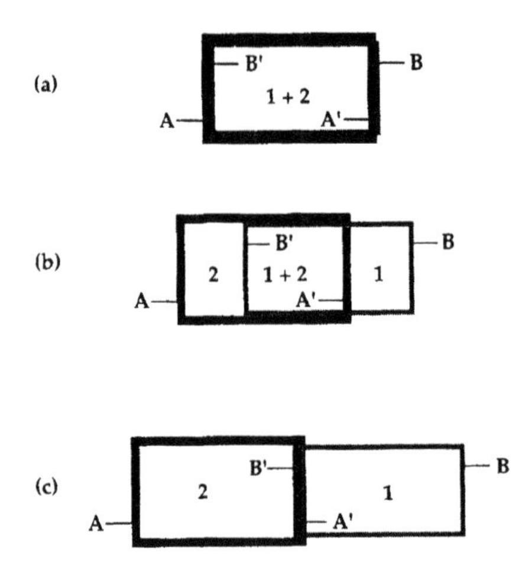


Fig. 1. (a) Gas species 1 and 2 before step 2. (b) Two containers telescoping outward during step 2, with membrane $B'$ permeable only to species 2 and membrane $A'$ permeable only to species 1. (c) The configuration upon completion of step 2, which separates species 1 and 2.

## III. SZILARD'S CYCLIC PROCESS

### A. Step 1: Measurement

Each molecule of type 1 is assigned an $x$ value of 1, and each one of type 2 is assigned an $x$ value of 2. Szilard envisioned "intervention" by an intelligent being, who measures the $x$ values of each molecule *at some initial time*, and assigns the value $y = 1$ to all molecules with $x = 1$ and the value $y = 2$ to all molecules with $x = 2$. The set $\{y\}$ is remembered by the intelligent being, and does not change even if a molecule is later transformed by interactions with other molecules and/or with the container walls, from type 1 to type 2 or *vice versa*. In such events the molecule's $x$ value changes but its $y$ value does not. The main point is: Each molecule's $y$ value represents its *initial* species type.

It is assumed that these measurements of the $y$ variables do not change the thermodynamic state of the gas; i.e., the entropy change of the gas is

$$[\Delta S_{\text{gas}}]_{\text{step 1}} = 0. \tag{5a}$$

On the other hand, Szilard proposed that the intervention needed to determine the set $\{y\}$ generates a positive entropy change,

$$[\Delta S_{\text{intervention}}]_{\text{step 1}} > 0. \tag{5b}$$

Szilard did not discuss the mechanism for this entropy increase, but assumed that there were no other entropy changes in the environment (the parts of the universe other than the gas and intervention mechanism); i.e.,

$$[\Delta S_{\text{env}}]_{\text{step 1}} = 0. \tag{5c}$$

Given Eqs. (5a) and (5c), evidently the intervention entropy in Eq. (5b) must be an entropy increase of the intervening intelligent being itself.

### B. Step 2: Separation

After the measurement intervention, the two container sections are telescoped outward, leading from the configuration in Fig. 1(a) to that in Fig. 1(c). The total volume increases from $V$ to $2V$. We assume this is done slowly enough to keep the process quasistatic, but fast enough that $N_1$ and $N_2$ are unchanged. Because of the semipermeable membranes $A$ and $B$, all type-1 molecules are moved into the resulting right chamber and all type-2 molecules go into the left chamber [Fig. 1(c)]. The ideal membranes are assumed to sort perfectly and to do zero work on the species they transmit. Zero *net* external work is done on *both* species because $A$ and $A'$ do equal and opposite work on species 2 and $B'$ and $B$ do equal and opposite work on species 1.

The internal energy of each component is unchanged because $U_i = w_i N e_i(T)$ for $i=1, 2$; the temperature $T$ does not vary; and during the time interval of step 2, neither do $N_1$ and $N_2$. Thus, there is zero net heat transfer between the gas and reservoir. Telescoping is a reversible process, so inward telescoping will recover the original thermodynamic state as long as $N_1$ and $N_2$ do not vary. It follows from Eqs. (3) and (4) that

$$[\Delta S_{\text{gas}}]_{\text{step 2}} = 0. \tag{6a}$$

Szilard regarded both the intervention mechanism and the environment as having no thermodynamic involvement during separation, i.e.,

$$[\Delta S_{\text{intervention}}]_{\text{step 2}} = 0, \tag{6b}$$

and

$$[\Delta S_{\text{env}}]_{\text{step 2}} = 0, \tag{6c}$$

though he did not show all these entropy changes explicitly.

### C. Step 3: Equilibration

After separation, neither of the gases in the two storage chambers is in thermal equilibrium. That is, the $\{x\}$ set in each chamber changes over time while the $\{y\}$ set remains fixed. Specifically, just after separation, $N_1 = w_1 N$ and $N_2=0$ in the right chamber, and $N_2 = w_2 N$ and $N_1 = 0$ in the left chamber. After a sufficiently long time interval, $(N_1)_{\text{right}} \to w_1^2 N$ and $(N_2)_{\text{right}} \to w_2 w_1 N$, while $(N_1)_{\text{left}} \to w_1 w_2 N$ and $(N_2)_{\text{left}} \to w_2^2 N$. This equilibration constitutes the third step of the cycle.

Szilard assumed that equilibration occurs at constant temperature, and thus the internal energy per molecule is unchanged for each species. Further, because the overall molecular fractions of the species do not change during equilibration, neither do $U_1$ and $U_2$ even though the energy in each chamber changes (if $e_1 \neq e_2$). Therefore after equilibration, $(N_1)_{\text{total}} = (N_1)_{\text{left}} + (N_1)_{\text{right}} \to w_1 w_2 N + w_1^2 N = w_1 N$, $(N_2)_{\text{total}} = (N_2)_{\text{left}} + (N_2)_{\text{right}} \to w_2^2 N + w_1 w_2 N = w_2 N$, and $U_i = w_i N e_i(T)$ for $i=1, 2$. Equation (4) implies that during equilibration the entropy change in the left chamber is $\Delta S_{\text{left equil}} = w_2 Nk[w_1 \ln(f_1/f_2) - w_1 \ln w_1 - w_2 \ln w_2]$; the entropy change in the right chamber is $\Delta S_{\text{right equil}} = w_1 Nk[-w_2 \ln(f_1/f_2) - w_1 \ln w_1 - w_2 \ln w_2]$. Adding these gives

$$[\Delta S_{\text{gas}}]_{\text{step 3}} = -Nk\{w_1 \ln w_1 + w_2 \ln w_2\} > 0. \tag{7a}$$

Because the intervention mechanism is not involved,

$$[\Delta S_{\text{intervention}}]_{\text{step 3}} = 0. \tag{7b}$$

Szilard wrote, "If we allow this equilibrium to be achieved in both containers independently and at constant volume and temperature, then the entropy of the system certainly has increased." He went on to say, "If we accomplish the equilibrium distribution in both containers in a reversible fashion then the entropy of the rest of the world will decrease by the same amount." Accordingly, he found an entropy decrease for the environment of

$$[\Delta S_{\text{env}}]_{\text{step 3}} = -[\Delta S_{\text{gas}}]_{\text{step 3}} = Nk\{w_1 \ln w_1 + w_2 \ln w_2\} < 0. \tag{7c}$$

Szilard wrote down an equation equivalent to Eq. (7c), but his assumption of reversibility is curious, for it is difficult to envision reversal of the equilibration process. We return to this point later.

### D. Step 4: Cycle completion

Seeking a zero-work process to remix the gases in a single volume $V$, Szilard specified that in step 4, membranes $A'$ and $B'$ are to be replaced with ones permeable only to molecules with $y=1$ and $y=2$, respectively. This is an interesting concept, for it requires an ability to sort molecules based upon remembrance of a previous physical state. It transcends the abilities of real membranes, which sort on the basis of the *present* molecular states. One might interpret the required Szilard membranes to be collections of gates that are operated by the intervener of step 1, who remembers the $y$ value for each molecule. With this view we cautiously follow Szilard's lead.

With the $y$-selector membranes in place, the sections are telescoped inward, reducing the total volume from $2V$ to $V$ isothermally, doing zero work (as before) in the process. Equation (2) implies zero internal energy change for the gas because $N_1$ and $N_2$ do not change, and $T$ is constant. Equations (3) and (4) imply the entropy change

$$[\Delta S_{\text{gas}}]_{\text{step 4}} = +Nk\{w_1 \ln w_1 + w_2 \ln w_2\} < 0. \tag{8a}$$

Because the intervention mechanism is used here, it appears that

$$[\Delta S_{\text{intervention}}]_{\text{step 4}} > -Nk\{w_1 \ln w_1 + w_2 \ln w_2\}. \tag{8b}$$

If, as implied by Szilard's discussion, the entropy of the environment is unaltered,

$$[\Delta S_{\text{env}}]_{\text{step 4}} = 0. \tag{8c}$$

Szilard expounded on what step 4 accomplishes: "The distribution of the $y$-coordinate ... now has become statistically independent of the $x$-values and besides we are able to reestablish the original distribution ... (of the $y$ values). Thus we would have gone through a complete cycle. The only change that we have to register is the resulting decrease of entropy ... [given by Eq. (8a)]." Szilard seemed to be acknowledging two things: a cyclic process for the working fluid *and* memory erasure—namely, re-establishment of the original $y$ distribution.

### E. Implications of the whole cycle

Szilard utilized the fact that for the gas cycle, $[\Delta S_{\text{gas}}]_{\text{cycle}} = 0$, and the net entropy change of the environment is the sum of Eqs. (5c), (6c), (7c), and (8c):

$$[\Delta S_{\text{env}}]_{\text{cycle}} = Nk\{w_1 \ln w_1 + w_2 \ln w_2\} < 0. \tag{9a}$$

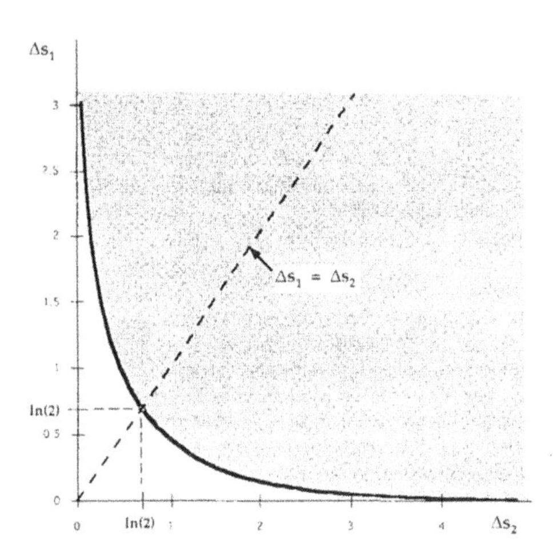


Fig. 2. A graphical depiction of relation (12), $\exp(-\Delta s_1/k)+\exp(-\Delta s_2/k)\leq 1$. For any $\Delta s_2$, the value of $\Delta s_1$ must lie above the solid curve, in the shaded region.

Recalling that Szilard did not account for intervention in step 4, Eqs. (5b), (6b), (7b), and (9a) along with the second law imply the total entropy change,

$$[\Delta S_{\text{intervention}}]_{\text{step 1}}+[\Delta S_{\text{env}}]_{\text{cycle}}$$
$$=[\Delta S_{\text{intervention}}]_{\text{step 1}}+Nk\{w_1 \ln w_1+w_2 \ln w_2\}\geq 0. \tag{9b}$$

This is his first major result for the membrane model.

Szilard then assumed that the intervention entropy could be written

$$[\Delta S_{\text{intervention}}]_{\text{step 1}}=N_1\Delta s_1+N_2\Delta s_2, \tag{10}$$

where $\Delta s_i$, the entropy change associated with the measurement of a species $i$ molecule, is independent of the probabilities $w_1$ and $w_2$. Combining Eqs. (9b) and (10),

$$w_1\Delta s_1+w_2\Delta s_2+k\{w_1 \ln w_1+w_2 \ln w_2\}\geq 0. \tag{11}$$

Minimizing the left-hand side with respect to $w_1(=1-w_2)$ for constant $\Delta s_1$ and $\Delta s_2$ gives[17]

$$\exp(-\Delta s_1/k)+\exp(-\Delta s_2/k)\leq 1, \tag{12}$$

which is Szilard's second major result. The inequality (12) means that the intervention entropy pairs $(\Delta s_1,\Delta s_2)$ must lie in the shaded region of Fig. 2.

The above description of Szilard's membrane model suggests two places where the analysis can be improved. The first is in step 3, where Szilard used the thermodynamic reversibility condition $\Delta S_{\text{gas}}+\Delta S_{\text{env}}=0$. There is no compelling reason to impose this reversibility condition because the irreversible equilibration does not require any entropy change in the environment. The analysis can be improved by replacing the reversibility assumption (7c) by the condition,

$$[\Delta S_{\text{env}}]_{\text{step 3}}=0. \tag{13}$$

Second, Szilard's omission of the intervention device in step 4 is questionable. The gas mixtures in the right and left chambers must be combined into the final volume $V$, incurring the negative entropy change for the gas given by Eq. (8a), which was *not* shown in Szilard's paper. To satisfy the second law a compensating entropy increase must exist somewhere. This cannot be in the environment because: $N_1$ and $N_2$ are unaltered, the process is isothermal, there is zero change in the internal energies $U_1$ and $U_2$, zero work is done on the gas, and thus the first law of thermodynamics implies zero heat transfer to the environment; i.e., Eq. (8c) holds. Evidently step 4 cannot be accomplished without an entropy increase associated with intervention, as given in Eq. (8b).

For completeness, we observe that one might imagine attempting the inward telescoping in step 4 using membranes $A'$ and $B'$ that are one-way filters, which allow molecules to cross membrane $B'$ from left to right only and membrane $A'$ from right to left only. However, these would presumably be subject to thermal fluctuations associated with Smoluchowski-type trapdoors, enabling molecules to actually move in both directions.[18,19] This again supports the conclusion that the use of the intervener's memory is unavoidable for Szilard's step 4, and that the entropy increase in Eq. (8b) is unavoidable.

Incorporating our two improvements, we replace Eq. (7c) with Eq. (13), add the entropy changes in parts (a), (b), and (c) of Eqs. (5)–(8), and apply the second law to get

$$[\Delta S_{\text{intervention}}]_{\text{step 1}}+[\Delta S_{\text{intervention}}]_{\text{step 4}}$$
$$>-Nk\{w_1 \ln w_1+w_2 \ln w_2\}. \tag{14}$$

The entropy increase in steps 1 and 4 can be understood as follows. In step 1, molecules are examined by the Maxwell's demon and their $y$ values are recorded in a memory. In step 4, as molecules approach the set of gates described earlier, the Maxwell's demon must again examine each molecule and check against its memory to see if its $y$ variable is 1 or 2. This again entails recording results in a memory, even if only temporarily.

These acts increase the entropy of the demon's memory as its memory addresses take on values other than their initial standardized values (say all zeros). In essence the memory goes from all zeros to a mixture of zeros and ones. If we repeat this experiment with many such similar systems and memories, the resulting ensemble of memories will have a nonzero statistical entropy. Presumably, this entropy increases with the number of measurements made until the memories are full and information must be erased in order to continue. We conclude that the results of the cyclic process are: (i) an environment with a lowered entropy, and (ii) a (jumbled) demon's memory whose entropy is higher than it was initially. Erasure of this memory would reduce its entropy to zero (in its standard state) but would send a compensating entropy to the environment in accordance with Landauer's principle.[10]

Our two modifications of Szilard's cycle are intended to clarify how that cycle can be interpreted in terms of our current understanding of measurement, memory, and memory erasure. In Sec. IV, we consider a variant of Szilard's membrane model that illustrates these same features in an even simpler and more direct way.

## IV. VARIATION ON SZILARD'S THEME

Reflection on the Szilard cycle described above suggests the use of an even simpler cycle that clearly illustrates four important elements: reversible measurement, storage in a memory, decoupling of the memory from the system, and

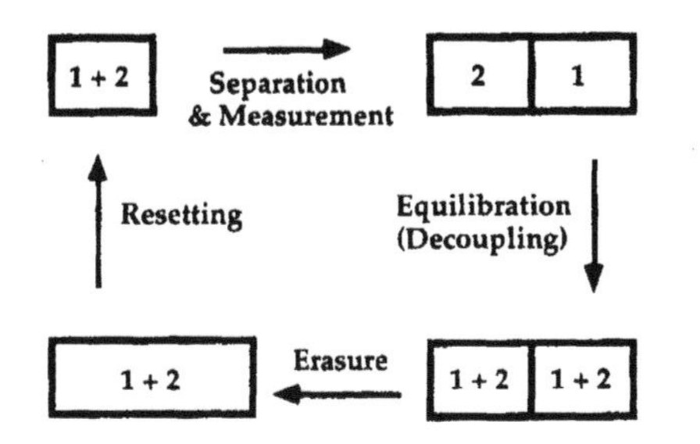


Fig. 3. Modified Szilard membrane cycle, which consists of: separation and measurement (step 2′), equilibration (and decoupling, step 3′), and erasure + resetting (step 4′). Note that after separation the total number of molecules in the left and right chambers are different if $w_1 \neq w_2$.

entropy increase associated with memory erasure and resetting. We describe this variant in terms of the modified steps 1′, 2′, 3′, and 4′, as illustrated in Fig. 3.

### A. Step 1′: Nothing

This is no step at all because we omit intervention by the intelligent being that is specified in Szilard's step 1. This omission is possible because measurement is accomplished in step 2′.

### B. Step 2′: Separation and measurement

This is the same as Szilard's step 2, namely a slow outward telescoping of the containers. However we now *interpret* the membrane actions as constituting "measurements" of the $x$ values of the molecules. That is, if the $i$th molecule passes through $A'$, it has $x_i=1$, and if the $j$th molecule passes through $B'$, it has $x_j=2$. The separation process stores molecules with initial values of $x=1$ and 2 in separate chambers. Together, these chambers can be regarded as a memory register that stores information about the original $x$ variable for each molecule. In essence the separation process results in the measurement of $(x_1,x_2,\ldots,x_N)$ and the correspondence: $(x_1,x_2,\ldots,x_N)\Rightarrow(y_1,y_2,\ldots,y_N)$ during the separation process: If molecule $i$ is in the right (or left) chamber, then $y_i=1$ (or 2). Of course

$$[\Delta S_{\text{gas}}]_{\text{step }2'}=[\Delta S_{\text{env}}]_{\text{step }2'}=0. \tag{15}$$

### C. Step 3′: Equilibration

Equilibration of the species is irreversible, and work can be done by neighboring molecules on one another to induce changes, say, in nuclear spin states. We assume however that zero net work is done on the gas by the container walls during equilibration. The entropy change of the gas in step 3′ is identical with Eq. (7a):

$$[\Delta S_{\text{gas}}]_{\text{step }3'}=-Nk\{w_1 \ln w_1+w_2 \ln w_2\}>0. \tag{16a}$$

Constancy of the internal energy and zero net external work imply zero net heat transfer and zero entropy change for the environment. Therefore, Eq. (7c) must be replaced by the equivalent of Eq. (13),

$$[\Delta S_{\text{env}}]_{\text{step }3'}=0. \tag{16b}$$

### D. Step 4′: Memory erasure and resetting

It is not possible to use a real membrane to bring both gases into the single volume $V$ with a zero external work process. Therefore we define step 4′ to consist of two parts: first, removal of the central partitions $A'$ and $B'$, so that the gases mix in volume $2V$; and second, reversible, isothermal compression of the mixed gas from volume $2V$ to volume $V$. In the first part there is no internal energy change, but there is an entropy change,

$$[\Delta S_{\text{gas}}]_{\text{erasure}}=Nk[w_1 \ln w_1+w_2 \ln w_2+\ln 2], \tag{17a}$$

which is positive if $w_1\neq w_2$ and zero if $w_1=w_2$. Equation (17a) is the entropy of erasure, which can be given a useful interpretation based upon Landauer's[20] view of erasure: the separate phase spaces of the gases in chambers 1 and 2 "diffuse into one another" irreversibly (unless $w_1=w_2$, in which case partition removal gives zero entropy change and partition replacement recovers the original state). During the erasure there is zero entropy change in the environment:

$$[\Delta S_{\text{env}}]_{\text{erasure}}=0. \tag{17b}$$

Resetting the memory by a reversible isothermal compression requires an external work $W>0$ *on* the gas and a concomitant heat transfer $Q<0$ (*to* the gas). The first law of thermodynamics demands that $\Delta U=Q+W=0$, or $Q=-W=NkT\ln 2$, and the corresponding entropy change of the constant-temperature environment is

$$[\Delta S_{\text{env}}]_{\text{resetting}}=-[\Delta S_{\text{gas}}]_{\text{resetting}}=Nk\ln 2. \tag{17c}$$

Adding Eqs. (17a) and (17c) gives the overall entropy change for the gas during step 4′,

$$[\Delta S_{\text{gas}}]_{\text{step }4'}=Nk[w_1 \ln w_1+w_2 \ln w_2]<0. \tag{17d}$$

Adding the entropy changes in Eqs. (17b)–(17d), the total entropy change for the erasure/resetting combination is seen to be

$$\begin{aligned}[\Delta S_{\text{gas+env}}]_{\text{erasure/resetting}}&=[\Delta S_{\text{gas}}]_{\text{step }4'}+[\Delta S_{\text{env}}]_{\text{step }4'}\\&=Nk[w_1 \ln w_1+w_2 \ln w_2+Nk\ln 2]\geqslant 0.\end{aligned} \tag{18}$$

The evident non-negativity of Eq. (18) is consistent with the second law.

### E. Implications of the whole cycle

The foregoing enables verification that $[\Delta S_{\text{gas}}]_{\text{cycle}}=0$ (as it must) and

$$[\Delta S_{\text{env}}]_{\text{cycle}}=Nk\ln 2. \tag{19}$$

If we make the association

$$[\Delta S_{\text{env}}]_{\text{cycle}}\equiv N[w_1\Delta s_1+w_2\Delta s_2]=Nk\ln 2, \tag{20}$$

it is clear that *if* $\Delta s_1=\Delta s_2\equiv\Delta s$, then $\Delta s=k\ln 2$, *independent of the values of* $w_1$ *and* $w_2$. Here $\Delta s_1$ and $\Delta s_2$ should be viewed as the entropies of erasure (per molecule) rather than entropies of measurement. Further if we adopt Szilard's assumption that $\Delta s_1$ and $\Delta s_2$ are in fact independent of $w_1$ and $w_2$, then the *only* possible entropies of erasure are those for which $\Delta s_1=\Delta s_2=k\ln 2$ because in Eq. (20), $w_1\Delta s_1+w_2\Delta s_2=w_1(\Delta s_1-\Delta s_2)+\Delta s_2$ must equal $k\ln 2$ independent of $w_1$. Thus the strict equality in Eq. (12) holds: $\exp(-\Delta s_1/k)+\exp(-\Delta s_2/k)=1$. Evidently this means that the erasure of each $y$ value is accompanied by an entropy increase $k \ln 2$ in the environment. This can be associated with the loss of information available about the (binary) $y$ values.

## V. DISCUSSION

In Sec. IV we have simplified Szilard's membrane model by omitting use of an intelligent being to make measurements, recognizing that the action of the semipermeable membranes effects both reversible measurement and memory storage, and observing that memory erasure and resetting requires work by an external agent. Prior to the erasure step, there exists a record of the initial $y$ variables—namely, the numbers of molecules in the left and right chambers. After erasure no record of the initial $y$ distribution remains.

An important characteristic of the membrane model is the irreversible equilibration that occurs after separation of the two species. This process destroys the correlation between the sets $\{x\}$ and $\{y\}$.[21] We can verify this by examining the joint probability $P(x,y)$ of finding a molecule in species state $x$ and with the memory value $y$, where $x$ and $y$ can take the values 1 or 2. We do this both before and after the equilibration following the separation step. Just after separation and *before* equilibration, there is perfect correlation between the $x$ and $y$ variables on both sides. In the left side, $P_{\text{left}}(x,y) = \delta_{xy}\delta_{y2}$ and on the right side, $P_{\text{right}}(x,y) = \delta_{xy}\delta_{y1}$, where $\delta_{xy}$ is the Kronecker delta, which=1 for $x=y$, and 0 for $x \neq y$. Information entropies associated with these distributions can be defined via $I_q = -k\Sigma_x\Sigma_y P_q(x,y)\ln P_q(x,y)$, where $q$=left or right. Because of the perfect $x-y$ correlation and the restriction that $x=y=(1,2)$ on the (right, left) prior to equilibration, it follows that $I_q=0$ for $q$=left and right just after separation. This means there is zero missing information. The species type of each molecule is fixed, and also known from the chamber (left or right) in which it resides.

After equilibration is complete, the joint probabilities are $P_{\text{left}}(x,y) = w_x\delta_{y2}$ and $P_{\text{right}}(x,y) = w_x\delta_{y1}$. The corresponding information entropies are $I_{\text{left}} = I_{\text{right}} = -k[w_1 \ln w_1 + w_2 \ln w_2]$. This can be interpreted as the entropy per molecule associated with the $x$ and $y$ variables after equilibration. The information entropy differences for step 3 are $\Delta I_{\text{left}} = \Delta I_{\text{right}} = -k[w_1 \ln w_1 + w_2 \ln w_2]$. Notice that $\Delta I_{\text{left}}$ and $\Delta I_{\text{right}}$ are not quite the same as $\Delta S_{\text{left equil}}$ and $\Delta S_{\text{right equil}}$ in the discussion preceding Eq. (7a) for two reasons. First, $\Delta I$ does not include temperature-dependent effects, but is restricted to configurational changes and second, $\Delta I$ is a per-molecule information entropy change. Because the fraction of molecules on the left is $w_2$ and that on the right is $w_1$, we may write the *total* information entropy change associated with the $x$ and $y$ variables as

$$\Delta I_{x-y \text{ equilibration}} = N w_1 \Delta I_{\text{right}} + N w_2 \Delta I_{\text{left}}$$
$$= -Nk[w_1 \ln w_1 + w_2 \ln w_2], \qquad (21)$$

which is identical with Eqs. (7a) and (16a). This equivalence shows that the irreversible equilibration process generates entropy solely because of the destruction of the correlations between the $x$ and $y$ variables. In other words, decoupling the memory from the system is an entropy-producing process that brings uncertainty about the species type of each molecule.

In our discussion of Szilard's original cycle, we observed that the memory gains entropy when information is recorded within it. A similar entropy increase occurs in the memory of a Maxwell's demon as it gathers information. Such entropy increases are best viewed in an ensemble sense: If many similar demons measure many similar systems, the resulting ensemble of memories has an entropy that is greater than the zero initial entropy when each ensemble member's memory was in a known reference state. The modified membrane model is very different because *every* species 1 (or 2) molecule is in the right (or left) chamber after the measurement and separation step. Separation yields no uncertainties other than those associated with fluctuations in the fractions $w_1$ and $w_2$. Indeed we showed above that the entropy change of the gas and environment were both zero during separation.

Although there is something troubling about filling a memory without generating a concomitant entropy change, this situation is only temporary. During the slow process of species equilibration, entropy is generated as the memory variables $\{y\}$ become decoupled from the gas variables $\{x\}$. Of course the memory remains intact while this happens. The entropy of decoupling, Eq. (16a), plays the role of an "entropy of memory" in that it results from the dual process of recording measurements of certain system variables in a memory and then decoupling that memory from those variables. It is certainly *not* an entropy of acquisition because it is generated long after the measurement step. The final resetting process transfers both the entropy of decoupling, Eq. (16a), and the entropy of erasure, Eq. (17a), to the environment. In essence the cycle transfers energy from one part of the environment (work source) to another part (constant-temperature reservoir). Its value lies not in what it accomplishes as a physical device, but in how it can help us learn about information acquisition, storage, and erasure.

We close with a remark about Leo Szilard. It is a tribute to his ingenuity that after 65 years, the clever models he invented are still providing food for thought and tools for understanding.

## ACKNOWLEDGMENT

We thank Rolf Landauer for his helpful comments on a first draft of this article.

[a]E-mail: hsleff@csupomona.edu
[b]E-mail: rex@ups.edu

[1]J. C. Maxwell, *Theory of Heat* (Longmans, London, 1871), Chap. 12.
[2]W. Thomson, "The kinetic theory of the dissipation of energy," Nature **9**, 441–444 (1874). Reprinted in *Lord Kelvin's Mathematical and Physical Papers* (Cambridge University Press, Cambridge, 1911), Vol. 5, pp. 11–20. Reprinted in Ref. 2.
[3]W. Thomson, "The sorting demon of Maxwell," R. Inst. Proc. **9**, 113–114 (1879). Reprinted in *Lord Kelvin's Mathematical and Physical Papers* in Ref. 2, pp. 21–23
[4]H. S. Leff and A. F. Rex, *Maxwell's Demon: Entropy, Information, Computing* (Princeton University Press, Princeton, and Adam Hilger/IOP, Great Britain, 1990).
[5]H. S. Leff and A. F. Rex, "Resource letter: Maxwell's demon," Am. J. Phys. **58**, 201–209 (1990).
[6]L. Szilard, "On the decrease of entropy in a thermodynamic system by the intervention of intelligent beings," Z. Phys. **53**, 840–856 (1929); in B. T. Feld and G. Weiss Szilard, *The Collected Works of Leo Szilard: Scientific Papers* (MIT Press, Cambridge, 1972), pp. 103–129, English translation is by A. Rapoport and M. Knoller; reprinted in J. A. Wheeler and W. H. Zurek, *Quantum Theory and Measurement* (Princeton University Press, Princeton, 1983), pp. 539–548, reprinted also in Ref. 4.
[7]L. Brillouin, "Maxwell's demon cannot operate: Information and entropy. I," J. Appl. Phys. **22**, 334–337 (1951). Reprinted in Ref. 4.
[8]L. Brillouin, *Science and Information Theory* (Academic Press, New York, 1956), Chap. 13.

[9]C. H. Bennett, "The thermodynamics of computation—a review," Int. J. Theor. Phys. **21**, 905–940 (1982). Reprinted in Ref. 4.

[10]R. Landauer, "Irreversibility and heat generation in the computing process," IBM J. Res. Dev. **5**, 183–191 (1961). Reprinted in Ref. 4.

[11]O. Penrose, *Foundations of Statistical Mechanics* (Pergamon Press, Oxford, 1970), pp. 221–238.

[12]R. Landauer, "Information is physical," Phys. Today **44** (5), 23–29 (May 1991).

[13]J. D. Barrow, *The World Within the World* (Oxford University Press, Oxford, 1988), pp. 127–130.

[14]W. H. Zurek, "Maxwell's demon, Szilard's engine and quantum measurements," in G. T. Moore and M. O. Scully, *Frontiers of Nonequilibrium Statistical Physics* (Plenum Press, New York, 1984), pp. 1 and 161. Reprinted in Ref. 4.

[15]W. Ehrenberg, "Maxwell's demon," Sci. Am. **217** (5), 103–110 (November 1967).

[16]This follows from the fact that the canonical partition function for classical ideal gases has the form $Z=[gV(2\pi mkT/h^2)^{3/2}q_{\rm rot}q_{\rm vib}q_{\rm el}]^N/N!$, where $m$ is the molecular mass, $h$ is Planck's constant, $g$ is the spin degeneracy, and the $q$ factors are the single-molecule partition functions for rotational, vibrational, and electronic transitions, respectively. Each $q$ depends only on $T$ and internal properties of the molecules; see for example, D. S. Betts and R. E. Turner, *Introductory Statistical Mechanics* (Addison-Wesley, Reading, MA, 1992), pp. 125–129.

[17]Setting the derivative of the left side of Eq. (11) equal to zero gives $w_2=w_1\exp([\Delta s_1-\Delta s_2]/k)$ and $w_1=1/[1+\exp([\Delta s_1-\Delta s_2]/k)$. Substituting back into Eq. (11) yields, $w_1[\Delta s_1/k+\ln w_1]\times[1+\exp([\Delta s_1-\Delta s_2]/k)]\geqslant 0$, which reduces to Eq. (12).

[18]P. A. Skordos and W. H. Zurek, "Maxwell's demon, rectifiers, and the second law: Computer simulation of Smoluchowski's trapdoor," Am. J. Phys. **60**, 876–882 (1992).

[19]A. F. Rex and R. Larsen, "Entropy and information for an automated Maxwell's demon," in *Workshop on Physics and Computation PhysComp '92* (IEEE Computer Society Press, Los Alamitos, 1993), pp. 93–101.

[20]R. Landauer, "Information is physical," in Ref. 19, pp. 1–4.

[21]A related analysis is given by M. J. Klein, "Note on a problem concerning the Gibbs paradox," Am. J. Phys. **26**, 80–81 (1958). A gas with its nuclei in an excited metastable state and another with the same type nuclei in the ground state, are mixed. As one species decays into the other, the entropy of mixing "disappears" but there is a compensating entropy increase in a constant temperature reservoir. In contrast, our present example unmixes (separates) the gases first and the subsequent equilibration induces new mixing because the single species gases each become two-species gases.

## Heat generation required by information erasure

Kousuke Shizume
*University of Library and Information Science, 1-2 Kasuga, Tsukuba, Ibaraki 305, Japan*
(Received 18 April 1995)

Landauer argued that the erasure of 1 bit of information stored in a memory device requires a minimal heat generation of $k_BT \ln 2$ [IBM J. Res. Dev. **5**, 183 (1961)], but recently several articles have been written to dispute the validity of his argument. In this paper, we deal with a basic model of the memory, that is, a system including a particle making the Brownian motion in a time-dependent potential well, and show that Landauer's claim holds rigorously if the random force acting on the particle is white and Gaussian. Our proof is based on the fact that the analogue of the second law of thermodynamics $dQ \leq k_BTdS$ holds rigorously by virtue of the Fokker-Planck equation, even if the potential is not static. Using the above result, we also discuss the counterargument of Goto *et al.* to Landauer's claim based on the quantum flux parametron.

PACS number(s): 05.40.+j, 05.20.−y, 89.70.+c

### I. INTRODUCTION

In 1961, Landauer discussed the limitation of the efficiency of computers imposed by physical laws [1]. He argued that the erasure of 1 bit of information requires a minimal heat generation of $k_BT \ln 2$ based on the second law of thermodynamics. To be more precise, the argument is as follows. Consider a memory device that can hold one of the two values ONE and ZERO. Physically it is a system which has two stable states. When it is in one of the two states, it can be regarded as holding the value ONE, and when it is in another state, ZERO. The erasure of information stored in the memory means the operation RESTORE TO ONE (RTO), which sets the value to ONE, regardless of its initial value. Physically, RTO forces the system into the state corresponding to ONE regardless of its initial state. Moreover, this operation must not leave any trace of the initial value (in other words, any trace of the initial state) anywhere in the system. Landauer's claim means that RTO is inevitably accompanied by the heat generation of at least $k_BT \ln 2$.

For the sake of concreteness, let us introduce a basic model of a memory [1,2]. It is a binary device in which a particle makes the Brownian motion in a bistable potential well (see Fig. 1). When the particle is in the right-hand-side well, one may regard the device as taking the value ONE, and when it is in the left-hand-side well ZERO. In this model one can perform RTO by varying the shape of the potential well with time so that the particle ends in the right-hand-side well, regardless of its initial position. In this model Landauer's claim means that, in whatever way one varies the shape of the potential well with time, the work dissipated into the environment due to the friction cannot be less than $k_BT \ln 2$.

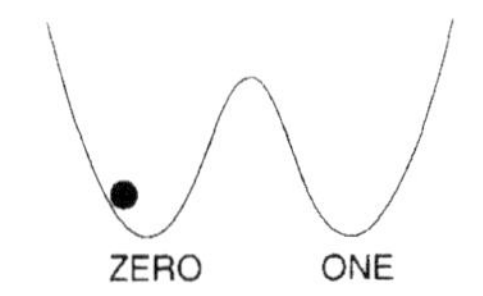


FIG. 1. The model of the memory. The particle is making a Brownian motion in the potential well.

Inspired by his study, a considerable number of studies have been made on the thermodynamics of information processing, which include Maxwell's demon problem [2,3], reversible computation [2,4], the proposal of the algorithmic entropy [5], etc. The thermal cost of more general information processes have also been investigated [6].

On the other hand, however, objections have been raised to his claim. Some authors indicate that his claim is based only on the second law of thermodynamics, and, although plausible, not very rigorous [7,8]. Other authors argued that the information has nothing to do with thermodynamical entropy at all [9]. One of the most interesting counterarguments was advanced by Goto *et al.* [9,10]. They argue that it is possible to erase information with infinitesimal heat generation by using the quantum flux parametron (QFP) [9–13] developed by themselves (see Sec. III). Their counterargument has special importance because one may be able to realize the basic model of the memory introduced above by using the QFP. Although some discussions have appeared thereafter [14,15], they are not quantitative or rigorous.

A major drawback in Landauer's discussion that makes these objections possible is that it is based only on the second law. Although the second law is very general, these objections seems to suggest that applying it to information processing requires more careful consideration. Since Landauer's claim has become a part of the foundations of the thermodynamics of information processing, more rigorous and quantitative discussion is desired.

In this paper we present a sufficient condition for Landauer's claim. Our discussion is based on the Fokker-Planck equation, so that it is less general than Landauer's discussion but more definite and instructive. We show that Landauer's claim holds rigorously in the basic model of the memory introduced above if the random force acting on the particle is a white and Gaussian noise force.

More specifically, we show rigorously that, no matter how one changes the potential well with time to perform RTO, the average of the total heat generated during the operation will be greater than $k_B T \ln 2$ if the random force acting on the particle is white and Gaussian. In our proof, which is given in Sec. II, we *do not presume* the second law of thermodynamics as Landauer did. Because of our assumption about the nature of the random force, the time dependence of the distribution function of the particle is described by the Fokker-Planck equation (FPE) [16] as is well known. We show that the analogue of the second law of thermodynamics $dQ \leq k_B T dS$ *is derived from the FPE, even if the potential is time dependent.* The minimal heat generation caused by RTO can be obtained as a direct result of the analogue of the second law.

Based on this result, Goto *et al.*'s counterargument is discussed in Sec. III and it is shown that, if the behavior of the Josephson junction used in the QFP can be described by the FPE, it is impossible to erase information with infinitesimal heat generation even by using the QFP. We also comment on some other discussions relating to Landauer's claim. We summarize our main results in Sec. IV. We use the units $k_B = 1$ below.

## II. THE MINIMUM HEAT GENERATION

We study the model of the memory involving a particle making the Brownian motion in a time dependent potential well. The motion of the particle is described by the Langevin equation [16]

$$m\frac{d^2x}{dt^2} + m\gamma\frac{dx}{dt} = -\frac{\partial V(x,t)}{\partial x} + F_R(t), \tag{2.1}$$

where $m$ is the mass of the particle, $\gamma$ is the friction constant, $V(x,t)$ is the potential that is assumed to be time dependent, and $F_R(t)$ is the thermal random force. We assume the random force $F_R(t)$ to be a white and Gaussian noise force satisfying

$$\langle F_R(t_1)F_R(t_2)\rangle = 2m\gamma T\delta(t_1 - t_2), \tag{2.2}$$

where $T$ is the temperature of the environment of the memory. Because of this assumption, the motion of the distribution function $f(x,u,t)$ of the particle in position and velocity space is described by the FPE [16]

$$\frac{\partial}{\partial t}f(x,u,t) = \left[-\frac{\partial}{\partial x}u + \frac{\partial}{\partial u}\left(\gamma u + \frac{1}{m}\frac{\partial V(x,t)}{\partial x}\right) + \frac{\gamma T}{m}\frac{\partial^2}{\partial u^2}\right]f(x,u,t). \tag{2.3}$$

If the potential $V$ is static, the well-known $H$ theorem is derived from (2.3), and the analog of the second law of thermodynamics,

$$dQ \leq TdS, \tag{2.4}$$

is derived based on the $H$ theorem. [The exact meaning of Eq. (2.4) is given by Eq. (2.6) below.] Moreover, several important thermodynamic concepts are incorporated into the stochastic process framework [17,18]. In our case, however, the $H$ theorem *does not hold* because $V$ is time dependent. (It is easy to show it. One can change the distribution $f$ cyclically by changing the potential $V$ cyclically. In this case any functional of $f$ such as $H$ should also change cyclically, so that it should increase at some period.) Nevertheless Eq. (2.4) is valid even in this case. In fact, letting $\dot{Q}$ be the ensemble average of the energy given to the particle by the environment per unit time (the dot over $Q$ means "per unit time") and $S$ be the Shannon–von Neumann entropy

$$S \equiv -\int_{-\infty}^{\infty} dx du f \ln f \tag{2.5}$$

of the distribution function at time $t$, the inequality

$$\dot{Q} \leq T\frac{dS}{dt} \tag{2.6}$$

holds at any time.

The proof of Eq. (2.6) is straightforward. By virtue of the energy conservation law, $\dot{Q}$ is given by

$$\dot{Q} = \frac{d\langle E\rangle}{dt} - \dot{W}, \tag{2.7}$$

where

$$E \equiv \frac{mu^2}{2} + V(x,t), \tag{2.8}$$

$\langle X\rangle$ denotes the ensemble average of any function $X$,

$$\langle X\rangle \equiv \int_{-\infty}^{\infty} dx du f(x,u,t)X(x,u,t), \tag{2.9}$$

and $\dot{W}$ is the average work done by the potential $V$ on the particle per unit time. Since the work given to the particle at a position $x$ by $V$ per unit time is $\partial V(x,t)/\partial t$ [19], $\dot{W}$ is given by

$$\dot{W} = \left\langle\frac{\partial V(x,t)}{\partial t}\right\rangle = \int_{-\infty}^{\infty} dx du f(x,u,t)\frac{\partial V(x,t)}{\partial t}. \tag{2.10}$$

By inserting Eqs. (2.8)–(2.10) into (2.7) and using the FPE, we obtain

$$\dot{Q} = \int_{-\infty}^{\infty} dx du \frac{\partial f(x,u,t)}{\partial t}V(x,u,t) = \gamma(T - \langle mu^2\rangle). \tag{2.11}$$

The time derivative of $S$ is given by differentiating Eq. (2.5) with respect to $t$ and using the FPE as follows:

$$\frac{dS}{dt} = \gamma\left[\frac{T}{m}\left\langle\left(\frac{\partial \ln f}{\partial u}\right)^2\right\rangle - 1\right]. \tag{2.12}$$

Therefore, by means of (2.11) and (2.12),

$$\dot{Q} - T\frac{dS}{dt} = -\frac{\gamma}{m}\left\langle\left(T\frac{\partial \ln f}{\partial u} + mu\right)^2\right\rangle \leq 0. \tag{2.13}$$

Thus we obtain Eq. (2.6).

Equation (2.6) allows us to use reasoning formally identical to that used in thermodynamics. Especially, letting $\Delta Q_{out}(t_i, t_f)$ be the average energy dissipated into the environment between any times $t_i$ and $t_f$, the lower bound of $\Delta Q_{out}(t_i, t_f)$ is determined by the value of Shannon–von Neumann entropy of the distribution function at those times as follows:

$$\Delta Q_{out}(t_i, t_f) = \int_{t_i}^{t_f} (-\dot{Q})dt \geq T[S(t_i) - S(t_f)]. \tag{2.14}$$

Now we can calculate the lower bound of the heat generation caused by the erasure of 1 bit of information by using (2.14). The following argument may be regarded as a refinement of Landauer's original discussion [1]. Consider an ensemble consisting of $N(\gg 1)$ memories. Let us assume that at time $t_i$ every memory in the ensemble stores 1 bit of information. This means that the potential well of each memory forms the double well form (Fig. 1), and the distribution of each particle is localized in either the right-hand-side well (when the stored value is ONE) or the left-hand-side well (when it is ZERO). Let $f_1(x,u)$ and $f_0(x,u)$ be the distribution functions of the particle when the memory takes the values ONE and ZERO, respectively, and also $p_1 N$ and $p_0 N$ be the number of memories whose values are ONE and ZERO, respectively. Then the number of particles whose positions and velocities are within $x \sim x + dx$ and $u \sim u + du$, respectively, is given by

$$N p_0 f_0(x,u) dx du + N p_1 f_1(x,u) dx du$$

$$= N[p_0 f_0(x,u) + p_1 f_1(x,u)] dx du. \tag{2.15}$$

Therefore the Shannon–von Neumann entropy $S_{init}$ of the ensemble per memory at time $t_i$ is given by

$$S_{init} = -\int_{-\infty}^{\infty} du dx (p_0 f_0 + p_1 f_1) \ln(p_0 f_0 + p_1 f_1). \tag{2.16}$$

The overlapping of $f_1(x,u)$ and $f_0(x,u)$ should be negligible, because if not it is impossible to decide the value of the memory by the measurement of the position of the particle. Thus

$$S_{init} \approx -\int_{-\infty}^{\infty} du dx [p_0 f_0 \ln(p_0 f_0) + p_1 f_1 \ln(p_1 f_1)]$$

$$= p_0 S[f_0] + p_1 S[f_1] + S[p_0, p_1], \tag{2.17}$$

where

$$S[f_k] \equiv -\int_{-\infty}^{\infty} dx du f_k \ln f_k \quad (k = 0 \text{ or } 1), \tag{2.18}$$

and

$$S[p_0, p_1] \equiv -\sum_{i=0,1} p_i \ln p_i. \tag{2.19}$$

$S[f_k]$ is the entropy due to the distribution of the particle in the phase space when the memory stores a definite value $k$, and $S[p_0, p_1]$ is that due to the distribution of the values of the memories. If RTO is performed, and then the values of all memories become ONE at time $t_f$, the entropy $S_{final}$ of the ensemble per memory at time $t_f$ is given by

$$S_{final} = S[f_1]. \tag{2.20}$$

By inserting Eqs. (2.17) and (2.20) into (2.14), we obtain the lower bound of the heat generated by RTO as follows:

$$\Delta Q^{\text{RTO}}_{Lower\ bound} = T(S[p_0, p_1] + p_0 S[f_0] + p_1 S[f_1] - S[f_1]). \tag{2.21}$$

Then if

$$S[f_0] = S[f_1] \tag{2.22}$$

holds, Eq. (2.21) is reduced to

$$\Delta Q^{\text{RTO}}_{Lower\ bound} = T S[p_0, p_1]. \tag{2.23}$$

The right-hand side of Eq. (2.23) is exactly the product of the temperature and the amounts of erased information, and gives $T \ln 2$ if $p_0 = p_1 = 1/2$. Thus we have derived Landauer's claim rigorously in our model.

In the above discussion, we assumed Eq. (2.22) to derive Eq. (2.23). But, in fact, the assumption can be dispensed with, if we take into account the thermal cost required to *set* a definite value for the memory after the erasure. (Landauer seems to have noted this point in Ref. [1]. However, because he gave no detailed explanation, we discuss it here for completeness.) The entropy of the ensemble per memory after setting a definite value for each memory is given by

$$S_{set} = p_0 S[f_0] + p_1 S[f_1], \tag{2.24}$$

where it is assumed that the ratio between the memories whose values are ZERO and those whose values are ONE is the same as that before the erasure. [$S[p_0, p_1]$ does not appear in the right-hand side of Eq. (2.24) because each memory has a definite value.] Therefore the lower bound of the work dissipated into the environment during the erasure-setting process is given by

$$T(S_{init} - S_{set}) = T S[p_0, p_1] \tag{2.25}$$

*without* the assumption (2.22). This result can be generalized to the case where each memory can take $M$ ($\geq 2$) values in a straightforward manner.

## III. DISCUSSION

In this section we first discuss Goto *et al.*'s counterargument based on the QFP, and then comment on some interesting discussions relating to Landauer's claim.

The QFP is a Josephson logic device developed by Goto *et al.*, which uses magnetic flux to hold and transfer information. The output flux $\Phi$ from the QFP can be controlled by the "input flux" $\Phi_s$ and the "activation flux" $\Phi_a$, because it is governed by the Langevin-type equation of motion

$$2C\frac{d^2\Phi}{dt^2} + \frac{2}{R}\frac{d\Phi}{dt} + \frac{\partial V}{\partial \Phi} = -I_R , \tag{3.1}$$

where

$$V = -2E_J \cos\left(\frac{2\pi\Phi}{\Phi_0}\right)\cos\left(\frac{2\pi\Phi_a}{\Phi_0}\right) + \frac{(\Phi - \Phi_s)^2}{2L_L} \tag{3.2}$$

($\Phi_0 = h/2e$), $I_R$ is the resistive thermal noise current, and $C, R, L_L$, and $E_J$ are constants. See Refs. [9–13] for details. The essential point is that one can make $V$ either a single well or double well in form by controlling the activation flux $\Phi_a$. When $\Phi_a = 0$, $V$ forms a single well, and when $\Phi_a = \Phi_0/2$ it forms a double well. Therefore one may be able to realize the basic model of the memory introduced in Sec. I by using the QFP.

Goto *et al.* [9,10] argued that the heat generated by QFP per clock period is given by

$$P \equiv \int \frac{\dot{\Phi}^2}{R} dt = H_0 f_c, \tag{3.3}$$

where $H_0$ is a constant and $f_c$ is the clock cycle, and that by making $f_c$ small one can make $P$ as small as one wishes in contradiction to Landauer's claim. However, since Eq. (3.1) is mathematically the same as Eq. (2.1), we can apply our result obtained in Sec. II. Therefore we can conclude that, if the noise current can be regarded as white and Gaussian, it is impossible even for the QFP to erase 1 bit of information with less heat generation than $k_BT \ln 2$.

We should keep in mind, however, that we need the assumption about the nature of the noise current to derive the above conclusion. The assumption that the noise current is white and Gaussian, or, in other words, the motion of the Josephson junction used in the QFP can be described by the FPE, is known to be valid in many cases [16,20,21]. Nevertheless, it may be possible to construct a QFP with a Josephson junction whose current noise is not white and Gaussian. In this case, our discussion is not enough to contradict Goto *et al.*'s counterargument and we must investigate whether the discussion given in the previous section can be generalized to the case where the random force is not white and Gaussian. It needs further investigation.

There is one more point to be discussed concerning their discussion. As stated above, they take the total heat $P$ generated by the resistors and discuss whether it is possible or not to make it lower than $k_BT \ln 2$. In the model discussed in Sec. II the quantity corresponding to $P$ is

$$P' \equiv \int_{t_i}^{t_f} \gamma m u^2 dt. \tag{3.4}$$

However, the resistors not only generate but also absorb heat from the environment. Then "heat generation required by information erasure" means the difference between the heat generated and that absorbed by the memory during the erasure process. In other words, it means the work done by the time-dependent potential and dissipated into the environment. It is for this reason that we should use, and have used, $\Delta Q_{out}(t_i, t_f)$ rather than $P'$ as the definition of the "heat generation required by information erasure." It may be worth pointing out that the relationship between the quantities $P'$ and $\Delta Q_{out}(t_i, t_f)$ is obtained, by using Eq. (2.11), as

$$P' = \Delta Q_{out}(t_i, t_f) + \gamma T(t_f - t_i). \tag{3.5}$$

As a result, the inequality

$$P' > \Delta Q_{out}(t_i, t_f) \tag{3.6}$$

always holds as expected.

Igeta [22] argued, against Landauer's claim, that "the physical entropy does not change because there is no thermodynamic difference between *Zero* and *One*. Each of them is definite and has no statistical factors. Also, the physical entropy of the states *Zero* and *One* can be the same by physical symmetry." His last statement corresponds to Eq. (2.22). However, our result supports Landauer's claim and contradicts Igeta's argument. The point is that the ensemble before the erasure contains both of memories whose values are ONE and those whose values are ZERO. As a result, $S_{init}$ given by Eq. (2.17) contains the term (2.19) which does not appear in $S_{final}$.

Fahn [23] argued, in connection with the analysis of Maxwell's demon (especially Szilard's engine), that "there is an entropy symmetry between the measurement and erasure steps, whereby the two steps additively share a constant entropy change, but the proportion that occurs during each of the two steps is arbitrary." His "measurement" process corresponds to our "setting process" in Sec. II, so that our result on the thermal cost of the erasure-setting process agrees with his argument.

Finally we comment on Schneider's note [24] on Landauer's claim. According to Schneider, the information erasure consists of two steps: priming and setting. The device is in one of the two stable states at the beginning, so that one must add some energy to the memory device to alter that state at first. He called this step the "priming step" and argued that information is lost at this step. On the next "setting step," the device is guided by preset inputs to fall into the standard state and the energy added at the priming step is dissipated. He argued that Landauer lumped these two steps into a single step.

However, the priming step is not essential for the information erasure. The information is assumed to be already lost before the erasure, that is, the value of the memory is already uncertain before the erasure. (If not so, one can set the value to ONE expending no energy [1].) In a typical erasure process energy is added to the particle and then dissipated almost simultaneously when the distribution of the particle is compressed.

## IV. SUMMARY

We have investigated the lower bound of the heat generation required by information erasure in the case of a basic model of the memory, that is, a Brownian particle in a moving potential well. We have shown that, if the random force acting on the particle can be regarded as white and Gaussian, Landauer's claim that the erasure of 1 bit of information is accompanied by the heat generation of at least $k_B T \ln 2$ holds rigorously. Next we have discussed Goto *et al.*'s counterargument and concluded that if the resistive thermal current noise involved in the QFP is white and Gaussian, or, in other words, if the behavior of the Josephson junction can be described by the Fokker-Planck equation, it is impossible to erase information with infinitesimal heat generation even by using the QFP.

## ACKNOWLEDGMENTS

The author would like to thank T. Arimitsu, H. Hayakawa, H. Hasegawa, T. Iwai, K. Sasaki, S. Takagi, and Y. Takahashi for helpful discussions.

---

[1] R. Landauer, IBM J. Res. Dev. **5**, 183 (1961); reprinted in Ref. [3].
[2] C.H. Bennet, Int. J. Theor. Phys. **21**, 905 (1982); reprinted in Ref. [3].
[3] H.S. Leff and A.F. Rex, *Maxwell's Demon: Entropy, Information, Computing* (Princeton University Press, Princeton, 1990).
[4] E. Fredkin and T. Toffoli, Int. J. Theor. Phys. **21** , 219 (1982).
[5] W.H. Zurek, Phys. Rev. A **40**, 4731(1989).
[6] R. Landauer and J.W.F. Woo, J. Appl. Phys. **42**, 2301 (1971).
[7] J. Berger, Int. J. Theor. Phys. **29**, 985 (1990).
[8] D. Wolpert, Phys. Today **45** (3), 98 (1992).
[9] E. Goto, N. Yoshida, K.F. Loe, and W. Hioe, in *Proceedings of the 3rd International Symposium on the Foundations of Quantum Mechanics, Tokyo*, edited by H. Ezawa, Y. Murayama, and S. Nomura (Phys. Soc. Jpn., Tokyo, 1990), p. 412.
[10] E. Goto, W. Hioe, and M. Hosoya, Physica C **185-189**, 385 (1991).
[11] E. Goto and K.F. Loe, *DC Flux Parametron* (World Scientific, Singapore, 1986).
[12] Y. Harada, H. Nanane, N. Miyamato, U. Kawabe, and E. Goto, IEEE Trans. Magn. **MAG-23**, 3801 (1987).
[13] W. Hioe and E. Goto, *Quantum Flux Parametron* (World Scientific, Singapore, 1991).
[14] R. Landauer, Physica C **208**, 205 (1993).
[15] E. Goto and N. Yoshida (unpublished).
[16] H. Risken, *The Fokker-Planck Equation*, 2nd. ed. (Springer-Verlag, New York, 1989).
[17] H. Hasegawa, Prog. Theor. Phys. **57**, 1523 (1977).
[18] H. Hasegawa, T. Nakagomi, M. Mabuchi, and K. Kondo, J. Stat. Phys. **23**, 281 (1980).
[19] L.D. Landau and E.M. Lifshitz, *Mechanics* (Pergamon, Oxford, 1969).
[20] A. Barone and G. Paterno, *Physics and Applications of the Josephson Effect* (Wiley, New York, 1982), and references therein.
[21] E. Ben-Jacobi, E. Mottola, and G. Schoen, Phys. Rev. Lett. **51**, 2064 (1983).
[22] K. Igeta, in *Proceedings of the Workshop on Physics and Computation PhysComp'92* (IEEE Computer Society Press, Los Alamitos, CA, 1992), p. 184.
[23] P.N. Fahn, in *Proceedings of the Workshop on Physics and Computation PhysComp'94* (IEEE Computer Society Press, Los Alamitos, CA, 1994), p. 217.
[24] T.D. Schneider, Nanotechnology **5**, 1 (1994).

# Information erasure

Barbara Piechocinska*
*Los Alamos National Laboratory, T-6, Los Alamos, New Mexico 87544*
(Received 27 September 1999; published 17 May 2000)

Landauer's principle states that in erasing one bit of information, on average, at least $k_B T \ln(2)$ energy is dissipated into the environment (where $k_B$ is Boltzmann's constant and $T$ is the temperature of the environment at which one erases). Here, Landauer's principle is microscopically derived without direct reference to the second law of thermodynamics. This is done for a classical system with continuous space and time, with discrete space and time, and for a quantum system. The assumption made in all three cases is that during erasure the bit is in contact with a thermal reservoir.

PACS number(s): 03.67.−a, 89.70.+c

## I. INTRODUCTION

The purpose of this paper is to show that Landauer's principle [1] [which states that in erasing one bit (binary digit) of information one dissipates, on average, at least $k_B T \ln(2)$ of energy into the environment, where $k_B$ is Boltzmann's constant and $T$ is the temperature at which one erases] can be derived from microscopic considerations. The intention is to show this for classical and quantum systems without direct reference to the second law of thermodynamics. This, however, does not imply that we will be proving the second law of thermodynamics. Since the beginning of this century it has been known that it is possible to derive many of the inequalities in thermodynamics from the canonical distribution [2].

The introductory part of the paper will try to give the reader a clear picture of what we mean by erasure, what Landauer's principle is about, and what has been done in this field.

In Sec. II of this paper we will attempt to derive Landauer's principle from microscopic considerations for three separate cases: the classical continuous case, the classical discrete case, and the quantum case. In Sec. III we will discuss nondegenerate energy levels for the states of the bits and probability distributions of the states of bits in ensembles of bits.

### A. Basic setup

The bit is a fundamental unit of information, the smallest item capable of indicating a choice. We will assume that all information is physically representable, and therefore all bits representing it are encoded in the states of physical systems [3]. The bits will have two distinguishable states that we will call "zero" and "one." They will be in contact with an environment that we will be modeling as a heat reservoir at a fixed temperature $T$. There will also be an external parameter that will let us do work on the bit. This external parameter will be the means with which we will erase the bit. In all cases we will assume to have a large number of bits but they will be erased individually, one by one. We will view the large number of bits as an ensemble. To have a clearer picture of the entire process one could for instance think of the bit as being a spin-1/2 particle and of the external parameter as being a magnetic field that we can alter. This shows us that all we need is one heat reservoir and one external parameter for erasure. We do not need additional heat reservoirs or additional external parameters. Erasure is a reset operation. It can be defined either as "restore to one" or as "restore to zero." In either case we are going from two possible states of the bit to one possible state.

### B. Landauer's argument

The relationship between physical entropy and information may have been mentioned first by Szilard [4,5]. Based on the second law of thermodynamics Szilard introduced the idea that a measurement, information gain, should at some point be accompanied by an increase in entropy. This has been further discussed in terms of quantum mechanics by Zurek [6] and Lloyd [7].

In his original paper from 1961 [1], Landauer argues that since erasure is a logical function that does not have a single-valued inverse it must be associated with physical irreversibility and therefore require heat dissipation. He argues that a bit has one degree of freedom and the heat dissipation should be of order $k_B T$. More precisely, that since before erasure a bit can be in any of the two possible states and after erasure it can only be in one state this implies a change in information entropy of $-k \ln(2)$. Since entropy cannot decrease it must appear, Landauer argues, somewhere else as heat. Implicit in this argument is the crucial assumption that information entropy translates into physical entropy.

### C. Later work on erasure

Bennett built on Landauer's principle in a paper on reversible computation [8] (see also [9]). He showed that every step in computation can be made reversible except for erasure (which also includes error correction). After these papers were published many other scientists wrote papers on Szilard's engines [10], different variations on Maxwell's demons [11,12], and on computations which all used Landauer's principle. In fact, their arguments often strongly depended on Landauer's principle. Later, papers were published which criticized Landauer's paper claiming that his proof is not rigorous enough. The reason for these objections was that Landauer's proof is only based on the second law of thermodynamics [10,13] and that it is not clear what

*Electronic address: bpiechocinska@hotmail.com

the connection between information and thermodynamic entropy is [14]. In response to the criticism Shizume showed that Landauer's principle holds for a model in which a particle having Brownian motion in a time-dependent double potential well in which a white and Gaussian force is acting on the particle [15]. This was shown using the Fokker-Planck equation. His derivation is restricted to a specific model and only works in classical mechanics. Therefore, it is not as general as Landauer's principle.

Even though computation can be made reversible, and therefore not generate heat (at least in theory), it is still desirable to have a rigorous derivation of Landauer's principle for demonic reasons and because computers always need error-correction. Error-correction is not a reversible operation for a cyclic process with a finite amount of memory and just like erasure it requires heat generation.

## II. MICROSCOPIC DERIVATION OF LANDAUER'S PRINCIPLE

In this section we will show the validity of Landauer's principle in three cases: for a classical system in continuous space and time, for a classical system in discrete space and time, and for a quantum system. In all three cases we will assume that the bits which are to be erased are in contact with a heat reservoir whose initial microstate is chosen from a canonical distribution. To make the microscopic derivation as general as possible we will not be using specific and detailed models of the bits and erasure. Instead we will be treating erasure as a thermodynamic process during which we can change an external parameter while the bit is in contact with a heat reservoir. To show Landauer's principle we will be using an ensemble of bits and averaging over the microscopic realizations of this process.

We will assume that the two states of the bit, the "zero" state and the "one" state, have equal energy. We make this assumption because it has been shown that for these systems all computational operations (except for erasure) can be made reversible and can, at least in theory, be performed using an arbitrarily small amount of work. In Sec. III B of this paper, we will generalize Landauer's principle to include the case where the two energy-states defining the "zero" and the "one" state are nondegenerate.

### A. The continuous classical case

We will be making the following assumptions:

(i) Our system is classical.

(ii) The memory state is a symmetric double potential well where the states "zero" and "one" have the same energy before and after the erasure.

(iii) The input is randomly distributed (the number of "zeros" and "ones" is equal and there are no correlations between the bits).

(iv) During erasure the system is in contact with a thermal reservoir with initial states chosen from a canonical distribution.

(v) The interaction term in the Hamiltonian is negligibly small.

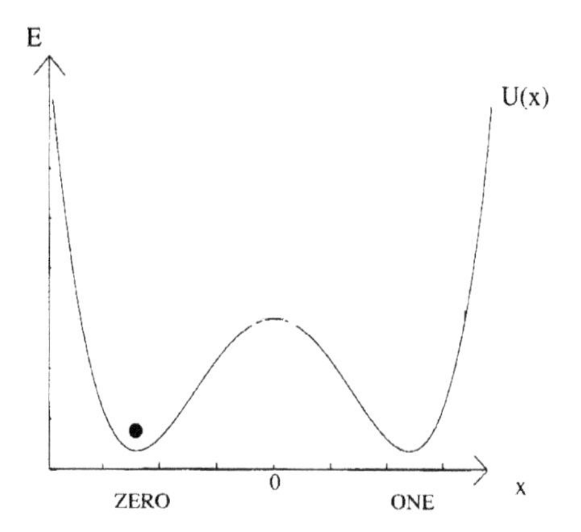


FIG. 1. Double potential well described by the function $U(x)$, where the state "one" is $x>0$ and the state "zero" is $x<0$. $E$ (in $J$) is the energy and $x$ (in $m$) is a distance.

We have an infinite ensemble of bits which we each model as a system with one continuous degree of freedom, $x$, subject to a symmetric double-well potential energy $E$, as shown in Fig. 1. The position of the particle in the double-well potential will determine the state of the bit. If the particle is found on the left-hand side of the potential ($x<0$), then we will say that the bit is in the state "zero." If it is found on the right-hand side of the well ($x>0$), then we will say that the bit is in the state "one." For it to be considered a useful bit one should also add that the energy barrier separating the two wells is much greater than $k_BT$. This way the bit is stable enough to store information for a longer period of time. If the energy barrier and $k_BT$ were comparable in size then the particle in the potential well would have enough energy to jump between the two distinct states and the initial information would not be stored. This is not a desirable situation for computational purposes. This assumption, however, is not necessary for the purpose of showing Landauer's principle.

If we were given a bit like the one described above and wanted to erase it by putting it into the "one" state it should be clear that this could be done by coupling it to a heat reservoir and changing an external parameter.

To show that erasure implies heat dissipation we will use some of the results presented by Jarzynski in his paper on Clausius-Duhem processes [16]. Before the erasure we want half of the bits to be in the "one" state and the other half to be in the "zero" state. We assume that the ensemble of bits is in contact with a thermal reservoir where the temperature of the reservoir is low enough not to change the state of the bits ($k_BT\leqslant\Delta U$). The system will instead reach a "local" thermal equilibrium in one of the half-wells. We therefore assume that the initial statistical state is described by the following distribution function for the bits before erasure:

$$\rho_{init}(x,p)=\frac{1}{Z}\exp\left\{-\beta\left[U(x)+\frac{p^2}{2m}\right]\right\} \tag{1}$$

and the statistical state after erasure can be described as

$$\rho_{final}(x,p)=\begin{cases}\frac{2}{Z}\exp\left(-\beta\left[U(x)+\frac{p^2}{2m}\right]\right) & \text{for } x>0\\ 0 & \text{for } x<0,\end{cases} \tag{2}$$

where $x$ is the position, $m$ the mass, $p$ the momentum, $\beta=1/(k_BT)$, and $Z=\int\exp\{-[U(x)+p^2/2m]/k_BT\}dx\,dp$ is the partition function.

Since the total system (the bit and the heat reservoir) is a classical and isolated system it will evolve according to the following Hamiltonian:

$$H(x,p,\mathbf{x}_T,\mathbf{p}_T,t)=H(x,p)+H_T(\mathbf{x}_T,\mathbf{p}_T)+H_{int}(x,p,\mathbf{x}_T,\mathbf{p}_T), \tag{3}$$

where $H(x,p)$ is the Hamiltonian of the system, $H_T(\mathbf{x}_T,\mathbf{p}_T)$ is the Hamiltonian of the heat reservoir, and $H_{int}$ is the Hamiltonian of interaction which we assume to be negligible in comparison with the other terms of the Hamiltonian. The $\mathbf{x}_T$ and $\mathbf{p}_T$ are the positions and momenta of the degrees of freedom that describe the heat reservoir. For easier notation let us use $\zeta=(x,p,\mathbf{x}_T,\mathbf{p}_T)$. Then the trajectory $\zeta(t)$, where $t$ is time, will describe the evolution of all degrees of freedom for one realization of the erasure process. Let us assume that the erasure process takes a time $\tau$ and use the shorthand $\zeta^0=\zeta(0)$ and $\zeta^\tau=\zeta(\tau)$.

Following [16] let us now define a function, $\Gamma(\zeta^0,\zeta^\tau)$. $\Gamma$ is defined for a given microscopic realization in the following way:

$$\Gamma(\zeta^0,\zeta^\tau)=-\ln[\rho_{final}(x^\tau,p^\tau)]+\ln[\rho_{init}(x^0,p^0)]+\beta\Delta E(\mathbf{x}_T^0,\mathbf{p}_T^0,\mathbf{x}_T^\tau,\mathbf{p}_T^\tau), \tag{4}$$

where $\Delta E=H_T(\mathbf{x}_T^\tau,\mathbf{p}_T^\tau)-H_T(\mathbf{x}_T^0,\mathbf{p}_T^0)$ is the change in the internal energy of the heat reservoir and $\beta$ is as defined previously using the temperature of the heat reservoir. $\Gamma$ is defined in this particular way because it has proven useful in the kind of calculations we want to perform. $\Gamma$ is explicitly a function of the initial and final microstates of the system and reservoir during the course of one realization. However, since the final state $\zeta^\tau$ can be viewed as a function of the initial state $\zeta^0$ [because the evolution $\zeta(t)$ is deterministic], $\Gamma$ can be viewed as a function of $\zeta^0$ alone:

$$\Gamma(\zeta^0,\zeta^\tau)=\Gamma(\zeta^0,\zeta^\tau(\zeta^0)). \tag{5}$$

For the purpose of Landauer's principle, which is a statement about the average heat released into the environment, we will be interested in averaging over the statistical ensemble of realizations. We will also be interested in finding an inequality relating these averages. For these purposes we will now compute $\langle\exp(-\Gamma)\rangle$, where the angular brackets denote the average over the statistical ensemble of realizations. Since the evolution [governed by the Hamiltonian written in Eq. (3)] is deterministic $\langle\exp(-\Gamma)\rangle$ can be written as an integral over initial conditions $\zeta^0$:

$$\langle\exp(-\Gamma)\rangle=\frac{1}{Z_T}\int\rho_{init}(x^0,p^0)\exp\left(-\frac{H_T(\mathbf{x}_T^0,\mathbf{p}_T^0)}{k_BT}\right)\times\exp(-\Gamma)d\zeta^0 \tag{6}$$

$$=\frac{1}{Z_T}\int\rho_{init}(x^0,p^0)\frac{\rho_{final}(x^\tau,p^\tau)}{\rho_{init}(x^0,p^0)}\times\exp\left(-\frac{H_T(\mathbf{x}_T^0,\mathbf{p}_T^0)}{k_BT}\right)\times\exp\left(\frac{H_T(\mathbf{x}_T^0,\mathbf{p}_T^0)}{k_BT}-\frac{H_T(\mathbf{x}_T^\tau,\mathbf{p}_T^\tau)}{k_BT}\right)d\zeta^0 \tag{7}$$

$$=\frac{1}{Z_T}\int\rho_{final}(x^\tau,p^\tau)\exp\left(-\frac{H_T(\mathbf{x}_T^\tau,\mathbf{p}_T^\tau)}{k_BT}\right)d\zeta^\tau$$

$$=\frac{Z_T}{Z_T}=1, \tag{8}$$

where $Z_T=\int\exp\{-[H_T(\mathbf{x}_T,\mathbf{p}_T)/k_BT]\}d\mathbf{x}_Td\mathbf{p}_T$. In the equations above we have changed the integration variables from $d\zeta^0$ to $d\zeta^\tau$. Since the evolution of our system is Hamiltonian the Jacobian associated with this change of variables is equal to 1. We thus have $\langle\exp[-\Gamma(\zeta^0,\zeta^\tau)]\rangle=1$, where the brackets indicated an average over an ensemble of realizations of the erasure process. Note that the equations above do not in any way imply that the final distribution is a canonical one. The function $\Gamma$ is just a function of $\zeta^0$ and $\zeta^\tau$ and it happens to be chosen in such as way that the terms involving $\zeta^0$ in the equations above cancel.

By the convexity of the exponential function $-\langle\Gamma\rangle\leq 0$. Written explicitly the inequality becomes

$$\langle\ln[\rho_{final}(x^\tau,p^\tau)]\rangle-\langle\ln[\rho_{init}(x^0,p^0)]\rangle\leq\langle\beta\Delta E\rangle. \tag{9}$$

Written even more explicitly, using the distribution functions in Eqs. (1) and (2) and the fact that for any function $A(x,p)$,

$$\langle A(x^0,p^0)\rangle=\int\rho_{init}(x,p)A(x,p)dx\,dp \tag{10}$$

as well as

$$\langle A(x^\tau,p^\tau)\rangle=\int\rho_{final}(x,p)A(x,p)dx\,dp, \tag{11}$$

the left-hand side of the inequality above becomes a sum of two contributions.

For $x>0$

$$\int_0^\infty\frac{2}{Z}\exp(-\beta H)\ln[(2/Z)\exp(-\beta H)]dx\,dp-\int_0^\infty\frac{1}{Z}\exp(-\beta H)\ln[(1/Z)\exp(-\beta H)]dx\,dp \tag{12}$$

$$=\int_0^\infty 2\alpha\ln(2\alpha)dx\,dp-\int_0^\infty\alpha\ln(\alpha)dx\,dp, \tag{13}$$

where $\alpha=(1/Z)\exp(-\beta H)$. For $x<0$

$$0\ln(0)-\int_{-\infty}^{0}(1/Z)\exp(-\beta H)\ln(1/Z)\exp(-\beta H)dx\,dp$$
$$=-\int_{-\infty}^{0}\alpha\ln(\alpha)\,dx\,dp. \quad (14)$$

The term 0 ln(0) is equal to 0 by l'Hôpital's rule. Since the initial distribution function is symmetric with respect to $x=0$, we have

$$\int_{0}^{\infty}\alpha\ln(\alpha)dx\,dp=\int_{-\infty}^{0}\alpha\ln(\alpha)dx\,dp. \quad (15)$$

Totally, for all $x$, the left-hand side of Eq. (9) becomes

$$\int_{0}^{\infty}2\alpha\ln(2\alpha)dx\,dp-2\int_{0}^{\infty}\alpha\ln(\alpha)dx\,dp$$
$$=\int_{0}^{\infty}2\alpha\ln\left(\frac{2\alpha}{\alpha}\right)dx\,dp=\int_{0}^{\infty}2\alpha\ln(2)dx\,dp. \quad (16)$$

Since the distribution function is normalized to unity Eq. (9) becomes

$$\ln(2)\leq\beta\langle\Delta E\rangle. \quad (17)$$

In defining $\Gamma$ we defined $\Delta E$ as the change in the internal energy of the heat reservoir. We recognize that the interaction term in the Hamiltonian is necessary for the heat reservoir and the system to be able to exchange energy. The size of this term depends on the nature of the heat reservoir as well as the nature of the bit. We will now use the approximation that the interaction term in the Hamiltonian of system and heat reservoir is negligible. To determine how good an approximation this is one would need to specify the physical systems. If we write the equation of conservation of energy for the system and the heat reservoir it gives:

$$W=\Delta E+\Delta E_{system}, \quad (18)$$

where $W$ is the work done on the system and heat reservoir, while $\Delta E_{system}$ is the change in the internal energy of the system. Due to the symmetry of $\rho_{init}$ and $\rho_{final}$ $\Delta E_{system}$ disappears when averaged over ($\langle\Delta E_{system}\rangle=0$). So, Eqs. (17) and (18) taken together give us

$$k_BT\ln(2)\leq\langle W\rangle. \quad (19)$$

This means that to erase one bit of information, on average, the work performed on the system has to be equal to or greater than $k_BT\ln(2)$,[1] or, equivalently, that the heat dissipation by the system into the heat reservoir has to be greater than or equal to $k_BT\ln(2)$.[2]

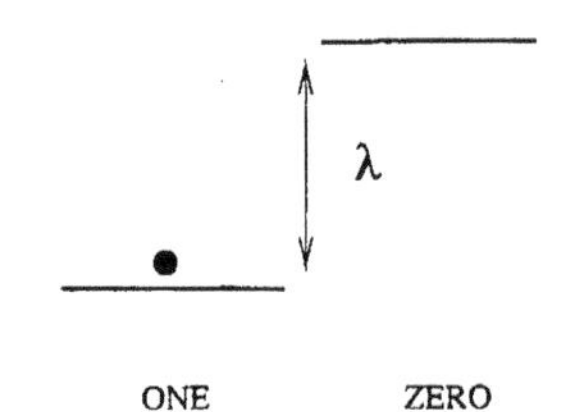


FIG. 2. Two-state system where the left state describes the "zero" state and the right state describes the "one" state. The difference in energy between the states is λ, which we treat as an external parameter.

#### B. The discrete classical case

We will now consider a case in which the evolution of the system (or bit) is modeled as a Markov process. In some sense the stochastic dynamics which we use here is less fundamental than the Hamiltonian evolution considered in the preceding section, or the quantum evolution studied in the next section. Nevertheless, Markov evolution is very often used to model a system in contact with a heat reservoir, and it is instructive to see that using this approach, we obtain exactly the same result (Landauer's principle) as when using more fundamental equations of motion. Both the discrete classical case and the quantum case are closely related to the information theoretic treatment of classical Markovian and quantum versions of Maxwell's demon presented by Lloyd [17].

Let us model our bit as a system with two classical states, analogous to quantum energy levels. One of the levels corresponds to the state "one" and the other is the state "zero" (see Fig. 2). We will use a discrete rather than a continuous time parameter: at each time step the system can either stay in its current state, or "jump" to the other state. In other words both our time and space will be discrete. To calculate the heat dissipation into the environment during erasure we will use the method of Crook described in [18].

We assume that in the beginning the energy difference between these levels, λ, is zero, and that half of the bits are in the state "one" while the other half are in the state "zero." Then, during the erasure procedure, we change the value of λ in discrete steps, thus performing work on the bit. During the time of erasure we couple the two-state system to a heat reservoir with a certain temperature, $T$. We could for instance imagine that we separate the energy levels in such a way as to make one of the energy levels have a much higher energy than $k_BT$. This would guarantee that the transition from the lower energy level to the higher energy level is

[1] Note that this is an approximation because we neglected the interaction term in the Hamiltonian.

[2] This, however, is a precise statement. We do not need to neglect the interaction term in the Hamiltonian for this to be true. It follows from the microscopic definition of "heat dissipated," which we have used.

highly improbable. If we wait for a sufficiently long period of time the probability of finding the system (bit) in the higher energy state will be extremely small. This is one way of seeing how erasure could work in this particular case. Note, however, that the discrete two-state system described in this section is not restricted to this particular scheme of erasure. After erasure the energy difference between the two levels is again set to zero. We will assume that during erasure every step in the process is independent of the other steps (the Markov approximation), and therefore we can write the probability of going from state $i_0$ to state $i_N$ with all the intermediate states and $\lambda$'s as:

$$P(i_0 \overset{\lambda_1}{\rightarrow} i_1 \overset{\lambda_2}{\rightarrow} \cdots \overset{\lambda_N}{\rightarrow} i_N) = P(i_0 \overset{\lambda_1}{\rightarrow} i_1) P(i_1 \overset{\lambda_2}{\rightarrow} i_2) \cdots P(i_{N-1} \overset{\lambda_N}{\rightarrow} i_N) \quad (20)$$

using the notation of [18]. Every step in the erasure process can be divided into two parts. The first part consists of changing $\lambda$ from $\lambda_t$ to $\lambda_{t+1}$. The subscript $t$ indicates the discrete time step. In changing $\lambda$, we perform external work on the bit, for instance, in the first time step *Work* $=E(i_0,\lambda_1)-E(i_0,\lambda_0)$, where $E()$ is the energy of the system. During the second part of the step in the erasure process the bit evolves from one state, $i_t$, to the next state, $i_{t+1}$. The corresponding heat absorbed by the system can be written as Heat$=E(i_1,\lambda_1)-E(i_0,\lambda_1)$ for the first time step. Looking at the entire process we can calculate the total work performed on the system, $W$, and the total heat absorbed by the system, $Q$,

$$W=\sum_{t=0}^{N-1} E(i_t,\lambda_{t+1})-E(i_t,\lambda_t), \quad (21)$$

$$Q=\sum_{t=1}^{N} E(i_t,\lambda_t)-E(i_{t-1},\lambda_t). \quad (22)$$

We assume that the transition probabilities obey detailed balance. Detailed balance in general can be written as ( [19,20])

$$\frac{P(i_0 \overset{\lambda_1}{\rightarrow} i_1)}{P(i_0 \overset{\lambda_1}{\leftarrow} i_1)} = \exp\{-\beta[E(i_1,\lambda_1)-E(i_0,\lambda_1)]\}, \quad (23)$$

where $P(i_t \overset{\lambda}{\rightarrow} i_{t+1})$ is the probability of transition from the state $i_t$ to the state $i_{t+1}$ with the external parameter $\lambda$. Using this definition for our particular case we can write

$$\frac{P(i_0 \overset{\lambda_1}{\rightarrow} i_1) P(i_1 \overset{\lambda_2}{\rightarrow} i_2) \cdots P(i_{N-1} \overset{\lambda_N}{\rightarrow} i_N)}{P(i_0 \overset{\lambda_1}{\leftarrow} i_1) P(i_1 \overset{\lambda_2}{\leftarrow} i_2) \cdots P(i_{N-1} \overset{\lambda_N}{\leftarrow} i_N)} \quad (24)$$

$$= \frac{\exp[-\beta E(i_1,\lambda_1)]\exp[-\beta E(i_2,\lambda_2)]\cdots\exp[-\beta E(i_N,\lambda_N)]}{\exp[-\beta E(i_0,\lambda_1)]\exp[-\beta E(i_1,\lambda_2)]\cdots\exp[-\beta E(i_{N-1},\lambda_N)]} \quad (25)$$

$$= \exp\left(-\beta \sum_{t=1}^{N-1} [E(i_t,\lambda_t)-E(i_{t-1},\lambda_t)]\right) = \exp(-\beta Q). \quad (26)$$

Let us now also take into consideration the initial and final probabilities and write

$$\frac{P_0(i_0)P(i_0 \rightarrow i_N)}{P_N(i_N)P(i_0 \leftarrow i_N)} = \frac{P_0(i_0)}{P_N(i_N)}\exp(-\beta Q) = \exp\{\ln[P_0(i_0)] - \ln[P_N(i_N)] - \beta Q\}. \quad (27)$$

Just like in the classical continuous case, we are interested in the average over all realizations. If the probabilities are normalized we can see that

$$\langle \exp\{-\ln[P_0(i_0)] + \ln[P_N(i_N)] + \beta Q\} \rangle \quad (28)$$

$$= \sum_{i_0,\ldots,i_N} P_0(i_0)P(i_0 \rightarrow i_N)\exp\{-\ln[P_0(i_0)] + \ln[P_N(i_N)] + \beta Q\} \quad (29)$$

$$= \sum_{i_0,\ldots,i_N} P_0(i_0)P(i_0 \rightarrow i_N)\frac{P_N(i_N)P(i_0 \leftarrow i_N)}{P_0(i_0)P(i_0 \rightarrow i_N)} \quad (30)$$

$$= \sum_{i_0,\ldots,i_N} P_N(i_N)P(i_0 \leftarrow i_N) = 1, \quad (31)$$

where $\langle\,\rangle$ denote the average value. To find the appropriate inequality we use the convexity of the exponential function and we write

$$-\langle \ln[P_0(i_0)] \rangle + \langle \ln[P_N(i_N)] \rangle + \langle \beta Q \rangle \leq 0. \quad (32)$$

In the case of erasure, if we erase by restoring to "one" then the initial probability will be $P_0(0)=P_0(1)=1/2$ and the final probabilities will be $P_N(1)=1$ and $P_N(0)=0$. Putting these values into Eq. (32) and using the fact that for an arbitrary function $A(i)$, $\langle A(i_0)\rangle = \Sigma_i P_0(i)A(i)$ and $\langle A(i_N)\rangle$

$=\Sigma_i P_N(i)A(i)$, and keeping in mind that the probabilities are normalized gives us the following inequality:

$$-\ln(1/2)+\ln(1)+\langle\beta Q\rangle\leq 0 \tag{33}$$

or just

$$\ln(2)\leq -\beta\langle Q\rangle. \tag{34}$$

In Eq. (21) we defined $Q$ as heat absorbed by the system. This means that $-Q$ is the heat dissipated into the heat reservoir by the system. We know that the total average work done on the system can be written as

$$\langle W\rangle=\langle\Delta E\rangle-\langle Q\rangle, \tag{35}$$

where $\Delta E$ is the change in the energy of the system. Since we start with $\lambda_0=0$ and end with $\lambda_N=0$ the change in the energy will be zero. This leaves us with $\langle W\rangle=-\langle Q\rangle$. Combining this with the inequality (34) give us

$$k_B T\ln(2)\leq\langle W\rangle, \tag{36}$$

which tells us that the average work we have to do on the system to erase one bit has to be greater than or equal to $k_B T\ln(2)$. Or, again, equivalently, we could say that the heat dissipated into the heat reservoir has to be equal to or greater than $k_B T\ln(2)$.

### C. The quantum case

First let us give a concrete example of what systems could be involved in erasure of a quantum system. The bit could be a two-level atom that initially has two degenerate energy states. The heat reservoir could be described as a photon reservoir with harmonic oscillators. We would couple it to the atom in the beginning of the erasure procedure and decouple it at the end. The external parameter could be a magnetic field that we can alter as we please. The field would be switched on in the beginning and, during erasure, it would split up the two initially degenerate energy states into two different energy states. If the energy difference between the two states became large enough so that the photon reservoir would not be able to excite the atom into the higher energy state, after a while the atom would find itself in the lower energy state. Then the field would be switched off and the energy levels of the two-state atom would become degenerate again. At this point the erasure would be complete. Another way to imagine "quantum erasure" could be to use a spin-1/2 particle as a quantum bit.

We will assume that once the erasure itself is complete, the reservoir becomes weakly coupled to some unspecified environment, causing it (the reservoir) to decohere. We will furthermore assume that this coupling is such that the energy eigenstates of the reservoir form the so-called preferred basis states [21], so that, once decoherence has set in, we can view the reservoir to be in a definite energy eigenstate. Effectively, the role of the environment in this situation is that of an "outside observer," who measures the final energy of the reservoir. Such a measurement is necessary if the heat absorbed by the reservoir is to be a well-defined quantity.

In our derivation we will be using a two-state quantum system, not necessarily in a pure state. Therefore we will use a density matrix, $\hat{\rho}$, to describe its statistical state. In the case where we have a $2\times 2$ density matrix we can always write it as

$$\hat{\rho}_{2\times 2}=a\hat{\rho}_a+b\hat{\rho}_b. \tag{37}$$

In other words we could say that the density matrix is diagonal in some basis (that does not have to be the energy basis). We can interpret the statistical state described by the density matrix of Eq. (37) as follows: the system is either in state $|a\rangle$ or in state $|b\rangle$, with probability $a$ and $b$, respectively. The statistical state of any two-state system can be described by a density matrix with the properties outlined above.

We will also be using a heat reservoir. As usual we assume it to be initially in thermal equilibrium. This allows us to write its density matrix as

$$\hat{\rho}_{\hat{H}}=\frac{\exp(-\beta\hat{H})}{tr[\exp(-\beta\hat{H})]}, \tag{38}$$

where $\hat{H}$ is the Hamiltonian of the heat reservoir. We can interpret the $\hat{\rho}_{\hat{H}}$ by imagining the reservoir to be in a definite energy eigenstate. The probability of finding the heat reservoir in its energy eigenstate $|E_n\rangle$ with the eigenvalue $E_n$ is

$$\text{Prob}(|E_n\rangle)=P_n=\frac{\exp(-\beta E_n)}{\sum_m \exp(-\beta E_m)}=\frac{\exp(-\beta E_n)}{Z}. \tag{39}$$

We also have an external parameter, $\lambda(t)$. This parameter serves the purpose of splitting up the two degenerate energy-eigenstates of the two-state system.

Let us imagine the erasure procedure as follows.

(1) At time $t=0$ the bit begins in some statistical state described by the $2\times 2$ density matrix

$$\hat{\rho}_{init}=\begin{pmatrix}1/2 & 0\\ 0 & 1/2\end{pmatrix} \tag{40}$$

in the energy eigenstate basis. (Therefore it can be viewed as starting in either the "zero" or the "one" state, with equal probability.) The reservoir begins in a definite eigenstate of $\hat{H}$, of energy $E_n$, with a thermal probability distribution [Eq. (39)]. The initial value of $\lambda$ is zero.

(2) At time $t=0^+$ we couple the bit to the reservoir.

(3) Between times $t=0^+$ and $t=\tau$ we change the value of the external parameter in some way which we believe will cause the bit to get erased. At the end, $t=\tau$, we make sure that the value of $\lambda$ is once again zero.

(4) At time $t=\tau^+$ we decouple the bit from the reservoir.

Assuming that the erasure was successful, the bit will now (with excellent probability) be in the pure state corresponding to "one." The reservoir will be in some statistical state, typically described by a density matrix which is not diagonal in the energy eigenbasis. This is where we invoke our assumption about the reservoir being weakly coupled to an external environment which causes it to decohere: the effect of the decoherence is to cause the off-diagonal elements of the density matrix to vanish, without changing the diagonal ones. Thus, at the end, the reservoir will again be in one of the energy eigenstates, with a probability determined by the diagonal density matrix elements.

We can therefore say that the bit begins in either the "zero" or the "one" state, and ends in the "one" state, whereas the reservoir begins in some state $|n\rangle$ and ends in a state $|m\rangle$. Then we can define

$$Q=E_n-E_m \tag{41}$$

as being the heat lost by the heat reservoir. Furthermore, let $|i\rangle$ and $|f\rangle$ denote the initial and final states of the bits, and let $P_{init}(i)=P_i$ and $P_{final}(f)=P_f$ denote the probability distributions of these bits. Then, by assumption, we have

$$P_{init}(i)=1/2 \quad \text{for} \quad i=0,1, \tag{42}$$

$$P_{final}(f)=\begin{cases}0 & \text{for} \quad f=0,\\ 1 & \text{for} \quad f=1.\end{cases} \tag{43}$$

Finally, let us define an observable $\Gamma$, as

$$\Gamma=\ln(P_i)-\ln(P_f)-\beta(E_n-E_m). \tag{44}$$

We can now calculate $\langle\exp(-\Gamma)\rangle$ where the angled brackets denote the average of the function written between them,

$$\langle\exp(-\Gamma)\rangle=\sum_{n,m,i,f}P_iP_n|U_{f,m,i,n}|^2 \times\exp[-\ln(P_i)+\ln(P_f)+\beta(E_n-E_m)] \tag{45}$$

$$=\sum_{n,m,i,f}P_i|U_{f,m,i,n}|^2\frac{P_f}{P_i}\frac{\exp(-\beta E_n)}{Z} \times\exp(-\beta E_m)\exp(\beta E_n) \tag{46}$$

$$=\frac{1}{Z}\sum_{f,m}P_f\exp(-\beta E_m)\sum_{i,n}|U_{f,m,i,n}|^2, \tag{47}$$

where $U_{f,m,i,n}$ corresponds to $\langle f,m|U(\tau)|i,n\rangle$ and it is the time evolution operator. At the same time it is a unitary matrix and therefore has the property that the sum of the absolute value squared of the elements in a column or a row is equal to 1. $|U_{f,m,i,n}|^2$ is the probability for finding the bit and reservoir in final states $|f\rangle$ and $|m\rangle$ given initial states $|i\rangle$ and $|n\rangle$. We then see that

$$\langle\exp(-\Gamma)\rangle=1. \tag{48}$$

Just like in the classical continuous case, Eqs. (45)–(47) do not in any way imply that the final distribution of reservoir states is canonical. The convexity of the exponential function gives us $-\langle\Gamma\rangle\leq 0$, which written more explicitly using Eq. (44) is

$$-\langle\ln(P_i)\rangle+\langle\ln(P_f)\rangle+\langle\beta Q\rangle\leq 0. \tag{49}$$

Putting in the assumed values of $P_{init}(i)$ and $P_{final}(f)$, we get

$$k_BT\ln(2)\leq-\langle Q\rangle. \tag{50}$$

From Eq. (41) we can deduce that $-Q$ is the heat dissipated into the heat reservoir.

Looking at both the system and heat reservoir we can define work the way it is defined in the classical case, namely,

$$W=\Delta E_{heat}+\Delta E_{system}. \tag{51}$$

Just like in the classical continuous case the above equation is valid under the assumption that the interaction energy between the heat reservoir and the two-state system is negligible. Again, the validity of this approximation depends on the physical systems used. As an example of a physical system we could look at nuclear magnetic resonance experiments for quantum computation where trichloroethylene was dissolved in chloroform. We will see that the interaction constant between the qubits and the chlorine is smaller than 1 Hz. This makes the interaction term negligible. Since the two energy-eigenstates of the two-state system are degenerate both before and after erasure we can say that the total change in its energy $\Delta E_{system}=0$. The change in the internal energy of the heat reservoir can be defined as $\Delta E_{heat}=E_m-E_n=-Q$. Putting these values into Eq. (51) gives us $W=-Q$. Putting them into Eq. (50) we see that

$$k_BT\ln(2)\leq\langle W\rangle. \tag{52}$$

This means that we can equally well[3] say that the work we have to do on the system in order for it to erase has to be at least $k_BT\ln(2)$.

## III. DISCUSSION

### A. Probability distributions

In all three cases we have assumed that the initial probability distributions have been 1/2. This is equivalent to saying that in the string of bits about to be erased half of the bits are in the "zero" state and half in the "one" state. This distribution happens to correspond to thermalized bits. However, in general, the initial string of bits can have any probability distribution. The amount of heat dissipated into the environment will then depend on that distribution. As an example we can take the case where all bits are already in one state only. The equations in the derivations of Landau-

[3]Again, just like in the classical continuous case, this is not an exact statement but an approximation because the interaction term in the Hamiltonian is neglected.

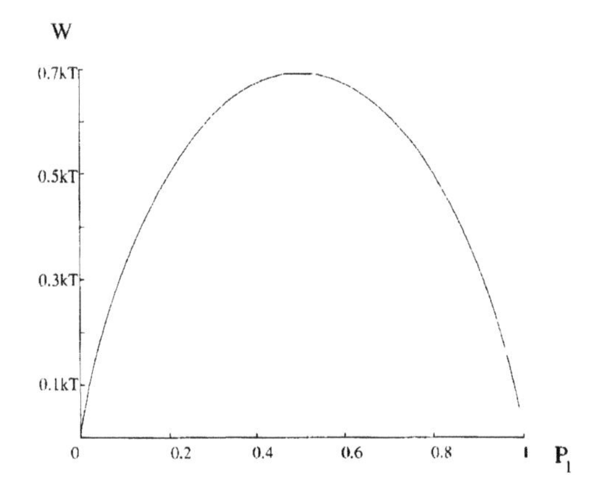


FIG. 3. Graph showing the average minimal amount of work, $P_{final}\ln P_{final} - P_{initial}\ln P_{initial} = \langle W\rangle$, required to erase a bit given the initial probability, $P_{initial}(1)$, of finding the bit in the state "one." $W$ (in $J$) is the work and $P_1$ is the probability to find the bit initially in state "one."

er's principle tell us that this kind of "erasure" can be done without any dissipation of heat. All other initial distributions will require some heat dissipation and the one which will require the largest dissipation will be the one where half of the states are in one state. More specifically, for both the classical cases considered above (Secs. II A and II B), we found that

$$\langle W\rangle \geq T\Delta S, \tag{53}$$

where $\Delta S$ is equal to minus $k_B$ times the change in the information entropy between the initial and final statistical states of the bit. In Fig. 3 we plot this lower bound on $\langle W\rangle$, as a function of the probability to find the bit initially in state "one."[4] We see that the distribution that requires the greatest amount of work (or heat dissipation) is the case $P_1 = 1/2$, in which the initial states are distributed equally between "zero" and "one."

In the quantum case we could have assumed that the initial density matrix for the bit, written in its energy eigenbasis, is some arbitrary matrix

$$\hat{\rho} = \begin{pmatrix} c & d \\ d^* & e \end{pmatrix}. \tag{54}$$

This would give us a different initial distribution from the case we considered ($c = e = 1/2$, $d = 0$). To see what the minimal heat dissipation would be for a particular density matrix we would have to diagonalize it first,

$$\hat{\rho} = \begin{pmatrix} a & 0 \\ 0 & b \end{pmatrix}. \tag{55}$$

[4]There is no particular reason for us having chosen the state "one" here. We could just as easily have chosen to write the state "zero."

The diagonal density matrix is expressed using basis states $|a\rangle$ and $|b\rangle$. We now interpret $a$ and $b$ as initial probabilities:

$$P_{init}(i') = \begin{cases} a & \text{for } i' = a, \\ b & \text{for } i' = b, \end{cases} \tag{56}$$

where the prime indicates that we are using a basis different from the energy eigenstate basis. For the final state of the bit, we still have

$$P_{final}(f) = \begin{cases} 0 & \text{for } f = 0, \\ 1 & \text{for } f = 1 \end{cases} \tag{57}$$

in the energy basis. Desipite the difference in the basis set used to describe the initial state of the bit, the calculation in Eqs. (45)–(47) goes through as before but with $i \rightarrow i'$ and we end with

$$\beta\langle W\rangle \geq -\sum_{i'} P_{init}(i')\ln P_{init}(i') = -a\ln a - b\ln b, \tag{58}$$

assuming perfect erasure. This result tells us that $\langle W\rangle$ is bounded from below by minus the change in the von Neumann entropy of the bit. As in the classical case, the greatest value of this lower bound occurs when $a = b = 1/2$, in which case $\langle W\rangle \geq k_B T \ln 2$.

Note, however, that if we are using an algorithm for erasure where we assume to receive a string with a random distribution and the string we actually receive has a different distribution, for instance, all "ones," this does not mean that we will automatically be erasing without heat dissipation. To do this we will need to change the algorithm used for erasure.

### B. Nondegenerate energy levels

All along we have assumed that the energy levels of the two states of the bits are equal before and after erasure. But what happens if we omit this assumption? Based on Landauer's argument one would have to say that his principle should apply in this case as well. To see in what form it still applies let us imagine the following system. We have an infinite ensemble of bits with degenerate energy values, like in the classical case with discrete space and time. We assume them to be populated so that half of them are in the "zero" state and half in the "one" state. We then use some external parameter to lower the energy of the "zero" state by $\Delta E$. This is our point of departure ($a$ in Fig. 4).

To do the erasure, which we will assume to be defined as restore to "zero," we start by raising state "zero" in all of the bits to the energy of state "one" ($b$ in Fig. 4). To do this we will on average have to do $1/2\Delta E$ work. Then we go through with the erasure procedure which we have shown to require $k_B T\ln(2)$ of work ($c$ in Fig. 4). Assuming that the erasure was perfect all the bits will be in the state "zero" ($d$ in Fig. 4). To go back to the original state we lower the energy of the state "zero" by $\Delta E$ ($e$ in Fig. 4). This will on average return the energy $\Delta E$. Summing up the energy put in and gotten out of the system, on average, we have

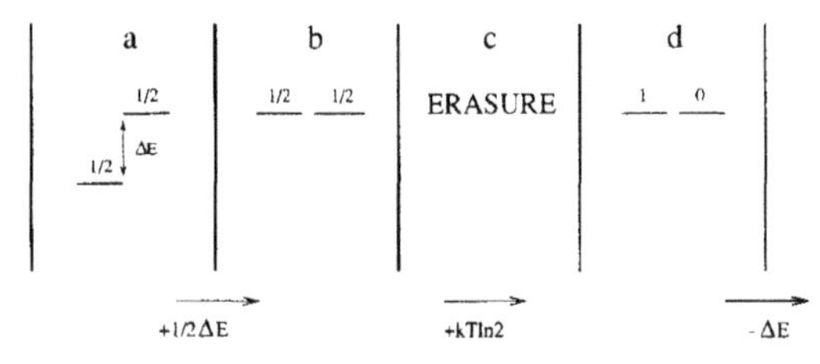


FIG. 4. Schematic picture of erasure for an ensemble of bits with different energy values for the ''zero'' state and the ''one'' state. The numbers above the lines showing the energy levels show the probability of finding the bit in that state. In going from a to b we need to add $1/2\Delta E$ amount of work on average. In c we are erasing the bit which requires $k_B T \ln(2)$ amount of work. In going from d to e we can recover $\Delta E$ of work. The average total amount of work we can recover in this kind of erasure is $\Delta E - k_B T \ln(2)$.

$$W_{out} = -1/2\Delta E - k_B T \ln(2) + \Delta E = 1/2\Delta E - k_B T \ln(2). \tag{59}$$

This means that in a system with a ground state and an excited state, where erasure is equivalent to restore to ''zero,'' with ''zero'' being the ground state, we can on average only hope to recover as much as $1/2\Delta E - k_B T \ln(2)$ of useful energy, $\Delta E$ being the difference in energies between the two states. The term ''useful energy'' is here used to denote energy that is not necessarily heat. We could have decided to define erasure as restore to ''one''. That, however, would not have given us any extra energy. We would have had to add as much as $1/2\Delta E + k_B T \ln(2)$ to the system in that procedure.

One could argue that it is not always physically possible to change the difference in energy between the levels so that one will have degeneracy. To show that even in this case one cannot get any more ''usable'' energy out of the system than $1/2\Delta E - k_B T \ln(2)$ we can use any of the above discussed models (the continuous classical case, the discrete classical case, or the quantum case) and perform the necessary calculations. The calculations will give us the same result as the thought experiment above.

In summary, we can say that for a system with equal energy levels the work required for erasure is equal to the heat dissipated into the environment, $k_B T \ln(2)$. For a system with different energy levels as the ''zero'' and ''one'' states we do not have that equality. At best we can get out half of the energy difference between the states minus the heat dissipated into the environment which will still be $k_B T \ln(2)$.

### C. Suggestions for future research

The microscopic derivation of Landauer's principle could perhaps be made more general if we could drop the assumption that the states of the thermal reservoir, with which the bits are in contact, are chosen from a canonical distribution. Perhaps it is enough to assume that the states of the thermal reservoir are chosen form a microcanonical distribution. The microcanonical ensemble would still provide us with a well defined temperature.

Landauer's principle gives us a fundamental lower bound on the amount of heat dissipated into the environment in the process of erasure. It would be interesting to see if this lower bound can actually be reached physically. One can certainly imagine a process where erasure is done on an infinite time scale where one reaches the lower bound. But is it physically realizable? If so, would it be practical for computational purposes?

## ACKNOWLEDGMENTS

This paper was written at Los Alamos National Laboratory. I would like to to thank Wojciech H. Zurek, N. Balazs, Christof Zalka, and Tanmoy Bhattacharya for new ideas and valuable and very interesting discussions about erasure. I would especially like to thank Chris Jarzynski for his guidance.

[1] R. Landauer, IBM J. Res. Dev. **3**, 183 (1961).
[2] J. W. Gibbs, *Elementary Principles in Statistical Mechanics* (Charles Scribner's Sons, New York, 1902), Chap. XIII.
[3] R. Landauer, in *Statistical Physics, Invited Papers from STATPHYS 20, 20th IUPAP International Conference on Statistical Physics UNESCO and Sorbonne Paris, 1998*, edited by A. Gervois, D. Iagolnitzer, M. Moreau, and Y. Pomeau (North-Holland, Elsevier, 1998).
[4] L. Szilard, Z. Phys. **53**, 840 (1929) [Translation in Wheeler and Zurek (Ref. [5])].
[5] J. A. Wheeler and W. H. Zurek, *Quantum Theory and Measurement* (Princeton University Press, Princeton, NJ, 1983).
[6] W. H. Zurek, in *Maxwell's Demon, Entropy, Information, Computing* (Princeton University Press, Princeton, NJ, 1990).
[7] S. Lloyd, Phys. Rev. A **56**, 3374 (1997).
[8] C. H. Bennett, IBM J. Res. Dev. **17**, 525 (1987).
[9] C. H. Bennett and R. Landauer, Sci. Am. **253**, 38 (1985).
[10] J. Berger, Int. J. Theor. Phys. **29**, 9 (1990).
[11] W. H. Zurek, e-print quant-ph/9807007.
[12] H. S. Leff and A. F. Rex, *Maxwell's Demon, Entropy, Information, Computing* (Ref. [6]).
[13] D. Wolpert, Phys. Today **45**, 98 (1992).
[14] E. Goto, N. Yoshida, K. F. Loe, and W. Hioe, in *Proceedings of the 3rd International Symposium on the Foundations of Quantum Mechanics, Tokyo*, edited by H. Ezawa, Y. Murayama, and S. Nomura (Physical Society Japan, Tokyo, 1990), p. 412.
[15] K. Shizume, Phys. Rev. E **52**, 3495 (1995).
[16] C. Jarzynski, e-print cond-mat/9802249.
[17] S. Lloyd, Phys. Rev. A **39**, 5378 (1989).
[18] G. E. Crooks, J. Stat. Phys. **90**, 1481 (1998).
[19] S. R. de Groot and P. Mazur, *Nonequilibrium Thermodynamics* (North-Holland, Amsterdam, 1962).
[20] D. Chandler, *Introduction to Modern Statistical Mechanics* (Oxford University Press, New York, 1987), p. 165.
[21] W. H. Zurek, Prog. Theor. Phys. **89**, 281 (1993).

# Chapter 5

# Quantum Nuances

# MAXWELL'S DEMON, SZILARD'S ENGINE AND QUANTUM MEASUREMENTS

W.H. Zurek

Theoretical Astrophysics
Los Alamos National Laboratory
Los Alamos, New Mexico 87545

and

Institute for Theoretical Physics
University of California
Santa Barbara, California 93106

ABSTRACT

We propose and analyze a quantum version of Szilard's "one-molecule engine." In particular, we recover, in the quantum context, Szilard's conclusion concerning the free energy "cost" of measurements: $\Delta F \geqslant k_B T \ln 2$ per bit of information.

## I. INTRODUCTION

In 1929 Leo Szilard wrote a path-breaking paper entitled "On the Decrease of Entropy in a Thermodynamic System by the Intervention of Intelligent Beings."[1] There, on the basis of a thermodynamic "gedanken experiment" involving "Maxwell's demon," he argued that an observer, in order to learn, through a measurement, which of the two equally probable alternatives is realized, must use up at least

$$\Delta F = k_B T \ln 2 \tag{1}$$

of free energy. Szilard's paper not only correctly defines the quantity known today as information, which has found a wide use in the work of Claude Shannon and others in the field of communication science.[3] It also formulates physical connection between thermodynamic entropy and information-theoretic entropy by establishing "Szilard's limit," the least price which must be paid in terms of free energy for the information gain.

The purpose of this paper is to translate Szilard's classical thought experiment into a quantum one, and to explore its consequences for quantum theory of measurements. A "one-molecule gas" is a "microscopic" system and one may wonder whether conclusions of Szilard's classical analysis remain valid in the quantum domain. In particular, one may argue, following Jauch and Baron,[4] that Szilard's analysis is inconsistent, because it employs two different, incompatible classical idealizations of the one-molecule gas--dynamical and thermodynamical--to arrive at Eq. (1). We shall show that the apparent inconsistency pointed out by Jauch and Baron is removed by quantum treatment. This is not too surprising, for, after all, thermodynamic entropy which is central in this discussion is incompatible with classical mechanics, as it becomes infinite in the limit $\hbar \to 0$. Indeed, information--theoretic analysis of the operation of Szilard's engine allows one to understand, in a very natural way, his thermodynamical conclusion, Eq.(1), as a consequence of the conservation of information in a closed quantum system.

The quantum version of Szilard's engine will be considered in the following section. Implications of Szilard's reasoning for quantum theory and for thermodynamics will be explored in Sec. III--where we shall assume that "Maxwell's Demon" is classical, and in Sec. IV, where it will be a quantum system.

## II. QUANTUM VERSION OF SZILARD'S ENGINE

A complete cycle of Szilard's classical engine is presented in Fig. 1. The work it can perform in the course of one cycle is

$$\Delta W = \int_{V/2}^{V} p(v)dv = k_B T \int_{V/2}^{V} dv/v = k_B T \ln 2 \tag{2}$$

Above, we have used the law of Gay-Lussac, $p=kT/V$, for one-molecule gas. This work gain per cycle can be maintained in spite of the fact that the whole system is at the same constant temperature T. If Szilard's model engine could indeed generate useful work, in the way described above at no additional expense of free energy, it would constitute a perpetuum mobile, as it delivers mechanical work from an (infinite) heat reservoir with no apparent temperature difference. To fend off this threat to the thermodynamic irreversibility, Szilard has noted that, "If we do not wish to admit that the Second Law has been violated, we must conclude that...the measurement...must be accompanied by a production of entropy." Szilard's conclusion has far-reaching consequences, which have not yet been fully explored. If it is indeed correct, it can provide an operational link between the concepts of "entropy" and "information." Moreover, it forces one to admit that a measuring apparatus can be used to gain information only if measurements are essentially irreversible.

Before accepting Szilard's conclusion one must realize that it is based on a very idealized model. In particular, two of the key issues have not been explored in the original paper. The first, obvious one concerns fluctuations. One may argue that the one-molecule engine cannot be analyzed by means of thermodynamics, because it is nowhere near the thermodynamic limit. This objection is overruled by noting that arbitrarily many "Szilard's engines" can be linked together to get a "many-cylinder" version of the original design. This will cut down fluctuations and allow one to apply thermodynamic concepts without difficulty.

A more subtle objection against the one-molecule engine has been advanced by Jauch and Baron.[4] They note that "Even the single-molecule gas is admissible as long as it satisfies the gas laws. However, at the exact moment when the piston is in the middle of the cylinder and the opening is closed, the gas violates the law of Gay-Lussac because gas is compressed to half its volume without expenditure of energy." Jauch and Baron "...therefore conclude that the idealizations in Szilard's experiment are inadmissible in their actual context..." This objection is not easy to refute for the classical one-molecule gas. Its molecule should behave as a billiard ball. Therefore, it is difficult to maintain that

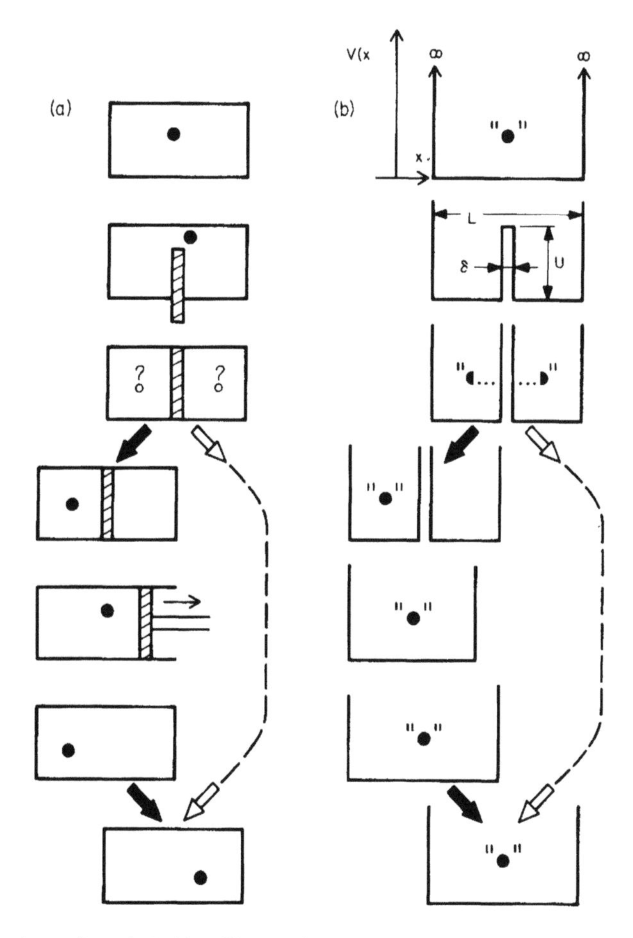


Figure 1. A cycle of Szilard's engine.
a) Original, classical version
b) Quantum version discussed here. The measurement of the location of the molecule is essential in the process of extracting work in both classical and quantum design.

after the piston has been inserted, the gas molecule still occupies the whole volume of the container. More defensible would be a claim that while the molecule is on a definite side of the partition, we do not know on which side, and this prevents extraction of the promised $k_BT \ln 2$ of energy. (This was more or less Szilard's position.) The suspicious feature of such argument is its subjectivity: Our (classical) intuition tells us that the gas molecule is on a definite side of a partition. Moreover, what we (or Maxwell's demon) know should have nothing to do with the objective state of the gas. It is this objective state of the gas which should allow one to extract energy. And, objectively, the gas has been compressed to half its volume by the insertion of the partition no matter on which side of the piston the molecule is. The eventual "observation" may help in making the engine work, but one may argue, as do Jauch and Baron, that the "potential to do work" seems to be present even before such a measurement is performed.

To re-examine arguments of Jauch and Baron, consider the quantum version of Szilard's engine, shown in Fig. 1. The role of the container is now played by the square potential well. A single gas molecule of mass m is described by the (nonrelativistic) Schrödinger equation with the boundary conditions $\psi(-L/2)= \psi(L/2)=0$. Energy levels and eigenfunctions are given by;

$$E_n=n^2\pi^2\hbar^2/(2mL^2)=\epsilon n^2 \tag{2a}$$

$$\langle x|\psi_n\rangle=\begin{cases}(2/L)^{\frac{1}{2}}\cos 2\pi nx/L \text{ for } n=2k+1 & \text{(2b)}\\ (2/L)^{\frac{1}{2}}\sin 2\pi nx/L \text{ for } n=2k & \text{(2c)}\end{cases}$$

At a finite temperature $T=\beta^{-1}k_B^{-1}$ the equilibrium state of the system will be completely described by the density matrix:

$$\rho=Z^{-1}\sum_n \exp(-\beta E_n)|\psi_n\rangle\langle\psi_n|. \tag{3}$$

Above, Z is the usual partition function:

$$Z=\sum_{n=1}^{\infty}\exp(-\beta\epsilon n^2)=\sum_{n=1}^{\infty}\zeta^{n^2} \tag{4}$$

For $1/2<\zeta< 1$, Z can be adequately approximated by:

$$Z=\frac{1}{2}\left(\sqrt{\pi/|\ln\zeta|}-1\right). \tag{5}$$

For our purposes a still simpler, high-temperature approximation

$$Z=(\pi/\epsilon\beta)^{\frac{1}{2}}/2=L(h^2/2\ mk_BT)^{\frac{1}{2}}\ . \tag{6}$$

valid for $\epsilon \ll k_BT$, will be sufficient most of the time. This is, of course, the familiar Boltzmann gas partition function. It can be readily generalized to the three-dimensional box

$$Z=L_xL_yL_z/(h^2/2\pi mk_BT)^{3/2}\ , \tag{7}$$

as well as to the case when there are N "classically" indistinguishable particles. The point of this elementary calculation is to demonstrate that, in the further analysis, we can rely on classical estimates of pressure, internal energy, entropy, etc...for one molecule gas which were used by Szilard[1]: A partition function completely determines thermodynamic behavior of the system.

Consider now a "piston," slowly inserted in the middle of the potential well "box". This can be done either (1) while the engine is coupled with the reservoir, or (2) when it is isolated. In either case, it must be done slowly, so that the process remains either (1) thermodynamically or (2) adiabatically reversible. We shall imagine the piston as a potential barrier of width $d \ll L$ and height that eventually attains $U \gg k_B T$. The presence of this barrier will alter the structure of the energy levels: these associated with even n will be shifted "upwards" so that the new eigenvalues are:

$$E'_{2k}=\varepsilon'(2k)^2+\Delta_k=E_k+\Delta_k \tag{8}$$

where

$$\varepsilon'=\varepsilon\, L^2/(L-d)^2 \tag{9}$$

and

$$\Delta_k \cong (4\varepsilon'/\pi)\exp(-d\sqrt{2m(U-E_k)}/\hbar). \tag{10}$$

Energy levels corresponding to the odd values of n are shifted upwards by $\Delta E_n \sim (2n+1)\varepsilon'$ so that

$$E'_{2k-1}=\varepsilon'(2k)^2-\Delta_k=E_k-\Delta_k \tag{11}$$

A pair of the eigenvalues $E'_{2k}$, $E'_{2k-1}$ can be alternatively regarded as the kth doubly degenerate eigenstate of the newly created two-well potential with degeneracy broken for finite values of U.

For $U\to\infty$ exact eigenfunctions can be given for each well separately. For finite U, for these levels where $\Delta_k \ll E_k$, eigenfunctions of the complete potential can be reconstructed from the kth eigenfuctions of the left ($|L_k\rangle$) and right ($|R_k\rangle$) wells:

$$E_k+\Delta_k \leftrightarrow |\psi_k^+\rangle=(|L_k\rangle-|R_k\rangle)/\sqrt{2}$$

$$E_k-\Delta_k \leftrightarrow |\psi_k^-\rangle=(|L_k\rangle+|R_k\rangle)/\sqrt{2}$$

Alternatively, eigenfunctions of the left and right wells can be expressed in terms of energy eigenfunctions of the complete Hamiltonian

$$|L_k\rangle=(|\psi_k^+\rangle+|\psi_k^-\rangle)/\sqrt{2} \tag{12c}$$

$$|R_k\rangle=(|\psi_k^-\rangle-|\psi_k^+\rangle)/\sqrt{2} \tag{12d}$$

## III. MEASUREMENTS BY THE CLASSICAL MAXWELL'S DEMON

Consider a measuring apparatus which, when inserted into Szilard's engine, determines on which side of the partition the molecule is. Formally, this can be accomplished by the measurement of the observable

$$\hat{\Pi}=\lambda(|L\rangle\langle L|-|R\rangle\langle R|). \tag{13}$$

Here $\lambda>0$ is an arbitrary eigenvalue, while

$$|L\rangle=\left(\sum_{k=1}^{N}|L_k\rangle\right)/N^{1/2} \tag{13a}$$

$$|R\rangle=\left(\sum_{k=1}^{N}|R_k\rangle\right)/N^{1/2} \tag{13b}$$

and N is sufficiently large, $N^2\varepsilon\beta>>1$.

The density matrix before the measurement, but after piston is inserted, is given by

$$\tilde{\rho}=\tilde{Z}^{-1}\sum_{k=1}^{\infty}\exp(-\beta E_k)\{\exp(-\beta\Delta_k)|\psi_k^+\rangle\langle\psi_k^+|+\exp(\beta\Delta_k)|\psi_k^-\rangle\langle\psi_k^-|\} \tag{14}$$

$$=\tilde{Z}^{-1}\sum_{k=1}^{\infty}\exp(-\beta E_k)\{\cosh\beta\Delta_k(|L_k\rangle\langle L_k|+|R_k\rangle\langle R_k|)+\sinh\beta\Delta_k(|L_k\rangle\langle R_k|+|R_k\rangle\langle L_k|)\}$$

Depending on the outcome of the observation, the density matrix becomes either $\rho_L$ or $\rho_R$ where;

$$\rho_L=Z_L^{-1}\sum_{k=1}^{\infty}\exp(-\beta E_k)\cosh\beta\Delta_k|L_k\rangle\langle L_k|, \tag{15a}$$

$$\rho_R=Z_R^{-1}\sum_{k=1}^{\infty}\exp(-\beta E_k)\cosh\beta\Delta_k|R_k\rangle\langle R_k|. \tag{15b}$$

Both of these options are chosen with the same probability. The classical "demon" "knows" the outcome of the measurement. This information can be used to extract energy in the way described by Szilard.

We are now in a position to return to the objection of Jauch and Baron. The "potential to do work" is measured by the free energy A(T,L) which can be readily calculated from the partition function

$$A(T,L) = -k_BT\ln Z(T,L) .$$

For the one-molecule engine this free energy is simply

$$A= -k_BT\ln[L/(h^2/2\pi mk_BT)^{\frac{1}{2}}] \tag{16}$$

before the partition is inserted. It becomes

$$\tilde{A}= -k_BT\ln[(L-d)/(h^2/2\pi mk_BT)^{\frac{1}{2}}] \tag{17}$$

after the insertion of the partition. Finally, as a result of the measurement, free energy increases regardless of the outcome

$$A_L=A_R= -k_BT\ln[((L-d)/2)/h^2/2\pi mk_BT)^{\frac{1}{2}}]$$

$$= -k_BT(\tilde{A}-\ln 2) \tag{18}$$

Let us also note that the change of A as a result of the insertion of the partition is negligible:

$$\tilde{A}-A=k_BT\ln(L/(L-d))\sim O(d/L). \tag{19}$$

However, the change of the free energy because of the measurement is precisely such as to account for the $W=k_BT\ln 2$ during the subsequent expansion of the piston:

$$A_L-\tilde{A}=k_BT\ln 2. \tag{20}$$

It is now difficult to argue against the conclusion of Szilard. The formalism of quantum mechanics confirms the key role of the measurement in the operation of the one-molecule engine. And the objection of Jauch and Baron, based on a classical intuition, is overruled.

The classical gas molecule, considered by Szilard, as well as by Jauch and Baron, may be on the unknown side of the piston, but cannot be on "both" sides of the piston. Therefore, intuitive arguments concerning the potential to do useful work could not be unambiguously settled in the context of classical dynamics and thermodynamics. Quantum molecule, on the other hand, can be on "both" sides of the potential barrier, even if its energy is far below the energy of the barrier top, and it will "collapse" to one of the two potential wells only if $\hat{\Pi}$ is "measured."

It is perhaps worth pointing out that the density matrix $\tilde{\rho}$ , Eq. (15), has a form which is consistent with the "classical" statement that " a molecule is on a definite, but unknown side of the piston" almost equally well as with the statement that "the molecule in in a thermal mixture of the energy eigenstates of the two-well potential." This second statement is rigorously true in the limit of weak coupling with the resevoir: gas is in contact with the heat bath and therefore is in thermal equilibrium. On the other hand, the off-diagonal terms of the very same density matrix in the $|L_k>, |R_k>$ representation are negligible ($\sim \sinh\beta\Delta_k$). Therefore, one can almost equally well maintain that this density matrix describes a molecule which is on an "unknown, but definite" side of the partition. The phase between $|R_n>$ and $|L_n>$ is lost.

The above discussion leads us to conclude that the key element needed to extract useful work is the correlation between the state of the gas and the state of the demon. This point can be analyzed further if we allow, in the next section, the "demon" to be a quantum system.

## IV. MEASUREMENT BY "QUANTUM MAXWELL'S DEMON"

Analysis of the measurement process in the previous section was very schematic. Measuring appartus, which has played the role of the "demon" simply acquired the information about the location of the gas molecule, and this was enough to describe the molecule by $\rho_L$ or $\rho_R$. The second law demanded the entropy of the apparatus to increase in the process of measurement, but, in the absence of a more concrete model for the apparatus it was hard to tell why this entropy increase was essential and how did it come about. The purpose of this section is to introduce a more detailed model of the apparatus, which makes such an analysis possible.

We shall use as an apparatus a two-state quantum system. We assume that it is initially prepared by some external agency -- we shall refer to it below as an "external observer" -- in the "ready to measure" state $|D_0\rangle$. The first stage of the measurement will be accomplished by the transition:

$$|L_n>|D_0> \rightarrow |L_n> |D_L> \tag{21a}$$

$$|R_n>|D_0> \rightarrow |R_n>|D_R> \tag{21b}$$

for all levels n. Here, $|D_R>$ and $|D_L>$ must be orthogonal if the measurement is to be reliable. This will be the case when the gas and the demon interact via an appropriate coupling Hamiltonian, e.g.:

$$H_{int}=i\delta(|L_n><L_n|-|R_n><R_n|)(|D_L><D_R|-|D_R><D_L|) \tag{22}$$

for the time interval $\Delta t=\Pi\hbar/(4\delta)^{5,6}$, and the initial state of the demon is:

$$|D_0\rangle = (|D_L\rangle + |D_R\rangle)/\sqrt{2} \tag{23}$$

For, in this case the complete density matrix becomes:

$$P = \exp(-iH_{int}\Delta t/\hbar)\tilde{\rho}|D_0\rangle\langle D_0| =$$
$$\cong(\rho_L|D_L\rangle\langle D_L| + \rho_R|D_R\rangle\langle D_R|)/2 \tag{24}$$

Here and below we have omitted small off-diagonal terms ($\sim\beta\Delta_n$) present in $\tilde{\rho}$, Eq. (14) and, by the same token, we shall drop corrections ($\sim\beta^2\Delta_n^2$) from the diagonal of $\rho_L$ and $\rho_R$. As it was pointed out at the end of the last section, for the equilibrium there is no need for the "reduction of the state vector." The measured system -- one molecule gas -- is already in the mixture of being on the left and right-hand side of the divided well.

To study the relation of the information gain and entropy increase in course of the measurement, we employ the usual definition of entropy in terms of the density matrix[3,6,7,8]

$$S(\rho) = -k_B \mathrm{Tr}\,\rho\ln\rho \tag{25}$$

The information is then:

$$I(\rho) = \ln(\mathrm{Dim}(\mathcal{H})) - S(\rho)/k_B, \tag{26}$$

where $\mathcal{H}$ is the Hilbert space of the system in question. Essential in our further discussion will be the mutual information $I_\mu(P_{AB})$ defined for two subsystems, A and B, which are described jointly by the density matrix $P_{AB}$, while their individual density matrices are $\rho_A$ and $\rho_B$:

$$I_\mu(P_{AB}) = I(P_{AB}) - (I(\rho_A) + I(\rho_B)). \tag{27}$$

In words, mutual information is the difference between the information needed to specify A and B are not correlated, $P_{AB} = \rho_A\rho_B$, then $I_\mu(P_{AB}) = 0$.

The readoff of the location of the gas molecule by the demon will be described by an external observer not aware of the outcome by the transition:

$$\tilde{\rho}|D_0\rangle\langle D_0| \to (\rho_L|D_L\rangle\langle D_L| + \rho_R|D_R\rangle\langle D_R|)/2 \tag{28}$$

The density matrix of the gas is then:

$$\tilde{\rho} = \tilde{Z}^{-1}\sum_{n=1}\exp(-\beta\epsilon n^2)(|L_n\rangle\langle L_n| + |R_n\rangle\langle R_n|) = (\rho_L + \rho_R)/2 \tag{29}$$

Thus, even after the measurement by the demon, the density matrix of the gas, $\rho_G$ will be still :

$$\rho_G = \langle D_L|P|D_L\rangle + \langle D_R|P|D_R\rangle \cong \tilde{\rho} \tag{30}$$

The state of the demon will, on the other hand, change from the initial pure state $|D_0\rangle\langle D_0|$ into a mixture:

$$\rho_D = \sum_n(\langle L_n|P|L_n\rangle + \langle R_n|P|R_n\rangle) = (|D_L\rangle\langle D_L| + |D_R\rangle\langle D_R|)/2 \tag{31}$$

Entropy of the gas viewed by the external observer remains constant:

$$S(\rho_G)=S(\tilde{\rho})= \partial(\beta\ln\tilde{Z})/\partial\beta \tag{32}$$

Entropy of the demon has however increased:

$$S(\rho_D)-S(|D_0\rangle\langle D_0|)=k_B\ln 2 \tag{33}$$

Nevertheless, the combined entropy of the gas-demon system could not have changed: In our model evolution during the read-off was dynamical, and the gas-demon system was isolated. Yet, the sum of the entropies of the two subsystems -- the gas and the demon -- has increased by $k_B\ln 2$. The obvious question is then: where is the "lost information" $\Delta I=I(P)-(I(\tilde{\rho})+I(\rho_D))$? The very form of this expression and its similarity to the right-hand side of Eq. (27) suggests the answer: The loss of the information by the demon is compensated for by an equal increase of the mutual information:

$$\Delta I_\mu=I_\mu(P)-I_\mu(\tilde{\rho}|D_0\rangle\langle D_0|). \tag{34}$$

Mutual infromation can be regarded as the information gained by the demon. From the conservation of entropy during dynamical evolutions it follows that an increase of the mutual information must be compensated for by an equal increase of the entropy.

$$\Delta I_\mu-\Delta S/k_B=0 \tag{35}$$

This last equation is the basis of Szilard's formula, Eq. (1).

At this stage readoff of the location of the gas molecule is reversible. To undo it, one can apply the inverse of the unitary operation which has produced the correlation. This would allow one to erase the increase of entropy of the demon, but only at a price of the loss of mutual information.

One can now easily picture further stages of the operation of Szilard's engine. The state of the demon can be used to decide which well is "empty" and can be "discarded."[6] The molecule in the other well contains twice as much energy as it would if the well were adiabatically (i.e. after decoupling it from the heat reservoir) expanded from its present size $\sim L/2$ to L. Adiabatic expansion conserves entropy. Therefore, one can immediately gain $\Delta W=\frac{1}{2}\langle E\rangle=k_BT/4$ without any additional entropy increases. One can also gain $\Delta W=k_BT\ln 2$, Eq. (2), by allowing adiabatic expansion to occur in small increments, inbetween which one molecule gas is reheated by the re-established contact with the reservoir. In the limit of infinitesimally small increments this leads, of course, to the isothermal expansion. At the end the state of the gas is precisely the same as it was at the beginning of the cycle, and we are $\Delta W$ "richer" in energy. If Szilard's engine could repeat this cycle, this would be a violation of the second law of thermodynamics. We would have a working model of a perpetuum mobile, which extracts energy in a cyclic process from an infinite reservoir without the temperature difference.

Fortunately for the second law, there is an essential gap in our above design: The demon is still in the mixed state, $\rho_D$, Eq. (31). To perform the next cycle, the observer must "reset" the state of the demon to the initial $|D_0\rangle$. This means that the entropy $dS=k_B\ln 2$ must be somehow removed from the system. If we were to leave demon in the mixed state, coupling it with the gas through $H_{int}$ will not result in the increase of their mutual information: The one-bit memory of the demon is still filled by the outcome of the past measurement of the gas. Moreover, this measurement cannot any longer be reversed by allowing the gas and the demon to interact. For, even though the density matrix of each of these two systems has remained the same, their joint density matrix is now very different:

$$\tilde{\rho}\rho_D=\{(\rho_L|D_L><D_L|+\rho_R|D_R><D_R|)/2+(\rho_L|D_R><D_R|+\rho_R|D_L><D_L|)/2\}/2 \quad (36)$$

One could, presumably, still accomplish the reversal using the work gained in course of the cycle. This would, however, defy the purpose of the engine. The only other way to get the demon into the working order must be executed by an "external observer" or by its automated equivalent: The demon must be reset to the "ready-to-measure" state $|D_0>$.

As it was pointed out by Bennett in his classic analysis, this operation of memory erasure is the true reason for irreversibility, and the ultimate reason why the free energy, Eq. (1), must be expended.[9] Short of reversing the cycle, any procedure which resets the state of the demon to $|D_0>$ must involve a measurement. For instance, a possible algorithm could begin with the measurement whether the demon is in the state $|D_L>$ or in the orthogonal state $|D_R>$. Once this is established then, depending on the outcome, the demon may be rotated either "rightward" from $|D_L>$, or "leftward" from $|D_R>$, so that its state returns to $|D_0>$. The measurement by some external agency -- some "environment" -- is an essential and unavoidable ingredient of any resetting algorithm. In its course entropy is "passed on" from the demon to the environment.

## V. SUMMARY

The prupose of our discussion was to analyze a process which converts thermodynamic energy into useful work. Our analysis was quantum, which removed ambiguities pointed out by Jauch and Baron in the original discussion by Szilard. We have concluded, that validity of the second law of thermodynamics can be satisfied only if the process of measurement is accompanied by the increase of entropy of the measuring apparatus by the amount no less than the amount of gained information. Our analysis confirms therefore conclusions of Szilard (and earlier discussion of Smoluchowski[10], which has inspired Szilard's paper). Moreover, we show that the ultimate reason for the entropy increase can be traced back to the necessity to "reset" the state of the measuring apparatus[9], which, in turn, must involve a measurement. This necessity of the "readoff" of the outcome by some external agency -- the "environment" of the measuring apparatus has been already invoked to settle some of the problems of quantum theory of measurements.[11] Its role in the context of the above example is to determine what was the outcome of the measurement, so that the apparatus can be reset. In the context of quantum measurements its role is very similar: it forces one of the outcomes of the measurement to be definite, and, therefore, causes the "collapse of the wavepacket."

In view of the title of Szilard's paper, it is perhaps worthwhile to point out that we did not have to invoke "intelligence" at any stage of our analysis: all the systems could be presumably unanimated. However, one can imagine that the ability to measure, and make the environment "pay" the entropic cost of the measurement is also one of the essential attributes of animated beings.

## ACKNOWLEDGMENTS

I would like to thank Charles Bennett and John Archibald Wheeler for many enjoyable and stimulating discussions on the subject of this paper. The research was supported in part by the National Science Foundation under Grant No. PHY77-27084, supplemented by funds from the National Aeronautics and Space Administration.

REFERENCES

1. L. Szilard, "On the Decrease of Entropy in a Thermodynamic System by the Intervention of Intelligent Beings," Z. Phys. 53:840 (1929); English translation reprinted in Behavioral Science 9:301 (1964), as well as in Ref. 2, p. 539.
2. J.A. Wheeler and W.H. Zurek, eds., Quantum Theory and Measurement (Princeton University Press, Princeton, 1983).
3. C.E. Shannon and W. Weaver, The Mathematical Theory of Communication, (University of Illinois Press, Urbana, 1949).
4. J.M. Jauch and J.G. Baron, "Entropy, Information, and Szilard's Paradox," Helv. Phys. Acta, 45:220 (1972).
5. W.H. Zurek, "Pointer Basis of Quantum Apparatus: Into What Mixture Does the Wavepacket Collapse?" Phys. Rev. D24:1516 (1981).
6. W.H. Zurek, "Information Transfer in Quantum Measurements: Irreversibility and Amplification," Quantum Optics, Experimental Gravitation and Measurement Theory, eds. P. Meystre and M.O. Scully (Plenum Press, New York, 1983).
7. J.R. Pierce and E.C. Rosner, Introduction to Communication Science and Systems (Plenum Press, New York, 1980).
8. H. Everett III, Dissertation, reprinted in B.S. DeWitt and N. Graham, eds., The Many -- Worlds Interpretation of Quantum Mechanics (Princeton University Press, Princeton, 1973).
9. C.H. Bennett, "The Thermodynamics of Computation" Int. J. Theor. Phys. 21:305 (1982).
10. M. Smoluchowski, "Vorträge über die Kinetische Theorie der Materie und Elktrizitat," Leipzig 1914.
11. W.H. Zurek, "Environment-Induced Superselection Rules," Phys. Rev. D26:]862 (1982).

# Keeping the Entropy of Measurement: Szilard Revisited

**Elihu Lubkin**[1]

*Received January 13, 1987*

What happens to von Neumann's entropy of measurement after we get the outcome? It becomes entropy of erasure. This is cribbed from Szilard (1929). Also, two errors in that celebrated paper are corrected.

## 1. INTRODUCTION

The Second Law of Thermodynamics forbids a net gain of information. Yet a measurement "provides information." Measurement itself thus becomes paradoxical, until one reflects that the gain in information about the system of interest might be offset by a gain in entropy of some "garbage can" gc. Indeed, it *must* be so offset to save the bookkeeping of the Second Law. This apparent and paradoxical gain in information attendant upon observation, presumably due to neglect of some dissipant gc, has long prompted aberrant speculation that intelligent beings and even life in general somehow indeed violate the Second Law, an erroneous view I cite only for perspective. For some time I have fallen prey to a version of this paradox, developed in the context of standard quantum theory of measurement as delineated by von Neumann (1955), a trap I have recently been able to escape with the help of Szilard (1929), the celebrated paper in which the related paradox of Maxwell's demon is broken. Here I describe a precise formulation, then resolution of my paradox of information through measurement.

## 2. REVIEW OF VON NEUMANN

Von Neumann finds *no* paradoxical loss of entropy through measurement, but rather a *gain* of entropy quite in conformity with the Second

[1]Department of Physics, University of Wisconsin-Milwaukee, Milwaukee, Wisconsin 53201.

Law. I go over this well-known ground to explore the paradox of the *missing* paradox!

Indeed, if we start with a pure state vector $|x\rangle = \sum a_i |e_i\rangle$ resolved on the orthonormal basis of states $|e_i\rangle$ separated by distinct outcomes of the measurement, then the $p_i = |a_i|^2$ are the probabilities of the various outcomes, or, in an early way of saying it, the probabilities of the different possible "quantum jumps." Interaction with the measuring device makes the system "jump" with probabilities $p_i$, which introduce nonnegative entropy

$$\sum p_i \ln \frac{1}{p_i} \equiv \left\langle \ln \frac{1}{p} \right\rangle \equiv vN \tag{1}$$

(in Boltzmen or $e$-folds). Perhaps I should attribute this notion of production of entropy to Dirac (1938–39), who noted that the quantum jumps not only introduce entropy, but might account for *all* production of entropy, a thesis with ancient roots (Lucretius ~55 B.C.; see Latham, 1951, p. 66) that I also long ago found support for in a calculation (Lubkin, 1978).

Von Neumann makes this already familiar production of entropy $vN$ unambiguous by dealing with very concrete ensembles. Thus, the pure state $|x\rangle$ before the measurement is presented as some large number $N$ of copies of an $|x\rangle$-prepared system. After measurement, $p_i N$ estimates the number of copies cast into state $|e_i\rangle$, hence the originally pure ensemble $|x\rangle\langle x| = P$ gets replaced by the mixture $\sum |e_i\rangle p_i \langle e_i| = \Upsilon$. The entropy per copy of $P$ is 0, the entropy per copy of $\Upsilon$ is quite unambiguously the $vN$ attributable to measurement, or to "quantum jumps."

The clarity of this mixing of the multicopy ensemble unfortunately reaches its resulting increase $vN$ of entropy by so well bypassing the phenomenon of *gain* of information through measurement that we simply do not learn enough about our paradox from it. Of course, more generally, the original ensemble $P$ need not be pure, and the mutually orthogonal outcome spaces $E_i$ [of a sharp test (Lubkin, 1979b), definition on p. 550; also Lubkin (1974)] need not be one-dimensional, yet the final ensemble $\Upsilon = \sum_i E_i P E_i$ if distinct from $P$ is of greater entropy than $P$, which happens if any $E_i$ fails to commute with $P$, this greater generality being, however, equally useless to us in regard to our paradox.

## 3. THE PARADOX

So, to formulate the paradox, as it were to look for trouble, I focus attention as before upon the *individual* trial, rather than on the statistical behavior of the very many trials dealt with by von Neumann. We still have the probabilities $p_i = |a_i|^2$ (in the case of initial pure state $x$) as predictions of the likelihoods of our single outcome for our individual trial. Indeed, after the trial is over, *but before we look* at the outcome, the original state matrix $P$ is replaced by the new state matrix $\Upsilon$. Yet when we do learn the outcome, $\Upsilon$ is in turn replaced by $E_i \Upsilon E_i$ (unnormalized), with $i$ the index of the single observed outcome, or more simply (and normalized) by $E_i$ itself if $E_i$ is one-dimensional (dimension of its image space). The threefold sequence

$$P \rightarrow \Upsilon \rightarrow E_i \qquad (2)$$

(to keep to the simple one-dimensional case, which illustrates the point nicely), with recognition of a state $\Upsilon$ after the essential interaction of measurement but before transition of the answer "$i$" to the observer, emphasizes the increase in entropy in the first step $P \rightarrow \Upsilon$, so that we may be properly shocked at the loss of this entropy in the second step, $\Upsilon \rightarrow E_i$. Also, the $P \rightarrow \Upsilon$ increase in entropy is of course precisely the $\langle \ln (1/p) \rangle$ entropy of measurement $vN$ of von Neumann, which we bet is the "right answer." We are now in the uncomfortable position of betting on a "right answer" for an entropy of measurement provided that we do not look at the outcome, yet getting zero if we do look, as if a proper measurement does not involve looking at the result, which is surely an unhappy state of affairs!

## 4. DEBATE ON THE SINGLE TRIAL

Some may fault the notion of probability without an ensemble as the root of my trouble; to relate the probabilities $p_i$ to experience, we must run many trials, not one, and so the probabilities, to be physically meaningful, *must* refer only to the very many trials. Yet let us run our $N$ trials one by one. If *each* time we generate no entropy, then we will not have generated any entropy in the *string* of $N$ trials, either. And the per-trial "entropy" $vN$ of the "$N$-ensemble" of $N$ trials in my above version of von Neumann appears only in describing the *nonuniformity* of the ensemble. Our knowledge of each outcome after it happens is replaced by the randomness of entropy in the amount $vN$ only if the state we produce for a next experiment is *randomly* chosen from that ensemble. To repeat, the generation of entropy by measurement, if nil in the particular trial, is nil in the $N$ trials together, and entropy of the ensemble for a subsequent test is produced only by the possible policy of disregarding the separation into cases provided by the particular outcomes, and using the whole ensemble in a future experiment *with no regard for the known sorting into cases of the earlier individual outcomes.*

Is there, then, no true entropy of measurement, but just entropy from a refusal to accept the sorting provided? No, there must be more to it than that, because of the $P \to \Upsilon \to E_i$ paradox; $\Upsilon$, the situation before we look, but after the measurement has taken place, already possesses entropy of measurement $vN$, which cannot be lost in the $\Upsilon \to E_i$ phase; *therefore, there must be a garbage can gc.*

## 5. BOHREAN DOUBTS

Or is the situation too vague? Increase of entropy is, after all, not universal. Define your system of interest to shrink to nothing, then its entropy of course goes to zero. If, as another but less trivial case, the system is so well isolated as to obey a law of unitary motion, its entropy remains constant, and frustratingly for Boltzmann, will not increase. The Second Law is for situations in which the system is imperfectly isolated, so that it tends to develop new correlations with the outside, yet where somehow the gross content of the system remains fixed, on the average, so that shrinking to nothing, for example, is forbidden. Then the density matrix of the system of interest, in having the increasing external correlations lopped off through Landau tracing over externalities (Lubkin 1978), becomes increasingly mixed. Another reason that having "the system" perfectly isolated is suspicious is that specification of a precise value of some observable forces a detailed correlation of the system with the outside in regard to complementary observables (e.g., Lubkin, 1979b, p. 537). Perhaps these various philosophical demands attending definition of the type of "system" that should indeed obey the Second Law engender an incompatibility with the "systems" in a paradigm of measurement; perhaps "systems" for clear Second-Law bookkeeping and "systems" in measurement are complementary. But I feel that these Bohrean misgivings anent the clarity of my paradox are mooted by finding gc with the help of Szilard's example. More strongly, it *must* be that the Second Law and measurement *go together*, because they both refer to the empiricism of everyday life, the particularization of definite outcomes in everyday measurement serving as the injector of randomness that drives the Second Law, as noted already by Lucretius and Dirac.

## 6. AMPLITUDES IN SUPPORT OF THE SINGLE TRIAL

Having drawn attention to the philosophical objection to probabilities for a single trial, I should make it clear why I nevertheless regard the $P \to \Upsilon$ phase to be meaningful for a single trial, again for the clear pure case $P = |x\rangle\langle x|$. We may regard state $x$ times the equipment's original state $y_0$ to evolve, $\sum a_i|e_i\rangle \otimes |y_0\rangle \to \sum a_i|e_i\rangle \otimes |y_i\rangle$, to a new *pure* state, with density matrix

$$\bar{\Upsilon} = \sum_{ij} a_i a_j^* |e_i\rangle \otimes |y_i\rangle\langle e_j| \otimes \langle y_j|$$

with $\Upsilon = \sum_i |a_i|^2 |e_i\rangle\langle e_i|$ obtained from $\bar{\Upsilon}$ by the Landau tracing-out of the equipment. That is, the various *amplitudes* for different "outcomes" are in principle capable of subsequent interference, and our option to disregard this by confining attention to the $x$-system subsequently is what reduces the amplitudes to probabilities and engenders entropy. So the branching into distinct channels is there, physically, in the Schrödinger equation, which gives us $\bar{\Upsilon}$, even before the neglect or approximation that limits our attention to $\Upsilon$. And this $\Upsilon$ already formally has entropy $vN = \mathrm{Tr}\, \Upsilon \ln(1/\Upsilon)$, even though operator $\Upsilon$ acts on the Hilbert space appropriate to *one* trial, *not* on a tensor product of $N$ copies of that Hilbert space. The fact that entropy $vN$ times $N$ is most easily *displayed* by the scatter in an $N$-copy ensemble *associated* to $\Upsilon$ should not be used to obscure the fact that the entropy $vN$ times 1 is already there without any such display. Of course, to emphasize the already physical qualities of this amplitudinous branching has become known as the "many-worlds" point of view; the reader will follow better if I say that for me, this is the *right* view.

## 7. PLURAL REALITIES

Indeed for clarity I restate the paradox in grossly multiworldly language.

*First Multiworldly Statement*: $vN = \sum p_i \ln(1/p_i)$ explicitly contemplates the multiplicity of branches issuing from a node of measurement, hence is obviously appropriate for a contemplator of that multiplicity, say, for an early observer who has not yet read the outcome, or for an outer observer or friend outside the laboratory. But in the reality relative to an inner or late observer who *has* read the outcome, the other branches are excluded, and the entropy expression $vN$ in seeming to yet take those other branches seriously seems no longer appropriate.

*Preliminary Resolution.* Now, the amount of entropy $vN$ or more must nevertheless yet be there, even in the bookkeeping of an inner observer, because the reality of experienece is the steady progress "inward" of an observer, through a series of particularized outcomes, and it is to this common experience or *stream* of consciousness that the Second Law and indeed the notion of time applies, else the Second Law could not have been found out in the 19th century. So while $vN$ for the *outer* or early observer is simply a feature of multitudinous branching, there must also be a $vN$ *inner* to each branch harbored in a gc in each branch.

*Second Multiworldly Statement.* Branching realities might be feared inconsistent with laws framed in a philosophy of one unique reality, hence inconsistent with all the traditional "*First*" laws of conservation, and inconsistent, too, with the Second Law. Since "each branch has all the baryons" (Lubkin, 1979a, p. 174), laws of conservation are fine—of course such laws are also tied to symmetry, and enough symmetries will remain. The present investigation may be taken to deal with the scare that the *Second* Law might fail.

*Resolution Again.* But it must not fail. When the probabilities are swept away by a specific outcome, their entropy must yet be swept into a gc, not actually annulled. This is the usual caution, in an entropic embarrassment, that one may not have been sufficiently careful in defining the problem's thermodynamically relevant "universe."

## 8. THE ANSWER, gc

Where, then, is the garbage can? Szilard (1929) finds it for me: gc is the damper on the register that receives the outcome, the damper that allows that register to get rid of its former configuration. I call this dissipation *entropy of erasure* (er), and argue that er $\geq vN$.

*Lemma.* Let the $k$-labeled orthogonal states of the register reg occur with Boltzmann's probabilities

$$q_k \propto \exp(-\varepsilon_k / T_0) \tag{3}$$

in the mixed state of reg at temperature $T_0$ that is to correspond to "erasure"; note that by choosing the energies $\varepsilon_k$ of the levels, one can adjust the $q_k$ at will. Outcome $k$ of a measurement is assumed to set reg instead to pure energy level $k$. Then, if the probability of this outcome is $p_k$, the entropy of erasure is given by the "cross entropy" expression

$$\mathrm{er} = -\sum p_k \ln q_k \tag{4}$$

*Proof.* Suppose that, due to the outcome having been $k$, the reg starts at pure state $E_k$ before erasure to $T_0$; to calculate the entropy of that erasure "$\mathrm{er}_k$" from such a start: $\mathrm{er}_k$ is a sum of two terms.

$$\mathrm{er}_k = \Delta S_{\mathrm{reg}} + \Delta S_{\mathrm{res}} \tag{5}$$

The part $\Delta S_{\mathrm{reg}}$ is $S_{\mathrm{reg}}$ at $T_0$ minus $S_{\mathrm{reg}}$ at $E_k$. As the latter is pure, its entropy $S_{\mathrm{reg}}$ is 0, whereas $S_{\mathrm{reg}}$ at $T_0 = -\sum q_j \ln q_j$. So also $\Delta S_{\mathrm{reg}} = -\sum q_j \ln q_j$. The other term $\Delta S_{\mathrm{res}}$ is the heat gained by the $T_0$ reservoir used for the quenching, divided by $T_0$. If the thermodynamic internal energy of reg after quenching is $U_0$, then the heat gained by reg is $U_0 - \varepsilon_k$, whence the heat gained by res is $-(U_0 - \varepsilon_k)$; so

$$\mathrm{er}_k = -\sum q_j \ln q_j - (U_0 - \varepsilon_k)/T_0 \tag{6}$$

is the result for the entropy $\text{er}_k$ of erasure from the definite outcome $k$. Of course,

$$U_0 = \sum q_j \varepsilon_j \tag{7}$$

Finally, the expected value er for the entropy of erasure if reg starts from an unknown setting $k$ but with probability $p_k$ is

$$\text{er} \equiv \sum p_k \cdot \text{er}_k \tag{8}$$

This convex combination alters (6) only in replacing $\varepsilon_k$ by $\sum p_k \varepsilon_k$, giving

$$\text{er} = -\sum q_j \ln q_j + \sum (p_j - q_j)\varepsilon_j / T_0 \tag{9}$$

Let

$$z \equiv \sum \exp(-\varepsilon_k / T_0) \tag{10}$$

Then $\varepsilon_k / T_0 = -\ln q_k - \ln z$ eliminates the $\varepsilon$'s, to give

$$\text{er} = -\sum q_j \ln q_j + \sum (p_j - q_j)(-\ln q_j - \ln z) \tag{11}$$

which indeed simplifies to $-\sum p_j \ln q_j$, hence to (4). ■

*Theorem*:

$$\text{er} \geq \upsilon N \tag{12}$$

*Proof.* It is to be shown that $-\sum p_k \ln q_k \geq -\sum p_k \ln p_k$, that is, that $-\sum p_k \ln q_k$ considered as a function of the arbitrarily assignable quenching probabilities $q_k$ for fixed values of the experiment's own probabilities $p_k$ is minimum at $q_k = p_k$. This is immediately verified, using a Lagrangian multiplier $\lambda$ for the constraint $\sum q_k = 1$. Thus, $d(-\sum p_k \ln q_k) = \lambda d \sum q_k$, $-\sum p_k \, dq_k / q_k = \lambda \sum dq_k$, $-p_k / q_k = \lambda$, hence $q_k \propto p_k$, hence $q_k = p_k$ since both $p$'s and $q$'s must sum to 1. It is also easy to check that this stationary point is indeed a minimum, e.g., $-\sum p_k \ln q_k \to +\infty$ at the $q_k = 0$ boundaries. ■

Varying instead on the $p$'s for fixed $q$'s produces no interesting result.

*Significance of the Theorem.* The reservoir that quenches reg, thereby erasing its old contents, indeed is an adquate gc for upholding the Second Law, in that quenching the outcome of a former trial of the same experiment in that gc produces enough entropy to compensate for the loss of entropy $\upsilon N$ upon learning the outcome of a new trial. Quenching indeed produces *more* than enough entropy, unless the quenched or erased mixed state of reg is selected to have the same probabilities as the experiment's own. In this way each trial of a long sequence needed to establish empirical probabilities will indeed contribute in the mean at least its proper share $\upsilon N$ to increase the entropy of the universe, and will do that by dissipation in the gc cribbed from Szilard.

## 9. CONTEMPLATION OF SZILARD (1929), WITH MINOR CORRECTIONS

*Erasure in Szilard.* I first explain wherein lies my debt to Szilard (1929). Szilard is exorcising the Maxwell's demon. This demon is an entity "who" by observing molecules can operate an engine in a way that reduces the entropy of the universe, or equivalently, extracts work from a single heat reservoir, in violation of Kelvin's principle. Szilard argues convincingly that the essence of the demon is the storage of information about one dynamical variable $x$ in another one $y$, and concentrates on cases where the set of possible values for $y$ has two elements: in modern jargon, $y$ is embodied in a one-bit register. Szilard gives examples to show that if such writing of $x$ on $y$ could be done without producing entropy, then the demon would work, and the Second Law would fail, but that if each such writing somehow entails the production of one bit of entropy, the demon fails. (In 1929, "amount $k_B \ln 2$ of entropy, where $k_B$ is Boltzmann's constant.") He concludes abstractly from the Second Law that such writing must produce the required bit of entropy to compensate the deficit, but he is not satisfied with that, and he accordingly builds a model of a one-bit register, to see precisely where entropy gets produced. He is so careful not to produce entropy unnecessarily that he frighteningly manages to write demonically on his register *without* producing entropy, if reg is in a known state before writing. This disaster is avoided when he does find the requisite production of entropy upon *erasing* back to a known state, for the next cycle. His known state is equilibrium at some temperature $T_0$; the erasure is effected by plunging reg into a reservoir at $T_0$. (Szilard does not use the word "erase," but that is the idea.) Since I also find my gc through entropy of erasure, I have now explained my debt to Szilard; I note that for a two-state register er $= -p_1 \ln q_1 - p_2 \ln q_2 \leq \ln 2$ and is ln 2, one bit, only when $p_1 = p_2 = q_1 = q_2$, and that the more general expression appears also in Szilard; I have simplified for readability.

I will now attempt to discharge my debts to Szilard and to the patient reader by correcting some mistakes; the excitement of discovery carried the brilliant author past some fine points.

*Degeneracy g.* Szilard uses a two-level quantum system or atom for his register. To record one answer "0," the atom is to be cooled to its ground state, which is unobjectionable; cool (near to) absolute zero, $1/T \to +\infty$. To record the other answer "1," the atom should be heated to its *nondegenerate* excited state. The easy way to do that is to heat to $1/T \to -\infty$, or $T \to 0^-$, the "other absolute zero" of negative temperature, but Szilard refuses to anticipate the discovery (Purcell and Pound, 1951) of negative temperature. He instead makes his excited state *highly degenerate*, giving it a multiplicity

$g \gg 1$. Then a large, positive temperature, $1/T \approx 0$, which is really only halfway to $-\infty$, seems to work, as the odds of occupation of the upper level are $g$ times the odds for the lower. Unfortunately, the hot mixed state has large entropy $\ln(g+1)$, and will produce entropy more than the one bit $\ln 2$ on being plunged into the resetting reservoir at $T_0$. (Or if $T_0$ is near $\infty$, resetting the *cold* state will produce excessive entropy.) As Szilard wishes to show that it is possible to *just* compensate the deficit of entropy, such excessive production of entropy would contradict his point about that. To see how the $g$-foldness of the upper level actually leads to the trouble of excessive heat/temperature upon cooling, note that although the energy ("heat") transferred is independent of $g$, the $T_0$ denominator does involve $g$: In the easy case, $q_k = p_k = 1/2$, and in particular is $1/2$ for the ground state, we have $e^0 = 1 = ge^{-\varepsilon/T_0}$, where $\varepsilon$ is the step between levels, hence $T_0 = \varepsilon/\ln g$ is indeed depressed by the largeness of $g$, and entropy $\sim \varepsilon/T_0$ is enhanced and excessive.

If one does not like my glib repair with negative temperature, one may instead write upon a nondegenerate upper level as follows: First cool to the lower level. Then apply a causal Hamiltonian motion to "rotate" that to the upper level.

I of course wish to extend optimal management of a two-level reg to an $n$-level reg. For writing, I must be able to set reg to one of these $n$ levels, not only to the ground state (by cooling to $0^+$) or to the top state (by heating to $0^-$). There are enough *other* absolute zeros available (Lubkin, 1984) through chemical potentials to indeed select any level, approximately, by direct Gibbsian equilibrium. Here the alternative method of causal Hamiltonian motions subsequent to cooling to the ground state is much plainer.

*Bits and Pieces.* Szilard tries so hard to avoid unnecessary production of entropy that he unwittingly lets his crucial bit slip by even in erasure, and so seemingly *creates* a demon: I erase by plunging reg directly into a $T_0$-bath, which may seem a gratuitous crudeness, to be replaced by a more quasistatic scheme . . . but which would, if possible, reinstate the demon! Szilard at first tries to avoid this seeming crudeness. His equipment is a cold $T_A$-bath, a hot $T_B$-bath, the erasing $T_0$-bath, body $K$ that is the register, but also extra pieces $A$ and $B$, and a large number of other reservoirs for implementing quasistatic non-entropy-producing processes. After having been written on, $K$ is either at $T_A$, signifying one of the two $y$ values, or at $T_B$, signifiying the other. If it were generally known which of these, $T_A$ or $T_B$, was $K$'s temperature, then $K$ could indeed be reset to $T_0$ quasistatistically, hence without producing entropy. Szilard wishes to convince us that when it is, however, *not* known which of $T_A$, $T_B$ is $K$'s temperature, *then* erasure does entail the production of entropy demanded by the Second

Law: *indeed this is his essential and correct contribution.* Unfortunately, he has used his pieces $A$ and $B$ "too well." One of $A$, $B$ is in contact with $K$. Piece $A$ is at $T_A$, $B$ is at $T_B$, and $K$ is at the temperature of the piece touching it, but we do not know which that is. Yet in order to bring $K$ to $T_0$ *without* producing entropy, we need only to move quasistatically *both* $A$ *and* $B$ to $T_0$ separately! Then $K$ will go to $T_0$ automatically and gradually, by conduction of heat through whichever piece it touches.

Having thus seemingly exploded Szilard's central point, I must somehow patch it up: It seems to me that the shifting of contact of $K$, sometimes with piece $A$ but sometimes $B$, *itself* requires a lever with two settings, and it is exclusion of this lever's budget of entropy from the discussion that allows a bit to escape scrutiny. Indeed, Szilard's *first* contraption instructs me about levers. It is a cylinder of volume $V_1 + V_2$ containing one ideal-gas molecule, the volume being then split into $V_1$, $V_2$ by slipping in a piston sideways. Then if the molecule is in $V_1$, its pressure will force the piston in one direction; if the molecule is in $V_2$, however, the force will be oppositely directed. *A lever* is provided, to in either case cause the force to *raise* a weight, thus seemingly achieving a demonic engine—if we forget to bookkeep entropy for the lever, which Szilard does not let us do in this case.

But adding *detailed* consideration of a lever, except to fend off Szilard's unwitting $A$-$B$-$K$ demon, is *not* instructive. The purpose of Szilard's body $K$ is to see the dissipation happen: If that dissipation instead happens elsewhere, in some extra lever, then that will involve another body $K'$, and we will have made no progress at all! So I got rid of pieces $A$ and $B$, and let $K$ (or reg) itself touch the dissipant entity. Indeed Szilard's *mathematics* pays no attention to his pieces $A$ and $B$. The strategy is to refuse to complicate with extra registers, and so to show the fallacy of simple demons. Then the Second Law itself, having survived Maxwell's assault, gains our confidence, and so causes us to lose interest in building other demons.

## 10. CLASSICAL DEMONS AND ORTHOGONALITY IN HILBERT SPACE

Is it possible to revive the paradoxical disappearance of entropy by changing the construction of reg to make er smaller? No; to have unambiguous separation of cases, the different outcomes must write reg into mutually *orthogonal* states, which already fixes the model. It would be silly to go on for my original problem, the statement of which stems from a quantum mechanical context. But the problem of Maxwell's demon antedates quantum mechanics, which may make us wonder whether Szilard's solution is as essentially quantal as it seems to be, from his use of a register with two energy levels. Indeed, if classical logic is allowed (Birkhoff and von Neumann, 1936; Jauch, 1968; Finkelstein *et al.*, 1962), er can be made arbitrarily small, thus breaking Szilard's solution: Just let the several pure recording states of reg all make arbitrarily *small* angles in Hilbert space with one common pure state vector $y_0$, and erase to $y_0$. E.g., have Szilard's two settings of $y$ be two linear polarizations of a photon, but separated by only a small angle. If we think about this classically, the electric vector will have slightly distinct directions, and that is classically enough to cause unambiguously distinct consequences. The demon does work classically. I leave conversion of my blend of Hilbert space with "classical logic" into a thoroughly engineered classically mechanical demon as an "exercise"!

*From the Second Law to Wigner's Principle.* Contrapositively, we may choose to assume the Second Law, which demands that $\text{er} \geq vN$, and so reach a denial of the usefulness of a set of nonorthogonal states as a register. This, then, is a *thermodynamic* foundation for Wigner's familiar principle (Wigner, 1952) that if a measurement unambiguously separates states in always leading to distinct settings of some dial, those states must not only be distinct, they must be orthogonal. Of course, Wigner's argument from unitarity of the overall process in time is undoubtedly clearer . . . unless you set out to *build* time from observation.

## 11. NOT LANDAU TRACING?

The entropy of any single mixed state Y may be imagined found from Landau tracing of an encompassing pure state: Diagonalize $Y = \sum p_i |x_i\rangle\langle x_i|$, and use $\psi = \sum p_i^{1/2} x_i \otimes y_i$ for an encompassing pure state's vector, where the $y_i$ are orthonormal and orthogonal to all the $x$'s. Is the entropy of erasure er also of this character?

It has *not* been so computed: The computations of separate $\text{er}_k$ were done first, *then* convexly combined to $\text{er} = \sum p_k \cdot \text{er}_k$. The nonlinearity of $x \to -x \ln x$ in Boltzmann's definition of entropy guarantees that if the convex combination were done first, the result would be wrong. In particular, for the optimum case $q_k = p_k$, one would "get" *no* production of entropy upon quenching to $T_0$ were the wrong order used. The computation of er did not investigate one single density matrix Y; indeed, naively replacing the separate $E_k$ by the single density matrix $Y = \sum p_k E_k$ just gave a wrong answer.

Yet there *is* in a sense an "encompassing" pure state $\psi$: the wave function of the system and reg, in interaction together. Nevertheless, the analysis of details within $\psi$ was *not* done by selecting some factor Hilbert space to be Landau-traced out. The physical analogue of such Landau-ignoring of a factor space is subdividing a system into a system of interest and a complementary part to be ignored. This Landau philosophy may, however, not be general, in that *reality* is not subdivided. In the calculation of er, we instead used a *different* simple reality for each outcome $k$, namely reg set at $E_k$, we $T_0$-quenched *that*, and then convexly combined the produced entropies $\mathrm{er}_k$ on the excuse of calculating a mean entropy over a long run: a time average rather than an ensemble average. This is also what Szilard does. Hence, since 1929 we have had a calculation of entropy production outside the scope of Landau tracing, and based upon relative reality, albeit disguised as an old-fashioned averaging over time.

It should be noted that Landau tracing does implicitly play its part here: If, in contemplating any single $E_k \rightarrow T_0$ quench, we imagine following the detailed unitary motion of reg in interaction with a $T_0$-reservoir, then no entropy will be produced until we Landau-neglect that reservoir; and that will get you $\mathrm{er}_k$. What I suspect may *not* be attainable by Landau tracing is a unified derivation of er, as distinct from $\mathrm{er}_k$.

A related trouble—for reviving my paradox, not Maxwell's—is the thought that you need never erase if you have enough "clean paper" to write on. My answer to this is that the entropic debt is then paid in advance, when you manufacture all that clean paper. It is roughly if not precisely analogous to "getting work from heat without a cold reservoir" by letting cylinders of ideal gas expand without restoring their original condition.

## ACKNOWLEDGMENTS

Thanks to Leonard Parker for pointing out the old-fashioned nature of the time-averaging, and the possible gap separating this from Landau tracing; to Thelma Lubkin for the objection of "clean paper"; to Atsushi Higuchi for the thought of a possibly *thoroughly* classical demon; and to Ming-Lap Chow for pressing me to look at Wheeler and Zurek (1983).

## REFERENCES

Birkhoff, G., and von Neumann, J. (1936). *Annals of Mathematics*, **37**, 823.

Dirac, P. A. M. (1938-39). *Proceedings of the Royal Society of Edinburgh*, **59**, 2, 122.

Finkelstein, D., Jauch, J. M., Schiminovich, S., and Speiser, D. (1962). *Journal of Mathematical Physics*, **3**, 207.

Jauch, J. M. (1968). *Foundations of Quantum Mechanics*. Addison-Wesley, Reading, Massachusetts.

Latham, R. E. (1951). *On the Nature of the Universe, A translation of Lucretius* (~*55* B.C.), Penguin Books, Baltimore, Maryland.

Lubkin, E. (1974). *Journal of Mathematical Physics*, **15**, 663.

Lubkin, E. (1978). *Journal of Mathematical Physics*, **19**, 1028.

Lubkin, E. (1979a). *International Journal of Theoretical Physics*, **18**, 165-177.

Lubkin, E. (1979b). *International Journal of Theoretical Physics*, **18**, 519-600.

Lubkin, E. (1984). Lie algebras of first laws of thermodynamics for physics without time, *XIIIth International Colloquium on Group Theoretical Methods in Physics*, W. W. Zachary, ed., pp. 275-278, World Scientific, Singapore.

Purcell, E. M., and Pound, R. V. (1951). *Physical Review*, **81**, 279-280.

Szilard, L. (1929). *Zeitschrift für Physik*, **53**, 840-56 [Translation in Wheeler and Zurek (1983).]

Von Neumann, J. (1955). *Mathematical Foundations of Quantum Mechanics*, Princeton University Press, Princeton, New Jersey [Reprinted in Wheeler and Zurek (1983).]

Wheeler, J. A., and Zurek, W. H. eds. (1983). *Quantum Theory and Measurement*, Princeton University Press, Princeton, New Jersey.

Wigner, E. P. (1952). *Zeitschrift für Physik*, **133**, 101.

# Use of mutual information to decrease entropy: Implications for the second law of thermodynamics

Seth Lloyd*
*The Rockefeller University, 1230 York Avenue, New York, New York 10021*
(Received 9 November 1987; revised manuscript received 5 December 1988)

Several theorems on the mechanics of gathering information are proved, and the possibility of violating the second law of thermodynamics by obtaining information is discussed in light of these theorems. Maxwell's demon can lower the entropy of his surroundings by an amount equal to the difference between the maximum entropy of his recording device and its initial entropy, without generating a compensating entropy increase. A demon with human-scale recording devices can reduce the entropy of a gas by a negligible amount only, but the proof of the demon's impracticability leaves open the possibility that systems highly correlated with their environment can reduce the environment's entropy by a substantial amount without increasing entropy elsewhere. In the event that a boundary condition for the universe requires it to be in a state of low entropy when small, the correlations induced between different particle modes during the expansion phase allow the modes to behave like Maxwell's demons during the contracting phase, reducing the entropy of the universe to a low value.

Maxwell[1] was the first to note the trade off between entropy and information: He pointed out that a being who could measure the velocity of individual molecules in a gas could operate a shutter between two containers of the gas, and by shunting fast molecules into one container and slow molecules into the other, create a difference in temperature between the two containers, in apparent contradiction to the second law of thermodynamics. William Thomson, Baron Kelvin, called this being a "demon." Such a demon takes information about the microscopic state of the gas and by acting judiciously, uses that information to reduce the entropy of the gas. Szilard[2] subsequently suggested that the act of acquiring information by its very nature generates entropy; he showed that one bit of information could be used to reduce the entropy of a one-molecule gas by $k_B \ln 2$ and gave an example of a simple measuring device that created least $k_B \ln 2$ of entropy for each bit of information that it acquired. Brillouin[3] elevated Szilard's result to the status of a "generalized Carnot principle:" the amount by which a demon can reduce entropy by putting his information to work is always less than or equal to the amount by which he increases entropy in the act of acquiring information in the first place. More recent authors have dealt with the question of minimum entropy increase in telecommunications[4,5] and in computation.[6,7,8]

The above authors investigate the trade off between information and entropy in specific devices. In this paper, the mechanics of the general process by which one system obtains information about another are examined. Almost any interaction generates correlation; not only does a drift chamber receive information about the trajectories of the particles that pass through it—an electron receives information about the particles from which it scatters, as well. Four theorems that exploit the measure-preserving properties of classical and quantum-mechanical dynamics are proved. The first two theorems put limits on the amount of information that a given system can gather and imply that a system such as a drift chamber or an electron that is to obtain information through interaction must be prepared in a state of low entropy beforehand. The third theorem confirms the well-known fact that a reduction of the entropy of one system by normal thermodynamic means requires and increase in entropy elsewhere; the fourth theorem shows that the existence of correlations can mitigate the amount of the required entropy increase. The generalized Carnot principle holds for drift chambers and electrons, photocells and eyes, because in order to gather information repeatedly such systems must repeatedly be placed in states of low entropy, at the cost of a counterbalancing increase in entropy elsewhere. However, the generalized Carnot principle does not necessarily hold for systems in preexisting states of low entropy, such as gravitationally induced clusters of matter, that interact once and for all to acquire information about other systems, such as nearby clusters. Maxwell's demon cannot function to reduce the entropy of a gas, but the existence of states of low entropy at early times together with the presence of long-range forces to induce correlations may imply a violation of the second law of thermodynamics for the universe as a whole.

In the event that a boundary condition for the universe requires that it be in a state of high order when small, the second law of thermodynamics must be violated during the recontracting phase. In the final section, the theorems proved here are put to use to show how in this event the different particle modes behave as Maxwell's demons, using information acquired during the expanding phase to decrease coarse-grained entropy during the recontracting phase.

We proceed as follows. First, we propose a version of the second law of thermodynamics based not on the traditional coarse graining over phase space, but on a coarse graining over evolutions. Second, we review some concepts from information theory and give natural generalizations to quantum-mechanical systems; we apply the notion of conditional information to give an information theoretic description of coarse graining. Third, we prove and apply four theorems on how entropy can be reduced through acquisition of mutual information. Finally, we discuss entropy reduction for the universe as a whole.

## ENTROPY

The Shannon entropy of a system with states $i$ is given by

$$S=-\sum_i p_i \ln p_i \;,$$

where $p_i$ is the probability that the system is in its $i$th state given macroscopic constraints such as total energy, volume, number of particles, etc. For a thermodynamic system, the Shannon entropy is equal to the fine-grained entropy divided by Boltzmann's constant. In what follows, the world "entropy" refers to the Shannon entropy unless otherwise noted. For a discrete system, the entropy is non-negative. For a continuous classical system with coordinates and momenta labeled by $a$, we have

$$S=-\int p(a)\ln p(a)da \;,$$

where $p(a)$ is now a probability density; $S$ for a continuous system can be negative. For a quantum-mechanical system, $S=-\mathrm{tr}\rho\ln\rho$, where $\rho$ is the density matrix for the system.[9] In each of these expressions for the entropy, $S$ gives a measure of how little we know about the system in question. If $p_i=1/W$, $i=1$ to $W$, so that the system can be in any one of $W$ states with equal probability, the usual expression for the entropy $S=\ln W$ is recovered.

An immediate problem with identifying the fine-grained entropy with the thermodynamic entropy is that $S=-\sum_i p_i \ln p_i$ has the property of remaining constant under volume-preserving evolution in phase space for classical systems and under unitary evolution for quantum-mechanical systems.[10] $S=$const is particularly easy to see in the quantum-mechanical case. Under unitary evolution $U$, $\rho\rightarrow\rho'=U\rho U^\dagger$, and

$$S\rightarrow S'=-\mathrm{tr}U\rho U^\dagger \ln U\rho U^\dagger \;.$$

Expanding the logarithm in a power series and using the cyclic property of the trace, we immediately get $S'=S$.

The problem arises as follows. If the probability density $p_0(a)$ or the density matrix $\rho_0$ reflects accurately our knowledge about the state of a given system at time $t=0$, then the probability density $p_t(a)=p_0(u_t^{-1}(a))$ or the density matrix $\rho_t=U_t\rho_0 U_t^\dagger$ reflects accurately our knowledge at time $t$ only if we know the exact form of the evolution $u_t$ in phase space or of the unitary evolution $U_t$. But a detailed knowledge of the evolution of a system such as a gas is out of the question. The equations of motion for such a system are rarely integrable, and if we constantly update our probability distributions and density matrices according to experimentally acquired knowledge of the evolution, the corresponding entropies will either remain constant or increase.

If the exact evolution of a system can be determined by integration of the equations of motion, or by experiment, the entropy remains constant. We prove that if the evolution of a system can be determined only inexactly by analytic or experimental means, the fine-grained entropy remains constant or increases. That is, we give a formal proof of the straightforward idea that if we know something about the state of a system to begin with, but know only approximately how the system evolves in time, then we will know less about the state of the system in the future. We thus prove a version of the second law of thermodynamics for a closed system based on a coarse graining over the space of the system's possible dynamical laws, rather than on a coarse graining over phase or Hilbert space, an idea suggested by the mathematician Borel.[11]

*Proposition:* Second law of thermodynamics for isolated systems. If experiment and analysis determine the time evolution of an isolated classical or quantum-mechanical system only inexactly, and if under such an imperfectly determined evolution the entropy $S$ goes to $S'$, then $S'\geq S$.

*Demonstration:* Take first the case of a classical, discrete system with $n$ states. A time evolution of such a system is given by a member of the group of permutations of $n$ objects, $S_n$. Let $p_i$ be the probability that the system is in its $i$th state. Let $\chi(\sigma)$ be the probability that the actual evolution of the system is $\sigma$. As the system evolves, $p_i\rightarrow p_i'$, where

$$p_i'=\sum_{\sigma\in S_n}\chi(\sigma)p_{\sigma^{-1}(i)}=\sum_j F_{ij}p_j \;,$$

and $F_{ij}=$the sum over all $\sigma$, such that $\sigma^{-1}(i)=j$, of $\chi(\sigma)$. Under such an inexact evolution, the entropy $S=-\sum_i p_i\ln p_i$ goes to

$$S'=-\sum_i p_i'\ln p_i'=-\sum_{i,j}F_{ij}p_j\ln\left[\sum_{j'}F_{ij'}p_{j'}\right]$$
$$\geq-\sum_j p_j\ln p_j=S \;,$$

since $F_{ij}$ is double stochastic

$$\sum_i F_{ij}=\sum_j F_{ij}=1 \;.$$

This demonstration holds also for any classical continuous system that can be suitably modeled by a discrete system, e.g., any system that we can model on a digital computer.

The proof that $S'\geq S$ for quantum-mechanical systems goes as follows. Let $\chi_i$ be the probability that the actual unitary evolution of the given system is $U_i$. A particular density matrix $\rho$ then transforms as

$$\rho\rightarrow\rho'=\sum_i \chi_i U_i\rho U_i^\dagger \;.$$

Look at a representation in which $\rho'$ is diagonal: $\rho'=\mathrm{diag}(p_1',\ldots,p_n')$. The entropy $S=-\mathrm{tr}\rho\ln\rho$ goes to

$$S' = -\mathrm{tr}\rho'\ln\rho' = -\sum_i p_i'\ln p_i'$$

$$= -\sum_j \sum_i \chi_i p_i^j \ln\left[\sum_{i'} \chi_{i'} p_{i'}^j\right],$$

where $p_i^j$ is the $j$th diagonal element of $U_i \rho U_i^\dagger$ in the representation in which $\rho'$ is diagonal. But $x \ln x$ is a convex function, and we have

$$S' \geq -\sum_i \chi_i S_i ,$$

where

$$S_i = -\sum_j p_i^j \ln p_i^j .$$

But if $\rho = \mathrm{diag}(p_1, \ldots, p_n)$, then

$$p_i^j = (U_i \rho U_i^\dagger)^{jj} = \sum_{k,k'} U_i^{jk} \rho_{kk'} \bar{U}_i^{jk'}$$

$$= \sum_k |U_i^{jk}|^2 p_k ,$$

where the matrix $F^{jk} = |U_i^{jk}|^2$ is double stochastic, $\sum_j F^{jk} = \sum_k F^{jk} = 1$, and so $S_i \geq S$ as in the discrete case. Hence $S' \geq \sum_i \chi_i S_i \geq S$.

We can also define a quantum-mechanical entropy with respect to a particular basis: $S^b(A) = -\mathrm{tr}\rho_A^b \ln\rho_A^b$, where

$$\rho_A^b = \sum_i |b_i\rangle\langle b_i|\rho_A|b_i\rangle\langle b_i| ,$$

and $\{|b_i\rangle\}$ is a basis for the Hilbert space of $A$. The entropy $S = -\mathrm{tr}\rho\ln\rho$ defined above is equal to the entropy defined with respect to a basis in which $\rho$ is diagonal, $\rho = \mathrm{diag}(p_1, \ldots, p_n)$. Note that

$$S^b = -\sum_i \sum_j |U^{ij}|^2 p_j \ln\left[\sum_{j'} |U^{ij'}|^2 p_{j'}\right],$$

where $U^{ij}$ are the components relating the basis in which $\rho$ is diagonal to $\{|b_i\rangle\}$. Hence $S^b \geq S$, as above. Any process, such as measurement, that replaces $\rho$ by $\rho^b$ also tends to increase entropy.

That the statistically defined entropy obeys a version of the second law of thermodynamics confirms the plausibility of identifying the entropy defined in terms of probabilities with the thermodynamic entropy. The proof of the second law given here differs from the proof of Gibbs in that we have introduced a coarse graining over the space of possible dynamical laws rather than a coarse graining over phase or Hilbert space. The form of a particular coarse graining for a system depends on the interactions that we can arrange between the system and the various devices that can record information about the system's state and evolution. We now turn to the mechanics of getting such information.

## INFORMATION

The relation $S = \ln W$ gives an explicit expression to the tradeoff between information and entropy. If we compress an ideal gas of $N$ molecules isothermally to half its original volume, we decrease its entropy by $N \ln 2$, and reduce the number of possible microstates of the gas by a factor of $2^N$—we gain information about the actual microscopic state of the gas. Conversely, the more we know about a system, the greater the number of constraints on its microscopic state, the lower its entropy. The amount of information obtained about a system during some process is defined to be equal to the amount by which the entropy of the system is reduced during the process:[12]

$$\Delta I = -\Delta S .$$

Two systems are correlated if when we look at one we get information about the other. Given a joint probability distribution $p(a_i b_j)$ for the states $a_i b_j$ of $AB$, the conditional entropy of $A$, given that $B$ is in the state $b_j$, is

$$S(A/b_j) = -\sum_i p(a_i/b_j)\ln p(a_i/b_j) ,$$

where

$$p(a_i/b_j) = p(a_i b_j)/p(b_j)$$

is the probability that $A$ is in the state $a_i$, given that $B$ is in the state $b_j$. The average entropy of $A$, given the state of $B$, is

$$S(A/B) = \sum_j p(b_j) S(A/b_j) = S(AB) - S(B) .$$

In the quantum-mechanical case, the average entropy of $A$, given the state of $B$, depends on which basis $\{|b_i\rangle\}$ we use to describe $B$: $S^b(A/B) = S^b(AB) - S^b(B)$.

The *mutual information,* or correlation, between $A$ and $B$ is defined to be the average amount that the entropy of $A$ is reduced, given knowledge of the state of $B$. $I(A:B)$ is the entropy of $A$ minus the average entropy of $A$, given the state of $B$:

$$I(A:B) = S(A) - S(A/B) = S(A) + S(B) - S(AB) .$$

There are two important facts about mutual information. The first is $I(A:B) = I(B:A)$—the average amount by which the entropy of $A$ is reduced if we know the state of $B$ is equal to the average amount by which the entropy of $B$ is reduced if we know the state of $A$. The second is $I(A:B) \geq 0$—mutual information is never negative.[12]

The quantum-mechanical version of mutual information between two systems $A$ and $B$ with joint density matrix $\rho_{AB}$ can be defined as

$$I(A:B) = S(A) + S(B) - S(AB)$$

$$= -\mathrm{tr}\rho_A \ln\rho_A - \mathrm{tr}\rho_B \ln\rho_B + \mathrm{tr}\rho_{AB}\ln\rho_{AB} ,$$

where $\rho_A$ is the density matrix for $A$ alone, obtained by taking the trace of $\rho_{AB}$ over $B$ degrees of freedom, similarly for $\rho_B$. Once again, $I(A:B) = I(B:A)$ and $I(A:B) \geq 0$. The superposition principle and the fact that a compound system is described by a Hilbert space that is the tensor product of the Hilbert spaces of its parts implies that $S(A)$ and $S(B)$ need not equal zero when $S(AB) = 0$, since the state for $AB$ as a whole need not be representable by the Cartesian product of any two states of $A$ and $B$. For example, two electrons in the state

$$(1/\sqrt{2})(|\uparrow_x\rangle|\downarrow_x\rangle - |\downarrow_x\rangle|\uparrow_x\rangle)$$

exhibit anticorrelation of spins not only along the $x$ axis, but along any other axis, as well. As a result, the quantum-mechanical mutual information is generally larger than the classical: the mutual information between the electrons is $2\ln 2$, twice the maximum value for the corresponding classical case. We can also define quantum-mechanical mutual information with respect to a particular choice of bases $\{|a_i\rangle\},\{|b_j\rangle\}$ for $A,B$:

$$I^{ab}(A:B) = S^a(A) + S^b(B) - S^{a:b}(AB)$$
$$= S^a(A) - S^{a:b}(A/B) \ .$$

Here,

$$S^{a:b}(AB) = -\mathrm{tr}\rho_{AB}^{a:b}\ln\rho_{AB}^{a:b} \ ,$$

where

$$\rho_{AB}^{a:b} = \sum_{i,j}(|a_i\rangle\langle a_i|)(|b_j\rangle\langle b_j|)$$
$$\times \rho_{AB}(|a_i\rangle\langle a_i|)(|b_j\rangle\langle b_j|) \ .$$

Note that $I^{ab}(A:B) \le I(A:B)$. For the two electrons in the state above, $I^{ab}$ is strictly less than $I$: No measurement with respect to a single basis can reveal all the mutual information that the two electrons possess.

## COARSE-GRAINED ENTROPY

Conditional entropies can be used to define thermodynamic or coarse-grained entropies in terms of interactions with measuring devices. Let $A$ be a system, such as a gas, with $n$ states $a_i$ initially described by a probability distribution $p(a_i) = 1/n$, i.e., nothing is known about the state of $A$. Let $B$ be a system, initially in state $b_0$, that interacts with $A$. If after the interaction $B$ is in the state $b_i$, then $A$ must have been in a state $a_j$ such that $a_j b_0 \to a_k b_i$; the transition $b_0 \to b_i$ defines a set of states that $A$ could have been in before the interaction. Such a set of states corresponds to a *macroscopic state* of $A$ defined by the interaction with $B$. In the continuous case, each transition for $B$ defines a volume in the phase space of $A$. In the quantum-mechanical case, each transition $|b_0\rangle \to |b_i\rangle$ defines a subspace of the Hilbert space for $A$. The *coarse-grained* entropy of the macroscopic state of $A$ given that $B$ went from $b_0$ to $b_i$ is $\bar{S}(A) = k_B S(A/b_i)$ after the interaction, where $k_B$ is Boltzmann's constant. Boltzmann's constant is brought in to make the connection between information theory and thermodynamics: Coarse-grained entropies such as $\bar{S}(A)$ are what is normally meant by "the entropy of a system." For example, if a thermometer $B$ is brought into contact with a gas $A$ confined to a certain volume, then $S(A/b_j)$ is the conditional entropy of the gas given that the thermometer is in a state $b_j$ in which the thermometer's scale registers a certain temperature: $\bar{S}(A)$ is the thermodynamic entropy of the gas at that temperature.

Note that the coarse-grained entropy of a macroscopic state is equal to the fine-grained entropy for a uniform distribution over the microscopic states corresponding to the macroscopic state. The coarse-grained entropy of a system in a particular state is thus always greater than or equal to the system's fine-grained entropy.

## REDUCING ENTROPY

The thermodynamic entropy of a system is a conditional entropy; it is Boltzmann's constant times the Shannon entropy of the system conditioned on the states of various measuring devices after the system and devices have undergone prescribed interactions. The following theorems put restrictions on how much one can reduce thermodynamic entropy of a system through interaction. Their method is to use the law of overall nondecrease of fine-grained entropy to put limits on the thermodynamically allowed reduction of conditional entropy.

*Theorem 1:* The amount by which the average entropy of $A$ given the state of $B$ can be reduced during an interaction is limited by the difference between the final value of $B$'s entropy, and its initial value:

$$-[S_t(A/B) - S_0(A/B)] \le S_t(B) - S_0(B) \ .$$

*Proof:*

$$S_t(AB) - S_0(AB) \ge 0$$
$$\to S_t(AB) - S_t(B) - S_0(AB) + S_0(B) \ge S_0(B) - S_t(B)$$
$$\to -[S_t(A/B) - S_0(A/B)] \le S_t(B) - S_0(B) \ .$$

In the proof for the quantum-mechanical case we must take into account the bases for $B$ with respect to which the average entropy of $A$ is given at time zero and at time $t$. If at time zero we look at $S(A/B)$ with respect to a basis $\{|b_j\rangle\}$ for $B$ and at time $t$ we look at $S(A/B)$ with respect to another basis $\{|b_j'\rangle\}$ for $B$, we have $S_t^{bb'}(AB) - S_0^b(AB) \ge 0$. Here

$$S_t^{bb'}(AB) = -\mathrm{tr}\rho_{AB}^{bb'}\ln\rho_{AB}^{bb} \ ,$$

where

$$\rho_{AB}^{bb'} = -\sum_j |b_j'\rangle\langle b_j'| U_t \rho_{AB}^b U_t^\dagger |b_j'\rangle\langle b_j'| \ .$$

The rest of the proof continues as after the first line above, and we have

$$-[S_t^{bc}(A/B)-S_0^b(A/B)]\le S_t^{bc}(B)-S_0^b(B)\ .$$

According to an outside observer who applies the second law of thermodynamics to $AB$ as an isolated system, a decrease in the entropy of $A$ given $B$ requires an increase in the entropy of $B$. But for an inside observer who has access to the actual state of $B$, the entropy of $B$ is always zero. Where the outside observer sees an increase in $S(B)$ in the course of $B$'s interaction with $A$, the insider sees not an increase in entropy, but an increase in the amount of *information* that $B$ has about $A$.

Mutual information allows us to give a measure to the amount of information that one system obtains about another during any interaction. Suppose $A$ and $B$ are initially described by a classical probability density $p_0(ab)$, where $a$ and $b$ label both coordinates and momenta for $A$ and $B$, respectively. Suppose that $A$ and $B$ undergo a volume-preserving evolution on phase space $u_t$ in which the state $ab$ goes to $u_t(ab)$ at time $t$; the probability density $p_0(ab)$ then goes to $p_t(ab)=p_0(u_t^{-1}(ab))$. At time $t$, $A$ and $B$ will in general be correlated: $I_t(A:B)>0$. The following theorem puts a limit on the amount of information that $B$ can obtain about $A$ during such an interaction.

*Theorem 2:*

$$I_t(A:B)\le S_{\max}(A)+S_{\max}(B)-S_0(AB)\ ,$$

where $S_{\max}(A)$, $S_{\max}(B)$ are the maximum possible entropies for $A$ and $B$ and

$$S_0(AB)=-\int p_0(ab)\ln p_0(ab)da\, db$$

is the initial entropy for $AB$.

*Proof:* The mutual information between $A$ and $B$ at time $t$ is

$$I_t(A:B)=S_t(A)+S_t(B)-S_t(AB)\ .$$

But $S_t(AB)\ge S_0(AB)$, as noted above, and the maximum value for

$$S_t(A)+S_t(B)=\ln V_A+\ln V_B\ ,$$

where $V_{A,B}$ are the volumes of phase space accessible to $A$ and $B$, respectively. The quantum-mechanical version of this proof is identical except that the maximum value for $S(A)+S(B)$ is equal to $\ln(n_A)+\ln(n_B)$, where $n_{A,B}$ are the dimensions of the Hilbert spaces for $A,B$.

Theorem 2 implies that if a system is to function as a measuring device, it must be in a state of less than maximum entropy to begin with, together with the system about which it is getting information. If $I(A:B)=0$ at time 0, and if both $A$ and $B$ are in states of maximum entropy at time 0, then $I(A:B)$ will remain zero for all future times $t$.

Theorem 2 puts limits on the amount by which the entropy of a system can be reduced by correlation. The following two theorems are elementary consequences of the definition of mutual information and put limits on the reduction of entropy by constraint.

*Theorem 3:* If $A$ and $B$ are initially uncorrelated, $I_0(A:B)=0$, then any interaction between the two that decreases $A$'s entropy by $\Delta S$ must increase $B$'s by at least $\Delta S$.

*Proof.* We have $S_0(A)+S_0(B)=S_0(AB)$ at time zero. At time $t$ we have $S_t(A)+S_t(B)\ge S_t(AB)$. But $S_t(A)=S_0(A)-\Delta S$, and $S_t(AB)\ge S_0(AB)$, so that

$$S_t(B)\ge S_0(AB)-S_0(A)+\Delta S=S_0(B)+\Delta S\ .$$

In the case of an ideal gas isothermally compressed to half its original volume with corresponding entropy decrease $\Delta S=N\ln 2$, the heat flowing out of the gas must increase the entropy elsewhere by at least that amount.

We now prove a theorem that puts a limit on the amount by which entropy can be reduced by a combination of correlation and constraint.

Theorem 4. Any interaction between $A$ and $B$ that decreases $A$'s entropy by $\Delta S$ must increase $B$'s entropy by at least $\Delta S-I_0(A:B)$, where $I_0(A:B)$ is the mutual information between $A$ and $B$ at the start of the interaction.

*Proof:* We have

$$S_0(A)+S_0(B)=S_0(AB)+I_0(A:B)$$

at time 0. At time $t$ we have

$$S_t(A)=S_0(A)-\Delta S,\quad S_t(AB)\ge S_0(AB)\ ,$$

and

$$S_t(A)+S_t(B)\ge S_t(AB)\ .$$

Now

$$\begin{aligned}S_t(B)\ge S_t(AB)-S_t(A)&\rightarrow S_t(B)\ge S_0(AB)-S_0(A)+\Delta S\\&\rightarrow S_t(B)\ge S_0(A)+S_0(B)-I_0(A:B)-S_0(A)+\Delta S\\&\rightarrow S_t(B)\ge S_0(B)+\Delta S-I_0(A:B)\ .\end{aligned}$$

Theorem 4, which includes theorem 3 as a special case, tells us that if $A$ and $B$ are correlated to begin with, a decrease in $A$'s entropy need not be fully compensated for by an increase in the entropy of $B$; correlation can be "traded in" to decrease entropy at less than the normal cost. So, when a gas undergoes a Poincaré recurrence cycle the correlations between the molecules conspire to produce a state of abnormally low entropy.

## ENTROPY AND INFORMATION

The theorems proved above allow us to determine exactly when the collection of information requires an increase in entropy and when it does not. Let us analyze how much information a system $B$ (a photocell, a drift chamber, an electron) can get about another system $A$ (a photon, an electron) by interacting with it, and by how much or how little the entropy of $A$ and $B$ increases during such a process.

Theorem 1 tells us that if the initial entropy of $B$ is less than its maximum value $S_0(B) < S_{max}(B)$, then the maximum amount of information that $B$ can get about $A$, i.e., the maximum amount by which the entropy of $A$ can be reduced by becoming correlated with $B$, is $\Delta I = S_{max}(B) - S_0(B)$. There is *no* reason why $S(AB)$ must *increase* during such a process. Nor is there any reason why the coarse-grained entropy must increase. When an electron scatters off of another electron their states become correlated; yet if the electrons begin in a pure state and scatter elastically, then they end up in a pure state, with entropy zero. Nor need entropy increase when $B$ is macroscopic.[8] Landauer[6] pointed out that a Brownian particle in a potential well that can be continuously modulated between bistability and biased monostability can be used to record one bit of information with vanishingly small increase in entropy as long as the initial configuration of the well is known. Such a system can be realized as a one-domain ferromagnet,[6,7] or as a Josephson junction with a controllable critical current.[13]

Even the prototypical measuring device of Szilard that creates ln2 of entropy for every bit of information acquired can be slightly modified to give as small an entropy increase as desired. Szilard's device consists of a piece of matter that is put in contact with one of two reservoirs at different temperatures, either the first at $T_1$ or the second at $T_2$ depending on the position of a pointer. The energy of the matter and the position of the pointer become correlated, but at the cost of an average increase of entropy of at least ln2 as heat flows from matter to reservoir and *vice versa*. If the initial temperature of the matter is known, however, one can make the increase in entropy as small as one likes simply by supplying reservoirs with incrementally increasing temperatures between $T_1$ and $T_2$, with which the piece of matter comes into equilibrium on its way from one reservoir to the other.

Theorem 1 tells us that the reason that these devices can get information without increasing entropy is that they have been prepared in states of less than maximum entropy: In each case above, the initial state of the device is known at least approximately. Suppose that $B$ has obtained a maximal amount of information about $A$, so that $S_0(B) = S_{max}(B)$. At this point it is certainly possible to arrange an interaction between $B$ and another system $C$, such that $B$ gets additional information about $C$, i.e., $S_t(C/B) - S_0(C/B) < 0$, but $B$ can only get information about $C$ by "forgetting" information about $A$,

$$S_t(AC/B) - S_0(AC/B) \geq S_0(B) - S_{max}(B) = 0 \ .$$

That is, once $S(B) = S_{max}(B)$, an interaction with $B$ that decreases the entropy of $C$ by $\Delta S$ must increase the entropy of $A$ by at least $\Delta S$. Once $B$ has been saturated as a recording device, any further information that $B$ gets about any system must be fully compensated for by an increase in entropy elsewhere.

Landauer[6] and Bennett[8] have pointed out that it is the act of erasure, rather than the act of recording information, that requires an increase in entropy. We apply the theorems proved above to treat the act of erasure in detail. Consider an interaction between $A$ and $B$ that "erases" $A$ by restoring $A$ to a particular "blank" state regardless of $A$'s initial state. If $S(A)$ is initially positive, the act of erasure reduces it to zero. But by theorem 3, if $A$ and $B$ are initially uncorrelated, a reduction of $\Delta S$ in the entropy of $A$ must be compensated for by an increase of $\Delta S$ in the entropy of $B$. If $A$ and $B$ are initially correlated, with mutual information $I_0$, then by theorem 4 a reduction of $\Delta S$ in the entropy of $A$ need be compensated for only by an increase of $\Delta S - I_0$ in the entropy of $B$. $B$'s initial information about $A$ mitigates somewhat the entropy increase required by erasure.

Let us apply these notions to a Maxwell's demon that has filled up his memory with junk information and desires to get rid of it. The second law of thermodynamics for isolated systems implies that the demon must arrange an interaction with some external system if he is to decrease the entropy of his recording device $B$ when $S(B) = S_{max}(B)$. Note that when $S(B) = S_{max}(B)$, the fine-grained entropy of $B$ equals the coarse-grained entropy: $k_B S(B) = \bar{S}(B)$. But this reduction in the entropy of $B$ cannot take place through correlation; even if $B$ becomes correlated with some other system $C$, this does the demon no good, since the demon can obtain no information about $C$ as long as $S(B) = S_{max}(B)$. The demon's memory is already filled to capacity. Hence the demon must arrange to decrease $S(B)$ by constraint. We can now apply theorems 3 and 4. If the demon arranges an interaction between $B$ and a system $C$ that has no initial correlation with $B$, theorem 3 requires that every decrease of $\Delta S$ in $B$'s entropy be compensated for by an increase of $\Delta S$ in the entropy of $C$. In addition, since $\bar{S}(B) = k_B S(B)$, a decrease of $k_B \Delta S$ in $\bar{S}(B)$ requires an increase of $k_B \Delta S$ in the fine-grained entropy of $C$. If $C$ is already correlated with $B$, theorem 4 requires that a decrease of $\Delta S$ in the entropy of $B$ be matched by an increase in the entropy of $C$ of $\Delta S - I_0(B:C)$, where $I_0(B:C)$ is the initial mutual information between $B$ and $C$. Theorem 4 thus holds out the hope of decreasing the entropy of $B$ without a fully compensating increase in entropy elsewhere; but by theorem 2, such a decrease can only be accomplished by also reducing the information that $B$ has about $C$.

Maxwell's demon can function up to a point. He can reduce the entropy of a gas without generating entropy elsewhere by $S_{max}(B) - S_0(B)$, where $S_0(B)$ is the initial entropy of the device that he uses to record information about the gas. He can reduce the coarse-grained entropy of the gas by $k_B$ times this amount. But that is all: Any further reductions in entropy can only be accomplished at the expense of generating entropy elsewhere as the demon erases the information that he has already gathered.

**DISCUSSION**

Maxwell's demon cannot function effectively because in order to decrease the entropy of a gas by a substantial amount without increasing entropy elsewhere, he needs to have a memory bank with at least as many bits as there are molecules in the gas, all at a state of low entropy to begin with. Even if the demon can use all the computers in the world to record his information, with, say, $10^{16}$ bits of memory, he will run out of memory space before he has reduced the entropy of a gram of gas by a factor of $10^{-10}$. In practice, therefore, clever devices that can violate the second law of thermodynamics by any substantial amount are out of the question. Systems in states of low entropy, however, can get a certain amount of information about other systems without increasing entropy overall. One such example is the case of two scattering electrons that are initially in a pure state. Gravitational systems give another example; it was the information encoded in the perturbations of the orbit of Uranus that allowed the location of Neptune to be pinpointed. Yet the interaction between the two planets that produced those perturbations did not produce any increase in entropy (apart from an irrelevant increase due to tidal action). In reducing the uncertainty about the position of Neptune by their calculations, though, Adams[14] and Le Verrier[15] produced more than a compensating amount of uncertainty in the positions and velocities of the molecules in the paper on which they recorded the results of those calculations.

A system in a state of zero entropy to begin with can record an amount of information equal to the logarithm of the number of states accessible to it in the course of interaction. The larger the system, the more information it can record, the more it can reduce the entropy of other systems without increasing entropy in turn. Maxwell's demon operating with human-scale resources can decrease entropy by only the smallest amount. The universe started out in a state of low entropy, however, and a demon that can put to use a substantial fraction of the matter in the universe as a recording device can reduce the entropy of the remaining part by a correspondingly substantial amount. The theorems proved here apply to the mechanics of getting information in general and do not single out specific processes. However, they suggest that by inducing correlations between the positions and velocities of interacting masses the long-range electromagnetic and gravitational forces could accomplish just such a large-scale uncompensated reduction in entropy.

A number of authors[16–21] have pointed out that if there is a boundary condition that requires that the universe be in a state of high order when small, then entropy must decrease as the universe recollapses. The theorems proved here suggest the mechanism by which such an entropy decrease can take place. The universe starts off with both fine- and coarse-grained entropy low. As the universe expands, the fine-grained entropy remains constant, while the coarse-grained entropy increases. In addition, as the universe expands, the states of different parts of the universe become correlated. As we will show, the mutual information between the various parts of the universe gives a lower bound on the increase in coarse-grained entropy. We will also show that mutual information between the particle modes of the fields in the universe is virtually certain to rise to its maximum value, driving the coarse-grained entropy up to *its* maximum value. As the universe contracts, this mutual information is used to decrease the coarse-grained entropy of the universe as by theorem 4 above. Each part of the universe behaves like Maxwell's demon, using the information that it has collected about the other parts of the universe during the expanding phase to decrease their coarse-grained entropy during the contracting phase.

We analyze this process in detail. First, we show that mutual information gives a lower bound to the increase in coarse-grained entropy.

*Theorem 5:* Given a system $B$ with degrees of freedom $l_1, \ldots, l_m$, a coarse graining for the states of $B$ such that $\bar{S}(B)=\bar{S}(l_1)+\cdots+\bar{S}(l_m)$, a macroscopic state $b$ of $B$ at time $t$ determined by the results of inexact measurements on each of the degrees of freedom $l_1, \ldots, l_m$ of $B$ corresponding to an initial distribution $p_0(l_1, \ldots, l_m)$ at time $t_0$, then we have

$$\bar{S}_t(b)-\bar{S}_0(l_1, \ldots, l_n)\geq k_B I_t(l_1, \ldots, l_m) ,$$

where

$$I_t(l_1, \ldots, l_m)=S_t(l_1)+\cdots+S_t(l_m) - S_t(l_1, \ldots, l_m)$$

is the mutual information shared between $l_1, \ldots, l_m$ at time $t$.

*Proof:* By assumption, the coarse-grained entropy of $b$ is equal to the sum of the coarse-grained entropies in the degrees of freedom of $B$ partially fixed by the measurement that determines $b$,

$$\bar{S}_t(b)=\bar{S}_t(l_1)+\cdots+\bar{S}_t(l_m) .$$

But $\bar{S}_t(l_i)\geq k_B S_t(l_i)$, where $S_t(l_i)$ is the Shannon entropy for $l_i$ at time $t$ given the underlying distribution, hence

$$\bar{S}_t(b)\geq k_B[S_t(l_1)+\cdots+S_t(l_m)] = k_B[I_t(l_1, \ldots, l_m)+S_t(l_1, \ldots, l_m)] ,$$

where

$$I_t(l_1, \ldots, l_m)=S_t(l_1)+\cdots+S_t(l_m) - S_t(l_1, \ldots, l_m)$$

is the mutual information shared between the degrees of freedom of $B$ at time $t$. But

$$S_t(l_1, \ldots, l_m)=S_0(l_1, \ldots, l_m) =(1/k_B)\bar{S}_0(l_1, \ldots, l_m) .$$

Thus,

$$\bar{S}_t(b)-\bar{S}_0(l_1, \ldots, l_m)=\bar{S}_t(b)-k_B S_t(l_1, \ldots, l_m) \geq k_B I_t(l_1, \ldots, l_m) .$$

Theorem 5 says that if the measurements that fix the

coarse graining for a system are unable to detect correlations between different degrees of freedom, then increases in the system's coarse-grained entropy are bounded below by increases in correlation. Dilute gases satisfy this coarse-graining requirement. Since most of the observed entropy of the universe is in the microwave background—a dilute gas of photons—theorem 5 holds for our universe at present and has held for it since the decoupling of matter and radiation. Bose and Fermi gases in the quantum regime and liquids for which excluded volume effects are important do not satisfy the coarse-graining requirement of the theorem. In such systems correlations give rise to macroscopic effects.

Next we show that the mutual information shared between the particle modes of the quantum fields in the universe is very likely to increase to its maximum value.

The modes of the quantum fields can be represented as a collection of harmonic oscillators. We first examine the case of $n$ oscillators, then take the thermodynamic limit $n \to \infty$. $n$ noninteracting harmonic oscillators are described by a Hamiltonian $H = H_1 + H_2 + \cdots + H_n$, where

$$H_k = \sum_{i=0}^{\infty} i\omega_k |i\rangle_k \langle i| ,$$

where $|i\rangle_k$ is the $i$th excited state of the $k$th oscillator, $\omega_k$ is its fundamental frequency, and we have set the zero-point energy to zero. Suppose that the oscillators start out in the state $|\chi_0\rangle = |i\rangle_1 |j\rangle_2 \cdots |k\rangle_n$, with total energy

$$E = (h/2\pi)(i\omega_1 + j\omega_2 + \cdots + k\omega_n) .$$

If there is no interaction between the oscillators, then they will stay in this state forever. Suppose now that one perturbs the system by adding a small, randomly selected interaction Hamiltonian to the original Hamiltonian $H' = H + H_{\text{int}}$. By *random* we mean that the eigenvalues $\lambda_i$ of $H_{\text{int}}$ are selected at random such that $\lambda_{\max} \geq \lambda_i \geq \lambda_{\min}$, $E >> |\lambda_{\max}|, |\lambda_{\min}|$, and the eigenvectors of $H_{\text{int}}$ confined to the subspace of energy $E$ are orthogonal linear combinations of the unperturbed states of the subspace with coefficients distributed at random subject to normalization constraints. Since $H_{\text{int}}$ is selected at random, its eigenvectors form a randomly selected basis for this subspace. If we wait an amount of time $t = h/(2\pi\Delta E)$, where $\Delta E$ is the average energy level spacing of $H_{\text{int}}$, the evolution of the system will take $|\chi_0\rangle$ to a state that is itself a random superposition of the original states $|i\rangle_1 |j\rangle_2 \cdots |k\rangle_n$ with total energy $E$, i.e.,

$$|\chi_0\rangle \to |\chi_1\rangle = \sum_{i,j,\ldots,k} \alpha_{i,j,\ldots,k} |i\rangle_1 |j\rangle_2 \cdots |k\rangle_n ,$$

where $\alpha_{i,j,\ldots,k}$ are distributed randomly subject to the normalization constraint

$$\sum_{i,j,\ldots,k} |\alpha_{i,j,\ldots,k}|^2 = 1 .$$

It is possible to show[22] that the density matrix for the first oscillator then takes the form

$$\rho_1 = (1/N) \sum_{i=0}^{\infty} |i\rangle_1 \langle i| d_{2,\ldots,n}(E - \hbar\omega_1 i) \times \{1 \pm 1/[d(E - \hbar\omega_1 i)^{1/2}]\} ,$$

where $d_{2,\ldots,n}(E - \hbar\omega_1 i)$ is the dimension of the space of states of the oscillators $2, \ldots, n$ with total energy $E - \hbar\omega_1 i$, $N$ is a normalization constant, and the $\pm 1/[d(E - \hbar\omega_1 i)^{1/2}]$ term expresses the uncertainty in the result due to the lack of knowledge of the exact form of $H_{\text{int}}$. If $E$ is large and $n >> 1$, we can write

$$d_{2,\ldots,n}(E - \hbar\omega_1 i) = e^{S_{2,\ldots,n}(E - \hbar\omega_1 i)} = e^{S_{2,\ldots,n}(E) - (\partial S_{2,\ldots,n}/\partial E)\hbar\omega_1 i} .$$

Taking the thermodynamic limit $n \to \infty$, we obtain

$$\rho_1 = (1/N') \sum_{i=0}^{\infty} |i\rangle_1 \langle i| e^{\hbar\omega_1 i/T} ,$$

where

$$1/T = \partial S_{2,\ldots,n}(E)/\partial E .$$

The expressions for $\rho_2, \ldots, \rho_n$ also take the exact thermal form. We may state these results as follows.

*Theorem 6:* In the thermodynamic limit, a randomly chosen $H_{\text{int}}$ between oscillators with total-energy $E$ causes the oscillators as a group to evolve into a state in which the density matrix for each of the oscillators is exactly that for the oscillators at thermal equilibrium. The set of $H_{\text{int}}$ that do not put the oscillators in a thermal state is of measure zero. The evolution drives both the mutual information and the coarse-grained entropy up to their maximum values $\bar{S} - S_0$ and $\bar{S}$, respectively, where $\bar{S}$ is the equilibrium thermodynamic entropy and $S_0$ is the fine-grained entropy.

As the universe expands, the different modes of the quantum fields collect exactly the right amount of information about each other to arrange the required decrease in coarse-grained entropy back to $S_0$ as the universe contracts. In general, this canceling out of coarse-grained entropy through mutual information shared between modes will not occur until the end of a Poincaré cycle for the modes. However, if the boundary condition for the universe requires that entropy be small at the big crunch, the interaction between the different modes causes them to behave like Maxwell's demons. Using existing information they decrease the coarse-grained entropy of the remaining modes to its original, low level. The effect of such boundary conditions is to pick out the trajectories for the modes whose Poincaré recurrence time equals the time required for the universe to expand and recollapse.

Classically, this trading in of mutual information cannot reduce the overall fine-grained entropy to zero. The fine-grained entropy does not decrease and cannot be zero initially if mutual information is to be generated in the first place. In contrast, mutual information can be generated quantum mechanically even when the universe as a whole is in a pure state with fine-grained entropy zero. Quantum-mechanical generation of mutual information occurs during measurement situations, for example, when the gravitational field couples to quantum fluctuations of the matter modes during inflation. When this mutual information is used to cancel coarse-grained entropy, quantum-mechanical "unmeasurement" takes place; as they become uncorrelated, the different components of the superposition interfere to produce a state of low entropy. In the case that the universe starts out in a pure state $S_0=0$, the interaction between the modes can actually reduce the coarse-grained entropy to zero.

## ACKNOWLEDGMENTS

I would like to thank Heinz Pagels, E. G. D. Cohen, Joel Cohen, and Wojciech Zurek for helpful discussions. Work supported in part by the Department of Energy under Contract Grant Number DE-AC02-87ER40325. TaskB.

---

*Permanent address: Department of Physics, California Institute of Technology, Pasadena, CA 91125.

[1]J. C. Maxwell, *Theory of Heat* (Appleton, London, 1871).
[2]L. Szilard, Z. Phys. **53**, 840 (1929).
[3]L. Brillouin, *Science and Information Theory,* 2nd ed. (Academic, New York, 1962).
[4]D. A. Bell, Am. Sci. **40**, 682 (1952).
[5]R. C. Raymond, Am. J. Phys. **19**, 109 (1951).
[6]R. Landauer, IBM J. Res. Dev. **5**, 183 (1961).
[7]R. W. Keyes and R. Landauer, IBM J. Res. Dev. **14**, 152 (1970).
[8]C. H. Bennett, IBM J. Res. Dev. **17**, 525 (1973); Int. J. Theor. Phys. **21**, 905 (1982); Sci. Am. **257**, 108 (1987).
[9]J. von Neumann, *Mathematical Foundations of Quantum Mechanics* (Princeton University Press, Princeton, 1955).
[10]R. C. Tolman, *The Principles of Statistical Mechanics* (Oxford University Press, Oxford, 1938).
[11]E. Borel, *Introduction Geometrique a Quelques Theories Physiques* (Gauthier-Villars, Paris, 1914). Borel makes the point that our knowledge of the exact conditions of evolution of a system such as a gas is acquired by experiment and suffers from statistical inexactness. A further development of this notion is given by J. M. Blatt in *An Alternative Approach to the Ergodic Problem* [Prog. Theor. Phys. **22**, 745 (1959)], who points out that the interaction between the molecules of a gas and the walls of their container causes the fine-grained entropy of the gas to increase.
[12]C. E. Shannon and W. Weaver, *The Mathematical Theory of Communication* (University of Illinois Press, Urbana, 1949).
[13]J. C. Adams, manuscript Nos. 1841–1846, St. John's College Library, Cambridge, England.
[14]U. J. J. LeVerrie, C.R. Acad. Sci. **21**, 1050 (1845); **22**, 907 (1846); **23**, 428 (1846); **23**, 657 (1846); **22**, 208 (1848); **27**, 273 (1848); **27**, 304 (1848); **27**, 325 (1848).
[15]K. K. Likharev, Int. J. Theor. Phys. **21**, 311 (1982).
[16]T. Gold, Am. J. Phys. **30**, 403 (1962).
[17]Y. Ne'eman, Int. J. Theor. Phys. **3**, 1 (1970).
[18]W. J. Cocke, Phys. Rev. **160**, 1165 (1967).
[19]H. Schmidt, J. Math. Phys. **7**, 495 (1966).
[20]S. W. Hawking, Phys. Rev. D **32**, 2489 (1985).
[21]D. Page, Phys. Rev. D **32**, 2496 (1985).
[22]S. Lloyd, Ph.D. thesis, The Rockefeller University, 1988.

# Quantum-mechanical Maxwell's demon

Seth Lloyd*
*d'Arbeloff Laboratory for Information Systems and Technology, Department of Mechanical Engineering, Massachusetts Institute of Technology, MIT 3-160, Cambridge, Massachusetts 02139*
(Received 8 November 1996; revised manuscript received 26 March 1997)

A Maxwell's demon is a device that gets information and trades it in for thermodynamic advantage, in apparent (but not actual) contradiction to the second law of thermodynamics. Quantum-mechanical versions of Maxwell's demon exhibit features that classical versions do not: in particular, a device that gets information about a quantum system disturbs it in the process. This paper proposes experimentally realizable models of quantum Maxwell's demons, explicates their thermodynamics, and shows how the information produced by quantum measurement and by decoherence acts as a source of thermodynamic inefficiency. [S1050-2947(97)02910-7]

PACS number(s): 03.65.Bz, 32.80.Pj, 33.80.Ps, 33.25.+k

## INTRODUCTION

In 1871, Maxwell noted that a being that could measure the velocities of individual molecules in a gas could shunt fast molecules into one container and slow molecules into another, thereby creating a difference in temperature between the two containers, in apparent violation of the second law of thermodynamics [1]. Kelvin called this being a "demon:" by getting information and being clever how it uses it, such a demon can in principle perform useful work. Maxwell's demon has been the subject of considerable discussion over the past century [2]. The contemporary view of the demon, spelled out in the past decade [3], is that a demon could indeed perform useful work $k_BT$ ln2 for each bit obtained, but must increase entropy by at least $k_B$ln2 for each bit erased (a result known as "Landauer's principle" [4]). As a result, a demon that operates in cyclic fashion, erasing bits after it exploits them, cannot violate the second law of thermodynamics.

Up to now, Maxwell's demon has functioned primarily as a thought experiment that allows the exploration of theoretical issues. This paper, in contrast, proposes a model of a Maxwell's demon that could be realized experimentally using magnetic or optical resonance techniques. Any experimentally realizable model of a "demon," like the molecules of Maxwell's original example, must be intrinsically quantum mechanical. The classic reference on quantum demons is Zurek's treatment of the quantum Szilard engine [5] (see also Refs. [6,7]). Zurek investigated a gedanken experiment consisting of a single quantum particle sitting in a classical cylinder, acted upon by a classical piston, and measured by a classical measuring device. The quantum nature of the particle allowed Zurek to give an accurate thermodynamic accounting of the heat taken in and work done by the particle as the piston expanded. Zurek's picture is semiclassical in that only the particle itself is taken to be quantum mechanical, while the remainder of the engine is classical; the semiclassical nature of this treatment does not allow Landauer's principle and the thermodynamics of the erasure process to be investigated in a quantum context.

This paper makes the following advances in treating Maxwell's demons.

(1) It uses well-established physics of the quantum electrodynamics of spin and optical resonance to present a fully quantum-mechanical model of a Maxwell's demon that unlike previous semiclassical models allows the thermodynamics of the demon's entire cycle of operation—heat absorption, measurement and decoherence, work generation, and erasure—to be treated within a unified quantum framework.

(2) Unlike other models of demons, the model presented here provides a detailed mechanism for erasing information and so allows Landauer's principle to be investigated in a realistic quantum context. In the proposed device, "waste" information is rejected by putting the spin that makes up the demon's quantum memory in contact with a low temperature mode of the electromagnetic field. Instead of violating the second law of thermodynamics, the demon operates as a quantum heat engine, doing work while pumping heat from hot modes to cold modes.

(3) The paper demonstrates the truth of Haus's conjecture [8] that processes such as quantum measurement and decoherence, by introducing information, effectively increase entropy and reduce the thermodynamic efficiency of the demon. As a result, quantum demons and quantum heat engines in general suffer from peculiarly quantum sources of inefficiency: each bit of information introduced by measurement or decoherence increases entropy by $k_B$ln2.

(4) Finally, the paper proposes experiments for realizing quantum demons using nuclear magnetic resonance on molecules.

*Electronic address: slloyd@mit.edu

The outline of the paper is as follows. Section I explains how a nuclear or electronic spin can function as the "working fluid" for a quantum demon, and shows how Landauer's principle prevents the demon from violating the second law of thermodynamics. Section II shows how quantum measurement and decoherence make the demon less thermodynamically efficient and provides formulas that quantify the resulting reduction in efficiency: not only does the operation of the quantum demon fail to violate the second law, it may actually increase entropy beyond what is required classically. In Sec. III a nuclear magnetic resonance (NMR) model of a demon is presented. The model can be realized by a variety of NMR techniques on a variety of molecules. Section IV provides a detailed analysis of the thermodynamics and quantum statistical mechanics of the NMR model, and goes through the cyclic operation of the demon step by step, identifying the thermodynamic costs of information gathering and erasure. The mechanisms by which information is obtained, exploited, and erased are explored in detail. Landauer's principle is confirmed: such a device cannot violate the second law of thermodynamics, but if supplied with heat reservoirs at different temperatures it can undergo a cycle analogous to the Carnot cycle and function as a heat engine. Section V analyzes the effects of decoherence on the NMR demon and shows that each bit of information introduced by quantum measurement and decoherence functions as an extra $k_B \ln 2$ of entropy, decreasing the engine's efficiency. A quantum-mechanical engine that processes information is limited not by the Carnot efficiency, but by the potentially lower quantum efficiency $\epsilon_Q$ defined in Sec. II. Section VI discusses a variety of potential experimental realizations in addition to NMR. Finally, Sec. VII concludes and discusses future work.

## I. A MAGNETIC RESONANCE MODEL OF A QUANTUM "DEMON"

To see how a device that gets information about a quantum system can use that information to perform useful work, consider a spin in a magnetic field. If the spin points in the same direction as the field it has energy $-\mu B$, where $\mu$ is the spin's magnetic dipole moment, and $B$ is the field strength. If it points in the opposite direction it has energy $+\mu B$. The state of the spin can be controlled by conventional magnetic resonance techniques: for example, the spin can be flipped from one energy state to the other by applying a $\pi$ pulse at the spin's Larmor precession frequency $\omega = 2\mu B/\hbar$ [9,10].

When the spin flips, it exchanges energy with the oscillatory field. If the spin flips from the lower energy state to the higher energy state it coherently absorbs one photon with energy $\hbar\omega$ from the field; if it flips from the higher energy state to the lower, it coherently emits one photon with energy $\hbar\omega$ to the field. Either the field does work on the spin, or the spin does work on the field.

When the quantum nature of the electromagnetic field is taken into account, one might worry that the interaction that exchanges energy between spin and field also induces an exchange of information that effectively entangles the quantum state of the spin with the state of the field. A detailed quantum electrodynamic treatment of this interaction [11] shows that this worry is indeed justified when the field is in a nonclassical state such as a highly squeezed state. When the field is in a coherent state, however, as is the case for fields normally produced by lasers, masers, or rf coils, the energy exchange involves no information exchange, entropy increase, or loss of quantum coherence [11]. As a result, even though the oscillatory field that flips the spin is in fact quantum mechanical, it may be treated as if it were classical for the purpose of the experiments described below.

It is clear how a device that acquires information about such a spin could use the information to make the spin do work. Suppose that some device can measure whether the spin is in the low-energy quantum state $|\downarrow\rangle$ that points in the same direction as the field, or in the high-energy quantum state $|\uparrow\rangle$ that points in the opposite direction to the field, and if the spin is in the high-energy state, send in a $\pi$ pulse to extract its energy. The device can then wait for the spin to come to equilibrium at temperature $T_1 \gg 2\mu B/k_B$ and repeat the operation. Each time it does so, it converts an average of $\mu B$ of heat into work. The device gets information and uses that information to convert heat into work. The amount of work done by such a device operating on a single spin is negligible; but many such devices operating in parallel could function as a "demon" maser, coherently amplifying the pulse that flips the spins.

Landauer's principle prevents such a device from violating the second law of thermodynamics. To operate in a cyclic fashion, the device must erase the information that it has gained about the state of the spin. When this information is erased, entropy $S_{\text{out}} \geq k_B \ln 2$ is pumped into the device's environment, compensating for the entropy $S_{\text{in}} \approx k_B \ln 2$ of entropy in the spin originally. If the environment is a heat bath at temperature $T_2$, which may be different from the temperature $T_1$ of the spins' heat bath, heat $k_B T_2 \ln 2$ flows to the bath along with the entropy, decreasing the energy available to convert into work. The detailed mechanism by which the "waste" information is converted to heat is described in Sec. IV below.

The overall accounting of energy and entropy in the course of the cycle is as follows: heat in $Q_{\text{in}} = T_1 S_{\text{in}}$, heat out $Q_{\text{out}} = T_2 S_{\text{out}}$, work out $W_{\text{out}} = Q_{\text{in}} - Q_{\text{out}}$, efficiency

$$\epsilon = W_{\text{out}}/Q_{\text{in}} = 1 - T_2 S_{\text{out}}/T_1 S_{\text{in}} \leq 1 - T_2/T_1 \equiv \epsilon_C, \tag{1.1}$$

where $\epsilon_C$ is the Carnot efficiency. Since $S_{\text{out}} \geq S_{\text{in}}$, $W_{\text{out}}$ can be greater than zero only if $T_1 > T_2$. That is, Landauer's principle implies that instead of violating the second law of thermodynamics, the device operates as a heat engine, pumping heat from a high-temperature reservoir to a low-temperature reservoir and doing work in the process.

## II. MEASUREMENT AND DECOHERENCE MAKE THE DEMON LESS EFFICIENT

Why quantum measurement introduces added inefficiency into the operation of such a device can be readily understood. Suppose that the spin is originally in the state $|\rightarrow\rangle = 1/\sqrt{2}(|\uparrow\rangle + |\downarrow\rangle)$. One way to extract energy from such a spin is to apply a $\pi/2$ pulse to rotate the spin to the state $|\downarrow\rangle$, extracting work $\mu B$ in the process. A second way to extract energy is to repeat the process described above: measure the spin to see if it is in the state $|\uparrow\rangle$, and if it is, apply a $\pi$ pulse to extract work $2\mu B$. Half the time the measurement will find the spin in the state $|\uparrow\rangle$ and half the time in the state $|\downarrow\rangle$; as a result, this process also generates work $\mu B$ on average, but in addition generates a "waste" bit of information that costs energy $k_B T_2 \ln 2$ to erase. Quantum measurement introduces added inefficiency to the process of getting information about a quantum system and exploiting that information to perform work.

More generally, suppose the spin is initially described by a density matrix $\rho$. Let the spin interact with a measuring device or other system that decoheres the spin by destroying off-diagonal terms in the density matrix. The new density matrix is then

$$\rho' = \sum_i P_i \rho P_i, \tag{2.1}$$

where $P_i$ are projection operators onto the eigenspaces of the operator corresponding to the measurement or onto the preferred subspaces of the decohering process [12]. The extra information generated by quantum measurement or decoherence is

$$\Delta I_Q = \Delta S_Q / k_B \ln 2 = -\mathrm{tr}\rho' \log_2 \rho' - (-\mathrm{tr}\rho \log_2 \rho) \geq 0. \tag{2.2}$$

The efficiency of the device in converting heat to work is limited not by the Carnot efficiency $\epsilon_C$ but by the quantum efficiency

$$\epsilon_Q = 1 - T_2(S_{\rm in} + \Delta S_Q)/T_1 S_{\rm in} = \epsilon_C - T_2 \Delta S_Q / T_1 S_{\rm in}. \tag{2.3}$$

Equation (2.3) quantifies the inefficiency due to any process, such as decoherence, that destroys off-diagonal terms of the density matrix $\rho$ [12–14].

The degree to which the quantum efficiency differs from the Carnot efficiency depends on the amount of extra entropy $\Delta S_Q$ or information $\Delta I_Q$ introduced by measurement or decoherence. If $\rho$ is already diagonal with respect to the projections $P_i$ or diagonal in the preferred basis of the decohering process then no new information is introduced and $\epsilon_Q = \epsilon_C$. At the other extreme, the measurement or decohering process can completely randomize the system, giving

$$\rho' = I/2, \quad S_{\rm in} + \Delta S_Q = k_B \ln 2, \tag{2.4}$$

$$\epsilon_Q = 1 - T_2 k_B \ln 2 / T_1 S_{\rm in}. \tag{2.5}$$

Note that $\epsilon_Q$ can be negative, corresponding either to a process that generates heat but does no net work, or (as discussed in Secs. IV–VI below) to a refrigerator that pumps heat from a cold reservoir to a hot reservoir.

Equation (2.3) applies not only to "demon"-like heat engines that operate by obtaining information about a quantum systems, thereby decohering them, it also applies to more conventional quantum heat engines such as lasers that undergo decoherence through interactions with their environment. For example, the reduction in quantum efficiency described by Eq. (2.3) could be observed by taking a laser in which a population inversion has been established between two levels, and "tipping" that inversion by an angle $\theta$ by applying a pulse at the resonant frequency between the levels with integrated intensity $\hbar\theta/2D$, where $D$ is the dipole moment between the levels. The amount by which the population of the higher level exceeds that of the lower level is proportional to $\cos\theta$, so that when the inversion has been tipped by $\pi/2$, there is no longer an effective inversion between the levels and further light will not be amplified. The maser analog of this effect is discussed in greater detail in Sec. V below.

## III. A QUANTUM DEMON CAN BE REALIZED USING NUCLEAR MAGNETIC RESONANCE

To investigate more thoroughly how quantum measurement and decoherence introduce thermodynamic inefficiency in a quantum information-processing "demon" requires a more detailed model of how such a device gets and gets rid of information. One of the simplest quantum systems that can function as a measuring device is another spin. Magnetic resonance affords a variety of techniques, called spin-coherence double resonance, whereby one spin can coherently acquire information about another spin with which it interacts [9,10]. The basic idea is to apply a sequence of pulses that makes spin 2 flip if and only if spin 1 is in the excited state $|\uparrow\rangle_1$, while leaving spin 1 unchanged. If spin 2 is originally in the ground state $|\downarrow\rangle_2$, then after the conditional spin-flipping operation, the two spins will either be in the state $|\uparrow\rangle_1|\uparrow\rangle_2$ if spin 1 was originally in the state $|\uparrow\rangle_1$, or in the state $|\downarrow\rangle_1|\downarrow\rangle_2$ if spin 1 was originally in the state $|\downarrow\rangle_1$. Spin 2 has acquired information about spin 1. A variety of spin coherence double resonance techniques (going under acronyms such as INEPT and INADEQUATE) can be used to perform this conditional flipping operation [9,10], which can be thought of as an experimentally realizable version of Zurek's treatment of the measurement process in Ref. [5]. Readers familiar with quantum computation will recognize the conditional spin flip as the quantum logic operation "controlled-NOT" [15–17].

How can this information be used to extract energy from spin 1? Simply apply a second pulse sequence to flip spin 1 if and only if spin 2 is in the state $|\uparrow\rangle_2$, while leaving spin 2 unchanged. The energy transfer from spins to field is as follows. If spin 1 was originally in the state $|\downarrow\rangle_1$, then spin 1 and spin 2 remain in the state $|\downarrow\rangle$ through both pulse sequences and no energy is transferred to the field. If spin 1 was originally in the state $|\uparrow\rangle$, first spin 2 flips, then spin 1, yielding a transfer of energy from spins to field of $\hbar(\omega_1 - \omega_2) = 2(\mu_1 - \mu_2)B$, which is $>0$ as long as $\mu_1 > \mu_2$. The average energy extracted is half this value. As long as the conditional spin flips are performed coherently, the amount of energy extracted depends only on overall conservation of energy, and is independent of the particular double resonance technique used. Note that the entire process maintains quantum coherence and can be reversed simply by repeating the conditional spin flips in reverse order.

The steps just described have been realized experimentally in a variety of systems. One of the first realizations of such a double resonance experiment is the Pound-Overhauser effect [9] in which spin 1 is a proton and spin 2 is an electron. Here the amount of energy required to flip the electron is three orders of magnitude higher than the amount of energy gained by flipping the proton: instead of operating as an engine the double resonance process functions as a refrigerator, pumping heat from the proton to the electron.

## IV. THERMODYNAMICS OF THE NMR DEMON

In the previous section it was shown how spins can get information about each other and put that information to use to do work. The spins are fulfilling their role as demon by extracting work without generating waste heat. They are not yet violating the second law of thermodynamics, however, which states that it is not possible for an engine to turn heat into work with no waste heat *while operating in a cyclic fashion.* To complete the cycle, the spins must be restored to their original state without generating waste heat. As will now be seen, this cannot happen.

To give a full treatment of the thermodynamics of this device and to understand the role of decoherence and quantum measurement in its functioning, we must investigate how the "demon" interacts with its thermal environment to take in heat and erase information. This section will show that a quantum device that interacts with a thermal environment can indeed get information and "cash it in" to do useful work, but not by violating the second law of thermodynamics: a detailed model of the erasure process confirms Landauer's principle (one bit of information "costs" entropy $k_B\ln 2$). As a result, instead of functioning as a perpetual motion machine, the device operates as a heat engine that undergoes a cycle analogous to a Carnot cycle.

The environment for our spins will be taken to consist of two sets of modes of the electromagnetic field, the first a set of modes at temperature $T_1$ with average frequency $\omega_1$ and with frequency spread greater than the coupling constant $|\kappa|$ between the spins but less than $\omega_1-\omega_2$, and the second a set of modes at temperature $T_2$ with average frequency $\omega_1$ and the same frequency spread. Such an environment can be obtained, for example, by bathing the spins in incoherent radiation with the given frequencies and temperatures. The purpose of such an environment is to provide effectively separate heat reservoirs for spin 1 and spin 2: spin 1 interacts strongly with the on-resonance radiation at frequency $\omega_1$, and weakly with the off-resonance radiation at frequency $\omega_2$ and vice versa for spin 2. Over short times, to a good approximation spin 1 can be regarded as interacting only with mode 1, and spin 2 as interacting only with mode 2. A spin can be put in and out of "contact" with its reservoir by isentropically altering the frequency of the reservoir mode to put the spin in and out of resonance.

With this approximation, the initial probabilities for the state of the $j$th spin are (ignoring for the moment the coupling between the spins)

$$p_j(\uparrow)=e^{-\mu_j B/k_B T_j}/Z_j,\quad p_j(\downarrow)=e^{\mu_j B/k_B T_j}/Z_j,\qquad (4.1)$$

yielding energy

$$E_j=-\mu_j B\tanh(\mu_j B/k_B T_j),\qquad (4.2)$$

and entropy

$$S_j=-k_B\sum_{i=\uparrow,\downarrow}p_j(i)\ln p_j(i)=E_j/T_j+k_B\ln Z_j,\qquad (4.3)$$

where $Z_j=e^{-\mu_j B/k_B T_j}+e^{\mu_j B/k_B T_j}=2\cosh(\mu_j B/k_B T_j)$. Even though it does not start out in a definite state, spin 2 can still acquire information about spin 1, and this information can be exploited to do electromagnetic work. The spins can function as a heat engine by going through the following cycle.

(1) Using spin coherence double resonance, flip spin 2 if spin 1 is in the state $|\uparrow\rangle_1$. This causes spin 2 to gain information $(\tilde{S}_2-S_2)/k_B\ln 2$ about spin 1 at the expense of work $W_1=p_1(\uparrow)2\mu_2 B\tanh(\mu_2 B/k_B T_2)$ supplied by the oscillating field. Here $\tilde{S}_2=-k_B\Sigma_{i=\uparrow,\downarrow}\tilde{p}_2(i)\ln\tilde{p}_2(i)$, where $\tilde{p}_2(\uparrow)=p_1(\uparrow)p_2(\downarrow)+p_1(\downarrow)p_2(\uparrow)$ and $\tilde{p}_2(\downarrow)=p_1(\downarrow)p_2(\downarrow)+p_1(\uparrow)p_2(\uparrow)$ are the probabilities for the states of spin 2 after the conditional spin flip.

(2) Flip spin 1 if spin 2 is in the state $|\uparrow\rangle_1$. This step allows spin 2 to "cash in" $(S_2-S_1)/k_B\ln 2$ of the information it has acquired, thereby performing work $-\mu_1 B[\tanh(\mu_1 B/k_B T_1)-\tanh(\mu_2 B/k_B T_2)]$ on the field.

(3) Spin 2 still possesses information $(\tilde{S}_2-S_1)/k_B\ln 2$ about spin 1, which can be converted into work by flipping spin 2 if spin 1 is in the state $|\uparrow\rangle_1$, thereby performing work $p_2(\uparrow)2\mu_2 B\tanh(\mu_1 B/k_B T_2)$ on the field.

It is straightforward to verify that after these three conditional spin flips, spin 1 has probabilities $p_1'(i)=p_2(i)$ while spin 2 has probabilities $p_2'(i)=p_1(i)$. That is, the sequence of pulses has "swapped" the information in spin 1 with the information in spin 2. As a result, $S_1'=S_2$, $S_2'=S_1$, and the new energies of the spins are $E_1'=-\mu_1 B\tanh(\mu_2 B/k_B T_2)$ and $E_2'=-\mu_2 B\tanh(\mu_1 B/k_B T_1)$. The total amount of work done by the spins on the electromagnetic field is

$$W=-(E_1'+E_2'-E_1-E_2)=-(\mu_1-\mu_2)B[\tanh(\mu_1 B/k_B T_1)-\tanh(\mu_2 B/k_B T_2)].\qquad (4.4)$$

At temperatures $T_i\gg\mu_i B/k_B$, Eq. (4.4) reduces to

$$W=-(\mu_1-\mu_2)(\mu_1/T_1-\mu_2/T_2)B^2/k_B.\qquad (4.5)$$

These formulas for the work done depend only on conservation of energy and do not depend on the specific set of coherent pulses that are used to "swap" the spins. Equations (4.4) and (4.5) shows that $W>0$ if either $\mu_1>\mu_2$, $\mu_1/T_1<\mu_2/T_2$ or $\mu_1<\mu_2$, $\mu_1/T_1>\mu_2/T_2$. If $T_1=T_2$, $W$ is zero or negative: no work can be extracted from the spins at equilibrium. The device cannot function as a *perpetuum mobile* of the second kind. The cycle can be completed by letting the spins reequilibrate with their respective reservoirs. Each time steps (1)–(3) are repeated, heat $Q_{\rm in}=E_1-E_1'$ flows from reservoir 1 to spin 1 and heat $Q_{\rm out}=E_2'-E_2$ flows from spin 2 into reservoir 2. The efficiency of this cycle is $W/Q_{\rm in}=1-\mu_2/\mu_1<1-T_2/T_1=\epsilon_C$: the efficiency is less than the Carnot efficiency because when the spins equilibrate with their respective reservoirs, heat flows but no work is done.

The following steps can be added to the cycle to allow the spins to reequilibrate isentropically.

(5) Return spin 1 to its original state: (i) Take the spin out of "contact" with its reservoir by varying the frequency of the reservoir modes as above. (ii) Alter the quasistatic field from $B\rightarrow B_1=BT_1/T_2$ adiabatically, with no heat flowing between spin and reservoir. (iii) Gradually change the

field from $B_1 \rightarrow B$ keeping the spin in "contact" with the reservoir at temperature $T_1$ so that heat flows isentropically between the spin and the reservoir. During this process, entropy $S_2 - S_1$ flows from the spin to the reservoir, while the spin does work $E_1 - E_1' - T_1(S_2 - S_1)$ on the field.

(6) Return spin 2 to its original state by the analogous set of steps.

The total work done by the spins on the electromagnetic field throughout the cycle is

$$W_C = (T_1 - T_2)(S_1 - S_2). \quad (4.6)$$

With the added steps (5) and (6) the demon undergoes a cycle analogous to the Carnot cycle and in principle operates at the Carnot efficiency $1 - T_2/T_1$. In practice, of course, the steps that go into operation of such an engine will be neither adiabatic nor isentropic, leading to an actual efficiency below the Carnot efficiency.

To attain the Carnot efficiency, it is important the demon "know" the temperatures $T_1$ and $T_2$. If the demon has incorrect values for these temperatures then after step [(5)(ii)] above spin 1 will be out of equilibrium with its reservoir and heat will flow with no work being done. An ordinary, macroscopic Carnot engine has a completely analogous source of inefficiency: altering the quasistatic field adiabatically for the spin is analogous to the adiabatic expansion-contraction stages for a Carnot engine. If the adiabatic expansion of the working fluid in a Carnot engine goes too far or not far enough, then the fluid will be at a temperature different from that of the low temperature reservoir, and when the fluid is put in contact with that reservoir the heat flow will not be isentropic but will increase entropy instead.

Equation (4.4) implies that when $\mu_1 > \mu_2$ and $\mu_1/T_1 > \mu_2/T_2$ the net work performed over the cycle is negative. In this case, the demon functions as a refrigerator, pumping net heat and entropy from the reservoir at temperature $T_2$ to the reservoir at temperature $T_1$ (note that $T_1$ need no longer be greater than $T_2$: indeed, $T_1$ and $T_2$ could now be equal). The efficiency of the refrigerator can be measured by the ratio between heat pumped and work done; for the idealized model given above this is equal to $T_2/(T_1 - T_2)$, which is just the usual Carnot coefficient of refrigerator performance. Naturally, any experimental realization of such a refrigerator will operate at an efficiency lower than the Carnot coefficient.

Such "demon refrigerators" have in fact been in operation since the 1950s. In the Pound-Overhauser effect [9], double resonance is used to "swap" information between an electron and a nucleon, thereby pumping entropy and energy from the nucleon to the electron. The Pound-Overhauser effect can be implemented by performing steps (1)–(3) above on an electron (spin 1) and a nucleon (spin 2). Swapping information between electron and nucleon interchanges their Boltzmann factors, reducing the effective temperature of the nucleon by a factor $g_e/g_n$ where $g_e, g_n$ are the gyromagnetic ratios of the electron and nucleon, respectively. In the Pound-Overhauser effect, entropy $S_2 - S_1$ is pumped from nucleon to electron at a cost in microscopic work given by Eq. (4.4). This entropy transfer is highly efficient: the work put into the transfer can be reextracted by swapping the information between the spins again. The only inefficiency arises from inaccuracies in the conditional spin flips. (At the macroscopic scale, of course, dissipation in the coils used to supply the pulses to flip the spins dwarfs the entropy transfers between spins.) Even at the microscopic level, when operated cyclically the Pound-Overhauser effect does not approach the Carnot coefficient of refrigerator performance, as the transfers of heat from lattice to nuclei and from electrons to electromagnetic field are not isentropic. Further examples of "demon" refrigerators will be discussed in Sec. VI.

## V. THERMODYNAMIC COST OF QUANTUM MEASUREMENT AND DECOHERENCE

So far, although the demon has functioned within the laws of quantum mechanics, quantum measurement and quantum information have not entered in any fundamental way. Now that the thermodynamics of the demon have been elucidated, however, it is possible to quantify precisely the effects both of measurement and of the introduction of quantum information on the demon's thermodynamic efficiency. The simple quantum information-processing engine of the previous section can in principle be operated at the Carnot efficiency: as will now be shown, when the engine introduces information by a process of measurement or decoherence, it cannot be operated even in principle (let alone in practice) above the lower, quantum efficiency $\epsilon_Q$ of Eq. (2.3).

To isolate the effects of quantum information, let us first look more closely at the simple model of Sec. I above, in which spin 1 is initially in the state

$$|\rightarrow\rangle_1 = 1/\sqrt{2}(|\uparrow\rangle_1 + |\downarrow\rangle_1). \quad (5.1)$$

This state has nonminimum free energy which is available for immediate conversion into work: simply apply a $\pi/2$ pulse to rotate spin 1 into the state $|\downarrow\rangle_1$, adding energy $\hbar\omega_2/2 = \mu_1 B$ to the oscillating field in the process. Suppose, however, that instead of extracting this energy directly, the demon operates in information-gathering mode as above, using magnetic resonance techniques to correlate the state of spin 2 with the state of spin 1 (cf. Ref. [5]). Suppose spin 2 is initially in the state $|\downarrow\rangle_2$. In this case, coherently flipping spin 2 if spin 1 is in the state $|\uparrow\rangle_1$ results in the state

$$1/\sqrt{2}(|\uparrow\rangle_1|\uparrow\rangle_2 + |\downarrow\rangle_1|\downarrow\rangle_2), \quad (5.2)$$

a quantum "entangled" state in which the state of spin 2 is perfectly correlated with the state of spin 1. Continuing the energy extraction process by flipping spin 1 if spin 2 is in the state $|\uparrow\rangle_2$ as before allows on average an amount of energy $(\mu_1 - \mu_2)B$ to be extracted from the spin. The resulting state of the spins is $1/\sqrt{2}|\downarrow\rangle_1(|\uparrow\rangle_2 + |\downarrow\rangle_2) = 1/\sqrt{2}|\downarrow\rangle_1|\rightarrow\rangle_2$. Up until this point, no extra thermodynamic cost has been incurred. Indeed, since the conditional spin flipping occurs coherently, the process can be reversed by repeating the steps in reverse order to return to the original state $|\rightarrow\rangle_1$, with a net energy and entropy change of zero.

When is the cost of quantum measurement realized? When decoherence occurs. In the original cycle, decoherence takes place when spin 2 is put in contact with the reservoir to "erase" it [13–14]: the exchange of energy between the spin and the reservoir is an incoherent process during which the pure state

$$|\rightarrow\rangle_2 = 1/\sqrt{2}(|\uparrow\rangle_2 + |\downarrow\rangle_2) \quad (5.3)$$

goes to the mixed state described by the density matrix

$$(1/2)(|\uparrow\rangle_2\langle\uparrow| + |\downarrow\rangle_2\langle\downarrow|). \quad (5.4)$$

The time scale for this process of decoherence is equal to spin 2's dephasing time $T_2^*$ and is typically much faster than the time scale for the transfer of energy [14]. In effect, interaction with the reservoir turns the process by which spin 2 coherently acquires quantum information about spin 1 into a decoherent process of measurement, during which a bit of information is created. This bit corresponds to an increase in entropy $k_B \ln 2$. During the process of erasure as in Sec. IV above, this entropy is transferred from spin 2 to the low-temperature reservoir, in accordance with Landauer's principle.

By decohering spin 2 and effectively measuring spin 1, the demon has increased entropy and introduced thermodynamic inefficiency. The amount of inefficiency can be quantified precisely by going to the Carnot cycle model of the demon above. In general, the state of spin 1 is described initially by a density matrix

$$\rho_1' = p_1(\uparrow')|\uparrow'\rangle_1\langle\uparrow'| + p_1(\downarrow')|\downarrow'\rangle_1\langle\downarrow'|, \quad (5.5)$$

where $|\uparrow'\rangle$ and $|\downarrow'\rangle$ are spin states along an axis at some angle $\theta$ from the $z$ axis. Without loss of generality, $T_1$ and $B$ can be taken to be such that

$$p_1(\uparrow') = e^{-\mu_1 B/k_B T_1}/Z_1, \quad p_1(\downarrow') = e^{\mu_1 B/k_B T_1}/Z_1. \quad (5.6)$$

This state is not at equilibrium, and possesses free energy that can be extracted by applying a tipping pulse that rotates the spin by $\theta$ and takes $|\uparrow'\rangle \rightarrow |\uparrow\rangle$ and $|\downarrow'\rangle \rightarrow |\downarrow\rangle$. The amount of work extracted is

$$W^* = E_1^* - E_1, \quad (5.7)$$

where

$$E_1 = \mu_1 B[p_1(\uparrow) - p_1(\downarrow)], \quad E_1^* = \mu_1 B[p_1^*(\uparrow) - p_1^*(\downarrow)] \quad (5.8)$$

and

$$p_1^*(\uparrow) = p_1(\uparrow)\cos^2\theta + p_1(\downarrow)\sin^2\theta,$$

$$p_1^*(\downarrow) = p_1(\downarrow)\cos^2\theta + p_1(\uparrow)\sin^2\theta. \quad (5.9)$$

Running the engine through a Carnot cycle by steps (1)–(5) above then extracts work $(T_1 - T_2)(S_1 - S_2)$. This process extracts the free energy of the spin isentropically, without increasing entropy.

By operating in this fashion—by only making measurements with respect to which the density matrix is diagonal—the demon reaches the upper limit of eq. (2.3) in which $\epsilon_Q = \epsilon_C$. Inefficiency due to measurement arises when instead of first applying the tipping pulse to extract spin 1's free energy, one simply operates the engine cyclically as before. In this case, the demon is making measurements with respect to which the density matrix has off-diagonal terms. As a result, the measurement introduces information and $\epsilon_Q$ is strictly less than $\epsilon_C$.

The steps are as above: three conditional flips swap the states of spin 1 and spin 2, so that spin 1 is in the state $\rho_2$ and spin 2 is in the state $\rho_2' = \rho_1'$. The interaction with the heat reservoir then decoheres spin 2, destroying the off-diagonal terms in the density matrix so that

$$\rho_2' \rightarrow p_1^*(\uparrow)|\uparrow\rangle\langle\uparrow| + p_1^*(\downarrow)|\downarrow\rangle\langle\downarrow|, \quad (5.10)$$

with entropy

$$S_1^* = -k_B \sum_{i=\uparrow,\downarrow} p_1^*(i)\ln p_1^*(i), \quad (5.11)$$

$$\Delta S_Q = (S_1^* - S_1), \quad \Delta I_Q = (S_1^* - S_1)/k_B \ln 2 \quad (5.12)$$

are the "extra" entropy and information introduced by decoherence. The entropy $S_1^* - S_2 = \Delta S_Q + S_{in}$ that flows out to reservoir 2 is greater than the entropy $S_1 - S_2 = S_{in}$ that flowed in from reservoir 1. The total amount of work done is

$$T_1(S_1 - S_2) - T_2(S_1^* - S_2) + W^*, \quad (5.13)$$

$T_2(S_1^* - S_1)$ less than the work $(T_1 - T_2)(S_1 - S_2) + W^*$ done by simply undoing the tipping pulse and operating the engine as before. The overall efficiency with decoherence and measurement included is

$$\epsilon_Q = 1 - T_2(S_1^* - S_2)/T_1(S_1 - S_2) = 1 - T_2(S_{in} + \Delta S_Q)/T_1 S_{in} \leqslant \epsilon_C, \quad (5.14)$$

in accordance with Eq. (2.3) above. The extra information introduced by quantum measurement and decoherence has decreased the efficiency of the demon. The quantum efficiency $\epsilon_Q$ rather than the Carnot efficiency $\epsilon_C$ provides the upper limit to the maximum efficiency of such an engine.

Decoherence also decreases efficiency when the quantum demon is operated as a refrigerator. If the demon is used to pump heat from a reservoir at a low temperature $T_2$ to a reservoir at a high temperature $T_1$ in the presence of decoherence as above, then the upper limit to the coefficient of performance for the refrigerator is no longer given by the Carnot coefficient $\gamma_C = T_2/(T_1 - T_2)$ but by the quantum coefficient of performance

$$\gamma_Q = T_2/[(1 + \Delta S_Q/S_{in})T_1 - T_2], \quad (5.15)$$

where $S_{in}$ is the net entropy pumped out of the low temperature reservoir, and $\Delta S_Q$ is the increase in entropy due to decoherence.

## VI. EXPERIMENTAL REALIZATIONS

This paper has proposed several effects that could be realized experimentally.

First, Secs. I–IV showed how a quantum system such as a spin that gets information about another spin can function as a Maxwell's demon, "cashing in" that information to do useful work. Nucleon-nucleon double resonance methods of the sort described in Sec. III could be used to construct a "demon maser" that performs a net amplification of the pulses that flip the spins. Equations (4.4) and (4.5) imply that for the maser to provide a net amplification the two different species of nucleons must start the cycle at significantly different temperatures, which could be accomplished by preparing one of the species in a low temperature state using electron-nucleon double resonance as in the Pound-Overhauser effect, or by optical pumping as in sideband cooling [18].

Such a maser, though unlikely to set any power records, could be used to verify Landauer's principle directly. Boltzmann factors for the different spin species can be verified at each stage of the cycle by looking at the induction signal for the different species. In addition, it may be possible to measure directly the heating of the lattice caused by the transfer of "waste" information from spins to lattice.

Although the detailed model given above treated the quantum demon in terms of interacting spins, exactly the same thermodynamics applies to any interacting two-level quantum systems. At bottom, a quantum "demon" consists of nothing more than an interaction between two quantum systems that allows the controlled transfer of information from one to the other. In particular, any system that can provide the coherent quantum logic operation controlled-NOT that flips one quantum bit conditioned on the state of another could form the basis for an information-processing quantum heat engine [18,19]. In Wineland's ion-trap quantum logic gate, for example, one of the quantum bits consists of two hyperfine states of a trapped ion, while the other bit consists of the lowest two vibrational states of the ion in the trap. Optical pumping techniques can be used to flip one of the bits conditioned on the state of the other.

A quantum demon can also be used as a refrigerator to pump heat from a low temperature reservoir to a high temperature reservoir, using up work in the process. As noted above, a "demon" refrigerator was realized as long ago as the 1950s in the Pound-Overhauser effect [9]. Wineland's sideband cooling technique [18] gives a second example of such a "Maxwell's refrigerator" [20]: in this technique, the same types of laser pulses used to perform the conditional quantum logic operations described in the previous paragraph are used to pump entropy from the vibrational mode of the ion to an internal electronic state which radiates heat to the environment via spontaneous emission. Both of these demons operate using microscopic, quantum degrees of freedom as the effective measuring apparatus whereby the demon gets information about the state of a quantum system and acts on that information to obtain work or to pump heat. It is also possible for the measuring apparatus to be macroscopic, as in Ref. [21]. Without a detailed model of the microscopic dynamics of the measurement apparatus and its interaction with the environment, however, it is not possible to verify Landauer's principle or the decrease in efficiency due to decoherence.

The second significant theoretical claim of this paper is that measurement and decoherence necessarily introduce inefficiency in the operation of a quantum heat engine or quantum refrigerator as described by Eqs. (2.3), (5.14), and (5.15). This inefficiency due to decoherence could be measured in a variety of ways. In any of the systems described in the previous paragraphs, decoherence can be introduced as described in Secs. II and V by introducing an extra pulse to "tip" the state of one of the systems before it interacts incoherently with its environment. In the case of the Pound-Overhauser experiment, for example, tipping the electron then performing the conditional spin-flipping pulse sequence results in an effective temperature for the nucleon that is higher than the effective temperature $T_1 g_n/g_e$ that results from performing the experiment without decoherence. This increase in the effective temperature could be measured by performing the cooling, waiting for a time greater than the decoherence time of the nucleon but less than its decay time, then tipping the nucleon and monitoring the strength of the nuclear induction signal: the strength of the induction signal is determined by the relative size of the Boltzmann factors in Eq. (4.1) and can be used to gauge the effective temperature of the nucleon. In the case of sideband cooling, the rate at which the vibrational mode is cooled will be diminished if the electronic state of the ion is tipped before the sideband pulse is administered: this diminished cooling rate could be verified by using the ion to monitor the effective temperature of the vibrational mode as the cooling progresses. In all cases, decoherence implies an additional dissipation $T\Delta S$ when the extra entropy $\Delta S$ created by decoherence is pumped into a reservoir at temperature $T$. For systems that would otherwise operate at the Carnot efficiency, this decrease in efficiency is quantified by Eq. (5.14) for a quantum heat engine and by Eq. (5.15) for a quantum refrigerator.

The decrease in efficiency due to measurement and decoherence is not confined to "demon" heat engines that operate by processing information. A variety of quantum heat engines exist [22]: the best-known examples are lasers and masers. These heat engines should not be considered "demons" because they do not operate by acquiring information and then acting on it. (All heat engines operate by transferring entropy from one degree of freedom to another. The appellation "demon" should be reserved for systems that acquire information and then perform an action conditioned on the value of that information.) Nonetheless, even in the absence of measurement, decoherence can be a source of inefficiency in a conventional quantum heat engine such as a laser, or in an "inverse laser" that uses anti-Stokes emissions to perform refrigeration [23,24]. As noted in Sec. II, if a tipping pulse is applied to such a system before the stage in its cycle in which it interacts incoherently with its environment then the efficiency of the device is decreased in accordance with Eqs. (2.3), (5.14), and (5.15).

The use of a tipping pulse is an experimental convenience to put one of the quantum systems that make up the demon in a state in which its density matrix has off-diagonal terms and is susceptible to decoherence. In fact, the operation of demon heat engines and refrigerators necessarily puts the demon in an off-diagonal state at some point during its operation. Any decoherence that occurs during this time will also increase entropy and reduce the demon's efficiency.

## CONCLUSION

This paper investigated Maxwell's demon in the context of realistic quantum models. As usual, such demons fail to violate the second law of thermodynamics. However, when operated between heat reservoirs at different temperatures, the demon can function either as a heat engine that pumps heat from a hot reservoir to a cold reservoir while doing work on the electromagnetic field, or as a refrigerator that pumps heat from a cold reservoir to a hot reservoir while absorbing work from the electromagnetic field. The paper derived general formulas that show how quantum measurement and decoherence decrease the efficiency of such heat engines.

The systems discussed here have several advantages over Zurek's quantum Szilard engine gedanken experiment [5]. First, the systems can be realized experimentally. Second, all aspects of the systems can be treated consistently within a unified quantum framework. The use of magnetic resonance techniques to describe a quantum Maxwell's demon was for the sake of convenience of exposition and potential experimental realizability: many other quantum systems could be suitable for performing the heat–information–energy conversion described above. Essentially any quantum system that can obtain information about another quantum system can form the basis for a quantum demon.

The work presented here suggests a variety of practical questions. First, we intend to investigate the circumstances under which a collection of many demons of the sort described could be used to pump macroscopic amounts of heat and do macroscopic amounts of work. In addition, we will examine the limitations to the thermodynamic efficiency of actual quantum devices that operate by getting and using information. Finally, having identified decoherence as a source of thermodynamic inefficiency in quantum devices, we can estimate the degree to which decoherence is responsible for low efficiency in existing quantum devices such as lasers, masers, and quantum cooling systems.

### ACKNOWLEDGMENTS

This paper originated in a series of discussions with Hermann Haus, who supplied the crucial insight that Landauer's principle implies that quantum demons suffered from additional sources of inefficiency. Without these discussions, and without Professor Haus's insistence that the author present a formal treatment of a quantum demon, this paper would not have been written. This work was supported in part by Grant No. N00014-95-1-0975 from the Office of Naval Research, by ARO and DARPA under Grant No. DAAH04-96-1-0386 to QUIC, the Quantum Information and Computation initiative, and by DARPA under a grant to NMRQC, the Nuclear Magnetic Resonance Quantum Computing program.

---

[1] J. C. Maxwell, *Theory of Heat* (Appleton, London, 1871).

[2] H. S. Leff and A. F. Rex, *Maxwell's Demon: Entropy, Information, Computing* (Princeton University Press, Princeton, NJ, 1990).

[3] See, e.g., C. H. Bennett, Sci. Am. **257**, 108 (1987); W. H. Zurek, Nature (London) **341**, 119 (1989).

[4] R. Landauer, IBM J. Res. Dev. **5**, 183 (1961).

[5] W. H. Zurek, H. S. Leff, and A. F. Rex (Ref. [2]), pp. 249–259, reprinted from *Frontiers of Nonequilibrium Statistical Physics*, edited by G. T. Moore and M. O. Scully (Plenum, New York, 1986), pp. 151–161.

[6] S. Lloyd, Phys. Rev. A **39**, 5378 (1989).

[7] G. J. Milburn (unpublished); see also E. P. Gyftopoulos (unpublished).

[8] H. Haus (private communication).

[9] C. P. Slichter, *Principles of Magnetic Resonance*, 3rd ed. (Springer-Verlag, New York, 1990). A $\pi$ pulse is a transversely polarized oscillatory pulse with integrated intensity $\hbar^{-1}\int \mu H(t) = \pi$, where $H(t)$ is the envelope function of the oscillatory field.

[10] O. R. Ernst, G. Bodenhausen, and A. Wokaun, *Principles of Nuclear Magnetic Resonance in One and Two Dimensions* (Oxford University Press, Oxford, 1987). There are a number of ways to flip one spin coherently conditioned on the state of another. For example, consider a second spin that interacts with the first in an Ising-like fashion, so that the Hamiltonian for the two spins is $2\mu_1 B\sigma_z^1 + 2\mu_2 B\sigma_z^2 + \hbar\kappa\sigma_z^1\sigma_z^2$, where $\mu_1 = \mu$ and $\mu_2$ are the magnetic dipole moments for the two spins and $\kappa$ is a coupling constant. To perform a controlled-NOT, (i) apply a $\pi/2$ pulse with frequency $\omega_2 = 2\mu_2 B/\hbar$ and width $\ll 2|\mu_1 - \mu_2|$ and $> 2\kappa$, (ii) wait for time $\pi/2\kappa$, and (iii) apply a second $\pi/2$ pulse with a phase delay of $3\pi/2$ from the first pulse. Step (i) rotates spin 2 by $\pi/2$, step (ii) allows spin 2 to acquire a phase of $\pm\pi/2$ conditioned on whether spin 1 is in the state $|\uparrow\rangle$ or $|\downarrow\rangle$, and step (iii) either rotates spin 2 back to its original state if spin $1 = |\downarrow\rangle$ or rotates spin 2 to an angle of $\pi$ from its original state is spin $1 = |\uparrow\rangle$. Another way to flip spin 2 if spin $1 = |\uparrow\rangle$ is to apply a highly selective $\pi$ pulse with frequency $\omega_2$ and width $\ll \kappa$. Spin 1 is off-resonance and does nothing, while spin 2 is on-resonance and flips if spin 1 $= |\uparrow\rangle$.

[11] S. Lloyd and W. H. Zurek, J. Stat. Phys. **62**, 819 (1991).

[12] See, e.g., J. A. Wheeler and W. H. Zurek, *Quantum Theory and Measurement* (Princeton University Press, Princeton, NJ, 1983).

[13] M. Gell-Mann and J. Hartle, in *Complexity, Entropy, and the Physics of Information*, edited by W. H. Zurek, Santa Fe Institute Studies in the Sciences of Complexity Vol. VIII (Addison-Wesley, Redwood City, CA, 1990).

[14] W. H. Zurek, Phys. Today **44** (10), 36 (1991); Phys. Rev. D **24**, 1516 (1981); **26**, 1516 (1981).

[15] D. Divincenzo, Science **270**, 255 (1995).

[16] D. Deutsch, A. Ekert, and R. Jozsa, Phys. Rev. Lett. **74**, 4083 (1995).

[17] T. Sleator and H. Weinfurter, Phys. Rev. Lett. **74**, 4087 (1995).

[18] C. Monroe, D. M. Meekhof, B. E. King, W. M. Itano, and D. J. Wineland, Phys. Rev. Lett. **75**, 4714 (1995).

[19] Q. A. Turchette, C. J. Hood, W. Lange, H. Mabuchi, and H. J. Kimble, Phys. Rev. Lett. **75**, 4710 (1995).

[20] J. Smith (unpublished).
[21] T. Sleator and M. Wilkens, Phys. Rev. A **48**, 3286 (1993).
[22] E. Geva and R. Kosloff, J. Chem. Phys. **96**, 3054 (1992); Phys. Rev. E **49**, 3903 (1994); J. Chem. Phys. **104**, 7681 (1996).
[23] For an example of a quantum-optical refrigerator, see R. I. Epstein, M. I. Buchwald, B. C. Edwards, T. R. Gosnell, and C. E. Mungan, Nature (London) **377**, 500 (1995). Inefficiency due to decoherence could be observed in their system and in the systems of Ref. [22] by applying intense laser pulses to ''tip'' the state of the ions during the operation of the refrigerator.
[24] S. Lloyd (unpublished).

# Landauer's erasure, error correction and entanglement

By Vlatko Vedral
*Centre for Quantum Computation, Clarendon Laboratory, University of Oxford, Parks Road, Oxford OX1 3PU, UK*

*Received 22 March 1999; revised 15 July 1999; accepted 10 November 1999*

Classical and quantum error correction are presented in the form of Maxwell's demon and their efficiency analysed from the thermodynamic point of view. We explain how Landauer's principle of information erasure applies to both cases. By then extending this principle to entanglement manipulations we rederive upper bounds on purification procedures, thereby linking the 'no local increase of entanglement' principle to the second law of thermodynamics.

Keywords: Landauer's erasure; thermodynamics; classical and quantum error correction; entanglement

## 1. Introduction

Landauer (1961) showed that any erasure of information is accompanied by an appropriate increase in entropy. This result was then used by Bennett (1982) to finally exorcize Maxwell's demon in a Szilard-like set-up (Szilard 1929). His main conclusion was that the increase in entropy is not necessarily a consequence of observations made by the demon, but accompanies the resetting of the final state of the demon to be able to start a new cycle. In other words, information gained has to eventually be erased, which leads to an increase of entropy in the environment and prevents the second law of thermodynamics from being violated. In fact, the entropy increase in erasure has to be at least as large as the initial information gain. Bennett's analysis was, however, completely classical. Soon after this, Zurek (1984) analysed the demon quantum mechanically confirming Bennett's results and since then there has been a number of other related works on this subject (see, for example, Lubkin 1987; Lloyd 1997). Here, however, we want to relate the notion of information erasure to the concept of quantum entanglement. Quantum theory tells us that a measurement process is, in fact, the creation of correlations (entanglement) between the system under observation and a measuring apparatus. Loosely speaking, the amount of entanglement tells us how much information is gained by the apparatus and therefore it seems natural to assume that Landauer's principle has implications for entanglement manipulations. In this paper we show that this indeed is the case; we will argue that Landauer's erasure entropy limits the increase in the amount of entanglement (more precisely, as will be seen later, it limits the average increase) between two quantum subsystems when each one is manipulated separately. This will then lead to an entirely new derivation of known entanglement measures such as the relative entropy of entanglement (Vedral & Plenio 1998) and the entanglement of creation

(Bennett *et al.* 1996*b*) (for a review see Plenio & Vedral 1998). However, in order to become more familiar with Landauer's principle, we first analyse classical error correction and compare it to quantum error correction using the concept of erasure of information.

The paper is organized as follows. In the next section we demonstrate formal analogies between classical error correction and a thermodynamical cycle in general. We show exactly how Landauer's principle is manifested in such a protocol. In § 3 we repeat this analysis using quantum error correction and derive the most general statement for information erasure. In § 4 we apply Landauer's principle to explain recent measures of entanglement, and link it to the principle of 'no increase of entanglement by local means'. This is then discussed from two different points of view, the individual and the ensemble, and a number of different questions is raised. Finally we discuss the implications of this work to other phenomena in quantum information theory and state other open problems implicated by our investigations.

## 2. Classical error correction and Maxwell's demon

We use a simple reversible cycle, which is a slight modification of Bennett's (1982) version of Maxwell's demon, to illustrate the process of error correction. In order to link information to thermodynamics we will use a box containing a single atom as a representation of a classical bit of information: the atom in the left-hand half (LHH) will represent a 0, and the atom in the right-hand half (RHH) will be a 1. Now, if the atom is already confined to one of these halves, and we expand it isothermally and reversibly to occupy the whole volume, then the entropy of the atom increases by $\Delta S = k \log 2$ and the free energy decreases by $\Delta F = -kT \log 2$. The atom does work $\Delta W = kT \log 2$. Suppose that initially we want to have our atom in one of the halves in order to be able to do some work as described. However, suppose also that there is a possibility of error, namely the atom has a chance of $\frac{1}{2}$ to jump to the other half. Once this happens we cannot extract any work until we return the atom to the initial state. But this itself requires an amount of work of $kT \log 2$ in an isothermal compression. We would thus like to be able to correct this error and so we introduce another atom in a box to monitor the first one. This is represented in figure 1 and the whole error correction protocol goes through five stages.

(1) Initially the atoms are in the LHH and RHH of respective boxes.

(2) Then an error happens to atom A so that it now has a 50:50 chance of being in the LHH or RHH.

(3) Atom B observes atom A and correlates itself to it, so that either both occupy the LHH or both occupy the RHH. We make no assumptions about how the observation is made.

(4) Depending on the state of B we now compress A to one of the two halves; this involves no work, but the state of B is now not known—it has a 50:50 chance of being in the LHH or RHH. Thus we have corrected A at the expense of randomizing B. It should be pointed out that by work, we always mean the work done by the atom (or on the atom) against the piston (or by the piston). As is usual in thermodynamical idealizations of this kind, all other work is

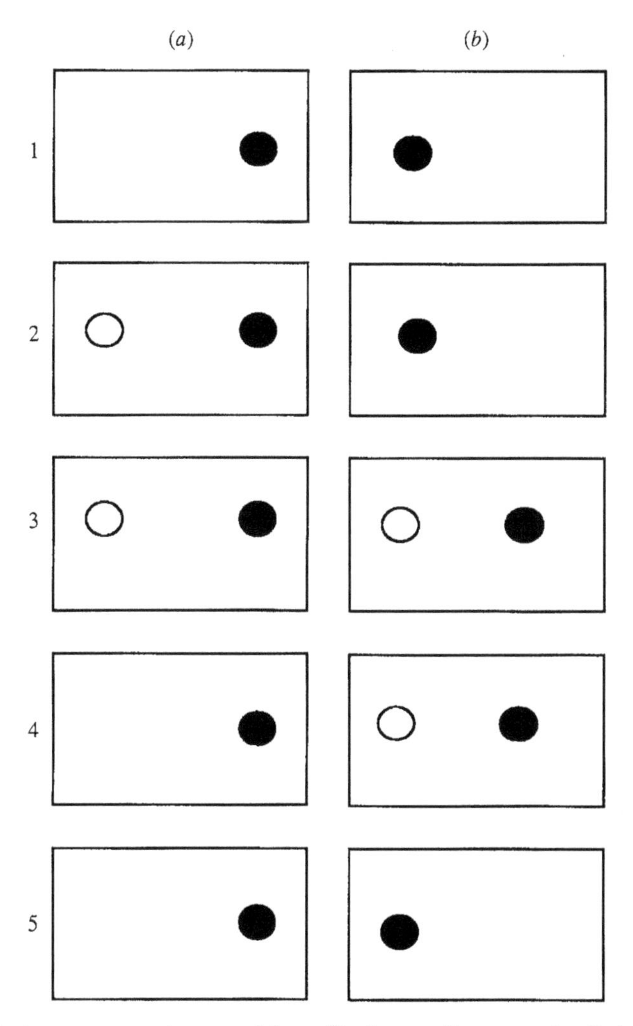


Figure 1. Classical error correction as a Maxwell's demon. Steps are detailed in the text and their significance explained: (1) states of atoms A and B are initially uncorrelated; (2) atom A undergoes an error; (3) atom B observes the atom A, and the atoms thereby become correlated; (4) atom A is corrected to its initial state and the atoms are now uncorrelated; (5) atom B is returned to its initial state and the whole cycle can start again.

neglected (or assumed negligible). For example, the partition itself is assumed to be very light (in fact, with zero weight), so that there is no work in pushing it. Here there is no work done by the atom, since it is not contained in that part of the box which is compressed (this information about the position of A is recorded by B).

(5) In order to be able to repeat the error correction we need to reset B to its initial state as in step 1. Thus we perform isothermal reversible compression of B.

Let us now analyse this process using entropy and free energy. In step 1 both of the atoms possess $kT \log 2$ of free energy. After A undergoes an error its free energy is decreased by $\Delta F_{\rm A} = -kT \log 2$, and nothing happens to B. The total free energy is now $\Delta F_{\rm AB} = kT \log 2$. In step 3 the total free energy is still $\Delta F_{\rm AB} = kT \log 2$, but the atoms are correlated. This means that atom B has information about A (and vice versa). The amount of information is $k \log 2$. This enables the error correction step to take place in step 4. This does not change the total free energy, but the atoms are now decorrelated. In step 5 a work of $kT \log 2$ is invested into resetting the state of atom B so that the initial state in step 1 is reached. This completes the cycle, which can now start again. What happens to entropy? The entropy of each atom is initially 0. Then error increases the entropy of A to $\Delta S_{\rm A} = k \log 2$. In step 3 atoms get correlated so that they both have the same entropy, i.e. $\Delta S_{\rm B} = k \log 2$. However, the crucial point is that the total entropy does not change from step 2 to step 3. This is the point of observation and the information gained by B about the state of A is $S_{\rm A} + S_{\rm B} - S_{\rm AB} = k \log 2$. In step 4 $\Delta S_{\rm A} = -k \log 2$ and there are no changes for atom B. In the resetting step $\Delta S_{\rm B} = -k \log 2$, so that now both of the atoms have zero entropy like at the beginning. Another change that took place, and this is the crux of Landauer's principle, is that in the compression of atom B, work was invested and the entropy of the environment increased by $k \log 2$. This final entropy increase is necessary for resetting and is in this case equal to the amount of information gained in step 3. Landauer's principle of erasure states that the entropy waste in resetting is at least as big as the information gain. If this were not so, we could use the above cycle to do work by extracting heat from the environment with no other changes and the second law of thermodynamics (Kelvin's form) would be violated. Thus, here an error meant that the ability of atom A to do work has been destroyed and in order to correct this we needed another atom B to transfer its free energy to A. In this process atom B loses its ability to do work and, in order to regain it, an amount of $k \log 2$ of entropy has to be wasted (thus 'saving' the second law). To gain more familiarity with these kinds of processes we will now analyse quantum error correction in general settings and then apply our reasoning to manipulations of entanglement.

### 3. Quantum error correction as Maxwell's demon

The aim of quantum error correction as presented in this section will be to preserve a given quantum state of a quantum mechanical system (Knill & Laflamme 1997), much as a refrigerator is meant to preserve the low temperature of food in a higher temperature environment (room). Some work is performed on the refrigerator, which then reduces the entropy of food by increasing the entropy of the surroundings. In accord with the second law, the entropy increase in the environment is at least as large as the entropy decrease of the food. Analogously, when there is an error in the state of a quantum system, then the entropy usually increases (this is, however, not always the case as we will see later), and the error correction reduces it back to the original state, thereby decreasing the entropy of the system, but increasing the entropy of the environment (or what we will call a garbage can). To quantify this precisely let us look at error correction process in detail (see also Nielsen *et al.* 1998).

Suppose we wish to protect a pure state $|\psi\rangle = \sum_i c_i |a_i\rangle$, where $\{|a_i\rangle\}$ form an orthonormal basis. This is usually done by introducing redundancy, i.e. encoding the state of a system in a larger Hilbert space according to some rule

$$|a_i\rangle \rightarrow |C_i\rangle,$$

where $\{|C_i\rangle\}$ are the so-called codewords. Note that this step was omitted in the classical case. This is because the very existence of system $B$ can be interpreted as encoding. The main difference between classical and quantum error correction is that errors in the classical case can always be distinguished. In quantum mechanics these can lead to non-orthogonal states so that the errors cannot always be distinguished and corrected. So it might be said that the encoding in quantum mechanics makes errors orthogonal and hence distinguishable (a precise mathematical statement of this is given in Knill & Laflamme (1997)). Of course, redundancy also exists in classical error correction (above we have another system, $B$, to protect $A$), but the states are already orthogonal and distinguishable by the very nature of being classical. In the quantum case we will introduce an additional system, called the apparatus, in order to detect different errors; this will play the role that $B$ plays in classical error correction. Now the error correction process can be viewed as a series of steps. First the initial state is

$$|\psi_c\rangle|m\rangle|e\rangle,$$

where $|\psi_c\rangle = \sum_i c_i|C_i\rangle$ is the encoded state, $|m\rangle$ is the initial state of the measuring apparatus and $|e\rangle$ is the initial state of the environment. Now, the second stage is the occurrence of errors, represented by the operators $\{E_i\}$ which act on the state of the system only, after which we have

$$\sum_i E_i|\psi_c\rangle|m\rangle|e_i\rangle.$$

Note that at this stage the measurement has not yet been made so that the state of the apparatus is still disentangled from the rest. In general, the states of the environment $|e_i\rangle$ need not be orthogonal (Vedral *et al.* 1997). If they are orthogonal, this leads to a specific form of decoherence, which we might call 'dephasing', and will be analysed later. However, the formalism we present here is completely general and applies to any form of errors. Now the measurement occurs and we obtain

$$\sum_i E_i|\psi_c\rangle|m_i\rangle|e_i\rangle.$$

The error correction is seen as an application of $E_i^{-1}$, conditional on the state $m_i$ (this, of course, cannot always be performed, but the codewords have to satisfy the conditions in Knill & Laflamme (1997) in order to be correctable. Here we need not worry about this; our aim is only to understand global features of error correction). After this the state becomes

$$|\psi_c\rangle \sum_i |m_i\rangle|e_i\rangle$$

and the state of the system returns to the initial encoded state; the error correction has worked. However, notice that the state of the apparatus and the environment is not equal to their initial state. This feature will be dealt with shortly. Before that let us note that the total state (system + apparatus + environment) is always pure.

Consequently, the Von Neumann entropy is always zero. Therefore it is difficult to see how this process will be compared with refrigeration where entropy is kept low at the expense of the environment's entropy increasing. However, in general, the environment is not accessible and we usually have no information about it (if we had this information, we would not need error correction!). Thus the relevant entropies are those of the system and apparatus. This means that we can *trace out* the state of the environment in the above picture; this leads to dealing with mixed states and increasing and decreasing entropies. In addition, the initial state of the system might be pure or mixed (above we assumed a pure state) and these two cases we now analyse separately.

(*a*) *Pure states*

We follow the above set of steps, but now the environment will be traced out of the picture after errors have occurred.

(1) Therefore, in the first step, the state is after errors $\sum_i E_i|\psi_c\rangle|m\rangle|e_i\rangle$, where we assume the 'perfect' decoherence, i.e. $\langle e_j|e_i\rangle = \delta_{ij}$, for simplicity.

(2) Now the environment is traced out leading to $\sum_i E_i|\psi_c\rangle\langle\psi_c|E_i^\dagger \otimes |m\rangle\langle m|$. Note that this is not a physical process, just a mathematical way of neglecting a part of the total state (we introduce the direct product sign just to indicate separation between the system and the apparatus; when there is no possibility of confusion we will omit it).

(3) Then the system is observed, thus creating correlations between the apparatus and the system $\sum_i E_i|\psi_c\rangle\langle\psi_c|E_i^\dagger \otimes |m_i\rangle\langle m_i|$. We assume that the observation is perfect so that $\langle m_j|m_i\rangle = \delta_{ij}$; we will deal with the imperfect observation in the following section. Note that we need different errors to lead to orthogonal states if we wish to be able to correct them. Here, also, if the observation is imperfect then error correction cannot be completely successful, since non-orthogonal states cannot be distinguished with perfect efficiency (see also the discussion at the end of this section).

(4) The correction step happens and the system is decorrelated from the apparatus so that we have $|\psi_c\rangle\langle\psi_c| \otimes \sum_i |m_i\rangle\langle m_i|$. As we remarked before this is not equal to the initial state of the system and apparatus. If we imagine that we have to perform correction a number of times in succession, then this state of apparatus would not be helpful at all. Somehow, we need to reset it back to the original state $|m\rangle$.

(5) This is done by introducing another system, called a garbage can (GC), which is in the right state $|m\rangle$, so that the total state is $|\psi_c\rangle\langle\psi_c| \otimes \sum_i |m_i\rangle\langle m_i| \otimes |m\rangle\langle m|$, and then swapping the state of the GC and the apparatus (this can be performed unitarily) so that we finally obtain $|\psi_c\rangle\langle\psi_c| \otimes |m\rangle\langle m| \otimes \sum_i |m_i\rangle\langle m_i|$. Only now are the system and the apparatus ready to undergo another cycle of error correction.

We can now apply the entropy analysis to this error correction cycle. In the first step the entropy of the system + apparatus has increased by $\Delta S_{\mathrm{S+A}} = S(\rho)$, where

$\rho = \sum_i E_i|\psi_c\rangle\langle\psi_c|E_i^\dagger$. Step 2 is not a physical operation and so there is no change in entropy. In the third step, there is also no change in entropy; it is only that the correlations between the system and the apparatus have been created. Step 4 is the same as step 2, so there is no change in entropy on the whole. In step 5, the entropy of the system + apparatus is zero since they are in the total pure state. Thus, $\Delta S_{\mathrm{S+A}} = -S(\rho)$, and now we see the formal analogy with the refrigeration process: the net change in entropy of the system + apparatus is zero, and the next error correction step can begin; however, the GC has at the end increased in entropy by $\Delta S_{\mathrm{GC}} = S(\rho)$. This is now exactly the manifestation of Landauer's erasure. The information gain in step 3 is equal to the mutual entropy between the system and the apparatus $I_{\mathrm{S+A}} = S_{\mathrm{S}} + S_{\mathrm{A}} - S_{\mathrm{S+A}} = S(\rho)$. The logic behind this formula is that before the observation the apparatus did not know anything about the system, therefore the state of the system was uncertain by $S(\rho)$, whereas after the observations it is zero—the apparatus knows everything about the system. We note that this information is the Shannon mutual information which exists between the two (Schmidt) observables pertaining to the system and the apparatus. This needs to be erased at the end to start a new cycle and the entropy increase is exactly (in this case) equal to the information gained. So from the entropic point of view we have performed the error correction in the most efficient way, since, in general, the GC entropy increase is larger than the information gained (as we will see in § 3 *c*).

Next we consider correction of mixed states. This might at first appear useless, because we might think that a mixed state is one that has already undergone an error. This is, however, not necessarily so, and this situation occurs when we are, for example, protecting a part of an entangled bipartite system (see Vedral *et al.* 1997). It might be thought that an analogous case does not exist in classical error correction. There an error was represented by a free expansion of the atom A from step 1 to step 2. However, we could have equally well started from A occupying the whole volume and treating an error as a 'spontaneous' compression of the atom to one of the halves. If A was correlated to some other atom C (so that they both occupied the LHH or RHH), then this compression would result in decorrelation, which really is an error. Thus classical and quantum error correction are in fact very closely related, which is also shown by their formal analogy to Maxwell's demon.

### (*b*) *Mixed states*

Now suppose that systems $A$ and $B$ are entangled and that we are only performing error correction on the system $A$. Then this is in our case the same as protecting a mixed state. We are not saying that protecting a mixed state is in general the same as protecting entanglement. For example, if a state $|00\rangle + |11\rangle$ flips to $|01\rangle + |10\rangle$ with probability $\frac{1}{2}$, then the entanglement is destroyed, but the reduced states of each subsystem are still preserved. What we mean is that quantum error correction is here developed to protect any pure state of a given system. In that case, any mixed state is also protected, and also any entanglement that it might have with some other systems. This also means that, using the standard quantum error correction (Knill & Laflamme 1997), an entangled pair can be preserved just by protecting each of the subsystems separately. Now, for simplicity say that we have a mixture of two orthogonal states $|\psi\rangle, |\phi\rangle$. The initial state is then, without normalization (and without the system $B$),

$$(|\psi\rangle\langle\psi| + |\phi\rangle\langle\phi|) \otimes |e\rangle\langle e| \otimes |m\rangle\langle m|.$$

Now we can go through all the above stages:
error,

$$\sum_i E_i(|\psi\rangle\langle\psi| + |\phi\rangle\langle\phi|)E_i^\dagger \otimes |e_i\rangle\langle e_i| \otimes |m\rangle\langle m|;$$

tracing out the environment,

$$\sum_i E_i(|\psi\rangle\langle\psi| + |\phi\rangle\langle\phi|)E_i^\dagger \otimes |m\rangle\langle m|;$$

observation,

$$\sum_i E_i(|\psi\rangle\langle\psi| + |\phi\rangle\langle\phi|)E_i^\dagger \otimes |m_i\rangle\langle m_i|;$$

correction,

$$(|\psi\rangle\langle\psi| + |\phi\rangle\langle\phi|) \otimes \sum_i |m_i\rangle\langle m_i|;$$

resetting,

$$(|\psi\rangle\langle\psi| + |\phi\rangle\langle\phi|) \otimes \sum_i |m_i\rangle\langle m_i| \otimes |m\rangle\langle m|$$
$$\rightarrow (|\psi\rangle\langle\psi| + |\phi\rangle\langle\phi|) \otimes |m\rangle\langle m| \otimes \sum_i |m_i\rangle\langle m_i|.$$

The entropy analysis is now as follows. In step 1, $\Delta S_{\text{S+A}} = S(\rho_f) - S(\rho_i)$, where $\rho_f = \sum_i E_i(|\psi\rangle\langle\psi| + |\phi\rangle\langle\phi|)E_i^\dagger$ and $\rho_i = |\psi\rangle\langle\psi| + |\phi\rangle\langle\phi|$ (not normalized). In steps 2 and 3 there is no change of entropy, although in step 3 an amount of $I = S(\rho_f)$ information was gained if the correlations between the system and the apparatus are perfect (i.e. $\langle m_j|m_i\rangle = \delta_{ij}$). In step 4, $\Delta S_{\text{S+A}} = S(\rho_i)$ as the system and the apparatus become decorrelated. In step 5, $\Delta S_{\text{S+A}} = -S(\rho_f)$, but the entropy of the GC increases by $S(\rho_f)$. Thus, altogether $\Delta S_{\text{S+A}} = 0$, and the entropy of the GC has increased by exactly the same amount as the information gained in step 3, thus confirming Landauer's principle again.

Now we want to analyse what happens if the observation in step 3 is imperfect. Suppose for simplicity that we only have two errors $E_1$ and $E_2$. Then there would be only two states of the apparatus $|m_1\rangle$ and $|m_2\rangle$; an imperfect observation would imply that $\langle m_1|m_2\rangle = a > 0$. Now the entropy of information erasure is $S(|m_1\rangle\langle m_1| + |m_2\rangle\langle m_2|)$, which is smaller than when $|m_1\rangle$ and $|m_2\rangle$ are orthogonal. This implies via Landauer's principle that the information gained in step 3 would be smaller than when the apparatus states are orthogonal, and this in turn leads to imperfect error correction. Thus, doing perfect error correction without perfect information gain is forbidden by the second law of thermodynamics via Landauer's principle. This is analogous to Von Neumann's (1952) proof that being able to distinguish perfectly between two non-orthogonal states would lead directly to violation of the second law of thermodynamics.

*(c) General erasure*

Previously we described erasure as a swap operation between the GC and the system. Now we will describe a more general way of erasing information, but which will be central to our understanding of entanglement manipulations in § 4. We follow Lubkin's (1987) analysis in somewhat more general settings.

A more general way of conducting erasure (resetting) of the apparatus is to assume that there is a reservoir which is in thermal equilibrium in a Gibbs state at certain temperature $T$. To erase the state of the apparatus we just throw it into the reservoir and bring in another pure state. The entropy increase of the operation now consists of two parts: the apparatus reaches the state of the reservoir and this entropy is now added to the reservoir entropy, and also the rest of the reservoir changes its entropy due to this interaction, which is the difference in the apparatus internal energy before and after the resetting (no work is done in this process). This quantum approach to equilibrium was also studied by Partovi (1989). A good model is obtained by imagining that the reservoir consists of a great number of systems (of the same 'size' as the apparatus) all in the same quantum equilibrium state $\omega$. Then the apparatus, which is in some state $\rho$, interacts with these reservoir systems one at a time. Each time there is an interaction, the state of the apparatus approaches more closely the state of the reservoir, while that single reservoir system also changes its state away from the equilibrium. However, the systems in the bath are numerous so that after a certain number of collisions the apparatus state will approach the state of the reservoir, while the reservoir will not change much since it is very large (this is equivalent to the so-called Born–Markov approximation that leads to irreversible dynamics of the apparatus described here).

Bearing all this in mind, we now reset the apparatus by plunging it into a reservoir in a thermal equilibrium at temperature $T$. Let the state of the reservoir be

$$\omega = \frac{e^{-\beta H}}{Z} = \sum_j q_j |\varepsilon_j\rangle\langle\varepsilon_j|,$$

where $H = \sum_i \varepsilon_i |\varepsilon_i\rangle\langle\varepsilon_i|$ is the Hamiltonian of the reservoir, $Z = \mathrm{tr}(e^{-\beta H})$ is the partition function and $\beta^{-1} = kT$, where $k$ is the Boltzmann constant. Now suppose that due to the measurement the entropy of the apparatus is $S(\rho)$ (and an amount $S(\rho)$ of information has been gained), where $\rho = \sum_i r_i |r_i\rangle\langle r_i|$ is the eigenexpansion of the apparatus state. Now the total entropy increase in the erasure is (there are two parts as we argued above: (1) change in the entropy of the apparatus; and (2) change in the entropy of the reservoir)

$$\Delta S_{\mathrm{er}} = \Delta S_{\mathrm{app}} + \Delta S_{\mathrm{res}}.$$

We immediately know that $\Delta S_{\mathrm{app}} = S(\omega)$, since the state of the apparatus (no matter what state it was before) is now erased to be the same as that of the reservoir. On the other hand, the entropy change in the reservoir is the average over all states $|r_i\rangle$ of heat received by the reservoir divided by the temperature. This is minus the heat received by the apparatus divided by the temperature; the heat received by the apparatus is the internal energy after the resetting minus initial internal energy

$\langle r_i|H|r_i\rangle$. Thus,

$$\begin{aligned}\Delta S_{\text{res}} &= -\sum_k r_k \frac{\text{tr}(\omega H) - \langle r_k|H|r_k\rangle}{T} \\ &= \sum_k (r_k \sum_j |\langle r_k|\varepsilon_j\rangle|^2 - q_k)(-\log q_k - \log Z) \\ &= -\,\text{tr}(\rho - \omega)(\log\omega - \log Z) \\ &= \text{tr}(\omega - \rho)\log\omega.\end{aligned}$$

Altogether we have that

$$\Delta S_{\text{er}} = -\,\text{tr}(\rho\log\omega)$$

(this result generalizes Lubkin's (1987) result, which applies only when $[\rho,\omega] = 0$). In general, however, the information gain is equal to $S(\rho)$, the entropy increase in the apparatus. Thus, we see that

$$\Delta S_{\text{er}} = -\,\text{tr}(\rho\log\omega) \geqslant S(\rho) = I$$

and Landauer's principle is confirmed (the inequality follows from the fact that the quantum relative entropy $S(\rho||\omega) = -\,\text{tr}(\rho\log\omega) - S(\rho)$ is non-negative). So the erasure is least wasteful when $\omega = \rho$, in which case the entropy of erasure is equal to $S(\rho)$, the information gain. This is when the reservoir is in the same state as the state of the apparatus we are trying to erase. In this case we just have a state swap between the new pure state of the apparatus, which is used to replace our old state $\rho$. This, in fact, was the case in all our examples of error correction above. However, sometimes it is impossible to meet this condition and it is this case we turn to next.

## 4. Entanglement purification

### (a) *General considerations*

Entanglement purification is a procedure whereby an ensemble of bipartite quantum systems all in a state $\rho$ is converted to a subensemble of pure maximally entangled states by local operation on the systems separately and with the aid of classical communications (Bennett *et al.* 1996*a*). The rest of the pairs end up completely separable (i.e. disentangled) and can be taken to be in a pure state of the form $|\psi\rangle|\phi\rangle$. For the sake of simplicity let us first assume that the initial ensemble is in a pure, but not maximally entangled, state. To link this with our previous analysis let us see how this situation might arise. Let a system $S$ be in the state

$$|\psi_{\text{S}}\rangle = \frac{1}{\sqrt{N}}\sum_{i=1}^{N}|s_i\rangle,$$

where $\{|s_i\rangle\}$ is an orthonormal basis. Let now the apparatus observe this superposition, after which the state is

$$|\psi_{\text{S+A}}\rangle = \frac{1}{\sqrt{N}}\sum_{i=1}^{N}|s_i\rangle|a_i\rangle$$

but let the observation be imperfect so that $\{|a_i\rangle\}$ is *not* an orthogonal set (which means that S and A are not maximally entangled). However, suppose that by acting on the apparatus we can transform the whole state $|\psi_{\mathrm{S+A}}\rangle$ into the maximally entangled state

$$|\phi_{\mathrm{S+A}}\rangle = \frac{1}{\sqrt{N}}\sum_{i=1}^{N}|s_i\rangle|b_i\rangle,$$

where $\{|b_i\rangle\}$ *is* an orthogonal set. This does not increase the information between the apparatus and the system since we are not interacting with the system at all. The crucial question is: what is the probability with which we can do this? Let

$$\mathrm{tr}_{\mathrm{S}}(|\psi_{\mathrm{S+A}}\rangle\langle\psi_{\mathrm{S+A}}|) = \rho_{\mathrm{A}}$$

be the state of the apparatus after the perfect measurement. Then Landauer's principle says that the entropy of erasure is $S(\rho_{\mathrm{A}})$ and this has to be greater than or equal to the information gain. But let us look at what the information gain is after we purified the state to $|\phi_{\mathrm{S+A}}\rangle$ with a probability $p$. First of all, we gained $p\log N$ information about the system since we now have maximal correlations. Secondly, the rest of the state contains no information (we assume it is completely disentangled) and this is with probability $(1-p)$. Thus writing Landauer's principle leads to

$$S(\rho_{\mathrm{A}}) \geqslant p\log N.$$

The upper bound to purification efficiency is therefore

$$p \leqslant S(\rho_{\mathrm{A}})/\log N.$$

That the upper bound is achievable was shown by Bennett *et al.* (1996*a*). We stress that in this reasoning we used the fact that the entropy erasure is greater than or equal to the information gain before purification (by Landauer), and this in turn is greater than or equal to the information after purification (since, in purification, the apparatus does not interact with the system).

Let us now analyse when the initial state of the system and the apparatus is mixed in a state $\rho$. Then by Lubkin's method the entropy of erasure is

$$\Delta S_{\mathrm{er}} = -\,\mathrm{tr}(\rho\log\omega).$$

The information gain after the purification is

$$I = S(\rho) + p\log N,$$

where $S(\rho)$ comes from the fact that all the states after purification are pure (being either maximally entangled or disentangled), so that the total information gain is the uncertainty before (i.e. $S(\rho)$) minus the uncertainty after (i.e. zero). We can view this in yet another way. The fact that $\rho$ is mixed means that it is entangled to another system, which we call ancilla. The total state of system + apparatus + ancilla can always be chosen to be pure. Now, system + apparatus are correlated to the ancilla, since the system + apparatus itself is in a mixed state (the ancilla is also mixed and has the same eigenspectrum as system + apparatus, which means that it also has the same Von Neumann entropy). After the purification this correlation disappears, and the resulting information is $S(\rho)$. Using Landauer's principle now gives

$$-\,\mathrm{tr}(\rho\log\omega) \geqslant S(\rho) + p\log N$$

or

$$p \leqslant S(\rho||\omega)/\log N,$$

where $S(\rho||\omega) = -\operatorname{tr}(\rho\log\omega) - S(\rho)$ is quantum relative entropy, which we met in the previous section. Now a tight upper bound on purification is found when $S(\rho||\omega)$ is minimal. However, remembering that we now acted on the system and the apparatus separately (i.e. all the operations were local), the state of the reservoir for resetting will also have to be local (i.e. separable or disentangled as in Vedral & Plenio (1998)). This means that the system is reset by plunging it into its own reservoir and the apparatus is reset by plunging it into its own, but separate, reservoir. By reservoir we will always mean two of these reservoirs together unless stated otherwise. We can always assume that they are in a separable, but classically correlated, state as will be shown later in this section. Let us call the set of all separable states of the system and the apparatus $D$. Then,

$$p \leqslant \min_{\omega\in D} S(\rho||\omega)/\log N.$$

The quantity $E_{\mathrm{RE}}(\rho) = \min_{\omega\in D} S(\rho||\omega)$ is known as the relative entropy of entanglement and has been recently argued by Vedral & Plenio (1998) and Rains (1998) to be an upper bound on the efficiency of purification procedures. For $\rho$ pure this reduces to our previous result. These results were originally derived from the principle that entanglement, being a non-local property, cannot increase under local operations and classical communications. This implied finding a mathematical form of local operations and then finding a measure which is non-increasing under them (Vedral & Plenio 1998). Now, we have derived the same result in a simpler and more physical way, but from at first sight a completely different direction, by taking Landauer's principle. We have thus related the 'no local increase of entanglement' principle to the second law of thermodynamics. Let us briefly summarize this link. Local increase of entanglement is in this context equivalent to *perfectly* distinguishing between non-orthogonal states. If we have this ability, we can then violate the second law with Maxwell's demon that Bennett (1982) used in his analysis (also Von Neumann (1952) showed this by a different argument). Therefore, local increase of entanglement is seen to be prohibited by the second law. The additional principle that we used is that information about some system can only be gained if we interact with it. We emphasize that this principle is not necessarily related only to entanglement. We saw in the previous section that, in error correction, the apparatus correlates itself with the system in order to correct errors and these correlations are purely classical in nature. It is here, however, that the link between 'no local increase of entanglement' and Landauer's principle is most clearly seen: entanglement between two subsystems cannot be increased unless one subsystem gains more information about the other, but this cannot be done locally without interaction.

It should be noted that the reservoirs in general have to be classically correlated (in order to obtain a tight upper bound on purification; otherwise, the bound still holds, but is in general too high). This poses a question as to how this could be achieved in practice. A way to do that is to remember that this result is applicable in the asymptotic case, meaning that we can average over a large number of different, but uncorrelated, reservoir states, which are certainly natural to consider. So, if $\omega = \sum_i p_i \omega_{\mathrm{S}}^i \otimes \omega_{\mathrm{A}}^i$ is the state of the reservoir achieving the minimum (where S refers to the reservoir into which the system is immersed and A to the reservoir into which the apparatus is immersed to erase their mutual information), then this implies that we would be using reservoirs of the form $\omega_S^i \otimes \omega_R^i$ with frequencies $p_i$ and this would on average produce the state $\omega$. Therefore if we consider at $n$ initial mixed states of system + apparatus, then to delete these correlations, we use $p_1 n$ times the uncorrelated reservoir state $\omega_S^1 \otimes \omega_A^1$, $p_2 n$ times the uncorrelated state $\omega_S^2 \otimes \omega_A^2$ and so on. Note that while each of these reservoirs is in a thermal state with a well-defined temperature, the total (classically correlated) state does not have a well-defined temperature. In general, however, the bound is universal and holds no matter how this purification is performed as long as it is local in character. Also note that if we are allowed to have a common reservoir for the system and the apparatus, then the erasure can be different to the above, i.e. it can be smaller than the above local erasure. However, if we allow non-local operations, then we can also create additional entanglement and the above analysis would not apply. An alternative way to view this local erasure is to allow a common reservoir for the system and the apparatus, but to restrict its thermal states to separable states only. Then we are sure that no additional entanglement is created to that already existing between the system and the apparatus and all our results remain true.

This now leads us to state another equivalent way of interpreting the relative entropy of entanglement. Suppose we again have a disentangled bipartite system + apparatus, but this time in a thermal state $\omega = \mathrm{e}^{-\beta H}/Z$, where $\beta^{-1} = kT$, $k$ is the Boltzmann constant, $T$ the temperature, $H$ is the Hamiltonian and $Z$ the partition function (note that this is now the state of system+apparatus which is the same as the state of the reservoir used for resetting. Here, however, there are no further locality requirements and we do have a well-defined temperature). The question that we wish to ask now is how much free energy would this system gain by going to the entangled state $\rho$ (see Donald (1987) for applications of quantum relative entropy to statistical mechanics, and Partovi (1989) for the quantum basis of thermodynamics)? Note that here we allow any operations, and are not restricted by locality requirements. First of all, the free energy is given by

$$F = U - TS$$

so that

$$F(\rho) = \mathrm{tr}(\rho H) - \beta^{-1} S(\rho),$$
$$F(\omega) = -\beta^{-1} \log Z.$$

Therefore, the difference in free energies of $\rho$ and $\omega$ is

$$F(\rho) - F(\omega) = \beta^{-1} S(\rho||\omega).$$

In the light of this, *entanglement as given by the relative entropy would be proportional to the amount of free energy lost in deleting all the quantum correlations and creating solely classical correlations* by, for example, plunging the system+apparatus into a reservoir whose thermal state is given by that classically correlated state (now we do allow a common reservoir with a well-defined thermal state and temperature). The constant of proportionality is the temperature times the Boltzmann constant $k$.

We stress that entanglement of creation, another measure of quantum correlations introduced by Bennett *et al.* (1996*b*), can also be interpreted using the above methods. It is also an upper bound to the efficiency of purification procedures although it is not tight as it is larger than the relative entropy of entanglement (Vedral & Plenio 1998). It arises from applying Landauer's principle to each pure state in the decomposition of $\rho = \sum_i p_i|\psi_i\rangle\langle\psi_i|$ (note that this is not necessarily the eigendecomposition). In fact, we can define the entanglement of creation via the free energy as *the average decrease in free energy due to resetting each of* $|\psi_i\rangle = \sum_j c_{ij}|a_j\rangle|b_j\rangle$ *to a completely separable state* $\omega_i = \sum_i |c_{ij}|^2|a_j\rangle|b_j\rangle\langle a_j|\langle b_j|$ (strictly speaking, the entanglement of creation is actually the minimum of this quantity over all possible decompositions of $\rho$). Mathematically this can be expressed as

$$E_C = \frac{\min_{\text{decomp of }\rho} \sum_i \Delta F_i}{T}$$

where $\Delta F_i = \beta^{-1} S(|\psi_i\rangle||\omega_i)$ (which is equal to the Von Neumann reduced entropy of $|\psi_i\rangle$ (Vedral & Plenio 1998)).

At the end of this section we emphasize again why local operations were important when we wanted to derive a bound on purification of entanglement. First of all, it is believed that entanglement does not increase under local operations and classical communication. Here, however, we did not use this fact to derive bounds on the efficiency of purification procedures. We only used the fact that local operations on one subsystem (e.g. apparatus) do not tell us anything more about some other subsystem than is already known; in other words, if the operations were not local the above analysis would be wrong. In fact, the upper bounds we derived on entanglement purification can now be restated: the most successful purification is the one which wastes no information, i.e. the free energy needed to reset the state of the ensemble before purification should be equal to the free energy needed to reset the state of the ensemble after purification (i.e. Landauer's erasure bound is saturated). At the end of this section we show an equivalent way of interpreting the relative entropy of entanglement where the operations are not restricted to the local ones only, but the reservoir state still has to be disentangled.

### (*b*) *Ensemble versus single trial view*

All the results we considered above are actually appropriate from the ensemble point of view. To explain what we mean by this consider the following problem. Suppose that Alice and Bob share a *single* pair of entangled systems in a state

$$|\psi\rangle = a|0\rangle|0\rangle + b|1\rangle|1\rangle.$$

Then it can be proven (Lo & Popescu 1997) that the best efficiency (i.e. the highest probability) with which they can purify this state by acting locally on their own systems is $2b^2$ if $b < a$. However, $2b^2 < -a^2 \log a^2 - b^2 \log b^2$ (the equality is achieved only when $b^2 = \frac{1}{2}$, and the state is already maximally entangled), which is the efficiency we derived above from Landauer's erasure principle. Thus Landauer's efficiency limit is only reached asymptotically when Alice and Bob share an infinite ensemble of entangled systems and they operate on all of their particles at the same time. So, although Landauer's erasure holds true even if we operate on single pairs (a 'single-shot' measurement), it gives an overestimate of the efficiency of this process. We note that, strictly speaking, both of these problems involve ensembles, but in the single-shot view we are allowed to act on only one pair at a time, whereas otherwise we can act on all of the ensembles simultaneously. It is therefore no surprise that the former method, which is a special case of the latter method, is less efficient. This is the reason why the bound we presented in this paper is too high for the single-shot purification. On the whole, however, by performing erasure the way we imagined, where we use thermal reservoirs to delete information, we have derived a universal upper bound no matter how the purification is performed. Thus an open question would be whether it is more appropriate to use a different measure for erasure in the single-shot case since the amount of entanglement as measured by the purification efficiency is different for a single pair measurement and for an ensemble measurement. This would be important to consider, since most of the practical manipulations at present involve only a few entangled particles and, as we said, the entropic measures overestimate various efficiencies of entanglement processing. So, in our bounds on entanglement purification, $\rho$ should actually stand for $\rho^n = \rho \otimes \rho \cdots \otimes \rho$ ($n$ times), where each $\rho$ now refers to a single pair of quantum systems (also $\log N$ would become $n \log N$). This brings us to the question of whether $\min_{\omega \in D} S(\rho^n || \omega) = n \min_{\omega \in D} S(\rho || \omega)$, i.e. whether the relative entropy of entanglement is additive, which is still open (although see Rains (1998)). A reasonable conjecture is that in all the quantum information manipulations that involve large ensembles the above reasoning will be suitable. Examples of this are quantum data compression (Schumacher 1995) and capacity of a quantum channel (see, for example, Feynman's derivation of Shannon's coding for classical binary symmetric channels using Landauer's principle in Hey & Allen (1996)). In quantum data compression, for instance, the free energy lost in deleting before compression is $n\beta^{-1}S(\rho)$ and after the compression is $m\beta^{-1} \log N$. These two free energies should be equal if no information is lost (i.e. if we wish to have maximum efficiency) in compressing, and therefore $m/n = S(\rho)/\log N$ as shown by Schumacher (1995).

## 5. Conclusions

We have analysed general classical and quantum error correcting procedures from the entropic perspective. We have shown that the amount of information gained in the observation step needed to perform a correction is then turned into an equal amount of wasted entropy in a GC. This GC is needed to reset the apparatus to its initial state so that the next cycle of error correction can then be performed. This fact is equally true both in the classical and the quantum case and is known as Landauer's principle of information erasure. We then analysed purification procedures using the same principle. Surprisingly, Landauer's principle when applied appropriately yields the correct upper bounds to the efficiency of purification procedures. Whether these bounds can be achieved in general remains an open question. Landauer's principle therefore provides a physical basis for several entanglement measures, notably relative entropy of entanglement and the entanglement of creation. In addition, this provides a link between the principle of 'no local increase of entanglement' and the second law of thermodynamics. Further open questions are the implications of Landauer's erasure to other forms of quantum information manipulations such as quantum cloning.

I thank Guido Bacciagaluppi for inviting me to give a talk at the Department of Philosophy in Oxford, which stimulated me to think more about the relationship between thermodynamics and entanglement, thus resulting in this exposition. I also thank M. Plenio and B. Schumacher for useful comments and discussions on the subject of this paper.

## References

Bennett, C. H. 1982 The thermodynamics of computation. *Int. J. Theor. Phys.* **21**, 905–940.

Bennett, C. H., Bernstein, H., Popescu, S. & Schumacher, B. 1996*a* Concentrating partial entanglement by local operations. *Phys. Rev.* A **53**, 2046–2684.

Bennett, C. H., DiVincenzo, D. P., Smolin, J. A. & Wootters, W. K. 1996*b* Mixed state entanglement and error correction. *Phys. Rev.* A **54**, 3824–3851.

Donald, M. J. 1987 Free energy and the relative entropy. *J. Stat. Phys.* **49**, 81–87.

Hey, A. J. G. & Allen, R. W. (eds) 1996 *Feynman lectures on computation.* Addison-Wesley.

Knill, E. & Laflamme, R. 1997 A theory of quantum error-correcting codes. *Phys. Rev.* A **55**, 900–911.

Landauer, R. 1961 Irreversibility and heat generation in the computing process. *IBM Jl Res. Develop.* **5**, 183–191.

Lloyd, S. 1997 Quantum mechanical Maxwell's demon. *Phys. Rev.* A **56**, 3374–3382.

Lo, H. & Popescu, S. 1997 Concentrating entanglement by local actions—beyond mean values. Preprint, quant-ph/9707038.

Lubkin, E. 1987 Keeping the entropy of measurement: Szilard revisited. *Int. J. Theor. Phys.* **26**, 523–535.

Nielsen, M. A., Caves, C., Schumacher, B. & Barnum, H. 1998 Information-theoretic approach to quantum error correction and reversible measurement. *Proc. R. Soc. Lond.* A **454**, 277–304.

Partovi, H. M. 1989 Quantum thermodynamics. *Phys. Lett.* A **137**, 440–444.

Plenio, M. B. & Vedral, V. 1998 Teleportation, entanglement and thermodynamics in the quantum world. *Cont. Phys.* **39**, 431–446.

Rains, E. 1998 An improved bound on distilable entanglement. Preprint, quant-ph/9809082.

Schumacher, B. 1995 Quantum coding. *Phys. Rev.* A **51**, 2738–2747.

Szilard, L. 1929 On the decrease of entropy in a thermodynamic system by the intervention of intelligent beings. *Z. Phys.* **53**, 840–856.

Vedral, V. & Plenio, M. B. 1998 Entanglement measures and purification procedures. *Phys. Rev.* A **57**, 1619–1633.

Vedral, V., Rippin, M. A. & Plenio, M. B. 1997 Quantum correlations, local interactions and error correction. *J. Mod. Opt.* **44**, 2185–2205.

Von Neumann, J. 1952 *Mathematical foundations of quantum mechanics.* Princeton University Press.

Zurek, W. H. 1984 Maxwell's demon, Szilard's engine and quantum measurements. In *Frontiers of non-equilibrium statistical physics* (ed. G. T. Moore & M. O. Scully), pp. 151–161. Plenum.

# Chapter 6

## Algorithmic Information

# Information and entropy

Carlton M. Caves
*Santa Fe Institute, 1660 Old Pecos Trail, Suite A, Santa Fe, New Mexico 87501*
*and Department of Physics and Astronomy, University of New Mexico, Albuquerque, New Mexico 87131-1156*
(Received 4 January 1993)

The Landauer cost for erasing information demands that information about a physical system be included in the total entropy, as proposed by Zurek [Nature **341**, 119 (1989); Phys. Rev. A **40**, 4731 (1989)]. A consequence is that most system states—either classical phase-space distributions or quantum pure states—have total entropy much larger than thermal equilibrium. If total entropy is to be a useful concept, this must imply that work can be extracted in going from equilibrium to a typical system state. The work comes from randomization of a "memory" that holds a record of the system state.

PACS number(s): 89.70.+c, 03.65.−w

## I. INTRODUCTION

To say that a system occupies a certain state implies that one has the information necessary to generate a complete description of that state. This apparently innocuous statement acquires physical significance from *Landauer's principle* [1, 2]: *to erase a bit of information in an environment at temperature $T$ requires dissipation of energy $\geq k_B T \ln 2$.* Landauer's principle demands that information be granted a physical status as a negative contribution to *free* energy. This leads to a *total free energy*

$$\mathcal{F} = F - k_B T \ln 2\, I = E - k_B T \ln 2\,(H + I)\,, \qquad (1.1)$$

where $E$, $H$, and $F$ are conventional energy, entropy (in bits), and free energy, and $I$ is the information (in bits) required to describe the state of interest. The absolute information measure $I$, called *algorithmic information* [3], is defined as the length (in bits) of the shortest program on a universal computer that can generate a complete description of the state of interest. Defining total free energy is equivalent to defining a *total entropy*

$$S = H + I\,, \qquad (1.2)$$

as proposed by Zurek [4, 5]. Throughout this paper I refer to the conventional entropy $H$ as the *statistical entropy*, and I call the entity that stores information about the system a "memory."

This paper explores consequences of including information in free energy and entropy. The bottom line is that *it takes an enormous amount of information to describe most system states—either classical phase-space distributions or quantum pure states—much more information, in fact, than the statistical entropy of thermal equilibrium* [6]. As a result, most phase-space distributions and most pure states have total entropy much larger than equilibrium and total free energy much lower than equilibrium. If total entropy and total free energy are to be useful concepts, this means that work—indeed, an enormous amount of work—can be extracted in going from equilibrium to a typical phase-space distribution or a typical pure state.

This initially puzzling conclusion is illustrated by two *gedanken* examples—one classical and one quantum-mechanical—which are aimed at removing the puzzlement. They show that the work comes from randomization of the memory that holds a record of the system state. In the language of conventional statistical physics, the memory goes from a low-entropy, high-free-energy state, when it stores a record of the system in equilibrium, to a high-entropy, low-free-energy state, when it stores a record of a typical phase-space distribution or a typical pure state.

Far from just demystifying a theoretical puzzle, however, this paper establishes a framework for more ambitious investigations. Having established that typical phase-space distributions and typical pure states are algorithmically extremely complex and, furthermore, that it makes sense to attribute to them a high total entropy, one can proceed to the following question: can Hamiltonian evolution—either classical or quantum-mechanical—take an algorithmically simple initial state to one of the typical algorithmically complex states? Investigation of this question has begun [6–8], with interesting consequences for the second law of thermodynamics and the nature of irreversibility and for the connection between chaos and quantum mechanics.

Section II reviews pertinent elements of algorithmic information theory. Section III considers the total entropy of the conventional *distinct* states of a physical system and reconciles the memory's "inside view" [5], taken in this paper, with the "outside view" of conventional statistical physics. Section IV introduces the phase-space patterns and pure states that are the subject of this paper. Section V presents the gedanken examples that elucidate the meaning of total entropy and total free energy. Section VI concludes with a brief discussion.

## II. ALGORITHMIC INFORMATION

Consider a set of $\mathcal{K}$ alternatives, labeled by an index $k$ and occupied with probabilities $p_k$. In the remaining sections the alternatives become states of a physical system.

Three kinds of information arise in describing these alternatives. The first is the background information needed to give the statistical description. This background information, denoted by $I_0$, is the algorithmic information required to generate an overall description of the alternatives, including a list of all of them and their probabilities. The second kind of information, conventionally denoted $I_{k,0}$, but denoted just by $I_k$ here, is the *joint* algorithmic information needed both to give the background information *and* to specify the particular alternative $k$. The third kind of information, conventionally denoted $I_{k|0}$, but denoted by $\Delta I_k$ here, is the *conditional* algorithmic information needed to specify alternative $k$, given the background information (more precisely, given the minimal program for the background information). The background information $I_0$, which is sufficient to generate a list of *all* the states, is often very much smaller than the the additional information $\Delta I_k$ needed to specify a particular *typical* alternative $k$.

Two results from algorithmic information theory are needed. The first result is the sensible one that

$$I_k = I_0 + \Delta I_k + O(1)\ , \tag{2.1}$$

where $O(1)$ denotes a computer-dependent constant, which arises from defining algorithmic information in terms of computer programs and which is bounded in absolute value for any particular universal computer. Equation (2.1) reveals the reason for my unconventional notation for algorithmic information: it is aimed at harmonizing with the conventional thermodynamic notation for differences between states—a reasonable aim when $H$ and $I$ are included on an equal footing in the total entropy.

The second result from algorithmic information theory is a double inequality [5, 9–11],

$$H \le \sum_{k=1}^{\mathcal{K}} p_k \Delta I_k \le H + O(1)\ , \tag{2.2}$$

which relates the *average* conditional information to the usual statistical information [12]

$$H = -\sum_{k=1}^{\mathcal{K}} p_k \log_2 p_k\ , \tag{2.3}$$

which is stored in $\mathcal{K}$ alternatives that have probabilities $p_k$ ($\log_2$ denotes a base-2 logarithm, so $H$ is in bits). The left-hand inequality in Eq. (2.2) is strict, whereas the right-hand inequality is soft because of the computer-dependent constant. For *typical* alternatives, $\Delta I_k$ is very close to the code-word length for alternative $k$ in an optimal instantaneous code [12] for all $\mathcal{K}$ alternatives.

For the remainder of this paper, I specialize to equally likely alternatives ($p_k = 1/\mathcal{K}$), as in the microcanonical ensemble. In this situation the double inequality (2.2) becomes

$$\log_2 \mathcal{K} \le \frac{1}{\mathcal{K}} \sum_{k=1}^{\mathcal{K}} \Delta I_k \le \log_2 \mathcal{K} + O(1)\ . \tag{2.4}$$

It reduces the remainder of this paper to counting alternatives: the conditional information $\Delta I_k$ to specify a *typical* alternative $k$ is very nearly $\log_2 \mathcal{K}$; there are simple alternatives for which $\Delta I_k \ll \log_2 \mathcal{K}$, but they are atypical.

## III. CONVENTIONAL STATES AND THE INSIDE AND OUTSIDE VIEWS

Consider now an isolated physical system that has $\mathcal{J}$ *distinct* states, labeled by an index $j$ and all of energy $E_0$. Classically these states are *nonoverlapping* cells in phase space, defined by a phase-space coarse graining; quantum mechanically they are *orthonormal* basis states in a $\mathcal{J}$-dimensional Hilbert space. These states are distinguished from the phase-space patterns and pure states introduced in Sec. IV precisely because they are distinct in the usual sense. To make a semantic distinction, I refer to a set of distinct states as conventional states.

Thermal equilibrium for this system is described by the microcanonical ensemble, in which all states are equally likely. Equilibrium has energy $E_0$, statistical entropy

$$H_0 = \log_2 \mathcal{J}\ , \tag{3.1}$$

and free energy

$$F_0 = E_0 - k_B T \ln 2\, H_0 = E_0 - k_B T \ln 2\, \log_2 \mathcal{J}\ . \tag{3.2}$$

The background information $I_0$ needed to describe equilibrium is small ($I_0 \simeq \log_2 H_0$ [6, 11]) and can be neglected. Hence, the total entropy and total free energy of equilibrium are essentially the same as their conventional counterparts.

Each conventional state $j$ has statistical entropy $H_j = 0$, corresponding to a statistical-entropy reduction

$$\Delta H_j = H_j - H_0 = -\log_2 \mathcal{J} \tag{3.3}$$

and a free-energy increase

$$\Delta F_j = -k_B T \ln 2\, \Delta H_j = k_B T \ln 2\, \log_2 \mathcal{J}\ , \tag{3.4}$$

relative to equilibrium. The additional information required to describe a *typical* conventional state $j$, however, is

$$\Delta I_j \simeq \log_2 \mathcal{J}\ . \tag{3.5}$$

Thus a typical conventional state has the same total entropy as equilibrium, i.e.,

$$\Delta S_j = S_j - S_0 = \Delta H_j + \Delta I_j \simeq 0 \tag{3.6}$$

and, likewise, the same total free energy, i.e.,

$$\Delta\mathcal{F}_j = \mathcal{F}_j - \mathcal{F}_0 = \Delta F_j - k_B T \ln 2\, \Delta I_j$$
$$= -k_B T \ln 2\, (\Delta H_j + \Delta I_j) \simeq 0\,. \quad (3.7)$$

Physically, this means that *once the Landauer erasure cost is recognized, no work is available in going from a typical conventional state to equilibrium.*

Equation (3.7) is the contemporary formula for exorcising Maxwell demons [4–6, 10, 11]. A demon-memory can operate an engine cycle in which it first observes the system in equilibrium, finding it in state $j$ (with probability $1/\mathcal{J}$), then extracts work $\Delta F_j$ as the system returns to equilibrium, and finally pays a Landauer erasure cost $k_B T \ln 2\, \Delta I_j$ to return to its background state of knowledge. The *net* work extracted *on the average*,

$$\frac{1}{\mathcal{J}} \sum_{j=1}^{\mathcal{J}} (\Delta F_j - k_B T \ln 2\, \Delta I_j)$$
$$= -\frac{1}{\mathcal{J}} \sum_{j=1}^{\mathcal{J}} \Delta\mathcal{F}_j$$
$$= k_B T \ln 2 \left( \log_2 \mathcal{J} - \frac{1}{\mathcal{J}} \sum_{j=1}^{\mathcal{J}} \Delta I_j \right) \le 0\,, \quad (3.8)$$

is guaranteed not to be positive by the strict left-hand inequality of the double inequality (2.4). Though the demon-memory cannot win, the soft right-hand inequality implies that it can come close to breaking even, at least in principle. Equation (3.8) illustrates how the Landauer erasure cost leads naturally to the notion of total free energy.

This analysis—and this paper—use the memory's own inside view [5], which assigns to the memory the specific, *minimal* amount of information (relative to a particular universal computer) needed to describe the system state—an amount of information that varies from one system state to another. The memory is also a physical system. Viewed from the outside, the memory should be treated by conventional statistical physics, like any other physical system. Once the demon-memory has observed the system's state, the outside view assigns equal probabilities to the $\mathcal{J}$ different memory configurations that record the system states and thus attributes a statistical entropy $\log_2 \mathcal{J}$ to the memory. Thus, from the outside view, the demon-memory cannot win because its conventional statistical entropy increases after it observes the system and must be reduced to return to its background state of knowledge. The double inequality (2.4) is a consistency condition: it ensures that an average over the inside view is equivalent to the conventional outside view.

The import of the first sentence of this paper is that a system state *implies* the existence of a memory that contains its description. Aside from the computer dependence, *the amount of information $I$ contained in the memory and, hence, the total entropy $S$ are properties of the system state.* From the inside view, it is a convenient shorthand to say that a system state *has* total entropy $S$, even though from the outside view the information part of the total entropy is the conventional statistical entropy of another physical system, the memory.

In contrast to typical conventional states, a *simple* conventional state $j$ has $\Delta I_j \ll H_0$, so the information contribution to the total entropy and total free energy can be neglected. A simple conventional state has lower total entropy and higher total free energy than equilibrium by the conventional amounts. Work

$$\Delta\mathcal{F}_j \simeq \Delta F_j = k_B T \ln 2\, \log_2 \mathcal{J} \quad (3.9)$$

must be supplied to transform the system reliably from equilibrium to a simple state and can be extracted as the system returns to equilibrium from a simple state. In conventional statistical physics, simple conventional states and equilibrium constitute the entire subject, precisely because the information needed to describe them can be neglected; typical conventional states are dealt with not individually, but only statistically within the equilibrium ensemble.

## IV. TYPICAL PHASE-SPACE PATTERNS AND TYPICAL PURE STATES

Typical conventional states are already too complex algorithmically to be dealt with individually in conventional statistical physics, but they are by no means the most algorithmically complex states that have a statistical-entropy reduction $\Delta H = -\log_2 \mathcal{J}$ relative to equilibrium. Classically, any *pattern* of *fine-grained* phase-space cells whose total phase-space volume is the same as the volume of a coarse-grained cell has $\Delta H = -\log_2 \mathcal{J}$. Quantum mechanically, any pure state—i.e., any linear combination of the orthonormal basis states—has $\Delta H = -\log_2 \mathcal{J}$. The number of fine-grained phase-space patterns is much greater than the number $\mathcal{J}$ of coarse-grained cells and is limited only by the scale of the fine graining; the number of pure states is much greater than the dimension $\mathcal{J}$ of Hilbert space and is limited only by the number of significant digits in the basis-state amplitudes—i.e., in $\mathcal{J}$ probabilities and $\mathcal{J}$ phases [6, 13]. Hence, the information $\Delta I$ needed to specify a *typical* fine-grained phase-space pattern or a *typical* pure state, beyond the background information (which must be supplemented by a small amount to specify the scale of the fine graining or the number of digits in the quantum amplitudes), is much greater than the equilibrium entropy $\log_2 \mathcal{J}$ [6]. Thus emerges the key result of this paper: *a typical fine-grained phase-space pattern or a typical pure state has total entropy much greater than equilibrium.*

Although this key result has far-reaching consequences, it is by no means difficult to understand. It is a consequence of dropping the usual insistence that states be distinct. The fine-grained phase-space patterns can overlap, so one is freed from the constraint set by the number of nonoverlapping patterns that can be fitted into the available volume of phase space. The pure states need not be orthogonal, so one is freed from the constraint set by the number of Hilbert-space dimensions.

Figure 1 depicts the resulting entropy and free-energy relationships. The two systems of Fig. 1—one classical and one quantum-mechanical—focus the remainder

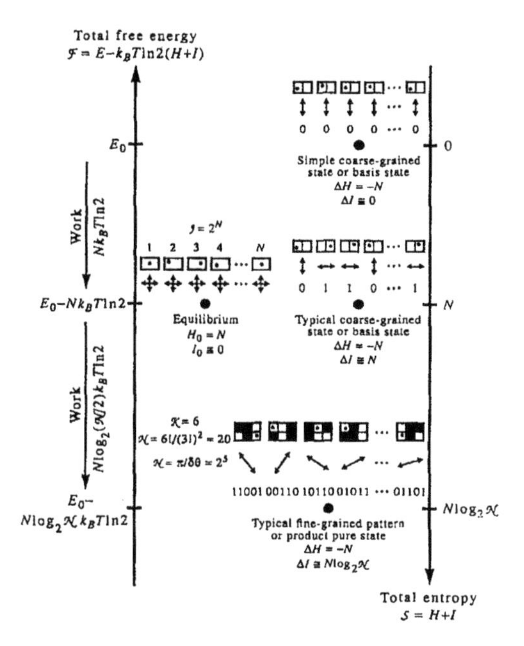

FIG. 1. Total free energy (upward) and total entropy (downward) for a classical system—$N$ molecules enclosed in separate containers—and a quantum system—$N$ linear polarizations (double-ended arrows). There are $\mathcal{J} = 2^N$ *distinct* states: coarse graining divides each container into two equal volumes; basis states are vertical and horizontal polarizations. In equilibrium each molecule explores its entire container, and both polarizations are equally likely. A classical coarse-grained state or quantum basis state has $\Delta H = -N$ and is represented by an $N$-digit binary string—0 for left or vertical and 1 for right or horizontal. *Typical* classical coarse-grained states or *typical* quantum basis states are represented by *random* strings, whose specification requires $\Delta I \simeq N$ bits; typical coarse-grained states or typical basis states have the same total entropy as equilibrium. *Simple* classical coarse-grained states or *simple* quantum basis states, illustrated by extreme cases in which all molecules are on the left or all polarizations are vertical, are represented by strings that can be specified by $\Delta I \lll N$ bits—in the cases here, essentially one bit, for 0 or 1; simple coarse-grained states or simple basis states have total entropy lower than equilibrium by nearly $N$ bits. Classical fine-grained patterns and quantum pure states, both of which have $\Delta H = -N$, are depicted at the bottom. Fine graining divides each container into $\mathcal{K}$ identical boxes; for each container there are $\mathcal{N} = \mathcal{K}!/[(\mathcal{K}/2)!]^2$ fine-grained patterns, each made up of half the boxes (white in the drawing) and each represented by a ($\log_2 \mathcal{N}$)-digit binary code word. For each photon a pure linear-polarization state is represented by an angle $\theta$, defined relative to vertical and given modulo $\pi$ to $\log_2 \mathcal{N}$ binary digits. A *typical* fine-grained pattern or a *typical* product pure state for all $N$ constituents requires $\Delta I \simeq N \log_2 \mathcal{N}$ bits for its description and thus has total entropy much larger than equilibrium. Work can be extracted in going from equilibrium to a typical fine-grained pattern or a typical pure state.

of the paper. The classical system has $N$ containers, each enclosing a single molecule; the choice of coarse graining divides each container into two equal volumes, which can be separated by a partition. The quantum-mechanical system is made up of the polarizations of $N$ photons; the chosen orthogonal basis states are vertical and horizontal linear polarization. For both systems there are $\mathcal{J} = 2^N$ distinct states.

For the classical system, suppose the fine graining divides each container into $\mathcal{K}$ boxes of equal volume. A fine-grained pattern is formed by selecting half the boxes in each container. These boxes (white in Fig. 1) can be separated physically from the other half of the boxes (black in Fig. 1) by a suitable partition. The number of patterns is $\mathcal{N}^N$, where

$$\mathcal{N} = \frac{\mathcal{K}!}{[(\mathcal{K}/2)!]^2} \tag{4.1}$$

is the number of patterns per container. Suppose the quantum system is restricted to *product* states of *linear* polarizations [14]. Then a pure state is a product of $N$ linear polarizations, each specified by an angle $\theta$ relative to vertical. If this polarization angle is given to accuracy $\delta\theta$, the number of pure states is $\mathcal{N}^N$, where

$$\mathcal{N} = \pi/\delta\theta \tag{4.2}$$

is the number of linear polarizations per photon. A fine-grained pattern or a pure state has a statistical-entropy reduction

$$\Delta H = -\log_2 \mathcal{J} = -N\ , \tag{4.3}$$

but the information needed to specify a *typical* pattern or a *typical* pure state is

$$\Delta I \simeq \log_2 \mathcal{N}^N = N \log_2 \mathcal{N}\ . \tag{4.4}$$

For the classical system $\Delta I \sim N\mathcal{K}$ for $\mathcal{K} \gg 1$.

A typical fine-grained pattern or typical pure state has total entropy higher than equilibrium by

$$\Delta \mathcal{S} = \Delta H + \Delta I \simeq N \log_2(\mathcal{N}/2) \tag{4.5}$$

and total free energy lower by

$$\Delta \mathcal{F} = -k_B T \ln 2\, \Delta \mathcal{S} \simeq -k_B T \ln 2\, N \log_2(\mathcal{N}/2)\ . \tag{4.6}$$

To erase the memory's record of a typical fine-grained pattern or a typical pure state, energy $k_B T \ln 2\, N \log_2 \mathcal{N}$ must be dissipated, of which $k_B T \ln 2\, N$ can be recovered in work as the system "expands" to equilibrium. Thus, to go from a typical fine-grained pattern or a typical pure state to equilibrium, energy $k_B T \ln 2\, N \log_2(\mathcal{N}/2) = -\Delta\mathcal{F}$ must be dissipated. There should be a reverse transformation, in which work $-\Delta\mathcal{F}$ is extracted in going from equilibrium to a typical fine-grained pattern or a typical pure state. Section V spells out such a reverse transformation.

## V. GEDANKEN EXAMPLES

To describe the transformation requires just one molecule or one photon, so henceforth I specialize to the case $N = 1$. I aim for clarity by presenting the gedanken examples in terms of typical fine-grained patterns or typical pure states, thereby neglecting the possibility of compressing records for simple states. A more rigorous account, particularly for the engine cycles discussed below, uses the double inequality (2.4) to constrain averages, but reaches the same conclusions.

Consider first the classical system, a single molecule enclosed in a container (Fig. 2). The objective is to transform the molecule from equilibrium, where it roams the entire container, to occupation of a fine-grained pattern that is recorded in the memory, while extracting work $-\Delta\mathcal{F} = k_BT\ln 2\,\log_2(\mathcal{N}/2)$. To store the $(\log_2\mathcal{N})$-bit record for which pattern—i.e., the additional information beyond the background information—the "memory" uses $\log_2\mathcal{N}$ binary registers, which are initially in a "standard" state, storing no information. The fine-grained patterns come in complementary pairs (white and black in Fig. 2), for each of which the memory has a partition that separates white and black boxes. There being $\mathcal{N}/2$ pattern pairs, it takes $\log_2(\mathcal{N}/2)$ bits to specify a pair or its partition. Figure 2 shows a three-step transformation.

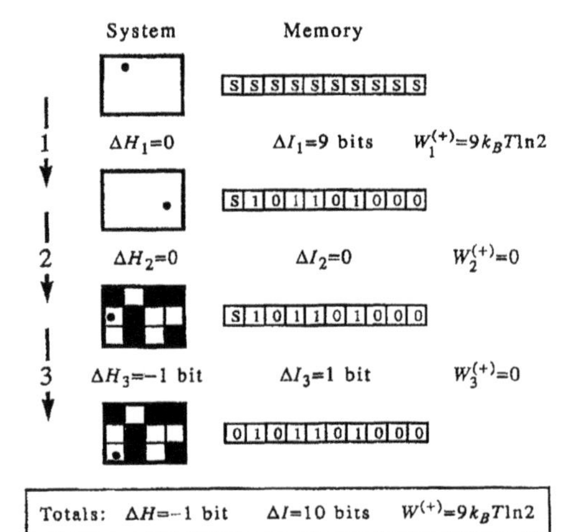


FIG. 2. A single molecule, initially free to explore its entire container, is transformed in three steps to occupy a fine-grained pattern consisting of 6 out of $\mathcal{K} = 12$ boxes. A memory with 10 binary registers, each initially in a standard state (denoted by "$S$"), stores the 10 bits ($\log_2\mathcal{N} = \log_2[12!/(6!)^2] = \log_2 924 = 9.85$) needed to specify the pattern. Given at each step are the change in the statistical entropy, the change in the number of bits of memory used, and the work extracted. After the transformation, the statistical entropy has changed by $\Delta H = -1$ bit, the memory stores $\Delta I = 10$ bits, and work $W^{(+)} = k_BT\ln 2\,(\Delta H + \Delta I) = 9k_BT\ln 2$ has been extracted.

*Step 1*: The $\log_2(\mathcal{N}/2)$ memory registers after the first are allowed to randomize, with extraction of work (inverse of erasing [10])

$$W_1^{(+)} = k_BT\ln 2\,\log_2(\mathcal{N}/2)\,, \tag{5.1}$$

after which the memory stores

$$\Delta I_1 = \log_2(\mathcal{N}/2)\ \text{bits}\,. \tag{5.2}$$

An explicit way for the memory to perform this step is to use an auxiliary system that has $\mathcal{N}/2$ distinct states, one for each record that can be stored in the $\log_2(\mathcal{N}/2)$ memory registers after the first. This auxiliary system is initially in equilibrium, with equal probability to occupy each of its distinct states. The memory first observes the state of the auxiliary system, storing a record of the observed state in its $\log_2(\mathcal{N}/2)$ registers after the first and thereby reducing the statistical entropy of the auxiliary system by $\log_2(\mathcal{N}/2)$ bits. The memory then uses its record to extract work $W_1^{(+)}$ as the auxiliary system returns to equilibrium, after which the auxiliary system is irrelevant to the further discussion.

*Step 2*: The memory uses its $[\log_2(\mathcal{N}/2)]$-bit record to select a partition and applies it to the container.

*Step 3*: The memory observes whether the molecule is in the white or the black part of the container and records

$$\Delta I_3 = 1\ \text{bit} \tag{5.3}$$

for white (0) or black (1) in its first register, thereby changing the molecule's statistical entropy by

$$\Delta H_3 = -1\ \text{bit}\,. \tag{5.4}$$

After step 3 the molecule occupies a particular pattern, with statistical-entropy reduction

$$\Delta H = \Delta H_3 = -1\ \text{bit}\,; \tag{5.5}$$

the memory stores

$$\Delta I = \Delta I_1 + \Delta I_3 = \log_2\mathcal{N}\ \text{bits}\,, \tag{5.6}$$

which record which pattern the molecule occupies; and work

$$W^{(+)} = W_1^{(+)} = k_BT\ln 2\,\log_2(\mathcal{N}/2) \tag{5.7}$$

has been extracted, as promised. An alternative transformation lets all $\log_2\mathcal{N}$ memory registers randomize, with extraction of work $k_BT\ln 2\,\log_2\mathcal{N}$, of which $k_BT\ln 2$ is used to "compress" the system into the pattern stored in the memory's record.

It is instructive to consider two engine cycles based on this example. The two cycles differ in what is regarded as the memory's background state of knowledge. The first cycle has the same initial state (situation before step 1): the molecule is in equilibrium, and the memory stores no information. This cycle proceeds through steps 1–3 and adds two further steps to get back to the initial state.

*Step 4*: The molecule returns to equilibrium with extraction of work

$$W_4^{(+)} = k_B T \ln 2 \,. \tag{5.8}$$

One way to extract this work is to move all the black boxes to one side of the container and all the white boxes to the other side, separating the two by a partition down the middle; then, knowing which side the molecule is on, the memory inserts a piston into the empty side and extracts work $k_B T \ln 2$ as the molecule pushes the piston out of the container.

*Step 5*: The memory returns to its standard state (all registers empty), paying a Landauer erasure cost

$$W_5^{(-)} = k_B T \ln 2 \, \Delta I = k_B T \ln 2 \, \log_2 \mathcal{N} \,. \tag{5.9}$$

The net work extracted, $W_1^{(+)} + W_4^{(+)} - W_5^{(-)}$, is zero. In this cycle the background information is the information needed to generate a description of the system at the level of division into boxes. The information gathered by the memory during the cycle includes both the $\log_2(\mathcal{N}/2)$ bits for choosing a partition and the 1 black-or-white bit from observing the molecule.

The second kind of cycle uses the result of step 1 as the initial state: the molecule is in equilibrium, the memory's first register is in the standard state, and the remaining $\log_2(\mathcal{N}/2)$ memory registers store which-partition information. Step 4 is the same, but in step 5 only the first memory register needs to be erased—at cost $k_B T \ln 2$—to return the memory to its background state of knowledge. In this cycle the $\log_2(\mathcal{N}/2)$ bits for choosing a partition are background information, which tells the memory how to partition the container into two equal volumes. The cycle is just a fancy Szilard [15] engine, with the container partitioned in an unusual way instead of by inserting a partition down the middle.

These two cycles emphasize that the change in total entropy must be defined relative to some initial system state *and* to some initial background information. The difference between the two cycles is precisely whether the $\log_2(\mathcal{N}/2)$ which-partition bits are background information, to be carried forward from cycle to cycle, or foreground information, collected afresh during each cycle and erased to return to the initial background information. The total-entropy increase when the molecule is confined to a typical pattern is real, but it must be understood as an increase relative to physical equilibrium *and* to equilibrium background information.

Turn now to photon polarization. The objective is to take an initially unpolarized photon to linear polarization at some angle $\theta$ (within $\delta\theta = \pi/\mathcal{N}$) that is recorded in the memory, while extracting work $k_B T \ln 2 \, \log_2(\mathcal{N}/2)$. The memory has $\log_2 \mathcal{N}$ binary registers, initially in a standard state, to store the $(\log_2 \mathcal{N})$-bit polarization angle. The analog of the classical partitions is a polarizing beam splitter that, when set at angle $\theta_p$, separates orthogonal polarizations at angles $\theta_p$ and $\theta_p + \pi/2$. Figure 3 shows a transformation that mimics the three steps in the classical example.

*Step 1*: The $\log_2(\mathcal{N}/2)$ memory registers after the first are allowed to randomize, with extraction of work

$$W_1^{(+)} = k_B T \ln 2 \, \log_2(\mathcal{N}/2) \,; \tag{5.10}$$

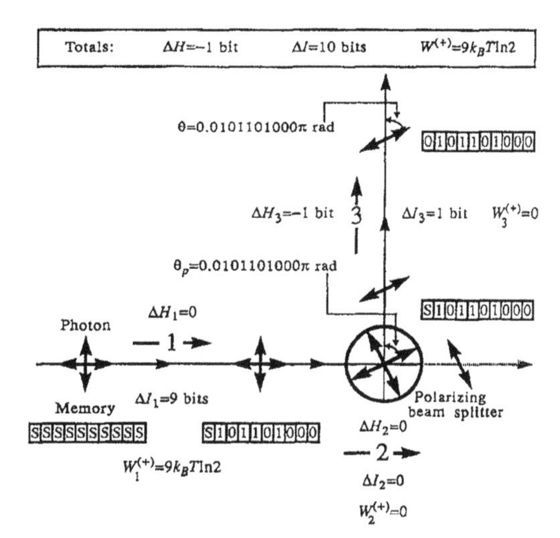


FIG. 3. An unpolarized photon is transformed in three steps to linear polarization at angle $\theta$ (relative to the vertical) given to 10 binary digits (accuracy $\delta\theta = 2^{-10}\pi$). The state of linear polarization is depicted in a plane rotated by 90° so that it lies in the plane of the paper. A memory with 10 binary registers, each initially in a standard state (denoted by "$S$"), stores the 10 angle bits. Given at each step are the change in the statistical entropy, the change in the number of bits of memory used, and the work extracted. After the transformation, the statistical entropy has changed by $\Delta H = -1$ bit, the memory stores $\Delta I = 10$ bits, and work $W^{(+)} = k_B T \ln 2 \, (\Delta H + \Delta I) = 9 k_B T \ln 2$ has been extracted.

after this step the memory stores a binary string $r$ of length

$$\Delta I_1 = \log_2(\mathcal{N}/2) \text{ bits} \,. \tag{5.11}$$

*Step 2*: The memory sets the polarizing beam splitter at angle $\theta_p = 0.0r\,\pi$ ($0.0r$ is the binary representation of $\theta_p/\pi$), after which the photon passes through the beam splitter.

*Step 3*: The memory observes the photon's output direction, recording in its first register the

$$\Delta I_3 = 1 \text{ bit} \tag{5.12}$$

for which orthogonal polarization—0 for angle $\theta_p$ and 1 for angle $\theta_p + \pi/2$; the observation changes the statistical entropy by

$$\Delta H_3 = -1 \text{ bit} \,. \tag{5.13}$$

After step 3 the photon has linear polarization at angle $\theta = \theta_p = 0.0r\,\pi$ or at angle $\theta = \theta_p + \pi/2 = 0.1r\,\pi$ and thus has statistical-entropy reduction

$$\Delta H = \Delta H_3 = -1 \text{ bit} \,; \tag{5.14}$$

the memory stores the polarization angle modulo $\pi$ as the binary string $0r$ or $1r$ of length

$$\Delta I = \Delta I_1 + \Delta I_3 = \log_2 \mathcal{N} \text{ bits} ; \tag{5.15}$$

and work

$$W^{(+)} = W_1^{(+)} = k_B T \ln 2 \log_2(\mathcal{N}/2) \tag{5.16}$$

has been extracted.

In both examples the memory's record splits naturally into two parts: $\log_2(\mathcal{N}/2)$ bits that tell the memory *how* to observe the system—which partition or which beam splitter angle—and 1 bit from the subsequent observation. The $\log_2(\mathcal{N}/2)$ bits of which-observation information are not stored in the system and cannot be obtained by observing it. The $\log_2(\mathcal{N}/2)$ which-observation registers are "called into existence" in a standard state when one contemplates storing a record of a typical fine-grained pattern or a typical pure state. The work $W^{(+)}$ comes wholly from randomizing these which-observation registers.

From the outside view of conventional statistical physics, the standard memory state is a low-entropy, high-free-energy state. Supplying memory registers in the standard state is the same as supplying fuel that can be turned into useful work [10]. Indeed, from the outside view, the work is extracted as the which-observation registers go from the low-entropy standard state, far from equilibrium, to the high-entropy equilibrium state, in which they store a record of how to observe the system.

The inside and outside views agree on the *average* work extracted or dissipated in steps 1–5. Thus both views agree on the *average* free-energy decrease in going from the system in equilibrium, with a record of equilibrium stored in the memory, to the system's occupying a fine-grained pattern or pure state, with a record of the pattern or pure state stored in the memory. The outside view achieves all this using only the conventional statistical entropies of the system and the memory. The inside view agrees with the conventional outside view *on the average*, but it has the advantage that it gives an account—indeed, it must give an account—of each memory state. This advantage is of little consequence in analyzing an engine cycle, since only average behavior is relevant for questions of thermodynamic efficiency, but it becomes important when one applies the concept of total entropy to system dynamics that begin with a particular initial state.

## VI. DISCUSSION

This paper provides background and motivation for addressing the following question: can Hamiltonian dynamics, either classical or quantum-mechanical, take a simple initial state—i.e., an algorithmically simple phase-space cell or an algorithmically simple quantum basis state—to one of the typical fine-grained phase-space patterns or typical quantum pure states, which are algorithmically extremely complex and thus have high total entropy? This question, foreshadowed by Zurek's [5] analysis of classical ergodic, but nonmixing systems, is only beginning to be investigated. Initial work [6] indicates that the answer is no, at least for reasonable times. The negative answer holds both for classical systems, either regular or chaotic, and for quantum systems.

Things get more interesting, however, when one realizes that the state that evolves from a simple initial state under classical chaotic—but not regular—evolution or under quantum-mechanical evolution, although itself algorithmically simple, can be easily perturbed into one of the typical algorithmically complex phase-space patterns or pure states [6]. Detailed analyses of a stochastically perturbed version of the baker's map [7], a prototype for classical chaos, and of a stochastically perturbed version of a quantized baker's map [8], indicate that the algorithmic information needed to keep track of the perturbed state grows extremely fast. Indeed, the growth of algorithmic information is much faster than the increase in statistical entropy that follows from averaging over the stochastic perturbation. These detailed analyses, together with the heuristic analyses [6] that motivated them, have deep implications for the second law of thermodynamics and for the nature of irreversibility and, moreover, hint at a previously unrecognized connection between classical chaos and quantum mechanics.

One further question can be addressed profitably to the gedanken examples in Sec. V: how are classical and quantum systems different? Successive observations of the classical molecule, using different partitions, eventually isolate the molecule in a single box, thus uncovering the fine-grained structure beneath the initial coarse graining. Successive observations of the photon, using different beam splitter angles, yield fresh information as long as there is memory space to store it, never revealing any structure beneath the pure states (no hidden variables). This ability of quantum systems to manufacture new information is the central mystery of quantum mechanics, stated in information-theoretic language.

## ACKNOWLEDGMENTS

I thank W. G. Unruh for challenging me to come up with the examples in Figs. 2 and 3, and I thank S. Lloyd, R. Schack, and W. H. Zurek for advising me how to revise this paper from its original, shorter form. This work was supported in part by the Complexity, Entropy, and Physics of Information program at the Santa Fe Institute.

[1] R. Landauer, IBM J. Res. Develop. **5**, 183 (1961).
[2] R. Landauer, Nature (London) **355**, 779 (1988).
[3] G. J. Chaitin, *Information, Randomness, and Incompleteness* (World Scientific, Singapore, 1987).
[4] W. H. Zurek, Nature (London) **341**, 119 (1989).
[5] W. H. Zurek, Phys. Rev. A **40**, 4731 (1989).
[6] C. M. Caves, in *Physical Origins of Time Asymmetry*, edited by J. J. Halliwell, J. Pérez-Mercader, and W. H. Zurek (Cambridge University, Cambridge, England, 1993).
[7] R. Schack and C. M. Caves, Phys. Rev. Lett. **69**, 3413 (1992).

[8] R. Schack and C. M. Caves (unpublished).
[9] A. K. Zvonkin and L. A. Levin, Usp. Mat. Nauk **25** (6), 85 (1970) [Russ. Math. Surveys **25** (6), 83 (1970)].
[10] C. H. Bennett, Int. J. Theor. Phys. **21**, 905 (1982).
[11] C. M. Caves, in *Complexity, Entropy and the Physics of Information*, edited by W. H. Zurek (Addison-Wesley, Redwood City, CA, 1990), p. 91.
[12] R. G. Gallager, *Information Theory and Reliable Communication* (Wiley, New York, 1968).
[13] I. C. Percival, in *Quantum Chaos, Quantum Measurement*, edited by P. Cvitanović, I. Percival, and A. Wirzba, (Kluwer, Dordrecht, Holland, 1992), p. 199. Percival emphasizes that so much information is required to specify a typical pure state that it is impossible in practice to prepare one.
[14] If one allows arbitrary pure states within the ($\mathcal{J} = 2^N$)-dimensional Hilbert space of the $N$ photon polarizations, then the information to specify a typical state is $\Delta I \simeq \mathcal{J} \times$(number of bits to specify a real amplitude and a phase for each Hilbert-space dimension). The number of pure states, at the Hilbert-space resolution defined by the number of digits in the amplitudes and phases, is $\mathcal{N} \simeq 2^{\Delta I}$—an enormous number even for Hilbert spaces with a modest number of dimensions. A more rigorous discussion of this point can be found in Ref. [6].
[15] L. Szilard, Z. Phys. **53**, 840 (1929) [English translation in *Quantum Theory and Measurement*, edited by J. A. Wheeler and W. H. Zurek (Princeton University, Princeton, NJ, 1983), p. 539].

# Algorithmic Information and Simplicity in Statistical Physics

**Rüdiger Schack**[1,2]

*Received April 2, 1996*

Applications of algorithmic information theory to statistical physics rely (a) on the fact that average conditional algorithmic information can be approximated by Shannon information and (b) on the existence of *simple states* described by short programs. More precisely, given a list of $N$ states with probabilities $0 < p_1 \leq \cdots \leq p_N$, the average conditional algorithmic information $\bar{I}$ to specify one of these states obeys the inequality $H \leq \bar{I} < H + O(1)$, where $H = -\Sigma\, p_j \log_2 p_j$ and $O(1)$ is a computer-dependent constant. We show how any universal computer can be slightly modified in such a way that (a) the inequality becomes $H \leq \bar{I} < H + 1$ and (b) states that are simple with respect to the original computer remain simple with respect to the modified computer, thereby eliminating the computer-dependent constant from statistical physics.

## 1. INTRODUCTION

Algorithmic information theory (Solomonoff, 1964; Kolmogoroff, 1965; Zvonkin and Levin, 1970; Chaitin, 1987), in combination with Landauer's principle (Landauer, 1961, 1988), which specifies the unavoidable energy cost $k_B T \ln 2$ for the erasure of a bit of information in the presence of a heat reservoir at temperature $T$, has been applied successfully to a range of problems: the Maxwell demon paradox (Bennett, 1982), a consistent Bayesian approach to statistical mechanics (Zurek, 1989a,b; Caves, 1993a,b), a treatment of irreversibility in classical Hamiltonian chaotic systems (Caves, 1993b; Schack and Caves, 1992), and a characterization of quantum chaos relevant to statistical physics (Caves, 1993b; Schack and Caves, 1993, 1996a,b; Schack *et al.*, 1994). The algorithmic information for a physical state is defined as

[1] Center for Advanced Studies, Department of Physics and Astronomy, University of New Mexico, Albuquerque, New Mexico 87131-1156.
[2] Present address: Department of Mathematics, Royal Holloway, University of London, Egham, Surrey TW20 0EX, U.K.; e-mail: r.schack@rhbnc.ac.uk.

the length in bits of the shortest self-delimiting program for a universal computer that generates a description of that state (Zurek, 1989b; Caves, 1990). Algorithmic information with respect to two different universal computers differs at most by a computer-dependent constant (Chaitin, 1987). Although typically the latter can be neglected in the context of statistical physics, the presence of an arbitrary constant in a physical theory is unsatisfactory and has led to criticism (Denker and leCun, 1993). In the present paper, we show how the computer-dependent constant can be eliminated from statistical physics.

In the following paragraphs we give a simplified account of the role of algorithmic information in classical statistical physics. A more complete exposition including the quantum case can be found in Zurek (1989b) and Caves (1993b). We adopt here the information-theoretic approach to statistical physics pioneered by Jaynes (1983). In this approach, the *state* of a system represents the observer's knowledge of the way the system was prepared. States are described by probability densities in phase space; observers with different knowledge assign different states to the system. Entropy measures the information missing toward a complete specification of the system.

Consider a set of $N$ states ($N \geq 2$) labeled by $j = 1, \ldots, N$, all having the same energy and entropy. The restriction to states of the same energy and entropy is not essential, but it simplifies the notation. Initially the system is assumed to be in a state in which state $j$ is occupied with probability $p_j > 0$. We assume throughout that the states $j$ are labeled such that $0 < p_1 \leq \cdots \leq p_N$. If an observation reveals that the system is in state $j$, the increased knowledge is reflected in an entropy decrease $\Delta S = -k_B \ln 2\, H$, where $H = -\Sigma\, p_j \log_2 p_j > 0$ is the original missing information measured in bits. To make the connection with thermodynamics, we assume that there is a heat reservoir at temperature $T$ to which all energy in the form of heat must eventually be transferred, possibly using intermediate steps such as storage at some lower temperature. In the presence of this fiducial heat reservoir, the entropy decrease $\Delta S$ corresponds to a free energy increase $\Delta F = -T\Delta S = +k_BT \ln 2\, H$. Each bit of missing information decreases the free energy by the amount $k_BT \ln 2$; if information is acquired about the system, free energy increases.

The fact that entropy can decrease through observation—which underlies most proposals for a Maxwell demon—does not conflict with the second law of thermodynamics, because the observer's physical state changes as a consequence of the interaction with the system. Szilard (1929) discovered that no matter how complicated the change in the observer's physical state, the associated irreducible thermodynamic cost can be described solely in terms of information. He found that in the presence of a heat reservoir at temperature $T$ each bit of information acquired by the observer has an energy cost at least as big as $k_B T \ln 2$. Total available work is reduced not only by missing information, but also by information the observer has acquired about the system. The physical nature of the cost of information was clarified by Bennett (1982), who applied Landauer's principle (Landauer, 1961, 1988) to the Maxwell demon problem and showed that the energy cost has to be paid when information is erased.

To keep the Landauer erasure cost of the observational record as low as possible, the information should be stored in maximally compressed form. The concept of a maximally compressed record is formalized in algorithmic information theory (Chaitin, 1987). Bennett (1982) and Zurek (1989a,b) gave Szilard's theory its present form by using algorithmic information to quantify the amount of information in an observational record. In particular, by exploiting Bennett's idea of a reversible computer (Bennett, 1982), Zurek (1989a) showed how an observational record can be replaced by a compressed form at no thermodynamic cost. This means that the energy cost of the observational record can be reduced to the Landauer erasure cost of the compressed form.

Let us denote by $s_j$ a binary string describing the $j$th state ($j = 1, \ldots, N$). A detailed discussion of how a description of a physical state can be encoded in a binary string is given in Zurek (1989b). The exact form of the strings $s_j$ is of no importance for the theory outlined here, however, because the information needed to generate a list of *all* the strings $s_j$ can be treated as *background information* (Caves, 1990, 1993b). Background information is the information needed to generate a list $s = ((s_1, p_1), \ldots, (s_N, p_N))$ of all $N$ states together with their probabilities; i.e., background information is the information the observer has before the observation. This formulation assumes that the probabilities $p_j$ are completely specified by the background information—a natural assumption in the Bayesian approach to probabilities (Jaynes, 1983) adopted in this paper. A generalization to approximately specified probabilities is discussed in Bennett (1982).

Algorithmic information is defined with respect to a specific universal computer $U$. We denote by $I_U(s_j|s)$ the conditional algorithmic information, with respect to the universal computer $U$, to specify the $j$th state, given the background information (Chaitin, 1987; Zurek, 1989b; Caves, 1990). More precisely, $I_U(s_j|s)$ is the length in bits of the shortest self-delimiting program for $U$ that generates the string $s_j$, given a minimal self-delimiting program to generate $s$. For a formal definition of a universal computer $U$ and of $I_U(s_j|s)$ see Section 2. It should be emphasized that a minimal program that generates the list $s$ of descriptions of all states and their probabilities can be short even when a minimal program that generates the description $s_j$ of a typical single state is very long (Zurek, 1989b).

Since total available work is reduced by $k_B T \ln 2$ by each bit of information the observer acquires about the system as well as by each bit of missing information, the change in *total free energy* or *available work* upon observing state $j$ can now be written as

$$\begin{aligned} \Delta F_{j,\text{tot}} &= -T[\Delta S + k_B \ln 2\, I_U(s_j \mid s)] \\ &= -k_B T \ln 2\, [-H + I_U(s_j \mid s)] \end{aligned} \tag{1}$$

This definition of total free energy is closely related to Zurek's definition of physical entropy (Zurek, 1989b). Average conditional algorithmic information $\overline{I_U(\cdot \mid s)} = \Sigma\, p_j I_U(s_j \mid s)$ obeys the double inequality (Zurek, 1989b; Caves, 1990)

$$H \leq \overline{I_U(\cdot \mid s)} < H + O(1) \tag{2}$$

where $O(1)$ denotes a positive computer-dependent constant (Chaitin, 1987). It follows immediately that the *average* change in total free energy, $\Delta F_{\text{tot}} = \Sigma\, p_j \Delta F_{j,\text{tot}}$, is zero or negative:

$$0 \geq \Delta F_{\text{tot}} > -O(1) k_B T \ln 2 \tag{3}$$

The left side of this double inequality establishes that acquiring information cannot increase available work on the average. For standard choices for the universal computer $U$, e.g., a Turing machine or Chaitin's LISP-based universal computer (Chaitin, 1987), the computer-dependent $O(1)$ constant on the right is completely negligible in comparison with thermodynamic entropies. Condition (3) therefore expresses that on the average, with respect to a standard universal computer, total free energy remains essentially unchanged upon observation. Despite the success of this theory, the presence of an arbitrary constant is disturbing. To understand the issues involved in removing the arbitrary constant, we must introduce the notions of simple and complex states.

Although the average information $\overline{I_U(\cdot \mid s)}$ is greater than or equal to $H$, there is a class of low-entropy states that can be prepared without gathering a large amount of information. For example, in order to compress a gas into a fraction of its original volume, free energy has to be spent, but the length in bits of written instructions to prepare the compressed state is negligible on the scale of thermodynamic entropies. States that can be prepared reliably in a laboratory experiment usually are *simple states*, which means that there is a short verbal description of how to prepare such a state.

The concept of a simple state is formalized in algorithmic information theory. A simple state is defined as a state for which $I_U(s_j \mid s) << H$; i.e., descriptions for simple states can be generated by short programs. The total free energy increases, in the sense of equation (1), upon observing the system to be in a simple state. Simplicity is a computer-dependent concept. Standard universal computers like Turing machines reflect our intuitive notion of simplicity. It is easy, however, to define a universal computer for which there are no short programs at all; such a computer would not recognize simplicity.

Intuitively, simplicity ought to be an intrinsic property of a state. A computer formalizing the intuitive concept of simplicity should reflect this. In particular, for such a computer a simple state should have a short program independent of the probability distribution $p_1, \ldots, p_N$. This is not true for all universal computers. In Section 2 we introduce a universal computer $U_\epsilon$ for with $I_{U_\epsilon}(s_j \mid s)$ is determined solely by the probabilities $p_1, \ldots, p_N$. For this computer, a short program for the $j$th state reflects a large probability $p_j$, not an intrinsic property of the state. We will say that such a computer does not recognize intrinsically simple states.

Simple states are rare—there are fewer than $2^n$ states $j$ for which $I_U(s_j \mid s) < n$ (Chaitin, 1987)—and thus arise rarely as the result of an observation, yet they are of great importance. Simple states are states for which the algorithmic contribution to total free energy is negligible. The concept of total free energy does not conflict with conventional thermodynamics because thermodynamic states are simple. If the theory does not have the notion of simple states, the connection with conventional thermodynamics is lost.

The opposite of a simple state, a *complex state*, is defined as a state for which $I_U(s_j \mid s)$ is of the same order as $H$. Complex states arise not just through Maxwell demon-like observations. We have shown (Caves, 1993b; Schack and Caves, 1992, 1993, 1996a,b; Schack *et al.*, 1994) that initially simple states of chaotic Hamiltonian systems in the presence of a perturbing environment rapidly evolve into extremely complex states (Caves, 1993a,b) for which the negative algorithmic contribution to total free energy is vastly bigger than $H$ and thus totally dominates conventional free energy. In addition to giving insight into the second law of thermodynamics, this result leads to a new approach to quantum chaos (Caves, 1993b; Schack and Caves, 1993, 1996a; Schack *et al.*, 1994).

In this paper, we show how the computer-dependent $O(1)$ constant can be eliminated from the theory summarized above. In Section 2 we construct an optimal universal computer for which the $O(1)$ constant is minimal. It turns out, however, that optimal universal computers do not recognize intrinsically simple states and thus are unsatisfactory in formulating the theory. This difficulty is solved in Section 3, where we show that any universal computer $U$ can be modified in a simple way such that (a) any state that is simple with respect to $U$ is also simple with respect to the modified universal computer $U_3$ and (b) average conditional information with respect to $U_3$ exceeds average conditional information with respect to an optimal universal computer by at most 0.5 bit. Moreover, conditional algorithmic information with respect to the modified computer $U_3$ obeys the inequality $H \leq \overline{I_{U_3}(\cdot \mid s)} < H + 1$. This double bound is the tightest possible in the sense that there is no tighter bound that is independent of the probabilities $p_j$.

## 2. AN OPTIMAL UNIVERSAL COMPUTER

The idea of an optimal universal computer is motivated by Zurek's discussion (Zurek, 1989b) of *Huffman coding* (Huffman, 1952) as an alternative way to quantify the information in an observational record. We consider only binary codes, for which the code words are binary strings. Before reviewing Huffman coding, we need to formalize the concept of a list consisting of descriptions of $N$ states together with their probabilities.

*Definition 1.* A *list of states* $s$ is a string of the form $s = ((s_1, p_1), \ldots, (s_N, p_N))$, where $N \geq 2$, $0 < p_1 \leq \cdots \leq p_N$, $\Sigma\, p_j = 1$, and $s_j$ is a binary string ($j = 1, \ldots, N$). More precisely, the list of states $s$ is the binary string obtained from the list $((s_1, p_1), \ldots, (s_N, p_N))$ by some definite translation scheme. One possible translation scheme is to represent parentheses, commas, and numbers (i.e., the probabilities $p_j$) in ascii code, and to precede each binary string $s_j$ by a number giving its length $|s_j|$ in bits. The entropy of a list of states is $H(s) = -\Sigma\, p_j \log_2 p_j$. Throughout this paper, $|t|$ denotes the length of the binary string $t$.

The Huffman code for a list of states $s = ((s_1, p_1), \ldots, (s_N, p_N))$ is a prefix-free or instantaneous code (Welsh, 1988)—i.e., no code word is a prefix of any other code word—and can, like all prefix-free codes, be represented by a binary tree as shown in Fig. 1. The number of links leading from the root

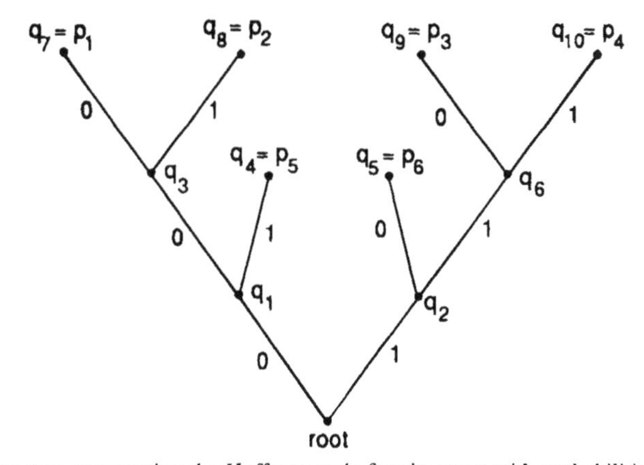


**Fig. 1.** Binary tree representing the Huffman code for six states with probabilities $p_1, \ldots, p_6$. The node probabilities $q_k$ are defined recursively, i.e., $q_7 = p_1$, $q_8 = p_2$, $q_3 = q_7 + q_8$, etc. Code words correspond to branch labels; e.g., the code word for the third state (probability $p_3$) is 110.

of the tree to a node is called the *level* of that node. If the level-$n$ node $a$ is connected to the level-$(n + 1)$ nodes $b$ and $c$, then $a$ is called the *parent* of $b$ and $c$; $a$'s *children* $b$ and $c$ are called *siblings*. There are exactly $N$ terminal nodes or *leaves*, each leaf corresponding to a state $j$. Each link connecting two nodes is labeled 0 or 1. The sequence of labels encountered on the path from the root to a leaf is the code word assigned to the corresponding state. The code-word length of a state is thus equal to the level of the corresponding leaf. Each node is assigned a probability $q_k$ such that the probability of a leaf is equal to the probability $p_j$ of the corresponding state and the probability of each nonterminal node is equal to the sum of the probabilities of its children.

A binary tree represents a Huffman code if and only if it has the *sibling property* (Gallager, 1978), i.e., if and only if each node except the root has a sibling, and the nodes can be listed in order of nonincreasing probability with each node being adjacent to its sibling in the list. The tree corresponding to a Huffman code and thus the Huffman code itself can be built recursively. Create a list of $N$ nodes corresponding to the $N$ states. These $N$ nodes will be the leaves of the tree that will now be constructed. Repeat the following procedure until the tree is complete: Take two nodes with smallest probabilities and make them siblings by generating a node that is their common parent; replace in the list the two nodes by their parent; label the two links branching from the new parent node by 0 and 1.

The procedure outlined above does not define a unique Huffman code for the list of states $s$, nor does it give generally a unique set of code-word lengths. In the following, we will assume that we are given some definite algorithm to assign a Huffman code where the freedom in the coding procedure is used to assign to the first state (the one with smallest probability) a code word of maximum length consisting only of zeros.

*Definition 2.* Given a list of states $s = ((s_1, p_1), \ldots, (s_N, p_N))$, the binary string $c_j(s)$ with length $l_j(s) \equiv |c_j(s)|$ denotes the Huffman code word assigned to the $j$th state using a definite algorithm with the property that $c_1(s) = 0 \cdots 0$ and $l_j(s) \leq l_1(s)$ for $j = 2, \ldots, N$. We denote the average Huffman code-word length by $\bar{l}(s) = \Sigma\, p_j l_j(s)$. The *redundancy* $r(s)$ of the Huffman code is defined by $r(s) = \bar{l}(s) - H(s)$.

The redundancy $r(s)$ obeys the bounds $0 \leq r(s) < 1$, corresponding to bounds

$$H(s) \leq \bar{l}(s) < H(s) + 1 \tag{4}$$

for the average code-word length. Huffman coding is optimal in the sense that there is no prefix-free binary code with an average code-word length less than $\bar{l}(s)$. There can be, however, optimal prefix-free codes that are not Huffman codes.

The length $l_j(s)$ of the Huffman code word $c_j(s)$ cannot be determined from the probability $p_j$ alone, but depends on the entire set of probabilities $p_1, \ldots, p_N$. The tightest general bounds for $l_j(s)$ are (Katona and Nemetz, 1976)

$$1 \leq l_j(s) < -\log_g p_j + 1 \tag{5}$$

where $g = (\sqrt{5} + 1)/2$ is the golden mean. The code-word length for some states $j$ thus can differ widely from the value $-\log_2 p_j$. For most states $j$, however, the Huffman code-word length is $l_j(s) \simeq -\log_2 p_j$. The following theorem (Schack, 1994) is a precise version of this statement.

*Theorem 1.* (a) $P_m^- = \sum_{j \in I_m^-} p_j < 2^{-m}$, where $I_m^- = \{i \mid l_i(s) < -\log_2 p_i - m\}$, i.e., the probability that a state with probability $p$ has Huffman code-word length smaller than $-\log_2 p - m$ is less than $2^{-m}$. (This is true for any prefix-free code.) (b) $P_m^+ = \sum_{j \in I_m^+} p_j < 2^{-c(m-2)+2}$, where $I_m^+ = \{i \mid l_i(s) > -\log_2 p_i + m\}$ and $c = (1 - \log_2 g)^{-1} - 1 \simeq 2.27$, i.e., the probability that a state with probability $p$ has Huffman code-word length greater than $-\log_2 p + m$ is less than $2^{-c(m-2)+2}$.

*Proof.* See Schack (1994). ■

The probability of encountering a state $j$ with a Huffman code word much longer than $-\log_2 p_j$ is therefore exponentially small. There are alternative coding schemes that avoid these long code words altogether at the cost of slightly increasing the average code-word length; one such scheme, Shannon–Fano coding, is discussed in Zurek (1989b). In the present paper, we use Huffman coding for specificity.

Suppose that one characterizes the information content of a state $j$ by its Huffman code-word length $l_j(s)$. Then in condition (2) average algorithmic information $\overline{I_U(\cdot \mid s)}$ is replaced by average code-word length $\bar{l}(s)$, the $O(1)$ constant is replaced by 1, and condition (3) assumes the concise form $0 \geq \Delta F_{\text{tot}} > -k_B T \ln 2$. This way of eliminating the $O(1)$ constant, however, has a high price. Since Huffman code-word lengths depend solely on the probabilities $p_1, \ldots, p_N$—states with high probability are assigned shorter code words than states with low probability—Huffman coding does not recognize intrinsically simple states. This means that one of the most appealing features of the theory is lost, namely that the Landauer erasure cost associated with states that can be prepared in a laboratory is negligible.

In the present paper we show that it is possible to retain this feature of the theory, yet still eliminate the computer-dependent constant. We first attempt to do this by constructing an optimal universal computer, i.e., a universal computer for which the $O(1)$ constant in condition (2) is minimal. We find, however, that optimal universal computers do not recognize intrinsically simple states either. A solution to this problem will be given in Section 3, where we discuss a class of nearly optimal universal computers.

We will need precise definitions of a computer and a universal computer, which we quote from Chapter 6.2 in Chaitin (1987).

*Definition 3.* A *computer C* is a computable partial function that carries a program string $p$ and a free data string $q$ into an output string $C(p, q)$ with the property that for each $q$ the domain of $C(., q)$ is a prefix-free set; i.e., if $C(p, q)$ is defined and $p$ is a proper prefix of $p'$, then $C(p', q)$ is not defined. In other words, programs must be self-delimiting. $U$ is a *universal computer* if and only if for each computer $C$ there is a constant sim($C$) with the following property: if $C(p, q)$ is defined, then there is a $p'$ such that $U(p', q) = C(p, q)$ and $|p'| \leq |p| + \text{sim}(C)$.

In this definition, all strings are binary strings, and $|p|$ denotes the length of the string $p$ as before. The self-delimiting or prefix-free property entails that for each free data string $q$, the set of all valid program strings can be represented by a binary tree.

For any binary string $t$ we denote by $t^*(U)$ (or just $t^*$ if no confusion is possible) the shortest string for which $U(t^*, \Lambda) = t$, where $\Lambda$ is the empty string; i.e., $t^*$ is the shortest program for the universal computer $U$ to calculate $t$. If there are several such programs, we pick the one that is first in lexicographic order. This allows us to define conditional algorithmic information.

*Definition 4.* The *conditional algorithmic information* $I_U(t_1 | t_2)$ to specify the binary string $t_1$, given the binary string $t_2$, is

$$I_U(t_1 | t_2) = \min_{p | U(p, t_2^*) = t_1} |p| \tag{6}$$

In words, $I_U(t_1 | t_2)$ is the length of a shortest program for $U$ that computes $t_1$ in the presence of the free data string $t_2^*$. In particular, the conditional algorithmic information $I_U(s_j | s)$ to specify the $j$th state, given a list of states $s = ((s_1, p_1), \ldots, (s_N, p_N))$, is

$$I_U(s_j | s) = \min_{p | U(p, s^*) = s_j} |p| \tag{7}$$

The average of $I_U(s_j | s)$ is denoted by $\overline{I_U(\cdot | s)} = \Sigma\, p_j I_U(s_j | s)$.

The next theorem puts a lower bound on the average information.

*Theorem 2.* For any universal computer $U$ and any list of states $s = ((s_1, p_1), \ldots, (s_N, p_N))$, the average conditional algorithmic information obeys the bound

$$\overline{I_U(\cdot | s)} \geq H(s) + r(s) + p_1 \tag{8}$$

*Proof.* We denote by $s_j'$ a shortest string for which $U(s_j', s^*) = s_j$. The $N$ strings $s_j'$ form a prefix-free code. If the $N$ strings $s_j'$ are represented by the leaves of a binary tree, then there is at least one node that has no sibling. Otherwise $U(p, s^*)$ would be defined only for a finite number $N$ of programs $p$, and $U$ would not be a universal computer. Let us denote by $\mathcal{Q}$ a sibling-free node and by $q$ its probability ($q \geq p_1$). Then a shorter prefix-free code $\{s_j''\}$ can be obtained by moving node $\mathcal{Q}$ down one level. More precisely, for states $j$ corresponding to leaves of the subtree branching from node $\mathcal{Q}$, $s_j''$ is obtained from $s_j'$ by removing the digit corresponding to the link between node $\mathcal{Q}$ and its parent; for all other states $j$, $s_j'' = s_j'$. The code-word lengths of the new code are $|s_j''| = |s_j'| - 1$ if state $j$ is a leaf of the subtree branching from node $\mathcal{Q}$ and $|s_j''| = |s_j'|$ otherwise. Since the new code is prefix-free, its average code-word length is greater than or equal to the Huffman code-word length $\bar{l}(s)$. It follows that

$$\overline{I_U(\cdot\,|s)} = \sum_j p_j|s_j'| = \sum_j p_j|s_j''| + q \geq \bar{l}(s) + p_1$$
$$= H(s) + r(s) + p_1 \qquad (9)$$

which proves the theorem. ■

We can now proceed to define an optimal universal computer.

*Definition 5.* $U$ is an *optimal universal computer* if there is a constant $\epsilon > 0$ such that for all lists of states $s = ((s_1, p_1), \ldots, (s_N, p_N))$ with $p_1 \geq \epsilon$ the average conditional algorithmic information has its minimum value

$$\overline{I_U(\cdot\,|s)} = H(s) + r(s) + p_1 \qquad (10)$$

*Theorem 3.* For any $\epsilon > 0$ there is an optimal universal computer $U_\epsilon$.

*Proof.* Let $U$ be an arbitrary universal computer and $\epsilon > 0$. For any list of states $s = ((s_1, p_1), \ldots, (s_N, p_N))$ with $p_1 \geq \epsilon$ we define $c_1'(s) = c_1(s) \circ 1 = 0\cdots01$ and $c_j'(s) = c_j(s)$ for $j = 2, \ldots, N$, where $\circ$ denotes concatenation of strings. The strings $c_j'(s)$ thus differ from the Huffman code $c_j(s)$ in that a 1 has been appended to the code word for the state $j = 1$. According to condition (5), $l_1(s) + 1 \leq N_0 \equiv \lfloor -\log_g \epsilon + 2 \rfloor$, where $g = (\sqrt{5} + 1)/2$ and $\lfloor x \rfloor$ denotes the largest integer less than or equal to $x$. We denote by $\sigma_0$ a string composed of $N_0$ zeros; none of the strings $c_j'(s)$ is longer than $\sigma_0$.

For the definition of $U_\epsilon(p, q)$ we distinguish two cases. If the binary string $q$ is of the form

$$q = \sigma_0 \circ q_s \qquad \text{with} \qquad U(q_s, \Lambda) = s \qquad (11)$$

for some list of states $s = ((s_1, p_1), \ldots, (s_N, p_N))$ with $p_1 \geq \epsilon$, then $U_\epsilon(p, q)$ is defined for

$$p \in D(q) \equiv \{\sigma_0 \circ p' \mid U(p', q) \text{ is defined}\}$$
$$\cup \{c_j'(s) \mid 1 \leq j \leq N\} \tag{12}$$

with

$$U_\epsilon(\sigma_0 \circ p', q)$$
$$= U(p', q) \qquad \text{whenever } U(p', q) \text{ is defined} \tag{13}$$

and

$$U_\epsilon(c_j'(s), q) = s_j \qquad \text{for } j = 1, \ldots, N \tag{14}$$

If the binary string $q$ is not of the form (11), then $U_\epsilon(p, q)$ is defined for

$$p \in D(q) \equiv \{\sigma_0 \circ p' \mid U(p', q) \text{ is defined}\} \tag{15}$$

with

$$U_\epsilon(\sigma_0 \circ p', q) = U(p', q) \qquad \text{whenever } U(p', q) \text{ is defined} \tag{16}$$

In both cases, the set $D(q)$, which is the domain of $U_\epsilon(\cdot, q)$, is clearly prefix-free. Moreover, since $U_\epsilon(\sigma_0 \circ p, q) = U(p, q)$ whenever $U(p, q)$ is defined and $U$ is a universal computer, $U_\epsilon$ is also a universal computer, with the simulation constant sim($C$) increased by $N_0$.

For any string $t$ the minimal program on $U_\epsilon$—i.e., the shortest program given an empty free data string—is $t^*(U_\epsilon) = \sigma_0 \circ t^*(U)$, where $t^*(U)$ is the minimal program for $t$ on $U$. In particular, the shortest program for $U_\epsilon$ to compute $s$ is $s^*(U_\epsilon) = \sigma_0 \circ s^*(U)$. Since $U_\epsilon(c_j'(s), s^*(U_\epsilon)) = s_j$ and $|c_j'(s)| \leq N_0$ for $j = 1, \ldots, N$, while $|p| \geq N_0$ for all other programs $p \in D(s^*(U_\epsilon))$, it follows immediately that

$$I_{U_\epsilon}(s_j \mid s) = |c_j'(s)| = |c_j(s)| + \delta_{1j} = l_j(s) + \delta_{1j} \tag{17}$$

and thus that

$$\overline{I_{U_\epsilon}(\cdot \mid s)} = \sum p_j I_{U_\epsilon}(s_j \mid s) = \sum p_j |c_j'(s)| = \sum p_j |c_j(s)| + p_1$$
$$= \bar{l}(s) + p_1 = H(s) + r(s) + p_1 \quad \blacksquare \tag{18}$$

If $U(q_s, \Lambda) = U_\epsilon(\sigma_0 \circ q_s, \Lambda) = s$, i.e., if $q_s$ is a program for $U$ generating a list of states $s$, the programs $p$ for which $U_\epsilon(p, \sigma_0 \circ q_s)$ is defined can be represented by a binary tree similar to Fig. 2. With respect to the binary tree representing the Huffman code (Fig. 1), the leaf for the $j = 1$ state has been moved up one level to make room for the new node labeled by $U$. This new node leads to a subtree representing all programs $p'$ for which $U(p', \sigma_0 \circ q_s)$ is defined.

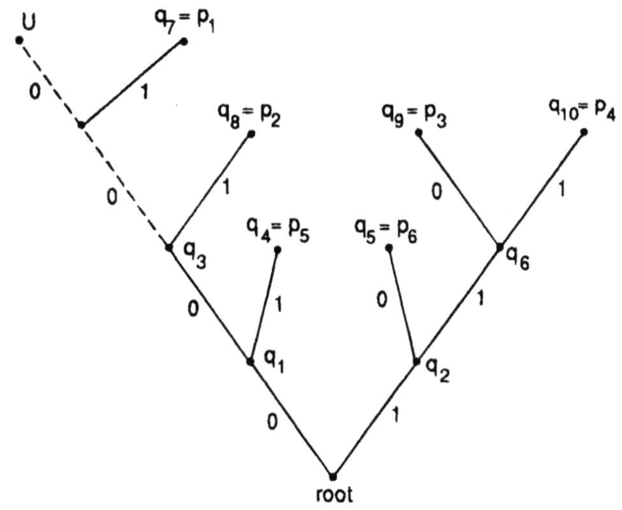


**Fig. 2.** Binary tree representing all valid programs for the optimal universal computer $U_\epsilon$ in the presence of a free data string generating a list of states $((s_1, p_1), \ldots, (s_6, p_6))$. With respect to the tree in Fig. 1, the node labeled $q_7 = p_1$ has been moved up one level to make room for the subtree representing programs for $U$.

The operation of the optimal universal computer $U_\epsilon$ can be described in the following way. When $U_\epsilon$ reads a string that begins with $N_0$ zeros from its program tape, $U_\epsilon$ disregards the $N_0$ zeros and interprets the rest of the string as a program for the universal computer $U$, executing it accordingly. If $U_\epsilon$ encounters the digit 1 while reading the first $N_0$ digits from its program tape, $U_\epsilon$ interrupts reading from the program tape, reads in the free data string, and executes it. If the result of executing the free data string is a list of states $s = ((s_1, p_1), \ldots, (s_N, p_N))$, $U_\epsilon$ establishes the modified Huffman code $\{c_j'(s)\}$ for $s$, continues reading digits from the program tape until the string read matches one of the code words, say $c_{j0}'(s)$, and then prints the string $s_{j_0}$. The output of $U_\epsilon$ is undefined in all other cases.

Since $r(s) + p_1 < 1$ (Gallager, 1978), $H(s) \leq \overline{I_U(\cdot \,|\, s)} < H(s) + 1$ for any optimal universal computer $U$. For the particular optimal universal computer $U_\epsilon$ defined in the proof of Theorem 3, however, the information $I_{U_\epsilon}(s_j \,|\, s)$ is completely determined by the Huffman code-word length for the $j$th state and therefore is completely determined by the probabilities $p_1, \ldots, p_N$. This optimal universal computer does not recognize intrinsically simple states. As an aside, note that $U_\epsilon$ cannot give a short description of the background information for any probability distribution, because a minimal program for computing the list of states $s$ on $U_\epsilon$ must begin with $N_0$ zeros. It turns out that all optimal universal computers, not just $U_\epsilon$, are unable to recognize intrinsically simple states. The following theorem formulates this inability for all optimal universal computers in a slightly weaker form than holds for $U_\epsilon$. As a consequence, the use of algorithmic information with respect to an optimal universal computer to quantify the information in an observational record presents no advantage over the use of Huffman coding.

*Theorem 4.* For any optimal universal computer $U$ and any list of states $s = ((s_1, p_1), \ldots, (s_N, p_N))$ for which $\overline{I_U(\cdot \mid s)} = H(s) + r(s) + p_1$, the following holds: If $p_i > p_j$, then $I_U(s_i \mid s) \leq I_U(s_j \mid s)$. Optimal universal computers therefore do not recognize intrinsically simple states.

*Proof.* To prove the theorem, we show that $\overline{I_U(\cdot \mid s)} > H(s) + r(s) + p_1$ for any universal computer $U$ and any list of states $s = ((s_1, p_1), \ldots, (s_N, p_N))$ for which there are indices $i$ and $j$ such that $p_i > p_j$ but $I_U(s_i \mid s) > I_U(s_j \mid s)$. We denote by $s_j'$ a shortest string for which $U(s_j', s^*) = s_j$. The strings $s_j'$ form a prefix-free code. Following an argument similar to the proof of Theorem 2, we can shorten that code on the average by moving a sibling-free node one level down and in addition by interchanging the code words for states $i$ and $j$. The resulting shorter code must obey the Huffman bound, from which the inequality $\overline{I_U(\cdot \mid s)} > \bar{l}(s) + p_1 = H(s) + r(s) + p_1$ follows. ■

## 3. PRESERVING SIMPLE STATES BY GIVING UP 1/2 BIT

Although the discussion in the last section shows that optimal universal computers present no advantages over Huffman coding, the main idea behind their construction can be further exploited. If the subtree representing the programs for the universal computer $U$ is not attached next to the $j = 1$ leaf as in Fig. 2, but instead is attached close to the root as in Fig. 3, the resulting universal computer $U_3$ combines the desirable properties of Huffman coding and the computer $U$. This is the content of the following theorem.

*Theorem 5.* For any universal computer $U$ there is a universal computer $U_3$ such that

$$I_{U_3}(t_1 \mid t_2) \leq I_U(t_1 \mid t_2) + 3 \tag{19}$$

for all binary strings $t_1$ and $t_2$, and that

$$H(s) \leq \overline{I_{U_3}(\cdot \mid s)} < H(s) + 1 \tag{20}$$

and

$$\overline{I_{U_3}(\cdot \mid s)} \leq H(s) + r(s) + 1/2 \tag{21}$$

for all lists of states $s = ((s_1, p_1), \ldots, (s_N, p_N))$.

*Proof.* Let $U$ be an arbitrary universal computer. For any list of states $s = ((s_1, p_1), \ldots, (s_N, p_N))$ we define the set of strings $c_j'(s)$ as follows. We

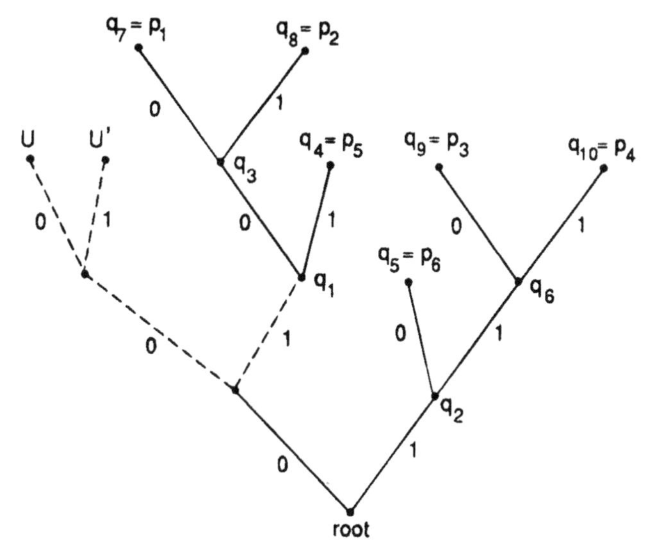


**Fig. 3.** Binary tree representing all valid programs for the universal computer $U_3$ in the presence of a free data string generating a list of states $s = ((s_1, p_1), \ldots, (s_6, p_6))$. With respect to the tree in Fig. 1, the level-1 node labeled $q_1$ has been moved up one level to make room for the subtrees representing programs for $U$. More precisely, the binary tree represents the programs $p$ for which $U_3(p, 000 \circ q_s)$ is defined if $U_3(000 \circ q_s, \Lambda) = s$. The node labeled $U$ is the root of a subtree corresponding to the programs $p'$ for which $U(p', 000 \circ q_s)$ is defined, and the node labeled $U'$ is the root of a subtree corresponding to the programs $p'$ for which $U(p', q_s)$ is defined.

start from the binary tree formed by the Huffman code words $c_j(s)$, where we denote by $q_1$ the probability of the level-1 node connected to the root by the link labeled 0 (see Fig. 1). According to the value of $q_1$, we distinguish two cases. In the case $q_1 \leq 1/2$, $c_j'(s) = 01 \circ c_j^+(s)$ if $c_j(s)$ is of the form $c_j(s) = 0 \circ c_j^+(s)$, and $c_j'(s) = c_j(s)$ if $c_j(s)$ is of the form $c_j(s) = 1 \circ c_j^+(s)$. In the case $q_1 > 1/2$, $c_j'(s) = 01 \circ c_j^+(s)$ if $c_j(s)$ is of the form $c_j(s) = 1 \circ c_j^+(s)$, and $c_j'(s) = 1 \circ c_j^+(s)$ if $c_j(s)$ is of the form $c_j(s) = 0 \circ c_j^+(s)$.

Figure 3 illustrates the binary tree formed by the code words $c_j'(s)$ for the case $q_1 \leq 1/2$. Of the two main subtrees emerging from the level-1 nodes in Fig. 1, the subtree having smaller probability is moved up one link and attached to the node labeled 01, and the subtree having larger probability is attached to the node labeled 1. In this way, the node labeled 00 is freed for the subtrees representing the valid programs for $U$.

For the definition of $U_3(p, q)$ we distinguish three cases. If the binary string $q$ is of the form

$$q = 000 \circ q_s \qquad \text{with} \quad U(q_s, \Lambda) = s \tag{22}$$

for some list of states $s = ((s_1, p_1), \ldots, (s_N, p_N))$, then $U_3(p, q)$ is defined for

$$p \in D(q) \equiv \{000 \circ p' \mid U(p', q) \text{ is defined}\}$$
$$\cup \{001 \circ p' \mid U(p', q_s) \text{ is defined}\}$$
$$\cup \{c_j'(s) \mid 1 \leq j \leq N\} \tag{23}$$

with

$$U_3(000 \circ p', q)$$
$$= U(p', q) \qquad \text{whenever } U(p', q) \text{ is defined} \tag{24}$$
$$U_3(001 \circ p', q)$$
$$= U(p', q_s) \qquad \text{whenever } U(p', q_s) \text{ is defined} \tag{25}$$

and

$$U_3(c_j'(s), q) = s_j \qquad \text{for } j = 1, \ldots, N \tag{26}$$

If the binary string $q$ is of the form

$$q = 000 \circ q' \tag{27}$$

but there is *no* list of states $s$ such that $U(q', \Lambda) = s$, then $U_3(p, q)$ is defined for

$$p \in D(q) \equiv \{000 \circ p' \mid U(p', q) \text{ is defined}\}$$
$$\cup \{001 \circ p' \mid U(p', q') \text{ is defined}\} \tag{28}$$

with

$$U_3(000 \circ p', q)$$
$$= U(p', q) \qquad \text{whenever } U(p', q) \text{ is defined} \tag{29}$$

and

$$U_3(001 \circ p', q)$$
$$= U(p', q') \qquad \text{whenever } U(p', q') \text{ is defined} \tag{30}$$

Finally, if $q$ is not of the form (27), then $U_3(p, q)$ is defined for

$$p \in D(q) \equiv \{000 \circ p' \mid U(p', q) \text{ is defined}\} \tag{31}$$

with

$$U_3(000 \circ p', q)$$
$$= U(p', q) \qquad \text{whenever } U(p', q) \text{ is defined} \tag{32}$$

In all three cases, the set $D(q)$, which is the domain of $U_3(\cdot, q)$, is clearly prefix-free. Moreover, since $U_3(000 \circ p, q) = U(p, q)$ whenever $U(p, q)$ is defined and $U$ is a universal computer, $U_3$ is also a universal computer, with the simulation constant sim($C$) increased by 3. Equation (19) holds because of the following. The minimal program for $t_2$ on $U_3$ in the presence of an empty free data string is $t_2^*(U_3) = 000 \circ t_2^*(U)$ since $U_3(p, \Lambda)$ is defined only if $p = 000 \circ p'$ and $U(p', \Lambda)$ is defined, in which case $U_3(p, \Lambda) = U(p', \Lambda)$. If $p$ is a minimal program for $t_1$ on $U$ in the presence of the minimal program for $t_2$, i.e., if

$$U(p, t_2^*(U)) = t_1, \qquad |p| = I_U(t_1 | t_2) \tag{33}$$

then

$$\begin{aligned} U_3(001 \circ p, t_2^*(U_3)) &= U_3(001 \circ p, 000 \circ t_2^*(U)) \\ &= U(p, t_2^*(U)) = t_1 \end{aligned} \tag{34}$$

and therefore

$$I_{U_3}(t_1 | t_2) \leq |001 \circ p| = |p| + 3 \tag{35}$$

The strings $c_j'(s)$ form a prefix-free code with an unused code word of length 2, for which $\Sigma\, p_j |c_j'(s)| < H(s) + 1$ according to Theorem 3 in Gallager (1978). [In Gallager (1978) the inequality appears with a $\leq$ sign, but equality can occur only if the smallest probability $p_1$ is equal to zero, a case we have excluded.] The shortest program for $U_3$ to compute $s$ is $s^*(U_3) = 000 \circ s^*(U)$, where $s^*(U)$ is the shortest program for $U$ to compute $s$. Since $U_3(c_j'(s), s^*(U_3)) = s_j$ for $j = 1, \ldots, N$, it follows immediately that $I_{U_3}(s_j | s) \leq |c_j'(s)|$ and thus that

$$\overline{I_{U_3}(\cdot | s)} = \sum p_j I_{U_3}(s_j | s) \leq \sum p_j |c_j'(s)| < H(s) + 1 \tag{36}$$

which establishes the upper bound in condition (20). The lower bound in (20) holds for all universal computers. Equation (21) follows from

$$\begin{aligned} \sum p_j |c_j'(s)| &= \sum p_j |c_j(s)| + \min(q_1, 1 - q_1) \\ &= \bar{l}(s) + \min(q_1, 1 - q_1) \\ &\leq H(s) + r(s) + 1/2 \quad \blacksquare \end{aligned} \tag{37}$$

If $U(q_s, \Lambda) = U_3(000 \circ q_s, \Lambda) = s$, i.e., if $q_s$ is a program for $U$ generating a list of states $s$, the programs $p$ for which $U_3(p, 000 \circ q_s)$ is defined can be represented by a binary tree similar to Fig. 3. The level-3 node labeled $U$ is the root of a subtree corresponding to the programs $p'$ for which $U(p', 000 \circ q_s)$ is defined, and the level-3 node labeled $U'$ is the root of a subtree corresponding to the programs $p'$ for which $U(p', q_s)$ is defined.

The operation of the universal computer $U_3$ can be described in the following way. When $U_3$ reads a string that begins with the prefix 000 from its program tape, $U_3$ disregards the prefix and interprets the rest of the string as a program for the universal computer $U$, executing it accordingly. When $U_3$ reads a string that begins with the prefix 001 from its program tape, the output is only defined if the free data string begins with 000, in which case $U_3$ disregards the first three digits of the program and free data strings and interprets the rest of the strings as program and free data strings for the universal computer $U$, executing it accordingly. If $U_3$ encounters the digit 1 while reading the first two digits from its program tape, $U_3$ interrupts reading from the program tape, reads in the free data string, and executes it. If the result of executing the free data string is a list of states $s = ((s_1, p_1), \ldots, (s_N, p_N))$, $U_3$ establishes the modified Huffman code $\{c_j'(s)\}$ for $s$, continues reading digits from the program tape until the string read matches one of the code words, say $c_{j0}'(s)$, and then prints the string $s_{j0}$. The output of $U_3$ is undefined in all other cases.

The computer $U_3$ compromises between the desirable properties of algorithmic information and Huffman coding. Since algorithmic information defined with respect to $U_3$ exceeds algorithmic information relative to $U$ by at most 3 bits, states that are simple with respect to $U$ are simple with respect to $U_3$. Those 3 bits are the price to pay for a small upper bound on average information. The average conditional algorithmic information $\overline{I_{U_3}(\cdot \mid s)}$ obeys the close double bound (20) and exceeds the Huffman bound $\bar{l}(s)$ by at most 0.5 bit. This half bit is the price to pay for the recognition of intrinsically simple states.

## 4. CONCLUSION

We have shown that any universal computer $U$ can be modified in such a way that (i) the modified universal computer $U_3$ recognizes the same intrinsically simple states as $U$ and (ii) average algorithmic information with respect to $U_3$ obeys the same close double bound as Huffman coding, $H(s) \leq \overline{I_{U_3}(\cdot \mid s)} < H(s) + 1$. If for any choice of a universal computer $U$, total free energy is defined with respect to the corresponding modified universal computer $U_3$, i.e., if the change of total free energy due to finding the system in the $j$th state is $\Delta F_{j,\mathrm{tot}} = -k_B T \ln 2\,[-H(s) + I_{U_3}(s_j \mid s)]$, then the bounds for the average change in total free energy are given by

$$0 \geq \overline{\Delta F_{\mathrm{tot}}} > -k_B T \ln 2 \tag{38}$$

instead of by (3).

This result effectively eliminates the undetermined computer-dependent constant from applications of algorithmic information theory to statistical physics. Except for an unavoidable loss due to the coding bounded by $k_B T \ln 2$, on the average available work is independent of the information the observer has acquired about the system, any decrease of the statistical entropy being balanced by an equal increase in algorithmic information.

## ACKNOWLEDGMENT

The author wishes to thank Carlton M. Caves for suggesting the problem and for many enlightening discussions.

## REFERENCES

Bennett, C. H. (1982). *International Journal of Theoretical Physics*, **21**, 905.

Caves, C. M. (1990). In *Complexity, Entropy, and the Physics of Information*, W. H. Zurek, ed., Addison-Wesley, Redwood City, California, p. 91.

Caves, C. M. (1993a). *Physical Review E*, **47**, 4010.

Caves, C. M. (1993b). In *Physical Origins of Time Asymmetry*, J. J. Halliwell, J. Pérez-Mercader, and W. H. Zurek, eds., Cambridge University Press, Cambridge, p. 47.

Chaitin, G. J. (1987). *Algorithmic Information Theory*, Cambridge University Press, Cambridge.

Denker, J. S., and leCun, Y. (1993). In *Workshop on Physics and Computation: PhysComp '92*, IEEE Computer Society Press, Los Alamitos, California, p. 122.

Gallager, R. G. (1978). *IEEE Transactions on Information Theory*, **IT-24**, 668.

Huffman, D. A. (1952). *Proceedings IRE*, **40**, 1098.

Jaynes, E. T. (1983). In *Papers on Probability, Statistics, and Statistical Physics*, R. D. Rosenkrantz, ed., Kluwer, Dordrecht, Holland.

Katona, G. O. H., and Nemetz, T. O. H. (1976). *IEEE Transactions on Information Theory*, **IT-22**, 337.

Kolmogoroff, A. N. (1965). *Problemy Peredachi Informatsii*, **1**, 3 [*Problems of Information Transmission*, **1**, 1 (1965)].

Landauer, R. (1961). *IBM Journal of Research and Development*, **5**, 183.

Landauer, R. (1988). *Nature*, **355**, 779.

Schack, R. (1994). *IEEE Transactions on Information Theory*, **IT-40**, 1246.

Schack, R., and Caves, C. M. (1992). *Physical Review Letters*, **69**, 3413.

Schack, R., and Caves, C. M. (1993). *Physical Review Letters*, **71**, 525.

Schack, R., and Caves, C. M. (1996a). *Physical Review E*, **53**, 3257.

Schack, R., and Caves, C. M. (1996b). *Physical Review E*, **53**, 3387.

Schack, R., D'Ariano, G. M., and Caves, C. M. (1994). *Physical Review E*, **50**, 972.

Solomonoff, R. J. (1964). *Information and Control*, **7**, 1.

Szilard, L. (1929). *Zeitschrift für Physik*, **53**, 840.

Welsh, D. (1988). *Codes and Cyptography*, Clarendon Press, Oxford.

Zurek, W. H. (1989a). *Nature*, **341**, 119.

Zurek, W. H. (1989b). *Physical Review A*, **40**, 4731.

Zvonkin, A. V., and Levin, L. A. (1970). *Uspekhi Matematicheskikh Nauk* **25**, 85 [*Russian Mathematical Surveys*, **25**, 83 (1970)].

# ALGORITHMIC RANDOMNESS, PHYSICAL ENTROPY, MEASUREMENTS, AND THE DEMON OF CHOICE

**W. H. Zurek**

## Abstract

Measurements — interactions which establish correlations between a system and a recording device — can be made thermodynamically reversible. One might be concerned that such reversibility will make the second law of thermodynamics vulnerable to the designs of the *demon of choice*, a selective version of Maxwell's demon. The strategy of the demon of choice is to take advantage of rare fluctuations to extract useful work, and to reversibly undo measurements which do not lead to such a favorable but unlikely outcomes. I show that this threat does not arise as the demon of choice cannot operate without recording (explicitly or implicitly) whether its measurement was a success (or a failure). Thermodynamic cost associated with such a record cannot be, on the average, made smaller than the gain of useful work derived from the fluctuations.

## 22.1 Feynman

When I was asked to write for a volume dedicated to Richard Feynman, I decided that I should select the subject in which I was influenced by him the most, and which would still be consistent with the overall theme of computation and physics. And these influences started well before I met him in person: I got Feynman's "Lectures on Physics" more than a quarter century ago, in Polish translation, from my father. As a finishing high school student I was accompanying him on a hunting expedition in the lake district of Poland — a remote corner of the country. Every few days we drove for supplies to the provincial capital, and there I noticed the volumes in the local bookstore. My father asked why (the expense was considerable), but surprisingly easily gave way to my arguments. I spent much of the rest of the hunting vacation (a couple of weeks altogether) getting through volume I.

Over the years I have developed a habit of treating the "Lectures" sort of like a collection of poems. I like some "poems" more than others, and I return to the favorites now and again. And when I am stuck with a physics problem, reading a few of the relevant "poems" is often the best way to get "unstuck". But there are a few chapters which have been read over and over again without any such an ulterior motive, for sheer pleasure. Amongst them, I would certainly include the discussion of the fluctuations and the second law (the famous "ratchet and pawl" argument [1]).

Thermodynamic concerns and arguments have often pre-saged the deepest developments in physics. I suspect this is because thermodynamics "knows" about the physical relevance of information, and hence, it knew about the Planck constant, stimulated emission, black hole entropy, and so on. When I met Feynman in person for the first time (at a small workshop organised near Austin, Texas, by John Archibald Wheeler in the Spring of 1981), I remember — amongst other things — a thermodynamic argument he used to great effect to prove that one cannot accelerate elementary particles by shaking them together with a bunch of heavier objects, so that they could acquire equipartition kinetic energies (and therefore, because of their small mass, enormous momenta). This idea (credible at first sight, as it is akin to the Fermi acceleration of cosmic rays) was brought up by one of the participants. It would not work — Feynman argued — because all sorts of other modes of the vacuum would have to get their fair share of energy, creating an equilibrium heat bath, with approximate equipartition between all the modes (rather then with the energy in the elementary particles one really wanted to accelerate in the first place).

But that was not the most vivid memory of that first encounter with the man whose "Lectures" I had acquired a decade or so earlier. Rather, I remember best that he showed up at the first lecture unshaved and uncombed, with dry grass in his hair. It turned out that he spent the night outside — apparently, he decided the accomodations for the speakers (which were in the posh tennis club) were too opulent, returned the key to his apartment at the reception, and decided to "camp out". During the morning coffee he has also reported in detail (and with great gusto) how he had trouble breaking the code to get into his briefcase (where he had the sweater — it got cold). He knew the code, of course, by heart, but it was middle of the night, so he somehow had to dial it in complete darkness. He clearly relished the challenge. I do not remember how did he solve the problem, but the flavor of the adventure and of his report was very much in the spirit of the "adventures of a curious character". And all of this was a few months after his (first) cancer operation.

I came to talk to Feynman regularly, more or less once a month, during my Tolman Fellowship at Caltech (which started in the Fall of 1981), and a bit less often for a few years afterwards. I have also sat occasionally in the class on physics and computation he taught with John Hopfield. And I remember discussing with him (among other subjects) the connection between physics, information and computation. In fact, this was a recurring theme. For me, it became somewhat of an obsession early on — I really liked the universality of Turing machines, the halting problem, and the algorithmic view of information. While I was in Austin the fascination with these ideas and their possible relevance for physics was reinforced under the influence of John Wheeler. Which brings me, at long last, to the *algorithmic information content*, *measurements*, and various *thermodynamic demons* which probe the utility of acquired information.

## 22.2 Algorithmic Information Content, Measurement and the Second Law

Maxwell's demon — a hypothetical intelligent entity capable of performing measurements on a thermodynamic system and using their outcomes to extract useful work — was considered a threat to the validity of the second law of thermodynamics for over a century [2, 3]. Feynman was fascinated with the subject, and his discussion of ratchet and pawl [1] banished forever the "unintelligent" trapdoor version of the demon by clarifying and updating the influential argument put forward by Smoluchowski [4] much earlier, and in a rather different setting [1]. However, Smoluchowski's trapdoor carries out no (explicit) measurements. Therefore, trapdoors and ratchets and pawls can be analysed without reference to information [1, 4].

The complete Maxwell's demon should be able to measure, and it (...?; he? she?!) should be of course intelligent. Smoluchowski's trapdoor does not fit this bill. Measurements were incorporated into the discussion by Szilard [6], Landauer [7], and Bennett [8] who have argued, in a setting involving ensembles of demons, that the acquisition of information is only possible when the demon's memory is repeatedly erased, to prepare it for the new data. The *cost of erasure* eventually offsets whatever thermodynamic advantages of the demon's information gain might offer. This point (which has come to be known as "Landauer's principle") is now widely recognised as a key ingredient of thermodynamic demonology. This originally classical reasoning has been since extended to quantum physics [9, 10], and may even be experimentally testable [11].

However, the widespread fascination with Maxwell's demon is ultimately due to its intelligence. A demon will record a specific outcome of the measurement, and — using its intelligence — will try to make an optimal decision about the best possible action, which would maximize the work extracted from a given recorded phase space configuration. This is very much the course of action we take (although, fortunately for us, in a far-from-equilibrium setting). How can one convince an intelligent demon that, all cleverness notwithstanding, its attempts at defeating the second law are doomed? This is hard to accomplish at the level of ensembles: Each demon knows

[1] Smoluchowski's original trapdoor was a hole surrounded by hairs combed so that they all come out on the same side of the partition between the two chambers (rather than a real trap*door*). Naively, this arrangement of hairs should favor molecules passing in the direction in which the hair is combed, and impede the reverse motion. Smoluchowski pointed out that thermal fluctuations will "ruffle the hair" and make this arrangement ineffective as a rectifier of fluctuations when the whole system is at the same fixed temperature. Numerical simulations of trapdoors confirm these conclusions [5]. They also show why our intuition based on far-from-equilibrium behavior of trapdoors can be easily misled.

nothing but its own record, and need not care about the other members of "its ensemble" that have found out something else in their measurements — it will find out its solution to its own problem.

The ultimate analysis of Maxwell's demon must involve a definition of intelligence, a characteristic which has been all too consistently banished from discussions of demons carried out by physicists. On the other hand, intelligence has been — since Turing and his famous test — often invoked in the discussions of computer scientists. To convince ourselves (and the intelligent demon) of the limits imposed by the second law we shall, following Ref. [12], adopt an operational definition of intelligence which arose in the context of the theory of computation. It is based on the so-called Church–Turing thesis [13] — which in effect formalizes Turing's expectations about the "mental" capabilities of computers and states that intelligence is equivalent to the same kind of information processing that is in principle implementable on a universal computer.

Using the Church–Turing thesis as a point of departure, the present author has demonstrated that even this intelligent threat to the second law can be eliminated — the original "smart" Maxwell's demon can be exorcized. This is easiest to establish when one recognizes that the net ability of demons to extract useful work from systems depends on the sum of measures of two distinct aspects of disorder [12]:

(i) The usual *statistical entropy* given by:

$$H(\rho) = -Tr \rho \lg \rho \tag{22.1}$$

where $\rho$ is the density matrix of the system, determines the ignorance of the observer.

(ii) The *algorithmic information content:* [14–20]

$$K(\rho) = |p^*_\rho| \tag{22.2}$$

is given by the size ("|...|"), in bits, of the shortest algorithm ($p^*$) which, for an "operating system" of a given Maxwell's demon, can reproduce the detailed description ($\rho$) of the state of the system. $K(\rho)$ quantifies the cost of storing of the acquired information, which is related to the randomness inherent in the state of the system revealed by the measurement.

The Church–Turing thesis enters in this second algorithmic ingredient, as it involves an assumption that the intellectual abilities of Maxwell's demons can be regarded as equivalent to those of a universal Turing machine: It is assumed that demons can execute programs (such as $p^*_\rho$) to reconstruct records of past measurements out of their optimally compressed versions, or to carry out other logical operations in optimizing performance. Algorithmic information content provides a well-defined measure of the storage space required to register the known characteristics of the system.

*Physical entropy* [12] is the sum of the statistical entropy and of the algorithmic information content:

$$\mathcal{Z}(\rho) = H(\rho) + K(\rho) \tag{22.3}$$

Above, it is assumed that the base for the logarithm in Eq. 22.1 is the same as the size of the alphabet used by the computer which constitutes the operating system of the Maxwell's demon. In practice, it is customary and convenient to employ a binary alphabet, so that both $H(\rho)$ and $K(\rho)$ are measured in bits.

In order to appreciate the physical significance of the algorithmic randomness contribution, it is useful to discuss the behavior of $H$, $K$ and $\mathcal{Z}$ in the course of measurements and to follow the operations of the engines controlled by demons. In short, the two measures turn out to be complementary — not in the quantum sense, but a bit like kinetic and potential energy — and their sum is, on the average, conserved under optimal measurements carried out on an equilibrium ensemble. Analysis which leads to this conclusion was carried out by this author [10, 12] and extended by Caves [21]. Below we offer only a brief summary of the salient points.

In the course of ideal measurement on an equilibrium ensemble the decrease of ignorance is, on the average, compensated for by the increase of the size of the minimal record [12]:

$$\Delta H \simeq - < \Delta K > . \tag{22.4}$$

Consequently, physical entropy $\mathcal{Z}$ plays a role analogous to a constant of motion. The transformation of the state of the system is now, however, brought about by a *demonical* (rather than *dynamical*) evolution, by the act of acquisition of information. This "conservation law" can be demonstrated within the context of the algorithmic theory of information [10, 12, 21, 22]. However, its validity can be traced to coding theory [12, 21–23]. According to the noiseless coding theorem of Shannon [23], the minimal size $\mathcal{L}$ of the message required to encode information which corresponds to a decrease of entropy by $\Delta H$ is, on the average over all of the messages, bounded by:

$$\Delta H \leq \mathcal{L} < \Delta H + 1$$

This inequality is used in the proof of Eq. 22.4 and is ultimately responsible for the constancy of the physical entropy $\mathcal{Z}$ in the course of the measurement [12, 21].

The role of $\mathcal{Z}$ in determining the efficiency of demon-operated engines is the ultimate reason for regarding $\mathcal{Z}$ as physical entropy. For the total amount of work which can be extracted from a physical system in contact with a heat reservoir of temperature $T$ in the course of a cycle which involves a measurement ($\rho \rightarrow \rho_i$) and isothermal expansion ($\rho_i \rightarrow \rho$) can be made as large as, but no larger than:

$$\Delta W = k_B T(\mathcal{Z}(\rho) - \mathcal{Z}(\rho_i)) \tag{22.5}$$

To justify this last assertion, I shall appeal to Landauer's principle [7] which formalizes earlier remarks of Szilard [6] and states that erasure of one bit of information from the memory carries a thermodynamic price of $k_BT$. Although Landauer's principle assigns a definite price to the storage of information, this price need not be paid right away: A demon with a large unused memory can continue to carry out measurements as long as it has room to store information. However, such a demon poses no threat to the second law: Its operation is not truly cyclic. In effect, it operates by employing its initially empty memory as a low temperature (zero entropy) heat sink.

Erasure of the results of used up measurements carries a price tag of

$$\Delta W^- = T < (K(\rho_i) - K(\rho) > , \qquad (22.6a)$$

which must be subtracted from the gain of useful work

$$\Delta W^+ = T(H(\rho) - H(\rho_i)) , \qquad (22.6b)$$

to obtain the net work extracted by the demon. This immediately justifies Eq. 22.5. The hybrid $\mathcal{Z}$ is the physical entropy which provides the demon with an individual, personal measure of the potential for thermodynamic gains due to the information in its possession. It also demonstrates that a demon operating on a system in thermodynamic equilibrium will never be able to threaten the second law, for the ensemble average of $\mathcal{Z}$ is at best conserved, so that $< \Delta \mathcal{Z} > \leq 0$ in course of the process of acquisition of information.

## 22.3 Physical Entropy and the Demon of Choice

This last assertion is, however, justified only if the demon is forced to complete each measurement-initiated cycle. One can, by contrast, imagine a *demon of choice*, an intelligent and selective version of Maxwell's demon, who carries out to completion only those cycles for which the initial state of the system is sufficiently nonrandom (concisely describable, or *algorithmically simple*) to allow for a brief compressed record (small $K(\rho)$). This strategy appears to allow the demon to extract a sizeable work ($\Delta W^+$) at a small expense ($\Delta W^-$). Moreover, if the measurements can be reversibly undone, then the ones with disappointing outcomes could be reversed at no cost. Such demons would still threaten the second law, even if the threat is somewhat more subtle than in the case of Smoluchowski's trapdoor.

Caves [22] has considered and partially exorcised such a demon of choice by demonstrating that in any case the net gain of work cannot exceed $k_BT$ per measurement. Thus, the demons would be, at best, limited to exploiting thermal fluctuations. Moreover, in a comment [24] on Ref. [22] it was noted that taking advantage of such fluctuations is not really possible. Here I shall demonstrate that the only decision-making process free of inconsistencies necessarily leaves in the observer's (demon's) memory a "residue" which requires eventual erasure. The least cost of erasure of this residue is just enough to restore the validity of the second law. The aim of this paper is to make this argument (first put forward by this author at the meeting of the *Complexity, Entropy, and the Physics of Computation* network of the Santa Fe Institute in April of 1990) more carefully and more precisely.

To focus on a specific example consider *Gabor's engine* [25] illustrated in Fig. 22.1. There, the unlikely but profitable fluctuation occurs whenever the gas molecule is found in the small compartment of the engine. The amount of extractable work is:

$$\Delta W_p^+ = k_B T \ \lg(L/\ell) \tag{22.7}$$

The expense (measured by the used up memory) is only:

$$\Delta W^- = k_B T, \tag{22.8}$$

so that the net gain of work per each successful cycle is:

$$\Delta W_p = k_B T \ (\lg(L/\ell) - 1) \tag{22.9}$$

The more likely "uneconomical" cycles would allow a gain of work:

$$\Delta W_u^+ = k_B T \ \lg L/(L-\ell) \ , \tag{22.10}$$

so that the cost of memory erasure (still given by Eq. 22.8) outweighs the profit, leaving the net gain of work:

$$\Delta W_u = -k_B T(1 - \lg L/(L-\ell)). \tag{22.11}$$

When each measurement is followed by the extraction and erasure routine, the averaged net work gain per cycle is negative (i.e., it becomes a loss):

$$< \Delta W >= \frac{\ell}{L}\Delta W_p + \frac{L-\ell}{L}\Delta W_u = -k_B T[1 + (\frac{\ell}{L}\lg\frac{\ell}{L} + \frac{L-\ell}{L}\lg\frac{L-\ell}{L})] \tag{22.12}$$

The break even point occurs for the case of Szilard's engine [6], where the partition divides the container in half. In the opposite limit, $\ell/L << 1$, almost every measurement leads to an unsuccessful case which results in a negligible amount of extracted work but undiminished cost of erasure per cycle.

The design of the demon of choice attempts to capitalize on precisely this otherwise unprofitable limit by *undoing* all of the likely (and unprofitable) measurements at no thermodynamic cost, thus avoiding the necessity for erasure of the unused outcomes. It is important to emphasize that a measurement of the thermodynamic quantities can be indeed undone at no cost: A prejudice that measurement must be thermodynamically expensive goes back at least to the ambiguities in the original paper of Szilard [6] (who has hinted at, but failed to clearly identify erasure as the only thermodynamically expensive part of the measuring process), and was further reinforced by the popular (but incorrect) discussion of Brillouin [26]. Fig. 22.2 demonstrates how to carry out a measurement on a particle in the Gabor's engine (such measurement becomes reversible when the operations indicated are carried out infinitesimally slowly).

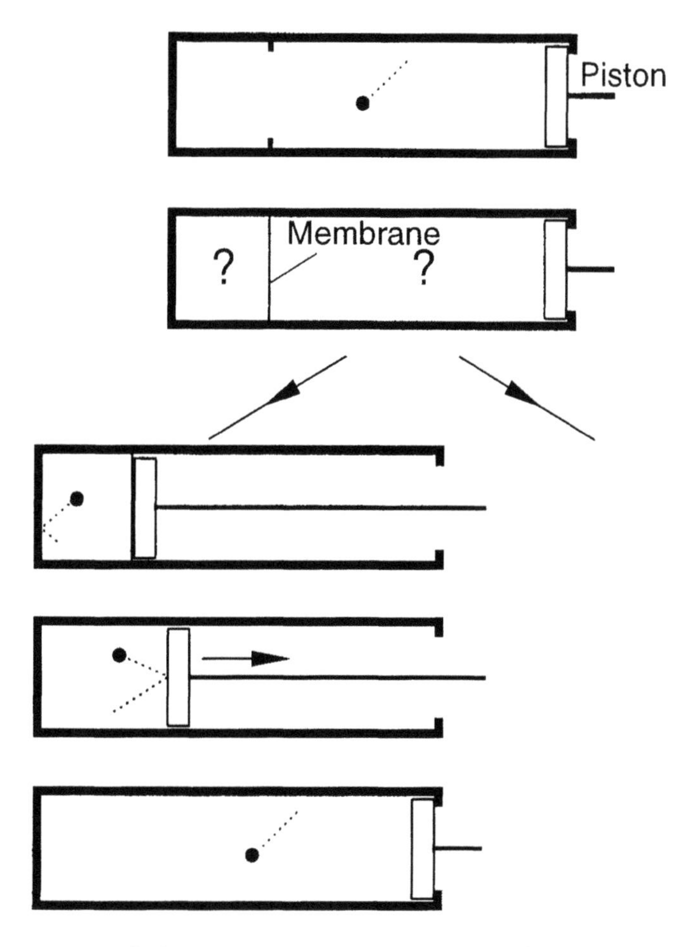


Fig. 22.1. Gabor's engine[25]. See text for the standard operating procedure. The decision between the two branches (of which only one — the profitable one — is shown) can be made reversibly with the help of the device shown in Fig. 22.2.

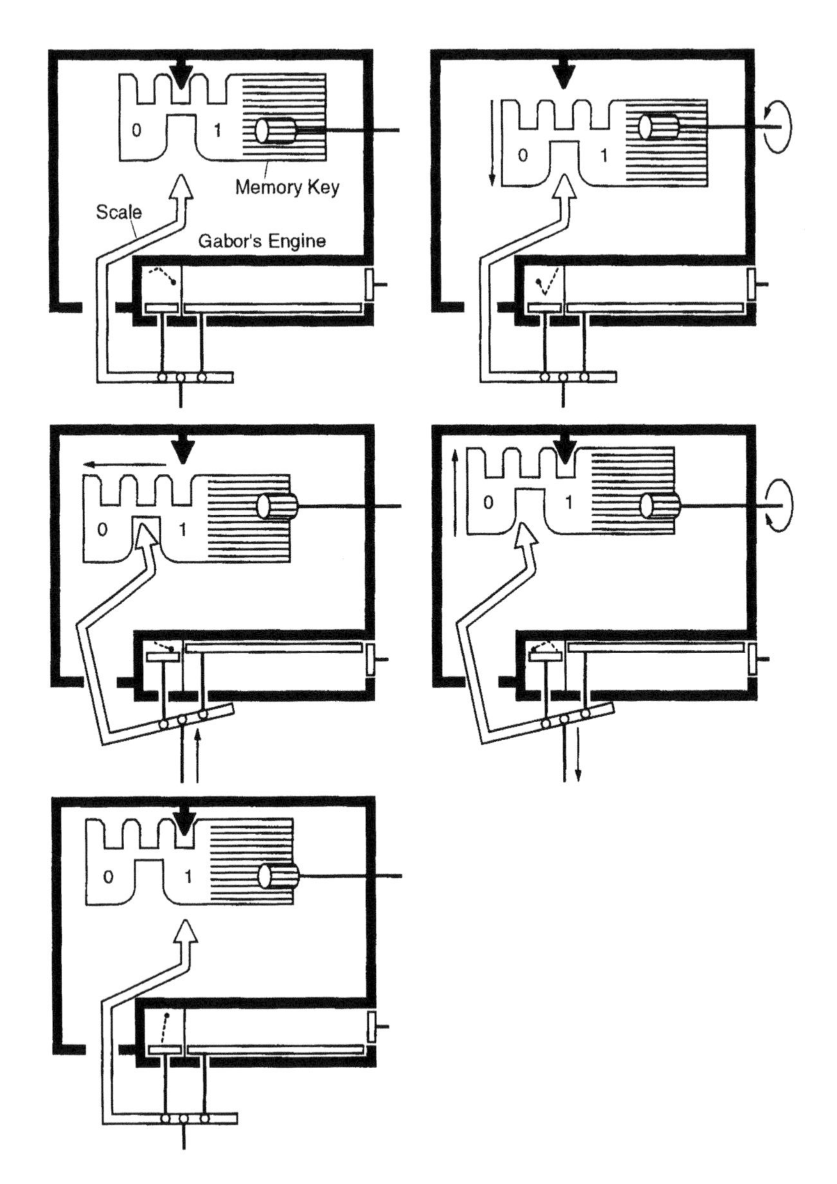


Fig. 22.2. Blueprint of a reversible measuring device for Gabor's engine. The measurements can be done (or undone) by turning the crank on the right in the appropriate direction and pushing in or pulling out the "scale". Thermodynamic reversibility is achieved in the limit of an infinitesimally slow operation. Faster controlled-not like measurements can be carried out on a dynamical timescale by implementing the unitary evolution given by Eq. 22.14. The design shown above is similar to the Szilard's engine contraption devised in Ref. [28].

## 22.4 Measurements and Decisions

The purpose of the measurement is to establish a correlation between the state of the system and the record — the state of the few relevant bits of memory. In the context of this paper we shall focus on the measurements which correlate memory with a cell in the phase space or a subspace of the Hilbert space of the system (corresponding to the projection operator $P_i$). In concert with the usual requirements I shall demand that the collection $\{P_i\}$ of all the measurements be mutually exclusive $(Tr(P_i, P_j) = 0)$, and exhaustive $(\Sigma_i P_i = 1)$. To avoid problems associated with quantum measurements we shall also demand that the measured observables should commute with the density matrix of the measured system $[P_i, \rho_S] = 0$. Thus, we shall allow for the best case [9] (from the demon's point of view), with no additional thermodynamic inefficiencies associated with the reduction of the state vector introduced into quantum measurement through decoherence [10, 11, 28–31].

A measurement performed by the demon, when viewed from the outside, results in the correlation between the state of the system (i.e. location of the particle in the Gabor's engine) and the state of the demon's memory. The total entropy can be prevented from increasing, as the only requirement for a successful measurement is to convert initial density matrix of the combined system-demon:

$$\rho_{SD}^{(o)} = \rho_S \times \rho_D^{(o)} = (\Sigma_i p_i P_i) \times \rho_D^{(o)} \tag{22.13a}$$

into the correlated [9, 10, 28–31]:

$$\rho_{SD} = \Sigma_i p_i (P_i \times \rho_D^{(i)}) \tag{22.13b}$$

Above, we have implicitly assumed that the measurement is exhaustive in the sense that the further refinements will reveal uniform probability distribution within the partitions defined by $P_i$. This need not be the case — it is straightforward to generalize the above formulae to the case when the different memory states of the demon are correlated with density matrices of the system. In any case, the entropies of $\rho_D^{(i)}$ and $\rho_D^{(o)}$ can, in principle, be the same: For, there exists a unitary *controlled-not* - like evolution operator:

$$U = \Sigma_i P_i \times (|\delta_i >< \delta_o| + |\delta_o >< \delta_i|) \tag{22.14}$$

with $|\delta_i >$ and $|\delta_o >$ defined by $\rho_D^{(i)} = |\delta_i >< \delta_o| \rho_D^{(o)}$, providing that $\rho_D^{(i)}$ correspond to distinguishable (orthogonal) memory states of the demon — a natural requirement for a successful measurement.

The statistical entropy of the system-demon combination is obviously the same before and after measurement, as, by construction of $U$, $H(\rho_D^{(i)}) = H(\rho_D^{(o)})$. Moreover, the measurement is obviously reversible: Applying the unitary evolution operator, Eq. 22.14, twice, will restore the pre-measurement situation.

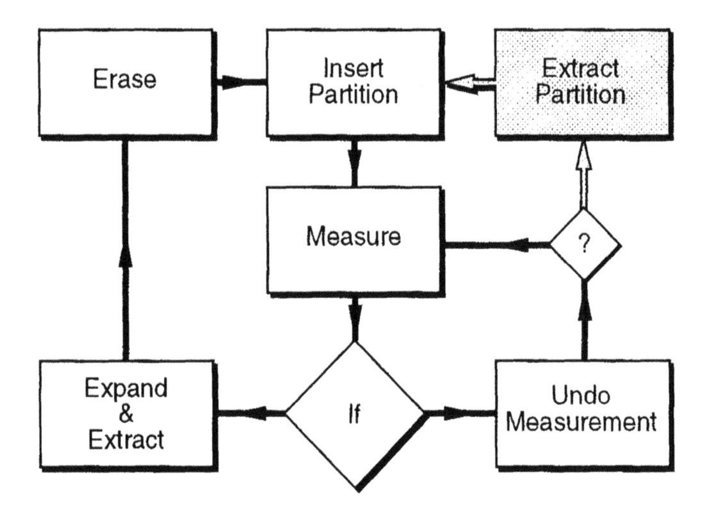


Fig. 22.3. Decision flowchart for the demon of choice. The branch on the left is profitable (and it is followed when the particle is "caught" in the small left chamber, see Fig. 22.1). The branch on the right is unprofitable, and as it is explained in the text in more detail, the demon of choice cannot be "saved" by reversing only the unprofitable measurements.

From the viewpoint of the outside observer, the measurement leads to a correlation between the system and the memory of the demon: The ensemble averaged increase of the ignorance about the content of demon's memory;

$$\Delta H_D = H(\rho_D) - H(\rho_D^{(o)}) = -\Sigma_i p_i \lg p_i \ , \qquad (22.15)$$

(where $\rho_D = Tr_S \rho_{SD}$ and $H(\rho) = -Tr \rho \lg \rho$) is compensated for by the increase of the mutual information defined as;

$$I_{SD} = H(\rho_D) + H(\rho_S) - H(\rho_{SD}), \qquad (22.16)$$

so that $\Delta H_D = \Delta I_{SD}$ (see Refs. [29] and [33] for the Shannon and algorithmic versions of this discussion in somewhat different settings).

From the viewpoint of the demon the acquired data are definite: The outcome is some definite demon state $\rho_D^{(n)}$ corresponding to the memory state $n$, and associated with the most concise record — increase of the algorithmic information content — given by some $\Delta K(n) = K(\rho_S^{(n)}) - K(\rho_S^{(o)})$.

The demon of choice would now either; (i) proceed with the expansion, extraction and erasure, providing that his estimate of the future gain:

$$\Delta W = k_B T(\Delta H - \Delta K) = k_B T \Delta \mathcal{Z} \qquad (22.17)$$

was positive, or, alternatively; (ii) undo the measurement at no cost, providing that $\Delta W < 0$. An algorithm that attempts to implement this strategy for the case of Gabor's engine is illustrated in Fig. 22.3. To see why this strategy will not work, we first note that the demon of choice threatens the second law only if its operation is cyclic — that is, it must be possible to implement the algorithm without it coming to an inevitable halt.

There is no need to comment on the left-hand side part of the cycle: it starts with the insertion of the partition. Detection of a particle in the left-hand side compartment is followed by the expansion of the partition (converted into a piston) and results in extraction of $\Delta W_p^+$, Eq. 22.7, of work. Since the partition was extracted, the results of the measurement must be erased (to prepare for the next measurement) which costs $k_B T$ of useful work, so that the gain per useful cycle is given by Eq. 22.9. The partition can be now reinserted and the whole cycle can start again.

There is, however, no decision procedure which can implement the goal of the right-hand side of the tree. The measurement can be of course undone. The demon — after undoing the correlation — no longer knows the location of the molecule inside the engine. Unfortunately for the demon, this does not imply that the state of the engine has also been undone. Moreover, the demon with empty memory will immediately proceed to do what demons with empty memory always do: It will measure. This action is an "unconditional reflex" of a demon with an empty memory. It is inevitable, as the actions of the demon must be completely determined by its internal state, including the state of its memory. (This is the same rule as for Turing machines.) But the particle in the Gabor's engine is still stuck on the unprofitable side of the partition. Therefore, when the measurement is repeated, it will yield the same disappointing result as before, and the demon will be locked forever into the measure - unmeasure "two-step" within the same unprofitable branch of the cycle by its algorithm, which compels it to repeat two controlled-not like actions, Eq. 22.14, which jointly amount to an identity.

This vicious cycle could be interrupted only if the decision process called for extraction and reinsertion of the partition *before* undoing the measurement (and thus causing the inevitable immediate re-measurement) in the unprofitable right branch of the decision tree. Extraction of the partition before the measurement is undone increases the entropy of the gas by $k_B[\lg(L-\ell)/L]$ and destroys the correlation with the demon's memory, thus decreasing the mutual information: The molecule now occupies the whole volume of the engine. Moreover it occurs with no gain of useful work. Consequently, reversibly undoing the measurement *after* the partition is extracted is no longer possible: The location on the decision tree (extracted partition, "full" memory) implicitly demonstrates that the measurement has been carried out and that it has revealed that the molecule was in the unprofitable compartment — it can occurr only in the right hand branch of the tree.

The opening of the partition has resulted in a free expansion of the gas, which squandered away the correlation between the state of the gas and the state of the memory of the demon. Absence of the correlation eliminates the possibility of undoing the measurement. Thus, now erasure is the only remaining option. It would have to be carried out before the next measurement, and the price of $k_BT$ per bit would have to be paid [6, 7].

One additional strategy should be explored before we conclude this discussion: The demon of choice can be assumed to have a large memory tape, so that it can put off erasures and temporarily store the results of its $\mathcal{N}$ measurements. The tape would then contain $\sim \mathcal{N}\cdot(\ell - L)/L$ 0's (which we shall take to signify an unprofitable outcome) and $\sim \mathcal{N}\ell/L$ 1's. In the limit of large $\mathcal{N}$ ($\mathcal{N}\ell/L >> 1$) the algorithmic information content of such a "sparse" binary sequence $s$ is given by [14–20]:

$$K(s) \simeq -\mathcal{N}[\frac{\ell}{L}\lg\frac{\ell}{L} + \frac{L-\ell}{L}\lg\frac{L-\ell}{L}] \tag{22.18}$$

Moreover, a binary string can be, at least in principle, compressed to its minimal record ($s^*$ such that $K(s) = |s^*|$) by a reversible computation [12]. Hence, it is possible to erase the record of the measurements carried out by the demon at a cost of no less than

$$< \Delta W^- >= k_BT[K(s)/\mathcal{N}] \ . \tag{22.19}$$

Thus, if the erasure is delayed so that the demon can attempt to minimize its cost before carrying it out, it can at best break even: The $-k_BT$ in Eq. 22.12 is substituted by the $< \Delta W^- >$, Eq. 22.19, which yields:

$$< \Delta W >=< \Delta W^+ > + < \Delta W^- >= 0. \tag{22.20}$$

It is straightforward to generalize this lesson derived on the example of Gabor's engine to other situations. The essential ingredient is the "noncommutativity" of the two operations: "undo the measurement" can be reversibly carried out only before "extract the partition." The actions of the demon are, by the assumption of the Church–Turing thesis, completely determined by its internal state, especially its memory content. Demons are forced to make useless re-measurements. Santayana's famous saying that "*those who forget their history are doomed to relive it*" applies to demons with a vengance! For, when the demon forgets the measurement outcome, it will repeat the measurement and remain stuck forever in the unprofitable cycle. One could consider more complicated algorithms, with additional bits and instructions on when to measure, and so on. The point is, however, that all such strategies must ultimately contain explicit or implicit information about the branch on which the demon has found itself as a result of the measurement. Erasure of this information carries a price which is on the average no less than the "illicit" gains which would violate the second law.

## 22.5 Conclusions

The aim of this paper was to exorcise the demon of choice — a selective version of Maxwell's demon which attempted to capitalize on large thermal fluctuations by reversibly undoing all of the measurements which did not reveal the system to be sufficiently far from equilibrium. I have demonstrated that a deterministic version of such a demon fails, as no decision procedure is capable of both (i) reversibly undoing the measurement, and, also, of (ii) opening the partitions inserted prior to the measurement to allow for energy extraction following readoff of the outcome.

Our discussion was phrased — save for an occassional reference to density matrices, Hilbert spaces, etc. — in a noncommital language, and it is indeed equally applicable in the classical and quantum contexts. As was pointed out already some time ago [9, 10], the only difference arises in the course of measurements. Quantum measurements are typically accompanied by a "reduction of the state vector". It occurs whenever an observer measures observables that are not co-diagonal with the density matrix of the system. It is a (near) instantaneous process [34], which is nowadays understood as a consequence of decoherence and einselection [19, 28, 30–34]. The implications of this difference are minor from the viewpoint of the threat to the second law posed by the demons (although decoherence is paramount for the discussion of the interpretation of quantum theory). It was noted already some time ago that decoherence (or, more generally, the increase of entropy associated with the reduction of the state vector) is not necessary to save the second law [9]. Soon after the algorithmic information content entered the discussion of demons [12, 21] it was also realised that the additional cost decoherence represents can be conveniently quantified using the "deficit" in what this author knew then as the 'Groenewold–Lindblad inequality' [35, 36], and what is now more often (and equally justifiably) called the 'Holevo quantity [37];

$$\chi = H(\rho) - \sum_i p_i H(\rho_S^{(i)}) , \tag{22.21}$$

which is a measure of the entropy increase due to the "reduction". The two proofs [36, 37] involving essentially the same quantity have appeared almost simultaneously, independently, and were motivated by — at least superficially — quite different considerations.

We shall not repeat these discussions here in detail. There are however several independently sufficient reasons not to worry about decoherence in the demonic context which deserve a brief review. To begin with, decoherence cannot help the demon as it only adds to the "cost of doing business". And the second law is apparently safe even without decoherence [9]. Moreover, especially in the context of Szilard's or Gabor's engines, decoherence is unlikely to hurt the demon either, since the obvious projection operators to use in Eq. 22.14 correspond to the particle being on the left (right) of the partition, and are likely to diagonalise the density matrix of the system in contact with a typical environment [9] (heat bath). (Superpositions of states corresponding to such obvious measurement outcomes are very Schrödinger cat-like, and, therefore, unstable on the decoherence timescale [34].) Last not least, even if demon for some odd reason started by measuring some observable which does not commute with the density matrix of the system decohering in contact with the heat bath environment, it should be able to figure out what's wrong and learn after a while what to measure to minimise the cost of erasure (demons are supposed to be intelligent, after all!).

So decoherence is of secondary importance in assuring validity of the second law in the setting involving engines and demons: Entropy cannot decrease already without it! But decoherence can (and often will) add to the *measurement* costs, and the cost of decoherence is paid "up front", during the measurement (and not really during the erasure, although there may be an ambiguity there — see a quantum calculation of erasure-like process of the consequences of decoherence in Ref. [38]). However, in the context of dynamics decoherence is the ultimate cause of entropy production, and, thus, the cause of the algorithmic arrow of time [33]. Moreover, there are intriguing quantum implication of the interplay of decoherence and (algorithmic) information that follow: Discussions of the interpretational issues of quantum theory are often conducted in a way which implicitly separates the information observers have about the state of the systems in the "rest of the Universe" from their own physical state — their identity. Yet, as the above analysis of the observer-like demons demonstrates, there can be *no information without representation.* The observer's state (or, for that matter, the state of its memory) determines its actions and should be regarded as an ultimate description of its identity. So, to end with one more "deep truth" *existence* (of the observers state, and, especially, of the state of its memory) *precedes the essence* (observer's information, and, hence their future actions).

## Acknowledgements

I have benefited from discussion on this subject with many, including Andreas Albrecht, Charles Bennett, Carlton Caves, Murray Gell-Mann, Chris Jarzynski, Rolf Landauer, Seth Lloyd, Michael Nielsen, Bill Unruh, and John Wheeler, who, in addition to stimulating the initial interest in matters concerning physics and information, insisted on my monthly dialogues with Feynman. This has led to one more "adventure with a curious character": In the Spring of 1984 I participated in the "Quantum Noise" program at the Institute for Theoretical Physics, UC Santa Barbara. It was to end with a one-week conference on various relevant quantum topics. One of the organisers (I think it was Tony Leggett), aware of my monthly escapades at Caltech, and of Feynman's (and mine) interests in quantum computation asked me whether I could ask him to speak. I did, and Feynman immediately agreed.

The lectures were held in a large conference room at the campus of the University of California at Santa Barbara. For the "regular speakers" and for most of the talks (such as my discussion of the decoherence timescale which was eventually published as Ref. [34]) the room was filled to perhaps a third of the capacity. However, when I walked in in the middle of the afternoon coffee break, well in advance of Feynman's talk, the room was already nearly full, and the air was thick with anticipation. A moment after I sat down in one of the few empty seats, I saw Feynman come in, and quietly take a seat somewhere in the midst of the audience. More people came in, including the organisers and the session chairman. The scheduled time of his talk came... and went. It was five minutes after. Ten minutes. Quarter of an hour. The chairman was nervous. I did not understand what was going on — I clearly saw Feynman's long grey hair and an occasional flash of an impish smile a few rows ahead.

Then it struck me: He was just being "a curious character", curious about what will happen... He did what he had promised — showed up for his talk on (or even before) time, and now he was going to see how the events unfold.

In the end I did the responsible thing: After a few more minutes I pointed out the speaker to the session chairman (who was greatly relieved, and who immediately and reverently led him to the speaker's podium). The talk (with the content, more or less, of Ref. [39]) started only moderately behind the schedule. And I was immediately sorry that I did not play along a while longer — I felt as if I had given away a high-school prank before it was fully consummated!

## References


[1] R. P. Feynman, R. B. Leighton, and M. Sands, *The Feynman Lectures on Physics*, vol. 1, pp 46.1 - 46.9 (Addison-Wesley, Reading, Massachussets, 1963).

[2] J. C. Maxwell, *Theory of Heat*, 4th ed., pp. 328-329 (Longman's, Green, & Co., London 1985).

[3] H. S. Leff and A. F. Rex, *Maxwell's Demon: Entropy, Information, Computing* (Princeton University Press, Princeton, 1990).

[4] M. Smoluchowski in *Vortgäge über die Kinetische Theorie der Materie und der Elektizität* (Teubner, Leipzig 1914).

[5] P. Skordos and W. H. Zurek, "Maxwell's Demons, Rectifiers, and the Second Law" *Am. J. Phys.* **60**, 876 (1992).

[6] L. Szilard, *Z. Phys.* **53** 840 (1929). English translation in Behav. Sci. **9**, 301 (1964), reprinted in *Quantum Theory and Measurement*, edited by J. A. Wheeler and W. H. Zurek (Princeton University Press, Princeton, 1983); Reprinted in Ref. 3.

[7] R. Landauer, *IBM J. Res. Dev.* **3**, 183 (1961); Reprinted in Ref. 3.

[8] C. H. Bennett, *IBM J. Res. Dev.* **17** 525 (1973); C. H. Bennett, *Int. J. Theor. Phys.* **21**, 905 (1982); C. H. Bennett, *IBM J. Res. Dev.*, **32**, 16-23 (1988); Reprinted in Ref. 3.

[9] W. H. Zurek, "Maxwell's Demons, Szilard's Engine's, and Quantum Measurements", Los Alamos Preprint LAUR 84-2751 (1984); pp. 151-161 in *Frontiers of Nonequilibrium Statistical Physics*, G. T. Moore and M. O. Scully, eds., (Plenum Press, New York, 1986); reprinted in Ref. 3.

[10] For a quantum treatement which employs the Groenewold-Lindblad/Holevo inequality and uses the "deficit" $\chi$ in that inequality to estimate of the price of decoherence, see W. H. Zurek, pp 115-123 in the *Proceedings of the $3^{rd}$ International Symposium on Foundations of Quantum Mechanics*, S. Kobayashi *et al.*, eds. (The Physical Society of Japan, Tokyo, 1990).

[11] S. Lloyd, *Phys. Rev.* **A56**, 3374-3382 (1997).

[12] W. H. Zurek, *Phys. Rev.* **A40**, 4731-4751 (1989); W. H. Zurek, *Nature* **347**, 119-124 (1989).

[13] For an accessible discussion of Church–Turing thesis, see D. R. Hofstadter, Gödel, Escher, Bach, chapter XVII (Vintage Books, New York, 1980).

[14] R. J. Solomonoff, *Inf. Control* **7**, 1 (1964).

[15] A. N. Kolmogorov, *Inf. Transmission* **1**, 3 (1965).

[16] G. J. Chaitin, *J. Assoc. Comput. Mach.* **13**, 547 (1966).

[17] A. N. Kolmogorov, *IEEE Trans. Inf. Theory* **14**, 662 (1968).

[18] G. J. Chaitin, *J. Assoc. Comput. Mach.* **22**, 329 (1975); G. J. Chaitin, *Sci. Am.* **23**(5), 47 (1975).

[19] A. K. Zvonkin and L. A.Levin, *Usp. Mat. Nauk.* **25**, 602 (1970).

[20] A. K. Zvonkin and L. A. Levin, *Usp. Mat. Nauk.* **25**, 602 (1970).

[21] C. M. Caves, "Entropy and Information", pp. 91-116 in *Complexity, Entropy, and Physics of Information*, W. H. Zurek, ed. (Addison-Wesley, Redwood City, CA, 1990).

[22] C. M. Caves, *Phys. Rev. Lett.* **64**, 2111-2114 (1990).

[23] W. Shannon and W. Weaver, *The Mathematical Theory of Communication* (University of Illinois Press, Urbana, 1949).

[24] C. M. Caves, W. G. Unruh, and W. H. Zurek, *Phys. Rev. Lett.*, **65**, 1387 (1990).

[25] D. Gabor, *Optics* **1**, 111-153 (1964).

[26] L. Brillouin, *Science and Information Theory*, 2nd ed. (Academic, London, 1962).

[27] C. H. Bennett, *Sci. Am.* **255** (11), 108 (1987).

[28] W. H. Zurek, *Phys. Rev.* **D24**, 1516 (1981); *ibid.* **D26**, 1862 (1982); *Physics Today* **44**, 36 (1991).

[29] W. H. Zurek, "Information Transfer in Quantum Measurements: Irreversibility and Amplification"; pp. 87-116 in *Quantum Optics, Experimental Gravitation, and Measurement Theory*, P. Meystre and M. O. Scully, eds. (Plenum, New York, 1983).

[30] E. Joos and H. D. Zeh, *Zeits. Phys.* **B59**, 223 (1985).

[31] D. Giulini, E. Joos, C. Kiefer, J. Kupsch and H. D. Zeh, *Decoherence and the Appearance of a Classical World in Quantum Theory* (Springer, Berlin, 1996).

[32] W. H. Zurek, *Progr. Theor. Phys.* **89**, 281-312 (1993).

[33] W. H. Zurek, in the *Proceedings of the Nobel Symposium 101 'Modern Studies in Basis Quantum Concepts and Phenomena'*, to appear in *Physica Scripta*, in press quant-ph/9802054.

[34] W. H. Zurek, "Reduction of the Wavepacket: How Long Does it Take?" Los Alamos preprint LAUR 84-2750 (1984); pp. 145-149 in the *Frontiers of Nonequilibrium Statistical Physics: Proceedings of a NATO ASI held June 3-16 in Santa Fe, New Mexico*, G. T. Moore and M. O. Scully, eds. (Plenum, New York, 1986).

[35] H. J. Groenewold, *Int. J. Theor. Phys.* **4**, 327 (1971).

[36] G. Lindblad, *Comm. Math. Phys.* **28**, 245 (1972).

[37] A. S. Holevo, *Problemy Peredachi Informatsii* **9**, 9-11 (1973).

[38] J. R. Anglin, R. Laflamme, W. H. Zurek, and J. P. Paz, *Phys. Rev.* **D52**, 2221-2231 (1995).

[39] R. P. Feynman, "Quantum Mechanical Computers", *Optics News*, reprinted in *Found. Phys.* **16**, 507-531 (1986).

# Chapter 7

# Computation: Thermodynamics and Limits

# The Thermodynamics of Computation—a Review

**Charles H. Bennett**

*IBM Watson Research Center, Yorktown Heights, New York 10598*

*Received May 8, 1981*

Computers may be thought of as engines for transforming free energy into waste heat and mathematical work. Existing electronic computers dissipate energy vastly in excess of the mean thermal energy $kT$, for purposes such as maintaining volatile storage devices in a bistable condition, synchronizing and standardizing signals, and maximizing switching speed. On the other hand, recent models due to Fredkin and Toffoli show that in principle a computer could compute at finite speed with zero energy dissipation and zero error. In these models, a simple assemblage of simple but idealized mechanical parts (e.g., hard spheres and flat plates) determines a ballistic trajectory isomorphic with the desired computation, a trajectory therefore not foreseen in detail by the builder of the computer. In a classical or semiclassical setting, ballistic models are unrealistic because they require the parts to be assembled with perfect precision and isolated from thermal noise, which would eventually randomize the trajectory and lead to errors. Possibly quantum effects could be exploited to prevent this undesired equipartition of the kinetic energy. Another family of models may be called Brownian computers, because they allow thermal noise to influence the trajectory so strongly that it becomes a random walk through the entire accessible (low-potential-energy) portion of the computer's configuration space. In these computers, a simple assemblage of simple parts determines a low-energy labyrinth isomorphic to the desired computation, through which the system executes its random walk, with a slight drift velocity due to a weak driving force in the direction of forward computation. In return for their greater realism, Brownian models are more dissipative than ballistic ones: the drift velocity is proportional to the driving force, and hence the energy dissipated approaches zero only in the limit of zero speed. In this regard Brownian models resemble the traditional apparatus of thermodynamic thought experiments, where reversibility is also typically only attainable in the limit of zero speed. The enzymatic apparatus of DNA replication, transcription, and translation appear to be nature's closest approach to a Brownian computer, dissipating $20-100kT$ per step. Both the ballistic and Brownian computers require a change in programming style: computations must be rendered *logically* reversible, so that no machine state has more than one logical predecessor. In a ballistic computer, the merging of two trajectories clearly cannot be brought about by purely conservative forces; in a Brownian computer, any extensive amount of merging of computation paths would cause the Brownian computer to spend most of its time bogged down in extraneous predecessors of states on the intended path, unless an extra driving force of $kT\ln 2$ were applied (and dissipated) at each merge point. The mathematical means of rendering a computation logically reversible (e.g., creation and annihilation of a history file) will be discussed. The old Maxwell's demon problem is discussed in the light of the relation between logical and thermodynamic reversibility: the essential irreversible step, which prevents the demon from breaking the second law, is not the making of a measurement (which in principle can be done reversibly) but rather the logically irreversible act of erasing the record of one measurement to make room for the next. Converse to the rule that logically irreversible operations on data require an entropy increase elsewhere in the computer is the fact that a tape full of zeros, or one containing some computable pseudorandom sequence such as pi, has fuel value and can be made to do useful thermodynamic work as it randomizes itself. A tape containing an algorithmically random sequence lacks this ability.

## 1. INTRODUCTION

The digital computer may be thought of as an engine that dissipates energy in order to perform mathematical work. Early workers naturally wondered whether there might be a fundamental thermodynamic limit to the efficiency of such engines, independent of hardware. Typical of early thinking in this area was the assertion by von Neumann, quoted from a 1949 lecture (von Neumann, 1966), that a computer operating at temperature $T$ must dissipate at least $kT\ln 2$ (about $3\times 10^{-21}$ J at room temperature), "per elementary act of information, that is per elementary decision of a two-way alternative and per elementary transmittal of one unit of information." Brillouin (1962) came to a similar conclusion by analyzing a thought experiment involving detection of holes in a punched tape by photons, and argued further that the energy dissipation must increase with the reliability of the measurement, being approximately $kT\ln(1/\eta)$ for a measurement with error probability $\eta$. These conjectures have a certain plausibility, in view of the quantitative relation between entropy and information exemplified by Maxwell's demon (Szilard, 1929), and the fact that each classical degree of freedom used to store a bit of information, e.g., the charge in a capacitor, suffers from $kT$ of thermal noise energy, which seemingly would have to be overcome in order to read or manipulate the bit reliably. However, it is now known that computers can in principle do an arbitrarily large amount of reliable computation per $kT$ of energy dissipated. In retrospect, it is hardly surprising that computation, like a complex, multistep industrial process, can in principle be accomplished with arbitrarily little waste, i.e., at thermodynamic cost only marginally greater than the difference in thermodynamic potential (if any) between its input and output. The belief that computation has an irreducible entropy cost per step may have been due to a failure to distinguish sufficiently between dissipation (an irreversible net increase in entropy) and reversible transfers of entropy.

Though they are several orders of magnitude more efficient than the first electronic computers, today's computers still dissipate vast amounts of energy compared to $kT$. Probably the most conspicuous waste occurs in volatile memory devices, such as TTL flip-flops, which dissipate energy continuously even when the information in them is not being used. Dissipative storage is a convenience rather than a necessity: magnetic cores, CMOS, and Josephson junctions exemplify devices that dissipate only or chiefly when they are being switched. A more basic reason for the inefficiency of existing computers is the macroscopic size and inertia of their components, which therefore require macroscopic amounts of energy to switch quickly. This energy (e.g., the energy in an electrical pulse sent from one component to another) could in principle be saved and reused, but in practice it is easier to dissipate it and form the next pulse from new energy, just as it is usually more practical to stop a moving vehicle with brakes than by saving its kinetic energy in a flywheel. Macroscopic size also explains the poor efficiency of neurons, which dissipate about $10^{11}kT$ per discharge. On the other hand, the molecular apparatus of DNA replication, transcription, and protein synthesis, whose components are truly microscopic, has a relatively high energy efficiency, dissipating 20–100$kT$ per nucleotide or amino acid inserted under physiological conditions.

Several models of thermodynamically reversible computation have been proposed. The most spectacular are the ballistic models of Fredkin and Toffoli (1982), which can compute at finite speed with zero energy dissipation. Less spectacular but perhaps more physically realistic are the Brownian models developed by Bennett (1973; see also below) following earlier work of Landauer and Keyes (1970), which approach zero dissipation only in the limit of zero speed. Likharev (1982) describes a scheme for reversible Brownian computing using Josephson devices.

Mathematically, the notion of a computer is well characterized. A large class of reasonable models of serial or parallel step-by-step computation, including Turing machines, random access machines, cellular automata, and tree automata, has been shown to be capable of simulating one another and therefore to define the same class of computable functions. In order to permit arbitrarily large computations, certain parts of these models (e.g., memory) are allowed to be infinite or indefinitely extendable, but the machine state must remain finitely describable throughout the computation. This requirement excludes "computers" whose memories contain prestored answers to infinitely many questions. An analogous requirement for a strictly finite computer, e.g, a logic net constructed of finitely many gates, would be that it be able to perform computations more complex than those that went into designing it. Models that are reasonable in the further sense of not allowing exponentially growing parallelism (e.g., in a $d$-dimensional cellular automaton, the effective degree of parallelism is bounded by the $d$th power of time) can generally simulate one another in polynomial time and linear space (in the jargon of computational complexity, time means number of machine cycles and space means number of bits of memory used). For this class of models, not only the computability of functions, but their rough level of difficulty (e.g., polynomial vs. exponential time in the size of the argument) are therefore model independent. Figure 1 reviews the Turing machine model of computation, on which several physical models of Section 3 will be based.

For time development of a physical system to be used for digital computation, there must be a reasonable mapping between the discrete logical states of the computation and the generally continuous mechanical states of the apparatus. In particular, as Toffoli suggests (1981), distinct logical variables describing the computer's mathematical state (i.e., the contents of a bit of memory, the location of a Turing machine's read/write head) ought to be embedded in distinct dynamical variables of the computer's physical state.

## 2. BALLISTIC COMPUTERS

The recent "ballistic" computation model of Fredkin and Toffoli (1982), shows that, in principle, a somewhat idealized apparatus can compute without dissipating the kinetic energy of its signals. In this model, a simple assemblage of simple but idealized mechanical parts (hard spheres colliding with each other and with fixed reflective barriers) determines a ballistic trajectory isomorphic with the desired computation. In more detail (Figure 2), the input end of a ballistic computer consists of a "starting line," like the starting line of a horse race, across which a number of hard spheres ("balls") are simultaneously fired straight forward into the computer with precisely equal velocity. There is a ball at each position in the starting line corresponding to a binary 1 in the input; at each position corresponding to a 0, no ball is fired. The computer itself has no moving parts, but contains a number of fixed barriers ("mirrors") with which the balls collide and which cause the balls to collide with each other. The collisions are elastic, and between collisions the balls travel in straight lines with constant velocity, in accord with Newton's second law. After a certain time, all the balls simultaneously emerge across a "finish line" similar to the starting line, with the presence or absence of balls again signifying the digits of the output. Within the computer, the mirrors perform the role of the logic gates of a conventional electronic computer, with the balls serving as signals.

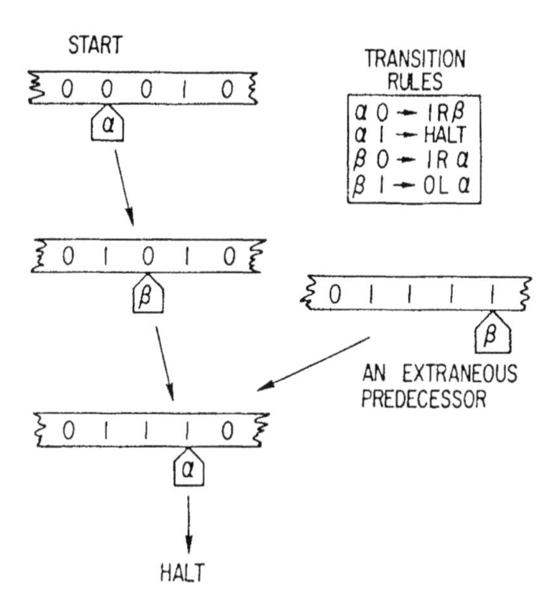


**Fig. 1.** An elementary mathematical model of computation, the Turing machine, consists of an infinite tape scanned by a movable finite automaton or "head," that can read or write one bit at a time, and shift a unit distance left or right along the tape. In order to remember what it is doing from one machine cycle to the next, the Turing machine head has a finite number of distinct internal states (here two: $\alpha$ and $\beta$). The Turing machine's behavior is governed by a fixed set of transition rules that indicate, for each combination of head state and scanned tape symbol, the new tape symbol to be written, the shift direction, and a new head state. The figure shows a short computation in which the machine has converted the input 00010, originally furnished on its tape, into the output 01110, and then halted. This Turing machine, because of its limited number of head states, can do only trivial computations; however, slightly more complicated machines, with a few dozen head states and correspondingly more transition rules, are "universal," i.e., capable of simulating any computer, even one much larger and more complicated than themselves. They do this by using the unlimited tape to store a coded representation of the larger machine's complete logical state, and breaking down each of the larger machine's machine cycles into many small steps, each simple enough to be performed by the Turing machine head. The configuration labeled "extraneous predecessor" is not part of the computation, but illustrates the fact that typical Turing machines, like other computers, often throw away information about their past, by making a transition into a logical state whose predecessor is ambiguous. This so-called "logical irreversibility" has an important bearing on the thermodynamics of computation, discussed in Section 4.

It is clear that such an apparatus cannot implement all Boolean functions: only functions that are conservative (the number of ones in the output equals the number of ones in the input) and bijective (to each output there corresponds one and only one input) can be implemented; but as Fredkin and Toffoli (1982) and Toffoli (1981) show, an arbitrarily Boolean function can be embedded in a conservative, bijective function without too much trouble.

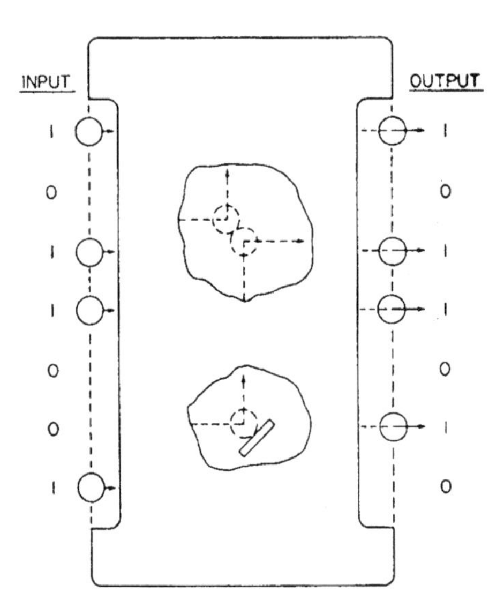


**Fig. 2.** Ballistic computer of Fredkin and Toffoli. In this example, the arrangement of mirrors inside the box is such that, when any five-bit number (here 13) is presented in the first five input positions, followed by 01 in the last two, the same five-bit number will appear in the first five output positions, followed by 01 if the number is composite, or 10 if the number is prime. The inclusion of the input as part of the output, and the use of two unlike bits to encode the desired answer, illustrate the embedding of an irreversible Boolean function into one that is reversible and conservative.

The two chief drawbacks of the ballistic computer are the sensitivity of its trajectory to small perturbations, and difficulty of making the collisions truly elastic. Because the balls are convex, small errors in their initial positions and velocities, or errors introduced later (e.g., by imperfect alignment of the mirrors) are amplified by roughly a factor of 2 at each collision between balls. Thus an initial random error of one part in $10^{15}$ in position and velocity, roughly what one would expect for billiard balls on the basis of the uncertainty principle, would cause the trajectory to become unpredictable after a few dozen collisions. Eventually the balls would degenerate into a gas, spread throughout the apparatus, with a Maxwell distribution of velocities. Even if classical balls could be shot with perfect accuracy into a perfect apparatus, fluctuating tidal forces from turbulence in the atmospheres of nearby stars would be enough to randomize their motion within a few hundred collisions. Needless to say, the trajectory would be spoiled much sooner if stronger nearby noise sources (e.g., thermal radiation and conduction) were not eliminated.

Practically, this dynamical instability means that the balls' velocities and positions would have to be corrected after every few collisions. The resulting computer, although no longer thermodynamically reversible, might still be of some practical interest, since energy cost per step of restoring the trajectory might be far less than the kinetic energy accounting for the computation's speed.

One way of making the trajectory insensitive to noise would be to use square balls, holonomically constrained to remain always parallel to each other and to the fixed walls. Errors would then no longer grow exponentially, and small perturbations could simply be ignored. Although this system is consistent with the laws of classical mechanics it is a bit unnatural, since there are no square atoms in nature. A macroscopic square particle would not do, because a fraction of its kinetic energy would be converted into heat at each collision. On the other hand, a square molecule might work, if it were stiff enough to require considerably more than $kT$ of energy to excite it out of its vibrational ground state. To prevent the molecule from rotating, it might be aligned in an external field strong enough to make the energy of the first librational excited state similarly greater than $kT$. One would still have to worry about losses when the molecules collided with the mirrors. A molecule scattering off even a very stiff crystal has sizable probability of exciting long-wavelength phonons, thereby transferring energy as well as momentum. This loss could be minimized by reflecting the particles from the mirrors by long-range electrostatic repulsion, but that would interfere with the use of short-range forces for collisions between molecules, not to mention spoiling the uniform electric field used to align the molecules.

Although quantum effects might possibly help stabilize a ballistic computer against external noise, they introduce a new source of internal instability in the form of wave-packet spreading. Benioff's discussion (1982) of quantum ballistic models shows how wave packet spreading can be prevented by employing a periodically varying Hamiltonian, but not apparently by any reasonably simple time-independent Hamiltonian.

In summary, although ballistic computation is consistent with the laws of classical and quantum mechanics, there is no evident way to prevent the signals' kinetic energy from spreading into the computer's other degrees of freedom. If this spread is combatted by restoring the signals, the computer becomes dissipative; if it is allowed to proceed unchecked, the initially ballistic trajectory degenerates into random Brownian motion.

## 3. BROWNIAN COMPUTERS

If thermal randomization of the kinetic energy cannot be avoided, perhaps it can be exploited. Another family of models may be called Brownian computers, because they allow thermal noise to influence the trajectory so strongly that all moving parts have nearly Maxwellian velocities, and the trajectory becomes a random walk. Despite this lack of discipline, the Brownian computer can still perform useful computations because its parts interlock in such a way as to create a labyrinth in configuration space, isomorphic to the desired computation, from which the trajectory is prevented from escaping by high-potential-energy barriers on all sides. Within this labyrinth the system executes a random walk, with a slight drift velocity in the intended direction of forward computation imparted by coupling the system to a weak external driving force.

In more concrete terms, the Brownian computer makes logical state transitions only as the accidental result of the random thermal jiggling of its information-bearing parts, and is about as likely to proceed backward along the computation path, undoing the most recent transition, as to proceed forward. The chaotic, asynchronous operation of a Brownian computer is unlike anything in the macroscopic world, and it may at first appear inconceivable that such an apparatus could work; however, this style of operation is quite common in the microscopic world of chemical reactions, were the trial and error action of Brownian motion suffices to bring reactant molecules into contact, orient and bend them into a specific conformation ("transition state") that may be required for reaction, and separate the product molecules after reaction. It is well known that all chemical reactions are in principle reversible: the same Brownian motion that accomplishes the forward reaction also sometimes brings product molecules together, pushes them backward through the transition state, and lets them emerge as reactant molecules. Though Brownian motion is scarcely noticeable in macroscopic bodies (e.g., $(kT/m)^{1/2} \approx 10^{-6}$ cm/sec for a 1-g mass at room temperature), it enables even rather large molecules, in a fraction of a second, to accomplish quite complicated chemical reactions, involving a great deal of trial and error and the surmounting of potential energy barriers of several $kT$ in order to arrive at the transition state. On the other hand, potential energy barriers of order 100 $kT$, the typical strength of covalent bonds, effectively obstruct chemical reactions. Such barriers, for example, prevent DNA from undergoing random rearrangements of its base sequence at room temperature.

To see how a molecular Brownian computer might work, we first consider a simpler apparatus: a Brownian tape-copying machine. Such an apparatus already exists in nature, in the form of RNA polymerase, the enzyme that synthesizes a complementary RNA copy of one or more genes of a DNA molecule. The RNA then serves to direct the synthesis of the proteins encoded by those genes (Watson, 1970). A schematic snapshot of RNA polymerase in action is given in Figure 3. In each cycle of operation, the enzyme takes a small molecule (one of the four nucleotide pyrophos-

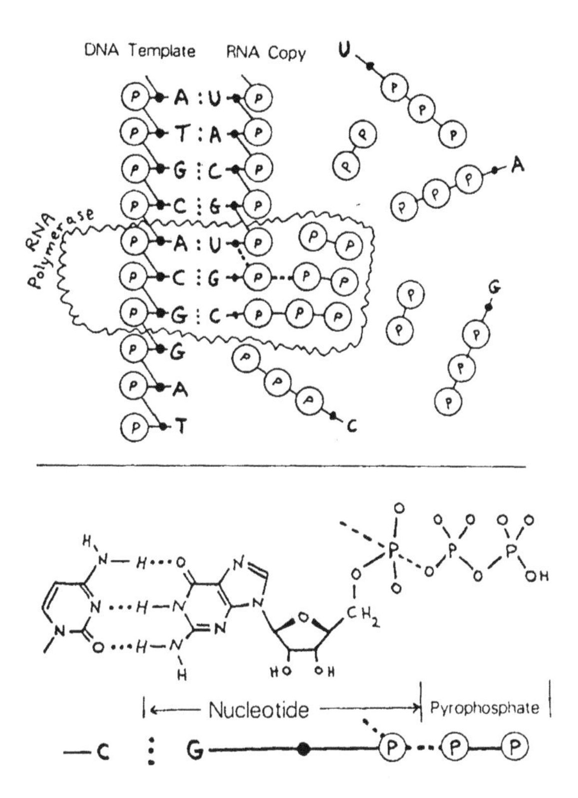


Fig. 3. RNA Polymerase synthesizing a complementary RNA strand on a single-strand DNA "template." Double and triple dots between DNA and RNA bases indicate base-pairing interaction; dashed lines indicate covalent bonds being formed and broken by RNA polymerase. Below, in more detail, the arriving GTP monomer about to lose its pyrophosphate group and be attached to the growing RNA strand.

phates, ATP, GTP, CTP, or UTP, whose base is complementary to the base about to be copied on the DNA strand) from the surrounding solution, forms a covalent bond between the nucleotide part of the small molecule and the existing uncompleted RNA strand, and releases the pyrophosphate part into the surrounding solution as a free pyrophosphate molecule (PP). The enzyme then shifts forward one notch along the DNA in preparation for copying the next nucleotide. In the absence of the enzyme, this reaction would occur with a negligible rate and with very poor specificity for selecting bases correctly complementary to those on the DNA strand. Assuming RNA polymerase to be similar to other enzymes whose mechanisms have been studied in detail, the enzyme works by forming many weak (e.g., van der Waals and hydrogen) bonds to the DNA, RNA, and incoming nucleotide pyrophosphate, in such a way that if the incoming nucleotide is correctly base-paired with the DNA, it is held in the correct transition state conformation for forming a covalent bond to the end of the RNA strand, while breaking the covalent bond to its own pyrophosphate group. The transition state is presumably further stabilized (its potential energy lowered) by favorable electrostatic interaction with strategically placed charged groups on the enzyme.

The reaction catalyzed by RNA polymerase is reversible: sometimes the enzyme takes up a free pyrophosphate molecule, combines it with the end nucleotide of the RNA, and releases the resulting nucleotide pyrophosphate into the surrounding solution, meanwhile backing up one notch along the DNA strand. The operation of the enzyme thus resembles a one-dimensional random walk (Figure 4), in which both forward and backward steps are possible, and would indeed occur equally often at equilibrium. Under biological conditions, RNA polymerase is kept away from equilibrium by other metabolic processes, which continually supply ATP, GTP, UTP, and CTP and remove PP, thereby driving the chain of reactions strongly in the direction of RNA synthesis. In domesticated bacteria, RNA polymerase runs forward at about 30 nucleotides per second, dissipating about $20kT$ per nucleotide, and making less than one mistake per ten thousand nucleotides.

In the laboratory, the speed and direction of operation of RNA polymerase can be varied by adjusting the reactant concentrations. The closer these are to equilibrium, the slower and the less dissipatively the enzyme works. For example, if ATP, GTP, UTP, and CTP were each present in 10% excess over the concentration that would be in equilibrium with a given ambient PP concentration, RNA synthesis would drift slowly forward, the enzyme on average making 11 forward steps for each 10 backward steps. These backward steps do not constitute errors, since they are undone by subsequent forward steps. The energy dissipation would be $kT\ln(11/10) \approx 0.1kT$ per (net) forward step, the difference in chemical potential between reactants and products under the given conditions. More

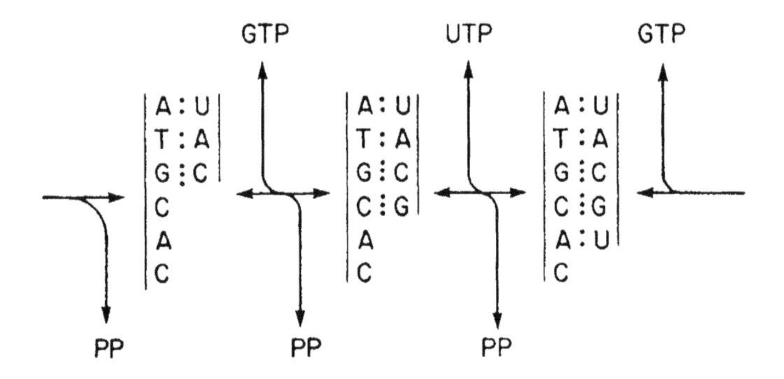


Fig. 4. RNA polymerase reaction viewed as a one-dimensional random walk.

generally, a dissipation of $\epsilon$ per step results in forward/backward step ratio of $e^{+\epsilon/kT}$, and for small $\epsilon$, a net copying speed proportional to $\epsilon$.

The analysis so far has ignored true errors, due to uncatalyzed reactions. Because these occur in some fixed, hardware-dependent ratio $\eta_0$ to the gross (rather than the net) number of catalyzed transitions, they set a limit on how slowly the copying system can be driven and still achieve reasonable accuracy. For example, if a copying system with an intrinsic error rate of $10^{-4}$ were driven at $0.1kT$ per step, its error rate would be about $10^{-3}$; but if it were driven at $10^{-4}kT$ or less, near total infidelity would result. Because the intrinsic error rate is determined by the difference in barriers opposing correct and incorrect transitions, it is a function of the particular chemical hardware, and does not represent a fundamental thermodynamic limit. In principle it can be made arbitrarily small by increasing the size and complexity of the recognition sites (to increase the potential energy difference $\Delta E$ between correct and incorrect reaction paths), by lowering the temperature (to increase the Boltzmann ratio $e^{\Delta E/kT}$ of correct to error transitions without changing $\Delta E$) and by making the apparatus larger and more massive (to reduce tunneling). In situations calling for very high accuracy (e.g., DNA copying), the genetic apparatus apparently uses another strategem for reducing errors: dissipative error correction or proofreading (Hopfield, 1974; Ninio, 1975), depicted in Figure 5. The dissipation-error tradeoff for model nonproofreading and proofreading copying systems is discussed by Bennett (1979). An amusing if impractical feature of this tradeoff is that when a copying system is operated at very low speed (and therefore high error rate), the errors themselves serve as a thermodynamic driving force, and can push the copying slowly forward even in the presence of a small reverse bias in the driving reaction. Of course, in obediance to the second law, the entropy of the incorporated errors more than makes up for the work done against the external reactants.

A true chemical Turing machine is not difficult to imagine (Figure 6). The tape might be a linear informational macromolecule analogous to RNA, with an additional chemical group attached at one site to encode the head state ($\alpha$) and location. Several hypothetical enzymes (one for each of the Turing machine's transition rules) would catalyze reactions of the macromolecule with small molecules in the surrounding solution, transforming the macromolecule into its logical successor. The transition $\alpha 0 \rightarrow 1R\beta$, for example, would be carried out by an enzyme that brings with it the groups 1 and $\beta$ that must be added during the transition, and has additional specific affinities allowing it to temporarily bind to groups 0 and $\alpha$ that must be removed. (Real enzymes with multiple, highly specific binding sites are well known, e.g., the acylating enzymes of protein synthesis.) In Figure 6, the hypothetical enzyme binds on the right, since its transition rule calls

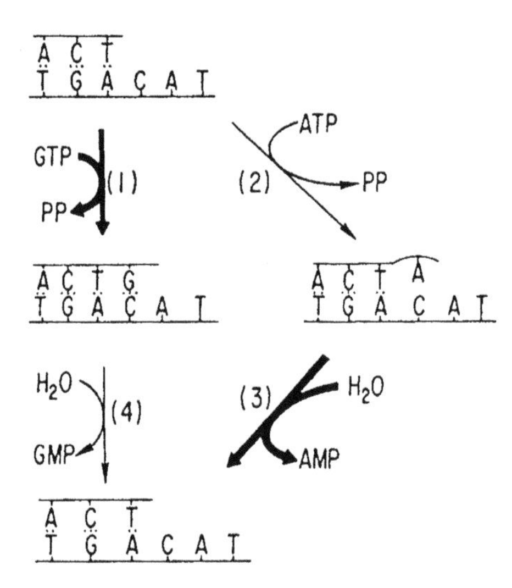


Fig. 5. Proofreading in DNA replication. The enzyme responsible for copying DNA usually inserts the correct nucleotide (1), but occasionally inserts an incorrect one (2). To counter this, another enzyme (or another active site on the same enzyme) catalyzes a proofreading reaction (3), which preferentially removes incorrectly paired nucleotides from the end of an uncompleted DNA strand. The proofreading enzyme also occasionally makes mistakes, removing a nucleotide even though it is correct (4). After either a correct or an incorrect nucleotide has been removed, the copying enzyme gets another chance to try to insert it correctly, and the proofreading enzyme gets another chance to proofread, etc. It is important to note that the proofreading reaction is not simply the thermodynamic reverse of the copying reaction: it uses different reactants, and has a different specificity (favoring incorrect nucleotides, while the copying reaction favors correct nucleotides). The minimum error rate (equal to the product of the error rates of the writing and proofreading steps) is obtained when both reactions are driven strongly forward, as they are under physiological conditions. Proofreading is an interesting example of the use of thermodynamic irreversibility to perform the logically irreversible operation of error correction.

for a right shift. After the requisite changes have been made, it drops off and is readied for reuse. At some point in their cycle of use the hypothetical enzymes are made to catalyze a reaction involving external reactants (here ATP:ADP), whose concentrations can be adjusted to provide a variable driving force.

Assume for the moment that the enzymatic Turing machine is *logically reversible*, i.e., that no whole-machine state has more than one logical predecessor (this mathematical requirement on the structure of the computation will be discussed in the next section). Then the net computation speed will be linear in the driving force, as was the case with RNA polymerase, because the logically accessible states (i.e., configurations of the macromolecule accessible by forward and backward operation of the enzymes) form a one-dimensional chain, along which the executes a random walk with drift velocity proportional to $\epsilon/kT$.

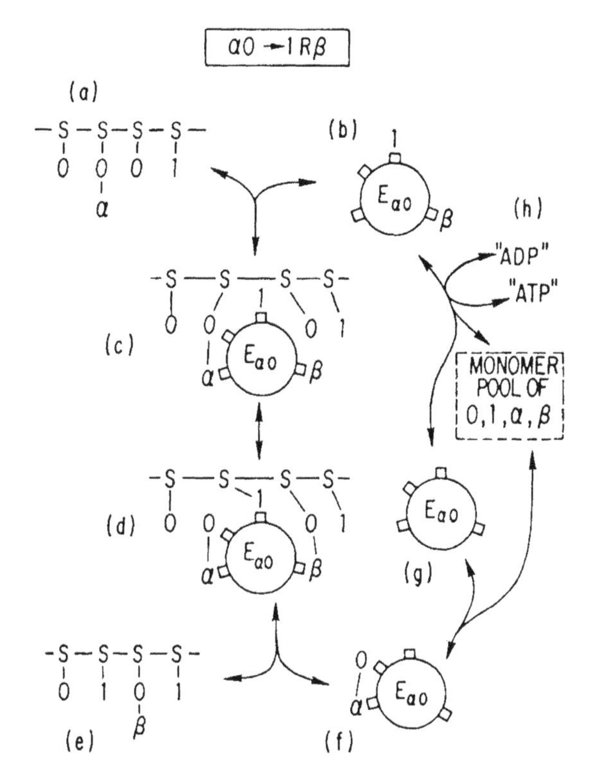


**Fig. 6.** Hypothetical enzymatic Turing machine. Macromolecular tape (a) consists of a structural backbone S–S–S bearing tape symbols 1,0 and head marker $\alpha$. Macromolecule reacts (c,d) with enzyme (b) that catalyzes the transition $\alpha 0 \rightarrow 1R\beta$, via specific binding sites (tabs), thereby resulting in logical successor configuration (e). Enzyme is then prepared for reuse (f,g,h). Coupling to external reaction (h) drives the reactions, which would otherwise drift indifferently forward and backward, in the intended forward direction.

It is also possible to imagine an error-free Brownian Turing machine made of rigid, frictionless clockwork. This model (Figure 7) lies between the billiard-ball computer and the enzymatic computer in realism because, on the one hand, no material body is perfectly hard; but on the other hand, the clockwork model's parts need not be machined perfectly, they may be fit together with some backlash, and they will function reliably even in the presence of environmental noise. A similar model has been considered by Reif (1979) in connection with the **P = PSPACE** question in computational complexity.

The baroque appearance of the clockwork Turing machine reflects the need to make all its parts interlock in such a way that, although they are free to jiggle locally at all times, no part can move an appreciable distance except when it is supposed to be making a logical transition. In this respect it resembles a well worn one of those wooden cube puzzles that must be

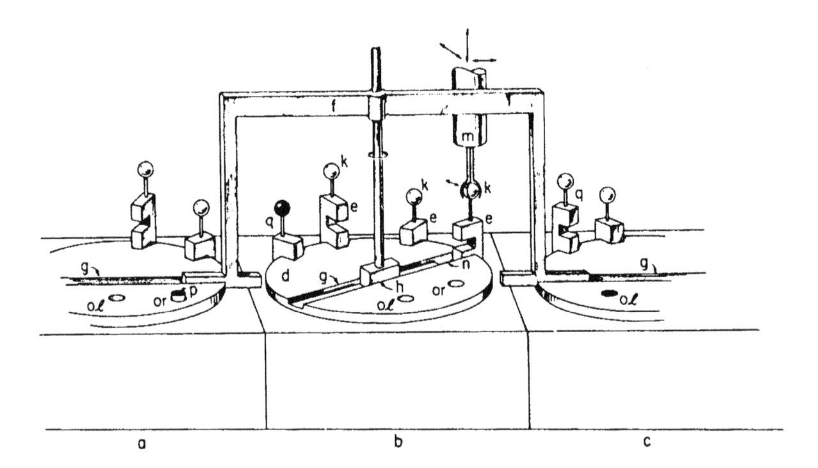


**Fig. 7.** Brownian Turing machine made of rigid, frictionless, loosely fitting clockwork. This figure shows the Turing machine tape (a,b,c) and the read–write–shift equipment. The machine is scanning square b. Each tape square has a disk (d) which interlocks with several E-shaped bit-storage blocks (e), holding them in the up (1) or down (0) position. A framework (f) fits over the scanned tape square, engaging the disks of the two adjacent squares (via their grooves g), to keep them from rotating when they are not supposed to. After the bits are read (cf. next figure) they must in general be changed. In order to change a bit, its knob (k) is first grasped by the manipulator (m), then the notch (n) is rotated into alignment by the screwdriver (h) and the bit storage block (e) is slid up or down. The block is then locked into place by further rotating the disk, after which the manipulator can safely let go and proceed to grasp the next bit's knob. Each tape square has a special knob (q) that is used to help constrain the disks on nonscanned tape squares. In principle these might all be constrained by the framework (f), but that would require making it infinitely large and aligning it with perfect angular accuracy. To avoid this, the framework (f) is used only to constrain the two adjacent tape squares. All the remaining tape squares are indirectly constrained by pegs (p) coupled to the special knob (q) of an adjacent square. The coupling (a lever arrangement hidden under the disk) is such that, when any square's q knob is down, a peg (p) engages the rightmost of two openings (o r) on the next tape square to the left, and another peg disengages the leftmost (o l) of two openings on the next tape square to the right. A q knob in the up position does the opposite: it frees the tape square to its left and locks the tape square to its right. To provide an outward-propagating chain of constraints on each side of the scanned square, all the q knobs to its right must be up, and all the q knobs to its left must be down. The q knob on the scanned square can be in either position, but just before a right shift it is lowered, and just before a left shift it is raised. To perform the shift, the screwdriver rotates the scanned square's groove (g) into alignment with the framework, then the manipulator (m), by grasping some convenient knob, pulls the whole head apparatus (including m itself, as well as f, h, and parts not shown) one square to the left or right.

solved by moving one part a small distance, which allows another to move in such a way as to free a third part, etc. The design differs from conventional clockwork in that parts are held in place only by the hard constraints of their loosely fitting neighbors, never by friction or by spring pressure. Therefore, when any part is about to be moved (e.g., when one of the E-shaped blocks used to store a bit of information is moved by the manipulator **m**), it must be grasped in its current position before local constraints on its motion are removed, and, after the move, local constraints must be reimposed before the part can safely be let go of.

Perhaps the most noteworthy feature of the machine's operation is the "obstructive read" depicted in Figure 8. In general, the coordinated motion of the screwdriver **h** and manipulator **m**, by which the Turing machine control unit acts on the tape, can be described as a deterministic unbranched path in the five-dimensional configuration space spanning the screwdriver's rotation and the manipulator's translation and grasp (because of backlash, this path is not single trajectory, but a zero-potential-energy channel of finite width surrounded by infinite potential energy barriers, within which the system performs a Brownian random walk, with a slight forward drift velocity). However, before it can know what to write or which way to shift, the control unit must ascertain the current contents of the scanned tape square. To do this, during the read stage of the machine cycle, the path of the manipulator branches nondeterministically into two paths, one of which is obstructed (due to collision with the knob **k**) by a bit in the up position, the other by the same bit in the down position. This bifurcation followed by obstruction is repeated for each additional bit stored on the tape square, so that, by the time the manipulator has negotiated all the

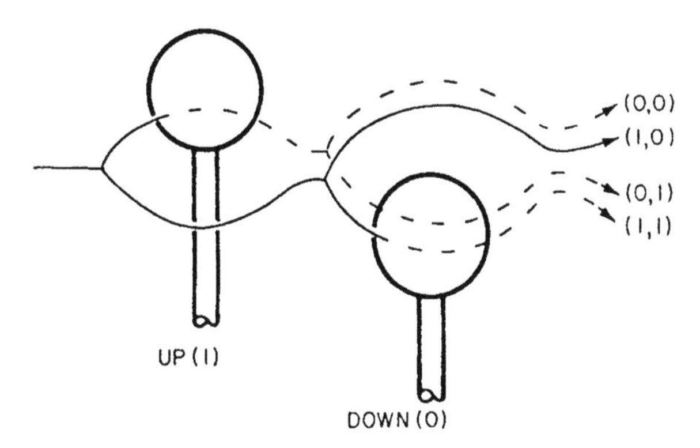


**Fig. 8.** Obstructive read. The clockwork Turing machine's control unit guides the manipulator along a branching path, one of whose branches is obstructed by a knob in the up position, the other by the same knob in the down position.

bifurcations and obstructions, it is again on a single path determined by the contents of the scanned tape square. If the manipulator by chance wanders into the wrong path at a bifurcation and encounters an obstruction, the forward progress of the computation is delayed until Brownian motion jiggles the manipulator back to the bifurcation and forward again along the right path.

Figure 9 suggests how the manipulator and screwdriver might be driven through their paces by the main control unit. A master camshaft, similar to a stereophonic phonograph record, would contain a network of tunnels isomorphic with the Turing machine's finite state transition graph. A weak spring (the only spring in the whole apparatus) biases the camshaft's Brownian motion, causing it to revolve on average in the direction corresponding to forward computation. Revolution of the camshaft imposes certain motions on a captive cam follower, which in turn are translated by appropriate mechanical linkages into synchronous motions of the manipulator and screwdriver in Figure 7, required to perform read, write, and shift operations on the Turing machine tape. During the read phase, the cam follower passes through a bifurcation in its tunnel for each bit to be read, causing the manipulator to perform an obstructive read, and delaying the

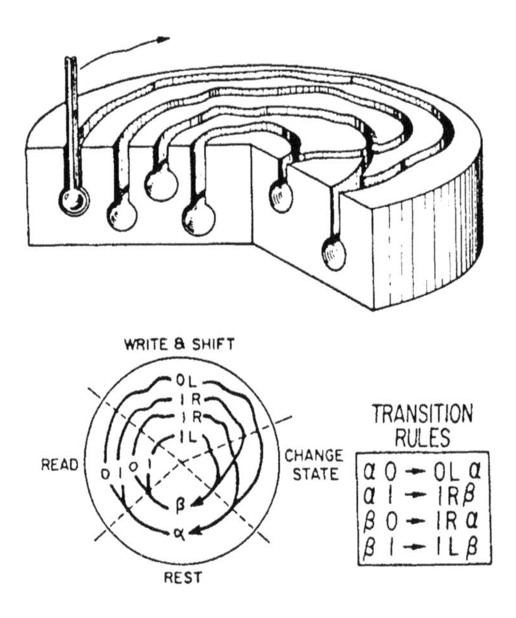


**Fig. 9.** Above: master camshaft of the clockwork Turing machine's control unit. Below: top view of camshaft, and the transition rules to which its tunnels are isomorphic.

computation until the cam follower, by Brownian trial and error, chooses the right tunnel. The obstructive read places the cam follower at the beginning of a specific tunnel segment corresponding to the current head state and tape symbol, and the rest of this tunnel segment drives the manipulator and screwdriver through the coordinated motions necessary to accomplish the write and shift operations. The obstructive read also serves another, less obvious purpose: it prevents the machine from wandering backward into states that are not logical predecessors of the present state. If the machine is logically reversible as supposed, this means that the only unobstructed path for the camshaft to rotate backwards one full revolution is the path leading to the present state's unique logical predecessor.

As in the case of the enzymatic Turing machine, the drift velocity is linear in the dissipation per step. Because the clockwork Turing machine cannot make illogical transitions, the only kind of error it is susceptible to is failure to be in the final logical state of its computation. Indeed, if the driving force $\epsilon$ is less than $kT$, any Brownian computer will at equilibrium spend most of its time in the last few predecessors of the final state, spending only about $\epsilon/kT$ of its time in the final state itself. However, the final state occupation probability can be made arbitrarily large, independent of the number of steps in the computation, by dissipating a little extra energy during the final step, a "latching energy" of $kT\ln(kT/\epsilon)+kT\ln(1/\eta)$ sufficing to raise the equilibrium final state occupation probability to $1-\eta$.

Quantum mechanics probably does not have a major qualitative effect on Brownian computers: in an enzymatic computer, tunneling and zero-point effects would modify transition rates for both catalyzed and uncatalyzed reactions; a quantum clockwork computer could be viewed as a particle propagating in a multidimensional labyrinth in configuration space (since the clockwork computer's parts are assumed to be perfectly hard, the wave function could not escape from this labyrinth by tunneling). Both models would exhibit the same sort of diffusive behavior as their classical versions.

[It should perhaps be remarked that, although energy transfers between a quantum system and its environment occur via quanta (e.g., photons of black body radiation) of typical magnitude about $kT$, this fact does not by itself imply any corresponding coarseness in the energy cost per step: a net energy transfer of $0.01kT$ between a Brownian computer and its environment could for example be achieved by emitting a thermal photon of $1.00kT$ and absorbing one of $0.99kT$. The only limitation on this kind of spontaneous thermal fine tuning of a quantum system's energy comes from its energy level spacing, which is less than $kT$ except for systems so cold that the system as a whole is frozen into its quantum ground state.]

## 4. LOGICAL REVERSIBILITY

Both the ballistic and Brownian computers require a change in programming style: *logically irreversible* operations (Landauer, 1961) such as erasure, which throw away information about the computer's preceding logical state, must be avoided. These operations are quite numerous in computer programs as ordinarily written; besides erasure, they include overwriting of data by other data, and entry into a portion of the program addressed by several different transfer instructions. In the case of Turing machines, although the individual transition rules (quintuples) are reversible, they often have overlapping ranges, so that from a given instantaneous description it is not generally possible to infer the immediately preceding instantaneous description (Figure 1). In the case of combinational logic, the very gates out of which logic functions are traditionally constructed are for the most part logically irreversible (e.g., AND, OR, NAND), though NOT is reversible.

Logically irreversible operations must be avoided entirely in a ballistic computer, and for a very simple reason: the merging of two trajectories into one cannot be brought about by conservative forces. In a Brownian computer, a small amount of logical irreversibility can be tolerated (Figure 10), but a large amount will greatly retard the computation or cause it to fail completely, unless a finite driving force (approximately $kT \ln 2$ per bit of information thrown away) is applied to combat the computer's tendency to drift backward into extraneous branches of the computation. Thus driven, the Brownian computer is no longer thermodynamically reversible, since its dissipation per step no longer approaches zero in the limit of zero speed.

In spite of their ubiquity, logically irreversible operations can be avoided without seriously limiting the power of computers. A means of simulating arbitrary irreversible computations reversibly is given by Bennett (1973) using Turing machines, was independently discovered by Fredkin, using reversible Boolean logic (Toffoli, 1980), and is outlined below.

We begin by noting that it is easy to render any computer reversible in a rather trivial sense, by having it save all the information it would otherwise have thrown away. For example the computer could be given an extra "history" tape, initially blank, on which to record enough about each transition (e.g., for a Turing machine, which quintuple was being used) that the preceding state would be uniquely determined by the present state and the last record on the history tape. From a practical viewpoint this does not look like much of an improvement, since the computer has only postponed the problem of throwing away unwanted information by using the extra tape as a garbage dump. To be usefully reversible, a computer ought to be required to clean up after itself, so that at the end of the computation the

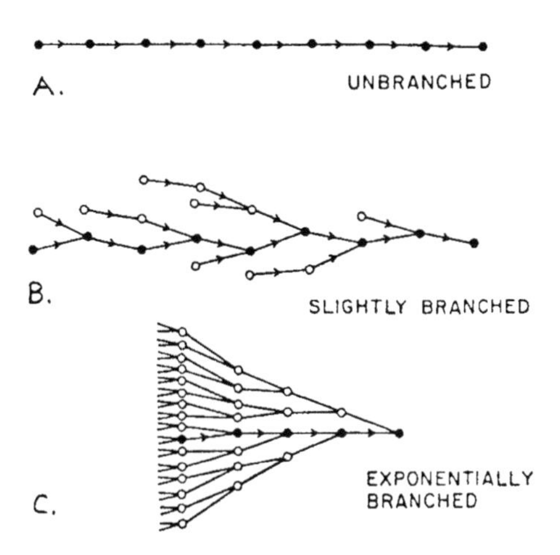


**Fig. 10.** Various kinds of computation graph, with black nodes denoting logical states (instantaneous descriptions, in Turing machine terminology) on the intended computation path, and open nodes denoting extraneous predecessors. Arrows indicate intended direction of transitions, but if the driving force is weak, backward transitions occur nearly as often as forward ones. In a strictly reversible computation (A), the graph is unbranched, and an arbitrarily small driving force $\epsilon$ suffices to drive the computation forward with drift velocity proportional to $\epsilon$. An arbitrarily small driving force still suffices for a slightly branched graph (B), with a few extraneous predecessors per state on the intended path, but the computation proceeds more slowly, due to temporary detours onto the extraneous branches. Slightly branching trees occur, for example when the same variable is assigned several times in succession: at any point in the computation, the most recent assignment can be randomly undone (giving the variable any value at all), but the result is typically a "garden-of-Eden" state with no predecessor, because this random value is inconsistent with the forward result of the previous assignment. Exponentially branching trees (C) also occur, for example, in loops that assign a succession of different variables. If such a tree is infinite, then a small driving force is insufficient to keep the computation from wandering onto an extraneous branch never to return; to drive the computation forward in such a tree, the dissipation per step must exceed $kT$ times the mean number of immediate predecessors per state. Even if the exponential backward branching is eventually stopped by garden-of-Eden states, as is commonly the case, extraneous states may outnumber states on the intended path by factors of $2^{100}$ or so, slowing down the computation by that much unless backward excursions are again suppressed by dissipating about $kT$ times the logarithm of the mean number of immediate predecessors of states near the intended computation path.

only data remaining are the desired output and the originally furnished input. [A general-purpose reversible computer must be allowed to save its input, or some equivalent information; otherwise it could not perform computations in which the input was not uniquely determined by the output. Formally this amounts to embedding the partial recursive function $x \to \varphi(x)$, computed by the original irreversible machine, in a 1 : 1 partial recursive function $x \to \langle x, \varphi(x) \rangle$, to be computed by the reversible machine;

since 1 : 1 functions are the only kind that can be computed by reversible machines.]

A tape full of random data can only be erased by an irreversible process. However, the history produced by the above untidy reversible computer is not random, and it can be gotten rid of reversibly, by exploiting the redundancy between it and the computation that produced it. If, at the end of the untidy computer's computation, another stage of computation were begun, using the inverse of the untidy machine's transition function, the inverse, or "cleanup" machine would begin undoing the untidy machine's computation step by step, eventually returning the history tape to its original blank condition. Since the untidy machine is reversible and deterministic, its cleanup machine is also reversible and deterministic. [Bennett (1973) gives detailed constructions of the reversible untidy and cleanup machines for an arbitrary irreversible Turing machine.] The cleanup machine, of course, is not a general-purpose garbage disposer: the only garbage it can erase is the garbage produced by the untidy machine of which it is the inverse.

Putting the untidy and cleanup machines together, one obtains a machine that reversibly does, then undoes, the original irreversible computation. This machine is still useless because the desired output, produced during the untidy first stage of computation, will have been eaten up along with the undesired history during the second cleanup stage, leaving behind only a reconstructed copy of the original input. Destruction of the desired output can be easily prevented, by introducing still another stage of computation, after the untidy stage but before the cleanup stage. During this stage the computer makes an extra copy of the output on a separate, formerly blank tape. No additional history is recorded during this stage, and none needs to be, since copying onto blank tape is already a 1 : 1 operation. The subsequent cleanup stage therefore destroys only the original of the output but not the copy. At the end of its three-stage computation, the computer contains the (reconstructed) original input plus the intact copy of the output. All other storage will have been restored to its original blank condition. Even though no history remains, the computation is reversible and deterministic, because each of its stages has been so. The use of separate tapes for output and history is not necessary; blank portions of the work tape may be used instead, at the cost of making the simulation run more slowly (quadratic rather than linear time). Figure 11A summarizes the three-stage computation.

The argument just outlined shows how an arbitrary Turing machine can be simulated in not much more time by a reversible Turing machine, at the cost of including the input as part of the output, and perhaps using a lot of temporary storage for the history. By cleaning up the history more often

A

| Computation Stage | Contents of Work area | Contents of History area | Contents of Output area |
|---|---|---|---|
| | _INPUT | _ | _ |
| Untidy | WORK | HISTO_ | _ |
| | _OUTPUT | HISTORY_ | _ |
| | OUTPUT | HISTORY_ | OUT_ |
| Copy output | | | |
| | OUTPUT_ | HISTORY_ | OUTPUT_ |
| | _OUTPUT | HISTORY_ | _OUTPUT |
| Cleanup | WORK | HISTO_ | _OUTPUT |
| | _INPUT | _ | _OUTPUT |

B

| Computation Stage | Contents of Work area | Contents of Hist. area | Contents of Output area |
|---|---|---|---|
| | INPUT | - | - |
| 1. Untidy $\varphi$ comp. | | | |
| | OUTPUT | $\varphi$ HISTORY | - |
| 2. Copy output | | | |
| | OUTPUT | $\varphi$ HISTORY | OUTPUT |
| 3. Cleanup $\varphi$ comp. | | | |
| | INPUT | - | OUTPUT |
| 4. Interchange data | | | |
| | OUTPUT | - | INPUT |
| 5. Untidy $\varphi^{-1}$ comp. | | | |
| | INPUT | $\varphi^{-1}$HISTORY | INPUT |
| 6. Cancel extra input | | | |
| | INPUT | $\varphi^{-1}$HISTORY | - |
| 7. Cleanup $\varphi^{-1}$ comp. | | | |
| | OUTPUT | - | - |

**Fig. 11.** (A) Reversible simulation of an irreversible computation. The reversible computer has three storage areas (e.g., tapes): a work area in which the simulation takes place; a history area in which garbage generated by the irreversible computation is saved temporarily, thereby rendering the computation reversible; and an output area. The work area initially contains the input; the history and output areas are initially blank. The computation takes place in three stages, representative snapshots of which are shown. The underbars represent the locations of read/write heads for a three-tape machine, or analogous place markers for other machines. (B) Reversible computation of an arbitrary 1 : 1 function $\varphi$ with no extra output, given irreversible algorithms for $\varphi$ and $\varphi^{-1}$. The computation proceeds by seven stages as shown. Stage 5 has the sole purpose of producing the $\varphi^{-1}$ history, which, after the extra input has been reversibly erased in stage 6, serves in stage 7 to destroy itself and the remaining copy of the input, leaving only the desired output.

(Bennett, 1973), space on the history tape may be traded off against time or additional garbage output. This tradeoff, the details of which remain to be worked out, provides an upper bound on the cost of making computations reversible. However, it has been observed in practice that many computations (e.g., numerical integration of differential equations) can be performed by reversible algorithms with *no* penalty in time, storage, or extra output.

In cases where the original irreversible Turing machine computes a 1 : 1 function, it can be simulated reversibly with no additional output, but with perhaps an exponential increase in run time. This is done by combining McCarthy's (1956) trial-and-error procedure for effectively computing the inverse of a 1 : 1 partial recursive function [given an algorithm $\varphi$ for the original function and an argument $y$, $\varphi^{-1}(y)$ is defined as the first element of the first ordered pair $\langle x, s\rangle$ such that $\varphi(x) = y$ in less than $s$ steps] with Bennett's procedure (1973; Figure 11B] for synthesizing a reversible Turing machine with no extra output from two mutually inverse irreversible Turing machines. These results imply that the reversible Turing machines provide a Gödel numbering of 1 : 1 partial recursive functions (i.e., every 1 : 1 partially reversible function is computable by a reversible TM and vice versa). The construction of a reversible machine from an irreversible machine and its inverse implies that the open question, of whether there exists a 1 : 1 function much easier to compute by an irreversible machine than by any reversible machine, is equivalent to the question of whether there is an easy 1 : 1 function with a hard inverse.

Simulation of irreversible logic functions by reversible gates is analogous (Toffoli, 1980) to the constructions for Turing machines. The desired function is first embedded in a one-to-one function (e.g., by extending the output to include a copy of the input) which is then computed by reversible gates, such as the three-input, three-output AND/NAND gate. The first two inputs of this gate simply pass through unchanged to become the first two outputs; the third output is obtained by taking the EXCLUSIVE-OR of the third input with the AND of the first two inputs. The resulting mapping from three bits onto three bits (which is its own inverse) can be used together with constant inputs to compute any logic function computable by conventional gates, but in so doing, may produce extra "garbage" bits analogous to the reversible Turing machine's history. If the function being computed is 1 : 1, these bits need not appear in the final output, because they can be disposed of by a process analogous to the reversible Turing machine's cleanup stage, e.g., by feeding them into a mirror image of the reversible logic net that produced them in the first place. The "interaction gate" used in the ballistic computer (Fredkin and Toffoli, 1981) may be regarded as having two inputs and four outputs (respectively, **x** & **y**, **y** & ¬ **x**, **x** & ¬ **y**, and **x** & **y**) for the four possible exit paths from the collision zone between balls **x** and **y**. The interaction gate is about as simple as can be imagined in its physical realization, yet it suffices (with the help of mirrors to redirect and synchronize the balls) to synthesize all conservative Boolean functions, within which all Boolean functions can be easily embedded. The rather late realization that logical irreversibility is not an essential feature of computation is probably due in part to the fact that reversible gates require somewhat more inputs and outputs (e.g., 3 : 3 or 2 : 4) to provide a basis for nontrivial computation than irreversible gates do (e.g., 2 : 1 for NAND).

## 5. REVERSIBLE MEASUREMENT AND MAXWELL'S DEMON

This section further explores the relation between logical and thermodynamic irreversibility and points out the connection between logical irreversibility and the Maxwell's demon problem.

Various forms of Maxwell's demon have been described; a typical one would be an organism or apparatus that, by opening a door between two equal gas cylinders whenever a molecule approached from the right, and closing it whenever a molecule approached from the left, would effortlessly concentrate all the gas on the left, thereby reducing the gas's entropy by $Nk \ln 2$. The second law forbids any apparatus from doing this reliably, even for a gas consisting of a single molecule, without producing a corresponding entropy increase elsewhere in the universe.

It is often supposed that *measurement* (e.g., the measurement the demon must make to determine whether the molecule is approaching from the left or the right) is an unavoidably irreversible act, requiring an entropy generation of at least $k \ln 2$ per bit of information obtained, and that this is what prevents the demon from violating the second law. In fact, as will be shown below, measurements of the sort required by Maxwell's demon can be made reversibly, provided the measuring apparatus (e.g., the demon's internal mechanism) is in a standard state before the measurement, so that measurement, like the copying of a bit onto previously blank tape, does not overwrite information previously stored there. Under these conditions, the essential irreversible act, which prevents the demon from violating the second law, is not the measurement itself but rather the subsequent restoration of the measuring apparatus to a standard state in preparation for the next measurement. This forgetting of a previous logical state, like the erasure or overwriting of a bit of intermediate data generated in the course of a computation, entails a many-to-one mapping of the demon's physical state, which cannot be accomplished without a corresponding entropy increase elsewhere.

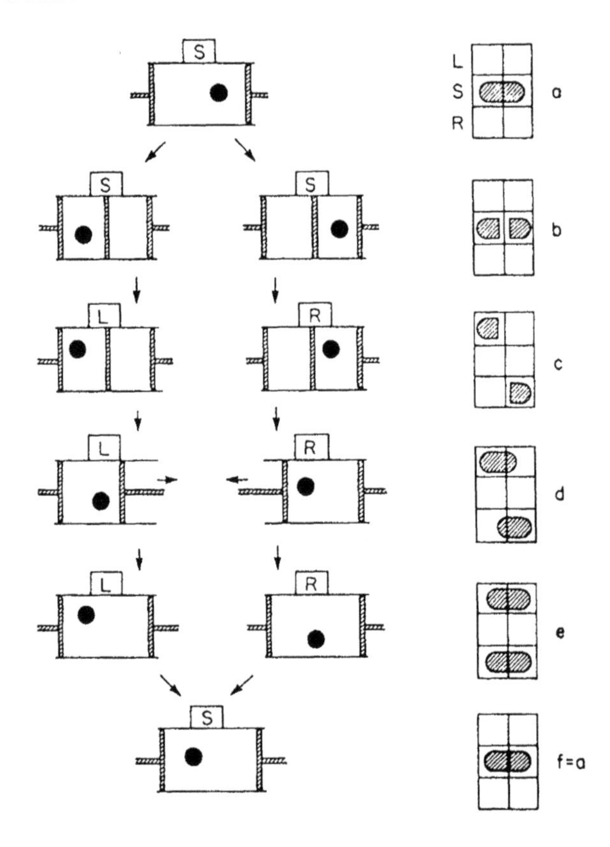


Fig. 12. A one-molecule Maxwell's demon apparatus.

Figure 12 shows the cycle of operation of a one-molecule Maxwell's demon apparatus. The left side of the figure shows the apparatus, and the right side shows the sequence changes in its phase space, depicted schematically as a product of a horizontal coordinate representing the location of the molecule and a vertical coordinate representing the physical state of the demon's "mind." The demon's mind has three states: its standard state *S* before a measurement, and two states **L** and **R** denoting the result of a measurement in which the molecule has been found on the left or right, respectively. At first (a) the molecule wanders freely throughout the apparatus and the demon is in the standard state *S*, indicating that it does not know where the molecule is. In (b) the demon has inserted a thin partition trapping the molecule on one side or the other. Next the demon performs a reversible measurement to learn (c) whether the molecule is on the left or the right. The demon then uses this information to extract $kT \ln 2$ of isothermal work from the molecule, by inserting a piston on the side not containing the molecule and allowing the molecule to expand (d) against the piston to fill the whole apparatus again (e). Notice that a different manipulation is required to extract work from the molecule depending on which side it is on; this is why the demon must make a measurement, and why at (d) the demon will be in one of two distinct parts of its own phase space depending on the result of that measurement. At (e) the molecule again fills the whole apparatus and the piston is in its original position. The only record of which side the molecule came from is the demon's record of the measurement, which must be erased to put the demon back into a standard state. This erasure (e–f) entails a twofold compression of the occupied volume of the demon's phase space, and therefore cannot be made to occur spontaneously except in conjunction with a corresponding entropy increase elsewhere. In other words, all the work obtained by letting the molecule expand in stage (d) must be converted into heat again in order to compress the demon's mind back into its standard state.

In the case of a measuring apparatus which is *not* in a standard state before the measurement, the compression of the demon's state, and the compensating entropy increase elsewhere, occur at the same time as the measurement. However, I feel it important even in this case to attribute the entropy cost to logical irreversibility, rather than to measurement, because in doing the latter one is apt to jump to the erroneous conclusion that all transfers of information, e.g., the synthesis of RNA, or reversible copying onto blank tape, have an irreducible entropy cost of order $kT \ln 2$ per bit.

As a further example of reversible copying and measurement, it may be instructive to consider a system simpler and more familiar than RNA polymerase, viz., a one-bit memory element consisting of a Brownian particle in a potential well that can be continuously modulated between bistability (two minima separated by a barrier considerably higher than $kT$) and monostability (one minimum), as well as being able to be biased to favor one well or the other. Such systems have been analyzed in detail in Landauer (1961) and Keyes and Landauer (1970); a physical example (Figure 13) would be an ellipsoidal piece of ferromagnetic material so small that in the absence of a field it consists of a single domain, magnetized parallel or antiparallel to the ellipse axis (alternatively, a round piece of magnetically anisotropic material could be used). Such a system can be modulated between bistability and monostability by a transverse magnetic field, and biased in favor of one minimum or the other by a longitudinal magnetic field. When the transverse field just barely abolishes the central minimum, the longitudinal magnetization becomes a "soft mode," very sensitive to small longitudinal components of the magnetic field.

This sensitivity can be exploited to copy information reversibly from one memory element to another, provided the memory element destined to

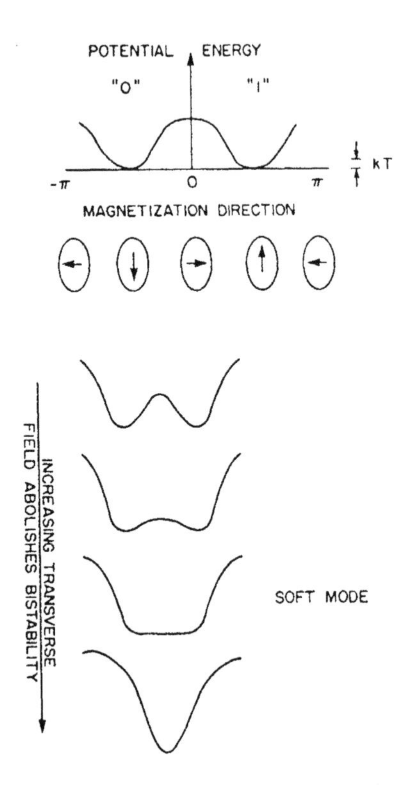


**Fig. 13.** A bistable potential well, realized by a one-domain ferromagnet. Increasing transverse magnetic field abolishes the bistability, resulting in a "soft mode," very sensitive to small longitudinal fields.

receive the information is in a standard state initially, so that the copying is logically reversible. Figure 14 shows how this reversible copying might be done. The apparatus contains two fixed memory elements, one (top) containing a zero for reference and the other (bottom) containing the data bit to be copied. A third memory element is movable and will be used to receive the copy. It is initially located next to the reference element and also contains a zero. To begin the copying operation, this element is slowly moved away from the reference bit and into a transverse field strong enough to abolish its bistability. This manipulation serves to smoothly and continuously change the element from its initial zero state to a unistable state. The element is then gradually moved out of the transverse field toward the data bit. As the movable bit leaves the region of strong transverse field, its soft mode is biased by the weak longitudinal field from the data bit, thereby making it choose the same direction of magnetization as the data bit. [The region of strong transverse field is assumed to be wide enough that by the time the movable bit reaches its bottom edge, the longitudinal bias field is

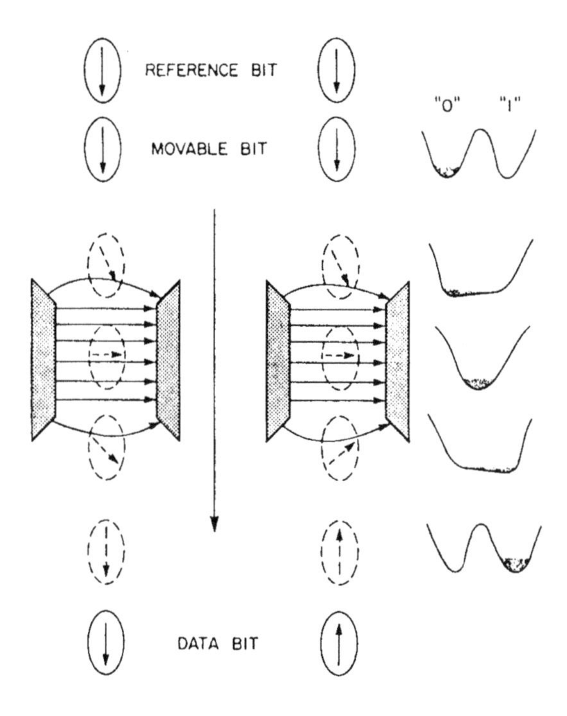


**Fig. 14.** Reversible copying using a one-domain ferromagnet. The movable bit, initially zero, is mapped into the same state as the data bit (zero in left column; one in center column). Right column shows how the probability density of the movable bit's magnetization, initially concentrated in the "**0**" minimum, is deformed continuously until it occupies the "**1**" minimum, in agreement with a "1" data bit.

due almost entirely to the data bit, with only a negligible perturbation (small compared to $kT$) from the more distant reference bit.] The overall effect has been to smoothly change the movable element's magnetization direction from agreeing with the reference bit at the top (i.e., zero) to agreeing with the data bit on the bottom. Throughout the manipulation, the movable element's magnetization remains a continuous, single-valued function of its position, and the forces exerted by the various fields on the movable element during the second half of the manipulation are equal and opposite to those exerted during the first half, except for the negligible long-range perturbation of the bias field mentioned earlier, and a viscous damping force (reflecting the finite relaxation time of spontaneous magnetization fluctuations) proportional to the speed of motion. The copying operation can therefore be performed with arbitrarily small dissipation. If carried out in reverse, the manipulation would serve to reversibly erase one of two bits known to be identical, which is the logical inverse of copying onto blank tape.

Like a copying enzyme, this apparatus is susceptible to degradation of its stored information by thermal fluctuations and tunneling. These phenomena, together with the damping coefficient, determine the minimum error probability of which any given apparatus of this type, operating at a given temperature, is capable, and determine a minimum dissipation per step required to operate the apparatus with approximately this error probability. However, as with copying enzymes, there is no fundamental thermodynamic obstacle to making the error probability $\eta$ and the dissipation $\epsilon/kT$ both arbitrarily small, and no practical obstacle to making them both much less than unity.

Bistable magnetic elements of this sort could also, in principle, be used to perform the reversible measurement required by Maxwell's demon. In Figure 15 a diamagnetic particle trapped in the right side of a Maxwell's demon apparatus introduces a weak upward bias in the mode-softening horizontal magnetic field, which yields an upward magnetization of the bistable element when the horizontal field is gradually turned off.

What would happen if a similar manipulation were used to perform a logically *irreversible* operation, e.g., restoring a bistable element that might initially be in either of two states to a standard state? An apparatus for doing this is shown in Figure 16: a memory element which might be magnetized either up or down is moved gradually into a transverse field, rotated slightly counterclockwise to bias the magnetization downward, moved out of the field, and rotated back again, leaving it in the down or zero state.

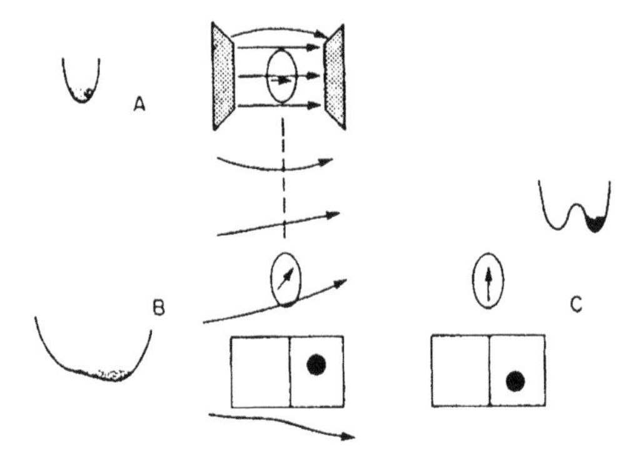


**Fig. 15.** A one-domain ferromagnet used to make measurements for Maxwell's demon: The bistable element has been moved out of strong transverse field (A) just far enough to create a soft mode (B), which is biased by the perturbation of the field by a diamagnetic Brownian particle in the right side of a Maxwell's demon apparatus. Slowly turning the field off (C) completes the measurement. Peripheral diagrams show potential energy and probability density of the ferromagnetic element's magnetization direction.

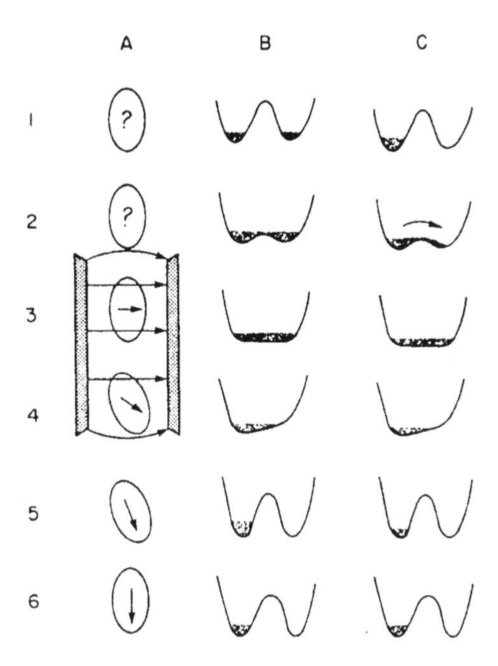


Fig. 16. Erasure of a bistable one-domain ferromagnet. Column A: the bistable element, which may be magnetized either up or down (1), is moved gradually (2) into a transverse field that abolishes its bistability symmetrically (3). It is then rotated slightly counterclockwise (4), to bias the soft mode downward, removed from the field (5), and rotated back again, leaving it in the down or zero state (6). Column B: Evolution of the probability density when the manipulation just described is used to erase a random, unknown bit. Column C: Behavior of the probability density when the manipulation is applied to a known bit (here zero). An irreversible entropy increase of $k \ln 2$ occurs at stage 2, when the probability density leaks out of the initially occupied minimum.

Although such many-to-one mappings have been called logically irreversible, in a subtler sense they may be reversible or not depending on the data to which they are applied. If the initial state in Figure 16 is truly unknown, and properly describable by a probability equidistributed between the two minima (Figure 16B), then undoing the manipulation of Figure 16 exactly restores the initial state of ignorance, and the operation can be viewed as reversible. It is also thermodynamically reversible: the $kT \ln 2$ of work required to carry out the manipulation is compensated by a decrease of $k \ln 2$ in the bistable element's entropy. This expenditure of external work to decrease a system's entropy is analogous to the isothermal compression of a gas to half its volume.

If, on the other hand, the initial state of the memory element were known (e.g., by virtue of its having been set during some intermediate stage of computation with known initial data) the operation would be irreversible:

undoing it would not restore the initial state, and the conversion of $kT \ln 2$ of work into heat in the surroundings would not be compensated by a decrease in the bistable element's entropy. The irreversible entropy increase occurs at the point indicated by the arrow in Figure 16C, when the system's probability density leaks from the minimum it was originally in to fill both minima (as it could have done all along had the initial data been unknown). This is analogous to the free expansion of gas into a previously evacuated container, in which the gas increases its own entropy without doing any work on its environment.

The normal biochemical mechanism by which RNA is destroyed when it is no longer needed (Watson, 1970) provides another example of logical irreversibility. As indicated before, the synthesis of RNA by RNA polymerase is a logically reversible copying operation, and under appropriate (nonphysiological) conditions, it could be carried out at an energy cost of less than $kT$ per nucleotide. The thermodynamically efficient way to get rid of an RNA molecule would be to reverse this process, i.e., to take the RNA back to the DNA from which it was made, and use an enzyme such as RNA polymerase with a slight excess of pyrophosphate to perform a sequence-specific degradation, checking each RNA nucleotide against the corresponding DNA nucleotide before splitting it off. This process does not occur in nature; instead RNA is degraded in a nonspecific and logically irreversible manner by other enzymes, such as polynucleotide phosphorylase. This enzyme catalyzes a reversible reaction between an RNA strand and free phosphate (maintained at high concentration) to split off successive nucleotides of the RNA as nucleotide phosphate monomers. Because the enzyme functions in the absence of a complementary DNA strand, the removal of each nucleotide is logically irreversible, and when running backwards (in the direction of RNA synthesis) the enzyme is about as likely to insert any of the three incorrect nucelotides as it is to reinsert the correct one. This logical irreversibility means that a fourfold higher phosphate concentration is needed to drive the reaction forward than would be required by a sequence-specific degradation. The excess phosphate keeps the enzyme from running backwards and synthesizing random RNA, but it also means that the cycle of specific synthesis followed by nonspecific degradation must waste about $kT \ln 4$ per nucleotide even in the limit of zero speed. For an organism that has already spent around $20kT$ per nucleotide to produce the RNA with near maximal speed and accuracy, the extra $1.4kT$ is obviously a small price to pay for being able to dispose of the RNA in a summary manner, without taking it back to its birthplace.

Corollary to the principle that erasing a random tape entails an entropy increase in the environment is the fact that a physical system of low entropy (e.g., a gas at high pressure, a collection of magnetic domains or individual atomic spins, all pointing "down," or more symbolically, a tape full of zeros) can act as a "fuel," doing useful thermodynamic or mechanical work as it randomizes itself. For example, a compressed gas expanding isothermally against a piston increases its own entropy, meanwhile converting waste heat from its surroundings into an equivalent amount of mechanical work. In an adiabatic demagnetization experiment, the randomization of a set of aligned spins allows them to pump heat from a cold body into warmer surroundings. A tape containing $N$ zeros would similarly serve as a fuel to do $NkT \ln 2$ of useful work as it randomized itself.

What about a tape containing a computable pseudorandom sequence like the first $N$ digits of the binary expansion of pi? Although the pattern of digits in pi looks random to begin with, and would not accomplish any useful work if allowed to truly randomize itself in an adiabatic demagnetization apparatus of the usual sort, it could be exploited by a slightly more complicated apparatus. Since the mapping ($N$ zeros)$\leftrightarrow$(first $N$ bits of pi) is one-to-one, it is possible in principle to construct a reversible computer that would execute this mapping physically, in either direction, at an arbitrarily small thermodynamic cost per digit. Thus, given a pi tape, we could convert it into a tape full of zeros, then use that as fuel by allowing it to truly randomize itself at the expense of waste heat in the surroundings.

What about a tape containing $N$ bits generated by coin tossing? Of course coin tossing might accidentally yield a tape of $N$ consecutive zeros, which could be used as fuel; however, the typical result would be a sequence with no such unusual features. Since the tape, after all, contains a specific sequence, $NkT \ln 2$ of work ought to be made available when it is mapped, in a one-to-many manner, onto a uniform distribution over all $2^N$ $N$-bit sequences. However, because there is no concise description of the way in which this specific sequence differs from the bulk of other sequences, no simple apparatus could extract this work. A complicated apparatus, containing a verbose description or verbatim copy of the specific sequence, could extract the work, but in doing so it would be doing no more than canceling the specific random sequence against the apparatus's copy of it to produce a blank tape, then using that as fuel. Of course there is a concise procedure for converting a random tape into a blank tape: erase it. But this procedure is not logically reversible, and so would require an input of energy equal to the fuel value of the blank tape it produced.

Finally, suppose we had seven identical copies of a typical random tape. Six of these could be converted into blank tapes by a logically and thermodynamically reversible process, e.g., subtracting one copy from another, digit by digit, to produce a tape full of zeros. The last copy could not be so canceled, because there would be nothing left to cancel it against. Thus the seven identical random tapes are interconvertible to six blank tapes and one random tape, and would have a fuel value equal to that of six blank tapes. The result of their exploitation as fuel, of course, would be seven *different* random tapes.

## 6. ALGORITHMIC ENTROPY AND THERMODYNAMICS

The above ideas may be generalized and expressed more precisely using algorithmic information theory, described in an introductory article by Chaitin (1975a), and review articles by Zvonkin and Levin (1970), and Chaitin (1977).

Ordinary information theory offers no solid grounds for calling one $N$-bit string "more random" than another, since all are equally likely to be produced by a random process such as coin tossing. Thus, in particular, it cannot distinguish bit strings with fuel value from those without. This inability reflects the fact that entropy is an inherently statistical concept, applicable to ensembles but not to the individual events comprising them. In equilibrium statistical mechanics, this fact manifests itself as the well-known impossibility of expressing macroscopic entropy as the ensemble average of some microscopic variable, the way temperature can be expressed as the average of $mv^2/2$: since the entropy of a distribution $\mathbf{p}$ is defined as

$$S[\mathbf{p}] = \sum_x p(x) \log[1/p(x)]$$

obviously there can be no one function $f(x)$, such that for all $\mathbf{p}$,

$$S[\mathbf{p}] = \sum_x p(x) f(x)$$

The absence of a microscopic quantity corresponding to entropy is a nuisance both practically and conceptually, requiring that entropy always be measured by indirect calorimetric methods, both in the laboratory and in Monte Carlo and molecular-dynamics "computer experiments" (Bennett, 1975), and frustrating the natural desire of molecular model builders to regard an individual molecular configuration as having an entropy.

Though it does not help the practical problem, the notion of algorithmic entropy resolves the conceptual problem by providing a microstate function, $H(x)$, whose average is very nearly equal to the macroscopic entropy $S[\mathbf{p}]$, not for all distributions $\mathbf{p}$ (which would be impossible) but rather for a large class of distributions including most of those relevant to statistical mechanics. Algorithmic entropy is a measure, not of any obvious physical property of the microstate $x$, but rather of the number of bits required to describe $x$ in an absolute mathematical sense, as the output of a universal computer. Algorithmic entropy is small for sequences such as the first million digits of pi, which can be computed from a small description, but large for typical sequences produced by coin tossing, which have no concise description. Sequences with large algorithmic entropy cannot be erased except by an irreversible process; conversely, those with small algorithmic entropy can do thermodynamic work as they randomize themselves.

Several slightly different definitions of algorithmic entropy have been proposed; for definiteness we adopt the definition of Levin (Gacs, 1974; Levin, 1976) and Chaitin (1975b) in terms of self-delimiting program size: the algorithmic entropy $H(x)$ of a binary string $x$ is the number of bits in the smallest self-delimiting program causing a standard computer to embark on a halting computation with $x$ as its output. (A program is self-delimiting if it needs no special symbol, other than the digits 0 and 1 of which it is composed, to mark its end.) The algorithmic entropy of discrete objects other than binary strings, e.g., integers, or coarse-grained cells in a continuous phase space, may be defined by indexing them by binary strings in a standard way. A string is called "algorithmically random" if it is not expressible as the output of a program much shorter than the string itself. A simple counting argument shows that, for any length $N$, most $N$-bit strings are algorithmically random (e.g., there are only enough $N-10$ bit programs to describe at most $1/1024$ of all the $N$-bit strings).

At first it might appear that the definition of $H$ is extremely machine dependent. However, it is well known that there exist computers which are universal in the strong sense of being able to simulate any other computer with at most an additive constant increase in program size. The algorithmic entropy functions defined by any two such machines therefore differ by at most $O(1)$, and algorithmic entropy, like classical thermodynamic entropy, may be regarded as well defined up to an additive constant.

A further noteworthy fact about $H(x)$ is that it is not an effectively computable function of $x$: there is no uniform procedure for computing $H(x)$ given $x$. Although this means that $H(x)$ cannot be routinely evaluated the way $x!$ and $\sin(x)$ can, it is not a severe limitation in the present context, where we wish chiefly to prove theorems about the relation between $H$ and other entropy functions. From a more practical viewpoint, the molecular model builder, formerly told that the question "what is its entropy?" was meaningless, is now told that the question is meaningful but its answer cannot generally be determined by looking at the model.

It is easy to see that simply describable deterministic transformations cannot increase a system's algorithmic entropy very much (i.e., by more than the number of bits required to describe the transformation). This follows because the final state can always be described indirectly by describing the initial state and the transformation. Therefore, for a system to increase its algorithmic entropy, it must behave probabilistically, increasing its statistical entropy at the same time. By the same token, simply describable reversible transformations (1 : 1 mappings) leave algorithmic entropy approximately unchanged.

Algorithmic entropy is a microscopic analog of ordinary statistical entropy in the following sense: if a macrostate **p** is *concisely describable*, e.g., if it is determined by equations of motion and boundary conditions describable in a small number of bits, then its statistical entropy is nearly equal to the ensemble average of the microstates' algorithmic entropy. In more detail, the relation between algorithmic and statistical entropy of a macrostate is as follows:

$$S_2[\mathbf{p}] < \sum_x p(x)H(x) \leqslant S_2[\mathbf{p}] + H(\mathbf{p}) + O(1) \qquad (1)$$

Here $S_2[\mathbf{p}] = \Sigma_x p(x)\log_2[1/p(x)]$, the macrostate's statistical entropy in binary units, measures the extent to which the distribution **p** is spread out over many microstates; $H(x)$, the microstate's algorithmic entropy, measures the extent to which a particular microstate $x$ is not concisely describable; and $H(\mathbf{p})$, the algorithmic entropy of **p**, is the number of bits required to describe the distribution **p**.

We need to say in more detail what it means to describe a distribution. Strictly construed, a description of **p** would be a program to compute a tabulation of the components of the vector **p**, to arbitrarily great precision in case any of the components were irrational numbers. In fact, equation (1) remains true for a much weaker kind of description: a Monte Carlo program for sampling some distribution **q** not too different from **p**. In this context, a Monte Carlo program means a fixed program, which, when given to the machine on its input tape, causes the machine to ask from time to time for additional bits of input; and if these are furnished probabilistically by tossing a fair coin, the machine eventually halts, yielding an output distributed according to the distribution **q**. For **q** not to be too different from **p** means that $\Sigma_x p(x)\log_2[p(x)/q(x)]$ is of order unity.

For typical macrostates considered in statistical mechanics, the equality between statistical entropy and the ensemble average of algorithmic entropy holds with negligible error, since the statistical entropy is typically of order $10^{23}$ bits, while the error term $H(\mathbf{p})$ is typically only the few thousand bits required to describe the macrostate's determining equations of motion, boundary conditions, etc. Macrostates occurring in nature (e.g., a gas in a box with irregular walls) may not be so concisely describable, but the traditional approach of statistical mechanics has been to approximate nature by simple models.

We now sketch the proof of equation (1). The left inequality follows from the convexity of the log function and the fact that, when algorithmic entropy is defined by self-delimiting programs, $\Sigma_x 2^{-H(x)}$ is less than 1. The right inequality is obtained from the following inequality, which holds for all $x$:

$$H(x) \leqslant H(\mathbf{q}) + \log_2[1/q(x)] + O(1) \qquad (2)$$

Here $\mathbf{q}$ is a distribution approximating $\mathbf{p}$ that can be exactly sampled by a Monte Carlo program of size $H(\mathbf{q})$. The additive $O(1)$ constant again depends on the universal computer but not on $\mathbf{q}$ or $x$. Equation (2) follows from the fact that one way of algorithmically describing $x$ is to first describe the distribution $\mathbf{q}$ and then describe how to locate $x$ within this distribution. The proof that a self-delimiting program of size $\log_2[1/q(x)] + O(1)$ bits suffices to compute $x$, given a Monte Carlo routine for sampling $\mathbf{q}$, is given by Chaitin (1975b, Theorem 3.2); slightly weaker versions of the idea are easy to understand intuitively. Equation (1) is finally obtained by summing equation (2) over the distribution $\mathbf{p}$, and applying the criterion of closeness of approximation of $\mathbf{p}$ by $\mathbf{q}$.

In conjunction with the second law, equation (1) implies that when a physical system increases its algorithmic entropy by $N$ bits (which it can only do by behaving probabilistically), it has the capacity to convert about $NkT \ln 2$ of waste heat into useful work in its surroundings. Conversely, the conversion of about $NkT \ln 2$ of work into heat in the surroundings is necessary to decrease a system's algorithmic entropy by $N$ bits. These statements are classical truisms when entropy is interpreted statistically, as a property of ensembles. The novelty here is in using algorithmic entropy, a property of microstates. No property of the ensembles need be assumed, beyond that they be concisely describable.

## ACKNOWLEDGMENTS

The work on enzymatic and clockwork Turing machines was done in 1970–72, at Argonne National Laboratory (Solid State Science Division), under the auspices of the U.S. Atomic Energy Commission. I wish to thank Rolf Landauer and Gregory Chaitin for years of stimulating discussions of reversibility and entropy, Michael Chamberlain for background information on polymerases, and John Slonczewski for background information on ferromagnets.

## REFERENCES


Benioff, Paul (1982) to appear in *Journal of Statistical Mechanics.*

Bennett, C. H. (1973). "Logical Reversibility of Computation", *IBM Journal of Research and Development*, **17**, 525–532.

Bennett, C. H. (1975). "Efficient Estimation of Free Energy Differences from Monte Carlo Data," *Journal of Computational Physics*, **22**, 245–268.

Bennett, C. H. (1979). "Dissipation-Error Tradeoff in Proofreading," *BioSystems*, **11**, 85–90.

Chaitin, G. (1975a). "Randomness and Mathematical Proof," *Scientific American*, **232**, No. 5, 46–52.

Chaitin, G. (1975b). "A Theory of Program Size Formally Identical to Information Theory," *Journal of the Association for Computing Machinery*, **22**, 329–340.

Chaitin, G. (1977). "Algorithmic Information Theory," *IBM Journal of Research and Development*, **21**, 350–359, 496.

Brillouin, L. (1956). *Science and Information Theory* (2nd edition, 1962), pp. 261–264, 194–196. Academic Press, London.

Fredkin, Edward, and Toffoli, Tommaso, (1982). "Conservative Logic," MIT Report MIT/LCS/TM-197; *International Journal of Theoretical Physics*, **21**, 219.

Gacs, P. (1974). "On the Symmetry of Algorithmic Information," *Soviet Mathematics Doklady*, **15**, 1477.

Hopfield, J. J. (1974). *Proceedings of the National Academy of Science USA*, **71**, 4135–4139.

Keyes, R. W., and Landuer, R. (1970). *IBM Journal of Research and Development*, **14**, 152.

Landauer, R. (1961). "Irreversibility and Heat Generation in the Computing Process," *IBM Journal of Research and Development*, **3**, 183–191.

Levin, L. A. (1976). "Various Measures of Complexity for Finite Objects (Axiomatic Description)," *Soviet Mathematics Doklady*, **17**, 522–526.

Likharev, K. (1982). "Classical and Quantum Limitations on Energy Consumption in Computation," *International Journal of Theoretical Physics*, **21**, 311.

McCarthy, John (1956). "The Inversion of Functions Defined by Turing Machines," in *Automata Studies*, C. E. Shannon and J. McCarthy, eds. Princeton Univ. Press, New Jersey.

Ninio, J. (1975). *Biochimie*, **57**, 587–595.

Reif, John H. (1979). "Complexity of the Mover's Problem and Generalizations," Proc. 20'th IEEE Symp. Found. Comp. Sci., San Juan, Puerto Rico, pp. 421–427.

Szilard, L. (1929). *Zeitschrift für Physik*, **53**, 840–856.

Toffoli, Tommaso (1980). "Reversible Computing," MIT Report MIT/LCS/TM-151.

Toffoli, Tommaso (1981). "Bicontinuous Extensions of Invertible Combinatorial Functions," *Mathematical and Systems Theory*, **14**, 13–23.

von Neumann, J. (1966). Fourth University of Illinois lecture, in *Theory of Self-Reproducing Automata*, A. W. Burks, ed., p. 66. Univ. of Illinois Press, Urbana.

Watson, J. D. (1970). *Molecular Biology of the Gene* (2nd edition). W. A. Benjamin, New York.

Zvonkin, A. K., and Levin, L. A. (1970). "The Complexity of Finite Objects and the Development of the Concepts of Information and Randomness by Means of the Theory of Algorithms," *Russian Mathematical Surveys*, **25**, 83–124.

# Computation: A Fundamental Physical View

Rolf Landauer

IBM Thomas J. Watson Research Center, P.O. Box 218, Yorktown Heights, New York 10598, U.S.A.

*Received September 3, 1986; accepted September 12, 1986*

### Abstract

Attempts to understand the fundamental physical limits of computation have been under way for over a quarter century. We discuss this field, with emphasis on the central notion of reversible computation, and emphasis on the relationship to the ultimate nature of physical law. A brief discussion of the generation of information is included. In ordinary computation, noise is a source of error, to be offset to the maximum possible extent. This can be done via reversible computation. Alternatively, there are situations in which noise controls the transitions in a system between many competing states of local stability, and can be used to explore this manifold. In the general case, where the noise depends on the state of the system, relative stability can only be determined by the kinetics along the *whole* pathway from one state of local stability to another one. Examination of the two terminal states to be compared cannot tell us which is the more likely state.

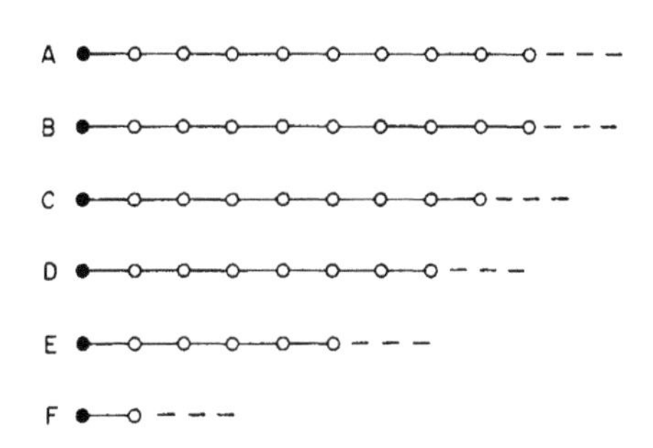


*Fig. 1.* The left-hand end of a horizontal chain represents the initial state, and forward computation represents motion to the right, through a sequence of states represented by successive circles. Different letters correspond to different initial states, i.e., different programs.

### 1. Introduction; reversible computation

The search for the fundamental physical limits of the computational process has been under way for more than a quarter century. The important notion of reversible computation, which came out of that search, was published in 1973 [1]. It is not our purpose, here, to review and explain this whole field. For an introduction to the area, and clues to the citation trail, see Refs. [2] and [3]. We will, instead, take the opportunity to make or emphasize some subsidiary points, which may not have had the same frequent exposure to public view.

First of all, let us explain why this field deserves attention. It is not for technological reasons. Reversible computation, which consists of a sequence of logical 1 : 1 mappings, is illustrated in Fig. 1. Each possible initial program defines a separate track in Fig. 1, with no merging or splitting. Such a computation resembles the motion of a particle through a periodic lattice, which – under a small driving force – will move diffusively over a short time scale, but will exhibit a drift velocity and predictable progress, over a longer time scale. Reversible computation can be carried out with energy dissipation proportional to the driving force and, therefore, to the computational velocity. That means that there is no minimal energy dissipation per step. But even if we ask for predictable computation, which does not act diffusively, a driving force corresponding to an expenditure of a few $kT$ per step will accomplish that. Note that this will be $kT$ per step of the total computer, and may correspond to the action of many simultaneous logic events. By comparison, low powered real circuits consume of the order of $10^9\,kT$ per logic event. When reality is so far from the fundamental limits, the limits do not serve as guide to the technologist.

Then why are we in this business? *Because it is at the very core of science.* Science, and physics most particularly, is expressed in terms of mathematics, i.e. in terms of rules for handling numbers. Information, numerical or otherwise, is not an abstraction, but is inevitably tied to a physical representation. The representation can be a hole in a punched card, an electron spin up or down, the state of a neuron, or a DNA configuration. There really is no software, in the strict sense of disembodied information, but only inactive and relatively static hardware. Thus, the handling of information is inevitably tied to the physical universe, its content and its laws. This is illustrated by the right hand arrow of Fig. 2. We have all been indoctrinated by the mathematicians and their sense of values. Given $\varepsilon$, $\exists N$, such that -------. Now we come back and ask, "Does this arbitrarily large $N$, this implied unlimited sequence of *infallible* operations, *really exist*?" If not, then the continuum mathematics of our early education can have no relationship to executable algorithms and physical

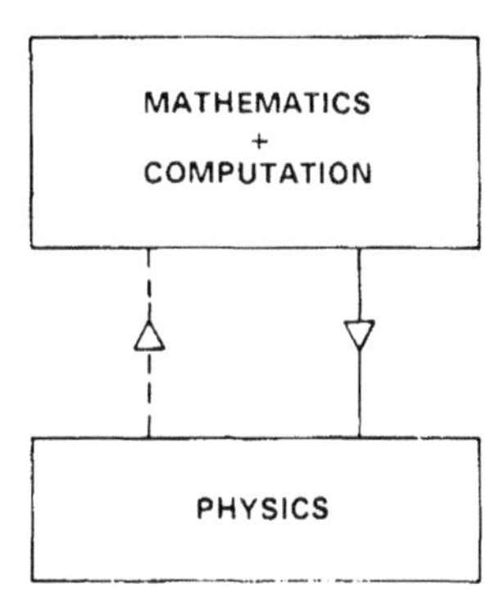


*Fig. 2.* Information handling, shown at the top, is dependent on that which the real physical universe permits, through available "parts," and through the laws of physics. This dependency is indicated by the right arrow. But the laws of physics are directions for processing information and, therefore, dependent on executable algorithms. This is indicated by the dashed arrow at left.

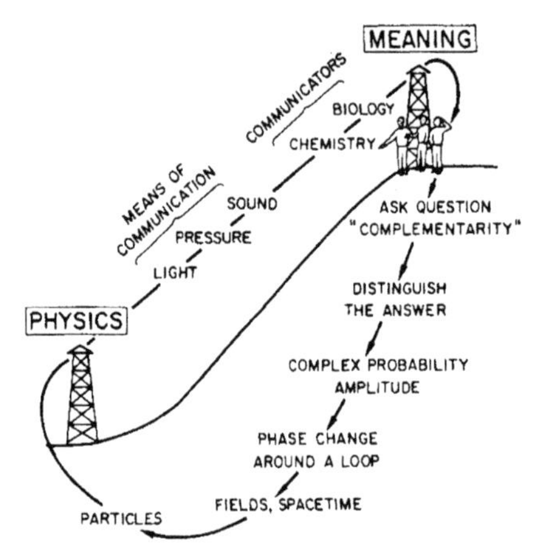


*Fig. 3.* Wheeler's "meaning circuit". According to Wheeler, Physics gives light and sound and pressure, tools of communication. It gives biology and chemistry and, through them, communicators. Communication between communicators gives meaning. Meaning calls for the asking of questions, but the asking of one question stands in a complementary relation to the asking of another.

law. This brings us to the left hand arrow of Fig. 2, which states that physical law requires operations which, at least in principle, are executable. Thus, ultimately, we need a self-consistent formulation of the laws of physics, as suggested in Fig. 2. The laws describing our universe must be restricted to algorithms which are executable in our actual universe; what is executable will, in turn, depend on the contents of that universe (i.e. the storehouse of possible parts) and to the laws describing the interaction of these parts.

The self-consistency invoked in Fig. 2 is reminiscent of Wheeler's Meaning Circuit, shown in Fig. 3, and adapted from Ref. [4]. John Wheeler has pointed us toward Era III of physics, in which, ". . . we have to seek nothing less than the foundation of physical law itself." Wheeler contrasts this to Era II, which commenced with Newton, where the laws of physics were to be discovered, but their origin was not a fit subject for discussion. Wheeler has given us a specific proposal for Era III, the "meaning circuit". In this, physics gives rise to all the phenomena whose study has usually occupied us. These phenomena, in turn, give rise to biology, communicators, and their questions, and these lead to measurement and the laws of physics. Our typical view in science has been one under which the laws of physics were there at the beginning; the laws are a process control program which steers the evolution of the universe. Wheeler reminds us that the discovery of laws is, or involves, quantum mechanical measurement. The results of measurement are not independent of the act of measurement, and the outcome of the measurement was not there, all along, to be revealed in its finally measured form. If I can be presumptuous enough to differ from Wheeler, it is related to his stress on human observers and their posing of questions. Measurement does not require intelligent beings, the environment is continually acting on a system, and making measurements [5]. Incidentally, that is the most elementary way in which Schrödinger's famous cat paradox is misleading. Long before the human observer looks at the cat, the air molecules bouncing into the cat have measured its temperature, or its breathing motion. All along, parts of the cat had made measurements on each other, e.g., the state of the brain cells depends on the continued blood flow. In fact, long before the bullet has left its barrel, there have been interactions of that sort. *I* do not really know what constitutes a *measurement*, though *others* are likely to have the answer to that [6, 7]. Most particularly, I do not know what constitutes a set of measurements leading to a physical law. It is not clear that this has to involve complex organisms that publish conference papers. Perhaps the universe, like Schrödinger's cat, is asking questions about the laws of physics all along. I do not assert that Wheeler, in Fig. 3, is wrong. Rather: I am not sure he is right, *perhaps* the circuit can close at a lower level.

While this author admits that he is uncertain about the exact character of a "measurement", others are less squeamish. Some of the critics of reversible computation [8–12] invoke a simple recipe which goes like this: Computation continually requires measurement on the existing state of the computer. Measurement requires energy dissipation, as is known from Szilard's analysis of Maxwell's demon [13], and requires energy dissipation related to the number of possibilities between which the measurement discriminates, even if done slowly. This criticism of reversible computation, based on a *perception* of *applicable principles*, has been refuted [14], in general terms. But there is a more cogent counterargument. Reversible computation has been described in terms of a number of detailed embodiments, listed in Ref. [3]. One of these [15] invokes Josephson junction circuitry, and could actually be made. The critics of reversible computation need to show us where such proposals err; otherwise, we are dependent upon their particular notion of what constitutes a measurement. Simply coupling two systems together, for a period, so that there is a cross-influence in their time evolution, does not require energy dissipation. This was already stated, in connection with the computational process, in 1961 [16]. Then why does measurement require energy dissipation? Because, after the measurement (or before the first measurement) the measurement apparatus has to be reset. This represents the erasure of information, and it is the erasure of information which in measurement [17], just as in computation [16], requires energy dissipation. Unfortunately, the literature on the energy requirements of the classical measurement process, while correct, manages to obscure this simple and central point. Reversible computation is, of course, a process that avoids erasure; Fig. 1 consists of a sequence of 1 : 1 mappings. It was the brilliant insight of Bennett [1] that universal computation can actually be carried out reversibly.

In connection with the continuing occasional objections to reversible computation [8–12], I take the liberty of including some editorial remarks, more personal and subjective in character than is typical for a scientific paper. When I was first exposed to Bennett's notion of reversible computation I was very skeptical, despite the fact that my own work had, in some ways, come very close. It took me months to become fully convinced that Bennett was right. Since then I have seen many colleagues, in a first exposure, struggle with this perhaps counter-intuitive idea. But it is now many years and many papers later. Reversible computation has been explored and expounded by a good many authors from a variety of veiwpoints, and it should not longer be that hard to understand. Remarkably enough, some of the critics [11, 12] refer to my work, but do not actually cite me.

The study of the physical limits of the computational process, so far, has served primarily to dispel limits that might have been expected on the basis of superficial speculation. $kT$ is not a minimal energy dissipation, per step, nor does the uncertainty principle help to set energy dissipation requirements. Noise immunity, in the intentionally used degrees of freedom (in contrast to fluctuations which actually cause equipment deterioration) is available to any required degree. Then where are the real limits? They are, presumably, related to more cosmological questions, and unlikely to be answered by those of us who started the field. How many degrees of freedom, for example, can be kept together effectively, to constitute a computer? If we cannot have an unlimited memory, then can we really calculate $\pi$ to as many placed as desired. A start toward such cosmological questions has been made by several investigators [18]. We will not try to evaluate or summarize these attempts; the important point: They are a thrust in the right general direction.

There are, of course, others who are concerned with the fact that our actual universe is not really characterized by a continuum. T. D. Lee [19] has proposed a "discrete mechanics". Feynman has stated [20], ". . . *everything* that happens in a finite volume of space and time would have to be exactly analyzable with a finite number of logical operations. The present theory of physics is not that way, apparently. It allows space to go down into infinitesimal distances, wavelengths to get infinitely great, terms to be summed in infinite order, and so forth; and, therefore, if this proposition is right, physical law is wrong". That statement clearly has a resemblance to our exposition. Ross [21] has invoked a model of the real number line based on probabilistic considerations. Woo [22] discusses the relation between discrete field theory and Turing machines. None of these, however, seem to go as far as we do, and none of them point out that the correct physical law must be based on algorithms whose nature is still to be found, rather than on classical mathematical analysis [23].

Mathematicians declared their independence of the real physical universe, about a century ago, and explained that they were really describing abstract objects and spaces. If some of these modelled real events, all right, but . . . . They ignored the question which we have emphasized. Mathematicians may be describing abstract spaces, but the actual information manipulation is still in the real physical universe. By and large, mathematicians do not welcome the points we have made, and scorn questions about physical executability, even though the mathematics community occasionally arrives at its own reasons for questioning the continuum. There are exceptions, however. One is found in the work of Bremermann [24]. A particularly clear version occurs in the work of Herman Weyl. Wheeler [25] quotes Weyl, . . . *belief in this transcendental world [of mathematical ideals, of propositions of infinite length, and of a continuum of natural numbers] taxes the strength of our faith hardly less than the doctrines of the early Fathers of the Church or of the scholastic philosophers of the Middle Ages.* Wheeler then goes on with his own statement: *Then how can physics in good conscience go on using in its description of existence a number system that does not even exist?*

The noise immunity in reversible computation is obtained by invoking large potentials, which act as barriers to noise in the undesired processes. In Bennett's original springless clockwork Turing machine [1] "hard" parts are assumed, which are not penetrable by each other, and which do not fall apart under the influence of thermal fluctuations. Similar assumptions underlie other reversible computer proposals. It is important to understand that the immunity is bought at the expense of the size of the potentials involved, or – mor or less equivalently – the size of the parts, and not at the expense of energy dissipation. In the real universe, with a limited selection of parts, arbitrary immunity may not be available. Noise immunity, however, which lasts for the lifetime of the universe should be adequate. The many real pressures, however, for continued miniaturization in computers, are likely to take us in the opposite direction. Techniques for improving the reliability of computation have been discussed in Ref. [3]. We stress, here, the undesirability of noise induced disturbances in computation, in anticipation of a subsequent section in which noise is viewed in a more constructive way.

## 2. Generation of information

Reference [16] pointed to the association between information destruction and physical irreversibility. The prevalence of friction and information loss leads to the questions: Are there counterbalancing processes? Does information get generated, or does it only disappear? We will try to give a very partial answer adapted from Ref. [26]. The generation of information must be the opposite of its destruction. Systems that are initially indistinguishable must develop a recognizable difference. There are several possible causes for the development of such differences. The most commonly cited one will be invoked later in this paper; noise that reaches our system from external sources (or from any degrees of freedom not explicitly described in the system's equations of motion) causes ensemble members, which were initially together, to diffuse apart. One can then go on, of course, to argue that the information did not really arise, but reflects information originally in the noise sources, and is transferred to the system of concern. In a universe with limited precision, however, the noise cannot be followed back in time and distance to any desired accuracy. Thus, the distinctions in behaviour produced by noise can be considered to be "generated".

In the last decade it has become acceptable to invoke chaotic processes, instead [27]. These are processes in deterministic nonlinear systems, possibly systems with only a few degrees of freedom, which are not periodic or quasi-periodic. Nearby trajectories diverge exponentially in time, instead of the typical diffusive $t^{1/2}$ divergence caused by noise [28]. Once again, that exponential divergence, by itself, does not really produce information. We must also claim that two systems that originally were very close, but not identical, are nevertheless in some "in principle" sense indistinguishable. Furthermore it is possible, as discussed in the paper by Swinney in this set of proceedings, to distinguish between the effects of external noise and the effects of deterministic chaos.

Amplifiers can also serve to push signals apart that are close together, initially. Amplifiers need not have the form of transistor circuits with a spatially separate input and output. In parametric amplifiers, to cite one example [28], we can utilize a departure from the neighbourhood of an unstable state and can achieve exponential growth in time. The growing separation of these signals will be accompanied by dissipative effects, i.e. by a compression of phase space for other degrees of freedom.

It is not clear whether one of the mechanisms cited above has to be regarded as the most fundamental. If a choice is necessary, then our preference lies with the direct effects of noise. After all noise clearly exists, and is pervasive, whereas noiseless dynamic systems do not exist. We add a cautionary note. Quantum mechanics, despite an intimate relationship to probabilities, contains no built in noise sources. It does not provide for a growing spread in nearby systems; entropy is conserved in quantum mechanics just as in the classical case.

Arbitrarily reliable computation, with unlimited precision, is unlikely to be available in our real worl. Thus, the very question discussed in the preceding paragraph, whether two initially identical states become separated with time, does not really have a very clear meaning. After all, in a world with limited precision, we can never be absolutely certain about the initial identity of two very close-lying states. Thus, the likely limited precision of our number handling introduces something like noise into our basic laws of dynamics. It is not necessarily an artifact which requires supplementary explanation.

### 3. What is computation?

We have equated the computational process with a universal Turing machine, essentially that which a real and practical computer can do if it had access to unlimited memory? This is a process in which branching is possible. Intermediate results are invoked as the basis for decisions about subsequent steps. As pointed out by Keyes [29], many otherwise interesting physical processes proposed for computational purposes do not allow such an unlimited chain of steps, but only a limited sequence chosen by the apparatus designer.

A good many physical processes, involving cellular automata rules, or other ways of coupling large arrays of elements in a relatively uncomplicated geometrical way, and without a great deal of detailed design, have been proposed as potentially useful tools for pattern recognition, as associative memories, and for other similar purposes. We can cite only a small proportion of this extensive literature [30]. It is not our purpose here to evaluate or explain these proposals. Our key point: Many of these proposals are unrelated to universal computation. Cellular automata can, of course, be chosen so as to be capable of universal computation, and even to do computation which is logically reversible [31], but many of the cellular automata discussed in the literature are more limited than that. Motion toward one of many attractors, in a system with competing states of local stability, is *not* universal computation. That does not mean that the processes are uninteresting, or that they have no utility. Nevertheless, I will insert here, a somewhat conservative reaction. Proposals for computer simulations of networks which learn, are unstructured or randomly structured initially, and mimic neuron interaction, date from the early 1950's and the beginning of large electronic computers. At IBM, this project carried out by N. Rochester and collaborators, on the engineering model of the 701, was called the "Conceptor". Such proposals have become more sophisticated with time, but have never achieved major practical success. Some of the enthusiasm, in the current audience, comes from physical scientists who have not asked: How can (and have) some of these goals been achieved through more conventional software techniques, and more conventionally organized hardware? For example, we do know how to build content-addressable memories out of silicon technology. Pattern recognition is a hard process for digital computers; nevertheless, it has been done in *real* applications. Over the decades, for example, a good deal of progress has been made toward the recognition of continuous speech. The hard part is not the motion toward an attractor, within a basis, but is the *definition* of the basin of attraction. In an associative memory, which can respond to slightly faulty patterns, one has to learn about the likely errors. "Hopfield," entered by keystroke, is quite likely to turn out "opfield", it is unlikely to be inverted to "Dleifpoh". A successful recognition system must take such statistics into account. If it is a system that learns, then it must be able to apply the experience with the name "Hopfield" to the name "Smith". Some of the recent physical literature on neural networks can, perhaps, be justified as a model of biological functions, rather than as a useful technique for man-made devices. This author is not qualified to judge that realm. I cannot help, however, noticing the unqualified assertions in the physics literature, exemplified by [32]:

> It is generally expected that the essential characteristics of the dynamics [of a neural network] are captured by a Hamiltonian of the form
>
> $$H_N = -\tfrac{1}{2}\sum_{i,j} J_{ij}S(i)S(j). \qquad (1)$$
>
> The $N$ neurons are described by Ising spin variables . . .

Others [33], publishing in the same journal of neurophysiology and computer science, are a little more cautious and point out that they are analyzing a particular *model*. The biologists, as exemplified by a recent review of neuronal circuits [34], have paid little attention, though that – by itself – leads us to no clear conclusion.

The relationship between physical processes and computation has become widely appreciated in recent years [35], and once again we can only cite some illustrative items. A computer, however, even a special purpose analog computer, needs places where input is entered and output collected, in a standardized representation. An electronic analog computer had dials to set the input. A slide rule had numbers and a scale printed on its surface; in fact, it was the very choice of representation which gave the slide rule its utility for multiplication. To call any physical evolution with time an analog computation seems a little strained, though we cannot say it is clearly wrong.

In connection with our reference to input and output, it is appropriate to discuss this in connection with reversible computation. It is sometimes assumed that, even if information transfer within the computer can be done without a mimimal dissipation [11, 12], this does not apply to the input and output; a *measurement* at that point becomes essential. It is true, of course, that these input-output processes can be made more dissipative than the computational steps but they do not have to be more dissipative. Why should information transfer into the system or out of the system *have* to be more dissipative than that within the system? Bennett [17] has pointed out that a magnetic tape can be copied slowly without minimal dissipation. The same point is already implicit in the work of Keyes and Landauer [36].

Do we need a dissipative measurement to signal the completion of a program? That, most certainly, is one possibility. We can, also, give ourselves lots of time to do whatever it is we want to do upon program completion, if we dissipate several $kT$ so that the computer cannot fluctuate right back out of the final state into an earlier state, as is likely in the diffusive behavior of reversible computers. But once again we must ask if the transfer of information out of whatever it is we call the computer, to another set of digital registers, is really very different from transfer within the so-called computer? It is, because reversible computation has to be undone, and if we also reverse the information transfer out of the computer prematurely, our computation will have been useless. The easiest answer to that, but not necessarily the optimum answer, is to perform occasonal dissipative measurements on the computer, to see if the "Task Completion Bit" is in its 1 status. When it is, the reversible computer can go into a long idling cycle, in which only an internal computer clock advances to give us a time to copy the output register with very little dissipation. After that the computer reverses itself, clearing out its own output register and restoring its initial state.

## 4. Fluctuations

In the preceding discussion noise has been a potential source of error in otherwise predictable computations. This is in accord with the historical interest in noise as a source of confusion and error in measurement and communication. In more recent times there has been a growing awareness that noise can be *a controlling qualitative influence on the development of systems*. In systems which are multistable, i.e., have two or more competing states of local stability, noise determines the motion of the system between these competing states. Small systems, as found in molecular biology, and in the drive toward miniaturization of computer components, can show a strong susceptibility to noise. Furthermore such noise induced transitions may be of utility, as invoked in the simulated annealing technique for optimization in the presence of complex constraints [37]. A similar application occurs in a model for the mechanism of perception and visual recognition [38] in which fluctuations are invoked to allow escape from premature fixation. First order phase transitions, in ordinary thermodynamic systems, can occur by homogeneous nucleation, and the formation of a critical nucleus is a fluctuation which takes the system from one state of local stability (i.e., the original metastable phase) to a new state of local stability. Evolution and the origin of life can be viewed as an optimization process, in which fluctuations (e.g., mutations) take us from one metastable ecology, to a new one [39, 40]. In contrast to the discussions in Ref. [40], some of which represent a longstanding orientation in ecology, physical scientists have been inclined to use the word *self-organization* very loosely. To apply this expression to the Benard instability, for example, to the Zhabotinskii reaction, or to the oscillations of a laser, under conditions where these temporal and spatial patterns are the only allowed behavior, seems strained. We might, with equal justice, refer to the revolution of an electron around a hydrogen nucleus, or the rotation of a water wheel, as *self-organization* [41].

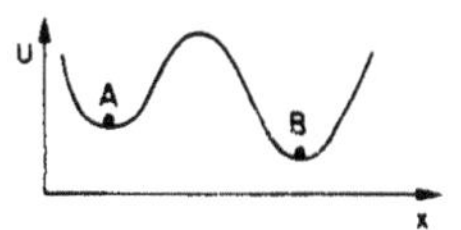


*Fig. 4.* Damped potential with metastable state A and state of lowest energy at B.

Questions about relative stability, e.g., is A or B of Fig. 4, the more likely state, are statistical questions. In the absence of fluctuations a particle will simply stay at a local minimum; it can escape from a well only with the assistance of noise. Let us discuss the type of situation shown in Fig. 4, but invoke greater generality than suggested by Fig. 4. Consider a law of motion along $x$ which is not necessarily that of a particle in a potential, but can represent a broader class of systems. $x$ could, for example, be the charge on a capacitor in an electrically excited bistable circuit, or represent the chemical composition in an open and driven chemical reactor. Without noise, the flux of probability $\varrho$

$$j = \varrho v(x) \tag{4.1}$$

is determined by the noiseless macroscopic velocity $v(x)$, determined by the system's kinetics. Equation (4.1) would cause the density to accumulate at the points of local stability, where $v(x) = 0$. In the presence of noise different ensemble members, which start together, subsequently diffuse apart from each other and also away from the deterministic trajectory defined by $\dot{x} = v(x)$. In the simplest cases this can be represented by an additional diffusion term in eq. (4.1), leading to

$$j = \varrho v(x) - D\frac{\partial p}{\partial x}. \tag{4.2}$$

In the steady state $j$ is independent of $x$. Unless current is introduced at $x = \pm\infty$ we will have $j = 0$ for all $x$. In that case, eq. (4.2) can readily be integrated to yield

$$\varrho(x) \simeq \exp\int (v/D)\,\mathrm{d}x, \tag{4.3}$$

showing that the steady state represents a balance between the deterministic velocity $v$, restoring the system to points of local stability, and $D$ which represents the noise, permitting departure from such points of local stability. As pointed out elsewhere [28], eq. (4.3) is a generalization of the Boltzmann factor, $\exp(-U/kT)$. If we compare the population between two points of local stability, we find

$$\varrho(B)/\varrho(A) = \exp\int_A^B (v/D)\,\mathrm{d}x. \tag{4.4}$$

Equation (4.4) shows the dependence on $D(x)$ between A and B. The kinetics along the connecting path matters, we cannot just inspect the end points A and B. We cannot expect, in general, some analog to Maxwell's equal area rule, which will predict relative stability on the basis of $\int_A^B v(x)\,\mathrm{d}x$. This was pointed out by this author [42] in 1962, is implicit in the work of R. L. Stratonovich [43] and was again emphasized [44] in 1975. The first version of the argument to be given subsequently in this paper [45] appeared in 1975, and was elaborated and repeated on a number of subsequent occasions. The point is frequently ascribed to later authors, and associated

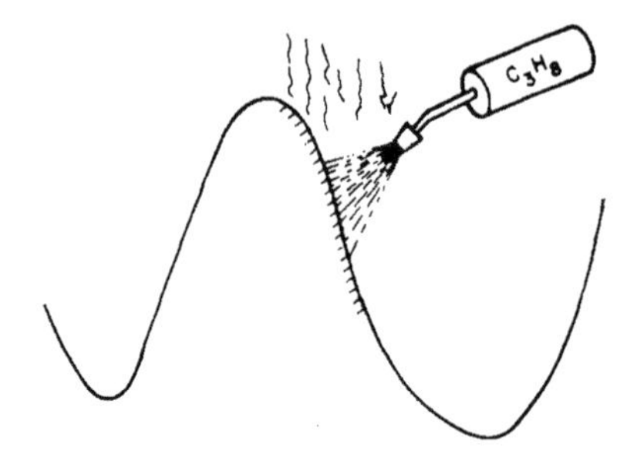


*Fig. 5.* A bistable potential in which the top right-hand side of the barrier has been brought to a temperature well above the adjacent portions. Particles escaping from the right-hand well can now get through the heated zone much more easily, leading to a smaller effective barrier.

with the labels *multiplicative noise, noise induced phase transitions*, and *external noise*.

In equilibrium, noise is defined by the temperature, and cannot be made an arbitrary function of the state of the system, i.e., of $x$ in our one-dimensional examples. It is this property which leads to the Maxwell equal area rule. More generally, however, the noise sources can be tailored to be almost any function of the state of the system. Our subsequent example will emphasize this. Consider a potential, as shown in Fig. 4. The right hand well is lower; in thermal equilibrium $\varrho(x)$ will be larger there. Let us now deviate from equilibrium in a very simple way. We heat up part of the right hand well, as illustrated in Fig. 5. We assume this is a heavily damped well; particles crossing the temperature boundary come to their new temperature very quickly.

To clarify the physical situation sketched in Fig. 5, we provide one possible embodiment in Fig. 6. This shows a sequence of insulating tubes with a particle. The particle is charged and the force-field is maintained by charges deposited on the outside of the tube. Different sections of the tube are maintained at different temperatures. The particle incident on a tube wall is assumed to bounce off without change in charge, but bounce off with the local tube temperature.

In Fig. 5 the particles entering the hot section from the right gain the energy needed for escape much more easily than without the elevated temperature. The particles in the right hand well, therefore, escape much more easily than without the blow-torch. The particles in the left hand well have an unmodified escape problem. In the steady state the two escape rates must balance. Therefore, the population in the right-hand well is depleted by the blow torch. If the temperature elevation is made large enough, most of the effect of the barrier in the hot section is eliminated. If the hot section is also made to extend over enough of the right-hand barrier shoulder, then the right-hand well which originally was more densely populated, becomes less densely occupied. The key point of this example: Relative stability can be controlled through the kinetics in parts of the system space which are rarely occupied. Therefore no amount of concern related to the neighborhood of the states of local stability can tell us which is the more likely neighborhood. This tells us that many of the favored principles in the physical chemistry literature, related to entropy, entropy production, excess entropy production, etc., cannot be applied to multistable systems. There are no short cuts for such systems; the detailed

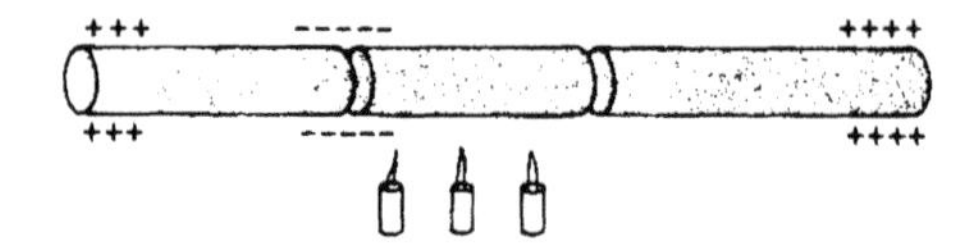


*Fig. 6.* Particle in an insulating tube, or series of tubes, subject to force field maintained by charges on the outside of the tube. The tube is heated nonuniformly.

kinetics along the transition paths must be taken into account. The relevance of this to discussions of evolution and the origin of life have been emphasized elsewhere [39, 46].

The lack of available short cuts, in predicting the noise-activated exploration of many competing states of local stability, relates to questions of complexity which have been discussed in the literature. As an introduction to this part of our discussion, let us first consider chaos. Multiply periodic motion, e.g., $y = \cos[\omega_0 t + (\delta \sin \omega_m t) t]$ may look a little complex, but does not present any unusual difficulty in the calculation of $y$ at a much later time; we do not need to follow the motion, in detail, over the intervening period. Chaos is usually more complex than that, in a genuine way. To predict a position a great many "cycles" later on, in the typical chaotic case, requires that we follow the motion very precisely. The program that directs such a calculation need not be complex, it is not a matter of great algorithmic complexity [47]. It is rather the detailed execution of the calculation which is long, i.e., the number of steps involved. This is a complexity measure which Bennett has proposed, discussed, and called "depth" [48].

The noise-activated search for likely states of local stability presents us with a similar dichotomy. If we are dealing with a multistable potential, and thermal equilibrium noise, then the depth of a set of wells allows us to compute their eventual (long-term) relative probability densities. (If we are concerned with the integrated probability of finding a particle *near* a minimum, rather than the probability *density*, then we must also take the well potential curvature into account, or equivalently compare free energies, rather than energies. That is not a significant distinction for our purposes.) If, however, we are dealing with a system which is not in thermal equilibrium, then it becomes more difficult. The additional difficulty relates to our discussion of the kinetics of Fig. 5, but of also the fact that in this more general case we can have circulation present in the steady state. The kinetics along the various pathways must then be taken into account. To predict the probability distribution at a much later time, from a given initial state, we *must* follow the motion in detail. (The italicized word *must* in the preceding sentence represents an intuitive assessment, and is not a result of a proof.) Such a measure of complexity, counting the number of states of local stability which have to be explored along the way, has been proposed by Kuhn [49]. He has called it *knowledge gained*, in connection with his discussions of the origin of life and the time-development of biological evolution. In general we can expect that there will be situations in which both of the above mentioned complexities are present simultaneously, and are not separable. We can start with a deterministic law of system motion, which is already chaotic, and the modify the system to make it stochastic.

The analogy between complexity in chaos and that in stochastic multistability is not meant to be carried too far.

Chaos was only used as an introduction to a complexity measure which characterizes program execution time. Chaotic motion, for example, does not settle down. The solution to a master equation, however, typically approaches a steady solution. In a space of unlimited extent, however, containing always more remote states of local stability, population changes *can* continue indefinitely.

This discussion of complexity in the presence of multistability is a first rough thrust in a direction which seems to deserve more thought, but is likely to demand more formal skills than this author can easily supply. The preceding discussion, for example, has only mentioned the possibility of circulation, without detailed consideration of it. Circulation can be present in purely deterministic motion; it can also be introduced by noise terms added to a deterministic equation.

## References


1. Bennett, C. H., IBM J. Res. Dev. **17**, 525 (1973).
2. Bennett, C. H. and Landauer, R., Sci. Am. **253**, No. 1, 48 (1985); **253**, No. 4, 9 (1985); Landauer, R. and Büttiker, M., Physica Scripta **T9**, 155 (1985); Landauer, R., in Computer Culture, The Scientific Intellectual and Social Impact of the Computer (Edited by H. R. Pagels), Ann. N. Y. Acad. Sci. **426**, 161 (1984); Landauer, R., Found. Phys. **16**, 551 (1986).
3. Landauer, R., in Der Informationsbegriff in Technik und Wissenschaft (Edited by O. G. Folberth and C. Hackl), p. 139, Oldenbourg, Munich (1986).
4. Wheeler, J. A., in Problems in Theoretical Physics (Edited by A. Giovannini, F. Mancini and M. Marinaro), p. 121, Univ. Salerno Press (1984); Wheeler, J. A., in Frontiers of Nonequilibrium Statistical Physics (Edited by G. T. Moore and M. O. Scully), Plenum, New York (1986).
5. Zurek, W. H., in Foundations of Quantum Mechanics in the Light of New Technology (Edited by S. Kamefuchi), p. 181, Physical Soc., Japan (1983). See also a number of papers in New Techniques and Ideas in Quantum Measurement Theory (Edited by D. Greenberger), N. Y. Acad. Sci. (in press).
6. Wheeler, J. A. and Zurek, W. H., Quantum Theory and Measurement, Princeton University Press (1983).
7. New Techniques and Ideas in Quantum Measurement Theory (Edited by D. Greenberger), N. Y. Acad. Sci. (in press).
8. Hastings, H. M. and Waner, S., Bio Systems **17**, 241 (1985).
9. Robinson, A. L., Science **223**, 1164 (1984).
10. Porod, W., Grondin, R. O., Ferry, D. K. and Porod, G., Phys. Rev. Lett. **52**, 232 (1984); Phys. Rev. Lett. **53**, 1206 (1984).
11. Rothstein, J., On Ultimate Thermodynamic Limitations in Communication and Computation, in Proceedings of the conference on Performance Limits in Communication Theory and Practice (Edited by J. K. Skwirzynski) Martinus Nijhoff, Dordrecht (in press).
12. Drogin, E. M., Defense Electronics, p. 31 (March, 1986).
13. Szilard, L., in Z. Phys. **53**, 840 (1929), English translation in Ref. [6], p. 539.
14. Benioff, P., Phys. Rev. Lett. **53**, 1203 (1984); ibid. Bennett, C. H., p. 1202, Landauer, R., p. 1205; Toffoli, T., p. 1204.
15. Likharev, K. K., Int. J. Theor. Phys. **21**, 311 (1982); Likharev, K. K., Tylov, S. V. and Semenov, V. K., IEEE Trans. Magn. **21**, 947 (1985).
16. Landauer, R., IBM J. Res. Dev. **5**, 183 (1961).
17. Bennett, C. H., Int. J. Theor. Phys. **21**, 905 (1982); Sci. Am. (to be published).
18. Barrow, J. D. and Tipler, F. J., The Anthropic Cosmological Principle, Ch. 10, p. 613, Clarendon, Oxford (1986); Dyson, F. J., Rev. Mod. Phys. **51**, 447 (1979); Tippler, F. J., Int. J. Theor. Phys. **25**, 617 (1986).
19. Friedberg, R. and Lee, T. D., Nucl. Phys. **B225**, 1 (1983); Lee, T. D., Difference Equations as the Basis of Fundamental Physical Theories, given at symposium, Old and New Problems in Fundamental Physics, held in honor of Wick, G. C. (Pisa, Italy, October 25, 1984).
20. Feynman, R. P., Int. J. Theor. Phys. **21**, 467 (1982).
21. Ross, D. K., Int. J. Theor. Phys. **23**, 1207 (1984).
22. Woo, C. H., Phys. Lett. **168B**, 376 (1986).
23. Landauer, R., IEEE Spectrum **4**, 105 (1967).
24. Bremermann, H. J., in The Encyclopedia of Ignorance (Edited by R. Duncan and M. Weston-Smith), p. 167, Pergamon, New York (1977).
25. Wheeler, J. A., Am. Sci. **74**, 366 (1986).
26. Landauer, R., in Computer Culture: The Scientific, Intellectual and Social Impact of the Computer, (Edited by H. R. Pagels), p. 167, Ann. N. Y. Acad. Sci. **426**, 161 (1984).
27. Shaw, R., Z. Naturforsch. **36a**, 80 (1981); Ford, J., Phys. Today **36**, 40 (1983).
28. Landauer, R., in Nonlinearity in Condensed Matter (Edited by A. R. Bishop, D. K. Campbell, P. Kumar and S. E. Trullinger), Springer (in press).
29. Keyes, R. W., Science **230**, 138 (1985).
30. Hopfield, J. J. and Tank, D. W., Science **233**, 625 (1986), includes an extensive list of earlier citations; Choi, M. Y. and Huberman, B. A., Phys. Rev. **B31**, 2862 (1985); Cooper, L. N., in J. C. Maxwell, the Sesquicentennial Symposium (Edited by M. S. Berger), Elsevier, Lausanne (1984); Ebeling, W., Pattern Processing and Optimization by Reaction Diffusion Systems (preprint); Edelman, G. M. and Reeke, G. N., Proc. Natl. Acad. Sci. USA **79**, 2091 (1982); Frumkin, A. and Moses, E., Phys. Rev. **A34**, 714 (1986); Huberman, B. A. and Hogg, T., Physica Scripta **32**, 271 (1985); Little, W. A. and Shaw, G. L., Behavioral Biology **14**, 115 (1975); Parisi, G., J. Phys. **A19**, L617 (1986); See also Ref. [38], and Kinzel, W. (in this issue).
31. Margolus, N., in Physica D, Cellular Automata. (Edited by D. Farmer, T. Toffoli and S. Wolfram), p. 81, North-Holland, Amsterdam (1984).
32. van Hemmen, J. L. and Kühn, R., Phys. Rev. Lett. **57**, 913 (1986).
33. Amit, D. J., Gutfreund, H. and Sompolinsky, H., Phys. Rev. Lett. **55**, 1530 (1985).
34. Dumont, J. P. C. and Robertson, R. Meldrum, Science **233**, 849 (1986); See also Thompson, R. F., Science **233**, 941 (1986) for more generous citation practices. Even here, however, the physics based papers are included in a peripheral and supplementary citation.
35. Albert, D. Z., Phys. Lett. **98A**, 249 (1983); Geroch, R. and Hartle, J. B., Found. Phys. **16**, 533 1986; Kantor, F. W., Information Mechanics, Wiley, New York (1977); Kantor, F. W., Int. J. Theor. Phys. **21**, 525 (1982); Manthey, M. J. and Moret, B. M. E., Commun. ACM **26**, 137 (1983); Noyes, H. P., Gefwert, C. and Manthey, M. J., in New Techniques and Ideas in Quantum Measurement Theory (Edited by D. Greenberger), N. Y. Acad. Sci. (in press); Steiglitz, K., Two Non-Standard Paradigms for Computation: Analog Machines and Cellular Automata, in Performance Limits in Communication Theory and Practice (Edited by J. K. Skwirzynski) Martinus Nijhoff, Dordrecht (in press); Vichniac, G. Y., Physica **10D**, 96 (1984); Wolfram, S., Phys. Rev. Lett. **54**, 735 (1985); Yates, F. E., Am. J. Physiol. **238**, R277 (1980); Marcer, P. J., The Church-Turing Hypothesis, Artificial Intelligence and Cybernetics, presented at International Cybernetics Congress, Namur, Aug. 1986.
36. Keyes, R. W. and Landauer, R., IBM J. Res. Dev. **14**, 152 (1970).
37. Kirkpatrick, S. and Toulouse, G., J. Physique **46**, 1277 (1985).
38. Ackley, D. H., Hinton, G. E. and Sejnowski, T. J., Cognitive Science **9**, 147 (1985); Kienker, P. K., Sejnowski, T. J., Hinton, G. E. and Schumacher, L. E., Separating figure from ground with a parallel network, Perception (in press); Sejnowski, T. J., Kienker, P. K. and Hinton, G. E., Learning symmetry groups with hidden units: Beyond the perceptron, Physica D (in press).
39. Landauer, R., Helv. Phys. Acta **56**, 847 (1983).
40. Anderson, P. W., Proc. Nat. Acad. Sci. **80**, 3386 (1983); Anderson, P. W. and Stein, D. L., in Self-Organizing Systems: The Emergence of Order (Edited by F. E. Yates, D. O. Walter and G. B. Yates) Plenum, New York (in press); Ebeling, W., in Strukturbildung Bei Irreversiblen Prozèssen, p. 168, Teubner, Leipzig (1976); see also Ebeling, W. and Feistel, R., Physik der Selbstorganisation und Evolution, Akademie-Verlag, Berlin (1982); Kirkpatrick, M., Am. Nat. **119**, 833 (1982); Kuhn, H. and Kuhn, C., Origins of Life **9**, 137 (1978); Küppers, B.-O., in Der Informationsbegriff in Technik und Wissenschaft (Edited by O. G. Folberth and C. Hackl), p. 181, Oldenbourg, Munich (1986); Lande, R., Proc. Natl. Acad. Sci. **82**, 7641 (1985), Levinton, J., Science **231**, 1490 (1986); Lewin, R., Science **231**, 672 (1986); Newman, C. M., Cohen, J. E. and Kipnis, C., Nature **315**, 399 (1985); Schuster, P., Physica Scripta, this issue; Wright, S., in Evolution and the Genetics of Populations, Vol. 3, p. 443, Univ. Chicago Press (1977).
41. Remarks adapted from Landauer, R., Am. J. Physiol. **241**, R107

(1981).
42. Landauer, R., J. Appl. Phys. **33**, 2209 (1962).
43. Stratonovich, R. L., Topics in the Theory of Random Noise, Gordon and Breach, New York, Vol. I (1963), Vol. II (1967).
44. Landauer, R., J. Stat. Phys. **13**, 1 (1975).
45. Landauer, R., Phys. Rev. **A12**, 636 (1975).
46. Landauer, R., in Self-Organizing Systems: The Emergence of Order (Edited by F. E. Yates, D. O. Walter and G. B. Yates), Plenum, N. Y. (in press).
47. Chaitin, G. J., IBM J. Res. Dev. **21**, 350 (1977).
48. Wright, R., The Sciences **25**, 10, No. 2 (1985). Bennett, C. H., Found. Phys. **16**, 585 (1986).
49. Kuhn, H. and Waser, J., in Biophysics (Edited by W. Hoppe, W. Lohmann, H. Markl and H. Ziegler), p. 830, Springer, Heidelberg (1983).

# Notes on the history of reversible computation

by Charles H. Bennett

**We review the history of the thermodynamics of information processing, beginning with the paradox of Maxwell's demon; continuing through the efforts of Szilard, Brillouin, and others to demonstrate a thermodynamic cost of information acquisition; the discovery by Landauer of the thermodynamic cost of information destruction; the development of the theory of and classical models for reversible computation; and ending with a brief survey of recent work on quantum reversible computation.**

Concern with the thermodynamic limits of computation was preceded historically by the paradox of Maxwell's demon [1] and the realization that one bit of information is somehow equivalent to $k$ ln 2 units of entropy, or about $2.3 \times 10^{-24}$ cal/Kelvin. This equivalence was implicit in the work of Szilard [2] and became explicit in Shannon's use [3] of the term "entropy" and the formula

$$H = -\sum_i P_i \log P_i$$

to describe the self-information of a message source.

The history of this subject is noteworthy because it offers an example of how ideas that are strikingly successful in one area of science (in this case the uncertainty principle and the theory of black-body radiation) can stimulate unconscious false analogies, and so impede progress in other areas of science (thermodynamics of measurement and computation).



In the nineteenth century, despite the vision of Babbage, computation was thought of as a mental process, not a mechanical one. Accordingly, the thermodynamics of computation, if anyone had stopped to wonder about it, would probably have seemed no more urgent as a topic of scientific inquiry than, say, the thermodynamics of love. However, the need to think seriously about the thermodynamics of perceptual and mental processes was thrust upon science by the famous paradox of "Maxwell's demon," described as follows by its inventor, in a passage of admirable clarity and foresight [1]:

"One of the best established facts in thermodynamics is that it is impossible in a system enclosed in an envelope which permits neither change of volume nor passage of heat, and in which both the temperature and the pressure are everywhere the same, to produce any inequality of temperature or pressure without the expenditure of work. This is the second law of thermodynamics, and it is undoubtedly true as long as we can deal with bodies only in mass, and have no power of perceiving or handling the separate molecules of which they are made up. But if we conceive a being whose faculties are so sharpened that he can follow every molecule in its course, such a being, whose attributes are still as essentially finite as our own, would be able to do what is at present impossible to us. For we have seen that the molecules in a vessel full of air at uniform temperature are moving with velocities by no means uniform, though the mean velocity of any great number of them, arbitrarily selected, is almost exactly uniform. Now let us suppose that such a vessel is divided into two portions, A and B, by a division in which there is a small hole, and that a being, who can see the individual molecules, opens and closes this hole, so as to allow only the swifter molecules to pass from A to B, and only the slower ones to pass from B to A. He will thus, without expenditure of work, raise the temperature of B and lower that of A, in contradiction to the second law of thermodynamics.

"This is only one of the instances in which conclusions we have drawn from our experience of bodies consisting of an immense number of molecules may be found not to be applicable to the more delicate observations and experiments which we may suppose made by one who can perceive and handle the individual molecules which we deal with only in large masses.

"In dealing with masses of matter, while we do not perceive the individual molecules, we are compelled to adopt what I have described as the statistical method of calculation, and to abandon the strict dynamical method, in which we follow every motion by the calculus.

"It would be interesting to enquire how far those ideas about the nature and methods of science which have been derived from examples of scientific investigation in which the dynamical method is followed are applicable to our actual knowledge of concrete things, which, as we have seen, is of an essentially statistical nature, because no one has yet discovered any practical method of tracing the path of a molecule, or of identifying it at different times.

"I do not think, however, that the perfect identity which we observe between different portions of the same kind of matter can be explained on the statistical principle of the stability of averages of large numbers of quantities each of which may differ from the mean. For if of the molecules of some substance such as hydrogen, some were of sensibly greater mass than others, we have the means of producing a separation between molecules of different masses, and in this way we should be able to produce two kinds of hydrogen, one of which would be somewhat denser than the other. As this cannot be done, we must admit that the equality which we assert to exist between the molecules of hydrogen applies to each individual molecule, and not merely to the average of groups of millions of molecules."

Maxwell offered no definitive refutation of the demon, beyond saying that we lack its ability to see and handle individual molecules. In subsequent years Smoluchowski [4] partly solved the problem by pointing out that a simple automatic apparatus, such as a trap door, would be prevented by its own Brownian motion from functioning as an effective demon. He also remarked [5],

"As far as we know today, there is no automatic, permanently effective perpetual motion machine, in spite of molecular fluctuations, but such a device might, perhaps, function regularly if it were appropriately operated by intelligent beings. . . ."

This apparent ability of intelligent beings to violate the second law called into question the accepted belief that such beings obey the same laws as other material systems. Szilard, in his famous paper [2], "On the Decrease of Entropy in a Thermodynamic System by the Intervention of Intelligent Beings," attempted to escape from this predicament by arguing that the act of measurement, by which the demon determines the molecule's speed (or, in Szilard's version of the apparatus, determines which side of the partition it is on) is necessarily accompanied by an entropy increase sufficient to compensate the entropy decrease obtained later by exploiting the result of the measurement. Szilard was somewhat vague about the nature and location of this entropy increase, but a widely held interpretation of the situation, ever since his paper appeared, has been that measurement is an inevitably irreversible process, attended by an increase of entropy in the universe as a whole by at least $k \ln 2$ per bit of information acquired by the measurement. Later we shall see this is not quite correct: The measurement itself can be performed reversibly, but an unavoidable entropy increase, which prevents the demon from violating the second law, occurs when the demon erases the result of one measurement to make room for the next. The existence of an irreducible thermodynamic cost for information destruction (as opposed to information acquisition) was only clearly recognized three decades later by Landauer [6], and another two decades elapsed before Landauer's insight was applied to explain the demon without invoking any thermodynamic cost of measurement [7–9].

Ironically, Szilard came quite close to understanding the thermodynamic cost of information destruction. At the end of his paper, where he followed one version of his demon apparatus through a complete cycle of operation, he found that resetting the demon in preparation for the next measurement generated $k \ln 2$ of entropy. Unfortunately, he did not pursue this finding to the point of recognizing that information destruction is always thermodynamically costly, and that therefore no thermodynamic cost need be postulated for information acquisition.

Szilard's partial insight was lost as subsequent workers neglected resetting, and instead attempted to prove in detail the irreversibility of various measurement processes, particularly those in which the demon observes the molecule with light. The emphasis on measurement and neglect of resetting probably represented unconscious biases from everyday experience, where information is thought of as valuable or at worst neutral, and from quantum mechanics, which strikingly demonstrated the nontriviality of the measurement process. The influence of quantum mechanics, particularly the quantum theory of black-body radiation, can be seen in a discussion of Maxwell's demon in Brillouin's influential 1956 book *Science and Information Theory* [10]:

"The essential question is . . . *Is it actually possible for the demon to see the individual atoms?* . . . The demon is in an enclosure at equilibrium at constant temperature, where the radiation must be black body radiation, and it is impossible to see anything in the interior of a black body. . . . The demon would see thermal radiation and its fluctuations, but he would never see the molecules.

"It is not surprising that Maxwell did not think of including radiation in the system in equilibrium at temperature $T$. Black body radiation was hardly known in 1871, and it was thirty years before the thermodynamics of radiation was clearly understood and Planck's theory was developed."

Brillouin goes on to consider a dissipative measurement scheme in which the demon observes the molecules by photons from a non-equilibrium source such as a hot lamp filament, concluding that to see the molecule, the demon must use at least one photon more energetic than the photons comprising the thermal background, thereby dissipating an energy of order $kT$ in the process of measurement.

By the 1950s the development of the theory of computation by Turing and others had made it commonplace to think of computation as a mechanical process. Meanwhile the development of electronic digital computers had naturally raised the question of the ultimate thermodynamic cost of computation, especially since heat removal has always been a major engineering consideration in the design of computers, limiting the density with which active components can be packed.

The general folklore belief at this time, descended from Szilard's and Brillouin's analyses, is expressed in a remark [11] from a 1949 lecture by von Neumann, to the effect that a computer operating at temperature $T$ must dissipate at least $kT \ln 2$ of energy "per elementary act of information, that is, per elementary decision of a two-way alternative and per elementary transmittal of one unit of information."

A major turning point in understanding the thermodynamics of computation took place when Landauer [6] attempted to prove this folklore belief and found he couldn't. He was able to prove a lower bound of order $kT$ for some data operations, but not for others. Specifically, he showed that "logically irreversible" operations—those that throw away information about the previous logical state of the computer—necessarily generate in the surroundings an amount of entropy equal to the information thrown away. The essence of Landauer's argument was that such operations compress the phase space spanned by the computer's information-bearing degrees of freedom, and so, in order to occur spontaneously, they must allow a corresponding expansion, in other words, an entropy increase, in other degrees of freedom.

[This argument is not without its subtleties; for example, a many-to-one operation such as erasure may be thermodynamically reversible or not, depending on the data to which it is applied. When truly *random* data (e.g., a bit equally likely to be 0 or 1) is erased, the entropy increase of the surroundings is compensated by an entropy decrease of the data, so the operation as a whole is thermodynamically reversible. This is the case in resetting Maxwell's demon, where two equiprobable states of the demon's mind must be compressed onto one. By contrast, in computations, logically irreversible operations are usually applied to nonrandom data deterministically generated by the computation. When erasure is applied to such data, the entropy increase of the environment is not compensated by an entropy decrease of the data, and the operation is thermodynamically irreversible [7].]

About 1970, having read Landauer's paper and heard him talk, I began thinking about the thermodynamics of computation. Initially I assumed, as he, that at least some logically irreversible operations were necessary to nontrivial computation. However, as a side project, I experimented with simple computations that could be done without them. For example, I wrote a reversible program that used repeated subtraction to test whether one integer is divisible by another. Such experiments revealed a common pattern: The computation consisted of two halves, the second of which almost exactly undid the work of the first. The first half would generate the desired answer (e.g., divisible or not) as well as, typically, some other information (e.g., remainder and quotient). The second half would dispose of the extraneous information by reversing the process that generated it, but would keep the desired answer. This led me to realize [12] that any computation could be rendered into this reversible format by accumulating a history of all information that would normally be thrown away, then disposing of this history by the reverse of the process that created it. To prevent the reverse stage from destroying the desired output along with the undesired history, it suffices, before beginning the reverse stage, to copy the output on blank tape. No history is recorded during this copying operation, and none needs to be, since copying onto blank tape is already logically reversible; the reverse stage of computation then destroys only the original of the output, leaving the copy intact. My technique for performing an arbitrary computation reversibly is illustrated in **Table 1**, with underbars indicating the positions of the tape heads.

A proof of the thermodynamic reversibility of computation requires not only showing that logically irreversible operations can be avoided, but also showing that, once the computation has been rendered into the logically

**Table 1** Scheme for reversible computation.

| *Stage* | *Contents of* | | |
|---|---|---|---|
| | *Work tape* | *History tape* | *Output tape* |
| Forward | _INPUT | _ | _ |
| | WO$\underline{\text{R}}$K | HIST_ | _ |
| | _OUT$\bar{\text{P}}$UT | HISTORY_ | _ |
| Copy output | _OUTPUT | HISTORY_ | _ |
| | OUT$\underline{\text{P}}$UT | HISTORY_ | OUT_ |
| | _OUT$\bar{\text{P}}$UT | HISTORY_ | _OUTPUT |
| Reverse | _OUTPUT | HISTORY_ | _OUTPUT |
| | WO$\underline{\text{R}}$K | HIST_ | _OUTPUT |
| | _INP$\bar{\text{U}}$T | _ | _OUTPUT |

reversible format, some actual hardware, or some physically reasonable theoretical model, can perform the resulting chain of logically reversible operations in a thermodynamically reversible fashion. Approaching the problem with a background of prior interests in biochemistry and computability theory, I saw an analogy between DNA and RNA and the tapes of a Turing machine. The notion of an informational macromolecule, undergoing transitions of its logical state by highly specific (e.g., enzyme-catalyzed) reversible chemical reactions, offered a felicitous model within which thermodynamic questions about information processing could be asked and rigorously answered. Within this theoretical framework it is easy to design an "enzymatic Turing machine" [7, 12] which would execute logically reversible computations with a dissipation per step proportional to the speed of computation. Near equilibrium, the machine would execute a slightly biased random walk, making backward steps nearly as often as forward ones. The backward steps would not result in errors, since they would be undone by subsequent forward steps. True errors—transitions to logically unrelated states—would also occur in any system with finite potential energy barriers, but their rate could be made small (in principle arbitrarily small) compared to the rate of logically correct forward and backward transitions. The enzymatic Turing machine is an example of a "Brownian" reversible computer, in which the non-information-bearing degrees of freedom are strongly coupled to, and exert a viscous drag on, the information-bearing ones, resulting in a dissipation per step proportional to the speed of computation.

Although there are no known general-purpose (i.e., universal) enzymatic Turing machines in nature, there are enzymes analogous to special-purpose Turing machines, notably RNA polymerase. This enzyme, whose function is to make an RNA transcript of the genetic information in one or more DNA genes, may be viewed as a special-purpose tape-copying Turing machine. Under physiological conditions the enzyme is driven hard forward, and dissipates about 20 $kT$ per step; however, the operation of RNA polymerase is both logically and thermodynamically reversible, and it is routinely operated both forward and backward in the laboratory by varying the relative concentrations of reactants (nucleoside triphosphates) and product (pyrophosphate) [13, 14]. When operating backward the enzyme performs the logical inverse of copying: It removes bases one by one from the RNA strand, checking each one for complementarity with the DNA before removing it.

Edward Fredkin, at MIT, independently arrived at similar conclusions concerning reversible computation. Fredkin was motivated by a conviction that computers and physics should be more like each other. On one hand he was dissatisfied with a theoretical physics based on partial differential equations and continuous space-time. He felt it unreasonable to invoke an infinite number of bits of information to encode the state of one cubic centimeter of nature, and an infinite number of digital operations to exactly simulate one second of its evolution. By the same token he felt it wrong to base the theory of computation on irreversible primitives, not found in physics. To remedy this he found a reversible three-input three-output logic function, the "conservative logic gate" able to simulate all other logic operations, including the standard ones AND, OR, and NOT [15, 16]. He showed that conservative logic circuits can perform arbitrary computations by essentially the same programming trick I had used with reversible Turing machines: Do the computation, temporarily saving the extra information generated in the course of obtaining the desired answer, then dispose of this information by the reverse of the process by which it was created.

Fredkin's displeasure with continuum models resembles Landauer's well-known displeasure [17] with mathematical operations that have no physical way of being performed, e.g., calculating the $10^{100}$th digit of pi. These doubts, however, led Fredkin to pursue the radical goal of finding a fully discrete basis for physics, whereas in Landauer they merely inspired a certain aesthetic indifference toward nonconstructive mathematics.

Fredkin was joined by T. Toffoli (who in his doctoral thesis [18] had refuted, by counterexample, an accepted but erroneous proof that reversible cellular automata cannot be computationally universal), and later by Gerard Vichniac and Norman Margolus to form the Information Mechanics group at MIT. The activities of this group are largely responsible for stimulating the current interest in reversible cellular automata with direct physical significance, notably deterministic Ising models [19–21] and momentum-conserving lattice gases that support a macroscopic hydrodynamics [22].

A major step toward Fredkin's goal of finding a reversible physical basis for computation was his discovery of the billiard-ball model of computation [16]. This takes

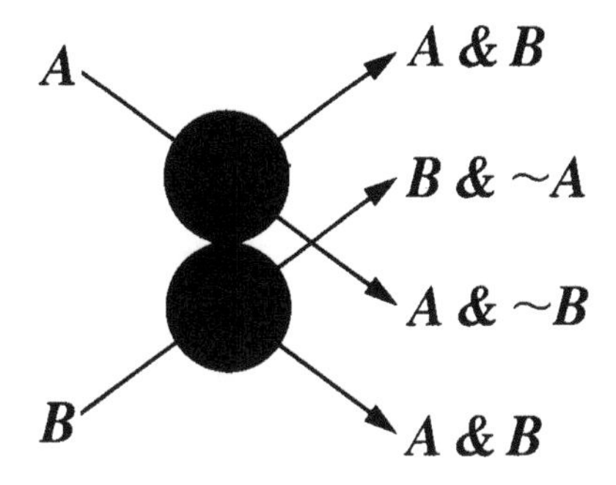


Use of a billiard-ball collision to realize a two-input, four-output logic function; data (1 or 0) represented by the presence or absence of a billiard ball on a given trajectory.

advantage of the fact that a collision between two classical hard spheres ("balls") diverts each one from the path it would have followed had the other been absent; thus a collision can be thought of as a two-input, four-output logic function whose outputs, for inputs $A$ and $B$, are, respectively (cf. **Figure 1**),

$A$ and $B$,
$B$ and not $A$,
$A$ and not $B$,
$A$ and $B$.

Fredkin showed that, with the addition of "mirrors" to redirect the balls, such collisions can simulate any conservative logic function, and therefore any ordinary logic function. This implies that an infinite two-dimensional hard sphere gas, in an appropriate periodic potential (i.e., a periodic array of mirrors), is *computationally universal*—capable of being programmed through its initial condition to simulate any digital computation.

The billiard-ball computer is the prime example of a ballistic reversible computer. In contrast to the Brownian computers described earlier, ballistic computers operate with zero dissipation at finite speed, but they depend on isolating the information-bearing degrees of freedom from all sources of thermal noise, such as internal degrees of freedom of the balls or mirrors. Another way of characterizing the difference between Brownian and ballistic computers is to say that the former work by creating a low-potential energy labyrinth in configuration space, isomorphic to the desired computation, through which the system drifts despite thermal noise; the latter instead work by creating a dynamical trajectory isomorphic to the desired computation, which the system follows exactly in the absence of noise.

A number of other classical-mechanical models of reversible computation can be characterized as clocked Brownian models: The information-bearing degrees of freedom are locked to and driven by a master "clock" degree of freedom, with dissipation proportional to speed. These include the early coupled-potential-well models of Landauer and Keyes [6, 23], which were invented before the trick of reversible programming was known, but would function as Brownian reversible computers if reversibly programmed; the author's clockwork Turing machine [7], which invokes infinitely hard potentials to achieve zero error in a Brownian setting; Likharev's reversible computer based on Josephson junctions [24], which could probably be built, and Landauer's ball-and-pipe model [15, 25].

Returning to the question of Maxwell's demon, we can now give a detailed entropy accounting of the demon's cycle of operation. We refer to Szilard's [2] version of the demon, which uses a gas consisting of a single molecule. The demon first inserts a partition trapping the molecule on one side or the other, next performs a measurement to learn which side the molecule is on, then extracts $kT \ln 2$ of work by allowing the molecule to expand isothermally to fill the whole container again, and finally clears its mind in preparation for the next measurement. The discussion below of the classical Szilard engine follows [7]; an analogous quantum analysis has been given by Zurek [26].

According to our current understanding, each step of the cycle is thermodynamically reversible if we make the usual idealization that operations are carried out quasistatically. In particular, the measurement is reversible and does not increase the entropy of the universe. What the measurement does do, however, is to establish a correlation between the state of the demon's mind and the position of the molecule. This correlation means that after the measurement the entropy of the combined system (demon + molecule) is no longer equal to the sum of the entropies of its parts. Adopting a convenient origin for the entropy scale, the entropy of the molecule is one bit (since it may be, equiprobably, on either side of the partition), the entropy of the demon's mind is one bit (since it may think, equiprobably, that the molecule is on either side of the partition), but the entropy of the combined system is only one bit, because the system as a whole, owing to the correlation, has only two equiprobable states, not four.

The next phase of the cycle, the isothermal expansion, reduces the entropy of the environment by one bit while increasing the entropy of the demon + molecule system from one bit to two bits. Because the expansion destroys the correlation between demon and molecule (rendering the information obtained by the measurement obsolete), the entropy of the demon + molecule system is now equal to the sum of the entropies of its parts, one bit each.

The last phase of the cycle, resetting the demon's mind, reduces the entropy of the demon from one bit to zero, and accordingly, by Landauer's argument, must increase the entropy of the environment by one bit. This increase cancels the decrease brought about during the expansion phase, bringing the cycle to a close with no net entropy change of demon, molecule, or environment.

One may wonder how, in view of the arguments of Brillouin and others, the demon can make its measurement without dissipation. Though plausible, these arguments only demonstrated the dissipativeness of certain particular mechanisms of measurement, not of all measurements. In a sense, the existence of copying mechanisms such as RNA polymerase demonstrates the reversibility of measurement, if one is willing to call RNA synthesis a measurement of the DNA. More traditional reversible-measurement schemes can also be devised which are ideal in the sense of having no other effect than to establish the desired correlation between the measuring apparatus and the system being measured. Such a measurement begins with the measuring apparatus in a standard dynamical or thermodynamic state and ends with it in one of several states depending on the initial state of the system being measured, meanwhile having produced no change either in the environment or in the system being measured. **Figure 2**, for example, shows a classical billiard-ball mechanism based on the ideas of Fredkin that uses one billiard ball (dark) to test the presence of another (light) without disturbing the dynamical state of the latter. The apparatus consists of a number of fixed mirrors (dark rectangles) which reflect the billiard balls. First assume that the dark ball is absent. Then a light ball injected into the apparatus at $X$ will follow the closed diamond-shaped trajectory $ABCDEFA$ forever, representing the value 1; conversely, the absence of the light ball (i.e., no balls in the apparatus at all) represents the value 0. The goal of the measurement is to inject another ball (dark color) into the apparatus in such a way that it tests whether the light ball is present without altering the light ball's state. By injecting the dark ball at $Y$ at the appropriate time, the light ball (if present) is diverted from, but then returned to, its original path (following $BGD$ instead of $BCD$), while the dark ball leaves the apparatus at $M$ if the light ball was present and at $N$ if it was absent.

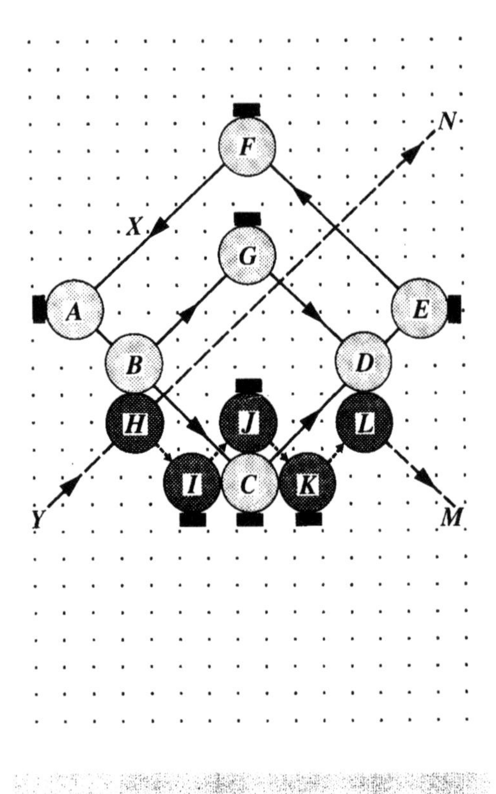


Reversible measurement in the billiard-ball model of computation.

One can design analogous mechanisms [7, 8] for reversibly measuring which side of Szilard's engine the molecule is on without otherwise disturbing the thermodynamic state of the engine or the environment. Such reversible nondemolition measurement schemes in general exist for classical systems, and for quantum systems in which the goal of the measurement is to distinguish among orthogonal states of the system, since these states may in principle be made eigenstates of an appropriate observable. Of course a quantum measurement cannot avoid disturbing a system which is presented to it in a superposition of eigenstates the measuring apparatus is designed to measure. The relation of irreversibility to quantum measurement has been considered by many authors (cf. the concise discussion in [27] and references therein).

An active research area recently has been the theory of quantum reversible computation. Chemical Brownian computers such as RNA polymerase are of course quantum systems, but because of the high temperature and short thermal de Broglie wavelength, quantum effects are subtle and quantitative (e.g., zero-point and tunneling corrections to reaction rates) rather than qualitative.

More distinctively quantum models have been considered by a number of authors [28–35]. These models are somewhat abstract by comparison with classical models, consisting typically of an array of two-state spins (each representing one bit) and a time evolution operator or Hamiltonian designed to make the spins pass through a sequence of states corresponding to the desired computation. The computationally relevant states are generally a subset of a set of orthonormal "basis states," in which each spin is either up or down.

One of the earliest quantum models, by Benioff [28], used a Hamiltonian such that a basis state corresponding to the initial logical state of a reversible Turing machine would be transformed, at integer times, to basis states corresponding to successive logical successors. In other words, the Hamiltonian $H$ was chosen so that the unitary operator $U$, representing evolution under the Hamiltonian for unit time, mapped each computationally relevant basis state onto its logical successor. Casting the logical time evolution of a Turing machine into the form of a unitary operator requires that all basis states have successors. Thus there can be no halt states, and all computations must be either infinite or cyclic. Since the Hamiltonian represents, albeit in an abstract form, the actual interactions among parts of the quantum computer that the designer is able to choose and control, Benioff considered it important for the Hamiltonian to be simple, and in particular not to depend explicitly on the global structure of the computation being performed. In order to achieve this, he found it necessary to make $H$ time-dependent, in effect using a three-phase clock (two phases would also have sufficed) to turn on three Hamiltonians one after another, and making $U$ the product of three non-commuting unitary operators $U = U_3U_2U_1$. In each of the clock phases, some of the spins (bits) in the computer flip conditionally on the state of others.

Feynman [29] found a way to define a simple, time-independent Hamiltonian for quantum computations: Instead of incorporating the direction of the computation (forward as opposed to backward) in the Hamiltonian, he incorporated it into the initial condition, which was now not a single basis state but rather a wave-packet-like superposition of basis states. The Feynman Hamiltonian for a given unitary transition operator $U$ is of the form $H = U + U^{+}$, analogous to the Hamiltonian for a one-dimensional crystal in which spin waves can propagate either forward or backward according to their initial momentum. Feynman also noted that quantum computers can exhibit behavior intermediate between Brownian and ballistic: Thermal fluctuations in the Hamiltonian scatter the propagating computation wave like phonons in a crystal, so that under appropriate conditions the mean free path between scattering events is finite but much larger than one computation step. The computation then proceeds with net velocity proportional to the driving force, as in a Brownian computer, but with a proportionality constant that varies inversely with the mean free path, like electrical conductivity.

Zurek [30] compares the dynamical stability of quantum and classical ballistic computers with respect to errors in the initial condition ("software") and the Hamiltonian ("hardware"). In the billiard-ball model either type of error produces an exponentially growing error in the trajectory, whereas for quantum computers hardware-induced errors increase only quadratically with time and software errors do not increase at all.

Margolus [33] and Benioff [34] considered the problem of finding a universal quantum computer (with infinite memory) whose Feynman-type Hamiltonian nevertheless would have a finite range of interaction. For a serial computer such as a Turing machine, in which only one part is active at a time, this is not difficult; but when an analogous construction is attempted for a parallel machine such as a cellular automaton, Margolus found, on the one hand, that the finite range of interaction forbade synchronous updating of all the sites, and, on the other hand, that with asynchronous updating the computation no longer proceeded ballistically.

Deutsch [32] considered a more general kind of quantum computer that could be programmed to perform distinctively quantum operations such as generating two bits in an Einstein–Podolsky–Rosen superposition state. With Deutsch's computer it is possible to split a computation into two (or more) subtasks, perform the subtasks simultaneously in different Everett worlds, and then allow the results of the subtasks to interfere. An appropriate measurement on the final superposed state of the computer produces a probabilistic behavior of the output, which sometimes yields the desired answer (e.g., the exclusive-or, or some other linear function of the results of the two subtasks), and sometimes yields a "failure," an eigenstate of the output operator which says nothing about the results of the subtasks. Because of the probability of failure, quantum parallelization does not reduce the average time required to complete a parallelizable computation.

Landauer [35] has reviewed several quantum computation models in more detail than given here, pointing out some of the unphysical idealizations in existing models and the importance of specifying a quantum computer more concretely than by merely inventing a Hamiltonian.

## Acknowledgments

I would like to acknowledge my early mentors Berni Alder and the late Aneesur Rahman, who had no direct interest in the theory of computation, but who taught me to think clearly about reversibility in statistical physics, resisting deception by common sense and discouragement by apparent paradoxes. I would also like to thank Rolf Landauer, under whose kind influence my interest in this subject took form nearly twenty years ago. Since then I have been inspired and assisted continually by many, especially Landauer, Gregory Chaitin, Edward Fredkin, Tom Toffoli, Paul Benioff, Norman Margolus, and David Deutsch.

## References

1. J. C. Maxwell, *Theory of Heat*, 4th Ed., Longmans, Green & Co., London, 1875 (1st Ed. 1871), pp. 328–329.

2. L. Szilard, *Z. Phys.* **53,** 840–856 (1929).
3. C. E. Shannon and W. Weaver, *The Mathematical Theory of Communication*, University of Illinois Press, Urbana-Champaign, IL, 1949.
4. M. von Smoluchowski, *Z. Phys.* (1912).
5. M. von Smoluchowski, lecture notes, Leipzig, 1914 (as quoted by Szilard [2]).
6. R. Landauer, *IBM J. Res. Develop.* **3,** 183–191 (1961).
7. C. H. Bennett, *Int. J. Theor. Phys.* **21,** 905–940 (1982).
8. C. H. Bennett, *Research Report RC-12526*, IBM Thomas J. Watson Research Center, Yorktown Heights, NY 10598; edited version in *Sci. Amer.* (November 1987).
9. E. Lubkin, *Int. J. Theor. Phys.* **16,** 523–535 (1987).
10. L. Brillouin, *Science and Information Theory*, 2nd Ed., Academic Press, London, 1962.
11. J. von Neumann, *Theory of Self-Reproducing Automata*, Arthur Burks, Ed., University of Illinois Press, Urbana-Champaign, IL, 1966, p. 66.
12. C. H. Bennett, *IBM J. Res. Develop.* **17,** 525–532 (1973).
13. Judith Levin, Doctoral Dissertation, Biochemistry Department, University of California, Berkeley, CA, 1985.
14. George Kassaveties, *J. Biol. Chem.* **261,** 14256–14265 (1986).
15. R. Landauer, *Ber. Bunsenges.* **80,** 1048 (1976).
16. E. Fredkin and T. Toffoli, *Int. J. Theor. Phys.* **21,** 219–253 (1982).
17. R. L. Landauer, *IEEE Spectrum* **4,** 105 (1967).
18. T. Toffoli, *J. Comp. Syst. Sci.* **15,** 213–231 (1977).
19. G. Vichniac, *Physica* **10D,** 96 (1984).
20. Y. Pomeau, *J. Phys. A* **17,** 415 (1984).
21. M. Creutz, *Ann. Phys.* **167,** 62 (1986).
22. L. Kadanoff, *Physics Today* **39,** 7 (September 1986).
23. R. W. Keyes and R. Landauer, *IBM J. Res. Develop.* **14,** 152 (1970).
24. K. K. Likharev, *Int. J. Theor. Phys.* **21,** 311 (1982).
25. R. L. Landauer, *Int. J. Theor. Phys.* **21,** 283 (1982).
26. W. Zurek, *Frontiers of Nonequilibrium Statistical Physics*, G. T. Moore and M. O. Scully, Eds., Plenum Publishing Co., New York, 1986, pp. 15–161.
27. W. Zurek, *Ann. N.Y. Acad. Sci.* **480,** 89–97 (1986).
28. P. Benioff, *Phys. Rev. Lett.* **48,** 1581–1585 (1982); *J. Stat. Phys.* **29,** 515–545 (1982).
29. R. P. Feynman, *Opt. News* **11,** 11–20 (1985).
30. W. H. Zurek, *Phys. Rev. Lett.* **53,** 391–394 (1984).
31. A. Peres, *Phys. Rev.* **32A,** 3266–3276 (1985).
32. D. Deutsch, *Proc. Roy. Soc. Lond. A* **400,** 97–117 (1985).
33. N. Margolus, *Ann. N.Y. Acad. Sci.* **480,** 487–497 (1986).
34. P. Benioff, *Ann. N.Y. Acad. Sci.* **480,** 475–486 (1986).
35. R. Landauer, *Found. Phys.* **16,** 551–564 (1986).

*Received April 8, 1987; accepted for publication October 5, 1987*

**Charles H. Bennett** *IBM Thomas J. Watson Research Center, P.O. Box 218, Yorktown Heights, New York 10598.* Dr. Bennett earned his Ph.D. from Harvard University in 1970 for molecular dynamics studies (computer simulation of molecular motion) under David Turnbull and Berni Alder. For the next two years he continued this research under the late Aneesur Rahman at Argonne Laboratories, Argonne, Illinois, coming to IBM Research in 1972. In 1973, building on the work of IBM Fellow Rolf Landauer, Dr. Bennett showed that general-purpose computation can be performed by a logically and thermodynamically reversible apparatus, one which is able to operate with arbitrarily little energy dissipation per step because it avoids throwing away information about past logical states. In 1982 he proposed a reinterpretation of Maxwell's demon, attributing its inability to break the second law to an irreducible thermodynamic cost of destroying, rather than acquiring, information. Aside from the thermodynamics of information processing, Dr. Bennett's research interests include the mathematical theory of randomness, probabilistic computation, and error-correction; the cryptographic applications of the uncertainty principle; and the characterization of the conditions under which statistical-mechanical systems evolve spontaneously toward states of high computational complexity. In 1983–85, as visiting professor of computer science at Boston University, he taught courses on cryptography and the physics of computation, and in 1986 was co-organizer of a conference on cellular automata held at MIT.

# The physical nature of information

Rolf Landauer [1]
IBM T.J. Watson Research Center, P.O. Box 218, Yorktown Heights, NY 10598, USA

Received 9 May 1996
Communicated by V.M. Agranovich

**Abstract**

Information is inevitably tied to a physical representation and therefore to restrictions and possibilities related to the laws of physics and the parts available in the universe. Quantum mechanical superpositions of information bearing states can be used, and the real utility of that needs to be understood. Quantum parallelism in computation is one possibility and will be assessed pessimistically. The energy dissipation requirements of computation, of measurement and of the communications link are discussed. The insights gained from the analysis of computation has caused a reappraisal of the perceived wisdom in the other two fields. A concluding section speculates about the nature of the laws of physics, which are algorithms for the handling of information, and must be executable in our real physical universe.

## 1. Information is physical

Information is not a disembodied abstract entity; it is always tied to a physical representation. It is represented by engraving on a stone tablet, a spin, a charge, a hole in a punched card, a mark on paper, or some other equivalent. This ties the handling of information to all the possibilities and restrictions of our real physical word, its laws of physics and its storehouse of available parts.

This view was implicit in Szilard's discussion of Maxwell's demon [1]. Szilard's discussion, while a major milestone in the elucidation of the demon, was by no means an unambiguous resolution. The history of that can be found in Refs. [2,3]. The acceptance of the view, however, that information is a physical entity, has been slow. Penrose [4], for example, argues for the Platonic reality of mathematics, independent of any manipulation. He tells us "... devices can yield only approximations to a structure that has a deep and 'computer-independent' existence of its own." Indeed, our assertion that information is physical amounts to an assertion that mathematics and computer science are a part of physics. We cannot expect our colleagues in mathematics and in computer science to be cheerful about surrendering their independence. Mathematicians, in particular, have long assumed that mathematics was there first, and that physics needed that to describe the universe. We will, instead, ask for a more self-consistent framework in Sec. V.

P.W. Bridgman, recognized as Nobel laureate for his work in high pressure physics, published a remarkable paper [5] in 1934. That was his attempt to wrestle with the paradoxes of set theory. His solution: Mathematics must be confined to that which can be handled by a succession of unambiguous executable operations. Bridgman's paper is essentially a want ad for a Turing machine, which came a few years later. In a remarkable coincidence Bridgman even uses the word *program* for a succession of executable instructions. Bridgman

[1] E-Mail: landaue@watson.ibm.com.

did not go on to discuss the *physical* executability of the successive instructions, but that is the additional requirement we emphasize here [6,7].

## 2. Quantum information

When we first learned to count on our very classical and sticky little fingers we were misled into thinking about information as a classical entity. In the binary case that that requires either a *0* or *1* state. But nature allows a quantum mechanically coherent superposition of *0* and *1*. That is a degree of freedom appreciated only in recent years, and its impact still needs to be fully understood [8]. This new possibility has been investigated in three different scenarios so far; undoubtedly the list will grow.

One of these scenarios deals with *quantum teleportation* [9]. Here we use two previously prepared and correlated quantum objects; an EPR pair. One is shipped to the transmitting end, the other to the receiving end. An interaction between this prepared object at the transmitting end and the source object whose state is to be transmitted is used to generate a classical signal. At the receiving end this classical signal interacts with the prepared object located there to generate a copy of the teleported source state. The state of the source object is changed in the interaction that generates the classical signal. Quantum teleportation is a subject of serious conceptual interest, but it is not clear that it is a practical recipe for something that really needs to be done, and will not receive further discussion.

The most developed and obviously useful of the three scenarios is quantum cryptography [10], which has been demonstrated successfully in real systems, though its eventual realm of applicability is still uncertain. This application makes direct practical use of the uncertainty principle. A stream of quantum information cannot be examined by an eavesdropper without leaving a mark on the measured stream. A communication link, in contrast to computation, subjects each bit to limited handling, and this minimizes the need for each operation to do *exactly* what it is supposed to do. Furthermore, it is well known [11] that in a communications link (or memory) occasional rare errors are easily remedied by redundancy. Quantum cryptography, therefore, typifies a promising direction for the use of quantum information, in contrast to quantum parallelism, which will be discussed next.

Paul Benioff [12] first understood that a purely quantum mechanical time evolution can cause interacting bits (or spins) to change with time just as we would want them to do in a computer. David Deutsch later [13] realized that such a computer does not have to be confined to executing a single program, but can be following a quantum mechanically coherent superposition of different computational trajectories. At the end we can gain some kinds of information that depend on all of these parallel trajectories, much as the diffraction pattern in a two-slit experiment depends on both trajectories. Eventually Shor [14] showed that this form of parallelism provides tremendous gains for the factoring problem, finding the prime factors of a large number. That, in turn, is important for cryptography, and as a result quantum parallelism has gained widespread attention. We cite only some of the most elementary surveys [15].

## 3. Quantum parallelism: A return to analog computation

An analog computer can do much more per step than a digital computer. But an analog computer, in which a physical variable such as a voltage can take on any value within a permitted range, does not allow for easy error correction. Therefore, in the analog computer errors, due to unintentional imperfections in the machinery, build up quickly and the procedure can go through only a few successive steps before the errors accumulate prohibitively. A digital computer, by contrast, allows only a *0* or *1*. That permits us to restore signals toward their intended values, before they drift far away from that. In typical digital logic the signal is restored toward the power supply voltage or ground at every successive stage. This is what permits us to go through a tremendous number of successive digital steps, and this has given the digital computer its power. In quantum parallelism we do not just use *0* and *1*, but all their possible coherent superpositions. This continuum range, which gives quantum parallelism its power, also gives it the problems of analog computation, a point first explicitly stated by Peres [16]. If we have a state which is mostly a *1*, with a small admixture of *0*, we cannot simply eliminate that admixture;

the superposition may be the intended signal.

Imperfections in the quantum mechanically coherent computer generate several separate problems. First of all, interaction of the intentional information bearing degrees of freedom with the environment causes decoherence and spoils the quantum mechanical interference between alternative trajectories which constitute the basis of quantum parallelism. In the literature generated by the advocates this is the most widely recognized of the several problems faced by quantum parallelism. We can give only a small sampling of that recognition [17]. Note that in order to carry out logic the information bearing degrees of freedom must interact strongly with each other. At the same time, to preserve coherence, they cannot interact with anything else, including the physical framework that holds the bits in their positions. That is a tall order! A second problem arises from the fact that there are manufacturing flaws; the machinery will not do exactly what is intended. For example, interacting bits may not be spaced exactly as needed. Or the external radiative pulses, invoked in a number of the more detailed embodiments, may not meet their exact specification. A $\pi/2$ pulse may be a little too long, too short, or come too early. Flaws in the machinery can cause two separate problems. First of all they can, and generally will, introduce erroneous components into the states reached by the computation. Additionally, however, they can cause unintended reversals of the computational process; they can reflect the motion of the computer. This problem exists most clearly in the case of the Feynman computer [18]. In this computer the computation is launched as a wave-packet moving down its computational track, much as an electron can be sent down a one-dimensional periodic chain of atoms. The one-dimensional electron case is beset by *localization*: transmission diminishes exponentially with length due to irregularities in the supposedly periodic potential. We can expect the same for the computational trajectory. Most of the recent quantum-mechanical computer proposals are not Feynman computers; they do not propagate with their initial kinetic energy. Rather, they are clocked externally. One can then *hope* that if we push the computation forward in a sufficiently determined manner (strong arm and stiff crank?) reflections can be avoided. That is a hope, not yet backed by analysis. One clocked scheme using time-dependent Hamiltonians has been analyzed [19], and does exhibit a reversal problem. *Localization* is a condensed matter theory concept, whereas quantum computation has been studied by computer scientists and by physicists who (largely) do not have a condensed matter background. That is why unintended reversals are not even discussed by the advocates. For exceptions see Ref. [20].

Error recognition and error correction in quantum computation cannot follow the recipes we learned for classical digital computers. Error recognition requires the ability to distinguish a signal from its ideal value. But we cannot, in general, tell whether two arbitrary quantum states differ, or not. Even if we were able to recognize errors, we cannot throw away the description of the error. Discarding information is a dissipative event and will spoil the coherence needed for quantum parallelism. If we do keep a record of the error it must be led aside, so as not appear in the subsequent interference. Despite these difficulties progress has been made toward error reduction, and we can cite only a sample of the material on its way [21]. This is far more progress in fact than this author thought possible, but not enough to permit computation. An effective error correction approach must work for *all* logic steps, and must not rely on perfect supplementary apparatus nor on additional signals which have to be presumed to be perfect. Undoubtedly, further progress will be made, but victory is not yet in sight.

A particularly serious difficulty is caused by components which, even if they meet the ideal specification, still do not do exactly what they are supposed to do. Ref. [19] listed the difficulties faced by Lloyd's proposal [22]. Ref. [23] listed the subtle deviations from those that are actually required in the steps needed for a quantum communications scheme. We suspect that most proposals, if analyzed equally carefully, will show similar flaws. If the devices, even in their ideal state, do not do *exactly* what is needed, that becomes a particularly stubborn problem. At best, only interaction of the results with those obtained in a totally different way can give any hope of error reduction, and it is not clear how this can be achieved and whether it can provide adequate correction.

Finally there is a computer science problem, unrelated to the physics and technology we have stressed. Even if quantum parallelism can be made to work, what is its range of application? The world is unlikely to want to pay for the development of a difficult technology without a broad payoff. The suggestion has been made [24] that quantum computers are particularly good at simulating quantum systems. But for that to be true the system, which needs to be analyzed, must have a structure which maps easily and cleanly onto that of the computer. Would a one-dimensional chain of interacting two level systems [22] make it easy to follow electronic motion in a heavy atom?

Despite the pessimism I have expressed, the mere possibility of quantum parallelism has changed theoretical computer science permanently. Those concerned with theorems about the minimal number of steps required for the execution of an algorithm must allow for quantum parallelism.

## 4. Energy dissipation requirements

The history that led to our understanding of the energy requirements of the computer has been summarized elsewhere [2]. Reversible classical computation, in the presence of viscous friction proportional to velocity, can be accomplished with as little energy dissipation, per step, as desired. Occasional objections still appear to reversible computation and/or the need to dissipate energy when discarding information. Some of the objections are discussed and put aside by Shizume [26]. But most of the interested community long ago accepted these notions. It is, however, not widely recognized that the same body of work that led to reversible computing also demonstrated that any desired immunity to noise can be obtained, and that this is not limited to thermal noise. Similarly, in the measurement process, the reversible operations can be carried out with arbitrarily little dissipation. Resetting the meter, after it has become separated from the system to which it was coupled for measurement, is irreversible, and is the essential lossy step [3]. It is a sociological puzzle why this insight, achieved so long after Maxwell first posed the demon question, has not been celebrated more widely. As in the case of most scientific insights, a number of investigators contributed along the way as discussed in Ref. [27]; nevertheless, the clear, confident and complete resolution stems from the 1980's.

A comparable understanding of the energy needs of the communications channel came later. For many decades the general perception has been that it takes $kT \ln 2$ to send a bit in a classical channel, and more if $h\nu > kT$, where $\nu$ is a typical signalling frequency. This mode of thought has been described in more detail elsewhere [23,28]. It is based on the assumption that a linear boson channel is used and that the energy used in the transmission has to be dissipated. As stressed in Ref. [25], this whole field of concern with the physical fundamentals of information handling is characterized by first answers which have been wrong. Alternative communication methods have been proposed which, in the classical case [28,29] take arbitrarily little energy dissipation, if the bits are moved slowly. Slow motion does not imply a low bit rate; the bits can follow each other as densely as desired. Furthermore, as in the case of reversible computation, the classical communication schemes allow for any desired immunity to noise by suitable choice of the device parameters. The methods do not require perfect components. The classical conclusion should have been obvious once reversible computation is accepted. After all bits are moved around in a computer and therefore there can be no minimal unavoidable energy penalty for bit motion, if there is none for the overall computation. The quantum communication link does not allow an equally easy chain of reasoning. The existing theoretical treatments of the quantum computer allow only for the interacting information bearing degrees of freedom, and ignore all others degrees of freedom including those in internal communication links. Therefore the fact that the literature describes Hamiltonian quantum computers provides no insight into their internal communication links.

It has been shown, however, that if we optimistically assume that frictionless and quantum mechanically coherent machinery is available, then a nondissipative quantum communications link can be constructed [23]. It does depend, not surprisingly, on a sequence of operations each of which is reversible. The classical and quantum communication links proposed by this author are not practical recipes, but are only existence theorems. In particular they invoke active machinery all along the link, in contrast to the passive nature of an electromagnetic transmission line or an optical fiber. It is not clear, however, that this is an essential restriction [23]; there may be room for invention. Again there is a sociological mystery. When I first proposed low energy communication links, in the classical case, I did not expect ready acceptance for my refutation of the conventional wisdom. I expected angry rebuttal. In actual fact, with the exception of one dissenting publication [30], concerned really with reversible computation rather than the communications link, there was remarkable silence.

## 5. Impact on the laws of physics

This is by far the most speculative part of this discussion, but perhaps also its most significant aspect. I return here to notions first presented almost thirty years ago [6] and elaborated on a number of occasions after that, e.g. in Ref. [31]. The laws of physics are essentially algorithms for calculation. These algorithms are significant only to the extent that they are executable in our real physical world. Our usual laws of physics depend on the mathematician's real number system. With that comes the presumption that given any accuracy requirement, there exist a number of calculational steps, which if executed, will satisfy that accuracy requirement. But the real world is unlikely to supply us with unlimited memory or unlimited Turing machine tapes. Therefore, continuum mathematics is not executable, and physical laws which invoke that can not really be satisfactory. They are references to illusionary procedures. Can we not prove that $\cos^2\theta + \sin^2\theta = 1$ exactly, and not just to some very large number of bits? Yes, within a closed postulate system that can be demonstrated with a limited number of steps. But the laws of physics need to go beyond that and require actual number crunching. In a world with limited memory we simply cannot distinguish between $\pi$ and a terribly close neighbor.

In fact, the limit on memory size is not the only limitation likely to be imposed by the real world. Computer elements are continually beset by an endless variety of adverse influences. Cosmic rays, alpha particles, electromigration, corrosion, spilled cups of coffee and earthquakes are a partial list. Can we keep going without limit in the presence of this exposure? Can we invent hardware which can be made arbitrarily immune to degradation? If we simply make the equipment more massive, then we will exhaust the universe's supply of available parts more quickly. At this point the skeptic will respond: Our inability to calculate the exact evolution under a law of physics does not prevent the universe from following that law. The evolution is precisely determined; it is only our ability to simulate that on a computer which is limited. The answer to that: It is not an incorrect response; it is an unverifiable and therefore meaningless response. The skeptic of my proposed statement will suggest that we can divide the universe into two halves, set them into identical initial states lending to the same subsequent history. Quantum mechanics does not, of course, allow us to be sure we have set two systems into the same state [32]. But even if we ignore that, how can we possibly be sure that two such large entities, requiring remarkably complex description, are alike? There cannot be a meaningful wave function for the universe, and very likely not even for half of it.

The limited precision available for the execution of algorithms implies that there is some sort of uncertainty in the laws of physics. This is not necessarily exhibited in the form of a limit to a certain number of bits, it may be more stochastic in nature. This may in turn be the ultimate source of noise and of irreversibility in the universe [28,33].

Even if the speculation just presented is not acceptable to the reader, our argument that mathematics, or any information handling process, cannot reasonably invoke an unlimited number of successive steps, is broader and more compelling. We cannot, of course, expect mathematicians to give up their beautiful array of concepts without hesitation and struggle.

A reluctance to accept the continuum can be found in many places. See for example John Wheelers's frequent slogan "no continuum" [34], and his citation of others with that view [35]. John Wheeler also has a second and more significant relation to this discussion. Our scientific culture normally views the laws of physics as predating the actual physical universe. The laws are considered to be like a control program in a modern chemical plant; the plant is turned on after the program is installed. *In the beginning was the Word and the Word was with God, and the Word was God* (John I,1), attests to this belief. *Word* is a translation from the Greek *Logos* "thought of as constituting the controlling principle of the universe" [36]. Wheeler, in a number of discussion [37] has an adventurous view in which the laws of physics result from our observation of the universe. Wheeler's details are not my details, but we both depart from the notion that the laws were there at the beginning. The view I have expounded here makes the laws of physics dependent upon the apparatus and kinetics available in our universe, and that kinetics in turn depends on the laws of physics. Thus, this is a want ad for a self-consistent theory.

## References


[1] L. Szilard, Z. Phys. 53 (1929) 840.
[2] C.H. Bennett, IBM J. Res. Dev. 32 (1988) 16.
[3] H.S. Leff and A.F. Rex, Maxwell's demon: entropy, information, computing (Princeton Univ. Press, Princeton, 1990).
[4] R. Penrose, The emperor's new mind (Oxford University Press, Oxford, 1989).
[5] P.W. Bridgman, Scr. Math. 2 (1934) 3.
[6] R. Landauer, IEEE Spectrum 4(9) (1967) 105; Reprinted with added commentary in Speculations in Science and Technology 10 (1987) 292.
[7] A. Ekert, in: Proceedings of the 14th ICAP, eds. D. Wineland et al. (AIP, New York, 1995) p. 450.
[8] C.H. Bennett, Phys. Today 48 (October 1995) 24.
[9] C.H. Bennett, G. Brassard, C. Crépeau, R. Jozsa, A. Peres and W.K. Wootters, Phys. Rev. Lett. 70 (1993) 1895.
[10] I. Peterson, Science News 149 (10 February 1996) 90; A. Muller, H. Zbinden and N. Gisin, Europhys. Lett. 33 (1996) 335.
[11] C.E. Shannon, Bell Syst. Tech. J. 27 (1948) 379, 623.
[12] P. Benioff, J. Stat. Phys. 22 (1980) 563; J. Stat. Phys. 29 (1982) 515; Phys. Rev. Lett. 48 (1982) 1581.
[13] D. Deutsch, Proc. R. Soc. Lond. A 400 (1985) 97.
[14] P.W. Shor, in: Proc. 35th Annual Symposium on the Foundation of computer science (IEEE Computer Society, Los Alamitos, CA, 1994) p. 124.
[15] S. Lloyd, Sci. Am. 273 (1995) 140;
D. DiVincenzo, Science 270 (1995) 255;
J. Glanz, Science 269 (1995) 28;
B. Schwarzschild, Phys. Today 49 (1996) 21.
[16] A. Peres, Phys. Rev. A 32 (1985) 3266.
[17] I.L. Chuang, R. Laflamme, P.W. Shor and W.H. Zurek, Science 270 (1995) 1633;
M.B. Plenio and P.L. Knight, Phys. Rev. A 53 (1996) 2986;
W.G. Unruh, Phys. Rev. A 51 (1995) 992.
[18] R. Feynman, Opt. News 11 (2) (1985) 11; reprinted in: Found. Phys. 16 (1986) 507.
[19] R. Landauer, in: Proc. Drexel-4 Symposium on Quantum nonintegrability – quantum-classical correspondence, eds. D.H. Feng and B-L. Hu (International Press, Boston), in press.
[20] S.C. Kak, Found. Phys. 26 (1996) 127;
P.B. Benioff, Hamiltonian models of quantum computers which evolve quantum ballistically, preprint.
[21] C.H. Bennett, D.P. DiVincenzo, J.A. Smolin and W.K. Wootters, Mixed state entanglement and quantum error-correction, Phys. Rev. A., to be published;
R. Laflamme, C. Miquel, J.-P. Paz and W.H. Zurek, Perfect quantum error correction code, to be published;
P. Shor, Phys. Rev. A. 52 (1995) 2493.
[22] S. Lloyd, Science 261 (1993) 1569.
[23] R. Landauer, Minimal energy requirements in communication, Science, to be published.
[24] R.P. Feynman, Int. J. Theor. Phys. 21 (1982) 467;
S. Lloyd, Universal quantum simulators, preprint.
[25] R. Landauer, in: Proc. Workshop on Physics and computation PhysComp '94, (IEEE Computer Society Press, Los Alamitos, 1994) p. 54.
[26] K. Shizume, Phys. Rev. E 52 (1995) 3495.
[27] R. Landauer, Physica A 194 (1993) 551.
[28] R. Landauer, in: Selected topics in signal processing, ed. S. Haykin (Prentice Hall, Englewood Cliffs, 1989) p. 18 (printed version has some figures oriented incorrectly).
[29] R. Landauer, Appl. Phys. Lett. 51 (1987) 2056.
[30] W. Porod, Appl. Phys. Lett. 52 (1988) 2191.
[31] R. Landauer, in: Proc. 3rd Int. Symp. on Foundations of quantum mechanics – in the light of new technology, eds. S. Kobayashi, H. Ezawa, Y. Murayama and S. Nomura (The Physical Society of Japan, Tokyo, Japan, 1990) p. 407.
[32] W.K. Wootters and W.H. Zurek, Nature 299 (1982) 802.
[33] R. Landauer, Z. Phys. B Condensed Matter 68 (1987) 217.
[34] J.A. Wheeler, in: Physical origins of time asymmetry, eds. J.J. Halliwell, J. Perez-Mercader and W.H. Zurek (Cambridge Univ. Press, Cambridge, UK, 1994).
[35] J.A. Wheeler, IBM J. Res. Dev. 32 (1988).
[36] Webster's New World Dictionary of the American Language, ed. D.B. Guralnik (William Collins + World Publishing, 1978).
[37] J.A. Wheeler, in: Problems in theoretical physics, A. Giovanni, F. Mancini and M. Marinaro, Eds. (University of Salerno Press, Salerno, 1984) p. 121; in: Frontiers of nonequilibrium statistical physics, eds. G.T. Moore and M.O. Scully (Plenum, New York, 1986).

# Minimal Energy Requirements in Communication

Rolf Landauer

The literature describing the energy needs for a communications channel has been dominated by analyses of linear electromagnetic transmission, often without awareness that this is a special case. This case leads to the conclusion that an amount of energy equal to $kT\ln 2$, where $kT$ is the thermal noise per unit bandwidth, is needed to transmit a bit, and more if quantized channels are used with photon energies $h\nu > kT$. Alternative communication methods are proposed to show that there is no unavoidable minimal energy requirement per transmitted bit. These methods are invoked as part of an analysis of ultimate limits and not as practical procedures.

Information is inevitably tied to a physical representation, such as a mark on a paper, a hole in a punched card, an electron spin pointing up or down, or a charge present or absent on a capacitor. This representation leads us to ask whether the laws of physics restrict the handling of information and in particular whether there are minimal energy dissipation requirements associated with information handling. The subject has three distinct but interrelated branches dealing, respectively, with the measurement process, the communications channel, and computation. Concern with the measurement process can be dated back to Maxwell's demon (*1*). In the development of that subject, the notion that information is physical was introduced by Szilard (*2*), although it was not widely accepted for many decades. Concern with the communications channel became a subject of intense concern after Shannon's work (*3*). It is the newest of the three branches, computation, that has caused us to reexamine the perceived wisdom in the two earlier areas (*1*, *4–8*). It was pointed out long ago (*9*) that the steps in the computational process that inevitably demand an energy consumption with a known and specifiable lower bound are those that discard information. It was also understood long ago (*9*) that operations that do throw away information, such as the logical AND and the logical OR, can be imbedded in larger operations that perform a logical 1:1 mapping and do not discard information. Nevertheless, a real understanding of what is now called reversible computation came from the work of Bennett (*10*, *11*), who showed that computation can always be conducted through a series of logical 1:1 mappings. Bennett furthermore showed that physical implementations exist that allow this mapping to be utilized to perform computation with arbitrarily little dissipation per step, if done sufficiently slowly. Bennett's discussion envisioned classical machinery with viscous frictional forces proportional to the velocity of motion. It is these forces that can be made as small as desired, through slow computation.

The notion of logically reversible operations, which do not discard information, provides the unifying thread between the three fields of measurement, communications, and computation. In the measurement process, transfer of information from the system to be measured to the meter does not require any minimal and unavoidable dissi-

The author is with the IBM Thomas J. Watson Research Center, Yorktown Heights, NY 10598, USA.

pation (9). The dissipation comes later, if we reset the meter by itself for subsequent reuse and discard the information in it (1, 11).

Communication can also be done with logically and physically reversible operations. As a result, there is no unavoidable lower limit on the energy dissipation per transmitted bit. The term reversible is used in the physical chemist's sense in the discussion of thermodynamic cycles, assuming frictional forces that can be made as small as we wish. I do not use it in the physicist's sense, requiring a total absence of frictional forces.

Our concern is with theoretical limits, not necessarily related to serious technological promise. Our proposed methods for communicating with very little energy expenditure are not put forth as practical candidates. In that connection, however, it is appropriate to recall that reversible computers were originally conceptual proposals with an unrealistic flavor. In recent years, however, closely related schemes have been suggested to save power in real CMOS (complementary metal-oxide semiconductor) logic (12).

The prevailing view about the minimal energy required for bit transmission is based on Shannon's work (3). Shannon provided general expressions that relate channel capacity to the relative probabilities for various received messages, as a function of the transmitted message. These probabilities define the likelihood of erroneous transmission. Shannon's general expressions are not in question, but problems arise when specific models are used for the error probabilities and their generality is overinterpreted. Shannon understood these limitations; later workers have been less cautious. Quoting from Shannon, "An important special case occurs when the noise is added to the signal and is independent of it (in the probability sense)."

In that case, Shannon finds the well-known result

$$C = W \log_2 \frac{P+N}{N} \tag{1}$$

where $C$ is the channel capacity, $W$ is the bandwidth, $P$ is the average received power, and $N$ is the average noise power. Thermal noise, for a classical transmission line with additive equilibrium noise, is given by $N = kTW$, where $k$ is Boltzmann's constant and $T$ is temperature. Equation 1 yields a maximum for $C/P$, at small $P$, given by

$$C = P/kT\ln 2 \tag{2}$$

This equation suggests that at least $kT\ln 2$ energy per transmitted bit is required, although it is not clear that this energy has to be dissipated. This result views noise as an error that maps the intended signal into other nearby signals. However, the linearity of the system, reflected in the addition of $P$ and $N$ in Eq. 1, does not lead to the best way to handle digital signals. These signals are best handled in bistable system in which 0 and 1 are states of local stability. Small noise pulses will cause a temporary deviation from the ideal desired 0 or 1 state, followed by a restoration back to the state of local stability.

If the frequencies $\omega$ in the signal are such that the photon energies $\hbar\omega$ ($\hbar$ is Planck's constant $h$ divided by $2\pi$) are comparable to or larger than $kT$, then the quantization of the signal becomes important. The literature on quantum channel capacity [summarized in (13)], just like that dealing with the classical case, is focused on the linear electromagnetic transmission channel. The work of Caves and Drummond (13) represents a remarkable exception in this long history. They emphasize the limitations that lead to their results.

The rest of the literature is variable in the clarity with which it describes the limits of its results. Brillouin (14), for example, believes he is giving a general law of nature when he states ". . . one bit of information can never be obtained for less than $k\ln 2$ in negentropy costs." Levitin (15) reaffirmed Brillouin's conclusion in recent years but has somewhat more careful language in describing the range of applicability. Bremermann and Marko (16) are two further examples, of many, who give the $kT\ln 2$ result (and its quantum extensions) very broad interpretation.

**Classical channel.** The fact that there is no minimal energy requirement of $kT\ln 2$ per transmitted bit should be apparent from several different perspectives. First of all, computation can be carried out with arbitrarily little dissipation (4, 10, 11). Within a computer, bits are transmitted; therefore, there can be no minimal unavoidable energy penalty for the motion of bits within the computer. A second argument (17): The difference between memory and communication is only a matter of perspective. What is viewed as the rest frame? If there is no unavoidable lower bound to the energy cost of storing a bit, then there should not be any for communication. This statement is closely allied to my own version: we can communicate by shipping the memory (18). These general arguments may not satisfy the reader who asks about connecting into and out of the memory, so I describe a more specific apparatus. First, a classical link discussed in earlier work (5, 19) demonstrated that communication with low energy dissipation does not require physical long-range transport of matter. Therefore, a physical memory structure does not actually have to be shipped. The basic element is a time-modulated potential containing a particle

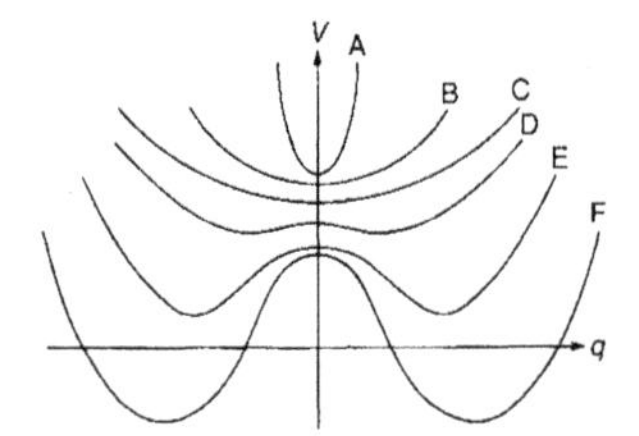


**Fig. 1.** Potential $V$ as a function of particle position $q$ changing with time. The potential starts with a single minimum (curve A), ends up in a deeply bistable state (curve F), and then returns to the single minimum. The relative vertical displacement of curves is selected for clarity and has no significance.

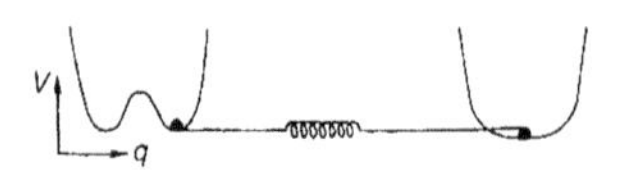


**Fig. 2.** Particle in deeply bistable potential well on the left, coupled to a particle on the right. The particle on the right is in a well about to undergo a transition to a bistable state. The spring is symbolic; it is the relative displacements from the center of the respective potentials that are coupled.

(Fig. 1). This scheme is an adaptation of a proposal for the use of subharmonic parametric excitation to carry out logic (20), which was subsequently elaborated by many investigators (21). The potential $V$ in Fig. 1 is assumed to be heavily damped as a result of viscous frictional forces proportional to particle velocity. The potential will change slowly with time, so as to minimize frictional forces on the moving particle. The mere time dependence of the potential, in the absence of motion of the particle, is not a source of energy dissipation. The time-dependent forces can be generated, for example, by moving controlled charges toward and away from a charged information-bearing particle in the well. Coupling particles in different wells with springs and choosing different phases for the time modulation of different wells can be used to accomplish all the logic functions of a computer (5, 22), but that will not be needed here. The basic unit of interest to us couples two wells (Fig. 2). The unit on the left, with the particle locked securely in its left or right pocket, exerts a biasing force on the particle on the right, in a well about to become bistable. Thus, the particle on the right will likely end up in the new, right-hand pocket.

Consider in more detail the particle in Fig. 1, subject to a biasing force pushing it to the right. Curve C in Fig. 1, at the onset of bifurcation, is relatively flat. The biasing force displaces the minimum to the right, and as the bifurcation proceeds, this minimum evolves into a deeper pocket, favored over the left pocket, which would start to form at a later stage of the bifurcation. When this metastable left pocket starts to form, the Boltzmann factor, $\exp(-V/kT)$, gives a nonvanishing probability for trapping the particle in this undesired left pocket. Without following the exact kinetics, it is still clear that the particle has some chance of eventually ending up in the undesired left pocket, after we have reached its maximum depth (curve F, Fig. 1). This error probability, however, can be made as small as desired by suitable design choices (*5, 22*). Such minimization requires a large enough biasing force and a sufficiently slow change of $V$ with time to allow the particle trapped in the incorrect metastable state to escape during the initial stages of the bifurcation.

After the transfer of information from the left well of Fig. 2 to the right one has been accomplished, the left well can then be restored to its monostable state to receive a new bit. In that connection, another precaution must be described. Consider Fig. 3, which shows the transferred old bit in the right well. The potential in the middle well, where the bit originated, has started to return to the monostable state, with a lowered barrier between its two pockets. Assume that immediately to its left there is a new bit to be transferred into the central well and that this new bit is of opposite polarity to the old one. The biasing forces exerted on the central well from its two neighbors cancel (Fig. 3). As the barrier in the central well is lowered, the particle has some chance of thermally activated escape into the other pocket of the well. The ability to control the escape velocity by use of a sufficiently slow potential modulation rate is lost; the escape velocity is controlled by the barrier height. More detailed considerations show that the energy dissipation is that of a particle released into a volume twice its original size, $kT\ln 2$. To eliminate such undesired losses, the far left well in Fig. 3 must be in its monostable state until the central well has been restored to its monostable state. Only after that can the left well receive a new bit. To prevent this error, as the left well (Fig. 2) is restored to its monostable state, the well on its left is in turn also brought to a monostable well, acting as a buffer. The bias force on the well being restored therefore comes primarily from its right neighbor and favors the occupied pocket. Only after the central well (Fig. 3) is restored to the monostable state can its left neighbor become bistable, allowing it to receive a new bit. Passage of a bit along such a chain clearly constitutes a communication link. It is not passive like a transmission line or optical fiber: Active machinery is required for the propagation. In order to minimize friction, the wells have to be modulated slowly; therefore, the information velocity is low. This restriction does not imply a low information rate; the linear density of particles can be as high as desired. It can be arbitrarily high, as long as we do not inquire as to how such time-dependent wells are actually constructed.

The time-dependent potentials and the springs do not have to be perfect; for example, the potential can be slightly asymmetric. The parameters involved can be chosen to let the machinery function as intended, as long as the deviations fall within designated small error bounds (*5, 22*).

**Quantum mechanical channel.** We now turn to a description of a quantum mechanical procedure, enlarging on an earlier suggestion (*23*). The procedure is quantum mechanical but assumes that the bits are classical and each is in a clearly defined 0 or 1 state. This formulation ignores the recently fashionable possibility of utilizing quantum mechanically coherent superpositions of 0 and 1 (*24, 25*). Consideration of this extra freedom could possibly lead to more optimistic results and certainly will not give more restrictive results. Even without that additional benefit, the conventional limits for quantized linear electromagnetic channels do not apply when we consider alternative communication possibilities. Caves and Drummond (*13*) point out that

$$C = \frac{\pi}{\ln 2}\sqrt{2P/3h} \qquad (3)$$

for the linear boson channel, where $C$ is the bit transmission rate and $P$ is the power. Thus, $P/C \sim C$. The minimal energy per bit is proportional to the bit rate, or roughly the quantum size. The following example will show, instead, that no matter how high $C$ is, no unavoidable minimal energy expenditure is needed. It is optimistically assumed, as in other quantum mechanical analyses in this field, that Hamiltonian and frictionless systems are available.

Again using symmetrical bistable wells, take a particle in the left pocket as a 0 and a particle in the right pocket as a 1. If the barrier between the pockets is high enough, the information can be stable against both tunneling and thermal agitation for a long time. Basically, the proposal is to ship the wells with their bits, which means that the source of the well potential is in motion. For example, if the information-bearing particle is charged and externally controlled charges produce the well field, then these external charges are in motion.

A deeply bistable symmetrical well has an even wave function for its ground state and an odd wave function for its first excited state (Fig. 4). A particle in the left pocket results from a superposition of the two exhibited wave functions, giving almost complete cancellation in the right pocket. The particle is not in the ground state, but we can make the elevation in energy above the ground state, proportional to the energy splitting of the two states, as small as we wish by making the barrier sufficiently high. (The two states become degenerate if the barrier is impenetrable.) Thus, information can be stored with as little energy elevation above the ground state as desired, and even this elevation is not necessarily an energy expenditure. After all, we do not need or even want a relaxation to the ground state. As is known from theories of totally quantum mechanical computation, the computation proceeds in excited states of the system, but that does not in itself require dissipation. Furthermore, if we do not ask how we can actually construct these wells, but stick to elementary quantum mechan-

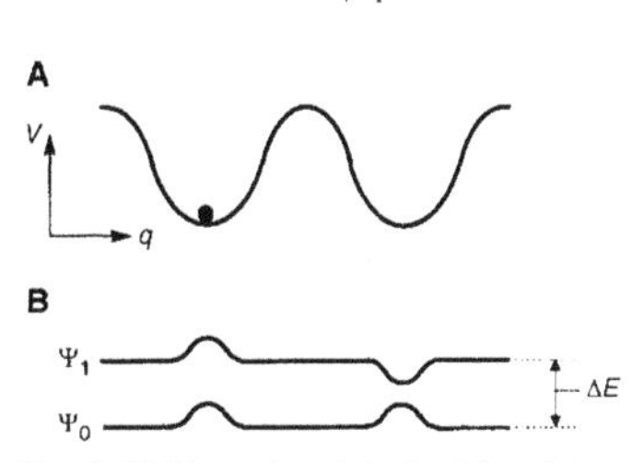


**Fig. 4.** (**A**) The system is in the left well, in a superposition of the symmetric ground state $\Psi_0$ and the antisymmetric first excited state $\Psi_1$. (**B**) The wave functions for the two lowest states, separated by energy $\Delta E$.

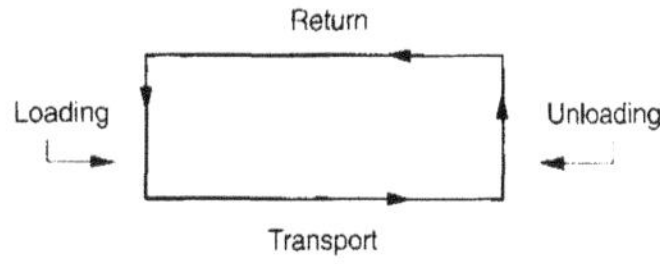


**Fig. 5.** Schematic characterization of communications link. Information is sent along bottom link. Standardized bits (say, all 0) are returned in the top link. On each side, the wells slow down (or even stop) for loading and unloading.

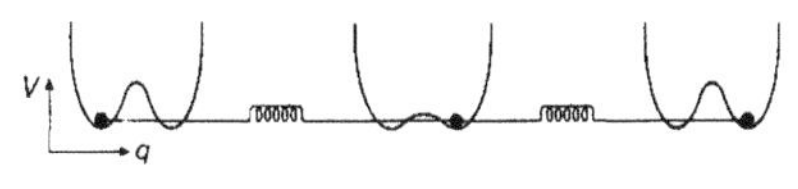


**Fig. 3.** Need to control influence of third well. Information in the central well has been transferred to right well, and the central well is returning to the monostable state. If the well on the far left is in a different information state, no bias will be exerted on the central well, allowing dissipative barrier crossing as the barrier disappears.

ics and statistical mechanics, then the wells can be as compact as desired. Thus, there is no restriction on the information density. Elementary quantum mechanics, per se, poses no restriction on bit density or on energy elevation in bit storage.

The communication link based on this theory resembles a ski lift (Fig. 5), but instead of chairs, it has moving bistable wells. There is a loading area where information is inserted into a well. As in a modern ski lift, the well is moved slowly (or even is stationary) during loading, and then is accelerated for transport. It slows down again, or halts, during unloading. Then it is left in a standardized state, say 0, and returned at higher speed to the loading area. To minimize the effects of acceleration and deceleration, narrow well pockets are used; thus, the states within a single pocket are separated by a large energy and are not very polarizable.

In the loading step (Fig. 6B), the incoming bit, which can be a 0 or 1, arrives in well α. The moving well on the communications link, well β, has just arrived (Fig. 5) in the 0 state. The state of α is copied into β , and later α is restored to the 0 state (Fig. 6C). It is the copying step (Fig. 6B) that requires the bit to be clearly in a 0 or 1 state rather than a quantum mechanical superposition; in the latter case, copying is impossible (*26*).

The copying is executed by a temporary modulation of the barrier in well β, which permits tunneling from the 0 to the 1 pocket if and only if α is in the 1 state. If the barrier in β is lowered so that tunneling can occur with appreciable probability, then the particle will oscillate back and forth between the two pockets with frequency $\Delta E/h$, where $\Delta E$ is the splitting between the two lowest states of the bistable well. If we allow the particle to tunnel for exactly half of a cycle, it will have moved to the other pocket; after that, the barrier is raised again. The potential during the copying step is a function of the position of both particles. The potential before any barrier lowering is symmetrical with respect to interchange of α and β (Fig. 7A). In that case, there is a barrier $V$ for one particle tunneling and $2V$ for both tunneling simultaneously. These barriers are presumed high enough to prevent tunneling. A lowered barrier for β permits the desired copying if α is in the 1 state (Fig. 7B). How might the barrier lowering be accomplished? The particle in α can be taken to have an attractive potential for the particle in β. (A similar process can be used for a repulsive potential.) The two wells are not oriented as in Fig. 6, with the two directions for barrier crossing aligned with each other; instead, the direction of barrier crossing in α is perpendicular to that in β (Fig. 8). The 1 valley in α is placed closest to the β barrier, the 0 valley farther away. Thus, if α is in the 1 state, then the β barrier is lowered. Meanwhile, the barrier in α is kept high enough to prevent unintended tunneling; it may even have to be raised temporarily. As the α well approaches the β well, the force needed by the deterministic (that is, heavy and essentially classical) apparatus controlling the motion depends on the information state of α, but such a force results in energy demands that are compensated in later parts of the cycle: it is not energy dissipation. Note also that as the α particle approaches its target, tunneling will commence before α reaches its final position. This early rise in tunneling probability must be taken into account in the total time allowed for tunneling.

To reset the α bit to 0, we can use a similar procedure (Fig. 7C). Both the α and β wells are turned 90° from that shown in Fig. 8, so that the two controlled α pockets are equal distances from the controlling particle in β. Turning the wells is a process similar to the motion of the wells along the link; the sources of the well field are turned in a process that does not depend on the information content of the wells. The wells must be turned so that the controlling β particle is closer or farther from the α barrier, depending on the state of β. Then, the tunneling in the α well is allowed to proceed, if the barrier has been lowered. The end result, involving two sequentially controlled tunneling events, is an exchange between the states of α and β. This exchange is a unitary operation, and if we permit ourselves the generosity characteristic of much of the literature on quantum computation [for example, (*27*)], which requires the specification of only a unitary time-evolution operator and not its physical embodiment, then all of the details related to the time-dependent wells with controlled tunneling can be skipped: We avoid the intermediate stage in which the content of α is copied to β. Therefore, the information state can be a quantum mechanical superposition of 0 and 1.

So that only the desired particles interact, rather than those further away, short-range forces are required, and we do not try to explain how this can be achieved. Coulomb forces screened by metallic electrodes are a possibility, but a complete analysis and design for that does not exist. Another possibility replaces the bistable wells with bistable quadrupole moments in arrays of quantum dots (*28*). The internal barriers to electron transfer in such arrays must be controlled and changed with time.

At the receiving end, a similar procedure can be used to unload and restore the moving bit to the 0 state. The communications link cannot be expected to act perfectly. There are errors due to the fact that the machinery is likely to be imperfect. The tunneling barrier may not be exactly as intended, the time allowed for tunneling transfer may be too short or too long, or the distance between two interacting particles invoked for controlled tunneling (Fig. 7)

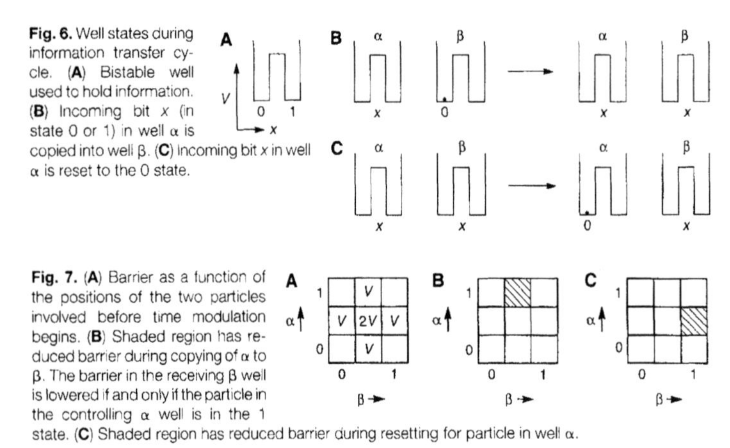


**Fig. 6.** Well states during information transfer cycle. **(A)** Bistable well used to hold information. **(B)** Incoming bit $x$ (in state 0 or 1) in well α is copied into well β. **(C)** Incoming bit $x$ in well α is reset to the 0 state.

**Fig. 7.** **(A)** Barrier as a function of the positions of the two particles involved before time modulation begins. **(B)** Shaded region has reduced barrier during copying of α to β. The barrier in the receiving β well is lowered if and only if the particle in the controlling α well is in the 1 state. **(C)** Shaded region has reduced barrier during resetting for particle in well α.

**Fig. 8.** Controlling α well is perpendicular to controlled β well. The position of α controls the barrier of β. The α barrier can be increased temporarily to lock the α particle firmly in its position.

may be incorrect. Even if the components act exactly as intended, there are still residual errors. For example, tunneling may occur unintentionally through the higher barrier. Another problem arises from the supposed cancellation of wave functions in realizing a 0 or 1 state (Fig. 4A). The two superposed eigenstates (Fig. 4B) have slightly different energies. Their exact wave functions within a pocket are slightly different; they cannot cancel exactly and cannot leave a totally unoccupied pocket. This difficulty can be minimized by using very narrow potential pockets, in which the next higher state within the pocket, having a node for the wave function within that pocket, is at a much higher energy. In the extreme limit, the behavior approaches that of a true two-level system, such as an electron spin, where no details beyond the spin-up and spin-down states exist.

Errors matter in two ways. First, an incorrect bit can be transmitted. This is not a serious problem, if the probability of error is small: a small error rate requires only a small amount of redundancy to regain error free transmission (*3*). The decoding that restores the original message is not dissipationless; after all, we are throwing away information about the error (*4*). However, the minimal required dissipation is proportional to the error rate. Additionally, if errors arise, then the slow accumulation of erroneous bits in the return link (Fig. 5) of the transport machinery must be prevented. The supposed 0-state bits returning to the loading stage (Fig. 5) may have become 1's. We can occasionally reset these bits on their return to the intended 0 state. One possible way to do this takes its clue from Lloyd's proposal (*25*) for error correction in quantum computation. The bit is biased so as to favor the 0 state, and then the barrier is slowly reduced. Eventually, a radiative decay from the metastable 1 state to the lower lying 0 state occurs. It can be assisted by coupling to oscillators (for example, a resonant cavity) at the emission frequency. The decay is a dissipative event, but once again, the dissipation is proportional to the error rate and can be far less than indicated in Eq. 3.

Incidentally, the methods of Figs. 6 through 8 can be adapted to carry out computer logic in arrangements where tunneling is controlled by more than one input well.

**Overview.** The two proposals discussed above use active channels with machinery all along the length of the channel. Such a setup makes these schemes unappealing as serious technological candidates. The most obvious alternative to an active link, for the classical case, is the use of particles with sufficiently well-defined speed and direction, as in the well-known colliding billiard ball computer (*29*). This scheme assumes that friction and noise can be completely eliminated. Furthermore, the chaotic nature of billiard ball collisions makes it extremely sensitive to flaws in the machinery. Using billiard ball motion for a communications link is most easily done within a framework where the net rate of billiard ball transfer does not depend on the information content of the link. For example, 0 is denoted by a ball followed by an empty slot, and a 1 is represented by an empty slot followed by a ball. This scheme permits use of a return link (Fig. 5). Alternatively, if we are optimistic about the lifetime of a photon, then conceivably there are possible inventions that make use of the polarization or timing of single photons. Or perhaps there could be nonlinear optical links.

Even if the limits discussed here are practically unachievable, it is still important to understand such limits. For example, I do not believe that a genuinely useful form of quantum parallelism (*24*) will be achieved; nevertheless, computer scientists concerned with the minimal number of steps required in the execution of an algorithm must take its possibility into account.

## REFERENCES


1. H. S. Leff and A. F. Rex, *Maxwell's Demon: Entropy, Information, Computing* (Princeton Univ. Press, Princeton, NJ, 1990).
2. L. Szilard, *Z. Phys.* **53**, 840 (1929).
3. C. E. Shannon, *Bell Syst. Tech. J.* **27**, 379 (1948); *ibid.*, p. 623.
4. C. H. Bennett, *IBM J. Res. Dev.* **32**, 1–6 (1988).
5. R. Landauer, in *Selected Topics in Signal Processing*, S. Haykin, Ed. (Prentice Hall, Englewood Cliffs, NJ, 1989), pp. 18–47. (The printed version has some figures oriented incorrectly.)
6. ———, *Physica A* **194**, 551 (1993).
7. ———, in *Proceedings of the Workshop on Physics and Computation: PhysComp '94* (IEEE Computer Society Press, Los Alamitos, CA, 1994), pp.54–59.
8. C. M. Caves, in *Physical Origins of Time Asymmetry*, J. J. Halliwell, J. Pérez-Mercader, W. H. Zurek, Eds. (Cambridge Univ. Press, Cambridge, 1994), pp. 47–89; W. Zurek, *Nature* **341**, 119 (1989).
9. R. Landauer, *IBM J. Res. Dev.* **5**, 183 (1961).
10. C. H. Bennett, *ibid.* **17**, 525 (1973).
11. ———, *Int. J. Theor. Phys.* **21**, 905 (1982).
12. D. Frank and P. Solomon, in *Proceedings of the International Symposium on Low Power Design* (Association for Computing Machinery, New York, 1995), pp. 197–202. See also *Proceedings of the Workshop on Physics and Computation: PhysComp '94* (IEEE Computer Society Press, Los Alamitos, CA, 1994), and citations therein.
13. C. M. Caves and P. D. Drummond, *Rev. Mod. Phys.* **66**, 481 (1994).
14. L. Brillouin, *Science and Information Theory* (Academic Press, New York, 1956), p. 162.
15. L. B. Levitin, in *Proceedings of the Workshop on Physics and Computation: PhysComp '92* (IEEE Computer Society Press, Los Alamitos, CA, 1993), pp. 210–214.
16. H. J. Bremermann, *Int. J. Theor. Phys.* **21**, 203 (1982); H. Marko, *Kybernetik* **2**, 274 (1965).
17. E. Fredkin, *Physica D* **45**, 254 (1990).
18. R. Landauer, in *Sixth International Conference on Noise in Physical Systems*, P. H. E. Meijer, R. D. Mountain, F. J. Soulen Jr., Eds. (Spec. Publ. 614, National Bureau of Standards, Washington, DC, 1981), pp. 12–17.
19. ———, *Appl. Phys. Lett.* **51**, 2056 (1987).
20. E. Goto, *J. Elec. Commun. Eng. Jpn.* **38**, 770 (1955); J. von Neumann, U.S. Patent 2,815,488 (1957).
21. For example, E. Goto *et al.*, *IRE Trans. Electron. Comput.* **9**, 25 (1960).
22. K. K. Likharev, *Int. J. Theor. Phys.* **21**, 311 (1982); R. C. Merkle, *Nanotechnology* **4**, 114 (1993).
23. R. Landauer, in *Proceedings of the Drexel-4 Symposium on Quantum Nonintegrability: Quantum-Classical Correspondence*, D. H. Feng and B.-L. Hu, Eds. (International Press, Boston, in press).
24. J. Glanz, *Science* **269**, 28 (1995); C. H. Bennett, *Phys. Today* **48**, 24 (October 1995); D. P. DiVincenzo, *Science* **270**, 255 (1995); B. Schumacher, *Phys. Rev. A.* **51**, 2738 (1995); P. Hauslanden, B. Schumacher, M. Westmoreland, W. K. Wootters, *Ann. N.Y. Acad. Sci.* **755**, 698 (1995); S. Lloyd, *Sci. Am.* **273**, 140 (October 1995).
25. S. Lloyd, *Science* **261**, 1569 (1993); *ibid.* **263**, 695 (1994).
26. W. K. Wootters and W. H. Zurek, *Nature* **299**, 802 (1982).
27. H. F. Chau and F. Wilczek, *Phys. Rev. Lett.* **75**, 748 (1995).
28. C. S. Lent, P. D. Tougaw, W. Porod, G. H. Bernstein, *Nanotechnology* **4**, 49 (1993).
29. E. Fredkin and T. Toffoli, *Int. J. Theor. Phys.* **21**, 219 (1982).


25 January 1996; accepted 22 April 1996

A minor variation of the scheme described in Figs. 6-8 can be used to load and unload qubits. Consult author for details.

# Information physics in cartoons

CHARLES H. BENNETT
*IBM Research Division, T. J. Watson Research Center, Yorktown Heights, NY 10598, U.S.A.*

*(Received 5 December 1997)*

Landauer's contributions to the physics of information transmission and processing are highlighted by figures and cartoons drawn by himself and others



**Key words:** fundamental limits, Maxwell's demon, Shannon, Landauer's principle.

## 1. Introduction

Rolf Landauer has done more than anyone else to establish the physics of information processing as a serious subject of scientific inquiry. Here I will take a less serious approach, highlighting his contributions to this field with the help of cartoons and diagrams drawn by himself and others. Landauer has always believed that physics and mathematics are profoundly interdependent (Fig. 1): on the one hand, the laws of physics are expressed in mathematical language; on the other, information can only be processed when it is embodied in a physical system. Perhaps because of its interdisciplinary nature, the physics of information has suffered more than its share of superficial thinking, exemplified by von Neumann's remark in a 1949 lecture, posthumously published and made famous, that any computing device, natural or artificial, must dissipate at least $kT \ln 2$ of energy 'per elementary act of information, that is, per elementary decision of a two-way alternative, and per elementary transmittal of 1 unit of information' [1]. Landauer's traditional response to such pronouncements has been a cartoon (Fig. 2) suggesting that the problem of fundamental limits is too complex to be solved in a dinnertime discussion. Even in the more formal setting of scholarly books and papers, von Neumann's principle has often been asserted as self-evident, or adduced from a few examples of particular mechanisms obeying it, without any effort to show that it must hold generally. Typically these mechanisms were optical, for example Brillouin's argument [2] that a Maxwell's demon, using light to observe molecules in flight, would need to expend at least one photon of energy at least $kT$ to see the molecule against the background of blackbody radiation, or Gabor's [3] transverse mode analysis of a light beam used to detect a molecule at one end of a long cylinder.

We now know that von Neumann's principle is too restrictive; it has been supplanted by Landauer's principle, according to which the data processing operations have an irreducible thermodynamic cost if and only if they are *logically irreversible*, i.e. throw away information about the computer's previous logical state. According to our current understanding [4–10], information acquisition or transmission need not be thermodynamically costly, but information destruction (e.g. when a Maxwell demon erases the result of one measurement to make room for the next) always is, and it is this cost which prevents the demon from violating the second law. The tendency of mid-twentieth century analyses of the demon to neglect the cost of information destruction, and concentrate instead on a supposed cost of information acquisition, probably represents a prejudice carried over from everyday life, where until recently information has been viewed as a valuable—or at worse worthless—commodity, not a waste product one would pay to have removed. It may also have been a side effect of the

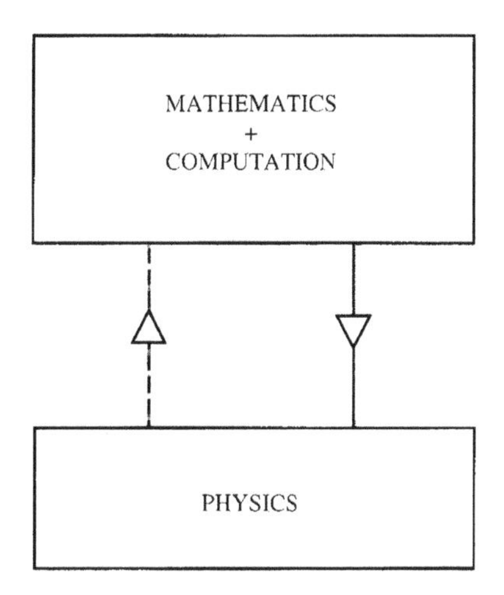


Fig. 1.

MAY WAH RESTAURANT

98 Triangle Shopping Center ( on the Mall ) Yorktown Heights, N. Y.

PRICE LIST FOR TAKE OUT SERVICE

TEL : ( 914 ) 245 - 3311

| SOUP | Pt. | Qt. |
|---|---|---|
| Wonton Soup | .75 | 1.40 |
| Egg Drop Soup | .70 | 1.30 |
| Chicken Noodle or Rice | | |
| Hot and Sour Soup | | |
| Yat Gow Mein | | |

CHOW MEIN
( with Noodles & R

| | | |
|---|---|---|
| Chicken Chow Mein | | |
| Chicken Chow Mein w. Mushroom | | |
| Shrimp Chow Mein | | |
| Roast Pork Chow Mein | | |
| Vegetable Chow Mein | | |
| Beef Chow Mein | 2.35 | 4.60 |
| Subgum Chicken Chow Mein | 2.80 | 4.95 |
| Subgum Shrimp Chow Mein | 3.05 | 5.95 |
| May Wah Special Chow Mein | 2.95 | 5.70 |
| Lobster Chow Mein | 3.75 | 7.25 |
| White Meat Chicken Chow Mein | 2.80 | 5.50 |

CHOP SUEY
( with Rice )

| | | |
|---|---|---|
| Fresh Pork Chop Suey | 2.35 | 4.55 |
| Chicken Chop Suey | 2.45 | 4.75 |
| Beef Chop Suey | 2.60 | 5.10 |
| Shrimp Chop Suey | 2.80 | 5.55 |
| Roast Pork Chop Suey | 2.50 | 4.85 |
| Vegetable Chop Suey | 2.20 | 4.35 |
| Lobster Chop Suey | 3.75 | 7.40 |

SEA FOOD

Portion

APPERIZERS

| | |
|---|---|
| Egg Rolls (2) | 1.40 |
| Shrimp Rolls (2) | 1.80 |
| ) 3.15 ( Lg. ) | 6.10 |
| ) 3.15 ( Lg. ) | 6.10 |
| | 2.50 |
| | 3.20 |
| ed | 3.15 |

( with Rice )

| | Portion |
|---|---|
| | 3.75 |
| de | 4.30 |
| | 5.45 |
| | 5.45 |
| Pepper Steak with Onions | 4.15 |
| Beef with Tomatoes | 5.30 |
| Beef with Gravy | 5.55 |
| Beef with Mushrooms | 5.55 |
| Chiang Po Beef | 5.55 |
| Shredded Spiced Beef | 5.70 |
| Beef Lo Mein | 3.75 |

PORK ( with Rice )

| | |
|---|---|
| Sweet and Sour Pork | 4.45 |
| Roast Pork with Bean Sprouts | 3.60 |
| Diced Roast Pork with Almond | 4.60 |
| Roast Pork with Vegetable | 3.85 |
| Roast Pork with Muhsrooms | 5.30 |
| Subgum Wonton | 6.75 |
| Roast Pork Lo Mein | 3.35 |
| Moo Shu Pork (Pancakes 30c Each) | 4.60 |

POULTRY ( with Rice )

$\Delta E \, \Delta t \sim \hbar$

Fig. 2.

great success of quantum mechanics, which focused attention on the nontriviality of measurement. Many of these ideas are summarized in Larry Gonick's cartoon history of Maxwell's demon (Fig. 3) [11].

Long after Landauer's principle was generally accepted in computation, the notion persisted among com-

**Fig. 3.**

munications theorists that at least $kT \ln 2$ of energy must be expended, or at least temporarily invested, in order to *transmit* a bit of information from one place to another. This belief arose from Shannon's proof [12] that this amount of energy is indeed needed to make signals in a linear electromagnetic (or other bosonic) channel distinguishable above the thermal background. Despite the practical importance of such channels,

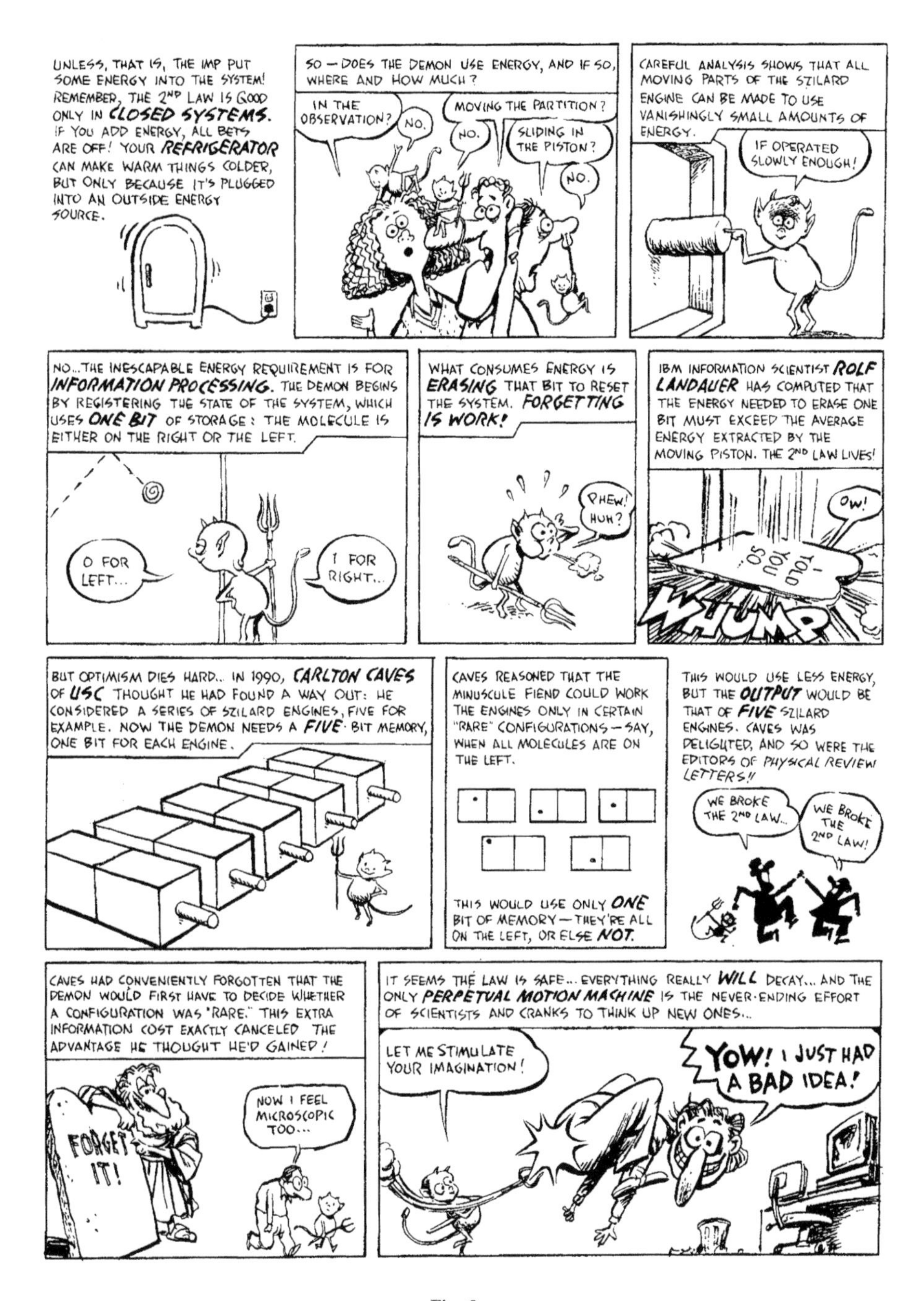


**Fig. 3.**

they are not the most energy-efficient: nonlinear channels, in which information is carried in the internal state of a material body (e.g. a molecule with two stable states separated by a high-energy barrier) can transport

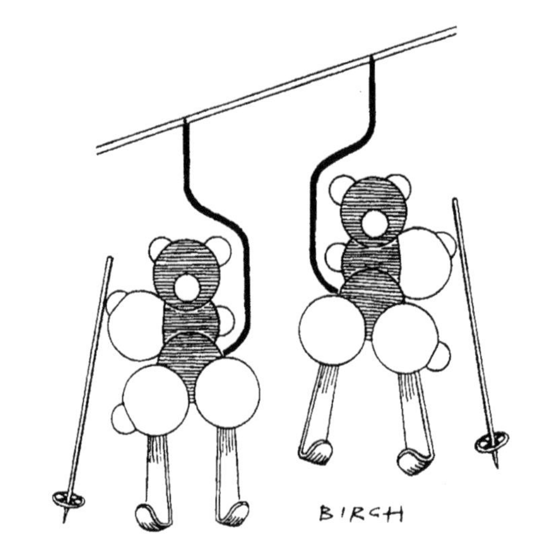


Fig. 4.

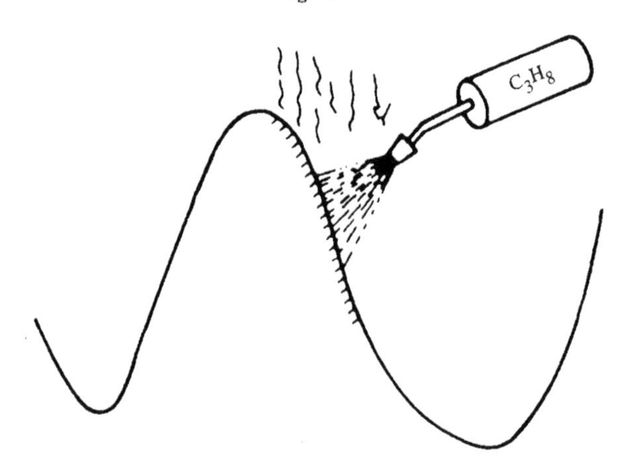


Fig. 5.

information at arbitrarily little cost per bit. Landauer's recent analysis of this type of channel as an information ski lift [10], which inspired a cartoon (Fig. 4) by Birch [13], appears finally to have gotten the message across.

Landauer has a good nose for ideas sexy enough to become fashionable but too grand and simple to be useful. Among his favorite examples from the past are catastrophe theory and all-optical computing. Another is the recurrent notion that there is some simple criterion for identifying preferred states of systems far from equilibrium, without reference to the systems' detailed dynamics. To help combat this idea he introduced the blowtorch model (Fig. 5), in which changes in the nonequilibrium dynamics of rarely populated states determine which of two locally stable energy minima is preferred. If the blowtorch is turned off, the energetically lower right minimum is preferred. If it is turned on, escape from this minimum is thermally facilitated, and the left minimum is preferred.

Recently Landauer has turned his skepticism toward the now fashionable field of quantum computation, doubting whether quantum computers can ever be made coherent and stable enough for the promised exponential speedups to materialize. Advocates of quantum computation responded with ingenious error-correction and stabilization schemes, earning, in his mind, a modicum of plausibility. A few years ago, when prospects for quantum computation looked far bleaker than they do today, Bill Unruh began a talk on the serious obstacles facing it with the words, 'I don't want to rain on your parade, but . . .'. Correctly foreseeing the sociological future of a field whose scientific future remains to be seen, Landauer remarked, 'It will take more than rain to stop this parade.'

## References


[1] J. von Neumann, *Theory of Self-Reproducing Automata*, edited and completed by Arthur W. Burks, p. 66, University Illinois Press, Urbana and London (1966).

[2] L. Brillouin, *Science and Information Theory*, 2nd edn, Academic Press, London (1962).

[3] D. Gabor, 'Light and Information', *Progress in Optics* **1**, 111–153 (1964).

[4] R. Landauer, *IBM J. Res. Develop* **3**, 183–191 (1961).

[5] C. H. Bennett, *IBM J. Res. Develop* **17**, 525–532 (1973).

[6] C. H. Bennett, *Int. J. Theor. Phys.* **21**, 905–940 (1982).

[7] E. Fredkin and T. Toffoli, *Int. J. Theor. Phys.* **21**, 219–253 (1982).

[8] E. Lubkin, *Int. J. Theor. Phys.* **16**, 523–535 (1987).

[9] R. Landauer, *Appl. Phys. Lett.* **51**, 2056 (1987).

[10] R. Landauer, 'Minimal Energy Requirements in Communication', *Science* **272**, 1914 (1996).

[11] 'Science Classics: Maxwell's Demon' *Discover* **12**, 82–83 (July 1991), reproduced by permission of Larry Gonick.

[12] C. Shannon, *Bell Syst. Tech. J.* **27**, 397, 623 (1948).

[13] *Nature* **382**, 669 (1996), reproduced by permission of Andrew Birch.

# Signal entropy and the thermodynamics of computation

by N. Gershenfeld

***Electronic computers currently have many orders of magnitude more thermodynamic degrees of freedom than information-bearing ones (bits). Because of this, these levels of description are usually considered separately as hardware and software, but as devices approach fundamental physical limits these will become comparable and must be understood together. Using some simple test problems, I explore the connection between the information in a computation and the thermodynamic properties of a system that can implement it, and outline the features of a unified theory of the degrees of freedom in a computer.***

Computers are unambiguously thermodynamic engines that do work and generate waste heat. It is hard to miss: Across the entire spectrum of machine sizes, power and heat are among the most severe limits on improving performance. Laptops can consume 10 watts (W), enough to run out of power midway during an airline flight. Desktop computers can consume on the order of 100 W, challenging the air-cooling capacity of the CPU, and adding up to a greater load in many buildings than the entire heating, ventilation, and air-conditioning system. Further, this load is poorly characterized; MIT would need to double the size of its electrical substation if it was designed to meet the listed consumption of all the computers being used on campus.[1] And supercomputers can consume 100 kilowatts (kW) in a small volume, pushing the very limits of heat transfer to prevent them from melting.

All of these problems point toward the need for computers that dissipate as little energy as possible. Current engineering practice is addressing the most obvious culprits (such as lowering the supply voltage, powering down subsystems, and developing more efficient backlights). In this paper I look ahead to consider the fundamental thermodynamic limitations that are intrinsic to the process of computation itself.

Consider a typical chip (Figure 1). Current is drawn from a power supply and returned to a ground, information enters and leaves the chip, and heat flows from the chip to a thermal reservoir. The question that I want to ask for this system is: What can be inferred about the possible thermodynamic performance of the chip from the analysis of the input and output signals? To clarify the important reasons to ask this question, this paper examines answers for some simple test cases:

- Estimating the fundamental physical limits for a given technology base, in order to understand how far it can be improved, requires insight into the expected workload.
- Short-term optimizations and long-term optimal design strategies must be based on this kind of system-dependent analysis.

The investigation of low-power computing is currently being done by two relatively disjoint camps: physicists who study the limits of single gates, but do not consider whole systems, and engineers who are making evolutionary improvements on existing

**Figure 1 Fluxes into and out of a chip**

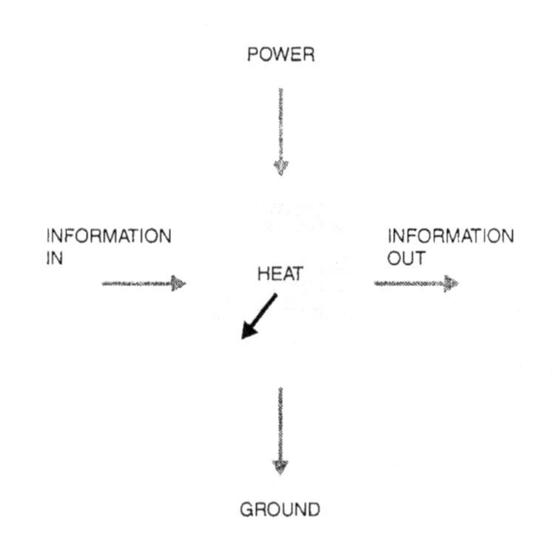


designs, without considering the fundamental limits. What has been missing is attention to basic limits for practical tasks; developing low-power computing is going to require as much attention to entropy as to energy.

## Bits

There has been a long-standing theoretical interest in the connection between physics and the thermodynamics of computation, reflecting the intimate connection between entropy and information. Although entropy first appeared as a way to measure heat engine efficiency, the search for a microscopic explanation helped create the fields of statistical mechanics and kinetic theory. The development of statistical mechanics raised a number of profound questions, including the paradox of Maxwell's Demon. This is a microscopic intelligent agent that apparently can violate the second law of thermodynamics by selectively opening and closing a partition between two chambers containing a gas, thereby changing the entropy of the system without doing any work on it.[2] Szilard reduced this concept to its essential features in a gas with a single molecule that can initially be on either side of the partition but then ends up on one known side.[3] Szilard's formulation introduced the notion of a bit of information, which provided the foundation for Shannon's theory of information[4] and, hence, modern coding theory. Through the study of the thermodynamics of computation, information theory is now returning to its roots in heat engines.

If all the available states of a system are equally likely (a micro-canonical ensemble) and there are $\Omega$ states, then the entropy is $k\log\Omega$. Reducing the entropy of a thermodynamic system by $dS$ generates a heat flow of $dQ = TdS$ out of the system. Landauer[5] realized that if the binary value of a bit is unknown, erasing the bit changes the logical entropy of the system from $k\log 2$ to $k\log 1 = 0$ (shrinking the phase space available to the system has decreased the entropy by $k\log 2$). If the physical representation of the bit is part of a thermalized distribution (so that thermodynamics does apply), then this bit erasure necessarily is associated with the dissipation of energy in a heat current of $kT\log 2$ per bit. This is due solely to the act of erasure of a distinguishable bit, and is independent of any other internal energy and entropy associated with the physical representation of the bit. Bennett[6] went further and showed that it is possible to compute reversibly with arbitrarily little dissipation per operation, and that the apparent paradox of Maxwell's Demon can be explained by the erasure in the Demon's brain of previous measurements of the state of the gas (that is when the combined Demon-gas system becomes irreversible).

For many years it was assumed that no fundamental questions about thermodynamics and computation remained, and that since present and foreseeable computers dissipate so much more than $kT\log 2$ per bit, these results would have little practical significance. More recently, a number of investigators have realized that there are immediate and very important implications. A bit stored in a CMOS (complementary metal-oxide semiconductor) capacitor contains an energy of $CV^2/2$ (where $V$ is the supply voltage and $C$ is the gate capacitance). Erasing a bit unnecessarily wastes this energy and so should be avoided unless it is absolutely necessary. A simple calculation also shows that charging a capacitor by instantaneously connecting it to the supply voltage dissipates an additional $CV^2/2$ through the resistance of the wire carrying the current (independent of the value of the resistance), whereas charging it by linearly ramping up the supply voltage at a rate $\tau$ reduces this to approximately $CV^2\ RC/\tau$.[7] Avoiding unnecessary erasure is the subject of charge recovery and reversible logic,[8,9] and making changes no faster than needed is the subject of adiabatic logic.[10,11] These ideas have been implemented quite successfully in conventional MOS processes, with power savings of at least an order of magnitude so far. The important point is that even though $CV^2/2$ is currently much greater than $kT$, by analogy many of the same concerns about erasing and moving bits apply at this much larger energy scale.

All of these results can be understood in the context of the first law of thermodynamics,

$$\begin{aligned} dU &= dW + dQ \\ &= dW + TdS \end{aligned} \tag{1}$$

The change in the internal energy of a system $dU$ is equal to the sum of the reversible work done on it $dW$ and the heat irreversibly exchanged with the environment $dQ = TdS$ (which is associated with a change in the entropy of the system). Integrating this gives the free energy

$$F = U - TS \tag{2}$$

which measures the fraction of the total internal energy that can reversibly be recovered to do work. Creating a bit by raising its potential energy $U$ (as is done in charging a capacitor) stores work that remains available in the free energy of the bit and that can be recovered. Erasing a bit consumes this free energy (which charge recovery logic seeks to save) and also changes the logical entropy $dS$ of the system; hence it is associated with dissipation (which is reduced in reversible logic).

Erasure changes the number of logical states available to the system, which has a very real thermodynamic cost. In addition, the second law of thermodynamics

$$dS \geq 0 \tag{3}$$

leads to the final principle of low-power computing. Entropy increases in spontaneous irreversible processes to the maximum value for the configuration of the system. If at any time the state of the bit is not close to being in its most probable state for the system (for example, instantaneously connecting a discharged capacitor to a power supply temporarily results in a very unlikely equilibrium state), then there will be extra dissipation as the entropy of the system increases to reach the local minimum of the free energy. Adiabatic logic reduces dissipation by always keeping the system near the maximum of the entropy. In the usual regime of linear nonequilibrium thermodynamics[12] (i.e., Ohm's law applies, and ballistic carriers can be ignored), this damping rate is given by the fluctuation-dissipation theorem in terms of the magnitude of the equilibrium fluctuations of the bit (and hence the error rate).

Fredkin and Toffoli introduced reversible gates that can be used as the basis of reversible logic families. These gates produce extra outputs, not present in ordinary logic, that enable the inputs to the gate to be deduced from the outputs (which is not possible with a conventional irreversible operation such as AND). Although this might make them appear to be useless in practice since all of these intermediate results must be stored or erased, Bennett used a pebbling argument to show that it is possible to use reversible primitives to perform an irreversible calculation with a modest space-time trade-off.[13,14] It is done by running the calculation as far forward as the intermediate results can be stored, copying the output and saving it, then reversing the calculation back to the beginning. The result is the inputs to the calculation plus the output; a longer calculation can be realized by hierarchically running subcalculations forward and then backward. This means that a steady-state calculation must pay the energetic and entropic cost of erasing the inputs and generating the outputs, but that there need be no intermediate erasure.

Therefore, an optimal computer should never need to erase its internal states. Further, the state transitions should occur no faster than they are needed, otherwise unnecessary irreversibility is incurred, and the answers should be no more reliable than they need be, otherwise the entropy is too sharply peaked and the system is over-damped. These are the principles that must guide the design of practical computers that can reach their fundamental thermodynamic limits.

To understand the relative magnitude of these terms, let us start with a bit stored in a 5-fF capacitor (a typical CMOS number, including the gate, drain, and charging line capacitances). It has

$$\frac{(5\text{fF})(3\text{V})}{1.6 \times 10^{-19} C/\text{electron}} \approx 10^5 \frac{\text{electrons}}{\text{bit}} \tag{4}$$

and at 3V this is an energy of

$$\frac{1}{2} \cdot 5\text{fF} \cdot (3\text{V})^2 = 2 \times 10^{-14} \frac{\text{J}}{\text{bit}} \approx 10^5 \frac{\text{eV}}{\text{bit}} \tag{5}$$

**Figure 2 A combinatorial gate that sorts its inputs, changing their entropy but not the number of zeros and ones**

| IN | OUT |
|---|---|
| 0 0 | 0 0 |
| 0 1 | 1 0 |
| 1 0 | 1 0 |
| 1 1 | 1 1 |

If $10^7$ transistors (typical of a large chip) dump this energy to ground at every clock cycle (as is done in a conventional CMOS gate) at 100 MHz (a typical clock rate), the energy being dissipated is

$$2 \times 10^{-14}\frac{\text{J}}{\text{bit}} \cdot 10^8\frac{\text{bits}}{\text{s}} \cdot 10^7\text{transistors} \approx 20\text{W} \qquad (6)$$

This calculation is an overestimate because not all transistors switch each cycle, but the answer is of the right order of magnitude for a "hot" CPU. The electrostatic energy per bit is still seven orders of magnitude greater than the room temperature thermal energy of

$$kT = 1.38 \times 10^{-23}\frac{\text{J}}{\text{K}} \cdot 300\text{K} \approx 0.02\text{eV} \qquad (7)$$

In addition to energy, entropy is associated with the distribution of electrons making up a bit because of the degrees of freedom associated with thermal excitations above the Fermi energy. It is a very good approximation to calculate this electronic entropy per bit in the Sommerfeld approximation for temperatures that are low compared to the Fermi temperature:[15]

$$S \approx \frac{N\pi^2 T}{2T_f}k = \frac{10^5\pi^2 300\text{K}}{2 \cdot 10^5\text{K}}k = 1480k \approx 10^{-20}\frac{\text{J}}{\text{K}} \qquad (8)$$

This entropy is not associated with a heat current if the bit is at the same temperature as the chip (remember that entropy is additive). But for comparison, if this configurational entropy was eliminated by cooling the bit, then the associated heat energy is

$$Q = TdS \approx 10^{-18}\frac{\text{J}}{\text{bit}} \qquad (9)$$

Finally, the dissipation associated with the logical erasure is

$$S = k\log 2 \approx 10^{-23}\frac{\text{J}}{\text{K}} \qquad (10)$$

or

$$Q = TdS \approx 10^{-21}\frac{\text{J}}{\text{bit}} \qquad (11)$$

We see that the heat from the thermodynamic and logical entropy associated with a bit are currently four and seven orders of magnitude lower than the electrostatic erasure energy.

## Gates

Now let us turn from bits to circuits, starting first with a simple combinatorial gate (one that has no memory) with the truth table shown in Figure 2. It sorts the input bits; the output has the same energy (the same number of 0s and 1s), but the entropy is reduced by half (assuming that all inputs are equally likely):

$$H_{in} = -\sum_{states} p_{state}\log_2 p_{state} = -4 \times \frac{1}{4}\log_2\frac{1}{4} = 2$$

$$H_{out} = -2 \times \frac{1}{4}\log_2\frac{1}{4} - \frac{1}{2}\log_2\frac{1}{2} = \frac{3}{2} \qquad (12)$$

(using the convention that $H$ measures logical entropy in bits, and $S$ measures thermodynamic entropy in J/K). On average, at each time step the gate consumes one-half bit of entropy and so must be dissipating energy at this rate.

Where does this dissipation appear in a circuit? Figure 3 shows a conventional CMOS implementation of the two-bit sorting task by an AND and an OR gate in parallel, each of which is a combination of parallel nMOS and series pMOS FETs (field-effect transistors) or *vice versa* followed by an inverter.

This circuit dissipates energy whenever its inputs change. Any time there is a $0 \rightarrow 1$ transition at the input, the input capacitance (the gate electrode and the

**Figure 3 A CMOS implementation of the combinatorial two-bit sort**

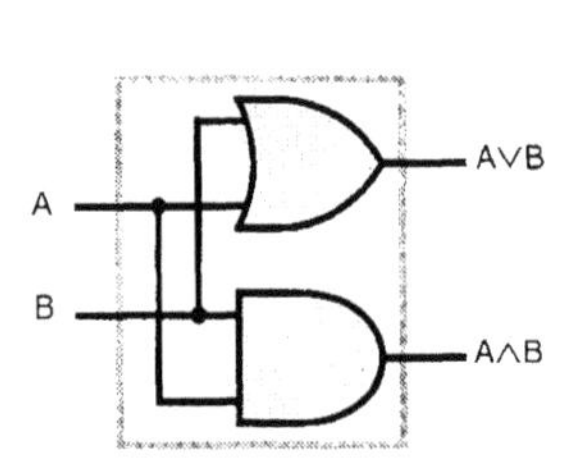


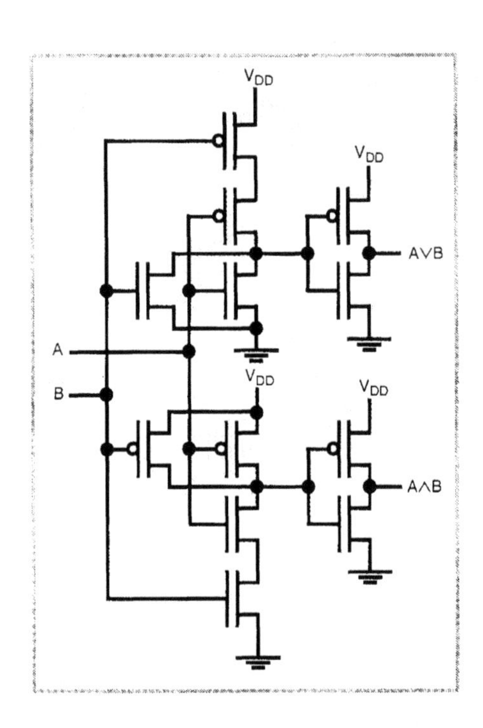


charging line) of each FET must be charged up, and when there is a $1 \rightarrow 0$ transition, this charge is dumped to ground. The same holds true for the outputs, which charge and discharge the inputs to the following gates. Therefore, the dissipation of this circuit is given solely by the expected number of bit transitions, and the one-half bit of entropic dissipation from the logical irreversibility appears to be missing. The issue is not that it is just a small perturbation, but that it is entirely absent.

To understand how this can be, consider the possible states of the system, shown in Figure 4. The system can physically represent four possible input states (00,01,10,11), and it can represent four output states (although 01 never actually appears). When the inputs are applied to this gate, the inputs and the outputs exist simultaneously and independently (as charge stored at the input and output gates). The system is capable of representing all 16 input-output pairs, even though some of them do not occur. When it erases the inputs, regardless of the output state, it must dissipate the knowledge of which input produced the output. This merging of four possible paths into one consumes $\log 4 = 2$ bits of entropy regardless of the state. That is why the half bit is missing: each step is always destroying all of the information possible.

The real problem is that we are using irreversible primitives to do a partially reversible task. Neither

**Figure 4 Possible states of the two-bit sort system**

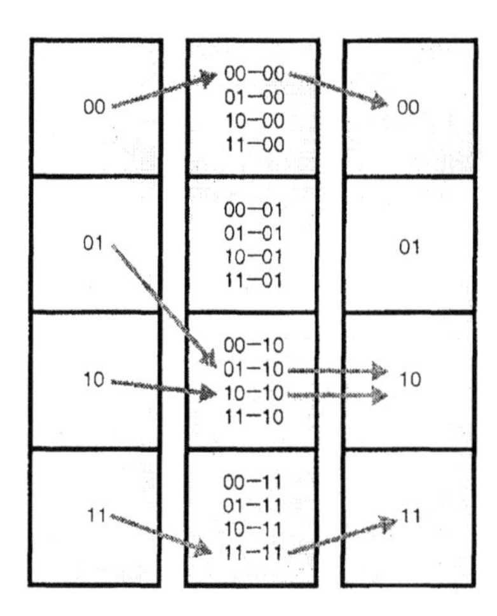


**Figure 5 A Fredkin gate: the control line flips the A and B lines**

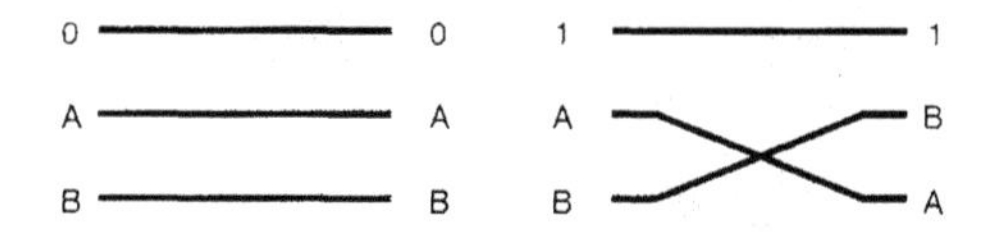


gate alone can recognize when their combined actions are reversible. This naturally suggests implementing this circuit with reversible gates in order to clarify where the half bit of entropic dissipation occurs. Figure 5 shows a simple universal reversible logic element, the Fredkin gate.[16,17] Inputs that are presented to the gate (which might be packets of charge, spins, or even billiard balls in a ballistic computer) are conserved and emerge at the output, but if the control input is present, then the output lines are reversed.

Figure 6 shows our two-bit sort implemented with Fredkin gates. In addition to the AND and OR operations, there are two extra gates that are needed for the reversible version of the FANOUT operation of copying one signal to two lines. For each of these operations to be reversible, extra lines bring out the information needed to run each gate backward. Here again, the half bit of entropy from the overall logical irreversibility is not apparent. It might appear that the situation is even worse than before because of all the extra unused information that this circuit generates. However, at each time step the outputs can be copied and then everything run backwards to return to the inputs, which can then be erased.[13] This means that, as with conventional CMOS, regardless of the state we must again erase the input bits and create the output bits. The problem is now that reversible primitives are being used to implement irreversible functions, which cannot recognize the reversibility of the overall system.

The lesson to draw from these examples is that there is an entropic cost to allocating degrees of freedom, whether or not they are used. It may be known that some are not needed, but unless this knowledge is explicitly built into the system, erasure must pay the full penalty of eliminating all the available configurations.

Let us look at one final implementation of the two-bit sort (Figure 7). In this one we have recoded in a new basis with separate symbols for 00, 01, 10, 11; consider these to be balls rolling down wells with curved sides. By precomputing the input-output pairs in this way it is now obvious that two symbols pass through unchanged, and two symbols are merged. The two that are merged (01 and 10) must emerge indistinguishably at the center of the well; therefore, lateral damping is needed in that irreversible well to erase the memory of the initial condition. In the other two wells lateral damping is not needed because only one symbol will pass through it. However, it is always possible to put the ball in away from the center of the well and then damp it so that it emerges from the center. This is what has been happening in the previous examples: the system is irreversibly computing a reversible operation. This unneeded dissipation has the same maximum value for all cases, and so the logical entropy is irrelevant.

This situation is analogous to the role of entropy in a communications channel. The measured entropy gives an estimate of the average number of bits needed to communicate a symbol. However, an *a posteriori* analysis of the number of bits that were needed has no impact on the number of bits that were actually sent. An inefficient code can, of course, be used that requires many more bits; the entropic minimum is only obtained if an optimal coding is used. An example is a Fano code, a variable length code in which each extra bit distinguishes between two groups of symbols that have equal probability until a particular symbol is uniquely specified.[18] By definition, each symbol adds approximately one bit of entropy, and bits are added only as needed. Similarly, thermodynamically optimal computing should add degrees of freedom in a calculation only as needed as the calculation progresses rather than always allowing for all possibilities (this is reminiscent of using asynchronous logic for low-power design).[19] This points to the need for a computational coding theory to reduce unnecessary degrees of freedom introduced into a computation, a generalization of the algorithms such as Quine-McCluskey that are used to reduce the size of combinatorial logic.[20]

**Figure 6 The two-bit sort implemented in Fredkin gates**

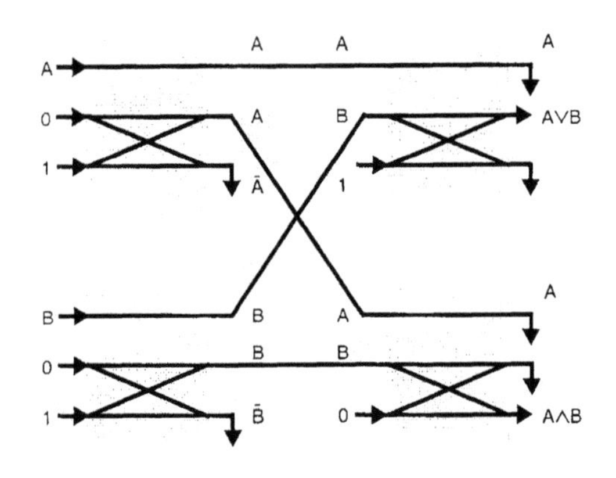


**Figure 7 Alternative coding for the two-bit sort, showing the trajectories for each input**

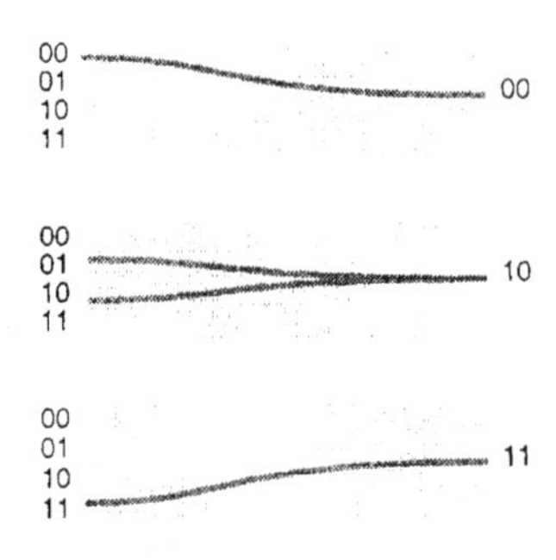


## Systems

Finally, let us look at a sequential gate (one that has memory) to see how that differs from the combinatorial case. Figure 8 shows our two-bit sort implemented as a clocked CMOS gate that sorts the temporal order of pairs of bits. Now it is necessary to study the predictability of the bit strings in time: If the input to the gate is random, then 0s and 1s are equally likely at the output, but their order is no longer simply random.

It is possible to extend statistical mechanics to include the algorithmic information content in a system as well as the logical information content.[21] This beautiful theory resolves a number of statistical mechanical paradoxes but unfortunately is uncomputable because the algorithmic information in a system cannot be calculated (otherwise the halting problem could be solved). Fortunately, a natural physical constraint can make this problem tractable: if the system has a finite memory depth $d$ over which past states can influence future ones, the conventional entropy can be estimated for a block of that length. In our example, $d = 2$.

Let us call the input at time $n$ to the gate $x_n$ and the output $y_n$. If the gate is irreversible, then there is information in $x_n$ that cannot be predicted from $y$, and the gate must dissipate this entropy decrease. Conversely, if in steady-state there is information in $y_n$ that cannot be predicted from $x$ or the internal initial conditions of the system, then the gate is actually a refrigerator, coupling internal thermal degrees of freedom from the heat bath to output information-bearing ones. The lat-

**Figure 8 A sequential two-bit sort, using a clock and latches**

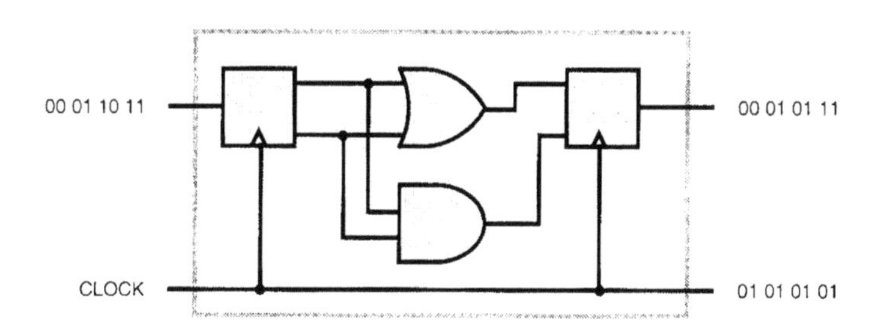


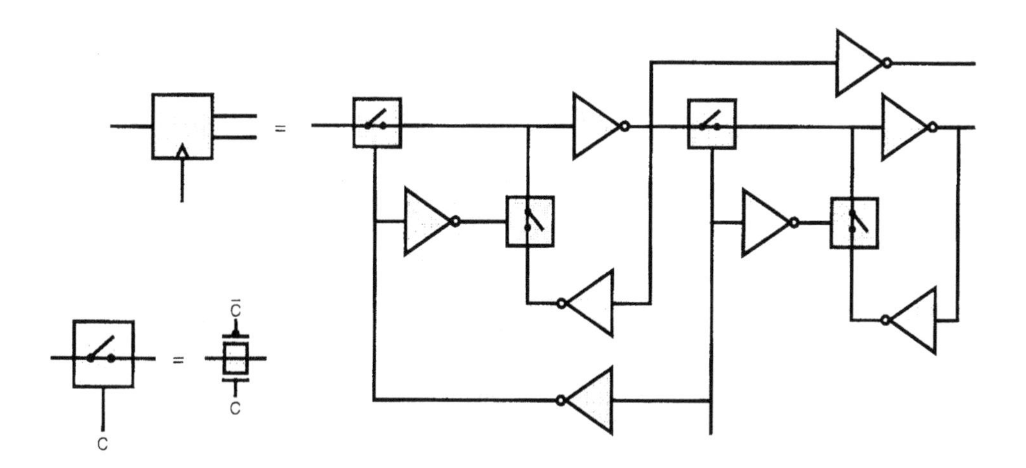


ter case never happens in deterministic logic but can occur in logic that has access to a physical source of random bits.

The average information $I_{reverse}$ irreversibly destroyed per input symbol $x_n$ can be measured by taking the difference between the entropy of a block of the $d$ input and output samples that $x_n$ can influence, and subtracting the entropy of the same block without $x_n$:

$$I_{reverse} = H(x_n, x_{n+1}, \ldots, x_{n+d}; y_n, y_{n+1}, \ldots, y_{n+d}; n) - H(x_{n+1}, \ldots, x_{n+d}; y_n, y_{n+1}, \ldots, y_{n+d}; n) \quad (13)$$

(I have also included $n$ in case the system has an explicit time dependence). If $x_n$ is completely predictable from the future, then these two numbers will be the same, and the mutual information $I_{reverse} = 0$. However, if $x_n$ cannot be predicted at all, then

$$H(x_n, x_{n+1}, \ldots, x_{n+d}; y_n, y_{n+1}, \ldots, y_{n+d}; n) = H(x_n) + H(x_{n+1}, \ldots, x_{n+d}; y_n, y_{n+1}, \ldots, y_{n+d}; n) \quad (14)$$

and so

$$I_{reverse} = H(x_n) \quad (15)$$

(all of the bits in $x_n$ are being consumed by the system). It is necessary to include the future of $x$ as well as $y$ in this calculation because $y$ may not determine $x$, but $x$ may not have any information content anyway (for example, the system clock is completely predictable and so carries no extra information per symbol).

In the opposite direction, the possible added information per output symbol is given by measuring its extra information relative to the symbols that can influence it:

$$I_{forward} = H(x_n, x_{n-1}, \ldots, x_{n-d}; y_n, y_{n-1}, \ldots, y_{n-d}; n) - H(x_n, x_{n-1}, \ldots, x_{n-d}; y_{n-1}, \ldots, y_{n-d}; n) \quad (16)$$

The difference between $I_{reverse}$ and $I_{forward}$ gives the net average flux of information being consumed or created by the system, and hence its informational heating (or cooling) load. It is bounded by the conventional assumption that all the input information must be destroyed. Once again, a measurement of this information difference provides only an empirical estimate of the limiting thermodynamic efficiency of a system that implements the observed transformation for the test workload and says nothing at all about any particular implementation. (The Chudnovskys[22] cannot determine if Pi is random by measuring the temperature of their computer as it prints out the digits.)

To get a feeling for these numbers, let us assume that our "hot" chip is driven by a 100 Mbit/sec network connection. RG58/U coaxial cable has a capacitance of approximately 30 pF/ft and a propagation velocity of approximately 1.5 nsec/ft, so the capacitance associated with the length of a bit is

$$\frac{1}{10^8 \frac{\text{bits}}{\text{sec}} \cdot 1.5 \times 10^{-9} \frac{\text{sec}}{\text{ft}}} \cdot 30 \frac{\text{pF}}{\text{ft}} \approx 200 \frac{\text{pF}}{\text{bit}} \quad (17)$$

and if we stay at 3V this is an energy of

$$\frac{1}{2} \cdot 200^{-12} \text{F} \cdot (3\text{V})^2 \approx 10^{-9} \frac{\text{J}}{\text{bit}} \quad (18)$$

Therefore, if the chip either consumes or generates completely random bits at this rate, the recoverable energy flux is

$$10^{-9} \frac{\text{J}}{\text{bit}} \cdot 10^8 \frac{\text{bits}}{\text{sec}} = 0.1\text{W} \quad (19)$$

for the electrostatic energy, and the irreversible heat is

$$10^{-21} \frac{\text{J}}{\text{bit}} \cdot 10^8 \frac{\text{bits}}{\text{sec}} = 10^{-13} \text{W} \quad (20)$$

for the logical entropy.

Once again, there is a huge difference between the erasure energy and entropy. However, just as the analogy between $kT$ and $CV^2/2$ has led to the development of techniques for low-power logic, the measurement of logical entropy in computers may prove to be useful for guiding subsystem optimization.[23] Although entropy is notoriously difficult to estimate reliably in practice, there are efficient algorithms for handling the required large data sets[24] (which are easy to collect for typical digital systems).

Taken together, we have seen three kinds of terms in the overall energy/entropy budget of a chip. The first is due to the free energy of the input and output bits, which can be recovered and reused. The second is due to the relaxation of the system back to the local minimum of the free energy after a state transition, which is proportional to how many times the bits must be moved and sets a bound on how quickly and how accurately the answers become available. The final term is due to the information difference between the inputs and the outputs, which sets a bound on the thermal properties of a device that can make a given transformation for a given workload.

### Concluding remarks

This paper has sketched the features of a theory of computation that can handle information-bearing and thermal degrees of freedom on an equal footing. The next step will be to extend the analysis from these simple examples to more complex systems, and to explicitly calculate the fluctuation-dissipation component associated with how sharply the entropy is peaked. This kind of explicit budgeting of entropy as well as energy is not done now but will become necessary as circuits approach their fundamental physical limits. The simple examples have shown that system-level predictability can be missed by gate-level designs, resulting in unnecessary dissipation. We have seen a very close analogy to the role of information measurements in characterizing a communications channel and then building optimal codes for it, pointing toward the possibility of "coders" that optimize chip layout for minimum dissipation over an observed workload. Although reaching this ambitious goal still lies in the future, and may not even be of much practical importance (since in many systems power consumption is already dominated by external I/O), the required elements of measuring, modeling, and predicting the information in a system will be much more broadly applicable throughout the optimization of information processing systems.

## Cited references


1. MIT Physical Plant (1994), personal communication.
2. *Maxwell's Demon: Entropy, Information, Computing*, H. S. Leff and A. F. Rex, Editors, Princeton University Press, Princeton, NJ (1990).
3. L. Szilard, *Zeitschrift für Physik* **53**, 840 (1929).
4. C. E. Shannon, "A Mathematical Theory of Communications," *Bell Systems Technical Journal* **27**, 379 (1948).
5. R. Landauer, "Irreversibility and Heat Generation in the Computing Process," *IBM Journal of Research and Development* **5**, No. 3, 183–191 (1961).
6. C. H. Bennett, "Notes on the History of Reversible Computation," *IBM Journal of Research and Development* **32**, No. 1, 16–23 (1988).
7. N. Gershenfeld, *The Physics of Information Technology*, Cambridge University Press, New York (1997), to be published.
8. R. C. Merkle, "Reversible Electronic Logic Using Switches," *Nanotechnology* **4**, 21–40 (1993).
9. S. Younis and T. Knight, "Practical Implementation of Charge Recovering Asymptotically Zero Power CMOS," *Proceedings of the 1993 Symposium on Integrated Systems*, MIT Press (1993), pp. 234–250.
10. W. C. Athas, L. J. Svensson, J. G. Koller, N. Tzartzanis, and E. Ying-Chin Chou, "Low-Power Digital Systems Based on Adiabatic-Switching Principles," *IEEE Transactions on VLSI Systems* **2**, 398–407 (1994).
11. A. G. Dickinson and J. S. Denker, "Adiabatic Dynamic Logic," *IEEE Journal of Solid-State Circuits* **30**, 311–315 (1995).
12. H. B. Callen, *Thermodynamics and an Introduction to Thermostatistics*, 2nd edition, John Wiley & Sons, Inc., New York (1985).
13. C. H. Bennett, "Time/Space Trade-offs for Reversible Computation," *SIAM Journal of Computation* **18**, 766–776 (1989).
14. R. Y. Levine and A. T. Sherman, "A Note on Bennett's Time-Space Tradeoff for Reversible Computation," *SIAM Journal of Computation* **19**, 673–677 (1990).
15. N. Aschcroft and N. D. Mermin, *Solid State Physics*, Holt, Rinehart and Winston, New York (1976).
16. R. Landauer, "Fundamental Limitations in the Computational Process," *Berichte der Bunsen Gesellschaft für Physikalische Chemie* **80**, 1048–1059 (1976).
17. E. Fredkin and T. Toffoli, "Conservative Logic," *International Journal of Theoretical Physics* **21**, 219–253 (1982).
18. R. E. Blahut, *Digital Transmission of Information*, Addison-Wesley Publishing Co., Reading MA (1990).
19. *Asynchronous Digital Circuit Design*, G. Birtwistle and A. Davis, Editors, Springer-Verlag, New York (1995).
20. F. J. Hill and G. R. Peterson, *Computer Aided Logical Design with Emphasis on VLSI*, 4th edition, John Wiley & Sons, Inc., New York (1993).
21. W. H. Zurek, "Algorithmic Randomness and Physical Entropy," *Physical Review Abstracts* **40**, 4731–4751 (1989).
22. D. Chudnovsky and G. Chudnovsky, "The Computation of Classical Constants," *Proceedings of the National Academy of Science* **86**, 8178–8182 (1989).
23. N. R. Shah and N. A. Gershenfeld, *Greedy Dynamic Page-Tuning* (1993), preprint.
24. N. A. Gershenfeld, "Information in Dynamics," Doug Matzke, Editor, *Proceedings of the Workshop on Physics of Computation*, IEEE Press, Piscataway, NJ (1993), pp. 276–280.


*Accepted for publication April 25, 1996.*

**Neil Gershenfeld** *MIT Media Laboratory, 20 Ames Street, Cambridge, Massachusetts 02139-4307 (electronic mail: gersh@media.mit.edu).* Dr. Gershenfeld directs the Physics and Media Group at the MIT Media Lab, and is co-Principal Investigator of the Things That Think research consortium. Before joining the Massachusetts Institute of Technology faculty he received a B.A. in physics with High Honors from Swarthmore College, did research at AT&T Bell Laboratories on the use of lasers in atomic and nuclear physics, received a Ph.D. in applied physics from Cornell University studying order-disorder transitions in condensed matter systems, and was a Junior Fellow of the Harvard Society of Fellows (where he directed a Santa Fe Institute/NATO study on nonlinear time series). His research group explores the boundary between the content of information and its physical representation, including the basic physics of computation, new devices and algorithms for interface transduction, and applications in collaborations that have ranged from furniture and footwear to Yo-Yo Ma and Penn & Teller.

Reprint Order No. G321-5625.

# Ultimate physical limits to computation

**Seth Lloyd**

*d'Arbeloff Laboratory for Information Systems and Technology, MIT Department of Mechanical Engineering, Massachusetts Institute of Technology 3-160, Cambridge, Massachusetts 02139, USA (slloyd@mit.edu)*

**Computers are physical systems: the laws of physics dictate what they can and cannot do. In particular, the speed with which a physical device can process information is limited by its energy and the amount of information that it can process is limited by the number of degrees of freedom it possesses. Here I explore the physical limits of computation as determined by the speed of light $c$, the quantum scale $\hbar$ and the gravitational constant $G$. As an example, I put quantitative bounds to the computational power of an 'ultimate laptop' with a mass of one kilogram confined to a volume of one litre.**

Over the past half century, the amount of information that computers are capable of processing and the rate at which they process it has doubled every 18 months, a phenomenon known as Moore's law. A variety of technologies — most recently, integrated circuits — have enabled this exponential increase in information processing power. But there is no particular reason why Moore's law should continue to hold: it is a law of human ingenuity, not of nature. At some point, Moore's law will break down. The question is, when?

The answer to this question will be found by applying the laws of physics to the process of computation[1–85]. Extrapolation of current exponential improvements over two more decades would result in computers that process information at the scale of individual atoms. Although an Avogadro-scale computer that can act on $10^{23}$ bits might seem implausible, prototype quantum computers that store and process information on individual atoms have already been demonstrated[64,65,76–80]. Existing quantum computers may be small and simple, and able to perform only a few hundred operations on fewer than ten quantum bits or 'qubits', but the fact that they work at all indicates that there is nothing in the laws of physics that forbids the construction of an Avogadro-scale computer.

The purpose of this article is to determine just what limits the laws of physics place on the power of computers. At first, this might seem a futile task: because we do not know the technologies by which computers 1,000, 100, or even 10 years in the future will be constructed, how can we determine the physical limits of those technologies? In fact, I will show that a great deal can be determined concerning the ultimate physical limits of computation simply from knowledge of the speed of light, $c = 2.9979 \times 10^8$ m s$^{-1}$, Planck's reduced constant, $\hbar = h/2\pi = 1.0545 \times 10^{-34}$ J s, and the gravitational constant, $G = 6.673 \times 10^{-11}$ m$^3$ kg$^{-1}$ s$^{-2}$. Boltzmann's constant, $k_B = 1.3805 \times 10^{-23}$ J K$^{-1}$, will also be crucial in translating between computational quantities such as memory space and operations per bit per second, and thermodynamic quantities such as entropy and temperature. In addition to reviewing previous work on how physics limits the speed and memory of computers, I present results — which are new except as noted — of the derivation of the ultimate speed limit to computation, of trade-offs between memory and speed, and of the analysis of the behaviour of computers at physical extremes of high temperatures and densities.

Before presenting methods for calculating these limits, it is important to note that there is no guarantee that these limits will ever be attained, no matter how ingenious computer designers become. Some extreme cases such as the black-hole computer described below are likely to prove extremely difficult or impossible to realize. Human ingenuity has proved great in the past, however, and before writing off physical limits as unattainable, we should realize that certain of these limits have already been attained within a circumscribed context in the construction of working quantum computers. The discussion below will note obstacles that must be sidestepped or overcome before various limits can be attained.

## Energy limits speed of computation

To explore the physical limits of computation, let us calculate the ultimate computational capacity of a computer with a mass of 1 kg occupying a volume of 1 litre, which is roughly the size of a conventional laptop computer. Such a computer, operating at the limits of speed and memory space allowed by physics, will be called the 'ultimate laptop' (Fig. 1).

First, ask what limits the laws of physics place on the speed of such a device. As I will now show, to perform an elementary logical operation in time $\Delta t$ requires an average amount of energy $E \geqslant \pi\hbar/2\Delta t$. As a consequence, a system with average energy $E$ can perform a maximum of $2E/\pi\hbar$ logical operations per second. A 1-kg computer has average energy $E = mc^2 = 8.9874 \times 10^{16}$ J. Accordingly, the ultimate laptop can perform a maximum of $5.4258 \times 10^{50}$ operations per second.

### Maximum speed per logical operation

For the sake of convenience, the ultimate laptop will be taken to be a digital computer. Computers that operate on non-binary or continuous variables obey similar limits to those that will be derived here. A digital computer performs computation by representing information in the terms of binary digits or bits, which can take the value 0 or 1, and then processes that information by performing simple logical operations such as AND, NOT and FANOUT. The operation, AND, for instance, takes two binary inputs $X$ and $Y$ and returns the output 1 if and only if both $X$ and $Y$ are 1; otherwise it returns the output 0. Similarly, NOT takes a single binary input $X$ and returns the output 1 if $X = 0$ and 0 if $X = 1$. FANOUT takes a single binary input $X$ and returns two binary outputs, each equal to $X$. Any boolean function can be constructed by repeated application of AND, NOT and FANOUT. A set of operations that allows the construction of arbitrary boolean functions is called universal. The actual physical device that performs a logical operation is called a logic gate.

How fast can a digital computer perform a logical operation? During such an operation, the bits in the computer on

**Figure 1** The ultimate laptop. The 'ultimate laptop' is a computer with a mass of 1 kg and a volume of 1 l, operating at the fundamental limits of speed and memory capacity fixed by physics. The ultimate laptop performs $2mc^2/\pi\hbar = 5.4258 \times 10^{50}$ logical operations per second on $\sim 10^{31}$ bits. Although its computational machinery is in fact in a highly specified physical state with zero entropy, while it performs a computation that uses all its resources of energy and memory space it appears to an outside observer to be in a thermal state at $\sim 10^{9}$ degrees Kelvin. The ultimate laptop looks like a small piece of the Big Bang.

which the operation is performed go from one state to another. The problem of how much energy is required for information processing was first investigated in the context of communications theory by Levitin[11–16], Bremermann[17–19], Beckenstein[20–22] and others, who showed that the laws of quantum mechanics determine the maximum rate at which a system with spread in energy $\Delta E$ can move from one distinguishable state to another. In particular, the correct interpretation of the time–energy Heisenberg uncertainty principle $\Delta E \Delta t \geq \hbar$ is not that it takes time $\Delta t$ to measure energy to an accuracy $\Delta E$ (a fallacy that was put to rest by Aharonov and Bohm[23,24]), but rather that a quantum state with spread in energy $\Delta E$ takes time at least $\Delta t = \pi\hbar/2\Delta E$ to evolve to an orthogonal (and hence distinguishable) state[23–26]. More recently, Margolus and Levitin[15,16] extended this result to show that a quantum system with average energy $E$ takes time at least $\Delta t = \pi\hbar/2E$ to evolve to an orthogonal state.

#### Performing quantum logic operations

As an example, consider the operation NOT performed on a qubit with logical states $|0\rangle$ and $|1\rangle$. (For readers unfamiliar with quantum mechanics, the 'bracket' notation $|\ \rangle$ signifies that whatever is contained in the bracket is a quantum-mechanical variable; $|0\rangle$ and $|1\rangle$ are vectors in a two-dimensional vector space over the complex numbers.) To flip the qubit, one can apply a potential $H = E_0|E_0\rangle\langle E_0| + E_1|E_1\rangle\langle E_1|$ with energy eigenstates $|E_0\rangle = (1/\sqrt{2})(|0\rangle + |1\rangle)$ and $|E_1\rangle = (1/\sqrt{2})(|0\rangle - |1\rangle)$. Because $|0\rangle = (1/\sqrt{2})(|E_0\rangle + |E_1\rangle)$ and $|1\rangle = (1/\sqrt{2})(|E_0\rangle - |E_1\rangle)$, each logical state $|0\rangle, |1\rangle$ has spread in energy $\Delta E = (E_1 - E_0)/2$. It is easy to verify that after a length of time $\Delta t = \pi\hbar/2\Delta E$ the qubit evolves so that $|0\rangle \to |1\rangle$ and $|1\rangle \to |0\rangle$. That is, applying the potential effects a NOT operation in a time that attains the limit given by quantum mechanics. Note that the average energy $E$ of the qubit in the course of the logical operation is $\langle 0|H|0\rangle = \langle 1|H|1\rangle = (E_0 + E_1)/2 = E_0 + \Delta E$. Taking the ground-state energy $E_0 = 0$ gives $E = \Delta E$. So the amount of time it takes to perform a NOT operation can also be written as $\Delta t = \pi\hbar/2E$. It is straightforward to show[15,16] that no quantum system with average energy $E$ can move to an orthogonal state in a time less than $\Delta t$. That is, the speed with which a logical operation can be performed is limited not only by the spread in energy, but also by the average energy. This result will prove to be a key component in deriving the speed limit for the ultimate laptop.

AND and FANOUT can be enacted in a way that is analogous to the NOT operation. A simple way to perform these operations in a quantum-mechanical context is to enact a so-called Toffoli or controlled-controlled-NOT operation[31]. This operation takes three binary inputs, $X$, $Y$ and $Z$, and returns three outputs, $X'$, $Y'$ and $Z'$. The first two inputs pass through unchanged, that is, $X' = X$, $Y' = Y$. The third input passes through unchanged unless both $X$ and $Y$ are 1, in which case it is flipped. This is universal in the sense that suitable choices of inputs allow the construction of AND, NOT and FANOUT. When the third input is set to zero, $Z = 0$, then the third output is the AND of the first two: $Z' = X$ AND $Y$. So AND can be constructed. When the first two inputs are 1, $X = Y = 1$, the third output is the NOT of the third input, $Z' =$ NOT $Z$. Finally, when the second input is set to 1, $Y = 1$, and the third to zero, $Z = 0$, the first and third output are the FANOUT of the first input, $X' = X$, $Z' = X$. So arbitrary boolean functions can be constructed from the Toffoli operation alone.

By embedding a controlled-controlled-NOT gate in a quantum context, it is straightforward to see that AND and FANOUT, like NOT, can be performed at a rate $2E/\pi\hbar$ times per second, where $E$ is the average energy of the logic gate that performs the operation. More complicated logic operations that cycle through a larger number of quantum states (such as those on non-binary or continuous quantum variables) can be performed at a rate $E/\pi\hbar$ — half as fast as the simpler operations[15,16]. Existing quantum logic gates in optical–atomic and nuclear magnetic resonance (NMR) quantum computers actually attain this limit. In the case of NOT, $E$ is the average energy of interaction of the qubit's dipole moment (electric dipole for optic–atomic qubits and nuclear magnetic dipole for NMR qubits) with the applied electromagnetic field. In the case of multiqubit operations such as the Toffoli operation, or the simpler two-bit controlled-NOT operation, which flips the second bit if and only if the first bit is 1, $E$ is the average energy in the interaction between the physical systems that register the qubits.

#### Ultimate limits to speed of computation

We are now in a position to derive the first physical limit to computation, that of energy. Suppose that one has a certain amount of energy $E$ to allocate to the logic gates of a computer. The more energy one allocates to a gate, the faster it can perform a logic operation. The total number of logic operations performed per second is equal to the sum over all logic gates of the operations per second per gate. That is, a computer can perform no more than

$$\sum_{\ell} 1/\Delta t_\ell \leq \sum_{\ell} 2E_\ell/\pi\hbar = 2E/\pi\hbar$$

operations per second. In other words, the rate at which a computer can compute is limited by its energy. (Similar limits have been proposed by Bremmerman in the context of the minimum energy required to communicate a bit[17–19], although these limits have been criticized for misinterpreting the energy–time uncertainty relation[21], and for failing to take into account the role of degeneracy of energy eigenvalues[13,14] and the role of nonlinearity in communications[7–9].) Applying this result to a 1-kg computer with energy $E = mc^2 = 8.9874 \times 10^{16}$ J shows that the ultimate laptop can perform a maximum of $5.4258 \times 10^{50}$ operations per second.

**Parallel and serial operation**

An interesting feature of this limit is that it is independent of computer architecture. It might have been thought that a computer could be speeded up by parallelization, that is, by taking the energy and dividing it up among many subsystems computing in parallel. But this is not the case. If the energy $E$ is spread among $N$ logic gates, each one operates at a rate $2E/\pi\hbar N$, and the total number of operations per second, $N2E/\pi\hbar N = 2E/\pi\hbar$, remains the same. If the energy is allocated to fewer logic gates (a more serial operation), the rate $1/\Delta t_\ell$ at which they operate and the spread in energy per gate $\Delta E_\ell$ increase. If the energy is allocated to more logic gates (a more parallel operation) then the rate at which they operate and the spread in energy per gate decrease. Note that in this parallel case, the overall spread in energy of the computer as a whole is considerably smaller than the average energy: in general $\Delta E = \sqrt{\Sigma_\ell \Delta E_\ell^2} \approx \sqrt{N}\Delta E_\ell$ whereas $E = \Sigma E_\ell \approx NE_\ell$. Parallelization can help perform certain computations more efficiently, but it does not alter the total number of operations per second. As I will show below, the degree of parallelizability of the computation to be performed determines the most efficient distribution of energy among the parts of the computer. Computers in which energy is relatively evenly distributed over a larger volume are better suited for performing parallel computations. More compact computers and computers with an uneven distribution of energy are better for performing serial computations.

**Comparison with existing computers**

Conventional laptops operate much more slowly than the ultimate laptop. There are two reasons for this inefficiency. First, most of the energy is locked up in the mass of the particles of which the computer is constructed, leaving only an infinitesimal fraction for performing logic. Second, a conventional computer uses many degrees of freedom (billions and billions of electrons) for registering a single bit. From the physical perspective, such a computer operates in a highly redundant fashion. There are, however, good technological reasons for such redundancy, with conventional designs depending on it for reliability and manufacturability. But in the present discussion, the subject is not what computers are but what they might be, and in this context the laws of physics do not require redundancy to perform logical operations — recently constructed quantum microcomputers use one quantum degree of freedom for each bit and operate at the Heisenberg limit $\Delta t = \pi\hbar/2\Delta E$ for the time needed to flip a bit[64,65,76–80]. Redundancy is, however, required for error correction, as will be discussed below.

In sum, quantum mechanics provides a simple answer to the question of how fast information can be processed using a given amount of energy. Now it will be shown that thermodynamics and statistical mechanics provide a fundamental limit to how many bits of information can be processed using a given amount of energy confined to a given volume. Available energy necessarily limits the rate at which a computer can process information. Similarly, the maximum entropy of a physical system determines the amount of information it can process. Energy limits speed. Entropy limits memory.

## Entropy limits memory space

The amount of information that a physical system can store and process is related to the number of distinct physical states that are accessible to the system. A collection of $m$ two-state systems has $2^m$ accessible states and can register $m$ bits of information. In general, a system with $N$ accessible states can register $\log_2 N$ bits of information. But it has been known for more than a century that the number of accessible states of a physical system, $W$, is related to its thermodynamic entropy by the formula $S = k_B \ln W$, where $k_B$ is Boltzmann's constant. (Although this formula is inscribed on Boltzmann's tomb, it is attributed originally to Planck; before the turn of the century, $k_B$ was often known as Planck's constant.)

The amount of information that can be registered by a physical system is $I = S(E)/k_B \ln 2$, where $S(E)$ is the thermodynamic entropy of a system with expectation value for the energy $E$. Combining this formula with the formula $2E/\pi\hbar$ for the number of logical operations that can be performed per second, we see that when it is using all its memory, the number of operations per bit per second that our ultimate laptop can perform is $k_B 2\ln(2)E/\pi\hbar S \propto k_B T/\hbar$, where $T = (\partial S/\partial E)^{-1}$ is the temperature of 1 kg of matter in a maximum entropy in a volume of 1 l. The entropy governs the amount of information the system can register and the temperature governs the number of operations per bit per second that it can perform.

Because thermodynamic entropy effectively counts the number of bits available to a physical system, the following derivation of the memory space available to the ultimate laptop is based on a thermodynamic treatment of 1 kg of matter confined to a volume of 1 l in a maximum entropy state. Throughout this derivation, it is important to remember that although the memory space available to the computer is given by the entropy of its thermal equilibrium state, the actual state of the ultimate laptop as it performs a computation is completely determined, so that its entropy remains always equal to zero. As above, I assume that we have complete control over the actual state of the ultimate laptop, and are able to guide it through its logical steps while insulating it from all uncontrolled degrees of freedom. But as the following discussion will make clear, such complete control will be difficult to attain (see Box 1).

**Entropy, energy and temperature**

To calculate the number of operations per second that could be performed by our ultimate laptop, I assume that the expectation value of the energy is $E$. Accordingly, the total number of bits of memory space available to the computer is $S(E, V)/k_B \ln 2$ where $S(E, V)$ is the thermodynamic entropy of a system with expectation value of the energy $E$ confined to volume $V$. The entropy of a closed system is usually given by the so-called microcanonical ensemble, which fixes both the average energy and the spread in energy $\Delta E$, and assigns equal probability to all states of the system within a range $[E, E + \Delta E]$. In the case of the ultimate laptop, however, I wish to fix only the average energy, while letting the spread in energy vary according to whether the computer is to be more serial (fewer, faster gates, with larger spread in energy) or parallel (more, slower gates, with smaller spread in energy). Accordingly, the ensemble that should be used to calculate the thermodynamic entropy and the memory space available is the canonical ensemble, which maximizes $S$ for fixed average energy with no constraint on the spread in energy $\Delta E$. The canonical ensemble shows how many bits of memory are available for all possible ways of programming the computer while keeping its average energy equal to $E$. In any given computation with average energy $E$, the ultimate laptop will be in a pure state with some fixed spread of energy, and will explore only a small fraction of its memory space.

In the canonical ensemble, a state with energy $E_i$ has probability $p_i = (1/Z(T))e^{-E_i/k_B T}$ where $Z(T) = \Sigma_i e^{-E_i/k_B T}$ is the partition function, and the temperature $T$ is chosen so that $E = \Sigma_i p_i E_i$. The entropy is $S = -k_B \Sigma_i p_i \ln p_i = E/T + k_B \ln Z$. The number of bits of memory space available to the computer is $S/k_B \ln 2$. The difference between the entropy as calculated using the canonical ensemble and that calculated using the microcanonical ensemble is minimal. But there is some subtlety involved in using the canonical ensemble rather than the more traditional microcanonical ensemble. The canonical ensemble is normally used for open systems that interact with a thermal bath at temperature $T$. In the case of the ultimate laptop, however, it is applied to a closed system to find the maximum entropy given a fixed expectation value for the energy. As a result, the temperature $T = (\partial S/\partial E)^{-1}$ has a somewhat different role in the context of physical limits of computation than it does in the case of an ordinary

Box 1
**The role of thermodynamics in computation**

The fact that entropy and information are intimately linked has been known since Maxwell introduced his famous 'demon' well over a century ago[1]. Maxwell's demon is a hypothetical being that uses its information-processing ability to reduce the entropy of a gas. The first results in the physics of information processing were derived in attempts to understand how Maxwell's demon could function[1–4]. The role of thermodynamics in computation has been examined repeatedly over the past half century. In the 1950s, von Neumann[10] speculated that each logical operation performed in a computer at temperature $T$ must dissipate energy $k_B T \ln 2$, thereby increasing entropy by $k_B \ln 2$. This speculation proved to be false. The precise, correct statement of the role of entropy in computation was attributed to Landauer[5], who showed that reversible, that is, one-to-one, logical operations such as NOT can be performed, in principle, without dissipation, but that irreversible, many-to-one operations such as AND or ERASE require dissipation of at least $k_B \ln 2$ for each bit of information lost. (ERASE is a one-bit logical operation that takes a bit, 0 or 1, and restores it to 0.) The argument behind Landauer's principle can be readily understood[37]. Essentially, the one-to-one dynamics of hamiltonian systems implies that when a bit is erased the information that it contains has to go somewhere. If the information goes into observable degrees of freedom of the computer, such as another bit, then it has not been erased but merely moved; but if it goes into unobservable degrees of freedom such as the microscopic motion of molecules it results in an increase of entropy of at least $k_B \ln 2$.

In 1973, Bennett[28–30] showed that all computations could be performed using only reversible logical operations. Consequently, by Landauer's principle, computation does not require dissipation. (Earlier work by Lecerf[27] had anticipated the possibility of reversible computation, but not its physical implications. Reversible computation was discovered independently by Fredkin and Toffoli[31].) The energy used to perform a logical operation can be 'borrowed' from a store of free energy such as a battery, 'invested' in the logic gate that performs the operation, and returned to storage after the operation has been performed, with a net 'profit' in the form of processed information. Electronic circuits based on reversible logic have been built and exhibit considerable reductions in dissipation over conventional reversible circuits[33–35].

Under many circumstances it may prove useful to perform irreversible operations such as erasure. If our ultimate laptop is subject to an error rate of $\epsilon$ bits per second, for example, then error-correcting codes can be used to detect those errors and reject them to the environment at a dissipative cost of $\epsilon k_B T_E \ln 2$ J s$^{-1}$, where $T_E$ is the temperature of the environment. ($k_B T \ln 2$ is the minimal amount of energy required to send a bit down an information channel with noise temperature $T$ (ref. 14).) Such error-correcting routines in our ultimate computer function as working analogues of Maxwell's demon, getting information and using it to reduce entropy at an exchange rate of $k_B T \ln 2$ joules per bit. In principle, computation does not require dissipation. In practice, however, any computer — even our ultimate laptop — will dissipate energy.

The ultimate laptop must reject errors to the environment at a high rate to maintain reliable operation. To estimate the rate at which it can reject errors to the environment, assume that the computer encodes erroneous bits in the form of black-body radiation at the characteristic temperature $5.87 \times 10^8$ K of the computer's memory[21]. The Stefan–Boltzmann law for black-body radiation then implies that the number of bits per unit area than can be sent out to the environment is $B = \pi^2 k_B^3 T^3/60 \ln(2)\hbar^2 c^2 = 7.195 \times 10^{42}$ bits per square meter per second. As the ultimate laptop has a surface area of $10^{-2}$ m$^2$ and is performing ~$10^{50}$ operations per second, it must have an error rate of less than $10^{-10}$ per operation in order to avoid over-heating. Even if it achieves such an error rate, it must have an energy throughput (free energy in and thermal energy out) of $4.04 \times 10^{26}$ W — turning over its own resting mass energy of $mc^2 \approx 10^{17}$ J in a nanosecond! The thermal load of correcting large numbers of errors clearly indicates the necessity of operating at a slower speed than the maximum allowed by the laws of physics.

thermodynamic system interacting with a thermal bath. Integrating the relationship $T = (\partial S/\partial E)^{-1}$ over $E$ yields $T = CE/S$, where $C$ is a constant of order unity (for example, $C = 4/3$ for black-body radiation, $C = 3/2$ for an ideal gas, and $C = 1/2$ for a black hole). Accordingly, the temperature governs the number of operations per bit per second, $k_B \ln(2)E/\hbar S \approx k_B T/\hbar$, that a system can perform. As I will show later, the relationship between temperature and operations per bit per second is useful in investigating computation under extreme physical conditions.

**Calculating the maximum memory space**

To calculate exactly the maximum entropy for a kilogram of matter in a litre volume would require complete knowledge of the dynamics of elementary particles, quantum gravity, and so on. Although we do not possess such knowledge, the maximum entropy can readily be estimated by a method reminiscent of that used to calculate thermodynamic quantities in the early Universe[86]. The idea is simple: model the volume occupied by the computer as a collection of modes of elementary particles with total average energy $E$. The maximum entropy is obtained by calculating the canonical ensemble over the modes. Here, I supply a simple derivation of the maximum memory space available to the ultimate laptop. A more detailed discussion of how to calculate the maximum amount of information that can be stored in a physical system can be found in the work of Bekenstein[19–21].

For this calculation, assume that the only conserved quantities other than the computer's energy are angular momentum and electric charge, which I take to be zero. (One might also ask that the number of baryons be conserved, but as will be seen below, one of the processes that could take place within the computer is black-hole formation and evaporation, which does not conserve baryon number.) At a particular temperature $T$, the entropy is dominated by the contributions from particles with mass less than $k_B T/2c^2$. The $\ell$'th such species of particle contributes energy $E = r_\ell \pi^2 V(k_B T)^4/30\hbar^3 c^3$ and entropy $S = 2r_\ell k_B \pi^2 V(k_B T)^3/45\hbar^3 c^3 = 4E/3T$, where $r_\ell$ is equal to the number of particles/antiparticles in the species (that is, 1 for photons, 2 for electrons/positrons) multiplied by the number of polarizations (2 for photons, 2 for electrons/positrons) multiplied by a factor that reflects particle statistics (1 for bosons, 7/8 for fermions). As the formula for $S$ in terms of $T$ shows, each species contributes $(2\pi)^5 r_\ell/90 \ln 2 \approx 10^2$ bits of memory space per cubic thermal wavelength $\lambda_T^3$, where $\lambda_T = 2\pi\hbar c/k_B T$. Re-expressing the formula for entropy as a function of energy, the estimate for the maximum entropy is

$$S = (4/3)k_B(\pi^2 rV/30\hbar^3 c^3)^{1/4} E^{3/4} = k_B \ln(2) I$$

where $r = \Sigma_\ell r_\ell$. Note that $S$ depends only insensitively on the total number of species with mass less than $k_B T/2c^2$.

A lower bound on the entropy can be obtained by assuming that energy and entropy are dominated by black-body radiation consisting of photons. In this case, $r = 2$, and for a 1-kg computer confined to a volume of a 1 l we have $k_B T = 8.10 \times 10^{-15}$ J, or $T = 5.87 \times 10^8$ K. The entropy is $S = 2.04 \times 10^8$ J K$^{-1}$, which corresponds to an amount of available memory space $I = S/k_B \ln 2 = 2.13 \times 10^{31}$ bits. When the ultimate laptop is using all its memory space it can perform $2\ln(2)k_B E/\pi\hbar S = 3\ln(2)k_B T/2\pi\hbar \approx 10^{19}$ operations per bit per second. As the number of operations per second $2E/\pi\hbar$ is independent of the number of bits available, the number of operations per bit per second can be increased by using a smaller number of bits. In keeping with the prescription that the ultimate laptop operates at the absolute limits given by physics, in what follows I assume that all available bits are used.

This estimate for the maximum entropy could be improved (and slightly increased) by adding more species of massless particles (neutrinos and gravitons) and by taking into effect the presence of electrons and positrons. Note that $k_B T/2c^2 = 4.51 \times 10^{-32}$ kg, compared with the electron mass of $9.1 \times 10^{-31}$ kg. That is, the ultimate laptop is close to a phase transition at which electrons and positrons are produced thermally. A more exact estimate of the maximum entropy and hence the available memory space would be straightforward to perform, but the details of such a calculation would detract from my general exposition, and could serve to alter $S$ only slightly. $S$ depends insensitively on the number of species of effectively massless particles: a change of $r$ by a factor of 10,000 serves to increase S by only a factor of 10.

#### Comparison with current computers

The amount of information that can be stored by the ultimate laptop, $\sim 10^{31}$ bits, is much higher than the $\sim 10^{10}$ bits stored on current laptops. This is because conventional laptops use many degrees of freedom to store a bit whereas the ultimate laptop uses just one. There are considerable advantages to using many degrees of freedom to store information, stability and controllability being perhaps the most important. Indeed, as the above calculation indicates, to take full advantage of the memory space available, the ultimate laptop must turn all its matter into energy. A typical state of the ultimate laptop's memory looks like a plasma at a billion degrees Kelvin — like a thermonuclear explosion or a little piece of the Big Bang! Clearly, packaging issues alone make it unlikely that this limit can be obtained, even setting aside the difficulties of stability and control.

Even though the ultimate physical limit to how much information can be stored in a kilogram of matter in a litre volume is unlikely to be attained, it may nonetheless be possible to progress some way towards such bit densities. In other words, the ultimate limits to memory space may prove easier to approach than the ultimate limits to speed. Following Moore's law, the density of bits in a computer has gone down from approximately one per square centimetre 50 years ago to one per square micrometre today, an improvement of a factor of $10^8$. It is not inconceivable that a similar improvement is possible over the course of the next 50 years. In particular, there is no physical reason why it should not be possible to store one bit of information per atom. Indeed, existing NMR and ion-trap quantum computers already store information on individual nuclei and atoms (typically in the states of individual nuclear spins or in hyperfine atomic states). Solid-state NMR with high gradient fields or quantum optical techniques such as spectral hole-burning provide potential technologies for storing large quantities of information at the atomic scale. A kilogram of ordinary matter holds on the order of $10^{25}$ nuclei. If a substantial fraction of these nuclei can be made to register a bit, then we could get close to the ultimate physical limit of memory without having to resort to thermonuclear explosions. If, in addition, we make use of the natural electromagnetic interactions between nuclei and electrons in the matter to perform logical operations, we are limited to a rate of $\sim 10^{15}$ operations per bit per second, yielding an overall information processing rate of $\sim 10^{40}$ operations per second in ordinary matter. Although less than the $\sim 10^{51}$ operations per second in the ultimate laptop, the maximum information processing rate in 'ordinary matter' is still quite respectable. Of course, even though such an 'ordinary matter' ultimate computer need not operate at nuclear energy levels, other problems remain — for example, the high number of bits still indicates substantial input/output problems. At an input/output rate of $10^{12}$ bits per second, an Avogadro-scale computer with $10^{23}$ bits would take about 10,000 years to perform a serial read/write operation on the entire memory. Higher throughput and parallel input/output schemes are clearly required to take advantage of the entire memory space that physics makes available.

#### Size limits parallelization

Up until this point, I have assumed that the ultimate laptop occupies a volume of 1 l. The previous discussion, however, indicates that benefits are to be obtained by varying the volume to which the computer is confined. Generally speaking, if the computation to be performed is highly parallelizable or requires many bits of memory, the volume of the computer should be greater and the energy available to perform the computation should be spread out evenly among the different parts of the computer. Conversely, if the computation to be performed is highly serial and requires fewer bits of memory, the energy should be concentrated in particular parts of the computer.

A good measure of the degree of parallelization in a computer is the ratio between the time it takes to communicate from one side of the computer to the other, and the average time it takes to perform a logical operation. The amount of time it takes to send a message from one side of a computer of radius $R$ to the other is $t_{com} = 2R/c$. The average time it takes a bit to flip in the ultimate laptop is the inverse of the number of operations per bit per second calculated above: $t_{flip} = \pi\hbar S/k_B 2\ln(2)E$. The measure of the degree of parallelization in the ultimate laptop is then

$$t_{com}/t_{flip} = k_B 4\ln(2)RE/\pi\hbar cS \propto k_B RT/\hbar c = 2\pi R/\lambda_T$$

That is, the amount of time it takes to communicate from one side of the computer to the other, divided by the amount of time it takes to flip a bit, is approximately equal to the ratio between the size of the

Box 2

**Can a black hole compute?**

No information can escape from a classical black hole: what goes in does not come out. But the quantum mechanical picture of a black hole is different. First of all, black holes are not quite black; they radiate at the Hawking temperature $T$ given above. In addition, the well-known statement that 'a black hole has no hair' — that is, from a distance all black holes with the same charge and angular momentum look essentially alike — is now known to be not always true[89–91]. Finally, research in string theory[92–94] indicates that black holes may not actually destroy the information about how they were formed, but instead process it and emit the processed information as part of the Hawking radiation as they evaporate: what goes in does come out, but in an altered form.

If this picture is correct, then black holes could in principle be 'programmed': one forms a black hole whose initial conditions encode the information to be processed, lets that information be processed by the planckian dynamics at the hole's horizon, and extracts the answer to the computation by examining the correlations in the Hawking radiation emitted when the hole evaporates. Despite our lack of knowledge of the precise details of what happens when a black hole forms and evaporates (a full account must await a more exact treatment using whatever theory of quantum gravity and matter turns out to be the correct one), we can still provide a rough estimate of how much information is processed during this computation[95–96]. Using Page's results on the rate of evaporation of a black hole[95], we obtain a lifetime for the hole $t_{life} = G^2m^3/3C\hbar c^4$, where $C$ is a constant that depends on the number of species of particles with a mass less than $k_B T$, where $T$ is the temperature of the hole. For $O(10^1–10^2)$ such species, $C$ is on the order of $10^{-3}–10^{-2}$, leading to a lifetime for a black hole of mass 1 kg of $\sim 10^{-19}$ s, during which time the hole can perform $\sim 10^{32}$ operations on its $\sim 10^{16}$ bits. As the actual number of effectively massless particles at the Hawking temperature of a 1-kg black hole is likely to be considerably larger than $10^2$, this number should be regarded as an upper bound on the actual number of operations that could be performed by the hole. Although this hypothetical computation is performed at ultra-high densities and speeds, the total number of bits available to be processed is not far from the number available to current computers operating in more familiar surroundings.

system and its thermal wavelength. For the ultimate laptop, with $2R = 10^{-1}$ m, $2E/\pi\hbar \approx 10^{51}$ operations per second, and $S/k_B \ln 2 \approx 10^{31}$ bits, $t_{com}/t_{flip} \approx 10^{10}$. The ultimate laptop is highly parallel. A greater degree of serial computation can be obtained at the cost of decreasing memory space by compressing the size of the computer or making the distribution of energy more uneven. As ordinary matter obeys the Beckenstein bound[20-22], $k_B RE/\hbar cS > 1/2\pi$, as the computer is compressed, $t_{com}/t_{flip} \approx k_B RE/\hbar cS$ will remain greater than one, that is, the operation will still be somewhat parallel. Only at the ultimate limit of compression — a black hole — is the computation entirely serial.

**Compressing the computer allows more serial computation**

Suppose that we want to perform a highly serial computation on a few bits. Then it is advantageous to compress the size of the computer so that it takes less time to send signals from one side of the computer to the other at the speed of light. As the computer gets smaller, keeping the energy fixed, the energy density inside the computer increases. As this happens, different regimes in high-energy physics are necessarily explored in the course of the computation. First the weak unification scale is reached, then the grand unification scale. Finally, as the linear size of the computer approaches its Schwarzchild radius, the Planck scale is reached (Fig. 2). (No known technology could possibly achieve such compression.) At the Planck scale, gravitational effects and quantum effects are both important: the Compton wavelength of a particle of mass $m$, $\lambda_C = 2\pi\hbar/mc$, is on the order of its Schwarzschild radius, $2Gm/c^2$. In other words, to describe behaviour at length scales of the size $\ell_P = \sqrt{\hbar G/c^3} = 1.616 \times 10^{-35}$ m, timescales $t_P = \sqrt{\hbar G/c^5} = 5.391 \times 10^{-44}$ s, and mass scales of $m_P = \sqrt{\hbar c/G} = 2.177 \times 10^{-8}$ kg, a unified theory of quantum gravity is required. We do not currently possess such a theory. Nonetheless, although we do not know the exact number of bits that can be registered by a 1-kg computer confined to a volume of 1 l, we do know the exact number of bits that can be registered by a 1-kg computer that has been compressed to the size of a black hole[87]. This is because the entropy of a black hole has a well-defined value.

In the following discussion, I use the properties of black holes to place limits on the speed, memory space and degree of serial computation that could be approached by compressing a computer to the smallest possible size. Whether or not these limits could be attained, even in principle, is a question whose answer will have to await a unified theory of quantum gravity (see Box 2).

The Schwarzschild radius of a 1-kg computer is $R_S = 2Gm/c^2 = 1.485 \times 10^{-27}$ m. The entropy of a black hole is Boltzmann's constant multiplied by its area divided by 4, as measured in Planck units. Accordingly, the amount of information that can be stored in a black hole is $I = 4\pi Gm^2/\ln(2)\hbar c = 4\pi m^2/\ln(2) m_P^2$. The amount of information that can be stored by the 1-kg computer in the black-hole limit is $3.827 \times 10^{16}$ bits. A computer compressed to the size of a black hole can perform $5.4258 \times 10^{50}$ operations per second, the same as the 1-l computer.

In a computer that has been compressed to its Schwarzschild radius, the energy per bit is $E/I = mc^2/I = \ln(2)\hbar c^3/4\pi mG = \ln(2) k_B T/2$, where $T = (\partial S/\partial E)^{-1} = \hbar c/4\pi k_B R_S$ is the temperature of the Hawking radiation emitted by the hole. As a result, the time it takes to flip a bit on average is $t_{flip} = \pi\hbar I/2E = \pi^2 RS/c \ln 2$. In other words, according to a distant observer, the amount of time it takes to flip a bit, $t_{flip}$, is on the same order as the amount of time $t_{com} = \pi R_S/c$ it takes to communicate from one side of the hole to the other by going around the horizon: $t_{com}/t_{flip} = \ln 2/\pi$. In contrast to computation at lesser densities, which is highly parallel, computation at the horizon of a black hole is highly serial: every bit is essentially connected to every other bit over the course of a single logic operation. As noted above, the serial nature of computation at the black-hole limit can be deduced from the fact that black holes attain the Beckenstein bound[20-22], $k_B RE/\hbar cS = 1/2\pi$.

## Constructing ultimate computers

Throughout this entire discussion of the physical limits to computation, no mention has been made of how to construct a computer that

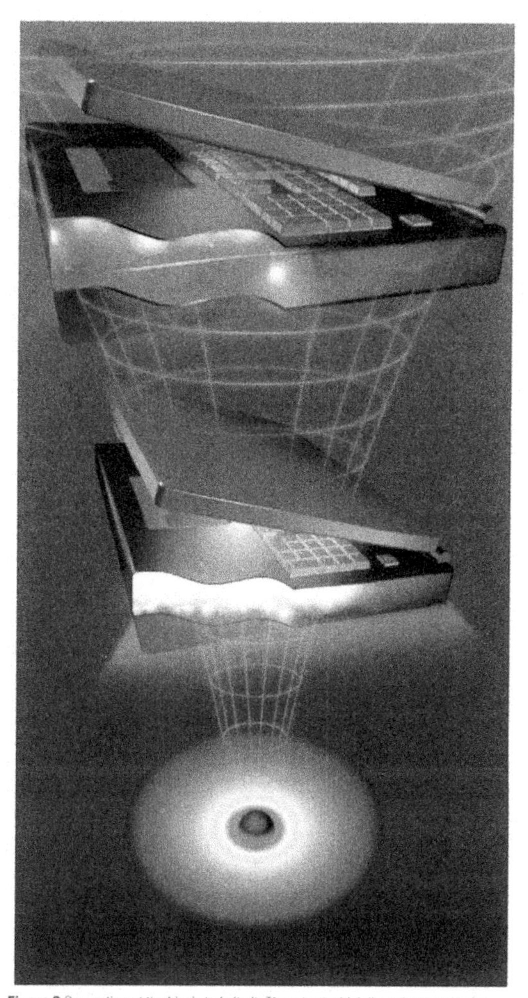

**Figure 2** Computing at the black-hole limit. The rate at which the components of a computer can communicate is limited by the speed of light. In the ultimate laptop, each bit can flip ~$10^{19}$ times per second, whereas the time taken to communicate from one side of the 1-l computer to the other is on the order of $10^{-9}$ s — the ultimate laptop is highly parallel. The computation can be speeded up and made more serial by compressing the computer. But no computer can be compressed to smaller than its Schwarzschild radius without becoming a black hole. A 1-kg computer that has been compressed to the black-hole limit of $R_S = 2Gm/c^2 = 1.485 \times 10^{-27}$ m can perform $5.4258 \times 10^{50}$ operations per second on its $I = 4\pi Gm^2/\ln(2)\hbar c = 3.827 \times 10^{16}$ bits. At the black-hole limit, computation is fully serial: the time it takes to flip a bit and the time it takes a signal to communicate around the horizon of the hole are the same.

operates at those limits. In fact, contemporary quantum 'microcomputers' such as those constructed using NMR[76-80] do indeed operate at the limits of speed and memory space described above. Information is stored on nuclear spins, with one spin registering one bit. The time it takes a bit to flip from a state $|\uparrow\rangle$ to an orthogonal state $|\downarrow\rangle$ is given by $\pi\hbar/2\mu B = \pi\hbar/2E$, where $\mu$ is the spin's magnetic moment, $B$ is the magnetic field, and $E = \mu B$ is the average energy of interaction between the spin and the magnetic field. To perform a quantum logic operation between two spins takes a time $\pi\hbar/2E_\gamma$, where $E_\gamma$ is the energy of interaction between the two spins.

Although NMR quantum computers already operate at the limits to computation set by physics, they are nonetheless much slower and process much less information than the ultimate laptop described above. This is because their energy is locked up largely in mass, thereby limiting both their speed and their memory. Unlocking this energy is of course possible, as a thermonuclear explosion indicates. But controlling such an 'unlocked' system is another question. In discussing the computational power of physical systems in which all energy is put to use, I assumed that such control is possible in principle, although it is certainly not possible in current practice. All current designs for quantum computers operate at low energy levels and temperatures, exactly so that precise control can be exerted on their parts.

As the above discussion of error correction indicates, the rate at which errors can be detected and rejected to the environment by error-correction routines places a fundamental limit on the rate at which errors can be committed. Suppose that each logical operation performed by the ultimate computer has a probability $\epsilon$ of being erroneous. The total number of errors committed by the ultimate computer per second is then $2\epsilon E/\pi\hbar$. The maximum rate at which information can be rejected to the environment is, up to a geometric factor, $\ln(2)cS/R$ (all bits in the computer moving outward at the speed of light). Accordingly, the maximum error rate that the ultimate computer can tolerate is $\epsilon \lesssim \pi\ln(2)\hbar cS/2ER = 2t_{flip}/t_{com}$. That is, the maximum error rate that can be tolerated by the ultimate computer is the inverse of its degree of parallelization.

Suppose that control of highly energetic systems were to become possible. Then how might these systems be made to compute? As an example of a 'computation' that might be performed at extreme conditions, consider a heavy-ion collision that takes place in the heavy-ion collider at Brookhaven (S. H. Kahana, personal communication). If one collides 100 nucleons on 100 nucleons (that is, two nuclei with 100 nucleons each) at 200 GeV per nucleon, the operation time is $\pi\hbar/2E \approx 10^{-29}$ s. The maximum entropy can be estimated to be $\sim 4k_B$ per relativistic pion (to within a factor of less than 2 associated with the overall entropy production rate per meson), and there are $\sim 10^4$ relativistic pions per collision. Accordingly, the total amount of memory space available is $S/k_B \ln 2 \approx 10^4$–$10^5$ bits. The collision time is short: in the centre-of-mass frame the two nuclei are Lorentz-contracted to $D/\gamma$ where $D = 12$–$13$ fermi and $\gamma = 100$, giving a total collision time of $\sim 10^{-25}$ s. During the collision, then, there is time to perform approximately $10^4$ operations on $10^4$ bits — a relatively simple computation. (The fact that only one operation per bit is performed indicates that there is insufficient time to reach thermal equilibrium, an observation that is confirmed by detailed simulations.) The heavy-ion system could be programmed by manipulating and preparing the initial momenta and internal nuclear states of the ions. Of course, we would not expect to be able do word processing on such a 'computer'. Rather it would be used to uncover basic knowledge about nuclear collisions and quark–gluon plasmas: in the words of Heinz Pagels, the plasma 'computes itself'[88].

At the greater extremes of a black-hole computer, I assumed that whatever theory (for example, string theory) turns out to be the correct theory of quantum matter and gravity, it is possible to prepare initial states of such systems that causes their natural time evolution to carry out a computation. What assurance do we have that such preparations exist, even in principle?

Physical systems that can be programmed to perform arbitrary digital computations are called computationally universal. Although computational universality might at first seem to be a stringent demand on a physical system, a wide variety of physical systems — ranging from nearest-neighbour Ising models[52] to quantum electrodynamics[84] and conformal field theories (M. Freedman, unpublished results) — are known to be computationally universal[51–53,55–65]. Indeed, computational universality seems to be the rule rather than the exception. Essentially any quantum system that admits controllable nonlinear interactions can be shown to be computationally universal[60,61]. For example, the ordinary electrostatic interaction between two charged particles can be used to perform universal quantum logic operations between two quantum bits. A bit is registered by the presence or absence of a particle in a mode. The strength of the interaction between the particles, $e^2/r$, determines the amount of time $t_{flip} = \pi\hbar r/2e^2$ it takes to perform a quantum logic operation such as a controlled-NOT on the two particles. The time it takes to perform such an operation divided by the amount of time it takes to send a signal at the speed of light between the bits $t_{com} = r/c$ is a universal constant, $t_{flip}/t_{com} = \pi\hbar c/2e^2 = \pi/2\alpha$, where $\alpha = e^2/\hbar c \approx 1/137$ is the fine structure constant. This example shows the degree to which the laws of physics and the limits to computation are entwined.

In addition to the theoretical evidence that most systems are computationally universal, the computer on which I am writing this article provides strong experimental evidence that whatever the correct underlying theory of physics is, it supports universal computation. Whether or not it is possible to make computation take place in the extreme regimes envisaged in this paper is an open question. The answer to this question lies in future technological development, which is difficult to predict. If, as seems highly unlikely, it is possible to extrapolate the exponential progress of Moore's law into the future, then it will take only 250 years to make up the 40 orders of magnitude in performance between current computers that perform $10^{10}$ operations per second on $10^{10}$ bits and our 1-kg ultimate laptop that performs $10^{51}$ operations per second on $10^{31}$ bits. □


1. Maxwell, J. C. *Theory of Heat* (Appleton, London, 1871).
2. Smoluchowski, F. *Vorträge über die kinetische Theorie der Materie u. Elektrizitat* (Leipzig, 1914).
3. Szilard, L. Über die Entropieverminderung in einem thermodynamischen System bei Eingriffen intelligenter Wesen. *Z. Physik* **53**, 840–856 (1929).
4. Brillouin, L. *Science and Information Theory* (Academic Press, New York, 1953).
5. Landauer, R. Irreversibility and heat generation in the computing process. *IBM J. Res. Dev.* **5**, 183–191 (1961).
6. Keyes, R. W. & Landauer, R. Minimal energy dissipation in logic. *IBM J. Res. Dev.* **14**, 152–157 (1970).
7. Landauer, R. Dissipation and noise-immunity in computation and communication. *Nature* **335**, 779–784 (1988).
8. Landauer, R. Information is physical. *Phys. Today* **44**, 23–29 (1991).
9. Landauer, R. The physical nature of information. *Phys. Lett. A* **217**, 188–193 (1996).
10. von Neumann, J. *Theory of Self-Reproducing Automata* Lect. 3 (Univ. Illinois Press, Urbana, IL, 1966).
11. Lebedev, D. S. & Levitin, L. B. Information transmission by electromagnetic field. *Inform. Control* **9**, 1–22 (1966).
12. Levitin, L. B. in *Proceedings of the 3rd International Symposium on Radio Electronics* part 3, 1–15 (Varna, Bulgaria, 1970).
13. Levitin, L. B. Physical limitations of rate, depth, and minimum energy in information processing. *Int. J. Theor. Phys.* **21**, 299–309 (1982).
14. Levitin, L. B. Energy cost of information transmission (along the path to understanding). *Physica D* **120**, 162–167 (1998).
15. Margolus, N. & Levitin, L. B. in *Proceedings of the Fourth Workshop on Physics and Computation—PhysComp96* (eds Toffoli, T., Biafore, M. & Leão, J.) (New England Complex Systems Institute, Boston, MA, 1996).
16. Margolus, N. & Levitin, L. B. The maximum speed of dynamical evolution. *Physica D* **120**, 188–195 (1998).
17. Bremermann, H. J. in *Self-Organizing Systems* (eds Yovits, M. C., Jacobi, G. T. & Goldstein, G. D.) 93–106 (Spartan Books, Washington DC, 1962).
18. Bremermann, H. J. in *Proceedings of the Fifth Berkeley Symposium on Mathematical Statistics and Probability* (eds LeCam, L. M. & Neymen, J.) Vol. 4, 15–20 (Univ. California Press, Berkeley, CA, 1967).
19. Bremermann, H. J. Minimum energy requirements of information transfer and computing. *Int. J. Theor. Phys.* **21**, 203–217 (1982).
20. Bekenstein, J. D. Universal upper bound on the entropy-to-energy ration for bounded systems. *Phys. Rev. D* **23**, 287–298 (1981).
21. Bekenstein, J. D. Energy cost of information transfer. *Phys. Rev. Lett.* **46**, 623–626 (1981).
22. Bekenstein, J. D. Entropy content and information flow in systems with limited energy. *Phys. Rev. D* **30**, 1669–1679 (1984).
23. Aharonov, Y. & Bohm, D. Time in the quantum theory and the uncertainty relation for the time and energy domain. *Phys. Rev.* **122**, 1649–1658 (1961).
24. Aharonov, Y. & Bohm, D. Answer to Fock concerning the time-energy indeterminancy relation. *Phys. Rev. B* **134**, 1417–1418 (1964).
25. Anandan, J. & Aharonov, Y. Geometry of quantum evolution. *Phys. Rev. Lett.* **65**, 1697–1700 (1990).
26. Peres, A. *Quantum Theory: Concepts and Methods* (Kluwer, Hingham, 1995).
27. Lecerf, Y. Machines de Turing réversibles. *C.R. Acad. Sci.* **257**, 2597–2600 (1963).
28. Bennett, C. H. Logical reversibility of computation. *IBM J. Res. Dev.* **17**, 525–532 (1973).
29. Bennett, C.H. Thermodynamics of computation—a review. *Int. J. Theor. Phys.* **21**, 905–940 (1982).
30. Bennett, C. H. Demons, engines and the second law. *Sci. Am.* **257**, 108 (1987).
31. Fredkin, E. & Toffoli, T. Conservative logic. *Int. J. Theor. Phys.* **21**, 219–253 (1982).
32. Likharev, K. K. Classical and quantum limitations on energy consumption in computation. *Int. J. Theor. Phys.* **21**, 311–325 (1982).
33. Seitz, C. L. *et al.* in *Proceedings of the 1985 Chapel Hill Conference on VLSI* (ed. Fuchs, H.) (Computer Science Press, Rockville, MD, 1985).

34. Merkle, R. C. Reversible electronic logic using switches. *Nanotechnology* **34,** 21–40 (1993).
35. Younis, S. G. & Knight, T. F. in *Proceedings of the 1993 Symposium on Integrated Systems*, Seattle, Washington (eds Berrielo, G. & Ebeling, C.) (MIT Press, Cambridge, MA, 1993).
36. Lloyd, S. & Pagels, H. Complexity as thermodynamic depth. *Ann. Phys.* **188,** 186–213 (1988).
37. Lloyd, S. Use of mutual information to decrease entropy—implications for the Second Law of Thermodynamics. *Phys. Rev. A* **39,** 5378–5386 (1989).
38. Zurek, W. H. Thermodynamic cost of computation, algorithmic complexity and the information metric. *Nature* **341,** 119–124 (1989).
39. Leff, H. S. & Rex, A. F. *Maxwell's Demon: Entropy, Information, Computing* (Princeton Univ. Press, Princeton, 1990).
40. Lloyd, S. Quantum mechanical Maxwell's demon. *Phys. Rev. A* **56,** 3374–3382 (1997).
41. Benioff, P. The computer as a physical system: a microscopic quantum mechanical Hamiltonian model of computers as represented by Turing machines. *J. Stat. Phys.* **22,** 563–591 (1980).
42. Benioff, P. Quantum mechanical models of Turing machines that dissipate no energy. *Phys. Rev. Lett.* **48,** 1581–1585 (1982).
43. Feynman, R. P. Simulating physics with computers. *Int. J. Theor. Phys.* **21,** 467 (1982).
44. Feynman, R. P. Quantum mechanical computers. *Optics News* **11,** 11 (1985); reprinted in *Found. Phys.* **16,** 507 (1986).
45. Zurek, W. H. Reversibility and stability of information-processing systems. *Phys. Rev. Lett.* **53,** 391–394 (1984).
46. Peres, A. Reversible logic and quantum computers. *Phys. Rev. A* **32,** 3266–3276 (1985).
47. Deutsch, D. Quantum-theory, the Church-Turing principle, and the universal quantum computer. *Proc. R. Soc. Lond.* **A 400,** 97–117 (1985).
48. Margolus, N. Quantum computation. *Ann. N.Y. Acad. Sci.* **480,** 487–497 (1986).
49. Deutsch, D. Quantum computational networks. *Proc. R. Soc. Lond. A* **425,** 73–90 (1989).
50. Margolus, N. in *Complexity, Entropy, and the Physics of Information, Santa Fe Institute Studies in the Sciences of Complexity* Vol. VIII (ed. Zurek, W. H.) 273–288 (Addison Wesley, Redwood City, 1991).
51. Lloyd, S. Quantum-mechanical computers and uncomputability. *Phys. Rev. Lett.* **71,** 943–946 (1993).
52. Lloyd, S. A potentially realizable quantum computer. *Science* **261,** 1569–1571 (1993).
53. Lloyd, S. Necessary and sufficient conditions for quantum computation. *J. Mod. Opt.* **41,** 2503–2520 (1994).
54. Shor, P. in *Proceedings of the 35th Annual Symposium on Foundations of Computer Science* (ed. Goldwasser, S.) 124–134 (IEEE Computer Society, Los Alamitos, CA, 1994).
55. Lloyd, S. Quantum-mechanical computers. *Sci. Am.* **273,** 140–145 (1995).
56. DiVincenzo, D. Quantum computation. *Science* **270,** 255–261 (1995).
57. DiVincenzo, D. P. 2-Bit gates are universal for quantum computation. *Phys. Rev. A* **51,** 1015–1022 (1995).
58. Sleator, T. & Weinfurter, H. Realizable universal quantum logic gates. *Phys. Rev. Lett.* **74,** 4087–4090 (1995).
59. Barenco, A. *et al.* Elementary gates for quantum computation. *Phys. Rev. A* **52,** 3457–3467 (1995).
60. Lloyd, S. Almost any quantum logic gate is universal. *Phys. Rev. Lett.* **75,** 346–349 (1995).
61. Deutsch, D., Barenco, A. & Ekert, A. Universality in quantum computation. *Proc. R. Soc. Lond. A* **449,** 669–677 (1995).
62. Cirac, J. I. & Zoller, P. Quantum computation with cold ion traps. *Phys. Rev. Lett.* **74,** 4091–4094 (1995).
63. Pellizzari, T., Gardiner, S. A., Cirac, J. I. & Zoller, P. Decoherence, continuous observation, and quantum computing—a cavity QED model. *Phys. Rev. Lett.* **75,** 3788–3791 (1995).
64. Turchette, Q. A., Hood, C. J., Lange, W., Mabuchi, H. & Kimble, H. J. Measurement of conditional phase-shifts for quantum logic. *Phys. Rev. Lett.* **75,** 4710–4713 (1995).
65. Monroe, C., Meekhof, D. M., King, B. E., Itano, W. M. & Wineland, D. J. Demonstration of a fundamental quantum logic gate. *Phys. Rev. Lett.* **75,** 4714–4717 (1995).
66. Grover, L. K. in *Proceedings of the 28th Annual ACM Symposium on the Theory of Computing* 212–218 (ACM Press, New York, 1996).
67. Lloyd, S. Universal quantum simulators. *Science* **273,** 1073–1078 (1996).
68. Zalka, C. Simulating quantum systems on a quantum computer. *Proc. R. Soc. Lond A* **454,** 313–322 (1998).
69. Shor, P. W. A scheme for reducing decoherence in quantum memory. *Phys. Rev. A* **52,** R2493–R2496 (1995).
70. Steane, A. M. Error correcting codes in quantum theory. *Phys. Rev. Lett.* **77,** 793–797 (1996).
71. Laflamme, R., Miquel, C., Paz, J. P. & Zurek, W. H. Perfect quantum error correcting code. *Phys. Rev. Lett.* **77,** 198–201 (1996).
72. DiVincenzo, D. P. & Shor, P. W. Fault-tolerant error correction with efficient quantum codes. *Phys. Rev. Lett.* **77,** 3260–3263 (1996).
73. Shor, P. in *Proceedings of the 37th Annual Symposium on the Foundations of Computer Science* 56–65 (IEEE Computer Society Press, Los Alamitos, CA, 1996).
74. Preskill, J. Reliable quantum computers. *Proc. R. Soc. Lond. A* **454,** 385–410 (1998).
75. Knill, E., Laflamme, R. & Zurek, W. H. Resilient quantum computation. *Science* **279,** 342–345 (1998).
76. Cory, D. G., Fahmy, A. F. & Havel, T. F. in *Proceedings of the Fourth Workshop on Physics and Computation—PhysComp96* (eds Toffoli, T., Biafore, M. & Leão, J.) 87–91 (New England Complex Systems Institute, Boston, MA, 1996).
77. Gershenfeld, N. A. & Chuang, I. L. Bulk spin-resonance quantum computation. *Science* **275,** 350–356 (1997).
78. Chuang, I. L., Vandersypen, L. M. K., Zhou, X., Leung, D. W. & Lloyd, S. Experimental realization of a quantum algorithm. *Nature* **393,** 143–146 (1998).
79. Jones, J. A., Mosca, M. & Hansen, R. H. Implementation of a quantum search algorithm on a quantum computer. *Nature* **393,** 344–346 (1998).
80. Chuang, I. L., Gershenfeld, N. & Kubinec, M. Experimental implementation of fast quantum searching. *Phys. Rev. Lett.* **80,** 3408–3411 (1998).
81. Kane, B. A silicon-based nuclear-spin quantum computer. *Nature* **393,** 133 (1998).
82. Nakamura, Y., Pashkin, Yu. A. & Tsai, J. S. Coherent control of macroscopic quantum states in a single-Cooper-pair box. *Nature* **398,** 786–788 (1999).
83. Mooij, J. E. *et al.* Josephson persistent-current qubit. *Science* **285,** 1036–1039 (1999).
84. Lloyd, S. & Braunstein, S. Quantum computation over continuous variables. *Phys. Rev. Lett.* **82,** 1784–1787 (1999).
85. Abrams, D. & Lloyd, S. Nonlinear quantum mechanics implies polynomial-time solution for NP-complete and P problems. *Phys. Rev. Lett.* **81,** 3992–3995 (1998).
86. Zel'dovich, Ya. B. & Novikov, I. D. *Relativistic Astrophysics* (Univ. of Chicago Press, Chicago, 1971).
87. Novikov, I. D. & Frolov, V. P. *Black Holes* (Springer, Berlin, 1986).
88. Pagels, H. *The Cosmic Code: Quantum Physics as the Language of Nature* (Simon and Schuster, New York, 1982).
89. Coleman, S., Preskill, J. & Wilczek, F. Growing hair on black-holes. *Phys. Rev. Lett.* **67,** 1975–1978 (1991).
90. Preskill, J. Quantum hair. *Phys. Scr. T* **36,** 258–264 (1991).
91. Fiola, T. M., Preskill, J. & Strominger A. Black-hole thermodynamics and information loss in 2 dimensions. *Phys. Rev. D* **50,** 3987–4014 (1994).
92. Susskind, L. & Uglum, J. Black-hole entropy in canonical quantum-gravity and superstring theory. *Phys. Rev. D* **50,** 2700–2711 (1994).
93. Strominger A. & Vafa, C. Microscopic origin of the Bekenstein-Hawking entropy. *Phys. Lett. B* 37, 99–104 (1996).
94. Das, S. R. & Mathur, S. D. Comparing decay rates for black holes and D-branes. *Nucl. Phys. B* **478,** 561–576 (1996).
95. Page, D. N. Particle emision rates from a black-hole: massless particles form an uncharged non-rotating black-hole. *Phys. Rev. D* **13,** 198 (1976).
96. Thorne, K. S., Zurek, W. H. & Price R. H. in *Black Holes: The Membrane Paradigm* Ch. VIII (eds Thorne, K. S., Price, R. H. & Macdonald, D. A.) 280–340 (Yale Univ. Press, New Haven, CT, 1986).

# Chronological Bibliography with Annotations and Selected Quotations

This chronological bibliography contains references dealing directly or indirectly with the Maxwell's demon puzzle. It includes background references on entropy, irreversibility, quantum theory, and related topics. Annotations and quotations illustrate the diverse viewpoints, proposed resolutions, and debates that have emerged regarding Maxwell's demon and related matters. Date, author and title are shown for each entry; full bibliographic information is in the separate Alphabetical Bibliography.

1867 Maxwell, J. C., 'Letter to P. G. Tait, 11 December 1867.' This contains the famous 1867 letter from J. C. Maxwell bearing the seminal idea for what has become known as Maxwell's demon.

1870 Maxwell, J. C., 'Letter to J. W. Strutt, 6 December 1870.' This letter to Lord Rayleigh is the second prepublication notice of the Maxwell's demon concept.

1871 Maxwell, J. C., *Theory of Heat*. Maxwell's public presentation of the demon in his classic thermodynamics book. See quotation in Section 1.2.1.

1874 Thomson, W., 'The Kinetic Theory of the Dissipation of Energy.' Reprinted here as Article 2.1.

1876 Loschmidt, J., 'Über den Zustand des Wärmegleichgewichtes eines System von Körpern mit Rücksicht auf die Schwerkraft.' The second law cannot be a purely mechanical principle because Newtonian mechanics allows the same sequence of motions backwards as forwards. Loschmidt's work challenged Boltzmann and helped him develop his statistical definition of entropy.

1878a Maxwell, J. C., 'Diffusion.'

*... the idea of dissipation of energy depends on the extent of our knowledge. Available energy is energy which we can direct into any desired channel. Dissipated energy is energy which we cannot lay hold of and direct at pleasure, such as the energy of the confused agitation of molecules ... the notion of dissipated energy could not occur to a being who could not turn any of the energies of nature to his own account, or to one who could trace the motion of every molecule and seize it*

*at the right moment. It is only to a being in the intermediate stage, who can lay hold of some forms of energy while others elude his grasp, that energy appears to be passing inevitably from the available to the dissipated state.*

1878b Maxwell, J. C., 'Tait's thermodynamics.'

*Hence we have only to suppose our senses sharpened to such a degree that we could trace the motions of molecules as easily as we now trace those of large bodies, and the distinction between work and heat would vanish, for the communication of heat would be seen to be a communication of energy of the same kind as that which we call work.*

1879 Thomson, W., 'The sorting demon of Maxwell.' Thomson gives a detailed description of the characteristics of a Maxwell demon. See Section 1.2.1.

*The conception of the 'sorting demon' is purely mechanical, and is of great value in purely physical science. It was not invented to help us deal with questions regarding the influence of life and of mind on the motions of matter, questions essentially beyond the range of mere dynamics.*

1885 Whiting, H., 'Maxwell's demons.' Whiting compares the sorting of a temperature-demon and the escape of high speed molecules from Earth's atmosphere.

*It seems to me of interest to point out that what, as Maxwell has shown, could be done by the agency of these imaginary beings, can be and often is actually accomplished by the aid of a sort of natural selection.*

1893 Poincaré, H., 'Mechanism and experience.' Poincaré contrasts the irreversibility of 'real experience' with the reversibility of mechanics. He refers to Maxwell's demon as an idea that gave rise to the kinetic theory of gases. Alluding to the famous recurrence theorem that bears his name:

*... to see heat pass from a cold body to a warm one, it will not be necessary to have the acute vision, the intelligence, and the dexterity of Maxwell's demon; it will suffice to have a little patience.*

1904 Boynton, W. P., *Applications of The Kinetic Theory*. One of earliest textbook references to Maxwell's demon, coming in a discussion of ideal gases, just after the Clausius and Kelvin statements of the second law.

*The validity of the second law is a matter of experience, and is not restricted to any particular substances. The reason seems to be that we are not able to deal individually with the motions of molecules, and discriminate between those with more and those with less energy, but have to deal with them in a lump. Hence it is that our treatment of the Kinetic Theory, dealing as it does with averages, presents the second law as a matter of course.*

1911 Knott, C. G., *Life and Scientific Work of Peter Guthrie Tait.* This contains Maxwell's 1867 letter and his 'Concerning demons' letter to Tait. See Section 1.2.1.

1912 Smoluchowski, M. von, 'Experimentell nachweisbare der üblichen Thermodynamik widersprechende Molekularphänomene.' A study of the validity of the second law of thermodynamics within the framework of kinetic theory and statistical mechanics.

1914 Smoluchowski, M. von, 'Gültigkeitsgrenzen des zweiten Hauptsatzes der Wärmtheorie.'

1922 Planck, M., *Treatise on Thermodynamics.* Planck discusses the impossibility of perpetual motion, referring only indirectly to the demon.

1923 Lewis, G. N. and Randall, M., *Thermodynamics and The Free Energy of Chemical Substances.* These authors focus on the entropy of the demon itself.

*Of course even in this hypothetical case one might maintain the law of entropy increase by asserting an increase of entropy within the demon, more than sufficient to compensate for the decrease in question. Before conceding this point it might be well to know something more of the demon's metabolism. Indeed a suggestion of Helmholtz raises a serious scientific question of this character. He inquires whether micro-organisms may not possess the faculty of choice which characterizes the hypothetical demon of Maxwell. If so, it is conceivable that systems might be found in which these micro-organisms would produce chemical reactions where the entropy of the whole system, including the substances of the organisms themselves, would diminish. Such systems have not as yet been discovered, but it would be dogmatic to assert that they do not exist.*

1924 Nyquist, H., 'Certain factors affecting telegraph speed.' The rate per character at which 'intelligence' can be transmitted is taken to be $K \log m$, where $m$ is the number of possible distinct characters and $K$ is a constant. Nyquist's use of 'intelligence' seems consistent with the modern term 'information.'

1925 Jeans, J. H., *Dynamical Theory of Gases.*

*Thus Maxwell's sorting demon could effect in a very short time what would probably take a very long time to come about if left to the play of chance. There would, however be nothing contrary to natural laws in the one case any more than in the other.*

1925 Szilard, L., 'On the extension of phenomenological thermodynamics to fluctuation phenomena.' This is Szilard's doctoral dissertation at the University of Berlin, a precursor to his seminal 1929 paper on Maxwell's demon.

1927 Costa, J. L., Smyth, H. D. and Compton, K. T., 'A mechanical Maxwell demon.' A 'mechanical' Maxwell's demon is used to effect an experimental measurement of Maxwell's speed distribution.

1928 Hartley, R. V. L., 'Transmission of information.' A very cogent discussion, proposing a quantitative measure of information, $H = n \log s$, where $s$ is the number of possible distinct symbols and $n$ is the number of symbols in a message.

1929 Clausing, P., 'Über die Entropieverminderung in einem thermodynamischen System bei Eingriffen intelligenter Wesen.' This early criticism of Szilard's 1929 paper contains an unsupported allegation that Szilard's analysis is incorrect.

1929 Preston, T., *The Theory of Heat.* The author discusses Maxwell's 'ingenious but illusory violation' of the second law.

*It must be remembered, however, that to this being the gas is by no means a uniformly heated mass. The faster-moving molecules are hot and the slower cold, and the whole mass to him is made up of discrete parts at very different temperatures, and this sifting of the molecules is no more a violation of the second law than would be the collection by an ordinary being of the warmer members of a system of bodies into one region of space and colder into another.*

1929 Szilard, L., 'On the decrease of entropy in a thermodynamic system by the intervention of intelligent beings.' Reprinted here as Article 3.3.

1930 Lewis, G. N., 'The symmetry of time in physics.' Maxwell's demon is introduced in connection with a simple model system of three distinguishable molecules in a cylinder with a partition that has a controllable shutter. Lewis finds that entropy increases when 'a known distribution goes over into an unknown distribution,' supporting the view that entropy is a subjective entity.

*Gain in entropy always means loss of information, and nothing more. It is a subjective concept, but we can express it in its least subjective form, as follows. If, on a page, we read the description of a physico-chemical system, together with certain data which help to specify the system, the entropy of the system is determined by these specifications. If any of the essential data are erased, the entropy becomes greater; if any essential data are added, the entropy becomes less. Nothing further is needed to show that the irreversible process neither implies one-way time, nor has any other temporal implications. Time is not one of the variables of pure thermodynamics.*

1932 von Neumann, J., *Mathematical Foundations of Quantum Mechanics.* In contrast with most books quantum theory, von Neumann devotes a significant effort to thermodynamic considerations and macroscopic applications of quantum mechanics, particularly to measurement theory and Szilard's heat engine. See quotation in Section 1.5.

1939 Allen, H. S. and Maxwell, R. S., *A Text-Book of Heat.* The authors seem to accept the view enunciated by Lewis (1930) that entropy is subjective.

1939 Slater, J. C., *Introduction to Chemical Physics.*

*Is it possible, we may well ask, to imagine demons of any desired degree of refinement? If it is, we can make any arbitrary process reversible, keep its entropy from increasing, and the second law of thermodynamics will cease to have any significance. The answer to this question given by the quantum theory is No. An improvement in technique can carry us only a certain distance, a distance practically reached in plenty of modern experiments with single atoms and electrons, and no conceivable demon, operating according to the laws of nature, could carry us further. The quantum theory gives us a fundamental size of cell in the phase space, such that we cannot regulate the initial conditions of an assembly on any smaller scale. And this fundamental cell furnishes us with a unique way of defining entropy and of judging whether a given process is reversible or irreversible.*

1939 Weber, H. C., *Thermodynamics for Chemical Engineers.* Weber alludes to a technique, which he attributes to Boltzmann, that avoids any necessity for intelligence in the demon's operation.

*Assume a cell containing an extremely dilute solution of colloidal iron. Place this between the poles of a permanent magnet having poles of dimensions approximating those of the iron particles. The magnet is provided with a coil from which a current developed by the Brownian movement of the iron particles past the magnet poles can be drawn. To maintain the colloidal motion (hold the temperature constant) the whole may be immersed in the ocean, which serves as a reservoir from which to draw heat energy. The sole net result of operating this system would be the conversion of a quantity of heat to work—an impossible result according to the second law. In all such cases we are either enlisting the aid of conscious choice to differentiate between particles of molecular dimensions or dealing with a number of particles so limited that the law of probability does not hold.*

1940 Jeans, J. H., *An Introduction to the Kinetic Theory of Gases.*

1941 Steiner, L. E., *Introduction to Chemical Thermodynamics.*

*The presence of fluctuations does not permit us to violate the second law of thermodynamics. If we wished to 'compress' the gas into $V_1$ ... without doing work, we should have to recognize the rare moments when all the molecules are in $V_1$ and place a retaining wall between $V_1$ and $V_1'$ before any of the molecules left $V_1$. Such recognition and action seem to be reserved for Maxwell's 'demon' although the possibility is not barred that living organisms can perform the necessary selectivity to produce separations which represent less probable states.*

1944 Darrow, K. K., 'The concept of entropy.' A cogent nonmathematical review of entropy.

1944 Demers, P., 'Les démons de Maxwell et le second principe de la thermodynamique.'

*(Maxwell's demon) … cannot be completely resolved by application of the uncertainty principle as proposed by Slater. In this regard, G. N. Lewis has made some thermodynamics remarks of interest that agree with our conclusions. One cannot however admit that an infinitesimal energy is sufficient to open a shutter between two compartments. It is necessary to expend an energy comparable to* $kT$*. The solution of this problem is based upon the existence of blackbody radiation and on the quantum nature of the emission. One might say that the kinetic theory and the second law demand blackbody radiation and quantum theory.*

1944 Dodge, B. F., *Chemical Engineering Thermodynamics.*

*Maxwell was one of the first to perceive that the reason for irreversibility is our inability to deal with individual molecules. He imagined a being, known commonly as 'Maxwell's demon,' who was small enough to deal with individual molecules and who could therefore readily violate the second law. If such a demon controlled the opening and closing of a very small aperture between two vessels containing gas at a uniform temperature, he could cause a temperature difference to develop by unshuffling the 'hot' and 'cold' molecules through proper manipulation of the shutter, or he could separate a mixture of two gases. Certain bacteria are able to assume a role approaching that of Maxwell's demon when they bring about the separation of racemic mixtures.*

1944 Gamow, G., *Mr. Tompkins Explores the Atom.* In a delightful introduction to Maxwell's demon, Gamow uses gambling schemes and fantasy to illustrate statistical physics. A human, temporarily shrunk to molecular size, witnesses a demon make part of an ice-filled drink boil!

1944 Schrödinger, E., *What is Life?.* Though somewhat outdated, this is a good general background reference on the role of thermodynamics in biology.

1945 Demers, P., 'Le second principe et la théorie des quanta.' This is a continuation of Demers' 1944 paper.

1948 Born, M., 'Die Quantenmchanik und der Zweite Hauptsatz der Thermodynamik.' This paper and the following one by Born and Green helped establish the connection between quantum theory and thermodynamics.

1948 Born, M. and Green, H. S., 'The kinetic basis of thermodynamics.'

1948 Wiener, N., *Cybernetics.*

*It is simpler to repel the question posed by the Maxwell demon than to answer it. Nothing is easier than to deny the possibility of such beings or structures. We shall actually find that Maxwell demons in the strictest sense cannot exist in a system in equilibrium, but if we accept this from the beginning, and so not try to demonstrate it, we shall miss an admirable opportunity to learn something about entropy and about possible physical, chemical, and biological systems. ... For a Maxwell demon to act, it must receive information from approaching particles concerning their velocity and point of impact on the wall. Whether these impulses involve a transfer of energy or not, they must involve a coupling of the demon and the gas. ... the only entropy which concerns us is that of the system gas-demon, and not that of the gas alone. ... The demon can only act on information received, and this information ... represents a negative entropy. However, under the quantum mechanics, it is impossible to obtain any information giving the position or the momentum of a particle, much less the two together, without a positive effect on the energy of the particle examined, exceeding a minimum dependent on the frequency of the light used for examination. In the long run, the Maxwell demon is itself subject to a random motion corresponding to the temperature of its environment.... In fact, it ceases to act as a Maxwell demon.*

1949 Brillouin, L., 'Life, thermodynamics, and cybernetics.' Reprinted here as Article 2.4.

1949 Jordan, P., 'On the process of measurement in quantum mechanics.'

*It is generally thought that the concepts of thermodynamics are strictly macrophysical ones—that a single atom never has a temperature, and never contains a definite amount of entropy. But the tendency of Szilard's views is to acknowledge also a microphysical applicability of thermodynamics. Let us again think of our photon—it may now be in such a state as can be represented by an unpolarised beam: The two wave functions $\phi$ and $\psi$ are incoherent and of the same intensity. According to Szilard this means that the photon possesses the entropy $k \ln 2$ ... According to V. Neumann this entropy is the result of a mental process of the observer: By forgetting the coherence between $\phi$ and $\psi$ ... he creates the entropy $k \ln 2$. Therefore for another observer, who did not forget, the photon has the entropy 0; the notion of entropy becomes a relative one, different for different observers or different memories. In my opinion—assuming the conversion of a polarised photon into an unpolarised one to be a real physical process—this entropy too has an objective meaning, independently of the mental processes of any observer. Therefore we are forced to acknowledge a new, hitherto not recognised type of physical uncertainty.*

1949 Shannon, C. E. and Weaver, W., *The Mathematical Theory of Communication*. This contains Claude Shannon's 1948 landmark paper on information theory, with an introduction by Warren Weaver.

1950 Raymond, R. C., 'Communication, entropy, and life.'

*It is the purpose of this paper to suggest that the entropy of a system may be defined quite generally as the sum of the positive thermodynamic entropy which the constituents of the system would have at thermodynamic equilibrium and a negative term proportional to the information necessary to build the actual system from its equilibrium state. This definition may be applied to open systems by closing them momentarily . . .*

1950 Wiener, N., 'Entropy and information.'

*The Maxwell demon is able to lower the entropy of the mechanical output compared with the mechanical input because the demon itself possesses a negative entropy of information. This negative entropy must give information of the momentum and position of particles approaching the gateway operated by the demon in such a way that when the particle collides with the gateway, the gateway is either open or locked shut, so that no absorption of energy appears. This involves a mode of transmission more rapid than the motion of the particles to be separated, and a mode which is probably light. This light itself has a negative entropy which decreases whenever the frequency of the light is lowered by a Compton effect or similar cause. Thus the Maxwell demon gives at least one way for comparing entropy of light with mechanical entropy . . . This whole point of view suggests that one place to look for Maxwell demons may be in the phenomena of photosynthesis.*

1950a Brillouin, L., 'Can the rectifier become a thermodynamical demon?' Brillouin shows that a resistor at temperature $T$, connected to a rectifier, generates zero rectified current and zero voltage across the rectifier. Thermal noise cannot be rectified to transform heat to electric work.

1950b Brillouin, L., 'Thermodynamics and information theory.' A relatively nonmathematical article in which Brillouin describes his ideas on information, thermodynamics, and computation.

*Whether he be a Maxwell's demon or a physicist, the observer can obtain a bit of information only when a certain quantity of negentropy is lost. Information gain means an increase of entropy in the laboratory. Vice versa, with the help of this information the observer may be in a position to make an appropriate decision, which decreases the entropy of the laboratory. In other words, he may use the information to recuperate part of the loss in negentropy, but the over-all balance is always an increase in entropy (a loss in negentropy).*

1951 Bohm, D., *Quantum Theory*. Brief discussion of the irreversibility of quantum measurement, with mention of Maxwell's demon.

1951a Brillouin, L., 'Maxwell's demon cannot operate: Information and entropy. I.' Reprinted here as Article 3.4.

1951b — Brillouin, L., 'Physical entropy and information. II.'

1951 — Jacobson, H., 'The role of information theory in the inactivation of Maxwell's demon.' Jacobson outlines how Brillouin's 'excellent, but not entirely general arguments' exorcising Maxwell's demon can be extended. He uses probability arguments to effect this extension for both 'pressure' and 'temperature' demons, obtaining estimates for: (1) the information needed to achieve a given entropy decrease; and (2) the entropy increase necessary to acquire that information. In (2) he assumes the demon has no need for a central nervous system, and avoids any discussion of how the demon gets its information. Rather, he focuses on the energy needed to operate the trap door reliably, given the existence of thermal fluctuations. Some of Jacobson's ideas were incorporated by Brillouin in his *Science and Information Theory*.

1951 — Raymond, R. C., 'The well-informed heat engine.' A heat engine is controlled by an outside observer (pressure demon) that takes advantage of density fluctuations in a gas. It is argued that the second law is not violated only if the demon 'creates in the system a negative information entropy equal to the negative entropy change involved in the operation of the engine.' Reprinted in MD1.

1951 — Rothstein, J., 'Information, measurement, and quantum mechanics.' Reprinted here as Article 2.5.

*... in a system like that of the Maxwell demon involving both light and matter, the supposed perpetual motion machine is not in reality perpetual. The light in it is degraded and exhausted by its use to operate the machine ....*

1952a — Rothstein, J., 'Information and Thermodynamics.' Rothstein suggests informational interpretations of the laws of thermodynamics.

*To sum up, the laws of thermodynamics can be stated as: (a) The conservation of energy. (b) The existence of modes of energy transfer incapable of mechanical description. In a sense (b) is implied in (a), for (a) without (b) is a mechanical theorem devoid of thermodynamic content. (c) The third law is true by definition, for in a perfectly ordered state at absolute zero there is no missing information, i.e., the entropy is zero (pure case).*

1952b — Rothstein, J., 'Organization and entropy.' Organization is characterized as a type of negative entropy that takes the form of encoded information. Theory is viewed as organization of observation.

1952c — Rothstein, J., 'A phenomenological uncertainty principle.'

*The existence of a thermodynamic limit to the precision of any measurement follows from the facts that: (a) information is conveyed by making choices from an ensemble of alternatives for which an 'entropy' is definable ...; (b) measurement chooses from an ensemble of possible results, thus yielding 'physical' information; (c) either the informational entropy of physical information multiplied by Boltzmann's constant must be a lower bound to the thermodynamic entropy generated in acquiring it or the second law can be violated. Example: A quantity equiprobably between 0 and u is measured to fall within* $\Delta u$. *Then choice from* $u/\Delta u$ *alternatives was made at entropy cost* $\Delta S$ *not less than* $k \ln(u/\Delta u)$. *For preassigned experimental conditions maximum* $\Delta S$ *is well defined and* $\Delta u \geq u \exp(-\Delta S/k) \ldots$

1952 Wiener, N., 'Cybernetics.' Reprinted in P. Masani, *Norbert Wiener: Collected Works with Commentaries* (MIT Press, Cambridge, Mass, 1985), p. 202. Wiener defines cybernetics as 'the employment of communication and of the notion of the quantity of information for the control of artificial and natural systems,' and uses these ideas to study Maxwell's demon.

1953a Balazs, N. L., 'Les relations d'incertitude d'Heisenberg empechent-elle le démon de Maxwell d'opérer?' It is found that Maxwell's demon is not restricted by the uncertainty principle if the system chosen is not degenerate. This agrees with, and was evidently done independently of, the work of Demers (1944).

1953b Balazs, N. L., 'L'effet des statistiques sur le démon de Maxwell.' If a Bose–Einstein gas is not degenerate, Maxwell's demon is found to be free of restrictions from the uncertainty principle. In contrast, no corresponding conditions whatsoever are imposed on Maxwell's demon for a gas obeying Fermi-Dirac statistics.

1953 Klein, M. J., 'Order, organisation, and entropy.'

*One can say that the entropy is low when the number of energy states, in which there is a non-negligible probability of finding the system is small. The entropy does not depend explicitly upon the nature of the wave functions for the system which are associated with these energy states. It is in the wave function, however, that the structure of the system is reflected, and it is this structure which is associated with the concept of organisation. ... We conclude that the degree or order in a system which is measured by the entropy (low entropy corresponding to high order) is not the same thing as the degree of organisation of the system in the sense used by the biologist.*

1955 Darling, L. and Hulburt, E. O., 'On Maxwell's demon.' The authors describe a situation encountered when a small dog entered a fenced area holding a hen turkey and her chicks. The resulting pandemonium and the authors' difficulties catching the dog and counting the chicks are likened to the challenge faced by a Maxwell's demon. They conclude, 'As a result of this experience we feel sure that the demon could not possibly do his molecule job, and that two demons would probably do worse than one.'

1956 Brillouin, L., *Science and Information Theory*. A clear, concise exposition of Brillouin's contributions to Maxwell's demon and information theory.

*Every physical measurement requires a corresponding entropy increase, and there is a lower limit, below which the measurement becomes impossible. This limit corresponds to change in entropy of ...* $k \ln 2$, *or approximately* 0.7 $k$ *for one bit of information obtained. ... It is very surprising that such a general result escaped attention until very recently. ... A general feature of all this discussion is that the quantum conditions were used in the reasoning, but Planck's constant h is eliminated in the final results, which depend only on Boltzmann's constant k. This proves that the results are independent of quanta and of the uncertainty principle, and, in fact, a discussion can be given along classical lines without the introduction of quantum conditions ...*

1956 Lotka, A. J., *Elements of Mathematical Biology (Originally published as Elements of Physical Biology).* Lotka cites Maxwell, 'who remarked that a demon capable of dealing with individual molecules would be able to cheat the second law of thermodynamics.'

1956 von Neumann, J., 'Probabilistic logics from unreliable components.' In a lengthy article dealing with automata, von Neumann gives a brief but notable reference to the similarity between the information theory and statistical mechanical entropy functions.

*That information theory should thus reveal itself as an essentially thermodynamical discipline, is not at all surprising: The closeness and the nature of the connection between information and entropy is inherent in L. Boltzmann's classical definition of entropy ...*

1957a Jaynes, E. T., 'Information theory and statistical mechanics.' Equilibrium statistical mechanics is developed using the maximum-entropy principle of information theory, and subjectivity in statistical mechanics is discussed.

*The essential point ... is that we accept the von Neumann–Shannon expression for entropy, very literally, as a measure of the amount of uncertainty represented by a probability distribution; thus entropy becomes the primitive concept with which we work, more fundamental even than energy. If in addition we reinterpret the prediction problem of statistical mechanics in the subjective sense, we can derive the usual relations in a very elementary way without any consideration of ensembles or appeal to the usual arguments concerning ergodicity or equal a priori probabilities.*

1957b Jaynes, E. T., 'Information theory and statistical mechanics. II.' Jaynes extends his development of statistical mechanics via information theory to time-dependent phenomena and irreversibility.

1957 Popper, K., 'Irreversibility; or, entropy since 1905.' Popper argues that the interpretation of the second law of thermodynamics as a statistical law became untenable once Brownian movement was observed, because the latter represents deviations from that law. He criticizes Szilard's work on entropy decrease via intervention by an intelligent being, arguing that the logic is circular.

*(Szilard) first gives four non-equivalent formulations of the entropy law clearly implying that they are all equivalent. ... (b) Szilard then assumes the axiomatic validity of the entropy law so defined. (c) He then proves that a Maxwell demon ... would have to pay for his information concerning the oncoming molecules by an increase of entropy; which is a trivial consequence if any form of the entropy law is to be preserved intact; and which therefore follows if the entropy law is assumed as a premiss. (d) He thereby shows that information or knowledge can be measured in terms of negative entropy, or entropy in terms of negative knowledge; which immediately leads to the interpretation of statistical probability in terms of subjective ignorance. ... But I think that hot air would continue to escape entropically even if there were no intelligent beings about to provide the corresponding amount of nescience.*

1957 Rothstein, J., 'Nuclear spin echo experiments and the foundations of statistical mechanics.'

*The problem of reconciling the irreversibility of thermodynamics with the completely reversible mechanics of the ultimate constituents of the thermodynamical system is examined from an operational viewpoint. The informational nature of entropy is demonstrated, and the famous paradoxes of statistical mechanics, due to Loschmidt and Zermelo, are resolved with its aid. Spin echo experiments are shown to realize the conditions of Loschmidt's reflection paradox, and used to illustrate how reversibility occurs only with perfect 'memory' or information storage, while 'forgetting' or loss of information implies irreversibility.*

1958 Elsasser, W. M., *The Physical Foundation of Biology*. A detailed nonmathematical discourse on entropy and information, including Maxwell's demon.

*...the observer must* **have** *knowledge of the approaching molecules in order to operate the shutter at the proper times. ... This process of seeing constitutes a* **physical interaction** *with the system observed. ... The simplest way to see a molecule is by means of ordinary light, and we may assume that in the idealized case one light quantum suffices to locate the molecule. ... It is hardly necessary to say that the use of light quanta to detect molecules is somewhat arbitrary and that the same general result would obtain if some other elementary particle was used for detection.*

1958 Haar (ter), D., *Elements of Statistical Mechanics*. A brief discussion of the work of Maxwell, Szilard, and Wiener, mainly useful as background information.

*...proofs of the general validity of the second law are different from the proofs of the general validity of the Heisenberg relations in quantum mechanics. In both cases one sets up an idealized experiment which should be suited to get around the theoretical limitations, but while in quantum mechanics one proves that it is* **never** *possible to break the limitations, in the thermodynamical case one can prove only that* **on the average** *it is not possible to violate the second law.*

1958 Rapoport, A., 'Remark.'

*It is noteworthy that the greatest discoveries of the physicists are stated in 'pessimistic' terms. They are statements about what cannot be done. For example the first law of thermodynamics ... says in effect that the perpetual motion machine cannot be constructed. But is also holds out a hope of a machine that will keep on working provided only that a large supply of heat is available—the so-called perpetual motion machine of the second kind. The second law of thermodynamics puts an end to that dream. ... Yet it would be a mistake to consider these discoveries as admissions of defeat only. Each has brought a broadened understanding; the first law of thermodynamics by revealing heat as a source of energy; the second law by revealing the role of entropy. Szilard's investigation rests on quantum-theoretical principles and so provides an important juncture between thermodynamics, information theory, and quantum theory. It appears, therefore, that the grand discoveries of physics have a sobering effect. I think the principles of information theory are of a similar kind. Typically they are statements of limitations. Their constructive side is in defining the framework in which the search for new knowledge or for new means of prediction and control must be confined.*

1958 Saha, M. N. and Srivastava, B. N., *A Treatise on Heat.*

*... to Maxwell's demon the gas does not appear as a homogeneous mass but as a system composed of discrete molecules. The law of entropy does not hold for individual molecules, but is a* **statistical law** *and is valid only when we are compelled to deal with matter in bulk, and are unable to direct of control the motion of individual molecules. Furthermore, ... there is always a chance, though a very meagre one, that the entropy may decrease.*

1959 Rothstein, J., 'Physical demonology.' The idea of demons such as Maxwell's and Laplace's is generalized, showing that any law can be formulated as a principle of impotence—i.e., in terms of the nonexistence of some type of demon.

1960 Finfgeld, C. and Machlup, S., 'Well-informed heat engine: efficiency and maximum power.' Raymond's well-informed heat engine is examined further to get estimates of its efficiency and power output. Reprinted in MD1.

1961 Bridgman, P. W., *The Nature of Thermodynamics.*

*If the Maxwell demon had been invented yesterday instead of in the last century I believe he would not have caused as much consternation. There are too many vital points that must be cleared up. In the first place, what is this 'intelligence' that must be presupposed? ... Another doubtful feature is the method by which the demon would learn of the approach of the individual molecules. The only method would appear to be by light signals; these must come in quanta and must react with the molecule. The reaction is uncontrollable, and may be sufficiently large to divert the molecule by so much as to vitiate the manipulations of the trap door. Then there is the question of the legitimacy of assuming that 'information' can be propagated in any purely thermodynamic system. ... Again, there is the question of the details of operation of the trap door; if the door is a mechanical system in the proper sense it must be perfectly elastic, and this means that its motion cannot be stopped, or is at most periodic. ... when the mechanism gets small, the mechanism itself and its controls (including in the controls the brain of the demon) become subject to temperature fluctuations which are proportionally large the smaller the mechanism. How do we know that this will not vitiate the entire program? These are serious questions, and so far as I know have received no adequate discussion.*

1961 Grad, H., 'The many faces of entropy.' Various 'faces' of entropy are examined in this review, including the $H$ function, irreversibility, and statistical and classical thermodynamics.

1961 Landauer, R., 'Irreversibility and heat generation in the computing process.' Reprinted here as Article 4.1.

1961 Pierce, J. R., *Symbols, Signals and Noise*. The author introduces Maxwell's demon as 'one of the most famous perpetual-motion machines of the second kind,' and analyzes a version of Szilard's one-molecule engine with suitable pulleys and weights.

1962 Bell, D. A., *Intelligent Machines: An Introduction to Cybernetics*. Maxwell's demon is described along the lines of Raymond and Brillouin.

1962 Rothstein, J., 'Discussion: Information and organization as the language of the operational viewpoint.' It is argued that the concepts and terminology of information theory correspond to physical measurement, and that the concept of organization corresponds to laws and to operations as used in physics.

1963 Feynman, R. P., Leighton, R. B. and Sands, M., *The Feynman Lectures on Physics–Vol. 1.* Feynman investigates whether the ratchet and pawl can be operated as a heat engine that violates the second law. Using statistical arguments, he shows that by carefully choosing the weight attached to the ratchet and pawl, the engine can either just barely be driven by the weight or it can just barely lift the weight; this approaches reversibility. [A critique can be found in Parrondo and Español (1996).] He estimates how the ratchet's angular velocity varies with torque, getting a curve reminiscent of an electrical rectifier. Like a rectifier, a ratchet works in reverse under certain temperature conditions. Feynman introduces Maxwell's demon, which:

*... is nothing but our ratchet and pawl in another form, and ultimately the mechanism will heat up. If we assume that the specific heat of the demon is not infinite, it must heat up. It has but a finite number of internal gears and wheels, so it cannot get rid of the extra heat that it gets from observing the molecules. Soon it is shaking from Brownian motion so much that it cannot tell whether it is coming or going, much less whether the molecules are coming or going, so it does not work.*

1963 Jaynes, E. T., 'Information Theory and Statistical Mechanics.' A concise, readable description of how statistical mechanics can be developed via information theory.

1964 Gabor, D., 'Light and information.' Gabor examines the use of information theory in optics, including an investigation of a variant of Szilard's model, using light to determine the molecule's location. The quantum theory of radiation is used to disprove the possibility of a perpetuum mobile; classical light theory is inadequate. Gabor's lecture on which this paper is based was given the same month that Brillouin's paper, 'Maxwell's Demon Cannot Operate: Information and Entropy I' was published, but evidently neither author was aware of the other's work. Reprinted in MD1.

1964 Rodd, P., 'Some comments on entropy and information.' This brief paper clarifies the work of Brillouin regarding the connections between entropy and information. Reprinted in MD1.

1965 Asimov, I., *Life and Energy*. A readable elementary discussion of molecules and molecular motion that introduces Maxwell's demon as a tool to dramatize the statistical nature of molecular motion.

1965 Bent, H. A., *The Second Law*.

*Obituary: Maxwell's Demon (1871–c.1949) The paradox posed by Maxwell's demon bothered generations of physicists. In 1912 Smoluchowski noted that Brownian agitation of the trap door, which would result in a random opening and closing of the door, would render ineffective the long range operation of any automatic device, such as a spring valve or a ratchet and pawl. In 1939 Slater suggested that the uncertainty principle might play a role in the problem. Later it was shown that this would not be the case for heavy atoms at low pressures. Not until 1944–51, however, did two physicists, Demers and Brillouin, call attention to the fact that in an isolated enclosure in internal thermal equilibrium* **it would be impossible for the demon to see the individual molecules.** *To make the molecules visible against the background black-body radiation, the demon would have to use a torch. Happily, as Demers and Brillouin showed, the entropy produced in the irreversible* *operation of the torch would always exceed the entropy destroyed bye the demon's sorting procedure. A real demon could not produce a violation of the second law.*

1965 Hatsopoulos, G. N. and Keenan, J. H., *Principles of General Thermodynamics*. The authors of this popular engineering thermodynamics text discuss Maxwell's demon in the book's Foreword, giving a good historical account of thermodynamics and statistical mechanics. The demon is discussed again later in connection with the definition of a thermodynamic system.

1965 Jaynes, E. T., 'Gibbs vs. Boltzmann entropies.' The Gibbs and Boltzmann entropies are compared, and the anthropomorphic nature of entropy is emphasized.

1965 Kubo, R., *Statistical Mechanics.*

*You may be able to find a demon who starts to do an extremely fine job of selecting molecules passing through the window. But he will never be able to continue the work indefinitely. Soon he will become dazzled and get sick and lose his control. Then the whole system, the gas molecules and the demon himself, will again approach a final equilibrium, where the temperature difference the demon once succeeded in building up disappears and the demon will run a fever at a temperature equal to that of the gas. A living organism may look like a Maxwell's demon, but it is not. A living organism is an open system, through which material, energy, and entropy are flowing. But life itself cannot violate thermodynamic laws.*

1965 Lehninger, A. L., *Bioenergetics: The Molecular Basis of Biological Energy Transformations*. Lehninger observes that that about $10^{23}$ bits of information are required to reduce the entropy of a system by 1 $\mathrm{cal}^{-1}$ $\mathrm{mol}^{-1}$ $\mathrm{K}^{-1}$, suggesting how energetically inexpensive information storage and communication are. Living cells store large amounts of information whose energy 'equivalent' is significant.

1966 Dugdale, J. S., *Entropy and Low Temperature Physics.*

*...Maxwell sought to demonstrate that the second law of thermodynamics has only statistical certainty. However, Brillouin, following earlier work by Szilard, has shown that if the demon himself is subject to the laws of physics (in particular the quantum theory) the operations necessary to detect the fast molecules cause an increase in entropy sufficient to offset the decrease in entropy which the demon is trying to bring about in the gas.*

1966 Feyerabend, P. K., 'On the possibility of a perpetuum mobile of the second kind.' The author provides an addendum to Popper's 1957 paper, 'Irreversibility; or, entropy since 1905.' He analyzes a variant of Szilard's one-molecule gas, finding (as did Popper) that it violates the second law without the need for information. Criticizing

Smoluchowski's and Szilard's efforts, he concludes 'the attempt to save the second law from systematic deviations, apart from being circular and based upon an ambiguous use of the term 'information,' is also ill-conceived, for such deviations are in principle possible.' See the discussion of memory resetting in Section 1.5.

1966 Frisch, D. H., 'The microscopic interpretation of entropy.' Boltzmann's entropy formula is deduced from the definition $dS = dQ/T$ and the quantum mechanical heat and work expressions.

1966 Singh, J., *Great Ideas in Information Theory, Language and Cybernetics.* Singh describes Maxwell's demon using Brillouin's information theory approach, following some of the ideas advanced by Wiener in his book Cybernetics (1961).

1967 Angrist, S. W. and Hepler, L. G., *Order & Chaos.* A literate, colorful treatise on thermodynamics for the lay science reader, with a chapter on 'Demons, poetry, and life.'

1967 Ehrenberg, W., 'Maxwell's demon.' A good historical overview through Brillouin's work.

*'How seriously Maxwell took his demon is hard to say. In any case, he neither carried out nor promoted experiments to test his hypothesis. His almost offhand remark has nonetheless intrigued many prominent physicists, because it holds out the possibility of a perpetual-motion machine deriving its mechanical effect from the temperature difference between the two portions of the vessel. ... Szilard's analysis of Smoluchowski's proposal that intelligent man could operate a perpetual-motion machine that violated the second law of thermodynamics led him neither to a working model nor to a proof that the proposal is unworkable but rather to a postulate relating entropy to information. One may therefore praise or blame Szilard for having opened the path leading to information theory and its mysteries. I am sure, however, that this was not his intention. He believed his paper put the final seal on half a century of argument.*

1967 Kauzmann, W., *Thermal Properties of Matter–Vol. II–Thermodynamics and Statistics: With Applications to Gases.* Kauzmann describes the demon's operation, giving a sorting strategy and a formula for the time it takes a demon to generate a temperature difference (time $\approx 4V/[u_{med}A]$, where $A$ = opening area, $V$ = container volume, and $u_{med}$ = median molecular speed). He introduces a model where a pressure demon lets molecules through the shutter from the right to left sides but not vice versa, lowering the gas entropy. He adopts Szilard's view that each demon decision requires conversion of nonthermal energy into heat, keeping the second law intact. No calculations are done with the above time formula, but for $V = 100$ m$^3$ (big room), $A = 10^{-16}$ m$^2$ (tiny trap door, to assure serial processing), and $u_{med} = 500$ m/s (room temperature), it implies a time $8 \times 10^{15}$ seconds, or $2.5 \times 10^8$ years.

1968 Bell, D. A., *Information Theory*. Bell reviews Maxwell's demon, combining Szilard's one-molecule heat engine and a variant of Brillouin's negentropy argument. He also uses an information-theoretic argument to show that the rate of entropy increase can be vanishingly small in an information channel when the power level approaches zero, in analogy with reversible heat transfer.

1968 Dutta, M., 'A hundred years of entropy.' The first one hundred years of entropy are reviewed, covering thermodynamics, statistical mechanics, communication theory, and other aspects.

1968 Morowitz, H. J., *Energy Flow in Biology*. The second law is related to life processes.

1969 Holman, J. P., *Thermodynamics*.

*The conclusion of a considerable body of opinion is that either (a) the demon will be unable to distinguish between fast and slow molecules because he is subjected to a random molecular bombardment or (b) the entropy of the demon must change to account for the 'information' he receives in the measurement process. This latter argument pertaining to information has led to the use of the term entropy in the science of information theory, and has even prompted some authors to develop the entire subject of thermodynamics on the basis of concepts from information theory. In this development entropy is taken to be a measure of our increasing lack of information.*

1969 Watanabe, S., *Knowing and Guessing: A Quantitative Study of Inference and Information*. A section on negentropy and thermodynamics contains a critical discussion of Brillouin's arguments. Watanabe finds that 'Brillouin's principle does not seem to be rigorously tenable although it reflects a great deal of valid insight into the difficult problem.'

1970 Daub, E. E., 'Maxwell's demon.' Reprinted here as Article 2.2.

1970 Heimann, P. M., 'Molecular forces, statistical representation and Maxwell's demon.' This paper concentrates on Maxwell's development of the demon from ideas found in his earlier work. Reprinted in MD1.

1970 Klein, M. J., 'Maxwell, his demon, and the second law of thermodynamics.' Reprinted here as Article 2.3.

1970 Morowitz, H. J., *Entropy for Biologists*. Morowitz calls entropy (regarding microstates) a measure of the system, while information is a measure that relates to the observer. He considers this an artificial distinction, because an observer manipulating constraints on the system is a necessary part of thermodynamics.

1970 Penrose, O., *Foundations of Statistical Mechanics*. See the quotation in Section 1.5 herein.

1970 Wright, P. G., 'Entropy and disorder.' Connections between entropy and intuitive qualitative ideas concerning disorder are explored. Wright warns that viewing entropy as a quantitative measure of disorder represents '... not the received doctrine of physical science, but ... a highly contentious opinion.'

1971 Brush, S. G., 'James Clerk Maxwell and the kinetic theory of gases: A review based on recent historical studies.' Brush emphasizes that if all molecules had the same speed for a given temperature, the separation process envisaged by Maxwell would be impossible. A demon requires a nonuniform speed distribution, such as that bearing Maxwell's name.

1971 Chambadal, P., *Paradoxes of Physics*. The author finds fault with various aspects of both Szilard's model and Brillouin's solution to the Maxwell's demon puzzle, and he calls the attempts to exorcise the demon 'vain and unnecessary.' Some of his provocative views are out of the mainstream, but they are interesting and worth reading.

1971 Rothstein, J., 'Informational Generalization of Entropy in Physics.'

*Informational generalization of entropy provides a language appropriate to the operational viewpoint. In that language many paradoxes simply dissolve, for the ambiguities inherent in previous ways of talking about them are clearly revealed. Included are famous paradoxes of statistical and quantum mechanics. The crucial role of irreversible phenomena in measurement and for the physical foundation of biology is also clearly shown. The historical importance of operational and thermodynamic considerations for both relativity and quantum mechanics suggests that informationally generalized thermodynamics may play a key role in constructing a satisfactory relativistic quantum theory.*

1971 Silver, R. S., *An Introduction to Thermodynamics*. Maxwell's demon is used to illustrate a contrast with normal engineering thermodynamics, which deals with interaction as a whole (e.g., in boilers) rather than with fluctuations.

1971 Tribus, M. and McIrvine, E. C., 'Energy and information.' Energy and entropy are scrutinized for a wide range of phenomena ranging from the hypothetical Maxwell's demon, to real computational, audio, and pictorial record activities.

*Maxwell's demon became an intellectual thorn in the side of thermodynamicists for almost a century. The challenge to the second law of thermodynamics was this: Is the principle of the increase of entropy in all spontaneous processes invalid where intelligence intervenes?*

1972 Bekenstein, J. D., 'Baryon number, entropy, and black hole thermodynamics.' A thought experiment of J. A. Wheeler envisages that a being drops entropy-bearing matter into a black hole, thereby destroying normal evidence of that entropy, in apparent violation of the second law. Bekenstein refers to the being who drops the matter into the black hole 'Wheeler's demon.' He recalls that Brillouin generalized entropy, relating it to information to show that a Maxwell's demon cannot violate the second law. Then he shows that entropy can be generalized in a manner appropriate for black holes to show that Wheeler's demon also cannot transcend the second law.

1972 Jauch, J. M. and Báron, J. G., 'Entropy, information and Szilard's paradox.' Reprinted here as Article 3.5.

1972 Zernike, J., *Entropy: The Devil on the Pillion.*

*... we might give (the demon) a torch, but then we are no longer dealing with a system at uniform temperature; and of course work can then be done ... Could the demon not set up signposts around his trapdoor, which warn him of the oncoming molecule? The whole thing might be automated; if the molecule passes two posts set in line with the door within a certain time, the door opens at the exact time of arrival. That indeed is possible in principle, but the transmission of a signal requires energy, more than a fast moving molecule could provide. We can surely think of as fast a molecule as we please with a correspondingly high amount of energy. But the signal has to be more rapid still, and by the same token would require more energy. The minute analysis of this case would lead us too far afield into the information theory ... So we have to restrict ourselves to writing down the conclusion: Maxwell's demon must remain incommunicado. ... he cannot operate, because the necessary information cannot reach him.*

1973 Bennett, C. H., 'Logical reversibility of computation.' A path-breaking paper, showing that a computing automaton can be made logically reversible at every step, allowing an in-principle thermodynamically reversible computer that: (1) saves all intermediate results, avoiding irreversible erasure; (2) prints out the desired output; and (3) reversibly disposes of all undesired intermediate results by retracing the steps of (1) in reverse order, restoring the machine to its original condition. Reprinted in MD1.

1973 Jammer, M., 'Entropy.' A comprehensive overview of entropy, with a section covering Maxwell's demon and other restrictions of the entropy concept.

1973 Jauch, J. M., *Are Quanta Real?*. Written in the style of a Galilean dialogue, this book introduces Maxwell's demon in a dream involving a roulette table.

*The demons cannot actually function because they are subject to the same fluctuations as the atoms which they are to control. The fact that they can function in the dream means that they are creatures from another level of reality than the dreamer, and thus carry the message of the existence of deeper levels of consciousness ...*

1973 Kelly, D. C., *Thermodynamics and Statistical Physics*. The demon is discussed prior to Kelly's introduction of the entropy concept.

*The demon sets out to pump heat from the cold reservoir to the hot reservoir. ... Unhappily, there is one minor problem—it will not work! Why not? Because we forgot the demon. We did not include the demon as part of the system. The demon cannot be regarded as part of the surroundings since he must communicate with the molecules in order to measure their speeds. The measurements made by the demon (or any instrumentalized version of the demon) require that he exchange energy with the molecules. Some of the energy being pumped uphill must be fed to the demon—he does not work free. Once the system is enlarged to include the demon one discovers that the heat pump has sprung a leak—a large leak. The demon is so hungry that he eats himself out of a job. His appetite completely wipes out the potential gain of his molecule-sorting action.*

1973 Lindblad, G., 'Entropy, information, and quantum measurements.' A formal mathematical treatment of the quantum measurement process, yielding a mathematical inequality that enables a comparison of entropy changes in an observed system and its measuring apparatus.

1974 Costa de Beauregard, O. and Tribus, M., 'Information theory and thermodynamics.' Reprinted here as Article 3.6.

1974 Davies, P. C. W., *The Physics of Time Asymmetry*.

*Everyday experience indicates that information only increases with time. Our own memories grow as we do, public libraries accumulate books, the moon accumulates craters from meteoric impacts. There is no incompatibility between the simultaneous growth of both entropy and information ... The law of entropy increase refers to closed systems, the law of information increase to open systems.*

1974 Gabor, D., 'Foreword.'

*An interesting new approach, but one which must be used with great caution is via the connection between thermodynamics and information theory, pioneered in 1928 by Leo Szilard. Were it not for the Second Principle, by asking questions and answering them by experiments, we could reduce the object the laboratory and ourselves to any improbable state. But in a time-reversed world, there would not be much point in asking questions; the answer would always precede the question! ... I believe that the Second Principle (like the Third) is meaningless in classical physics and received a physical meaning only by quantum mechanics, because with classical electromagnetic theory one can construct Maxwell demons and perpetuum mobili of the second kind. It is a hunch only, but I am in good company. Max Born believed in it, and Einstein made a remark to this effect to my friend Leo Szilard, around 1925.*

1974 Gasser, R. P. H. and Richards, W. G., *Entropy and Energy Levels.* Maxwell's demon is described, with quantitative connections for a crystal made up of independent two-state molecules.

1974 Grünbaum, A., 'Is the coarse-grained entropy of classical statistical mechanics an anthropomorphism?' The author argues that entropy depends on a human choice of cell size in phase space. He finds an invariance under cell size changes in the entropy behavior of a majority of the systems in an ensemble, even though individual systems generally display no such property. His answer to the title question is negative.

1974 Jammer, M., *The Philosophy of Quantum Mechanics.* In a chapter on the theory of measurement, Jammer refers to von Neumann's views, which were influenced by Smoluchowski and Szilard.

1974 Kivel, B., 'A relation between the second law of thermodynamics and quantum mechanics.' This short discussion attributes the demon's failure to beat the second law to energy quantization.

1974 Laing, R., 'Maxwell's demon and computation.' Maxwell's demon is treated as a computing automaton that can carry out arbitrary Turing computations. Reprinted in MD1.

1974 Lindblad, G., 'Measurements and information for thermodynamic quantities.' Lindblad shows that the reduction in entropy that an observer can obtain in a system described by a fluctuating thermodynamic parameter is less than the information possessed by the observer.

1974 Pekelis, V., *Cybernetics A to Z.* A lively, popularized account, where Maxwell's demon is used to impart an understanding of entropy. Relevant history from Clausius through Brillouin is covered qualitatively.

1974 Popper, K., *The Philosophy of Karl Popper.* Popper criticizes Szilard's 1929 paper, which establishes a link between entropy and information. He labels as 'spurious' Szilard's suggestion that knowledge and entropy are related. See the contrasting discussion in Section 1.5 herein.

*What I am* **not** *ready to accept is Szilard's more general argument by which he tries to establish the theorem that knowledge, or information, about the position of M (the molecule) can be converted into negentropy, and vice versa. This alleged theorem I regard, I am afraid, as sheer subjectivist nonsense. ... I assert, we do not need any knowledge regarding the location of M: all we need is to slide our piston into the cylinder. If M happens to be on the left, the piston will be driven to the right, and we can lift the weight. And if M is on the right, the piston will be driven to the left, and we can also lift a weight: nothing is easier than to fit the apparatus with some gear so that it lifts a weight in either case, without our having to know which of the two possible directions the impending movement will take. Thus no knowledge is needed here for the balancing of the entropy increase; and Szilard's analysis turns* *out to be a mistake: he has offered no valid argument whatever for the intrusion of knowledge into physics.*

1974 Rothstein, J., 'Loschmidt's and Zermelo's paradoxes do not exist.'

*Where reversal or recurrence are operationally realizable, no contradiction with the irreversible nature of macroscopic operations occurs. Paradox results either from neglecting irreversible phenomena in the means for preparing a reversed state, or from confusing elements or ensembles, which are meaningful in microstate language but meaningless operationally, with preparable macrostates, whose representation in microstate language is an ensemble whose very definition is incompatible with that of any paradox-generating element or ensemble.*

1974 Skagarstam, B., 'On the mathematical definition of entropy.'

1974 Sussman, M. V., 'Seeing entropy—the incompleat thermodynamics of the Maxwell demon bottle.' The Maxwell demon bottle is a demonstration device consisting of a long-necked, sealed flask with 5 black and 5 white spheres. The spheres can be mixed and unmixed, simulating the action of a Maxwell's demon. The author shows how the bottle can be used to 'see' entropy, and supports his discussion with calculations.

1975 Curzon, F. L. and Ahlborn, B., 'Efficiency of a Carnot engine at maximum power output.'

1975 Lerner, A. Y., *Fundamentals of Cybernetics.*

*Another interesting feature is that in disproving the possibility of a Maxwell demon, we are led to establishing a direct physical relationship between information and energy. It proves possible to calculate the minimum quantity of negentropy necessary for obtaining each unit of information.*

1975 Rosnay, J. de, *The Macroscope: A New World Scientific System.* Translated by R. Edwards. The author proposes a total systemic approach to scientific knowledge and its applications to human life and society. His treatment of information theory provides a unique sociological perspective.

1975 Skagarstam, B., 'On the notions of entropy and information.' This review focuses on the history of the information-theoretic and thermodynamic entropy concepts.

1975 Trincher, K. S., 'Information and Biological Thermodynamics.' Maxwell's demon is addressed, but the article contains some questionable statements.

1975 Zemansky, M. W., Abbott, M. M. and van Ness, H. C., *Basic Engineering Thermodynamics, 2nd Edition.* Maxwell's demon is introduced and discussed.

1976 Bhandari, R., 'Entropy, information and Maxwell's demon after quantum mechanics.' The subjective nature of entropy and its relation to information and irreversibility is examined in light of the quantum measurement problem. Bhandari asserts that wave function collapse during a measurement and concomitant entropy increase of the universe is seen by observers who are only able to observe a restricted manifold of states determined by their level of perception.

1976a Brush, S. G., 'Irreversibility and indeterminism: Fourier to Heisenberg.' Brush writes, 'Maxwell's immortal Demon proved that Victorian whimsy could relieve some of the gloom of the Germanic Heat Death.' He stresses that the demon 'gives us a new model for the fundamental irreversible process: he translates heat flow into molecular mixing.'

1976b Brush, S. G., *The Kind of Motion We Call Heat*. This detailed historical study of 'heat' contains discussions of Maxwell's demon in five different chapters.

*While Maxwell's demon is generally cited in connection with the possibility of violating the second law of thermodynamics, it seems equally important to note that by making the mixing of different molecules the fundamental irreversible process, Maxwell has really strengthened the concept of irreversibility, especially for those who seek molecular explanations for all phenomena.*

1976 Harney, R. C., 'Human perception and the uncertainty.' Order of magnitude calculations are given to illustrate the inadequacy of humans to directly observe the Heisenberg uncertainty principle. Even optically augmented vision is inadequate by three orders of magnitude. The intent is to help reconcile the uncertainty principle with students' own experiences. Maxwell's demon is not addressed, but an extension of the arguments shown for humans to a demon's scale is inviting.

1976 Sherwood, M., *Maxwell's Demon*. The following quotation shows the role of Maxwell's demon in this novel.

*That would be a fine turn-out, a touch of cosmic irony... That Maxwell's demon should be out to cause chaos. Sorry, didn't you ever learn any thermodynamics? I'll explain. About a hundred years ago, ... James Clerk Maxwell put forward an idea for getting around the second law of thermodynamics—you know, the one which Flanders and Swann explained as 'you can't pass heat from a colder to a hotter.' Maxwell postulated a two-part container... In the middle of the container was a trapdoor joining the two sides, and an imp, which opened the trapdoor whenever he saw a hot molecule moving from the cold side towards the hot, to let it through. In this way, the imp, which became known as Maxwell's demon, created order out of chaos... Until Leon Brillouin proved in the 1950s that even then you did not avoid the consequences of the second law of thermodynamics. Anyhow, that is why I liked your reverse Maxwell's demon—the one that creates chaos out of order...*

1976 Yu, F. T. S., *Optics and Information Theory*. An extensive treatment of Maxwell's demon, referring heavily to Brillouin's *Science and Information Theory* and Gabor's 'Light and Information.'

1977 Carnap, R., *Two Essays on Entropy*. See the quotation in Section 1.4.

1978 Brush, S. G., *The Temperature of History. Phases of Science and Culture in the Nineteenth Century*. A brief discussion of the demon, with reference to a suggestion by Stewart and Tait that 'the Maxwell demon might operate in an 'unseen universe,' linked by human thought to this one, so that we could look forward to a future state untroubled by dissipation of energy.'

1978 Wehrl, A., 'General properties of entropy.' An advanced mathematical review article.

*It is amusing to note that in practical applications of information theory ... the second law of thermodynamics has been adopted and there called the negentropy principle ... Thus we find a mutual interaction between physics and information theory rather than a perfect understanding of statistical mechanics on the grounds of information theory.*

1979 Jaynes, E. T., 'Where Do We Stand on Maximum Entropy?' The principle of maximum entropy is examined from historical, present, and potential future viewpoints.

1979 Penrose, O., 'Foundations of statistical mechanics.' An extensive review article that briefly covers various topics germane to Maxwell's demon. Discussing the nonuniqueness of entropy in nonequilibrium systems, he writes, 'Even in thermodynamics, where entropy is defined only for equilibrium states, the definition of entropy can depend on what problem we are interested in and on what experimental techniques are available ...'

1979 Poplavskii, R. P., 'Maxwell demon and the correspondence between information and entropy.' An entropic efficiency is defined, and is used to examine various order-producing control processes, including a traditional Maxwell's demon.

1979 Rothstein, J., 'Generalized Entropy, Boundary Conditions, and Biology.' Generalized entropy in the framework of a generalized thermodynamics is proposed to handle a wide variety of complex 'biotonic' laws.

1979 Yu, F. T. S., 'Super resolution and cost of entropy.'

1980 Bekenstein, J. D., 'Black-hole thermodynamics.' This is an update on ideas developed by the author in his Ph. D. dissertation (see Bekenstein, 1972).

1980 Chen, C. H., Hurst, G. S. and Payne, M. G., 'Direct counting of Xe atoms.' The authors refer to their resonance ionization spectroscopy apparatus as 'the modern analogue of Maxwell's demon.'

1980 Hurst, G. S., Payne, M. G., Kramer, S. D. and Chen, C. H., 'Counting the atoms.'

*Practically every element in the periodic table can be detected, down to single-atom sensitivity, by resonance-ionization-spectroscopy methods involving commercially available lasers.*

1981 Denbigh, K., 'How subjective is entropy?' Denbigh rejects the view that entropy is subjective. He argues that Brillouin's exorcism of Maxwell's demon can be accomplished without appeals to information theory or negentropy. Reprinted in MD1.

1981 Denur, J., 'The Doppler demon.' See also Motz (1983) and Chardin (1984).

*Equilibrium blackbody radiation, like all radiation, is, in general, Doppler shifted if it is emitted and/or reflected from a moving body. The relevance of this fact to the analysis of Maxwell's demon, which has been previously neglected, is revealed in this paper. In particular, we find that, by appropriately taking advantage of this fact, Maxwell's demon is, at least in principle, capable of operating, albeit very weakly, in violation of the second law of thermodynamics.*

1981 El'yashevich, M. A. and Prot'ko, T. S., 'Maxwell's contribution to the development of molecular physics and statistical methods.' The authors examine Maxwell's contributions to molecular physics during 1859–1879, including his ideas on the limitations of the second law: 'Maxwell's demon was essentially the first important illustration of the difference between microscopic and macroscopic processes.'

1981 Gordon, L. G. M., 'Brownian movement and microscopic irreversibility.' A microengine model that appears to violate the second law of thermodynamics is proposed and discussed.

1981 Kosloff, R., 'Thermodynamic aspects of the quantum-mechanical measuring process.' An advanced treatment of quantum measurement theory that uses Maxwell's demon and the work of Szilard and Brillouin as a starting point.

1981a Rossis, G., 'La logique des expériences de pensée et l'expérience de Szilard.' This article supports criticisms of Popper and Feyerabend regarding Szilard's one-molecule heat engine. The author argues that the model 'does not demonstrate the necessity of the identity between negentropy and information.'

| | |
|---|---|
| 1981b | Rossis, G., 'Sur la discussion par Brillouin de l'expérience de pensée de Maxwell.' Rossis argues that Brillouin's discussion of Maxwell's demon 'is limited to a mere examination of certain types of possible demons, which is methodologically incorrect.' He contends that 'Brillouin's discussion does not demonstrate in the least the necessity of the validity of the identity between entropy and information.' |
| 1982 | Benioff, P., 'Quantum mechanical models of Turing machines that dissipate no energy.' This paper demonstrates non-dissipative Hamiltonian models of Turing machines, constructed on a finite lattice of spin-1/2 systems. |
| 1982 | Bennett, C. H., 'The thermodynamics of computation—a review.' Reprinted here as Article 7.1 |
| 1982 | Campbell, J., *Grammatical Man: Information, Entropy, Language, and Life*. An excellent book for a general audience, with a chapter on 'The Demon Deposed.' The following quotation by Claude Shannon in 1979 is given: *I think the connection between information theory and thermodynamics will hold up in the long run, but it has not been fully explored and understood. There is more there than we know at present. Scientists have been investigating the atom for about a hundred years and they are continually finding more and more depth, more and more understanding. It may be that the same will be true of the relationship we are speaking about.* |
| 1982 | Feynman, R. P., 'Simulating physics with computers.' *I want to talk about the possibility that there is to be an exact simulation, that the computer will do exactly the same as nature. If this is to be proved ... then it's going to be necessary that everything that happens in a finite volume of space and time would have to be exactly analyzable with a finite number of logical operations. The present theory of physics is not that way, apparently. It allows space to go down into infinitesimal distances, wavelengths to get infinitely great, terms to be summed in infinite order, and so forth; and therefore, if this proposition is right, physical law is wrong.* |
| 1982 | Harman, P. M., *Energy, Force, and Matter: The Conceptual Development of Nineteenth-Century Physics*. An excellent discussion of the demon in the context of 19th century thermodynamics. |
| 1982 | Landauer, R., 'Uncertainty principle and minimal energy dissipation in the computer.' It is argued that Heisenberg's energy-time uncertainty principle, $\Delta E \Delta t \approx \hbar$, does not yield any information about dissipation in a computer system with switching time $\Delta t$. |
| 1982 | Landsberg, P. T., *The Enigma of Time*. Landsberg considers a pressure demon, and observes that Maxwell, who predated quantum theory, |

assumed implicitly that the energy required for any observation can be made negligibly small, but quantum physics does not allow this.

1982

Likharev, K. K., 'Classical and quantum limitations on energy consumption in computation.' Using a model (parametric quantron) of a real physical device based on the Josephson effect in superconductors, Likharev argues that logical operations can be physically reversible. This kind of device can be an elementary cell in a logically reversible computer. Energy dissipation per logical operation in the classical and quantum limits, respectively, is found to be $< kT$ and $< \hbar/\tau$ where $\tau$ is the process period.

1982

Pagels, H., *Cosmic Code*. This layman's book contains a readable account of statistical mechanics, including the statistical nature of the second law.

1982

Popper, K. R., *Quantum Theory and the Schism in Physics.*

*There is a further very real difficulty which has contributed to make the subjectivist interpretation of entropy more acceptable. I have in mind the problem of Maxwell's demon: since 1905, the success of the exorcism of the demon has become dubious, and this fact has led to attempts to solve the problem by invoking subjective probabilities, and other measures of subjective knowledge (or 'information'). No doubt some interesting analogies have been unearthed, but the soundness of the edifice, and especially the part played by Maxwell's demon, seems to me highly questionable.*

1982

Schweber, S., 'Demons, angels, and probability: Some Aspects of British Science in the Nineteenth Century.' Maxwell's demon is compared with a species-sorting 'being' proposed in 1844 by Charles Darwin, and genealogies of both beings are conjectured. Influences of the nineteenth century philosophers Quetelet and Buckle on Darwin, Maxwell, and Boltzmann are examined in detail.

1982

Silverman, M. P., 'The vortex tube: A violation of the second law?' The Hilsch vortex tube is considered and compared with Maxwell's demon. Analysis reveals that just as a Maxwell's demon cannot overthrow the second law, the vortex tube also fails to do so.

1982

Weinberg, A. M., 'On the relation between information and energy systems: A family of Maxwell's demons.' Weinberg uses Maxwell's demon to make some interesting connections between time, energy, and information. Reprinted in MD1.

1983

Davies, P. C. W., *God and the New Physics.* Davies asks: If a supreme being existed, constrained to act within the laws of physics, could it prevent the end of the universe? He concludes the answer is negative.

1983

Eigen, M. and Winkler, R., *Laws of the Game: How the Principles of Nature Govern Chance*. The authors examine an imagined cessation of irreversibility in nature, and discuss several thought-provoking hypothetical demons.

1983 Gordon, L. G. M., 'Maxwell's demon and detailed balancing.' Gordon analyzes an isothermal gas of molecules separated by a membrane, each pore of which is controlled by an independent molecular trap door that can exist in either of two states. Under certain conditions, this membrane acts like a Maxwell's demon, effecting a density and pressure gradient across the membrane, lowering the system's entropy—apparently violating the second law.

1983 Lindblad, G., *Non-equilibrium Entropy and Irreversibility.* The author gives a formal mathematical treatment of information and entropy, and applies it to the Maxwell's demon problem.

1983 Motz, H., 'The Doppler demon exorcised.' The author argues against the conclusion of Denur (1981) that a Doppler demon can defeat the second law. See also Chardin (1984).

1984 Atkins, P. W., *The Second Law.* Atkins discusses 'Boltzmann's demon,' who constantly rearranges microscopic states without changing macroscopically observed phenomena. Though apparently inspired by the Maxwell's demon, it does not violate or threaten any laws, but its random actions trap it in the future by assuring that return to an initial state is overwhelmingly improbable.

1984 Benioff, P., 'Comment on 'dissipation in computation.' Benioff responds to Porod et al. (1984).

*... the construction of the quantum mechanical models ... is an idealized construction only. It is an open question whether such models can also be physically constructed in the laboratory. However, the idealized existence of such models demonstrates that one cannot use purely quantum mechanical arguments ... to deny the existence of models which dissipate no energy.*

1984 Bennett, C. H., 'Thermodynamically reversible computation.' Bennett responds to Porod et al. (1984).

*Although it is easy to imagine dissipative ways of reading or logically transforming information, there also exist, in general, thermodynamically reversible ways, whenever the intended logical transformation conserves phase space. In such a transformation, although the individual parts of the computer (e.g., the reading head) may indeed evolve in a many-to-one or one-to-many manner, the state of the whole computer evolves in a one-to-one manner. A simple example of reversible information transfer is the use of a classical mass spectrometer to transcribe mass information into positional information. Even in quantum mechanics, where some irreversibility may be needed to fix the result of a measurement by destroying the phase coherence between measured and measuring systems, a single irreversible step is enough to fix the result of a long chain of reversible premeasurements.*

1984 Chardin, G., 'No free lunch for the Doppler demon.' The author argues against the conclusion of Denur (1981) that a Doppler demon can defeat the second law. See also Motz (1983).

1984 Hurst, G. S., *Resonance Ionization Spectroscopy and its Applications, Conference Series 71.*

1984a Landauer, R., 'Dissipation in computation.' Landauer responds to Porod et al. (1984).

*(Porod* et al*) is confined to general arguments and does not explain, in detail, where the specific reversible computer proposals, e.g., that of Likharev, go astray. …*

1984b Landauer, R., 'Fundamental physical limitations of the computational process.'

*This study of ultimate computer limits is in its infancy; it is easier to ask questions than answer them. To the extent that we have answers, the limits are terribly far away from reality. Therefore, we cannot claim to be guiding the technologist. Then why are the limits interesting? It is a matter of basic science; we are starting a physical theory of mathematics. Most mathematicians have little sympathy for the assertion that their subject requires a basis in physics. Information, however, whether it is in biological systems, in a digital computer, or handled by pencil and paper, inevitably has a physical form. Information is represented by a mark on a piece of paper, a hole in a punched card, a magnetic material magnetized one way or the other, a charge that is present or absent, a Josephson junction that is superconducting or not, etc. Information can come on a much smaller scale, e.g., in the molecular configuration of DNA or, conceivably, even through the spin of a single electron which can be pointed up or down. It is impossible to get away from this sort of physical embodiment. As a result, manipulation of information is inevitably subject to the laws of physics.*

1984a Porod, W., Grondin, R. O., Ferry, D. K. and Porod, G., 'Dissipation in computation.' The authors object to the view that computation with dissipation $< kT \ln 2$ per bit operation is possible. This drew strong responses from Benioff, Bennett, Landauer, and Toffoli, and a subsequent rejoinder from Porod *et al.*

*The question of the energy dissipation in the computational process is considered. Contrary to previous studies, dissipation is found to be an integral part of computation. A complementarity is suggested between systems that are describable in thermodynamic terms and systems that can be used for computation.*

1984b Porod, W., Grondin, R. O., Ferry, D. K. and Porod, G., 'Porod *et al.* respond.' This group defends its original position steadfastly, stating:

(1) physical and logical states of a system must be distinguished, and Landauer's 1961 paper fails to do so; (2) purely mechanical models of computers are unrealistic because noise is unavoidable for $T > 0$; (3) the zero speed limit is useless, for 'both dissipation and computation disappear'; (4) information erasure is a dissipative interaction with a heat bath, (rather than a noninvertible process); and (5) if reading can be dissipationless, then erasure can be too.

*Landauer is correct that merely coupling two systems together, especially in a closed environment, does not induce dissipation.* **It also does not constitute a measurement.** *Indeed, information initially in one of the systems will oscillate back and forth between the two systems. Measurement of the information must arise from opening the closed environment at some particular time (and thus subjecting it to the background heat bath). Erasure of information should be viewed not as a result of noninvertibility, but as a result of interactions with the heat bath.*

1984 — Prigogine, I. and Stengers, I., *Order out of Chaos.* This well-known book describes recent progress in nonequilibrium thermodynamics.

*...in the subjective interpretation of irreversibility (further reinforced by the ambiguous analogy with information theory), the observer is responsible for the temporal asymmetry characterizing the system's development. Since the observer cannot in a single glance determine the positions and velocities of all the particles composing a complex system, he cannot know the instantaneous state that simultaneously contains its past and its future, nor can he grasp the reversible law that would allow him to predict its developments from one moment to the next. Neither can he manipulate the system like the demon invented by Maxwell, who can separate fast- and slow-moving particles and impose on a system an antithermodynamic evolution toward an increasingly less uniform temperature distribution.*

1984 — Robinson, A. L., 'Computing without dissipating energy.' A status report on the minimum energy needed for computation.

1984 — Smorodinsky, Y. A., *Temperature.* In this semipopular book, the author devotes a section to 'Maxwell's Imp.' Although there are some interesting ideas here, the writing (or translation) is difficult to follow.

1984 — Toffoli, T., 'Comment on 'Dissipation in computation.' Toffoli responds to Porod et al. (1984).

*...the dissipation in a computer is bounded by the sum of* **two terms**. *Both are essentially proportional to $kT$; the first, through a coefficient $m \ln 2$, where $m$ is no greater than the number of input/output lines; the second, through a coefficient $nd$, where $n$ is no greater than the number of gates (each counted as many times as it is used ...), and $d$ is a 'technological factor' that, as far as one knows, can be made*

*as small as desired. Thus, the highest bound to the dissipation* **per elementary operation** *that is definitely known today is of the form* $[(m/n)\ln 2 + d]kT$. *For any* $T$, *this bound vanishes for computers that are designed to run long computations without displaying intermediate results (i.e.,* $n \gg m$*), and in the limit of perfect technology* ($d \to 0$). *This is what* **virtually nondissipative computation** *means.*

1984a Zurek, W. H., 'Maxwell's demon, Szilard's engine and quantum measurements.' Reprinted here as Article 5.1.

1984b Zurek, W. H., 'Reversibility and stability of information processing systems.'

*Instabilities in the evolution of the classical 'billiard ball computer' are analyzed and shown to result in a one-bit increase of entropy per step of computation. 'Quantum spin computers,' on the other hand, are not only microscopically, but also operationally reversible. Readoff of the output of quantum computation is shown not to interfere with this reversibility. Dissipation, while avoidable in principle, can be used in practice along with redundancy to prevent errors.*

1985 Bennett, C. H. and Landauer, R., 'The fundamental physical limits of computation.' A very readable article, intended for a sophisticated lay audience, on limiting factors in computation.

*In physical systems without friction, information can never be destroyed; whenever information is destroyed, some amount of energy must be dissipated ... As an example, imagine two easily distinguishable physical situations, such as a rubber ball held either one meter or two meters off the ground. If the ball is dropped, it will bounce. If there is no friction and the ball is perfectly elastic, an observer will always be able to tell what state the ball started out in (that is, what its initial height was) because a ball dropped from two meters will bounce higher than a ball dropped from one meter. If there is friction, however, the ball will dissipate a small amount of energy with each bounce, until it eventually stops bouncing and comes to rest on the ground. It will then be impossible to determine what the ball's initial state was; a ball dropped from two meters will be identical with a ball dropped from one meter. Information will have been lost as a result of energy dissipation.*

1985 Denbigh, K. G. and Denbigh, J. S., *Entropy in Relation to Incomplete Knowledge*. This critical analysis of the notion that entropy is subjective is an outgrowth of Denbigh's 1981 paper.

*Although information theory is more comprehensive than is statistical mechanics, this very comprehensiveness gives rise to objectionable consequences when it is applied in physics and chemistry. Since the subjective interpretation of probability, on which information theory has been based, is more general than is the objective interpretation, the language in which the theory is presented necessarily takes on a subjective character. A probability distribution is thus said to represent our state of knowledge, whereas in physico-chemical contexts it is usually justifiable to go much further in an objective direction and to regard a probability distribution as being characteristic of the nature of the system in question and of the physical constraints upon it. It remains true, nevertheless, that information theory can be of value in an heuristic sense. This was seen to be so in the Maxwell Demon problem ...*

1985 Hastings, H. M. and Waner, S., 'Low dissipation computing in biological systems.'

1985 Hurst, G. S., Payne, M. G., Kramer, S. D. and Chen, C. H., 'Method for counting noble gas atoms with isotopic selectivity.'

*A method has been developed for direct counting of noble gas atoms and has been demonstrated for selected isotopes of krypton. In principle, a few atoms of the noble gases argon, krypton, xenon and radon can now be counted with isotopic selectivity whether stable or radioactive. A concept was originated in which a laser method would be used to count noble gas atoms of a particular isotope that are moving freely in an enclosure. As the concept developed, a parallel with Maxwell's sorting demon became quite obvious since the plan was to sort out only atoms of a given type (Z selection), e.g. krypton atoms, from any other atom in the enclosure and then to sort the atom by isotope (A selection) before removing the atoms from the gas compartment. The plan was to count each atom as it was stored in a target until all atoms were counted.*

1985a Jaynes, E. T., 'Entropy and search theory.' The relation between entropy maximization and optimal strategy is explored for a model entailing a search for a hidden target.

1985b Jaynes, E. T., 'Generalized Scattering.' Inferential scattering is a process whereby inference is modified by added constraints based upon new knowledge. Remaximizing entropy using the new constraints, rather than using physical dynamics deductively, is examined, with emphasis on understanding why the MAXENT formalism works.

1985c Jaynes, E. T., 'Macroscopic Prediction.' Given information about a set A of macroscopic quantities in space-time region 1, how one can best predict the values of a set B of macroscopic quantities in space-time region 2? Jaynes addresses this, focusing on inference based on information, rather than on logical deduction via physical law.

1985 Poundstone, W., *The Recursive Universe.* This very readable book, designed for a general audience, contains a chapter on Maxwell's demon summarizing much of the work from Smoluchowski to Brillouin.

1985 Rosen, R., *Theoretical Biology and Complexity.*

*... historically, the relation between theoretical physics and biology has never been close. None of the great names of physical science, from*

*Newton to the present, have known or cared much about the properties of organisms, and therefore organic phenomena played no essential part in their science (leaving aside such diversions as Maxwell's demon, Schrödinger's informal essays, and the like).*

1985 Waldram, J. R., *The Theory of Thermodynamics.*

*... it is true in principle that we can reduce the entropy of an isolated system ... by taking measurements of the system (though the number of measurements needed to produce even a tiny reduction in entropy is formidable). But we do not reduce the entropy of the whole world in this process, for there is a compensating increase of entropy associated with the process of observation. There is therefore, alas, no prospect of providing for the energy needs of mankind by suitably marshalling a small horde of Maxwell demons.*

1985 Watanabe, S., 'Wiener on cybernetics, information theory, and entropy'. This contains commentary on six papers by Norbert Wiener on the title subjects, two of which (Wiener, 1950, 1952) are listed separately here.

*Maxwell's demon is of course closely related to the possible decrease of thermodynamical entropy. Indeed, Wiener's two major books ... start with a chapter on the Bergsonian concept of time. Wiener was fascinated by Bergson's idea that 'a living organism climbs the slope where the inert objects go down,' meaning that the former somehow behaves as if its entropy were decreasing. Wiener probably wanted to suggest that the cybernetical system somehow could reproduce such an entropy-decreasing phenomenon without violating the second law of thermodynamics.*

1986 Alekseev, G. N., *Energy and Entropy.* An interesting qualitative discourse concerned with the statistical nature of the second law, but with the nonstandard view that a Maxwell's demon that decreases a gas's entropy need not violate the second law.

1986 Barrow, J. D. and Tipler, F. J., *The Anthropic Cosmological Principle.* This contains a well written discussion of Maxwell's demon, emphasizing that the demon is the size of a molecule and sees things differently from macroscopic observers. The authors allude to existing controversies about Maxwell's demon, and side with the traditional view, based largely on Brillouin's work.

1986 Davies, P. C. W., *The Ghost in the Atom.* This excellent book deals with the mysteries of quantum physics. Although there is no discussion of Maxwell's demon *per se*, there is a discussion of a single electron in a box into which an impenetrable screen is inserted. See the related discussion herein, in Section 1.6.1.

*It is though, prior to the observation, there are two nebulous electron 'ghosts' each inhabiting one chamber waiting for an observation to turn one of them into a 'real' electron, and simultaneously to cause the other to vanish completely.*

1986 Drogin, E. M., 'Maxwell's demon lives.'

*According to Pierce you can feed a paper tape into a Turing machine and have it churn away madly for days and days, with no expenditure of energy required (provided that you don't have to know the initial state of the machine or reset it to get it going). But if a hammer comes out of the machine and hits you over the head to tell you the computation is complete, or if you use the teeniest, tiniest lit candle ... to read the scratches on the tape, you have broken the rules of the perpetual-motion machine game. Sorry about that.*

1986 Feynman, R. P., 'Quantum mechanical computers.' Feynman examines idealized quantum mechanical computers, and gives support to the concept that logical operations can in principle be physically reversible.

*... we are going to be even more ridiculous ... and consider bits written on one atom instead of the present* $10^{11}$ *atoms. Such nonsense is very entertaining to professors like me. I hope you will find it interesting and entertaining also. ... it seems that the laws of physics present no barrier to reducing the size of computers until bits are the size of atoms, and quantum behavior holds dominant sway.*

1986 Fine, A., *The Shaky Game: Einstein, Realism, and the Quantum Theory.* Fine describes Einstein's thought experiment discussed herein in Section 1.6.1.

1986a Landauer, R., 'Computation and physics: Wheeler's meaning circuit?'

*Computation is a physical process, inevitably utilizing physical degrees of freedom. Computation, therefore, is restricted by the laws of physics and also by the construction materials and operating environments available in our actual universe. These restrictions have been investigated for a quarter century. ... Physical law, in turn, consists of algorithms for information processing. Therefore, the ultimate form of physical laws must be consistent with the restrictions on the physical executability of algorithms, which is in turn dependent on physical law.*

1986b Landauer, R., 'Reversible computation.'

*Logically reversible functions can, of course, be carried out in ways which are very dissipative, e.g. via transistor circuitry. The logical reversibility, however, allows us to invent physical devices in which the process can actually be reversed, i.e., the information bearing degrees of freedom can be pushed back through the device to undo the operation. ... Energy dissipation, per step, and computing speed, are proportional to each other in these systems; if we are willing to compute slowly we have no minimal energy requirement per computer step. These are systems in which we allow friction, but we idealize and assume that the frictional forces, as in hydrodynamics and electricity, are proportional to velocity. We rule out static friction ...*

1986

Porter, T. M., *The Rise of Statistical Thinking: 1820–1900.* An interesting historical study of how statistical ideas became used in the social sciences, biology, and physics. Maxwell's strong influence on the development of kinetic theory and statistical mechanics are well documented:

*To Maxwell belongs the credit for first introducing the explicit consideration of probability distributions into physics. Indeed he was arguably the first to use distribution formulas in a significant and productive way in any science. ... Maxwell first pointed out that his dynamical theory of gases implied the possibility of violating the second law in 1867, in a playful letter to P. G. Tait. There he introduced the 'very observant and neat-fingered being,' later dubbed by William Thomson Maxwell's 'demon,' whose mission was to 'pick a hole' in the second law. ... Maxwell's invention implied that some physical principles, among them the second law, were really as much attributes of human perception as of nature itself.*

1986

Stasheff, C., *Her Majesty's Wizard.* Maxwell's demon is a character in this science fiction book, which includes the poem (p. 192): Long ago and far away, Maxwell felt the need one day For a Demon, scarce as high As the atoms going by. Over heat he gave it sway, Making warmth go either way From the vector Nature gave. Maxwell's Demon, come and save.

1986

Wang, Y. Z., 'Origin of the electric current in Fu's experiment—An analysis of Fu's work on realizing Maxwell's 'Demon.'

*...The current flowing in Fu's device is actually the initial velocity current of a vacuum diode tube. The initial velocity current is, in turn, a changed form of the Thomson thermoelectric effect, which requires both a hot tank and a cold tank, obeying the second law of thermodynamics.*

1987

Bennett, C. H., 'Demons, engines and the second law.' An excellent survey of Maxwell's demon history through Bennett's resolution of the puzzle via memory erasure.

*Another source of confusion is that we do not generally think of information as a liability. We pay to have newspapers delivered, not taken away. Intuitively, the demon's record of past actions seems to be a valuable (or at worst a useless) commodity. But for the demon 'yesterdays newspaper' (the result of a previous measurement) takes up valuable space, and the cost of clearing that space neutralizes the benefit the demon derived from the newspaper when it was fresh. Perhaps the increasing awareness of environmental pollution and the information explosion brought on by computers have made the idea that information can have a negative value seem more natural now than it would have seemed earlier in this century.*

1987 Hawking, S., 'The direction of time.' An expository article on the arrow of time. Hawking explicitly states that the process of writing to a computer memory is dissipative, putting him in conflict with proponents of the idea that in principle, computing can be done reversibly, and that only erasure of information is dissipative.

1987 Johnson, H. A., 'Thermal noise and biological information.' The steady loss of information due to thermal noise in information-handling biological systems is examined. The systems addressed include solutions, cell membranes, and regulatory organs. Pointing out that the renal tubules of the kidney perform a sorting function like that of Maxwell's demon, the author estimates a kidney's energy requirement and efficiency when thermal noise is taken into account.

1987a Landauer, R., 'Computation: A fundamental physical view.' Reprinted here as Article 7.2.

1987b Landauer, R., 'Computation: A fundamental physical view.' Note: Despite the title similarity, this article is distinct from Landauer, 1987a. See the Alphabetical Bibliography for citation details.

1987c Landauer, R., 'Energy requirements in communication.' This short paper uses physical transport of bistable systems as an example in which there is no minimal dissipation per transmitted bit of information, and in which dissipation occurs only when information is discarded.

1987d Landauer, R., 'Fundamental physical limitations of the computational process; an informal commentary.'

1987a Leff, H. S., 'Available work from a finite source and sink: How effective is a Maxwell's demon?' See Section 1.8.4 herein.

1987b Leff, H. S., 'Thermal efficiency at maximum work output: New results for old heat engines.' See Section 1.8.4 herein.

1987 Lubkin, E., 'Keeping the entropy of measurement: Szilard revisited.' Reprinted here as Article 5.2.

1987 Maddox, J., 'Quantum information storage.' A capsule summary of work by Obermayer, Mahler, and Haken (1987).

1987 Obermayer, K., Mahler, G. and Haken, H., 'Multistable quantum systems: Information processing at microscopic levels.' The authors propose building a semiconductor device on a nanometer scale that will allow them to measure directly the energy dissipation involved in certain processes.

1987 Rex, A. F., 'The operation of Maxwell's demon in a low entropy system.' The method of Brillouin is extended to the case where the two chamber temperatures are initially unequal. The demon still cannot effect a net reduction in entropy, no matter what the initial configuration. Reprinted in MD1.

1987 Spielberg, N. and Anderson, B. D., *Seven Ideas that Shook the Universe*. In a chapter on entropy and probability, one of the seven areas that comprise this book, there is a section, 'Entropy and Order: Maxwell's Demon.' Described as famous and fanciful, the demon is dismissed because the 'second law of thermodynamics states that elves and demons exist only in fairy tales, and Maxwell's demon was only a figment of his imagination ... '

1987 Wicken, J. S., 'Entropy and information: Suggestions for a common language.' This paper clarifies the differences between thermodynamic and information theory entropies, and suggests 'semantic house keeping,' whereby information entropy is renamed 'complexity' and 'entropy' is reserved for thermodynamics.

*There is no* **real** *information relevant to thermodynamics beyond that provided by the macroscopic state specification. Entropy is real in thermodynamics. And whereas microstates are probably real in some more-than-accounting sense, the concept of 'microscopic information' concerning their actual* **occupancy** *at any moment is an encumbering abstraction for thermodynamics. With information theory, on the other hand, entropy is the encumbering abstraction. Both abstractions are inimical to productive dialogue between thermodynamics and information theory.*

1988 Barrow, J. D., *The World within the World*. A good general interest book for the scientifically mature. Barrow gives credence both to the traditional (dissipative information acquisition) and modern (dissipative memory erasure) viewpoints, broadening the position in Barrow and Tipler's The Anthropic Cosmological Principle (1986).

1988 Bennett, C. H., 'Notes on the history of reversible computation.' Reprinted here as Article 7.3.

1988 Jaynes, E. T., 'Clearing up mysteries—The original goal.' The Bayesian inference viewpoint is used to understand 3 examples: diffusion, the EPR paradox, and the second law of thermodynamics in biology.

1988a Landauer, R., 'Dissipation and noise immunity in computation, measurement and communication.'

*... the word measurement is ambiguous, and remains so even in some sophisticated discussions of the concept. Some believe that involvement of a human observer is an essential ingredient. I believe that James Watt's invention of the centrifugal governor showed that the human mind did not have to participate. The thermostat in your home makes a measurement, even when you are not there. Other authors, although not arguing for human involvement do invoke the necessity of recording or registry in a macroscopic piece of apparatus, which I interpret as at least more reasonable. Nevertheless, at a time when we can manipulate and observe individual atoms and when we know how, in principle, to*

*use apparatus modelled on the genetic code to do computation, it is not clear that macroscopic apparatus is essential.*

1988b Landauer, R., 'Response.' A response to Porod's 1988 letter.

*Porod is not the only one who has voiced some disagreement with reversible computation. Porod, however is the only one to claim: 'There is, however, no reason to assume that, in general, logical irreversibility implies physical irreversibility.' Porod asserts this without providing examples of logical irreversibility unaccompanied by physical irreversibility. Enough has been published on this subject, including rebuttals to similar earlier arguments by Porod and collaborators, so that a further detailed rebuttal is not needed. A number of investigators ... have expanded our understanding of the energy requirements in computation.... We note that all of the critics invoke their perception of general principles supposedly violated by the discussions of reversible computation. The critics never point to flaws in the specific methods which are invoked to achieve low dissipation.*

1988 Layzer, D., 'Growth of order in the universe.' Layzer imagines a Maxwell's demon substitute, programmed to open and close the trap door according to the results of a calculation that predicts the positions and velocities of all molecules for times after an initial moment. Sometimes called 'Loschmidt's non-demon' (see Rothstein, 1959), this shows that the second law can fail if sufficient microscopic information about the initial state is present.

1988 Lindgren, K., 'Microscopic and macroscopic entropy.' Seeking a microscopic counterpart to entropy, Lindgren examines a 'measure entropy' for physical systems in which microstates can be represented as symbol sequences. In the thermodynamic limit, the ensemble average of the measure entropy equals the thermodynamic entropy if correlations decrease sufficiently fast with distance.

1988 Lloyd, S. and Pagels, H., 'Complexity as thermodynamic depth.'

*A measure of complexity for the macroscopic states of physical systems is defined. Called depth, the measure is universal: it applies to all physical systems. The form of the measure is uniquely fixed by the requirement that it be a continuous, additive function of the processes that can result in a state. Applied to a Hamiltonian system, the measure is equal to the difference between the system's coarse- and fine-grained entropy, a quantity that we call thermodynamic depth. The measure satisfies the intuitive requirements that wholly ordered and wholly random systems are not thermodynamically deep and that a complex object together with a copy is not much deeper than the object alone. Applied to systems capable of computation, the measure yields a conventional computational measure of complexity as a special case. The relation of depth and thermodynamic depth to previously proposed definitions of complexity is discussed and applications to physical, chemical, and mathematical problems are proposed.*

1988 Peres, A., 'Schrödinger's immortal cat.' A review of the quantum measurement problem with a brief reference to Szilard's 1929 paper and a section on the physical limitations of computing.

1988 Porod, W., 'Comment on 'Energy requirements in communication.' A critical letter, reacting to Landauer's 'Energy requirements in communication.' (1987c).

*...the flaw in Landauer's argument is the failure to distinguish between physical reversibility and logical reversibility. While physical irreversibility implies dissipation of energy, logical irreversibility (contrary to Landauer) does not, by itself, imply dissipation of energy. By the same token, logical reversibility does not imply physical reversibility. (This latter argument voids the basic premise on which all reversible computing schemes are based.) We emphasize that logical reversibility (or irreversibility) is irrelevant for the energy dissipation encountered in a process, be it computation or communication.*

1988a Rothstein, J., 'Entropy and the evolution of complexity and individuality.' A broad-based discussion of the concept of generalized entropy and its relevance to quantum theory, biology, and thermodynamics.

1988b Rothstein, J., 'On ultimate thermodynamic limitations in communication and computation.'

*Our attitude to reversible computing is thus much like that to Maxwell's demon—it is amusing, instructive, challenging and conceptually useful in helping to mark the boundary between what we would like to achieve and what can actually be achieved. We stress its value here to avoid misconstruing the intent this paper. Though it obviously denies that reversible computers hold any hope of being 'practical' (in addition to all the foregoing they are too slow), it affirms the value and necessity of scrutinizing scientific foundations carefully. Thought experiments are not idle amusements. Good ones confront accepted concepts and techniques with paradoxes and new challenges, clarifying new advances and pointing out the limits of the old. Reversible computers are valuable additions to the library of thought experiments, and we hope they help in pushing computers to their ultimate thermodynamic limits.*

1988 Tansjö, L., 'Comment on the discovery of the second law.' A brief historical paper discussing the ideas of Carnot, Thomson, Maxwell, and others.

1988 van Kampen, N. G., 'Ten theorems about quantum mechanical measurements.' A distinction is made between macroscopic measuring apparatus and microscopic objects being measured. It is argued that the Schrödinger equation fully describes the system + apparatus and that wave function collapse does not require an additional postulate. It is found that upon measurement, the entropies of the apparatus and system increase by the same amount, which is the difference between the metastable initial state and the stable final state of the apparatus.

1988 Wright, R., 'Did the universe just happen?' This article, on computer scientist Edward Fredkin, contains similar material about Fredkin and Maxwell's demon to that in *Three Scientists and Their Gods* (1988).

1988 Wright, R., *Three Scientists and Their Gods.* A popular account about three eccentric scientists concerned with information transfer in physics, computers, biology, and in animal societies.

*...the demon, all told, cannot reduce entropy... to create even the illusion of entropy reduction—to pile up lots of order in one place while sweeping disorder under the rug—the demon must process information. In the end, there is no escape from the second law, and it takes information even to buy time.*

1989 Costa de Beauregard, O., 'The computer and the heat engine.' A computer is viewed as an engine that delivers information. Both memory erasure and message duplication are argued to be dissipative.

1989 Einhorn, Richard, 'Maxwell's demon 4: for violin or violectra solo.' A recording by Mary Rowell, from the New York City Ballet production of Red Angels, can be heard on the Internet, at http://www.richardeinhorn.com. The composer, Richard Einhorn, can be contacted at richardein@richardeinhorn.com.

1989 Goto, E., Yoshida, N., Loe, K. F. and Hioe, W., 'A study on irreversible loss of information without heat generation.' The authors argue that the 'reversible logic scheme' seems to be an unnecessary complication and not necessarily relevant to the issue of null heat computing. The also question the validity of the following two premises: (1) An irreversible thermodynamical process necessarily generates heat; and (2) Irreversible loss of information necessarily generates at least $kT \ln 2$ heat for one bit of information loss.

1989a Landauer, R., 'Response to 'The computer and the heat engine.'

1989b Landauer, R., 'Computation, measurement, communication and energy dissipation.'

*...if we assume that we can characterize our initial system completely, can calculate its subsequent time evolution to any required degree of accuracy, and can build the equipment needed to extract the energy, then energy put into a finite Hamiltonian system can always be extracted, and there is not dissipation. All of the assumptions we have made, however, are open to question, if the system has a good many degrees of freedom and particularly if we have let a lot of time elapse since the energy was put into the system. ... We have, in a limited way come full circle. First*

*we learned that discarding information requires energy dissipation. Now we are suggesting that energy dissipation is the consequence (in part) of insufficient information to allow energy retrieval. Thus, there is almost an equivalence between dissipation and an information loss.*

1989c Landauer, R., 'Reversible computation: Implications for measurement, communication, and physical law.'

*...In all three areas (computation, communication, and measurement) dissipation can be made as small as desired, per step, as long as information is not discarded. Erasure of information is not essential in computation or communication, but is needed to reset the meter after measurement. We speculate on the implications for physical law, arising from the likely limited computational precision available in the universe. We become even more speculative, and suggest that the ultimate physical laws, as a result of their limited precision, have something akin to a built-in noise source, and that this is the ultimate source of irreversibility.*

1989 Landsberg, P. T. and Leff, H. S., 'Thermodynamic cycles with nearly-universal maximum-work efficiencies.' See Section 1.8.4 herein.

1989 Lloyd, S., 'Use of mutual information to decrease entropy: Implications for the second law of thermodynamics.' Reprinted here as Article 5.3.

1989 Smith, C. and Wise, M. N., *Energy and Empire: A Biographical Study of Lord Kelvin*. A thoughtful account of the early history of Maxwell's demon, with emphasis on William Thomson's contributions.

*So long as the demon had to employ material tools—cricket bats, valves, arms, trap doors, or switches—in order to direct molecules, any description of his action within abstract dynamics would require the transfer of finite amounts of energy from mind to the finite molecules of the directing material. The demon's action, therefore, could serve at best as a mere analogy for free will, limited to an idealized action that did not require mind to store energy. A number of authors in Thomson's circle employed Maxwell's analogy in this sense ...*

1989 van Kampen, N. G., 'A physicist looks at quantum mechanical measurement.'

*The problem of measurement in quantum mechanics is solved in practice. Our aim is to analyze how this works—not to indulge in interpretations and philosophy. To this end a model is constructed, which embodies the following essential features. The act of measurement is described by the Schrödinger equation for object + apparatus. The apparatus is macroscopic, i.e., a quantum system with so many degrees of freedom that it behaves classically. It is prepared in a metastable state, thereby registering the result permanently. The object system is left in a new subspace of the joint Hilbert space, in which it is described by a new wave function: that is the collapse of the wave function. The entropy of the object and the entropy of the apparatus both increase, and by the same amount. This amount is the entropy difference between the metastable and stable states of the apparatus.*

1989 Wheeler, J. A., 'Information, physics, quantum: The search for links.'

*... every physical quantity, every it, derives it ultimate significance from bits, binary yes-or-no indications, a conclusion which we epitomize in the phrase* it from bit.

1989a Zurek, W. H., 'Algorithmic randomness and physical entropy.' An ambitious study of physical entropy, using ideas of algorithmic information theory. Physical entropy is viewed as a sum of missing Shannon information and the algorithmic information content. This enables a reformulation of the first law of thermodynamics for engines operated by entities capable of information acquisition and processing.

*The definition of the physical entropy proposed here is made possible by the algorithmic definition of randomness. It is made necessary by the desire to discuss the function of the Maxwell's demon from its own perspective. Moreover, the demon's role can be successfully played by an automaton capable of (1) acquiring information through reversible measurements, (2) processing it in a manner analogous to the universal computer, and (3) adopting strategies which aim to optimize its behavior so that—for example—it can economically extract useful energy from the available sources.*

1989b Zurek, W. H., 'Algorithm randomness, physical entropy, measurements, and the second law.' See the related Article 6.3 published herein.

1989c Zurek, W. H., 'Thermodynamic cost of computation, algorithmic complexity and the information metric.' The author considers computation in which an input program is replaced by its output in a computer memory. He finds a resulting entropy increase equal to the difference between the algorighmic complexities of input and output. Replacement computation that does not decrease the algorithmic information content of a file can be carried out reversibly. This includes compression of information.

1990 Aspden, H., 'The law beats Maxwell's demon.'

1990a Bennett, C. H., 'How to Define Complexity in Physics, and Why.' Motivated by examples, from biology, Bennett analyzes various candidates for a satisfactory formal measure of complexity and discusses their usefulness.

1990b Bennett, C. H., 'Undecidable dynamics.' Motivated by examples, from biology, Bennett analyzes various candidates for a satisfactory formal measure of complexity and discusses their usefulness.

1990 Berger, J., 'Szilard's demon revisited.'

*We show that piston fluctuations are of crucial importance in the analysis of a Szilard engine. Some engines which do not require information in order to perform a Szilard cycle actually do not work. We pinpoint the mechanism and stages which require work investment when a measuring instrument is at rest.*

1990 Borg, F., 'The law beats Maxwell's demon.'

1990 Caves, C. M., Unruh, W. G. and Zurek, W. H., 'Comment on 'Quantitative limits on the ability of a Maxwell demon to extract work from heat.' Comment on the article by Caves (1990).

1990 Caves, C. M., 'Quantitative limits on the ability of a Maxwell Demon to extract work from heat.' Caves suggests that a Maxwell's demon can extract work only by waiting for rare thermal fluctuations. See also Caves, C. M., Unruh, W. G. and Zurek, W. H. (1990).

1990 Collier, J. D., 'Two faces of Maxwell's demon reveal the nature of irreversibility.'

1990 Exner, O., 'Perpetuum mobile once more, in fun and quite seriously.'

1990 Hayles, N. K., 'Self-reflexive metaphors in Maxwell's demon and Shannon's choice: Finding the passages.'

1990 Kronenberg, L. H., 'The law beats Maxwell's demon.'

1990 Landsberg, P. T., *Thermodynamics and Statistical Mechanics*. These pages contain a brief discussion of Maxwell's demon and a primer on information theory.

1990 Leff, H. S., 'Maxwell's demon, power, and time.' See Section 1.8.4 herein.

*What rate ($P_d$) of energy transfer is attainable by a Maxwell's demon who sorts gas molecules serially; and how much time ($t_d$) does it take to achieve a designated temperature difference $\Delta T$ across a partition? Two estimates are made, using (i) the energy-time form of Heisenberg's uncertainty principle and (ii) classical kinetic theory ...*

1990a Leff, H. S. and Rex, A. F., *Maxwell's Demon: Entropy, Information, Computing*. This first edition includes a historical background written by the editors, 25 reprints of seminal works on Maxwell's demon, going back to Maxwell and up through the 1980s, and an extensive bibliography.

1990b Leff, H. S. and Rex, A. F., 'Maxwell's Demon: Resource Letter MD-1.' This Resource Letter contains over 200 annotated references to Maxwell's demon and related subjects.

1990 Maddox, J., 'Maxwell's demon flourishes.'

1990 Samuelson, P. A., 'The law beats Maxwell's demon.'

1990 Yamamoto, Y. and Ueda, M., 'Quantum optics and new technology.'

*Quantum dynamics deals with observers' knowledge of the object and the observation results. It may give ultimate limits to data transmission (communication) and processing (computer) through the uncertainty principle. The authors describe several topics concerning quantum-dynamics-based communication capacity such as transmission path capacity involving quantum noise, minimum energy to carry 1-bit information, uncertainty relations between time and energy, and so forth. They also refer to the relation between contraction of wave flux and energy loss, taking examples of wave flux contraction in nondestructive quantum measurement, the reversible logic computer, Maxwell's demon, and Schrödinger's cat paradox.*

1990 Zurek, W. H., 'Algorithmic information content, Church-Turing thesis, physical entropy, and Maxwell's demon.'

1991 Brasher, J., 'Nonlinear-wave mechanics, information-theory, and thermodynamics.'

*A logarithmic nonlinear term is introduced in the Schrodinger wave equation, and a physical justification and interpretation are provided within the context of information theory and thermodynamics. From the resulting nonlinear Schrodinger equation for a system at absolute temperature $T > 0$, the energy equivalence, $kT \ln 2$, of a bit of information is derived.*

1991 Brown, J. R., *The Laboratory of the Mind.* Maxwell's demon is used as an example of a thought experiment that 'facilitates a conclusion drawn from a specific, well-articulated theory. Brown argues that the demon helps to make the concept of entropy decrease via fluctuations more plausible; i.e., to break down the barrier to understanding the statistical nature of entropy. A cartoon sketch of Maxwell's demon from this book was used to advertise Brown's book as an Alternate selection of the Library of Science.

1991 Gonick, L., 'Science classics.' This is a clever two-page color cartoon overview of the Maxwell's demon puzzle and its solution. Reprinted in Bennett, C. H. (1998).

1991 Gyftopoulos, E. P. and Beretta, G. P., *Thermodynamics: Foundations and Applications.*

1991 Landauer, R., 'Information is physical.'

1991 Landsberg, P. T., 'Maxwell's demon (book review).' This is a review of *Maxwell's Demon: Entropy, Information, Computing*, the first edition of this book, edited by Harvey S. Leff and Andrew F. Rex.

1991 Lloyd, S. and Zurek, W. H., 'Algorithmic treatment of the spin-echo effect.'

*We analyze the apparent increase in entropy in the course of the spin-echo effect using algorithmic information theory. We show that although the state of the spins quickly becomes algorithmically complex, then simple again during the echo, the overall complexity of spins together with the magnetic field grows slowly, as the logarithm of the elapsed time. This slow increase in complexity is reflected in an increased difficulty in taking advantage of the echo pulse. Our discussion illustrates the fundamental role of algorithmic information content in the formulation of statistical physics, including the second law of thermodynamics, from the viewpoint of the observer.*

1991 Lloyd, D. G., 'The Role Of Science And Technology In The Novels Of Thomas Pynchon.' Lloyd examines the use of scientific metaphors and allusions in the works of Thomas Pynchon. Maxwell's demon appears explicitly in 'The Crying of Lot 49.'

1991 Simons, J. P., 'Laser-beams and molecular dreams.'

*More than a century ago, Kekule dreamt of a chain of carbon atoms, first dancing in a line, then linking to form a ring-the central structure of the benzene molecule. James Clerk Maxwell dreamt of a supernatural demon, that could arrest the incessant motion of atoms and molecules, or view their restless activity like a 'fly on the wall'. A demon (physicist) might see the atoms or molecules collide, exchange energy and momentum, and separate. A demon (chemist) might, occasionally, see them emerge with a new identity or a new structural arrangement. A demon (photochemist) might see them pass through a stream of photons—absorb one—and be transformed or fragmented.*

1991 Zurek, W. H., 'Decoherence and the transition from quantum to classical.'

*The environment surrounding a quantum system can, in effect, monitor some of the system's observables. As a result, the eigenstates of those observables continuously decohere and can behave like classical states.*

1992 Bennett, C. H., Brassard, G. and Ekert, A. K., 'Quantum Cryptography.'

1992 De Meis, L., Montero-Lomeli, M. and Grieco, M. A. B., 'The Maxwell demon in biological systems. Use of glucose 6-phosphate and hexokinase as an ATP regenerating system by the $Ca^{2+}$-ATPase of sarcoplasmic reticulum and submitochondrial particles.'

1992 Hogenboom, D. L., 'Maxwell's Demon: Entropy, Information, Computing (book review).' Review of first edition of this book, edited by Harvey S. Leff and Andrew F. Rex.

1992a Landauer, R., 'Information is physical.'

1992b Landauer, R., 'Reversible computing and physical law—reply.'

1992 Matzke, D., *Proceedings of the Workshop on Physics and Computation.* This volume contains over 50 invited and contributed papers to the "PhysComp '92" conference in Dallas. Topics include entropy and information, quantum computing, Maxwell's demon, and applications in biology and neural networks.

1992 Rastogi, S. R., 'How a clever demon nearly blew up the second law of thermodynamics.'

1992 Rex, A. F. and Larsen, R., 'Entropy and information for an automated Maxwell's demon.' Reprinted here as Article 3.2.

1992 Schulman, A., 'Programmer's Bookshelf.' Contains a review of Leff/Rex's Maxwell's Demon: Entropy Information, Computing.

1992 Skordos, P. A. and Zurek, W. H., 'Maxwell's demon, rectifiers, and the second law: Computer simulation of Smoluchowski's trapdoor.' Reprinted here as Article 3.1.

1992 Skordos, P. A., 'Time-reversible Maxwell's Demon.'

*A time-reversible Maxwell's demon is demonstrated which creates a density difference between two chambers initialized to have equal density. The density difference is estimated theoretically and confirmed by computer simulations. It is found that the reversible Maxwell's demon compresses phase space volume even though its dynamics are time reversible. The significance of phase space volume compression in operating a microscopic heat engine is also discussed.*

1992 Wolpert, D. H., 'Memory systems, computation, and the second law of thermodynamics.'

1992 Zhang, K. and Zhang, K., 'Mechanical models of Maxwell's demon with noninvariant phase volume.'

*This paper is concerned with the dynamical basis of Maxwell's demon within the framework of classical mechanics. We show that the operation of the demon, whose effect is equivalent to exerting a velocity-dependent force on the gas molecules, can be modeled as a suitable force field without disobeying any laws in classical mechanics. An essential requirement for the models is that the phase-space volume should be noninvariant during time evolution. The necessity of the requirement can be established under general conditions by showing that (1) a mechanical device is able to violate the second law of thermodynamics if and only if it can be used to generate and sustain a robust momentum flow inside an isolated system, and (2) no systems with invariant phase volume are able to support such a flow. The invariance of phase volume appears as an independent factor responsible for the validity of the second law of thermodynamics. When this requirement is removed, explicit mechanical models of Maxwell's demon can exist.*

1992 Zimmerman, T. and Salamon, P., 'The demon algorithm.' An annealing algorithm, motivated by an information-theoretic analysis of simulated annealing, is constructed in analogy to the action of Maxwell's demon.

1993 Caves, C. M., 'Information and entropy.' Reprinted here as Article 6.1.

1993 Goldstein, M. and Goldstein, I., *The Refrigerator and the Universe.* Reviewed by H. S. Leff (1995b).

1993 Gyftopoulos, E. P., 'Can an omnipotent Maxwellian demon beat the laws of thermodynamics?'

1993a Landauer, R., 'Comment on 'Physical limits to quantum flux parametron operation.'

*Goto* et al *claim that for the quantum flux parametron 'there is no minimum dissipation per step even for logically irreversible computation.' In actual fact Goto et al embed an irreversible operation in a more complex reversible operation.*

1993b Landauer, R., 'Statistical physics of machinery - Forgotten middle-ground.'

*Statistical physics long concentrated on the equilibrium state in the thermodynamic limit, and small deviations from that state. Machinery is spatially inhomogeneous, non-linear, and operates far from equilibrium. After our field expanded to face these problems, they were soon displaced by still newer fashions. It is symptomatic of this situation that the real resolution of Maxwell's demon, in the 1980s, passed almost unnoticed. We revisit a number of subjects, emphasizing the need for concern with the detailed kinetics of the system at hand and the difficulties posed by magic short cuts. The maximum entropy formalism, minimal entropy production, and the difficulties in identifying the most likely state among competing states of local stability will be taken up. The blowtorch theorem demonstrates that relative stability cannot, in general, be decided by examining only the neighborhood of the competing states. The kinetics along the pathway connecting the states must be taken into account. The latter point will be connected to biological evolution. Circulating currents, set up by the simultaneous presence of force fields and temperature gradients, will be described.*

1993 Nardiello, G., 'Energy dissipation of computational processes.'

*We modify an ideal computer, Bennett's 'Brownian Turing Machine,' in order to show that logically irreversible operations can be executed with an arbitrarily small energy dissipation per bit.*

1993 Newton, R. G., *What Makes Nature Tick?*

1993 Robertson, H. S., *Statistical Thermophysics.*

1993 Schack, R. and Caves, C. M., 'Hypersensitivity to perturbations in the quantum baker's map.'

*We analyze a randomly perturbed quantum version of the baker's transformation, a prototype of an area-conserving chaotic map. By simulating the perturbed evolution, we estimate the information needed to follow a perturbed Hilbert-space vector in time. We find that the Landauer erasure cost associated with this grows very rapidly and becomes larger than the maximum statistical entropy given by the logarithm of the dimension of Hilbert space. The quantum baker's map displays a hypersensitivity to perturbations analogous to behavior found in the classical case. This hypersensitivity characterizes 'quantum chaos' in a way that is relevant to statistical physics.*

1993 Silverman, M. P., *And Yet It Moves: Strange Systems and Subtle Questions in Physics*. One of this book's sections is entitled, 'Exorcising a Maxwell demon.'

1993 Skordos, P. A., 'Compressible dynamics, time reversibility, Maxwell's demon, and the second law.'

*A tantalizing version of Maxwell's demon is presented which appears to operate reversibly. A container of hard-core disks is separated into two chambers of equal volume by a membrane that selects which disk can penetrate depending on the disk's angle of incidence. The analysis of this system confirms that the second law of thermodynamics requires the incompressibility of microscopic dynamics or an appropriate energy cost for compressible microscopic dynamics.*

1994 Beghian, L. E., 'Maxwell's demon revisited.'

*The problem of velocity selection by Maxwell's demon in a many-particle gas, as distinct from a one-particle gas, is addressed. For such a situation it is shown that it is impossible to effect a separation of the molecules into two separate partitions so as to bring about an overall decrease in entropy unless information entropy is made available to the system. The validity of this conclusion is discussed in terms of the Gibbs paradox and the principle of quantum indistinguishability of identical particles (6 Refs.)*

1994 Bennett, C. H., 'Quantum physics: Night thoughts, dark sight.'

1994 Berger, J., 'The fight against the second law of thermodynamics.'

*After presenting possible motives for fighting against the second law of thermodynamics, several attempts to beat this law are analyzed. The second law wins, but an interesting interpretation of it emerges. This interpretation uses the notion of 'encoded order' and claims that whether a system is or is not in thermodynamic equilibrium depends on the coordinates that the observer decides to measure. This interpretation may not be new, but most present-day physicists seem to be unaware of it. The question of subjectivity of entropy and the connection between the present interpretation and 'algorithmic randomness' are addressed.*

1994 Braginsky, V. B. and Khalili, F. Y., ' "Maxwell demon" in quantum-non-demolition measurements.'

1994 Caves, C. M., 'Information, entropy, and chaos.' Using algorithmic information theory, a tentative answer is given to the question: How does one lose the ability to extract work as a physical system evolves away from a simple initial state?

1994 Fahn, P. N., 'Entropy cost of information.'

*An entropy analysis of Szilard's (1929) one-molecule Maxwell's demon suggests a general theory of the entropy cost of information. The entropy of the demon increases due to the decoupling of the molecule from the measurement information. In general, neither measurement nor erasure is fundamentally a thermodynamically costly operation; however, the decorrelation of the system from the information must always increase entropy in the system-with-information. This causes a net entropy increase in the universe unless, as in the Szilard demon, the information is used to decrease entropy elsewhere before the correlation is lost. Thus information is thermodynamically costly precisely to the extent that it is not used to obtain work from the measured system.*

1994 Gordon, L. G. M., 'The molecular-kinetic theory and the second law.'

1994 Krus, D. J. and Webb, J. M., 'Maxwell's demons—Simulating probability-distributions with functions of propositional calculus.'

1994 Landauer, R., 'The zig-zag path to understanding.' Landauer provides a brief history of the development of Landauer's principle and some thoughts on quantum computers.

1994 Leff, H. S. and Rex, A. F., 'Entropy of Measurement and Erasure: Szilard's Membrane Model Revisited.' Reprinted here as Article 4.2.

1994 Leff, H. S., 'Thermodynamics of Crawford's energy equipartition journeys.'

*Energy equipartition 'journeys' proposed originally in the context of relativistic mechanics [F. S. Crawford, Am. J. Phys.* **61**, *317–326 (1993)] are examined thermodynamically. Equipartition is between ideal gases separated by an insulating piston. Each journey alternates slow adiabatic volume changes and zero-work adiabatic piston resettings effected by a benevolent Maxwell's demon until temperature and pressure equalities exist. For the systems considered, reversible adiabatic thermodynamic processes are equivalent to slow mechanically adiabatic, constant-action processes. Because phase space mixing is assumed implicitly in thermodynamics, the piston jiggling step needed in Crawford's mechanics-based treatment is unnecessary. Zero-work journeys maximize the gas system's entropy change, and maximum work output journeys leave that entropy unchanged. Thermodynamics confirms Crawford's result for the piston's effective spring constant,* *and statistical mechanics enables rigorous justification of his interesting recursion relation for the moments of $q$ = momentum $\times$ velocity.*

1994 Machado, A. C., 'Energy and Information in Society: A Study of the Relationships between Energy Use and Information Activities in Contemporary Society (Communications).' Maxwell's demon and information theory are related to social and economic processes.

1994 Maddox, J., 'Directed motion from random noise.'

*An interesting example has stilled fears that using the idea of a biochemical ratchet to explain muscle contraction would contradict the second law of thermodynamics. Such a ratchet mechanism is based on molecules having the ability to move relative to each other in one direction but unable to slip backward, a point that can be illustrated using Maxwell's demon.*

1994 Schneider, T. D., 'Sequence logos, machine channel capacity, Maxwell's demon, and molecular computers: A review of the theory of molecular machines.' An excellent review of how information theory can be applied to living cells.

1994 Schumacher, B., 'Demonic heat engines.'

*Thermal efficiency is discussed for a Szilard one-particle engine that operates with a READ demon connected to a high-temperature reservoir, and an ERASE demon connected to a low-temperature reservoir.*

1994 Sokirko, A. V., 'Does mass-action law breakdown occur in small thermodynamic systems?'

1994 Sporn, S. R., 'Air traffic controller and Maxwell's demon.'

*The controller functions in direct analogy with Maxwell's Demon. Faced with a disordered velocity and position distribution of aircraft in a control zone, the air traffic controller introduces order by supplying information so as to achieve a decrease in entropy. Controller workload is measured by the information (negentropy) he must supply.*

1994 Yamamoto, Y. and Chuang, I., 'Physical limits for computing and communication.'

*The standard and intrinsic physical limits of communication and computing are discussed in terms of a bit error rate, channel capacity, and minimum energy cost per one bit of information. Non-standard techniques for circumventing the standard physical limits, such as control of the wavefunction (squeezing), evasion of the measurement back action (quantum nondemolition measurement), and conservation of the logical reversibility (Fredkin gate), are reviewed.*

1995 Bennett, C. H., 'Quantum information and computation.' This paper is a good general introduction to the field, including definitions and explanations of some important terms in the context of quantum computing: superposition, interference, entanglement, nonlocality, uncertainty, qubits, and teleportation. The paper also contains some useful references to other work.

*A new quantum theory of communication and computation is emerging, in which the stuff transmitted or processed is not classical information, but arbitrary superpositions of quantum states.*

1995 — Biedenharn, L. C. and Solem, J. C., 'A quantum-mechanical treatment of Szilard's engine: Implication for the entropy of information.' Szilard's single-molecule engine is examined quantum mechanically and compared with the double-slit experiment. Distinctions are drawn between physical entropy and information entropy, and the authors '...conclude that information *per se* is a subjective, idealized, concept separated from the physical realm.'

1995 — Chernov, N. I. and Lebowitz, J. L., 'Stationary shear flow in boundary driven Hamiltonian systems.'

*We investigate stationary nonequilibrium states of particles moving according to Hamiltonian dynamics with Maxwell demon 'reflection rules' at the walls.*

1995 — Igeta, K. and Ogawa, T., 'Information dissipation in quantum-chaotic systems—computational view and measurement induction.'

1995 — Kyle, B. G., 'Exorcising Maxwell's demon. Entropy, information, and computing.'

1995a — Landauer, R., 'Is quantum mechanics useful?' Paper presented at NATO Advanced Research Workshop—Ultimate limits of fabrication and measurement, Cambridge—1994.

1995b — Landauer, R., 'Qualitative view of quantum shot noise.' Paper presented at New York Academy of Sciences Conference:—Fundamental problems in quantum theory—1995.

1995a — Leff, H. S., 'Thermodynamic insights from a one-particle gas.'

*Fundamentals of thermodynamics are introduced and illuminated using a model 'gas' consisting of a single particle of mass $m$ and kinetic energy $E$ in a one-dimensional box of length $L$. Constancy of the classical action $A = 2(2mE)^{1/2}L$ under slow, mechanically reversible changes in $L$ reflects the symmetry of reversibility. Heat is defined in terms of work using adiabatic paths to connect an arbitrary pair of states. The ensemble average of $A$ increases for many irreversible mechanical processes that break the reversibility symmetry, and entropy can be linked to $\ln(A)$. Although not strictly 'thermodynamic' this model helps clarify the statistical nature of the second law of thermodynamics and illuminates the very essence of thermodynamics, namely, the invisible mixing of energy within systems.*

1995b Leff, H. S., 'The Refrigerator and the Universe: Understanding the Laws of Energy (book review).' Review of book by Goldstein and Goldstein (1993).

1995 Letokhov, V. S., 'Laser Maxwell's demon.'

*A Maxwell's demon based on selective interaction between laser light and atomic particles is examined.*

1995 Lloyd, S., 'Quantum-mechanical computers.'

*Quantum-mechanical computers, if they can be constructed, will do things no ordinary computer can.*

1995 Malizia, D. and Tarsitani, C., 'Looking at the second law of thermodynamics through the eyes of Maxwell's demon.'

1995 Millonas, M. M., 'Self-consistent microscopic theory of fluctuation-induced transport.'

*A Maxwell's demon type 'information engine' that extracts work from a bath is constructed from a microscopic Hamiltonian for the whole system including a subsystem, a thermal bath, and a nonequilibrium bath of phonons or photons that represents an information source or sink. The kinetics of the engine is calculated self-consistently from the state of the nonequilibrium bath, and the relation of this kinetics to the underlying microscopic thermodynamics is established.*

1995 Morck, R. and Morowitz, H., 'Value and information: a profit maximizing strategy for Maxwells demon.' Analogies are suggested between economics and physics, including Maxwell's demon.

1995 Phipps, Jr., Thomas, E. and Brill, M. H., 'Bayesian entropy and interference.'

1995 Sheehan, D. P., 'A paradox involving the second law of thermodynamics.'

*A simple paradox is posed that appears to challenge the second law of thermodynamics in a blackbody plasma environment. Laboratory experiments approximating salient aspects of this system fail to resolve the paradox.*

1995 Shizume, K., 'Heat generation required by erasure.' Reprinted here as Article 4.3.

1995a Zaslavsky, G. M., 'From Hamiltonian chaos to Maxwell's demon.'

*The problem of the existence of Maxwell's Demon (MD) is formulated for systems with dynamical chaos. Property of stickiness of individual trajectories, anomalous distribution of the Poincaré recurrence time, and anomalous (non-Gaussian) transport for a typical system with Hamiltonian chaos results in a possibility to design a situation equivalent to the MD operation. A numerical example demonstrates a possibility to set without expenditure of work a thermodynamically non-equilibrium state between two contacted domains of the phase space lasting for an arbitrarily long time. This result offers a new view of the Hamiltonian chaos and its role in the foundation of statistical mechanics.*

1995b Zaslavsky, G. M., 'From Levy flights to the fractional kinetic equation for dynamical chaos.'

*Chaotic dynamics of Hamiltonian systems can be described by the random process which resembles the Levy-type flights and trappings in the phase space of a system. The probability distribution function satisfies the fractional in space and time generalization of the Fokker–Planck–Kolmogorov equation. Orders of the fractional derivatives in space and time can be connected to the Pesin's dimensions of the trajectories. A new look on the problem of Maxwell's Demon is discussed in the context of the anomalous ('strange') kinetics.*

1996 Bennett, C. H., 'Information theory—Freely communicating.'

1996 Birkhahn, R., 'The Return of Maxwell's Demon?'

1996 Clarke, B., 'Allegories of Victorian thermodynamics.'

1996 Dotzler, B. J., 'Demons—Magic—Cybernetics: On the introduction to natural magic as told by Heinz von Foerster.'

1996 Fahn, P. N., 'Maxwell's demon and the entropy cost of information.'

*We present an analysis of Szilard's one-molecule Maxwell's demon, including a detailed entropy accounting, that suggests a general theory of the entropy cost of information. It is shown that the entropy of the demon increases during the expansion step, due to the decoupling of the molecule from the measurement information. It is also shown that there is an entropy symmetry between the measurement and erasure steps, whereby the two steps additively share a constant entropy change, but the proportion that occurs during each of the two steps is arbitrary. Therefore the measurement step may be accompanied by an entropy increase, a decrease, or no change at all, and likewise for the erasure step. Generalizing beyond the demon, decorrelation between a physical system and information about that system always causes an entropy increase in the joint system comprised of both the original system and the information. Decorrelation causes a net entropy increase in the universe unless, as in the Szilard demon, the information is used to decrease entropy elsewhere before the correlation is lost. Thus, information is thermodynamically costly precisely to the extent that it is not used to obtain work from the measured system.*

1996 Feynman, R. P. (T. Hey and R. W. Allen, Editors), *Feynman Lectures on Computation.* This book provides a good overview of the physics of computation, along with Feynman's usual insights. See especially Chapter 5 'Reversible computation and the thermodynamics of computing' and Chapter 6 'Quantum Mechanical computers.'

1996 Gell-Mann, M. and Lloyd, S., 'Information measures, effective complexity, and total information.'

*This article defines the concept of an information measure and shows how common information measures such as entropy, Shannon information, and algorithmic information content can be combined to solve problems of characterization, inference, and learning for complex systems. Particularly useful quantities are the effective complexity, which is roughly the length of a compact description of the identified regularities of an entity, and total information, which is effective complexity plus an entropy term that measures the information required to describe the random aspects of the entity. Mathematical definitions are given for both quantities and some applications are discussed. In particular, it is pointed out that if one compares different sets of identified regularities of an entity, the 'best' set minimizes the total information, and then, subject to that constraint and to constraints on computation time, minimizes the effective complexity; the resulting effective complexity is then in many respects independent of the observer.*

1996 Gershenfeld, N., 'Signal entropy and the thermodynamics of computation.' Reprinted here as Article 7.7.

1996 Halliwell, J. J., Pérez-Mercader, J. and Zurek, W. H., Editors, *Physical Origins of Time Asymmetry*. This outstanding collection contains papers from many noted physicists, including Bennett, Caves, Gell-Mann, Hawking, Lloyd, Davies, Schumacher, Wheeler, and Zurek.

1996 Herron, N. and Farneth, W. E., 'The design and synthesis of heterogeneous catalyst systems.' A Maxwell's demon approach to catalysis, in which individual molecules can be distinguished, has come closer to reality in recent years. Two new approaches to catalyst synthesis are described that are leading the way to catalyst design. The Figure shows n-octane proceeding through the 8-ring window of zeolite A toward the iron active site for oxidation. The relative size of the window and alkane causes selectivity for oxidation towards the end of the chain.

1996 Jayannavar, A. M., 'Simple model for Maxwell's-demon-type information engine.' The authors develop 'a simple model of Maxwells-demon-type engine that extracts work out of a nonequilibrium bath by rectifying internal fluctuations.'

1996 Kioi, K., 'Reversible computing realizing ultimate low power LSI system.'

*It has been shown by Landauer and Bennett, through thermodynamics, that reversible computers can, in principle, dissipate arbitrarily small amounts of energy per logical operation. The quasi-reversible operating techniques using standard CMOS transistors presented in recent years are practical for ultimate low power LSI implementation. Extrapolating present trends, we should need reversible logic in the first decade of the 21st century. Theoretical background of reversible computing and its capability as a low power technology are given.*

1996a Landauer, R., 'The physical nature of information.' Reprinted here as Article 7.4.

1996b Landauer, R., 'Minimal energy requirements in communication.' Reprinted here as Article 7.5.

1996c Landauer, R., 'Reversible computation: Implications for measurement, communication, and physical law.' Paper presented at International symposium: Foundations of quantum mechanics—Tokyo—Aug. 1989.

1996 Li, M. and Vitanyi, P., 'Reversibility and adiabatic computation: Trading time and space for energy.'

*Future miniaturization and mobilization of computing devices requires energy parsimonious 'adiabatic' computation. This is contingent on logical reversibility of computation. An example is the idea of quantum computations which are reversible except for the irreversible observation steps. We propose to study quantitatively the exchange of computational resources like time and space for irreversibility in computations.*

1996 Likharev, K. K. and Korotkov, A. N., ' "Single-electron parametron": Reversible computation in a discrete-state system.'

*The energy dissipation in a proposed digital device in which discrete degrees of freedom are used to represent digital information (a 'single-electron parametron') was analyzed. If the switching speed is not too high, the device may operate reversibly (adiabatically), and the energy dissipation E per bit may be much less than the thermal energy scale kT ... The energy × time ($E\tau$) is, however, much greater than Planck's constant ħ, at least in the standard 'orthodox' model of single-electron tunneling that was used in these calculations.*

1996 LoSurdo, C., Guo, S. C. and Paccagnella, R., 'Self-consistent equilibria in a cylindrical reversed-field pinch.'

1996 Magnasco, M. O., 'Szilard's heat engine.'

*Szilard presented the first concrete embodyment of a Maxwell demon. We present a detailed kinematic analysis of his heat engine. We find that the phase space contains a branched manifold. After defining carefully the physical dynamics on such an object, we prove that the engine must operate at a loss.*

1996 Parrondo, J. M. R. and Español, P., 'Criticism of Feynman's analysis of the ratchet as an engine.'

*The well-known discussion on an engine consisting of a ratchet and pawl in (The Feynman Lectures on Physics) is shown to contain some misguided aspects: Since the engine is simultaneously in contact with reservoirs at different temperatures, it can never work in a reversible way. As a consequence, the engine can never achieve the efficiency of a Carnot cycle, not even in the limit of zero power (infinitely slow motion), in contradiction with the conclusion reached in the Lectures.*

1996 Reese, K. M., 'Odd device may be the work of Maxwell's demon.'

1996a Sheehan, D. P., 'Another paradox involving the second law of thermodynamics.'

*Recently a paradox has been posed that appears to challenge the second law of thermodynamics in a plasma blackbody environment [D. P. Sheehan, Phys. Plasmas* **2**, *1893 (1995)]. In this paper another, related paradox is posed in an unmagnetized Q plasma. Laboratory experiments simulating some necessary conditions for the paradoxical system corroborate theoretical predictions and fail to resolve the paradox in favor of the second law.*

1996b Sheehan, D. P., 'Response to comment on 'A paradox involving the second law of thermodynamics.'

1996 Thirion, J.-P., 'Non-rigid matching using demons.'

*We present the concept of non-rigid matching based on demons, by reference to Maxwell's demons. We contrast this concept with the more conventional viewpoint of attraction. We show that demons and attractive points are clearly distinct for large deformations, but also that they become similar for small displacements, encompassing techniques close to optical flow. We describe a general iterative matching method based on demons, and derive from it three different non-rigid matching algorithms, one using all the image intensities, one using only contours, and one for already segmented images. At last, we present results with synthesized and real deformations, with applications to Computer Vision and Medical Image Processing.*

1996 Trebbia, P., 'Maxwell's demon and data analysis.'

*The second principle of thermodynamics introduces a very important concept: the entropy of a macrosystem made of a very great number of microsystems. One of the aims of this work is to point out how, in data analysis, entropy and information, that Maxwell's demon should have to allow a macroscopic reversible process, are closely related. The measure of the information content of a data set allows the analyst to permanently keep under control the variation of information through the successive steps of data processing. On the other hand, multivariate statistics, and among them the factorial analysis of correspondence, can extract the independent and orthogonal sources of information contained in the experimental data. Through several examples dealing with very different experimental situations (stroboscopic observation of a mechanical stress, time-dependent spectroscopy, quantitative elemental mapping by X-ray radiography) we illustrate the usefulness of data analysis before and during data processing.*

1996 Tsuda, I., 'Linear demon.'

*(A) dynamical system may provide a mechanism of demon within proteins similar to Maxwell's demon. In order to obtain a model for demon, we pay attention to a dynamical mechanism of biological threshold such as seen in neuron's firing.*

1996 Walther, T. and Trebbia, P., 'Maxwell's demon and data analysis—Discussion.'

1997 Bennett, C. H., DiVincenzo, D. P. and Smolin, J. A., 'Capacities of quantum erasure channels.'

1997 Bier, M., 'A motor protein model and how it relates to stochastic resonance, Feynman's ratchet, and Maxwell's demon.'

*A motor protein turns chemical energy into motion, but it differs from an ordinary engine in that random Brownian kicks become important. Below we propose a description where the energy input is used to ratchet Brownian motion, i.e. to allow it in one direction and block it in the opposite direction. Our model relates to famous paradoxes like Feynman's Ratchet and Maxwell's Demon, but is thermodynamically consistent.*

1997 Brownlie, A. W., 'Private Colonies of the Imagination: Power and Possibility In Thomas Pynchon's "V.," "The Crying of Lot 49," and "Gravity's Rainbow".' Thomas Pynchon's use of science in his novels is explored. Maxwell's demon appears as an important metaphor in 'The Crying Of Lot 49.'

1997a Čápek, V., 'Isothermal Maxwell daemon.'

1997b Čápek, V., 'Isothermal Maxwell daemon and active binding of pairs of particles.'

1997 Caves, C. M. and Schack, R., 'Unpredictability, information, and chaos: Pursuing alternatives.'

1997 Grandy, Jr., W. T., 'Resource letter ITP-1: Information theory in physics.'

*This resource letter contains a historical discussion and well over 100 references related to the theory of information, including papers on the physics of computation, black holes, quantum cryptography, and quantum computing.*

1997 Grossmann, S. and Holthaus, M., 'Maxwell's Demon at work: two types of Bose condensate fluctuations in power-law traps.'

*After discussing the key idea underlying the Maxwell's Demon ensemble, we employ this idea for calculating fluctuations of ideal Bose gas condensates in traps with power-law single-particle energy spectra.*

1997 Harman, T., 'Maxwell's demon and a snowball's chance in hell.'

*Maxwell's demon can be used to show that entropy can decrease—it is just very unlikely to do so.*

1997 Kelly, T. R., Tellitu, I. and Sestelo, J. P., 'In search of molecular ratchets.' Kelly et al. synthesized the first working molecular ratchet. Their rotor had three benzene rings (acting like propeller blades) and another benzene ring acting as the pawl. The rotor was turned using random molecular motions and exhibited motion in both directions, as predicted by Feynman's analysis of the ratchet and pawl.

1997 Landauer, R., 'Fashions in science and technology.'

1997 Leff, H. S. and Rex, A. F., 'Maxwell's demon and the culture of entropy.'

*Maxwell's demon was conceived in Maxwell's mind two years after Clausius proposed the term entropy and has had a symbiotic relationship with entropy for the last 125 years. The demon has helped us understand entropy, and, in turn, the entropy concept has enabled clarification of the subtleties of Maxwell's demon. We trace the historic interplay of Maxwell's demon with entropy within the contexts of thermodynamics, statistical mechanics, information theory, and the theory of computing. We give evidence that Maxwell's demon has become an essential part of the culture of entropy.*

1997 Liboff, R. L., 'Maxwell's demon and the second law of thermodynamics.'

*An idealized, two-dimensional Maxwell demon is described which incorporates an irreversible process. The vertex of the device acts as a purely mechanical 'trap door.' This idealized mechanism is found to generate a violation of the second law of thermodynamics. These results indicate that the second law of thermodynamics is not valid in general for idealized, irreversible systems.*

1997 Lidorenko, N. S., 'The possibility of realization of 'Maxwell's demon.'

1997 Lloyd, S., 'Quantum-mechanical Maxwell's demon.' Reprinted here as Article 5.4.

1997 Machado, A. C. and Miller, R. E., 'Empirical relationships between the energy and information segments of the US economy—An input-output approach.'

1997 Markowsky, G., 'An introduction to algorithmic information theory: Its history and some examples.' A nice review of algorithmic information theory, with references, but no specific reference to Maxwell's demon.

1997 Navez, P., Bitouk, D., Gajda, M., Idziaszek, Z. and Rzazewski, K., 'Fourth statistical ensemble for the Bose-Einstein condensate.'

*We study the Bose–Einstein condensate (BEC) with a fixed number of particles. On the basis of conventional statistical mechanics we introduce the, so called, Maxwell's demon ensemble, where only particle transfer (without energy exchange) is allowed. We show that this new ensemble can be used for the microcanonical description of the system. We apply our formalism to the case of BEC in a harmonic trap and give the analytic expressions for the ground state fluctuations. We compare these expressions with the exact numerical results obtained for relatively small condensates.*

1997 Schack, R., 'Algorithmic information and simplicity in statistical physics.' Reprinted here as Article 6.2.

1997 Schlaffer, A. and Nossek, J. A., 'Is there a connection between adiabatic switching and reversible computing?

*In this paper it is shown that the predictions about the power dissipation of adiabatic switches for infinitely large charging times result in a circuit element, further denoted as the 'controlled ideal switch', that turns out to interfere with the second law of thermodynamics and is therefore an overidealization. The interference results from the fact that such a switch allows for the construction of Maxwell's demon. A smaller example hints that it is even possible to destroy information losslessly if an ideal switch is available.*

1997 Schulman, L. S., *Time's Arrow and Quantum Measurement.*

1997 Susskind, L., 'Black holes and the information paradox.'

1997 Weiss, C. and Wilkens, M., 'Particle number counting statistics in ideal Bose gases.' The 'Maxwell's demon ensemble' [see Navez *et al.* (1997)] is invoked to study the statistics of a degenerate boson system.

1997 Zaslavsky, G. M. and Edelman, M., 'Maxwell's demon as a dynamical model.'

*We consider a model of a billiard-type system, which consists of two chambers connected through a hole. One chamber has a circle-shaped scatterer inside (Sinai billiard with infinite horizon), and the other one has a Cassini oval with a concave border. The phase space of the Cassini billiard contains islands, and its parameters are taken in such a way as to produce a self-similar island hierarchy. Poincaré recurrences to the left and to the right chambers are considered. It is shown that the corresponding distribution function does not reach 'equipartition' even during the time $10^{10}$. The explanation is based on the existence of singularities in the phase space, which induces anomalous kinetics. The analogy to the Maxwell's Demon model is discussed.*

1997 Zemansky, M. W. and Dittman, R. H., *Heat and Thermodynamics*. One of the surprisingly few texts that includes a good description of the function of the demon. The arguments of Brillouin and Rodd (1964) are outlined and there is brief mention of advances in information theory by Wiener and Shannon and subsequent 'remarkable progress... deepening the understanding and applicability of the concepts of information storage and information erasure, which are fundamental to computer science and cybernetics.' Earlier, in the fourth edition (1957), Zemansky (sole author) alluded to Brillouin's work, stating 'information seems to be related to negative entropy or, as Brillouin puts it, 'negentropy'—but use of a light source is not mentioned. In the second edition (1943, prior to Brillouin's analysis), Zemansky presented the puzzle but did not attempt to solve it. He simply made Maxwell's point: 'If the trap door were left open and unattended by a Maxwell's demon, there would be a small probability of a separation of the fast from the slow molecules, but a much larger probability of an even distribution.'

1998 Anonymous, 'Profile.' Rolf Landauer seeks the physical limits of computation.

1998 Bennett, C. H., 'Information physics in cartoons.' Reprinted here as Article 7.6.

1998 Bonetto, F., Chernov, N. I. and Lebowitz, J. L., '(Global and local) fluctuations of phase space contraction in deterministic stationary nonequilibrium.'

1998 Čápek, V. and Bok, J., 'Isothermal Maxwell daemon: Numerical results in a simplified model.'

*A recently suggested simple model of an open microscopic system interacting with a bath and behaving like an isothermal Maxwell daemon is simplified and treated numerically. Results obtained confirm expectations that the system can, at the cost of thermal energy of the bath, spontaneously and in a cyclic process, pump particles from a particle reservoir to states with even higher energy. This contradicts the usual opinion based on the second law of macroscopic thermodynamics. For this effect, the interaction with the bath cannot be treated as weak. The role of dephasing owing to this interaction is illustrated.*

1998 Čápek, V., 'Isothermal Maxwell demon as a quantum 'sewing machine.'

*A model of an open microscopic quantum system interacting with an isothermal bath and able to bind actively particles from a reservoir to their even excited bound states at the cost of the bath energy is presented. The binding (potentially important in, e.g., chain reactions—hence 'sewing') is due to dynamic processes in a central part of the system accompanying the particle transfer. The outcome thus challenges the second law of thermodynamics.*

1998 DiVincenzo, D. P. and Loss, D., 'Quantum Information is Physical.'

*We discuss a few current developments in the use of quantum mechanically coherent systems for information processing. In each of these developments, Rolf Landauer has played a crucial role in nudging us, and other workers in the field, into asking the right questions, some of which we have been lucky enough to answer. A general overview of the key ideas of quantum error correction is given. We discuss how quantum entanglement is the key to protecting quantum states from decoherence in a manner which, in a theoretical sense, is as effective as the protection of digital data from bit noise. We also discuss five general criteria which must be satisfied to implement a quantum computer in the laboratory, and we illustrate the application of these criteria by discussing our ideas for creating a quantum computer out of the spin states of coupled quantum dots.*

1998 Earman, J. and Norton, J. D., 'Exorcist XIV: The wrath of Maxwell's demon. Part I. From Maxwell to Szilard.'

*(This paper) … will be devoted to understanding how a literature came to be that is so devoted to precarious constructions. Our answer will be provided in a historical review that traces how the role of Maxwell's Demon has changed since its earliest appearance.*

1998 Frank, M. P. and Knight, T., 'Ultimate theoretical models of nanocomputers.'

1998 Gorecki, H. and Zaczyk, M., 'Trade-off between energy and entropy of information in control systems.'

1998 Hatano, T. and Sasa, S., 'Numerical simulations on Szilard's engine and information erasure.' The Szilard heat engine is examined assuming a specific 'trapping potential' for Maxwell's demon and a stochastic white noise force on the piston. With these assumptions, the work done by the engine is no longer canceled by the heat generated by information erasure.

1998 Ivanitsky, G. R., Medvinsky, A. B., Deev, A. A. and Tsyganov, M. A., 'From Maxwell's demon to the self-organization of mass transfer processes in living systems.'

1998a Landauer, R., 'Energy needed to send a bit.'

*Earlier work demonstrated that quantum-mechanical machinery can be used to transmit classical bits without a minimal unavoidable energy cost per bit. It is shown that a very minor variation of the earlier proposal also works for qubits. In addition some casual remarks about error control in quantum parallelism are provided.*

1998b Landauer, R., 'The noise is the signal.'

1998 Landsberg, P. T., 'Maxwell's demon: Why warmth disperses and time passes (book review).' Review of book by von Baeyer (1998).

1998 Li, M., Tromp, J. and Vitany, I. P., 'Reversible simulation of irreversible computation.'

*Computer computations are generally irreversible while the laws of physics are reversible. This mismatch is penalized by among other things generating excess thermic entropy in the computation.*

1998 Milburn, G. J., 'Atom-optical Maxwell demon.'

*The separation, by an optical standing wave, of a beam of two-level atoms prepared in a thermal mixture of ground and excited states, is considered as an example of a Maxwell demon. By including the momentum exchanged with the cavity, it is shown how no violation of the second law is possible. A classical and quantum analysis is given which illustrates this principle in some detail.*

1998 Nielsen, M. A., Caves, C. M., Schumacher, B. and Barnum, H., 'Information-theoretic approach to quantum error correction and reversible measurement.'

*Quantum operations provide a general description of the state changes allowed by quantum mechanics. The reversal of quantum operations is important for quantum error-correcting codes, teleportation and reversing quantum measurements. We derive information-theoretic conditions and equivalent algebraic conditions that are necessary and sufficient for a general quantum operation to be reversible. We analyse the thermodynamic cost of error correction and show that error correction can be regarded as a kind of 'Maxwell demon', for which there is an entropy cost associated with information obtained from measurements performed during error correction. A prescription for thermodynamically efficient error correction is given.*

1998 Otsuka, J. and Nozawa, Y., 'Self-reproducing system can behave as Maxwell's demon; Theoretical illustration under prebiotic conditions.'

1998 Parellada, R. F. and Rojas, H. P., 'Competition between Maxwell's demons and some consequences.'

*We consider the problem of the competition between two Maxwell's demons. The presence of a more clever demon in one of two subsystems allows it to drain more order than the other. We suggest that this principle plays an important role in the evolution of organized systems. Examples from biological, social and economical sciences are proposed.*

1998 Scarani, V., 'Quantum computing.'

1998a Sheehan, D. P., 'Dynamically-maintained, steady-state pressure gradients.'

1998b Sheehan, D. P., 'Four paradoxes involving the second law of thermodynamics.'

*Recently four independent paradoxes have been proposed which appear to challenge the second law of thermodynamics. These paradoxes are briefly reviewed. It is shown that each paradox results from a synergism of two broken symmetries—one geometric, one thermodynamic.*

1998 Sheehan, D. P. and Means, J. D., 'Minimum requirement for second law violation: a paradox revisited.'

*A paradox has been posed that challenges the second law of thermodynamics in a plasma blackbody environment [D. P. Sheehan, Phys. Plasmas* **2***, 1893 (1995)]. Laboratory experiments testing critical aspects of the paradox have supported theoretical predictions and have failed to resolve the paradox in favor of the second law. In this article, the paradox is sharpened and expanded in scope by identifying the crux of the paradox and the general requirements for second law violation. It is found that for an electrically conducting probe immersed in a blackbody plasma and connected to ground through a load, the general requirement for second law violation is* $V_f \neq 0$*, where* $V_f$ *is the plasma floating potential. This requirement can be met by many natural and manmade plasmas.*

1998 Stix, G., 'Riding the back of electrons.' Stix offers a retrospective on the career of Rolf Landauer.

1998 Vologodskii, A., 'Maxwell demon and topology simplification by type II topoisomerases.'

1998 von Baeyer, H. C., *Maxwell's Demon: Why Warmth Disperses and Time Passes.* Designed for the lay reader, this book touches on many historical aspects of the Maxwell's demon puzzle. This includes the early development of the second law, Maxwell's kinetic theory and conception of the demon, and various incarnations of Maxwell's demon since then. The book concludes with some modern approaches, including a discussion of Zurek's algorithmic complexity.

1998 Williams, C. P. and Clearwater, S. H., *Explorations in Quantum Computing*. This book is intended to address developments in quantum computing and the technological hurdles that must be surmounted to achieve quantum computing in practice. The book comes with a CD-ROM containing simulations and tutorials on many of the topics covered in the text.

1999 Anonymous, 'Microrotors and Maxwell's demon.'

1999 Bennett, C. H., Shor, P. W., Smolin, J. A. and Thapliyal, A. V., 'Entanglement-assisted classical capacity of noisy quantum channels.'

*Prior entanglement between sender and receiver, which exactly doubles the classical capacity of a noiseless quantum channel, can increase the classical capacity of some noisy quantum channels by an arbitrarily large constant factor depending on the channel, relative to the best known classical capacity achievable without entanglement. The enhancement factor is greatest for very noisy channels, with positive classical capacity but zero quantum capacity. We obtain exact expressions for the entanglement-assisted capacity of depolarizing and erasure channels in d dimensions.*

1999a Bennett, C. H., 'Quantum information theory.'

*The new theory of coherent transmission and transformation of information in the form of intact quantum states represents a major extension and generalization of classical information and computation theory, and has a number of distinctive features, including dramatically faster algorithms for certain problems, more complex kinds of channel capacity and cryptography, and a second quantifiable kind of information—entanglement—which interacts with classical information in phenomena such as quantum teleportation.*

1999b Bennett, C. H., 'Retrospective: Rolf Landauer (1927–1999).'

1999 Čápek, V. and Mancal, T., 'Isothermal Maxwell daemon as a molecular rectifier.'

1999 Čápek, V. and Tributsch, H., 'Particle (electron, proton) transfer and self-organization in active thermodynamic reservoirs.' This article presents a 'new simplified model ... that is able to yield the previously reported isothermal Maxwell demon property ...'.

1999 Čápek, V. and Bok, J., 'A thought construction of working perpetuum mobile of the second kind.'

*The previously published model of the isothermal Maxwell's demon as one of models of open quantum systems endowed with the faculty of self-organization is reconstructed here. It describes an open quantum system interacting with a single thermodynamic bath but otherwise not aided from outside. Its activity is given by the standard linear Liouville equation for the system and bath. Owing to its self-organization property, the model then yields cyclic conversion of heat from the bath into mechanical work without compensation. Hence, it provides an explicit thought construction of perpetuum mobile of the second kind, contradicting thus the Thomson formulation of the second law of thermodynamics. No approximation is involved as a special scaling procedure is used which makes the employed kinetic equations exact.*

1999 Davis, B. R., Abbott, D. and Parrondo, J. M. R., 'The moving plate capacitor paradox.' The authors consider a moving plate capacitor driven by thermal noise from a resistor, with a demon restoring the capacitor to its original position, under specified conditions. This is compared with the case of the capacitor's position being restored by a simple spring.

1999 De Vos, A., 'Reversible computing.'

*In the present paper, we give a design methodology for reversible logic circuits. The implementation is based on complementary MOS (metal-oxide-semiconductor) transistor technology. Any Boolean function can be built from a 24-transistor building block. It follows that any combinatorial computer can thus be designed. The resulting digital computer is able to calculate in both directions, such that input and output are indistinguishable. Such reversible MOS circuits we call r-MOS systems. Because such systems have neither specific power lines nor clock lines, they are particularly suited for studying the fundamentals of digital computation. Endoreversible thermodynamics allow us to compare them to Carnot engines.*

1999 Earman, J. and Norton, J. D., 'Exorcist XIV: The Wrath of Maxwell's demon. Part II. From Szilard to Landauer and beyond.'

*We argue through a simple dilemma that these attempted exorcisms are ineffective, whether they follow Szilard ... or Landauer ... In so far as the Demon is a thermodynamic system already governed by the second law, no further supposition about information and entropy is needed to save the second law. In so far as the Demon fails to be such a system, no supposition about the entropy cost of information acquisition and processing can save the second law from the Demon.*

1999 Eggers, J., 'Sand as Maxwell's demon.'

*We consider a dilute gas of granular material inside a box, kept in a stationary state by vertical vibrations. A wall separates the box into two identical compartments, save for a small hole at some finite height h. As the gas is cooled, a second order phase transition occurs, in which the particles preferentially occupy one side of the box. We develop a quantitative theory of this clustering phenomenon and find good agreement with numerical simulations.*

1999 Fields, D. S., 'The spontaneous formation of inhomogeneity in a classical one-particle 'gas': An example in silico.'

*A single classical isoergic particle was placed in a cube and allowed to propagate for 100 ns to 10 ms. The interaction of the particle with the inner wall of the cube was modeled as a linear combination of specular and random reflection, the extent of the combination being governed by a user-defined 'roughness' parameter alpha. As a function of a, the particle's relative pressure and density spontaneously took on an inhomogenous distribution.*

1999 Fowler, A., Bennett, C. H., Keller, S. P. and Imry, Y., 'Obituary: Rolf William Landauer.'

*Rolf did more than anyone else to establish the physics of information processing as a serious subject for scientific inquiry ... It led him to pursue the thermodynamics of information processing more doggedly than any of his predecessors, who believed that each elementary information processing operation required an expenditure of work comparable to or larger than the mean thermal energy. In 1961, Rolf established that this generalization is true only for some information processing operations—those, like erasure, that cannot be undone—whereas other operations have no intrinsic, irreducible thermodynamic cost. The study of such operations led to the theory of reversible computers and communications channels, as well as to our modern understanding of Maxwell's demon, based on the thermodynamic cost of information destruction—a concept now called Landauer's principle.*

1999 Fuchikami, N., Iwata, H. and Ishioka, S., 'Thermodynamic entropy of computer devices.'

*To elucidate several issues relating to the thermodynamics of computations; we performed a simulation of a binary device using a Langevin equation. The simulation confirmed that the minimum energy consumption for a cycle of recording and erasing one-bit of information is* $kT \ln 2$ *as suggested by Landauer. On the basis of our numerical results, we consider how to estimate the thermodynamic entropy of computational devices. We then argue against the existence of the so-called residual entropy in frozen systems such as ice.*

1999 Goecke, J. F., *Maxwell's Demon Reads the Poetry of Emily Dickinson.* The author writes: 'Maxwell's demon was a scientific fiction that attempted to overcome entropy ... This scientific fiction had its origins in the literary fictions of the nineteenth-century as a variety of literary artists sought to overcome what they perceived as social uniformity brought about by evolution, democracy and capitalism. The common thread thee cultural fictions shared was the image of an agent, either the Demon or the artist, mediating between randomness and ideal order ... British authors such as Dickens, Tennyson, and Ruskin used the Maxwell's demon metaphor to maintain traditional order and values ... Americans like Emerson, Whitman, and Twain rejected the British model of order ... (Emily Dickinson) is a Maxwell's demon who seeks to increase the chaos around her.'

1999a Hey, A. J. G., *Feynman and Computation.* This reprint collection contains gems by Feynman, Landauer, Bennett, Wheeler, Fredkin, Zurek, and others. It is the companion volume for *Feynman Lectures on Computation*, Hey & Allen (1999a). Note: T. Hey and A. J. G. Hey are the same author.

1999b Hey, T., 'Quantum computing: an introduction.'

*The basic ideas of quantum computation are introduced by a brief discussion of Bennett and Fredkin's ideas of reversible computation. After some remarks about Deutsch's pioneering work on quantum complexity and Shor's factorisation algorithm, quantum logic gates, qubits and registers are discussed. The role of quantum entanglement is stressed and Grover's quantum search algorithm described in detail. The paper ends with a review of the current experimental status of quantum computers.*

1999 Ingarden, R. S., *Marian Smoluchowski: His Life and Scientific Work.* First published in 1986 by Panstwowe Wydawnictwo Naukow, Warszawe, this book contains a sketch of Marian Smoluchowski's life by his son Roman, plus articles about Smoluchowski's work by M. Kac and S. Chandrasekhar. Several lectures by Smoluchowski are reprinted in English. There is also a chronology of Smoluchowski's life and a bibliography of his original works and of works written about him. In his article, 'Marian Smoluchowski and the evolution of statistical thought in physics,' Mark Kac notes that Maxwell used his demon to show that 'the second law as it pertains to irreversible changes should be interpreted statistically.'

1999 Kelly, T. R., De Silva, H. and Silva, R. E., 'Unidirectional rotary motion in a molecular system.' The authors have built a (biological) molecular ratchet and pawl device, which turns in one direction with the input of chemical energy.

1999 Klein, J. P., Leete, T. H. and Rubin, H., 'A biomolecular implementation of logically reversible computation with minimal energy dissipation.'

*Energy dissipation associated with logic operations imposes a fundamental physical limit on computation and is generated by the entropic cost of information erasure, which is a consequence of irreversible logic elements. We show how to encode information in DNA and use DNA amplification to implement a logically reversible gate that comprises a complete set of operators capable of universal computation. We also propose a method using this design to connect, or 'wire', these gates together in a biochemical fashion to create a logic network, allowing complex parallel computations to be executed. The architecture of the system permits highly parallel operations and has properties that resemble well known genetic regulatory systems.*

1999 Koumura, N., Zijlstra, R. W. J., van Delden, R. A., Harada, N. and Feringa, B. L., 'Light-driven monodirectional molecular rotor.'

1999a Landauer, R., 'How molecules defy the demon (book review).' Review of *Maxwell's Demon: Why Warmth Disperses and Time Passes* by von Baeyer (1998).

1999b Landauer, R., 'Information is a physical entity.'

*This paper, associated with a broader conference talk on the fundamental physical limits of information handling, emphasizes the aspects still least appreciated. Information is not an abstract entity but exists only through a physical representation, thus tying it to all the restrictions and possibilities of our real physical universe. The mathematician's vision of an unlimited sequence of totally reliable operations is unlikely to be implementable in this real universe.*

*Speculative remarks about the possible impact of that, on the ultimate nature of the laws of physics are included.*

1999c Landauer, R., 'Information is inevitably physical.'

*Information is inevitably tied to a physical representation, ...*

1999 Lerner, E. J., 'Briefs (obituary for Rolf Landauer).'

1999 Lloyd, S., 'Obituary: Rolf Landauer (1927–99).'

1999 Machta, J., 'Entropy, information, and computation.'

*The relationship between entropy, information, and randomness is discussed. Algorithmic information theory is introduced and used to provide a fundamental definition of entropy. The relation between algorithmic entropy and the usual Gibbs–Shannon entropy is discussed.*

1999 Mahendra, A. K., Puranik, V. D. and Bhattacharjee, B., 'Separative power for a binary mixture of isotopes.'

*A more accurate expression for the separative power of a device separating a binary mixture into two streams with a large separation factor is worked out by taking recourse to the thermodynamic entropy function without any simplification. It is also shown that the expression reduces to the commonly used form under the assumption of close separation.*

1999 Majernik, V., 'A "realistic" Maxwell's demon is in fact a cognitive robot.'

*We show that a realistic device performing the activity of Maxwell's demon has to contain at least two units. An informer which gains information of speeds of molecules and an operator which passes the marked molecules through a microscopic door in the wall dividing a gas cylinder. Such a device represents in fact a cognitive robot, which is studied in robotics. We describe the functional interaction between both units with respect to the goal of Maxwell's to create a temperature difference in a gas cylinder and point out that in living objects there are many Maxwell-like devices, e.g. enzymes, which are organized in a fashion similar to the classical Maxwell device.*

1999 March, R. H., 'Maxwell's Demon: Why Warmth Disperses and Time Passes, (book review).' Review of book by von Baeyer (1998).

1999 Mityugov, V., 'Maxwell's Demon: On the quantum nature of heat.'

1999 Musser, G., 'Taming Maxwell's demon.' This article is about Brownian motors that use noise as a friend, rather than a foe.

1999 O'Connell, J., 'Maxwell's Demon: Why Warmth Disperses and Time Passes (book review).' Review of book by von Baeyer (1998).

1999 Plenio, M. B., 'The Holevo bound and Landauer's principle.'

*Landauer's principle states that the erasure of information generates a corresponding amount of entropy in the environment. We show that Landauer's principle provides an intuitive basis for Holevo bound on the classical capacity of a quantum channel.*

1999a Shenker, O. R., 'Is $-kTr(\rho \ln \rho)$ the entropy in quantum mechanics?'

*In quantum mechanics, the expression for entropy is usually taken to be $-kTr(\rho \ln \rho)$, where $\rho$ is the density matrix. The convention first appears in Von Neumann's Mathematical Foundations of Quantum Mechanics. The argument given there to justify this convention is the only one hitherto offered. All the arguments in the field refer to it at one point or another. Here this argument is shown to be invalid. Moreover, it is shown that, if entropy is $-kTr(\rho \ln \rho)$, then perpetual motion machines are possible. This and other considerations support the conclusion that this expression is not the quantum- mechanical correlate of thermodynamic entropy. Its usefulness in quantum-statistical mechanics can be explained by its being a convenient quantification of information, but information and entropy are not synonymous. As the present paper shows, one can change while the other is conserved.*

1999b Shenker, O. R., 'Maxwell's demon and Baron Munchausen: Free will as a perpetuum mobile.'

*... I propose to describe these goals and rules [of the Maxwell's demon enterprise] by analysing the logical structure of the argument underlying the thought experiment. While Szilard's followers believe the argument is sound, I suggest that we should see it as a reductio ad absurdum. This, I believe, is how Maxwell himself saw it. While returning to Maxwell's logical understanding of the experiment, though, we ought to take it one step farther. Maxwell found one contradiction in the argument's premisses, whereas I suggest that there are two. The argument is not only a reductio, but a double reductio. The demon has resisted solution so far because the second contradiction escaped Maxwell as well as Szilard and his followers. Finding this second contradiction and removing it is the project undertaken here.*

1999 Standish, R. K., 'Some techniques for the measurement of complexity in Tierra.'

*Adami (1998) and co-workers have been able to measure the information content of digital organisms living in the Avida artificial life system. They show that over time, the organisms behave like Maxwell's demon, accreting information as they evolve.*

1999 Wilczek, F., 'Maxwell's other demon.'

*Results from experiments and calculations relating to the interactions of elementary particles are proving difficult to reconcile. Experimenters at Fermilab, U.S., and CERN, Europe, involved in measuring CP violation—the asymmetry responsible for the dominance of matter over antimatter in the Universe—compared rare decay modes of exotic particles and found that r differs from zero in the standard model of particle physics. Meanwhile, a group of researchers at Columbia University and Brookhaven National Laboratory report that the standard model predicts r to be negative.*

1999 Williams, C. P., Editor, *Quantum Computing and Quantum Communications.* This book contains 43 papers from the First NASA International Conference on Quantum Computing and Quantum Communications, which was held in Palm Springs, California in February, 1998.

1999 Wootton, A., 'Physics community news (obituary for Rolf Landauer).'

1999 Zaslavsky, G. M., 'Chaotic dynamics and the origin of statistical laws.'

1999 Zurek, W. H., 'Algorithmic randomness, physical entropy, measurements, and the demon of choice.' Reprinted here as Article 6.3.

2000 Abbott, D., Davis, B. R. and Parrondo, J. M. R., 'The problem of detailed balance for the Feynman–Smoluchowski engine and the multiple pawl paradox.' Feynmann's treatment of detailed balance in *The Feynman Lectures on Physics, Vol. 1*, is scrutinized, a different approach is offered in its place, and several open questions are raised.

2000 Adami, C., Ofria, C. and Collier, T., 'Evolution of biological complexity.' The authors argue that 'because natural selection forces genomes to behave as a natural 'Maxwell Demon,' within a fixed environment, genomic complexity is forced to increase.'

2000a Allahverdyan, A. E. and Nieuwenhuizen, T. M., 'Extraction of work from a single thermal bath in the quantum regime.' Apparent violations of the second law are found to be the result of quantum coherence and a thermal bath slightly out of equilibrium.

2000b Allahverdyan, A. E. and Nieuwenhuizen, T. M., 'Optimizing the classical heat engine.'

2000 Bak, P. E. and Yoshino, R., 'A Dynamical Model of Maxwell's Demon and confinement systems.' A Dynamical Model of Maxwell's Demon (DMMD) is proposed as 'a contemporary version of Maxwell's Demon' and is used to study a magnetic confinement system not in equilibrium.

2000a Bennett, C. H., 'Notes on the history of reversible computation (Reprinted from IBM Journal of Research and Development, vol 32, 1988).' Reprinted as Article 7.3.

2000b Bennett, C. H., 'Rolf Landauer—in memoriam.'

2000a Binnewies, T., Sterr, U., Helmcke, J. and Riehle, F., 'Cooling by Maxwell's demon: Preparation of single-velocity atoms for matter-wave interferometry.' Laser cooling of atoms is associated with Maxwell's demon.

2000b Binnewies, T., Sterr, U., Helmcke, J. and Riehle, F., 'Preparation of single velocity atoms for matter wave interferometry.' Similar to Binnewies, et al. (2000a).

2000c Binnewies, T., Sterr, U., Helmcke, J. and Riehle, F., 'Sub-Doppler atomic velocities for an optical frequency standard by Maxwell's demon cooling.' A novel cooling scheme that allows sub-Doppler cooling of atoms previously subject to the Doppler limit.

2000 Birnbaum, J. and Williams, R. S., 'Physics and the information revolution.' The limits of CMOS technology are discussed, and the limits of Moore's law are seen as nanoscale devices are produced. Quantum computing is seen is one alternatives, but the prospects are assessed pessimistically.

2000 Čápek, V. and Frege, O., 'Dynamical trapping of particles as a challenge to statistical thermodynamics.'

*A model of a dynamical trap is considered. This type of trapping appears as a result of reorganization of the trap surroundings (relaxation to a new local configuration) upon particle arrival. This can dynamically break existing overlaps of localised particle orbitals, i.e. to interrupt the return path for the particle. Thus, e.g., trapping at even a local potential maximum appears. The effect has nothing to do with, e.g., renormalization of particle energies owing to the bath. Scaling arguments are also presented to make the used kinetic treatment unquestionable.*

2000 Duncan, T., 'Comment on "Dynamically maintained steady-state pressure gradients".'

*Sheehan [*Phys. Rev. *E* **57**, *6660 (1998)] recently discussed the possibility of establishing a dynamically maintained, steady-state pressure gradient in a gas filling a cavity. In this Comment it is pointed out that the pressure gradients in such a system, if attainable in the laboratory, could be used to violate the second law of thermodynamics.*

2000 Goychuk, I. and Hanggi, P., 'Directed current without dissipation: Reincarnation of a Maxwell–Loschmidt demon.'

2000 Ishioka, S. and Fuchikami, N., 'Entropy generation in computation and the second law of thermodynamics.' It is proposed that Landauer's principle that information erasure of one bit is accompanied by heat generation $kT \ln 2$ be replaced by the statement that such erasure is accompanied by entropy generation $k \ln 2$.

2000 Janzing, D., Wocjan, P., Zeier, R., Geiss, R. and Beth, T., 'The thermodynamic cost of reliability: Tightening Landauer's principle and the second law.'

*Landauer's principle states that the erasure of one bit of information requires the free energy* $kT \ln 2$. *We argue that the reliability of the bit erasure process is bounded by the accuracy inherent in the statistical state of the energy source ('the resources') driving the process.*

2000 Lieb, E. H. and Yngvason, J., 'A fresh look at entropy and the second law of thermodynamics.'

*The existence of entropy, and its increase, can be understood without reference to either statistical mechanics or heat engines.*

2000 Lloyd, S., 'Ultimate physical limits to computation.' Reprinted here as Article 7.8.

2000 Ng, Y. J., 'Physical limits to computation.'

*Computers and clocks are physical systems. As such, they must obey the laws of physics. Here we show that, according to the laws of quantum mechanics and gravitation, both the speed v with which a computer can process information and the amount of information I that it can process are limited by the input power. In particular, their product is bounded by a universal constant given by* $Iv^2 \leq t_p^{-2} \approx 10^{87}/sec^2$*, where* $t_p$ *is the Planck time. As a prelude, we show that the maximum time that a clock remains accurate is limited by the accuracy of the clock. All these and related bounds (including the holographic bound) originate from the same physics that governs the quantum fluctuations of space-time. (Hence they can be indirectly tested with future generations of gravitational-wave interferometers.) Furthermore, we show that these physical bounds are realized for black holes, which are thus poised to play an important role in linking together our concepts of information, gravity, and quantum uncertainty.*

2000 Nielsen, M. A. and Chuang, I. L., *Quantum Computing and Quantum Information*. This is a beautiful, up-to-date, detailed introduction to both quantum computing and, more generally, quantum information.

*...From the point of view of a physicist or hardware engineer worried about heat dissipation, the good news is that, in principle, it is possible to make computation dissipation-free by making it reversible, although in practice energy dissipation is required for system stability and immunity from noise. At an even more fundamental level, the ideas leading to reversible computation also lead to the resolution of a century-old problem in the foundations of physics, the famous problem of Maxwell's demon...*

2000 Parrondo, J. M. R., 'Entropy, macroscopic information, and phase transitions.' Generalizing from the Szilard heat engine example, it is argued that 'most problems of the relationship between entropy and information, embodied in a variety of Maxwell's demons, are present also in any symmetry breaking transition.'

2000 Pati, A. K. and Braunstein, S. L., 'Impossibility of deleting an unknown quantum state.'

*Here we show... that the linearity of quantum theory does not allow us to delete a copy of an arbitrary quantum state perfectly. Though in a classical computer information can be deleted (reversibly) against a copy, the analogous task cannot be accomplished, even irreversibly, with quantum information.*

2000 Piechocinska, B., 'Information erasure.' Reprinted here as Article 4.4.

2000 Sekimoto, K., Takagi, F. and Hondou, T., 'Carnot's cycle for small systems: Irreversibility and cost of operations.'

*(T)he so-called null-recurrence property of the cumulative efficiency of energy conversion over many cycles and the irreversible property of isolated, purely mechanical processes under external 'macroscopic' operations are discussed in relation to the impossibility of a perpetual machine, or Maxwell's demon. This analysis may serve as the basis for the design and analysis of mesoscopic energy converters in the near future.*

2000 Sheehan, D. P., 'Reply to 'Comment on 'Dynamically maintained steady-state pressure gradients ...'

2000 Sheehan, D. P., Glick, J. and Means, J. D., 'Steady-state work by an asymmetrically inelastic gravitator in a gas: A second law paradox.'

*A new member of a growing class of unresolved second law paradoxes is examined. In a sealed blackbody cavity, a spherical gravitator is suspended in a low density gas. Infalling gas suprathermally strikes the gravitator which is spherically asymmetric between its hemispheres with respect to surface trapping probability for the gas. In principle, this system can be made to perform steady-state work solely at the expense of heat from the heat bath, this in apparent violation of the second law of thermodynamics. Detailed three-dimensional test particle simulations of this system support this prediction. Standard resolutions to the paradox are discussed and found to be untenable. Experiments corroborating a central physical process of the paradox are discussed briefly. The paradox is discussed in the context of the Maxwell demon.*

2000 Shenker, O. R., 'Logic and entropy.'

*A remarkable thesis prevails in the physics of information, saying that the logical properties of operations that are carried out by computers determine their physical properties. More specifically, it says that logically irreversible operations are dissipative by* $k \log 2$ *per bit of lost information. (A function is logically irreversible if its input cannot be recovered from its output. An operation is dissipative if it turns useful forms of energy into useless ones, such as heat energy.) This is Landauer's dissipation thesis, hereafter LDT. LDT underlies and motivates numerous researches in physics and computer science. Nevertheless, this paper shows that is it plainly wrong. This conclusion is based on a detailed study of LDT in terms of the various notions of entropy used in main stream statistical mechanics. It is supported by a counter example for LDT. Further support is found in an analysis of the phase space representation on which LDT relies. This analysis emphasises the constraints placed on the choice of probability distribution by the fact that it has to be the basis for calculating phase averages corresponding to thermodynamic properties of individual systems. An alternative representation is offered, in which logical irreversibility has nothing to do with dissipation. The strong connection between logic and physics, that LDT implies, is thereby broken off.*

2000 Siegfried, T., *The Bit and the Pendulum.* This book examines classical and quantum information in matters ranging from computers to black holes. Chapter 3, entitled 'Information is Physical' contains a discussion of Maxwell's demon.

2000 Song, D. D. and Winstanley, E., 'Information erasure and the generalized second law of black hole thermodynamics.'

*We consider the generalized second law of black hole thermodynamics in the light of quantum information theory, in particular information erasure and Landauer's principle (namely, that erasure of information produces at least the equivalent amount of entropy). A small quantum system outside a black hole in the Hartle–Hawking state is studied, and the quantum system comes into thermal equilibrium with the radiation surrounding the black hole. For this scenario, we present a simple proof of the generalized second law based on quantum relative entropy. We then analyze the corresponding information erasure process, and confirm our proof of the generalized second law by applying Landauer's principle.*

2000 Svozil, K., 'Logic of reversible automata.' The empirical logic of reversible automata is studied.

2000 Touchette, H. and Lloyd, S., 'Information-Theoretic Limits of Control.'

*Fundamental limits on the controllability of physical systems are discussed in the light of information theory. It is shown that the second law of thermodynamics, when generalized to include information, sets absolute limits to the minimum amount of dissipation required by open-loop control. In addition, an information-theoretic analysis of control systems shows feedback control to be a zero sum game: each bit of information gathered from a dynamical system by a control device can serve to decrease the entropy of that system by at most one bit additional to the reduction of entropy attainable without such information. Consequences for the control of discrete state systems and chaotic maps are discussed.*

2000 van Enk, S. J. and Kimble, H. J., 'Single atom in free space as a quantum aperture.'

*We calculate exact three-dimensional solutions of Maxwell equations corresponding to strongly focused light beams, and study their interaction with a single atom in free space. We show how the naive picture of the atom as an absorber with a size given by its radiative cross section $\sigma = 3\lambda^2/(2\pi)$ must be modified. The implications of these results for quantum-information-processing capabilities of trapped atoms are discussed.*

2000 Vedral, V., 'Landauer's erasure, error correction and entanglement.' Reprinted here as Article 5.5.

2000a Zurek, W. H., 'Einselection and decoherence from an information theory perspective.'

*We introduce and investigate a simple model of conditional quantum dynamics. It allows for a discussion of the information-theoretic aspects of quantum measurements, decoherence, and environment-induced superselection (einselection).*

2000b Zurek, W. H., 'Quantum cloning—Schrödinger's sheep.'

*Classical and quantum error correction are presented in the form of Maxwell's demon and their effciency analysed from the thermodynamic point of view. We explain how Landauer's principle of information erasure applies to both cases. By then extending this principle to entanglement manipulations we rederive upper bounds on purification procedures, thereby linking the 'no local increase of entanglement' principle to the second law of thermodynamics.*

2001 Albert, D. Z., *Time and Chance.*

*This book is an attempt to get to the bottom of an acute and perennial tension between our best scientific pictures of the fundamental physical structure of the world and our everyday empirical experience of it. The trouble is about the direction of time. The situation (very briefly) is that it is a consequence of almost every one of those fundamental scientific pictures-and that it is at the same time radically at odds with our common sense-that whatever can happen can just as naturally happen backwards.*

2001 Allahverdyan, A. E. and Nieuwenhuizen, T. M., 'Breakdown of the Landauer bound for information erasure in the quantum regime.'

*A known aspect of the Clausius inequality is that an equilibrium system subjected to a squeezing $dS < 0$ of its entropy must release at least an amount $|Q| = T|dS|$ of heat. This serves as a basis for the Landauer principle, which puts a lower bound $kT \ln 2$ for the heat generated by erasure of one bit of information. Here we show that in the world of quantum entanglement this law is broken. A quantum Brownian particle interacting with its thermal bath can either generate less heat or even absorb heat during an analogous squeezing process, due to* *entanglement with the bath. The effect exists even for weak but fixed coupling with the bath, provided that temperature is low enough. This invalidates the Landauer bound in the quantum regime, and suggests that quantum carriers of information can be more efficient than assumed so far.*

2001 Aquino, G., Grigolini, P. and Scafetta, N., 'Sporadic randomness, Maxwell's demon and the Poincaré recurrence times.'

2001 Astumian, R. D., 'Making molecules into motors.' A brief, readable review of molecular ratchets and motors.

*Molecular turmoil, quantum craziness: microscopic machines must operate in a world gone mad. But if you can't beat the chaos, why not exploit it?*

2001 Bub, J., 'Maxwell's demon and the thermodynamics of computation.'

*It is generally accepted, following Landauer and Bennett, that the process of measurement involves no minimum entropy cost, but the erasure of information in resetting the memory register of a computer to zero requires dissipating heat into the environment. This thesis has been challenged recently in a two-part article by Earman and Norton. I review some relevant observations in the thermodynamics of computation and argue that Earman and Norton are mistaken: there is in principle no entropy cost to the acquisition of information, but the destruction of information does involve an irreducible entropy cost.*

2001 Callender, C., 'Taking thermodynamics too seriously.'

*This paper discusses the mistake of understanding the laws and concepts of thermodynamics too literally in the foundations of statistical mechanics. Arguing that this error is still pervasive (though slightly more subtle than before), we explore its consequences in three cases: explaining the approach to equilibrium, understanding equilibrium and defining phase transitions.*

2001 Crosignani, B. and DiPorto, P., 'On the validity of the second law of thermodynamics in the mesoscopic realm.' An ideal gas, in an insulated container partitioned by a movable piston, is found to suffer entropy decreases. The analysis suggests that if the system is mesoscopic, then second law violations would occur on a reasonable time scale.

2001 Čápek, V., 'Twilight of a dogma of statistical thermodynamics.'

*In very specific situations, processes are possible where particles get transferred from one subsystem to another one, even uphill in energy and at the cost of thermal energy of the bath only. This means that expected grand canonical equilibrium is not achieved in the course of the time development what imposes limitations on so far supposed universal validity of the standard statistical thermodynamics.*

2001 Čápek, V. and Bok, J., 'Violation of the second law of thermodynamics in the quantum microworld.'

*One of the previously reported linear models of open quantum systems (interacting with a single thermal bath but otherwise not aided from outside) endowed with the faculty of spontaneous self-organization challenging standard thermodynamics is reconstructed here. It is then able to produce, in a cyclic manner, a useful (this time mechanical) work at the cost of just thermal energy in the bath whose quanta get properly in-phased. This means perpetuum mobile of the second kind explicitly violating the second law in its Thomson formulation. No approximations can be made responsible for the effect as a special scaling procedure is used that makes the chosen kinetic theory exact. The effect is purely quantum and disappears in the classical limit.*

2001 Domokos, P., Kiss, T. and Janszky, J., 'Selecting molecules in the vibrational and rotational ground state by deflection.'

*A beam of diatomic molecules scattered off a standing wave laser mode splits according to the rovibrational quantum state of the molecules. Our numerical calculation shows that single state resolution can be achieved by properly tuned, monochromatic light. The proposed scheme allows for selecting non-vibrating and non-rotating molecules from a thermal beam, implementing a laser Maxwell's demon to prepare a rovibrationally cold molecular ensemble.*

2001 Gajda, M. and Rzazewski, K., 'Statistical physics of Bose–Einstein condensation.' A 'Maxwell demon ensemble' is used 'to compute fluctuations of the Bose–Einstein gas condensate of an ideal Bose gas according to the microcanonical ensemble.'

2001 Plenio, M. B. and Vitelli, V., 'The physics of forgetting: Landauer's erasure principle and information theory.'

*This article discusses the concept of information and its intimate relationship with physics. After an introduction of all the necessary quantum mechanical and information theoretical concepts we analyze Landauer's principle that states that the erasure of information is inevitably accompanied by the generation of heat. We employ this principle to rederive a number of results in classical and quantum information theory whose rigorous mathematical derivations are difficult. This demonstrates the usefulness of Landauer's principle and provides an introduction to the physical theory of information.*

2001 Scully, M. O., 'Extracting work from a single thermal bath via quantum negentropy.'

*Classical heat engines produce work by operating between a high temperature energy source and a low temperature entropy sink. The present quantum heat engine has no cooler reservoir acting as a sink of entropy but has instead an internal reservoir of negentropy which allows extraction of work from one thermal bath. The process is attended by constantly increasing entropy and does not violate the second law of thermodynamics.*

2001 Sheehan, D. P., 'The second law and chemically-induced, steady-state pressure gradients: controversy, corroboration and caveats.'

*Experimental results are reported supporting key aspects of a recently proposed chemical pressure gradient (D. P. Sheehan,* Phys. Rev. *E* **57** *(1998); 6660). This pressure gradient has been invoked in a paradox involving the second law of thermodynamics (T. L. Duncan,* Phys. Rev. *E* **61** *4661 (2000)).*

2001 Vedral, V., 'The role of relative entropy in quantum information theory.' Contains a proof of Landauer's principle.

*Quantum mechanics and information theory are among the most important scientific discoveries of the last century. Although these two areas initially developed separately it has emerged that they are in fact intimately related. In this review I will show how quantum information theory extends traditional information theory by exploring the limits imposed by quantum, rather than classical mechanics on information storage and transmission. The derivation of many key results uniquely differentiates this review from the 'usual' presentation in that they are shown to follow logically from one crucial property of relative entropy. Within the review optimal bounds on the speed-up that quantum computers can achieve over their classical counter-parts are outlined using information theoretic arguments. In addition important implications of quantum information theory to thermodynamics and quantum measurement are intermittently discussed. A number of simple examples and derivations including quantum superdense coding, quantum teleportation, Deutsch's and Grover's algorithms are also included.*

2002 Allahverdyan, A. E., Balian, R. and Nieuwenhuizen, T. M., 'Thomson's formulation of the second law: An exact theorem and limits of its validity.'

*Thomson's formulation of the second law—no work can be extracted from a system coupled to a bath through a cyclic process—is believed to be a fundamental principle of nature. For the equilibrium situation a simple proof is presented, valid for macroscopic sources of work. Thomson's formulation gets limited when the source of work is mesoscopic, i.e. when its number of degrees of freedom is large but finite. Here work-extraction from a single equilibrium thermal bath is possible when its temperature is large enough. This result is illustrated by means of exactly solvable models. Finally we consider the Clausius principle: heat goes from high temperature to low temperature. A theorem and some simple consequences are pointed out.*

2002a Allahverdyan, A. E. and Nieuwenhuizen, T. M., 'Statistical thermodynamics of quantum Brownian motion: Birth of perpetuum mobile of the second kind.'

*A known aspect of the Clausius inequality is that an equilibrium system subjected to a squeezing $dS < 0$ of its entropy must release at least an amount $|Q| = T|dS|$ of heat. This serves as a basis for the Landauer principle, which puts a lower bound $kT \ln 2$ for the heat generated by erasure of one bit of information. Here we show that in the world of quantum entanglement this law is broken. A quantum Brownian particle interacting with its thermal bath can either generate less heat or even absorb heat during an analogous squeezing process, due to entanglement with the bath. The effect exists even for weak but fixed coupling with the bath, provided that temperature is low enough. This invalidates the Landauer bound in the quantum regime, and suggests that quantum carriers of information can be more efficient than assumed so far.*

2002b Allahverdyan, A. E. and Nieuwenhuizen, T. M., 'Unmasking Maxwell's demon.'

*Maxwell's demon is a tiny but fine-fingered being, capable to achieve extraction of work from a system at instantaneous equilibrium, without needing energy input or information erasure. In the twentieth century many workers have claimed that the demon cannot operate. Here the point of view is taken that this exorcism of the demon never applied, since one did not consider Maxwell's original invention. For a Brownian particle coupled to a quantum bath it was shown by us that quantum entanglement can allow extraction of work from a non-equilibrium system coupled to a single bath. And mesoscopic work sources may establish work extraction cycles even when they are coupled to equilibrium mesoscopic systems immersed in a macroscopic thermal bath. Quantum entanglement and mesoscopicity are now identified with (true) Maxwell's demons.*

2002c Allahverdyan, A. E. and Nieuwenhuizen, T. M., 'Testing the violation of the Clausius inequality in nanoscale electric circuits.'

2002 Anonymous, 'June 1871: Maxwell and his demon.' This piece appears in the 'This Month in Physics History' column. It has some information on Maxwell, a description of his demon, and recent attempts to build Brownian ratchets. The centerpiece is a description of Eggers's sand experiment (see Eggers, 1999) which, it is explained, does not violate the second law.

2002 Bennett, C. H., 'Notes on Landauer's principle, reversible computation, and Maxwell's demon,'

*Landauer's principle, often regarded as the foundation of the thermodynamics of information processing, holds that any logically irreversible manipuoation of information, such as the erasure of a bit or the merging of two computation paths, must be accompanied by a corresponding entropy increase in non-information bearing degrees of freedom of the the information processing apparatus or its environment. Conversely, it is generally accepted that any logically reversible transformation of information can in principle be accomplished by an appropriate physical mechanism operating in a thermodynamically reversible fashion. These notions have sometimes been criticized either as being false, or as being trivial and obvious, and therefore unhelpful for purposes such as explaining why Maxwell's demon cannot violate the second law of thermodynamics. Here I attempt to refute some of the arguments against Landauer's principle, while arguing that although in a sense it is indeed a trivial and obvious restatement of the second law, it still has considerable pedagogic and explanatory power, especially in the context of other influential ideas in 18th and 20th century physics. Similar arguments have been given by Jeffrey Bub (2001).*

2002 Brey, J. J., Moreno, F., Garcia-Rojo, R. and Ruiz-Montero, M. J., 'Hydrodynamic Maxwell demon in granular systems.'

*Spontaneous symmetry breaking in a vibrated system confined into two connected compartments in the absence of external fields is reported. For a small number of particles, the grains are equipartitioned, but if it is increased beyond a critical value, the number of particles in each of the compartments becomes different in the steady state, and the number of particles in one of the compartments decreases monotonically tending to a given value. This phase transition is accurately described by the hydrodynamic equations for a granular gas. The relationship with previous phenomena of phase separation in vibrofluidized granular materials is discussed.*

2002 Callender, C., 'Who's afraid of Maxwell's demon—and which one?'

*Starting with Sir Karl Popper, who was born exactly 100 years ago today (July 27, 1902), philosophers of science have often found the physics literature surrounding Maxwell's demon deeply troubling. In my talk I explain Popper's arguments and why I believe they are still relevant today. Beginning with the historical context of Maxwell's demon, I briefly trace the evolution of the 'neat-fingered being' from a didactic tool of Maxwell's—a being whose existence Maxwell certainly didn't fear—to a physical mechanism that needs to be exorcized. I then explain why philosophers find exorcizing Maxwell's demon a very odd thing to do, connecting the worries here with other philosophically interesting issues such as the distinctions between the Gibbs and Boltzmann entropies and between objective and subjective probabilities. Of course, there are more versions of the second law besides the classical phenomenological one; for each of these there is a potential demon. I describe various distinct ways of conceiving the second law and its violation, providing a 'demonology' of sorts. Finally I propose some general challenges for new versions of the second law and would-be violators of these laws.*

2002 Čápek, V. and Bok, J., 'A model of spontaneous generation of population inversion in open quantum systems.'

2002 Čápek, V. and Mancal, T., 'Phonon mode cooperating with a particle serving as a Maxwell gate and rectifier.'

2002 Čápek, V. and Sheehan, D. P., 'Quantum mechanical model of a plasma system: A challenge to the second law of thermodynamics.'

*Within the framework of the theory of open quantum systems, a rigorous model is developed and applied to a previously proposed and experimentally corroborated plasma system that challenges the second law of thermodynamics (Sheehan, Phys. Plasmas* **2** *(1995) 1893; J. Scient. Explor.* **12** *(1998) 303). The results of the present model support the findings of the earlier challenge.*

2002 Čápek and Frege, O., 'Violation of the 2nd law of thermodynamics in the quantum microworld.'

*For one open quantum system recently reported to work as a perpetuum mobile of the second kind, basic equations providing basis for discussion of physics beyond the system activity are rederived in an appreciably simpler manner. The equations become exact in one specific scaling limit corresponding to the physical regime where internal processes (relaxations) in the system are commensurable or even slower than relaxation processes induced by bath. In the high-temperature (i.e. classical) limit, the system ceases to work, i.e., validity of the second law is reestablished.*

2002 Čápek, V. 'Zeroth and second laws of thermodynamics simultaneously questioned in the quantum microworld.' A possible violation of the Clausius form of the second law in the quantum regime is suggested.

2002 Deveraux, M., 'A modified Szilard's engine: Measurement, information, and Maxwell's demon.'

*Using an isolated measurement process, we calculate the effect measurement has on entropy for the multi-cylinder Szilard engine. We find that the system of cylinders possesses an entropy associated with cylinder total energy states, and that it records information transferred at measurement. Contrary to other's results, we find that the apparatus loses entropy due to measurement. The Second Law of Thermodynamics may be preserved if Maxwell's demon gains entropy moving the engine partition.*

2002 Devetak, I. and Staples, A. E., 'Towards a unification of physics and information theory.'

*A common framework for quantum mechanics, thermodynamics and information theory is presented... Within the model quantum thermodynamics is treated in such a way as to eliminate the problem of Maxwell's demon altogether, and a simple proof of the second law is given.*

2002 Gea-Banacloche, J., 'Splitting the wave function of a particle in a box.' The following problem is analyzed *without* reference to Maxwell's demon and the Szilard 1-particle gas model. A single particle is in a box, in its ground state, and a potential barrier is slowly placed off-center. The slowness is to assure that the particle remains in its ground state and, if so, the particle always ends up in the larger-volume side. When the analysis is repeated assuming the particle is in its first excited state, then the particle always ends up in the smaller-volume side (see T. Norsen, Am. J. Phys. **70**, 664–665 (2002)). These results hold even if the barrier is placed off center by an arbitrarily small distance. Although Gea-Banacloche does not mention it, no Maxwell's demon would be needed to know which side the particle is on in either case. A 'smart' demon could reduce the volume of the particle, to a known side, by nearly 50 percent. However, the analyses are for adiabatic, not isothermal, barrier placement, and the implications for isothermal off-center barrier placement (Szilard 1-particle gas model) are unclear.

2002 Gordon, L. G. M., 'Perpetual motion with Maxwell's demon.' This is a discussion of a hypothetical device that uses Brownian motion to produce a temperature gradient.

2002 Graeff, R. W., 'Measuring the temperature distribution in gas columns.' Experimental support is given for Loschmidt's notion that a vertical column of gas in a thermodynamically isolated system, but in a gravitational field, shows a temperature gradient.

2002a Gyftopoulos, E. P., 'Maxwell's demon. (I) A thermodynamic exorcism.' It is argued that a Maxwell's demon cannot function because of limitations imposed by the properties of the system on which it acts. The author claims that a demon would either reduce 'the nondestructible and nonstatistical entropy' of air without compensation by another system, or would extract only energy from air under conditions that require extraction of both energy and entropy.

2002b Gyftopoulos, E. P., 'Maxwell's demon. (II) A quantum-theoretic exorcism.' It is argued that within the context of an existing unified theory, for a gas in a thermodynamic equilibrium state, each molecule has zero momentum. The author states that with all molecules at a standstill, a Maxwell's demon cannot sort swift and slow molecules (relative to the average speed) because there are none.

2002 Hondou, T., 'Equation of state in a small system: Violation of an assumption of Maxwell's demon.'

*(It) is found that the equation of motions in small systems can be different from that expected by conventional thermodynamics, which is contrary to a basic and implicit assumption in studies of Maxwell's demon.*

2002 Hopfer, U., 'A Maxwell's demon type of membrane transport: Possibility for active transport by ABC-Type transporters?'

*(A) novel class of transport mechanisms is proposed based on Maxwell's demon idea.*

2002 Huchingson, J. E., 'Response to Stuart Kurtz and Ann Pederson.' The author responds to criticisms of his book. Issues include the relationship of theology to thermodynamics and information theory.

2002 Keefe, P. D., 'Coherent magneto-caloric effect heat engine process cycle.' An experiment using a type I superconductor is proposed to convert ambient heat fully into work, in violation of the second law. A patent has been obtained for the device.

2002 Kurtz, S. A., 'A computer scientist's perspective on chaos and mystery.'

*James E. Huchingson's Pandemonium Tremendum draws on a surprisingly fruitful analogy between metaphysics and thermodynamics, with the latter motivated through the more accessible language of communication theory. In Huchingson's model, God nurtures creation by the selective communication of bits of order that arise spontaneously in chaos.*

2002 Leff, H. S. and Rex, A. F., 'Maxwell's Demon and the Second Law.' A review of the history of Maxwell's demon is presented in light of the current status of the second law.

2002 Linke, H., Editor, *Special issue: Ratchets and Brownian motors: Basics, experiments and applications*. The editor writes:

*Operating far from thermal equilibrium, Brownian motors combine noise and asymmetry (the latter is often realized by a 'ratchet' potential) to generate useful work, for instance the transport of particles. This general concept was conceived in the present context in the late 1980s and early 1990s and has roots in several fields of physics and biology. Some of the most important stimuli came from theorists working on non-equilibrium statistical physics, and from scientists interested in the physical principles of biological motor proteins. This role of theoretical work as a driving force is documented by a substantial amount of theoretical literature published in journals on statistical and biophysics. In comparison, most experimental realizations of Brownian motors are relatively recent and are spread over many different fields of physics, chemistry and biology. It is therefore not always easy to appreciate the full range of existing experimental work. The present special issue aims at providing a representative overview of current experiments on Brownian motors and ratchets, and of possible applications. In order to introduce readers to the field, three reviews present the basic physics and history of Brownian motors, their energetics and efficiency, and the kinetics of biological motor proteins, respectively. Subsequent focussed articles describe experiments in a variety of systems and propose applications.*

2002 Maddox, J., 'Maxwell's demon: Slamming the door.' Maddox, a former editor of Nature, quotes Maxwell's famous passage describing the demon and mentions, correctly, that Maxwell's purpose was to 'pick some holes here and there' in the second law. However, he then goes on to question 'whether Maxwell's mechanism would have been effective.' Maddox alludes to the need for work to actually measure molecular velocities. He points out that if the trap door had mass, 'the demon would have to do work to move it.' If the trap door were massless, then he argues that the molecule would give up kinetic energy to the trap door, and he also worries about the tensile strength and heat conductivity of the membrane. He concludes that 'the functioning of Maxwell's demon must itself contradict the second law.' Maddox's argument circumvents the point that the effects he cites have no known finite, minimum energy or entropy costs. Thus they can in principle be made arbitrarily small. Most important, he ignores hundreds of articles published on the demon (some in Nature) that focus on information gathering and erasure, and the notion that erasure does have a known finite, minimum energy and entropy cost.

2002 Nieuwenhuizen, T. M. and Allahverdyan, A. E., 'Quantum Brownian motion and its conflict with the second law.' This is a summary of recent work by the authors on violations of the second law for a Brownian particle in a harmonic potential, in contact with a constant-temperature reservoir.

2002 Nikulov, A., 'About perpetuum mobile without emotions.' A violation of the second law for an electric ciruit near the superconducting transition is alleged on both theoretical and experimental grounds.

2002 Ollivier, H. and Zurek, W. H., 'Quantum discord: A measure of the quantumness of correlations.'

*Two classically identical expressions for the mutual information generally differ when the systems involved are quantum. This difference defines the quantum discord. It can be used as a measure of the quantumness of correlations. Separability of the density matrix describing a pair of systems does not guarantee vanishing of the discord, thus showing that absence of entanglement does not imply classicality. We relate this to the quantum superposition principle, and consider the vanishing of discord as a criterion for the preferred effectively classical states of a system, i.e., the pointer states.*

2002 Parrondo, J. M. R. and de Cisneros, B. J., 'Energetics of Brownian motors: A review.'

*We review the literature on the energetics of Brownian motors, distinguishing between forced ratchets, chemical motors—driven out of equilibrium by differences of chemical potential, and thermal motors—driven by temperature differences. The discussion is focused on the definition of efficiency and the compatibility between the models and the laws of thermodynamics.*

2002 Reimann, P., 'Brownian motors: Noisy transport far from equilibrium.' A comprehensive review of Brownian motors.

*Transport phenomena in spatially periodic systems far from thermal equilibrium are considered. The main emphasis is put on directed transport in so-called Brownian motors (ratchets), i.e. a dissipative dynamics in the presence of thermal noise and some prototypical perturbation that drives the system out of equilibrium without introducing a priori an obvious bias into one or the other direction of motion. Symmetry conditions for the appearance (or not) of directed current, its inversion upon variation of certain parameters, and quantitative theoretical predictions for specific models are reviewed as well as a wide variety of experimental realizations and biological applications, especially the modeling of molecular motors. Extensions include quantum mechanical and collective effects, Hamiltonian ratchets, the influence of spatial disorder, and diffusive transport.*

2002a Scully, M. O., 'The photo-Carnot quantum engine: I. Extracting work from a single thermal reservoir via quantum coherence.'

*We develop a new kind of heat engine based on cavity QED and the micromaser, together with the same phase coherence physics that gave us lasing without inversion . . . with properly chosen phase, it is possible to extract work even when* $T_h = T_c$. *Thus we find that quantum coherence allows us to extend Carnot's famous result and obtain work with a single reservoir. However the deep physics behind the second law of thermodynamics is not violated. The present Carnot quantum heat engine provides a simple model for studying various quantum effects in thermodynamics.*

2002b Scully, M. O., 'Quantum afterburner: Improving the efficiency of an ideal heat engine.'

*By using a laser and maser in tandem, it is possible to obtain laser action in the hot exhaust gases of a heat engine. Such a 'quantum afterburner' involves the internal quantum states of the working molecules as well as the techniques of cavity quantum electrodynamics and is therefore in the domain of quantum thermodynamics. It is shown that Otto cycle engine performance can be improved beyond that of the 'ideal' Otto heat engine. Furthermore, the present work demonstrates a new kind of lasing without initial inversion.*

2002c Scully, M. O., 'Quantum thermodynamics: From quantum heat engines to Maxwell's demon and beyond.' Scully discusses current research in: 1) Statistical mechanical resolution of a thermodynamic 'paradox' (ellipsoidal cavities); 2) Extracting work from a single thermal bath via quantum negentropy; and 3) Extracting work from a single thermal bath via quantum coherence.

*The second law of thermodynamics embodies deep truths and profound insights. Indeed thermodynamics was the main spring of quantum theory. Some think the time is ripe for quantum mechanics to return the favor by challenging the foundations of thermodynamics, and Norman Ramsey has shown how negative temperatures do just that. Aspects of current research into and controversy surrounding such issues will be discussed...*

2002a Sheehan, D. P., 'Experimentally-Testable Challenges to the Second Law.'

*Many equilibrium systems admit macroscopic potential gradients (e.g. gravitational, electrostatic, chemical). Over the last ten years these have been implicated in a series of second law challenges spanning plasma, gravitational, chemical, and solid-state physics. In this talk, the former will be briefly described, with emphasis on the experimental corroboration of their primary physical mechanisms. The solid-state challenge will be discussed in greater length. It taps the electrostatic potential inherent in a diffusion-frustrated depletion region of a p-n junction. In contrast to earlier challenges, it should operate at room temperature and at sizable power densities. Prospects for experimental tests will be discussed.*

2002b Sheehan, D. P., *First International Conference on Quantum Limits to the Second Law*. This volume contains over 60 invited and contributed papers to the First International Conference on Quantum Limits to the Second Law, held July 27–29, 2002 at the University of San Diego. A number of potential violations of the second law were suggested and discussed, particularly (but not exclusively) those that arise out of quantum considerations. Various kinds of Maxwell's demons surface in many of the papers.

2002 Sheehan, D. P., Glick, J., Duncan, T., Langton, J. A., Gagliardi, M. J. and Tobe, R., 'Phase space portraits of an unresolved gravitational Maxwell demon.'

*In 1885, during initial discussions of J. C. Maxwell's celebrated thermodynamic demon, Whiting observed that the demon-like velocity selection of molecules can occur in a gravitationally bound gas. Recently, a gravitational Maxwell demon has been proposed which makes use of this observation [D. P. Sheehan, J. Glick, and J. D. Means, Found. Phys.* **30***, 1227 (2000)]. Here we report on numerical simulations that detail its microscopic phase space structure. Results verify the previously hypothesized mechanism of its paradoxical behavior. This system appears to be the only example of a fully classical* *mechanical Maxwell demon that has not been resolved in favor of the second law of thermodynamics.*

2002 Sheehan, D. P., Putnam, A. R. and Wright, J. H., 'A Solid-state Maxwell demon.'

*A laboratory-testable, solid-state Maxwell's demon is proposed that utilizes the electric field energy of an open-gap p-n junction. Numerical results from a commercial semiconductor device simulator (Silvaco International-Atlas) verify primary results from a 1-d analytic model. Present day fabrication techniques appear adequate for laboratory tests of principle.*

2002 Trupp, A., 'Second law violations in the wake of the electrocaloric effect in liquid dielectrics.' Violation of the second law using the electrocaloric effect is claimed.

2002 Uffink, J., 'Irreversibility and the Second Law of Thermodynamics.'

*The aim of this paper is to analyse the relation between the second law of thermodynamics and the so-called arrow of time. For this purpose, a number of different aspects in this arrow of time are distinguished, in particular those of time-(a)symmetry and of (ir)reversibility. Next I review versions of the second law in the work of Carnot, Clausius, Kelvin, Planck, Gibbs, Cartheodory and Lieb and Yngvason, and investigate their connection with these aspects of the arrow of time. It is shown that this connection varies a great deal along with these formulations of the second law. According to the famous formulation by Planck, the second law expresses the irreversibility of natural processes. But in many other formulations irreversibility or even time-asymmetry plays no role. I therefore argue for the view that the second law has nothing to do with the arrow of time.*

2002 Weinstein, S., 'Objectivity, information, and Maxwell's demon.'

*This paper examines some common measures of complexity, structure, and information, with an eye toward understanding the extent to which complexity or information content may be regarded as objective properties of individual objects. A form of contextual objectivity is proposed which renders the measures objective, and which largely resolves the puzzle of Maxwell's Demon.*

2002 Wolfram, S., *A New Kind of Science*. From the publisher's summary:

*Wolfram uses his approach to tackle a remarkable array of fundamental problems in science, from the origins of apparent randomness in physical systems, to the development of complexity in biology, the ultimate scope and limitations of mathematics, the possibility of a truly fundamental theory of physics, the interplay between free will and determinism, and the character of intelligence in the universe.*

2002 Zurek, W. H., 'Quantum discord and Maxwell's demon.'

*Quantum discord was proposed as an information theoretic measure of the 'quantumness' of correlations. I show that discord determines the difference between the efficiency of quantum and classical Maxwell's demons in extracting work from collections of correlated systems.*

2003 Leff, H. S. and Rex, A. F., eds, *Maxwell's Demon 2: Entropy, Classical and Quantum Information, Computing*. This second edition includes a historical background written by the editors and includes 31 reprints of seminal works on Maxwell's demon, going back to Maxwell (1867) and up through 2002. It also includes an annotated bibliography with 570+ entries.

# Alphabetical Bibliography


Abbott, D., Davis, B. R. and Parrondo, J. M. R., 'The problem of detailed balance for the Feynman-Smoluchowski engine and the multiple pawl paradox.' In D. Abbott and L. B. Kish, Editors, *AIP Conf. Proc. (USA), No. 511—Unsolved Problems of Noise and Fluctuations. UPoN'99: Second International Conference* (American Institute of Physics, New York, 2000), pp. 213–218.

Adami, C., Ofria, C. and Collier, T., 'Evolution of biological complexity,' *Proc. Nat. Ac. Sci.* **97**, 4463–4468 (2000).

Albert, D. Z., *Time and Chance* (Harvard University Press, Cambridge, MA, 2001).

Alekseev, G. N., *Energy and Entropy* (Mir Publishers, Moscow, 1986), pp. 180–183.

Allahverdyan, A. E. and Nieuwenhuizen, T. M., 'Extraction of work from a single thermal bath in the quantum regime,' *Phys. Rev. Lett.* **85**, 1799–1802 (2000a).

Allahverdyan, A. E. and Nieuwenhuizen, T. M., 'Optimizing the classical heat engine,' *Phys. Rev. Lett.* **85**, 232–235 (2000b).

Allahverdyan, A. E. and Nieuwenhuizen, T. M., 'Breakdown of the Landauer bound for information erasure in the quantum regime,' *Phys. Rev.* E **64**, 0561171–0561179 (2001).

Allahverdyan, A. E. and Nieuwenhuizen, T. M., 'Statistical thermodynamics of quantum Brownian motion: Birth of perpetuum mobile of the second kind,' *Phys. Rev.* E **66**, 036102-1–036102-5 (2002a).

Allahverdyan, A. E. and Nieuwenhuizen, T. M., 'Unmasking Maxwell's demon.' In D. P. Sheehan, Editor, *First International Conference on Quantum Limits to the Second Law* (American Institute of Physics, New York, 2002b).

Allahverdyan, A. E. and Nieuwenhuizen, T. M., 'Testing the violation of the Clausius inequality in nanoscale electric circuits,' *Phys. Rev.* B **66**, 115309-1–115309-5 (2002c).

Allahverdyan, A. E., Balian, R. and Nieuwenhuizen, T. M., 'Thomson's formulation of the second law: An exact theorem and limits of its validity.' In D. P. Sheehan, Editor, *First International Conference on Quantum Limits to the Second Law* (American Institute of Physics, New York, 2002).

Allen, H. S. and Maxwell, R. S., *A Text-Book of Heat* (Macmillan and Co., Ltd., London, 1939), p. 629, 815.

Angrist, S. W. and Hepler, L. G., *Order and Chaos* (Basic Books, Inc., New York, 1967), pp. 193–199.

Anonymous, 'Profile,' *Sci. Am.* **279**, 32 (1998).

Anonymous, 'Microrotors and Maxwell's demon,' *Sci. Am.* (Science and the Citizen section) **280**, 24–25 (1999).

Anonymous, 'June 1871: Maxwell and his demon,' *APS News* **11**, 2 (2002).

Aquino, G., Grigolini, P. and Scafetta, N., 'Sporadic randomness, Maxwell's Demon and the Poincaré recurrence times,' *Chaos Solitons and Fractals* **12**, 2023–2038 (2001). See also arXiv: cond-mat/0006245.

Asimov, I., *Life and Energy* (Bantam Books, New York, 1965), pp. 65–76.

Aspden, H., 'The law beats Maxwell's demon,' *Nature* **347**, 25 (1990).

Astumian, R. D., 'Making molecules into motors,' *Sci. Am.* **285**, 56–64 (2001).

Atkins, P. W., *The Second Law* (Scientific American Books, New York, 1984), pp. 67–79.

Bak, P. E. and Yoshino, R., 'A Dynamical Model of Maxwell's Demon and confinement systems,' *Contrib. Plasma Phys.* **40**, 227–232 (2000).

Balazs, N. L., 'Les relations d'incertitude d'Heisenberg empechent-elle le démon de Maxwell d'opérer?,' *Comptes Rendus* **236**, 998–1000 (1953a).

Balazs, N. L., 'L'effet des statistiques sur le démon de Maxwell,' *Comptes Rendus* **236**, 2385–2386 (1953b).

Barrow, J. D. and Tipler, F. J., *The Anthropic Cosmological Principle* (Oxford University Press, New York, 1986), pp. 174, 179, 669.

Barrow, J. D., *The World within the World* (Oxford University Press, Oxford, 1988), pp. 127–130.

Beghian, L. E., 'Maxwell's demon revisited,' *Nuovo Cimento* B **109**, 611–616 (1994).

Bekenstein, J. D., 'Baryon number, entropy, and black hole thermodynamics,' *Ph. D. Dissertation*, Princeton University (1972).

Bekenstein, J. D., 'Black-hole thermodynamics,' *Phys. Today* **33**, 24–31 (1980).

Bell, D. A., *Intelligent Machines: An Introduction to Cybernetics* (Blaisdell Pub. Co., New York, 1962), pp. 7–10.

Bell, D. A., *Information Theory* (Sir Isaac Pitman and Sons, Ltd., London, 1968), pp. 20–21; 212–219.

Benioff, P., 'Quantum mechanical models of Turing machines that dissipate no energy,' *Phys. Rev. Lett.* **48**, 1581–1585 (1982).

Benioff, P., 'Comment on "dissipation in computation",' *Phys. Rev. Lett.* **53**, 1203 (1984).

Bennett, C. H., 'Logical reversibility of computation,' *IBM J. Res. Dev.* **17**, 525–532 (1973).

Bennett, C. H., 'The thermodynamics of computation—a review,' *Int. J. Theor. Phys.* **21**, 905–940 (1982).

Bennett, C. H., 'Thermodynamically reversible computation,' *Phys. Rev. Lett.* **53**, 1202 (1984).

Bennett, C. H. and Landauer, R., 'The fundamental physical limits of computation,' *Sci. Am.* **253**, 48–56 (July 1985).

Bennett, C. H., 'Demons, engines and the second law,' *Sci. Am.* **257**, 108–116 (1987).

Bennett, C. H., 'Notes on the history of reversible computation,' *IBM J. Res. Dev.* **32**, 16–23 (1988).

Bennett, C. H., 'How to Define Complexity in Physics, and Why.' In W. H. Zurek, *Complexity, Entropy, and the Physics of Information, SFI Studies in the Sciences of Complexity, Vol. VIII* (Addison-Wesley, Reading, 1990a), pp. 137–148.

Bennett, C. H., 'Undecidable dynamics,' *Nature* **346**, 606–607 (1990b).

Bennett, C. H., Brassard, G. and Ekert, A. K., 'Quantum Cryptography,' *Sci. Am.* **267**, 50–57 (1992).

Bennett, C. H., 'Quantum physics: Night thoughts, dark sight,' *Nature* **371**, 479 (1994).

Bennett, C. H., 'Quantum information and computation,' *Phys. Today* **48**, 24–30 (1995).

Bennett, C. H., 'Information theory—Freely communicating,' *Nature* **382**, 669–670 (1996).

Bennett, C. H., DiVincenzo, D. P. and Smolin, J. A., 'Capacities of quantum erasure channels,' *Phys. Rev. Lett.* **78**, 3217–3220 (1997).

Bennett, C. H., 'Information physics in cartoons,' *Superlattices and Microstructures* **23**, 367–372 (1998).

Bennett, C. H., Shor, P. W., Smolin, J. A. and Thapliyal, A. V., 'Entanglement-assisted classical capacity of noisy quantum channels,' *Phys. Rev. Lett.* **83**, 3081–3084 (1999).

Bennett, C. H., 'Quantum information theory.' In A. J. G. Hey, *Feynman and Computation: Exploring the Limits of Computers* (Perseus Books, Reading, 1999a), pp. 178–190.

Bennett, C. H., 'Retrospective: Rolf Landauer (1927-1999),' *Science* **284**, 1940 (1999b).

Bennett, C. H., 'Notes on the history of reversible computation (Reprinted from IBM Journal of Research and Development, vol 32, 1988),' *IBM J. Res. Dev.* **44**, 270–277 (2000a).

Bennett, C. H., 'Rolf Landauer in memoriam,' *Appl. Alg. Eng. Commun.* **10**, 273 (2000b).

Bennett, C. H., 'Notes on Landauer's principle, reversible computation, and Maxwell's demon,' ArXive: physics/0210005, v1 (2002).

Bent, H. A., *The Second Law* (Oxford University Press, New York, 1965), pp. 72–76.

Berger, J., 'Szilard's demon revisited,' *Int. J. Theor. Phys.* **29**, 985–995 (1990).

Berger, J., 'The fight against the second law of thermodynamics,' *Phys. Essays* **7**, 281–295 (1994).

Bhandari, R., 'Entropy, information and Maxwell's demon after quantum mechanics,' *Pramana* **6**, 135–145 (1976).

Biedenharn, L. C. and Solem, J. C., 'A quantum-mechanical treatment of Szilard's engine: Implication for the entropy of information,' *Found. Phys.* **25**, 1221–1229 (1995).

Bier, M., 'A motor protein model and how it relates to stochastic resonance, Feynman's ratchet, and Maxwell's demon.' In L. Schimansky-Geier and T. Poeschel, *Stochastic Dynamics* (Springer-Verlag, Berlin, 1997), pp. 81–87.

Binnewies, T., Sterr, U., Helmcke, J. and Riehle, F., 'Cooling by Maxwell's demon: Preparation of single-velocity atoms for matter-wave interferometry,' *Phys. Rev.* A **62**, 227–232 (2000a).

Binnewies, T., Sterr, U., Helmcke, J. and Riehle, F., 'Preparation of single velocity atoms for matter wave interferometry.' In *2000 International Quantum Electronics Conference* (IEEE, Piscatawy, NJ, 2000b), p. 242.

Binnewies, T., Sterr, U., Helmcke, J. and Riehle, F., 'Sub-Doppler atomic velocities for an optical frequency standard by Maxwell's demon cooling.' In J. Hunter and L. Johnson, *Proceedings - Conference on Precision Electromagnetic Measurements* (IEEE, Piscataway, NJ, 2000c), pp. 200–201.
Birkhahn, R., 'The Return of Maxwell's Demon?,' *J. Irreproducible Results* **41**, 14 (1996).
Birnbaum, J. and Williams, R. S., 'Physics and the information revolution,' *Phys. Today* **53**, 38–42 (January 2000).
Bohm, D., *Quantum Theory* (Prentice-Hall, Inc., Englewood Cliffs, New Jersey, 1951), pp. 608–609.
Bonetto, F., Chernov, N. I. and Lebowitz, J. L., '(Global and local) fluctuations of phase space contraction in deterministic stationary nonequilibrium,' *Chaos: An Interdisciplinary J. Nonlin. Sci.* **8**, 823–833 (1998).
Borg, F., 'The law beats Maxwell's demon,' *Nature* **347**, 24 (1990).
Born, M., 'Die Quantenmechanik und der Zweite Hauptsatz der Thermodynamik,' *Annalen der Physik* **3**, 107 (1948).
Born, M. and Green, H. S., 'The kinetic basis of thermodynamics,' *Proc. R. Soc. London* A **192**, 166–180 (1948).
Boynton, W. P., *Applications of The Kinetic Theory* (Macmillan Co., New York, 1904), pp. 53–54.
Braginsky, V. B. and Khalili, F. Y., ''Maxwell demon' in quantum-non-demolition measurements,' *Phys. Lett.* A **186**, 15 (1994).
Brasher, J., 'Nonlinear-wave mechanics, information-theory, and thermodynamics,' *Int. J. Theor. Phys.* **30**, 979–984 (1991).
Brey, J. J., Moreno, F., Garcia-Rojo, R. and Ruiz-Montero, M. J., 'Hydrodynamic Maxwell demon in granular systems,' *Phys. Rev.* E **6501**, art. no.-011305 (2002).
Bridgman, P. W., *The Nature of Thermodynamics* (Harper and Brothers, New York, 1961), pp. 155–159.
Brillouin, L., 'Life, thermodynamics, and cybernetics,' *Am. Sci.* **37**, 554–568 (1949).
Brillouin, L., 'Can the rectifier become a thermodynamical demon?,' *Phys. Rev.* **78**, 627–628 (1950a).
Brillouin, L., 'Thermodynamics and information theory,' *Am. Sci.* **38**, 594–599 (1950b).
Brillouin, L., 'Maxwell's demon cannot operate: Information and entropy. I,' *J. Appl. Phys.* **22**, 334–337 (1951a).
Brillouin, L., 'Physical entropy and information. II,' *J. Appl. Phys.* **22**, 338–343 (1951b).
Brillouin, L., *Science and Information Theory* (Academic Press Inc., New York, 1956), Ch. 13.
Brown, J. R., *The Laboratory of the Mind* (Routledge, London, 1991), pp. 36–38.
Brownlie, A. W., 'Private Colonies of the Imagination: Power and Possibility In Thomas Pynchon's "V.", "The Crying of Lot 49", and "Gravity's Rainbow",' *Ph. D Dissertation* University of Massachusetts (1997).
Brush, S. G., 'James Clerk Maxwell and the kinetic theory of gases: A review based on recent historical studies,' *Am. J. Phys.* **39**, 631–640 (1971).
Brush, S. G., 'Irreversibility and indeterminism: Fourier to Heisenberg,' *J. Hist. Ideas* **37**, 603–630 (1976a).
Brush, S. G., *The Kind of Motion We Call Heat* (North-Holland Publishing Co., New York, 1976b), Ch. 1, 4, 8, 14, 15.
Brush, S. G., *The Temperature of History. Phases of Science and Culture in the Nineteenth Century* (Burt Franklin and Co., Inc., New York, 1978), pp. 65–67.
Bub, J., 'Maxwell's demon and the thermodynamics of computation,' *Studies Hist. Phil. Mod. Phys.* **32**, 569-579 (2001). See also arXive: quant-ph/0203017.
Callender, C., 'Taking thermodynamics too seriously,' *Studies Hist. Phil. Mod. Phys.* **32**, 539–553 (2001).
Callender, C., 'Who's afraid of Maxwell's demon-and which one?' In D. P. Sheehan, Editor, *First International Conference on Quantum Limits to the Second Law* (American Institute of Physics, New York, 2002).
Campbell, J., *Grammatical Man: Information, Entropy, Language, and Life* (Simon and Schuster, Inc., New York, 1982), Ch. 3.
Čápek, V., 'Isothermal Maxwell daemon,' *Czech. J. Phys.* **47**, 845–849 (1997a).
Čápek, V., 'Isothermal Maxwell daemon and active binding of pairs of particles,' *J. Phys. A: Math. Gen.* **30**, 5245–5258 (1997b).
Čápek, V. and Bok, J., 'Isothermal Maxwell daemon: Numerical results in a simplified model,' *J. Phys. A: Math. Gen.* **31**, 1–12 (1998).
Čápek, V., 'Isothermal Maxwell demon as a quantum "sewing machine",' *Phys. Rev.* E **57**, 3846–3852 (1998).
Čápek, V. and Bok, J., 'A thought construction of working perpetuum mobile of the second kind,' *Czech. J. Phys.* **49**, 1645–1652 (1999).
Čápek, V. and Mancal, T., 'Isothermal Maxwell daemon as a molecular rectifier,' *Europhys. Lett.* **48**, 365–371 (1999).

Čápek, V. and Tributsch, H., 'Particle (electron, proton) transfer and self-organization in active thermodynamic reservoirs,' *J. Phys. Chem.* B **103**, 3711–3719 (1999).
Čápek, V. and Frege, O., 'Dynamical trapping of particles as a challenge to statistical thermodynamics,' *Czech. J. Phys.* **50**, 405–423 (2000).
Čápek, V.,'Twilight of a dogma of statistical thermodynamics,' *Mol. Cryst. Liquid Cryst.* **355**, 13–23 (2001).
Čápek, V. and Bok, J., 'Violation of the second law of thermodynamics in the quantum microworld,' *Physica* A **290**, 379–401 (2001).
Čápek, V. and Bok, J., 'A model of spontaneous generation of population inversion in open quantum systems,' *Chem. Phys.* **277**, 131–143 (2002).
Čápek, V. and Mancal, T., 'Phonon mode cooperating with a particle serving as a Maxwell gate and rectifier,' *J. Phys. A: Math. Gen.* **35**, 2111–2130 (2002).
Čápek, V. and Sheehan, D. P., 'Quantum mechanical model of a plasma system: A challenge to the second law of thermodynamics,' *Physica* A **304**, 461–479 (2002).
Čápek, V. and Frege, O., 'Violation of the 2nd law of thermodynamics in the quantum microworld,' *Czech. J. Phys.* **52**, 679–694 (2002).
Čápek, V., 'Zeroth and second laws of thermodynamics simultaneously questioned in the quantum microworld,' *Euro. Phys. J.* B **25**, 101–113 (2002).
Carnap, R., *Two Essays on Entropy* (University of California Press, Berkeley, 1977), pp. 72–73.
Caves, C. M., Unruh, W. G. and Zurek, W. H., 'Comment on "Quantitative limits on the ability of a Maxwell demon to extract work from heat",' *Phys. Rev. Lett.* **65**, 1387 (1990).
Caves, C. M., 'Quantitative limits on the ability of a Maxwell Demon to extract work from heat,' *Phys. Rev. Lett.* **64**, 2111–2114 (1990).
Caves, C. M., 'Information and entropy,' *Phys. Rev.* E **47**, 4010–4017 (1993).
Caves, C. M., 'Information, entropy, and chaos.' In J. Halliwell, J. Pérez-Mercader and W. H. Zurek, *The Physical Origins of Time Asymmetry* (Cambridge University Press, Cambridge, 1994), pp. 47–89.
Caves, C. M. and Schack, R., 'Unpredictability, information, and chaos: Pursuing alternatives,' *Complexity* **3**, 46–57 (1997).
Chambadal, P., *Paradoxes of Physics* (Transworld Publishers, London, 1971), Ch. IV and V, pp. 80–115.
Chardin, G., 'No free lunch for the Doppler demon,' *Am. J. Phys.* **52**, 252–253 (1984).
Chen, C. H., Hurst, G. S. and Payne, M. G., 'Direct counting of Xe atoms,' *Chem. Phys. Lett.* **75**, 473–477 (1980).
Chernov, N. I. and Lebowitz, J. L., 'Stationary shear flow in boundary driven Hamiltonian systems,' *Phys. Rev. Lett.* **75**, 2831–2834 (1995).
Clarke, B., 'Allegories of Victorian thermodynamics,' *Configurations* (The Johns Hopkins University Press) **4**, 67 (1996).
Clausing, P., 'Über die Entropieverminderung in einem thermodynamischen System bei Eingriffen intelligenter Wesen,' *Z. Physik* **56**, 671–672 (1929).
Collier, J. D., 'Two faces of Maxwell's demon reveal the nature of irreversibility,' *Studies Hist. Phil. Sci.* **21**, 257 (1990).
Costa, J. L., Smyth, H. D. and Compton, K. T., 'A mechanical Maxwell demon,' *Phys. Rev.* **30**, 349–353 (1927).
Costa de Beauregard, O. and Tribus, M., 'Information theory and thermodynamics,' *Helv. Phys. Acta* **47**, 238–247 (1974).
Costa de Beauregard, O., 'The computer and the heat engine,' *Found. Phys.* **19**, 725–727 (1989).
Crosignani, B. and DiPorto, P., 'On the validity of the second law of thermodynamics in the mesoscopic realm,' *Europhys. Lett.* **53**, 290–296 (2001).
Curzon, F. L. and Ahlborn, B., 'Efficiency of a Carnot engine at maximum power output,' *Am. J. Phys.* **43**, 22–24 (1975).
Darling, L. and Hulburt, E. O., 'On Maxwell's demon,' *Am. J. Phys.* **23**, 470–471 (1955).
Darrow, K. K., 'The concept of entropy,' *Am. J. Phys.* **12**, 183–196 (1944).
Daub, E. E., 'Maxwell's demon,' *Studies Hist. Phil. Sci.* **1**, 213–227 (1970).
Davies, P. C. W., *The Physics of Time Asymmetry* (University of California Press, Berkeley, 1974), pp. 53–54; 77.
Davies, P. C. W., *God and the New Physics*, (Simon and Schuster, New York, 1983), pp. 211–213.
Davies, P. C. W., *The Ghost in the Atom* (Cambridge University Press, Cambridge, 1986), pp. 20–22.

Davis, B. R., Abbott, D. and Parrondo, J. M. R., 'The moving plate capacitor paradox.' In D. Abbott and L. B. Kish, *AIP Conf. Proc. (USA), No. 511 - Unsolved Problems of Noise and Fluctuations. UPoN'99: Second International Conference* (American Institute of Physics, New York, 1999), pp. 553–558.

De Meis, L., Montero-Lomeli, M. and Grieco, M. A. B., 'The Maxwell demon in biological systems. Use of glucose 6-phosphate and hexokinase as an ATP regenerating system by the $Ca^{2+}$-ATPase of sarcoplasmic reticulum and submitochondrial particles,' *Annals* **671**, 19 (1992).

De Vos, A., 'Reversible computing,' *Prog. Quant. Elec.* **23**, 1–49 (1999).

Demers, P., 'Les démons de Maxwell et le second principe de la thermodynamique,' *Can. J. Research* **22**, 27–51 (1944).

Demers, P., 'Le second principe et la théorie des quanta,' *Can. J. Research* **23**, 47–55 (1945).

Denbigh, K., 'How subjective is entropy?,' *Chem. Brit.* **17**, 168–185 (1981).

Denbigh, K. G. and Denbigh, J. S., *Entropy in Relation to Incomplete Knowledge* (Cambridge University Press, London, 1985), pp. 1–5; 108–112.

Denur, J., 'The Doppler demon,' *Am. J. Phys.* **49**, 352–355 (1981).

Deveraux, M., 'A modified Szilard's engine: Measurement, information, and Maxwell's demon.' In D. P. Sheehan, Editor, *First International Conference on Quantum Limits to the Second Law* (American Institute of Physics, New York, 2002).

Devetak, I. and Staples, A. E., 'Towards a unification of physics and information theory,' arXiv: quant-ph/0112166 v4 (2002).

DiVincenzo, D. P. and Loss, D., 'Quantum Information is Physical,' *Superlattices Microstructures* **23**, 419–432 (1998).

Dodge, B. F., *Chemical Engineering Thermodynamics* (McGraw-Hill Book Co., Inc.., New York, 1944), pp. 48–49.

Domokos, P., Kiss, T. and Janszky, J., 'Selecting molecules in the vibrational and rotational ground state by deflection,' *Euro. Phys. J.* **14**, 49–53 (2001).

Dotzler, B. J., 'Demons-Magic-Cybernetics: On the introduction to natural magic as told by Heinz von Foerster,' *Syst. Res.* **13**, 245–250 (1996).

Drogin, E. M., 'Maxwell's demon lives,' *Defense Electron.* 31 (March 1986).

Dugdale, J. S., *Entropy and Low Temperature Physics* (Hutchinson University Library, London, 1966), pp. 151–153.

Duncan, T., 'Comment on "Dynamically maintained steady-state pressure gradients",' *Phys. Rev.* E **61**, 4661–4665 (2000).

Dutta, M., 'A hundred years of entropy,' *Physics Today* **21**, 75–79 (1968).

Earman, J. and Norton, J. D., 'Exorcist XIV: The wrath of Maxwell's demon. Part I. From Maxwell to Szilard.,' *Studies Hist. Phil. Mod. Phys.* **29**, 435–471 (1998).

Earman, J. and Norton, J. D., 'Exorcist XIV: The wrath of Maxwell's demon. Part II. From Szilard to Landauer and beyond,' *Studies Hist. Phil. Mod. Phys.* **30**, 1–40 (1999).

Eggers, J., 'Sand as Maxwell's demon,' *Phys. Rev. Lett.* **83**, 5322–5325 (1999).

Ehrenberg, W., 'Maxwell's demon,' *Sci. Am.* **217**, 103–110 (1967).

Eigen, M. and Winkler, R., *Laws of the Game: How the Principles of Nature Govern Chance*, (Harper and Row, Publishers, New York, 1983), pp. 157–160.

Einhorn, R., 'Maxwell's demon 4: for violin or violectra solo,' (1989). http://www.richardeinhorn.com.

El'yashevich, M. A. and Prot'ko, T. S., 'Maxwell's contribution to the development of molecular physics and statistical methods,' *Sov. Phys. Usp.* **24**, 876–903 (1981).

Elsasser, W. M., *The Physical Foundation of Biology* (Pergamon Press, New York, 1958), pp. 203–214.

Exner, O., 'Perpetuum mobile once more, in fun and quite seriously,' *Chem. Listy* **84**, 1233–1238 (1990).

Fahn, P. N., 'Entropy cost of information,' in *Proc. Workshop on Physics and Computation–Phys. Comp. '94*, (IEEE Press, Los Alamitos, 1994), pp. 217–226.

Fahn, P. N., 'Maxwell's demon and the entropy cost of information,' *Found. Phys.* **26**, 71–93 (1996).

Feyerabend, P. K., 'On the possibility of a perpetuum mobile of the second kind.' In P. K. Feyerabend and G. Maxwell, *Mind, Matter, and Method: Essays in Philosophy and Science in Honor of Herbert Feigl* (University of Minnesota Press, Minneapolis, 1966), pp. 409–412.

Feynman, R. P., Leighton, R. B. and Sands, M., *The Feynman Lectures on Physics* Vol. 1, (Addison-Wesley, Reading, Massachusetts, 1963), pp. 46.41–46.49.

Feynman, R. P., 'Simulating physics with computers,' *Int. J. Theor. Phys.* **21**, 467–488 (1982).
Feynman, R. P., 'Quantum mechanical computers,' *Found. Phys.* **16**, 507–531 (1986).
Feynman, R. P., *Feynman Lectures on Computation*, (Addison-Wesley, Reading, 1996). Listed also under T. Hey and R. W. Allen, Editors.
Fields, D. S., 'The spontaneous formation of inhomogeneity in a classical one-particle 'gas': An example in silico,' *Found. Phys. Lett.* **12**, 67–79 (1999).
Fine, A., *The Shaky Game: Einstein, Realism, and the Quantum Theory* (University of Chicago Press, Chicago, 1986), pp. 26–39.
Finfgeld, C. and Machlup, S., 'Well-informed heat engine: efficiency and maximum power,' *Am. J. Phys.* **28**, 324–326 (1960).
Fowler, A., Bennett, C. H., Keller, S. P. and Imry, Y., 'Obituary: Rolf William Landauer,' *Phys. Today* **52**, 104 (1999).
Frank, M. P. and Knight, T., 'Ultimate theoretical models of nanocomputers,' *Nanotechnology* **9**, 162–176 (1998).
Frisch, D. H., 'The microscopic interpretation of entropy,' *Am. J. Phys.* **34**, 1171–1173 (1966).
Fuchikami, N., Iwata, H. and Ishioka, S., 'Thermodynamic entropy of computer devices,' *J. Phys. Soc. Japan* **68**, 3751–3754 (1999).
Gabor, D., 'Light and information,' *Prog. Optics* **1**, 111–153 (1964).
Gabor, D., 'Foreword.' In B. Gal-or, *Modern Developments in Thermodynamics* (John Wiley, New York, 1974), p. x.
Gajda, M. and Rzazewski, K., 'Statistical physics of Bose–Einstein condensation,' *Acta Physica Polonica* A **100**, 7–28 (2001).
Gamow, G., *Mr. Tompkins Explores the Atom* (Macmillan, Cambridge, 1944). Reprinted in *Mr. Tompkins in Paperback* (Cambridge University Press, London, 1971), pp. 95–111. See section entitled 'Maxwell's Demon.'
Gasser, R. P. H. and Richards, W. G., *Entropy and Energy Levels* (Clarendon Press, Oxford, 1974), pp. 116–119.
Gea-Banacloche, J., 'Splitting the wave function of a particle in a box,' *Am. J. Phys.* **70**, 307–312 (2002).
Gell-Mann, M. and Lloyd, S., 'Information measures, effective complexity, and total information,' *Complexity* **2**, 44–52 (1996).
Gershenfeld, N., 'Signal entropy and the thermodynamics of computation,' *IBM Sys. J.* **35**, 577–586 (1996).
Goecke, J. F., *Maxwell's Demon Reads the Poetry of Emily Dickinson Ph. D. Thesis* University of Nebraska, Lincoln, NE, 1999.
Goldstein, M. and Goldstein, I., *The Refrigerator and the Universe* (Harvard University Press, Cambridge, 1993).
Gonick, L., 'Science classics,' *Discover* 82–83 (July 1991).
Gordon, L. G. M., 'Brownian movement and microscopic irreversibility,' *Found. Phys.* **11**, 103–113 (1981).
Gordon, L. G. M., 'Maxwell's demon and detailed balancing,' *Found. Phys.* **13**, 989–997 (1983).
Gordon, L. G. M., 'The molecular-kinetic theory and the second law,' *J. Colloid Interface Sci.* **162**, 512–513 (1994).
Gordon, L. G. M., 'Perpetual motion with Maxwell's demon.' In D. P. Sheehan, Editor, *First International Conference on Quantum Limits to the Second Law* (American Institute of Physics, New York, 2002).
Gorecki, H. and Zaczyk, M., 'Trade-off between energy and entropy of information in control systems.' In S. Domek, R. Kaszynski and L. Tarasiejski, *Proc. Fifth International Symposium on Methods and Models in Automation and Robotics* (Tech. Univ. Szczecin, Szczecin, Poland, 1998), pp. 359–364.
Goto, E., Yoshida, N., Loe, K. F. and Hioe, W., 'A study on irreversible loss of information without heat generation.' In S. Kobayashi, *et al Proc. 3rd Int. Symp. Foundations of Quantum Mechanics*, 1989), pp. 412–418.
Goychuk, I. and Hanggi, P., 'Directed current without dissipation: Reincarnation of a Maxwell–Loschmidt demon,' *Lecture notes in physics* **557**, 7–20 (2000).
Grad, H., 'The many faces of entropy,' *Commun. Pure Appl. Math.* **14**, 323–354 (1961).
Graeff, R. W., 'Measuring the temperature distribution in gas columns.' In D. P. Sheehan, Editor, *First International Conference on Quantum Limits to the Second Law* (American Institute of Physics, New York, 2002).
Grandy, Jr., W. T., 'Resource letter ITP-1: Information theory in physics,' *Am. J. Phys.* **65**, 466–476 (1997).
Grossmann, S. and Holthaus, M., 'Maxwell's Demon at work: Two types of Bose condensate fluctuations in power-law traps,' *Optics Express* **1**, (1997).
Grünbaum, A., 'Is the coarse-grained entropy of classical statistical mechanics an anthropomorphism?' In B. Gal-or, Editor, *Modern Developments in Thermodynamics* (John Wiley, New York, 1974), pp. 413–428.
Gyftopoulos, E. P. and Beretta, G. P., *Thermodynamics: Foundations and Applications* (Macmillan, New York, 1991).

Gyftopoulos, E. P., 'Can an omnipotent Maxwellian demon beat the laws of thermodynamics?,' *ASME Adv. Energy Sys.* **30**, 1–5 (1993).

Gyftopoulos, E. P., 'Maxwell's demon. (I) A thermodynamic exorcism,' *Physica* A **307**, 405–420 (2002a).

Gyftopoulos, E. P., 'Maxwell's demon. (II) A quantum-theoretic exorcism,' *Physica* A **307**, 421–436 (2002b).

Haar, D. ter, *Elements of Statistical Mechanics* (Rinehart and Co., Inc., New York, 1958), pp. 160–162.

Halliwell, J. J., Pérez-Mercader, J. and Zurek, W. H., *Physical Origins of Time Asymmetry* (Cambridge University Press, Cambridge, 1996).

Harman, P. M., *Energy, Force, and Matter: The Conceptual Development of Nineteenth-Century Physics* (Cambridge University Press, London, 1982), pp. 8, 139–143.

Harman, T., 'Maxwell's demon and a snowball's chance in hell,' *Phys. Ed.* **32**, 66–67 (1997).

Harney, R. C., 'Human perception and the uncertainty,' *Am. J. Phys.* **44**, 790–792 (1976).

Hartley, R. V. L., 'Transmission of information,' *Bell System Tech. J.* **7**, 535–563 (1928).

Hastings, H. M. and Waner, S., 'Low dissipation computing in biological systems,' *BioSystems* **17**, 241–244 (1985).

Hatano, T. and Sasa, S., 'Numerical simulations on Szilard's engine and information erasure,' *Prog. Theor. Phys.* **100**, 695–702 (1998).

Hatsopoulos, G. N. and Keenan, J. H., *Principles of General Thermodynamics* (John Wiley and Sons, Inc., New York, 1965), pp. xxxv–xl; 356–362.

Hawking, S., 'The direction of time,' *New Sci.* **115**, 46–49 (July 1987).

Hayles, N. K., 'Self-reflexive metaphors in Maxwell's demon and Shannon's choice: Finding the passages.' In S. Peterfreund, *Literature and science: Theory and Practice* (Northeastern University Press, Boston, 1990).

Heimann, P. M., 'Molecular forces, statistical representation and Maxwell's demon,' *Studies Hist. Phil. Sci.* **1**, 189–211 (1970).

Herron, N. and Farneth, W. E., 'The design and synthesis of heterogeneous catalyst systems,' *Adv. Materials* **8**, 959 (1996).

Hey, T. and Allen, R. W., *Feynman Lectures on Computation* (Perseus, Cambridge, MA, 1996). Listed also under R. P. Feynman. Note: Hey, T. and Hey, A. J. G. are the same author.

Hey, A. J. G., *Feynman and Computation* (Perseus Books, Reading, MA, 1999a).

Hey, T., 'Quantum computing: an introduction,' *Comp. Control Eng. J.* **10**, 105–112 (1999b).

Hogenboom, D. L., 'Maxwell's Demon: Entropy, Information, Computing (book review),' *Am. J. Phys.* **60**, 282–283 (1992).

Holman, J. P., *Thermodynamics* (McGraw-Hill Book Co., New York, 1969), pp. 144–145.

Hondou, T., 'Equation of state in a small system: Violation of an assumption of Maxwell's demon,' arXiv: cond-mat/0202417 v1 (2002).

Hopfer, U., 'A Maxwell's demon type of membrane transport: Possibility for active transport by ABC-Type transporters?,' *J. Theor. Biology* **214**, 539–547 (2002).

Huchingson, J. E., 'Response to Stuart Kurtz and Ann Pederson,' *Zygon* **37**, 433–441 (2002).

Hurst, G. S., Payne, M. G., Kramer, S. D. and Chen, C. H., 'Counting the atoms,' *Phys. Today* **33**, 24–29 (September 1980).

Hurst, G. S., *Resonance Ionization Spectroscopy and its Applications, Conference Series 71* (Institute of Physics, Bristol, 1984).

Hurst, G. S., Payne, M. G., Kramer, S. D. and Chen, C. H., 'Method for counting noble gas atoms with isotopic selectivity,' *Rep. Prog. Phys.* **48**, 1333–1370 (1985).

Igeta, K. and Ogawa, T., 'Information dissipation in quantum-chaotic systems—computational view and measurement induction,' *Chaos Solitons Fractals* **5**, 1365–1379 (1995).

Ingarden, R. S., *Marian Smoluchowski: His Life and Scientific Work* (Polish Scientific Publishers PWN, Warszawa, 1999).

Ishioka, S. and Fuchikami, N., 'Entropy generation in computation and the second law of thermodynamics.' In D. Abbott and L. B. Kish, *AIP Conf. Proc. (USA), No. 511 Unsolved Problems of Noise and Fluctuations. UPoN'99: Second International Conference* (American Institute of Physics, New York, 2000), pp. 329–340.

Ivanitsky, G. R., Medvinsky, A. B., Deev, A. A. and Tsyganov, M. A., 'From Maxwell's demon to the self-organization of mass transfer processes in living systems,' *Phys. Uspekhi* **41**, 1115 (1998). Original publication: *Uspekhi Fiz. Nauk* **168**, 11, 1221–1233 (1998).

Jacobson, H., 'The role of information theory in the inactivation of Maxwell's demon,' *Trans. N. Y. Acad. Sci.* **14**, 6–10 (1951).
Jammer, M., 'Entropy.' In P. Wiener, *Dictionary of the History of Ideas, Vol. 2* (Charles Scribner's Sons, New York, 1973), pp. 112–120.
Jammer, M., *The Philosophy of Quantum Mechanics* (John Wiley, New York, 1974), pp. 479–480.
Janzing, D., Wocjan, P., Zeier, R., Geiss, R. and Beth, T., 'The thermodynamic cost of reliability: Tightening Landauer's principle and the second law,' (2000). See also arXiv: quant-ph/0002048
Jauch, J. M. and Báron, J. G., 'Entropy, information and Szilard's paradox,' *Helv. Phys. Acta* **45**, 220–232 (1972).
Jauch, J. M., *Are Quanta Real?* (Indiana University Press, Bloomington, 1973), pp. 103–104.
Jayannavar, A. M., 'Simple model for Maxwell's-demon-type information engine,' *Phys. Rev.* E **53**, 2957–2959 (1996).
Jaynes, E. T., 'Information theory and statistical mechanics,' *Phys. Rev.* **106**, 620–630 (1957a).
Jaynes, E. T., 'Information theory and statistical mechanics. II,' *Phys. Rev.* **108**, 171–190 (1957b).
Jaynes, E. T., 'Information Theory and Statistical Mechanics.' In K. Ford, *Statistical Physics–1962 Brandeis Summer Institute Lectures in Theoretical Physics Vol. 3* (W. A. Benjamin, Inc., New York, 1963), pp. 181–218.
Jaynes, E. T., 'Gibbs vs. Boltzmann entropies,' *Am. J. Phys.* **33**, 391–398 (1965).
Jaynes, E. T., 'Where Do We Stand on Maximum Entropy?' In R. D. Levine and M. Tribus, *The Maximum Entropy Formalism* (The MIT Press, Cambridge, Mass., 1979), pp. 15–118.
Jaynes, E. T., 'Entropy and search theory,' in C. R. Smith and W. T. Grandy, Jr., *Maximum-Entropy and Bayesian Methods in Inverse Problems* (D. Reidel Publishing Co., Dordrecht, 1985a), pp. 443–454.
Jaynes, E. T., 'Generalized Scattering.' In C. R. Smith and W. T. Grandy, Jr., *Maximum-Entropy and Bayesian Methods in Inverse Problems* (D. Reidel Publishing Co., Dordrecht, 1985b), pp. 377–398.
Jaynes, E. T., 'Macroscopic Prediction.' In H. Haken, *Complex Systems—Operational Approaches* (Springer-Verlag, Berlin, 1985c), pp. 254–269.
Jaynes, E. T., 'Clearing up mysteries–The original goal.' In J. Skilling, *Proceedings of the MAXENT Workshop, St. John's College, Cambridge, England, Vol. J* (Kluwer Academic Publishers, Dordrecht, 1988), pp. 1–27.
Jeans, J. H., *Dynamical Theory of Gases* (Dover Publications, Inc., New York, 4th edition, 1954; original publication, 1925), p. 183.
Jeans, J. H., *An Introduction to the Kinetic Theory of Gases* (Macmillan Co., New York, 1940), p. 271.
Johnson, H. A., 'Thermal noise and biological information,' *Q. Rev. Biol.* **62**, 141–152 (1987).
Jordan, P., 'On the process of measurement in quantum mechanics,' *Phil. Sci.* **16**, 269–278 (1949).
Kauzmann, W., *Thermal Properties of Matter Vol. II Thermodynamics and Statistics: With Applications to Gases* (W. A. Benjamin, Inc., New York, 1967), pp. 209–211.
Keefe, P. D., 'Coherent magneto-caloric effect heat engine process cycle.' In D. P. Sheehan, Editor, *First International Conference on Quantum Limits to the Second Law* (American Institute of Physics, New York, 2002).
Kelly, D. C., *Thermodynamics and Statistical Physics* (Academic Press, New York, 1973), pp. 97–101.
Kelly, T. R., Tellitu, I. and Sestelo, J. P., 'In search of molecular ratchets,' *Angew. Chem. Int. Edn. Engl.* **36**, 1866–1868 (1997).
Kelly, T. R., De Silva, H. and Silva, R. E., 'Unidirectional rotary motion in a molecular system,' *Nature* **401**, 150–152 (1999).
Kioi, K., 'Reversible computing realizing ultimate low power LSI system,' *Sharp Tech. J.* **66**, 63068 (1996).
Kivel, B., 'A relation between the second law of thermodynamics and quantum mechanics,' *Am J. Phys.* **42**, 606–608 (1974).
Klein, M. J., 'Order, organisation, and entropy,' *Brit. J. Phil. Sci.* **4**, 158–160 (1953).
Klein, M. J., 'Maxwell, his demon, and the second law of thermodynamics,' *Am. Sci.* **58**, 84–97 (1970).
Klein, J. P., Leete, T. H. and Rubin, H., 'A biomolecular implementation of logically reversible computation with minimal energy dissipation,' *Biosystems* **52**, 15–23 (1999).
Knott, C. G., *Life and Scientific Work of Peter Guthrie Tait* (Cambridge University Press, London, 1911).
Kosloff, R., 'Thermodynamic aspects of the quantum-mechanical measuring process,' *Adv. Chem. Phys.* **46**, 153–193 (1981).
Koumura, N., Zijlstra, R. W. J., van Delden, R. A., Harada, N. and Feringa, B. L., 'Light-driven monodirectional molecular rotor,' *Nature* **401**, 152–155 (1999).

Kronenberg, L. H., 'The law beats Maxwell's demon,' *Nature* **347**, 25 (1990).

Krus, D. J. and Webb, J. M., 'Maxwell's demons—Simulating probability-distributions with functions of propositional calculus,' *J. Stat. Comput. Simul.* **51**, 71–77 (1994).

Kubo, R., *Statistical Mechanics* (North-Holland Publishing Co., Amsterdam, 1965), p. 13.

Kurtz, S. A., 'A computer scientist's perspective on chaos and mystery,' *Zygon* **37**, 415–420 (2002).

Kyle, B. G., 'Exorcising Maxwell's demon. Entropy, information, and computing,' *Chem. Eng. Ed.* **29**, 94 (1995).

Laing, R., 'Maxwell's demon and computation,' *Phil. Sci.* **41**, 171–178 (1974).

Landauer, R., 'Irreversibility and heat generation in the computing process,' *IBM J. Res. Dev.* **5**, 183–191 (1961).

Landauer, R., 'Uncertainty principle and minimal energy dissipation in the computer,' *Int. J. Theor. Phys.* **21**, 283–297 (1982).

Landauer, R., 'Dissipation in computation,' *Phys. Rev. Lett.* **53**, 1205 (1984a).

Landauer, R., 'Fundamental physical limitations of the computational process,' in *Computer Culture: The Scientific, Intellectual, and Social Impact of the Computer*, (Ann. New York Acad. Sci. Vol 426, New York, 1984b), pp. 161–171.

Landauer, R., 'Computation and physics: Wheeler's meaning circuit?,' *Found. Phys.* **16**, 551–564 (1986a).

Landauer, R., 'Reversible computation,' in O. G. Folberth and C. Hackl, *Der Informationsbegriff in Technik und Wissenschaft*, (R. Oldenbourg, München, 1986b), pp. 139–158.

Landauer, R., 'Computation: A fundamental physical view,' *Phys. Scr.* **35**, 88–95 (1987a).

Landauer, R., 'Computation: A fundamental physical view,' *Speculations in Science and Technology* **10**, 292–302 (1987b).

Landauer, R., 'Energy requirements in communication,' *Appl. Phys. Lett.* **51**, 2056–2058 (1987c).

Landauer, R., 'Fundamental physical limitations of the computational process; an informal commentary,' *Cybernetics Machine Group Newsheet (1/1/87)*, (1987d).

Landauer, R., 'Dissipation and noise immunity in computation, measurement and communication,' *Nature* **335**, (Oct. 27) 779–784 (1988a).

Landauer, R., 'Response,' *Appl. Phys. Lett.* **52**, 2191–2192 (1988b).

Landauer, R., 'Response to "The computer and the heat engine",' *Found. Phys.* **19**, 729–732 (1989a).

Landauer, R., 'Computation, measurement, communication and energy dissipation.' In S. Haykin, *Signal Processing* (Prentice-Hall, Englewood Cliffs, New Jersey, 1989b).

Landauer, R., 'Reversible computation: Implications for measurement, communication, and physical law,' in *Proc. 3rd International Symposium on Foundations of Quantum Mechanics, Tokyo. In the Light of New Technology* (Phys. Soc. Japan, Tokyo, 1989c), pp. 407–411.

Landauer, R., 'Information is physical,' *Phys. Today* 23–29 (May 1991).

Landauer, R., 'Information is physical,' in *Proc. Workshop on Physics and Computation* (IEEE Press, Los Alamitos, 1992a), pp. 1–4.

Landauer, R., 'Reply to "Reversible computing and physical law",' *Phys. Today* **45**, 100 (1992b).

Landauer, R., 'Comment on "Physical limits to quantum flux parametron operation",' *Physica* C **208**, 205–207 (1993a).

Landauer, R., 'Statistical physics of machinery—Forgotten middle-ground,' *Physica* A **194**, 551–562 (1993b).

Landauer, R., 'The zig-zag path to understanding,' in *Proc. Workshop on Physics and Computation (PhysComp '94)* (IEEE Press, Los Alamitos, 1994), pp. 54–59.

Landauer, R., 'Is quantum mechanics useful?,' *Phil. Trans. Phys. Sci. Eng.* **353**, (1703) 367–376 (1995a).

Landauer, R., 'Qualitative view of quantum shot noise,' *Ann. N. Y. Acad. Sci.* **755**, 417–428 (1995b).

Landauer, R., 'The physical nature of information,' *Phys. Lett.* A **217**, 188–193 (1996a).

Landauer, R., 'Minimal energy requirements in communication,' *Science* **272**, 1914–1918 (1996b).

Landauer, R., 'Reversible computation: Implications for measurement, communication, and physical law,' *Adv. Series in App. Phys.* **4**, 384–388 (1996c).

Landauer, R., 'Fashions in science and technology,' *Phys. Today* **50**, 61–62 (1997).

Landauer, R., 'Energy needed to send a bit,' *Proc. R. Soc. London* **454**, 305–311 (1998).

Landauer, R., 'The noise is the signal,' *Nature* **392**, 658–659 (1998).

Landauer, R., 'How molecules defy the demon (book review),' *Physics World* **12**, 37–38 (1999a).

Landauer, R., 'Information is a physical entity,' *Physica* A **263**, 63–67 (1999b).

Landauer, R., 'Information is inevitably physical.' In A. J. G. Hey, *Feynman and Computation: Exploring the Limits of Computers* (Perseus Books, Reading, 1999c).

Landsberg, P. T., *The Enigma of Time* (Adam Hilger Ltd., Bristol, G.B., 1982), pp. 15, 78, 81.

Landsberg, P. T. and Leff, H. S., 'Thermodynamic cycles with nearly-universal maximum-work efficiencies,' *J. Phys. A: Math. Gen.* **22**, 4019 (1989).

Landsberg, P. T., *Thermodynamics and Statistical Mechanics* (Dover, New York, 1990), pp. 131, 361–367.

Landsberg, P. T., 'Maxwell's demon (book review),' *Nature* **349**, (Jan. 31) 376 (1991).

Landsberg, P. T., 'Maxwell's demon: Why warmth disperses and time passes (book review),' *Nature* **393**, (6683) 324–324 (1998).

Layzer, D., 'Growth of order in the universe.' In B. H. Weber, *et al Entropy, Information, and Evolution: New Perspectives on Physical and Biological Evolution* (MIT Press, Cambridge, 1988), pp. 23–39.

Leff, H. S., 'Available work from a finite source and sink: How effective is a Maxwell's demon?,' *Am. J. Phys.* **55**, 701–705 (1987a).

Leff, H. S., 'Thermal efficiency at maximum work output: New results for old heat engines,' *Am. J. Phys.* **55**, 602–610 (1987b).

Leff, H. S., 'Maxwell's demon, power, and time,' *Am. J. Phys.* **58**, 135–142 (1990).

Leff, H. S. and Rex, A. F., *Maxwell's Demon: Entropy, Information, Computing* (Adam Hilger, Bristol, jointly with Princeton University Press, Princeton, 1990a).

Leff, H. S. and Rex, A. F., 'Maxwell's Demon: Resource Letter MD-1,' *Am. J. Phys.* **58**, 201–209 (1990b).

Leff, H. S. and Rex, A. F., 'Entropy of Measurement and Erasure: Szilard's Membrane Model Revisited,' *Am. J. Phys.* **63**, 994–1000 (1994a).

Leff, H. S., 'Thermodynamics of Crawford's energy equipartition journeys,' *Am. J. Phys.* **62**, 120–129 (1994b).

Leff, H. S., 'Thermodynamic insights from a one-particle gas,' *Am. J. Phys.* **63**, 895–905 (1995a).

Leff, H. S., 'The Refrigerator and the Universe: Understanding the Laws of Energy (book review),' *Am. J. Phys.* **63**, 282–283 (1995b).

Leff, H. S. and Rex, A. F., 'Maxwell's demon and the culture of entropy,' *Phys. Essays* **10**, 125–149 (1997).

Leff, H. S. and Rex, A. F., 'Maxwell's Demon and the Second Law.' In D. P. Sheehan, Editor, *First International Conference on Quantum Limits to the Second Law* (American Institute of Physics, New York, 2002).

Leff, H. S. and Rex, A. F., *Maxwell's Demon 2: Entropy, Classical and Quantum Information, Computing* (Institute of Physics, Bristol, 2003).

Lehninger, A. L., *Bioenergetics: The Molecular Basis of Biological Energy Transformations* (W. A. Benjamin, Inc., New York, 1965), pp. 225–227.

Lerner, A. Y., *Fundamentals of Cybernetics* (Plenum Pub. Corp., New York, 1975), pp. 256–258.

Lerner, E. J., 'Briefs (obituary for Rolf Landauer),' *The Industrial Physicist* **5**, 17 (1999).

Letokhov, V. S., 'Laser Maxwell's demon,' *Contemp. Phys.* **36**, 235–243 (1995).

Lewis, G. N. and Randall, M., *Thermodynamics and The Free Energy of Chemical Substances* (McGraw-Hill Book Co., Inc., New York, 1923), pp. 120–121.

Lewis, G. N., 'The symmetry of time in physics,' *Science* **71**, 569–577 (1930).

Li, M. and Vitanyi, P., 'Reversibility and adiabatic computation: Trading time and space for energy,' *Proc. Royal Soc. A-Math. Phys. Eng. Sci.* **452**, 769–789 (1996).

Li, M., Tromp, J. and Vitany, I. P., 'Reversible simulation of irreversible computation,' *Physica* D **120**, 168–176 (1998).

Liboff, R. L., 'Maxwell's demon and the second law of thermodynamics,' *Found. Phys. Lett.* **10**, 89–92 (1997).

Lidorenko, N. S., 'The possibility of realization of "Maxwell's demon",' *Appl. Energy* (Izvestiia RAN, Energetika) **35**, 91 (1997).

Lieb, E. H. and Yngvason, J., 'A fresh look at entropy and the second law of thermodynamics,' *Phys. Today* **53**, 32–37 (2000).

Likharev, K. K., 'Classical and quantum limitations on energy consumption in computation,' *Int. J. Theor. Phys.* **21**, 311–326 (1982).

Likharev, K. K. and Korotkov, A. N., ' "Single-electron parametron": Reversible computation in a discrete-state system,' *Science* **273**, 763–765 (1996).

Lindblad, G., 'Entropy, information, and quantum measurements,' *Commun. Math. Phys.* **33**, 305–322 (1973).

Lindblad, G., 'Measurements and information for thermodynamic quantities,' *J. Stat. Phys.* **11**, 231–255 (1974).
Lindblad, G., *Non-equilibrium Entropy and Irreversibility* (D. Reidel Publishing Co, Dordrecht, Holland, 1983), pp. 113–122.
Lindgren, K., 'Microscopic and macroscopic entropy,' *Phys. Rev.* A **38**, 4794–4798 (1988).
Linke, H., Editor, *Special Issue: Ratchets and Brownian Motors: Basics, Experiments and Applications, Appl. Phys.* A **75**(2) (2002).
Lloyd, D. G., 'The Role Of Science And Technology In The Novels Of Thomas Pynchon,' *Ph. D Dissertation* Texas Christian University (1991).
Lloyd, S. and Pagels, H., 'Complexity as thermodynamic depth,' *Ann. Phys.* **188**, 186–213 (1988).
Lloyd, S., 'Use of mutual information to decrease entropy: Implications for the second law of thermodynamics,' *Phys. Rev.* A **39**, 5378–5386 (1989).
Lloyd, S. and Zurek, W. H., 'Algorithmic treatment of the spin-echo effect,' *J. Stat. Phys.* **62**, 3–4 (1991).
Lloyd, S., 'Quantum-mechanical computers,' *Sci. Am.* **273**, 140–145 (October 1995).
Lloyd, S., 'Quantum-mechanical Maxwell's demon,' *Phys. Rev.* A **56**, 3374–3382 (1997).
Lloyd, S., 'Obituary: Rolf Landauer (1927–99),' *Nature* **400**, 720 (1999).
Lloyd, S., 'Ultimate physical limits to computation,' *Nature* **406**, 1047–1054 (2000). See also arXiv: quant-ph/9908043.
Loschmidt, J., 'Über den Zustand des Wärmegleichgewichtes eines System von Körpern mit Rücksicht auf die Schwerkraft,' *Wiener Berichte* **73**, 128–142 (1876).
LoSurdo, C., Guo, S. C. and Paccagnella, R., 'Self-consistent equilibria in a cylindrical reversed-field pinch,' *Nuovo. Cim.* **18**, 1425–1442 (1996).
Lotka, A. J., *Elements of Mathematical Biology (Originally published as Elements of Physical Biology)* (Dover Publications, Inc., New York, 1956), pp. 35–40; 121.
Lubkin, E., 'Keeping the entropy of measurement: Szilard revisited,' *Int J. Theor. Phys.* **26**, 523–535 (1987).
Machado, A. C., 'Energy and Information in Society: A Study of the Relationships between Energy Use and Information Activities in Contemporary Society (Communications),' *Ph. D. Dissertation* University Of Pennsylvania (1994).
Machado, A. C. and Miller, R. E., 'Empirical relationships between the energy and information segments of the US economy—An input-output approach,' *Energy Policy* **25**, 913–921 (1997).
Machta, J., 'Entropy, information, and computation,' *Am. J. Phys.* **67**, 1074–1077 (1999).
Maddox, J., 'Quantum information storage,' *Nature* **327**, 97 (1987).
Maddox, J., 'Maxwell's demon flourishes,' *Nature* **345**, 109 (May 1990).
Maddox, J., 'Directed motion from random noise,' *Nature* **369**, 181 (May 1994).
Maddox, J., 'Maxwell's demon: Slamming the door,' *Nature* **417**, 903–904 (2002).
Magnasco, M. O., 'Szilard's heat engine,' *Europhys. Lett.* **33**, 583–588 (1996).
Mahendra, A. K., Puranik, V. D. and Bhattacharjee, B., 'Separative power for a binary mixture of isotopes,' *Sep. Sci. Tech.* **34**, 3211–3225 (1999).
Majernik, V., 'A "realistic" Maxwell's demon is in fact a cognitive robot,' *Kybernetes* **28**, 1065–1071 (1999).
Malizia, D. and Tarsitani, C., 'Looking at the second law of thermodynamics through the eyes of Maxwell's demon.' In C. Bernardini, C. Tarsitani and M. Vicentini, *Proceedings: International Conference: Thinking Physics for Teaching–1994 (Rome)* (Plenum Press, New York, 1995), pp. 355–366.
March, R. H., 'Maxwell's Demon: Why Warmth Disperses and Time Passes, (book review),' *Phys. Today* **52**, 84 (March 1999).
Markowsky, G., 'An introduction to algorithmic information theory: Its history and some examples,' *Complexity* **2**, 14–22 (1997).
Matzke, D., Editor, *Proceedings of the Workshop on Physics and Computation* (IEEE Press, Los Alamitos, 1992).
Maxwell, J. C., 'Letter to P. G. Tait, 11 December 1867,' published in C. G. Knott, *Life and Scientific Work of Peter Guthrie Tait* (Cambridge University Press, London, 1911), pp. 213–215.
Maxwell, J. C., 'Letter to J. W. Strutt, 6 December 1870,' published in R. J. Strutt, *Life of John William Strutt, Third Baron Rayleigh* (E. Arnold, London, 1924), pp. 47–48.
Maxwell, J. C., *Theory of Heat* (Longmans, Green, and Co., London, 1871), Ch. 12.
Maxwell, J. C., 'Diffusion,' in *Encyclopedia Britannica, 9th Edition*, New York, 1878a), pp. 214–221.

Maxwell, J. C., 'Tait's thermodynamics,' *Nature* **17**, 250 ff (1878b). In *The Scientific Papers of James Clerk Maxwell, Vol. 2* (Cambridge University Press, Cambridge, 1890 (1878)), pp. 660–671.

Milburn, G. J., 'Atom-optical Maxwell demon,' *Aus. J. Phys.* **51**, 1–8 (1998).

Millonas, M. M., 'Self-consistent microscopic theory of fluctuation-induced transport,' *Phys. Rev. Lett.* **74**, 10–13 (1995).

Mityugov, V., 'Maxwell's Demon: On the quantum nature of heat,' *Quantum* **10**, 10 (1999).

Morck, R. and Morowitz, H., 'Value and information: a profit maximizing strategy for Maxwell's demon,' *Complexity* **1**, 58–63 (1995).

Morowitz, H. J., *Energy Flow in Biology* (Academic Press, Inc., New York, 1968), pp. 124–133.

Morowitz, H. J., *Entropy for Biologists* (Academic Press, New York, 1970), pp. 108–111.

Motz, H., 'The Doppler demon exorcised,' *Am. J. Phys.* **51**, 72–73 (1983).

Musser, G., 'Taming Maxwell's demon,' *Sci. Am.* **280**, 24 (February 1999).

Nardiello, G., 'Energy dissipation of computational processes,' *Nuovo Cim.* **108**, 943–946 (1993).

Navez, P., Bitouk, D., Gajda, M., Idziaszek, Z. and Rzazewski, K., 'Fourth statistical ensemble for the Bose–Einstein condensate,' *Phys. Rev. Lett.* **79**, 1789–1792 (1997).

Newton, R. G., *What Makes Nature Tick?* (Harvard University Press, Cambridge, 1993).

Ng, Y. J., 'Physical limits to computation,' arXiv: gr-qc/0006105 (2000).

Nielsen, M. A., Caves, C. M., Schumacher, B. and Barnum, H., 'Information-theoretic approach to quantum error correction and reversible measurement,' *Proc. Royal Soc. A-Math. Phys. Eng. Sci.* **454**, 277–304 (1998).

Nielsen, M. A. and Chuang, I. L., *Quantum Computing and Quantum Information* (Cambridge University Press, Cambridge, 2000).

Nieuwenhuizen, T. M. and Allahverdyan, A. E., 'Quantum Brownian motion and its conflict with the second law.' In D. P. Sheehan, Editor, *First International Conference on Quantum Limits to the Second Law* (American Institute of Physics, New York, 2002).

Nikulov, A., 'About perpetuum mobile without emotions.' In D. P. Sheehan, Editor, *First International Conference on Quantum Limits to the Second Law* (American Institute of Physics, New York, 2002).

Nyquist, H., 'Certain factors affecting telegraph speed,' *Bell Syst. Tech. J.* **3**, 324–346 (1924).

O'Connell, J., 'Maxwell's Demon: Why Warmth Disperses and Time Passes (book review),' *Am. J. Phys.* **67**, 552 (1999).

Obermayer, K., Mahler, G. and Haken, H., 'Multistable quantum systems: Information processing at microscopic levels,' *Phys. Rev. Lett.* **58**, 1792–1795 (1987).

Ollivier, H. and Zurek, W. H., 'Quantum discord: A measure of the quantumness of correlations,' *Phys. Rev. Lett.* **88**, 017901-1–017901-4 (2002).

Otsuka, J. and Nozawa, Y., 'Self-reproducing system can behave as Maxwell's demon; Theoretical illustration under prebiotic conditions,' *J. Theoret. Bio.* **194**, 205–221 (1998).

Pagels, H., *Cosmic Code* (Simon and Schuster, New York, 1982), pp. Part I, Sec. 8.

Parellada, R. F. and Rojas, H. P., 'Competition between Maxwell's demons and some consequences,' *Rev. Mex. Fis.* **44**, 128–132 (1998).

Parrondo, J. M. R. and Español, P., 'Criticism of Feynman's analysis of the ratchet as an engine,' *Am. J. Phys.* **64**, 1125–1130 (1996).

Parrondo, J. M. R. and de Cisneros, B. J., 'Energetics of Brownian motors: A review,' *Appl. Phys.* A **75**, 179–191 (2002).

Parrondo, J. M. R., 'Entropy, macroscopic information, and phase transitions.' In D. Abbott and L. B. Kish, *AIP Conf. Proc. (USA), No. 511—Unsolved Problems of Noise and Fluctuations. UPoN'99: Second International Conference* (American Institute of Physics, New York, 2000), pp. 314–325.

Pati, A. K. and Braunstein, S. L., 'Impossibility of deleting an unknown quantum state,' *Nature* **404**, 164–165 (2000).

Pekelis, V., *Cybernetics A to Z* (Mir Publishers, Moscow, 1974), pp. 104–110.

Penrose, O., *Foundations of Statistical Mechanics* (Pergamon Press, Oxford, 1970), pp. 221–238.

Penrose, O., 'Foundations of statistical mechanics,' *Rep. Prog. Phys.* **42**, 1937–2006 (1979).

Peres, A., 'Schrödinger's immortal cat,' *Found. Phys.* **18**, 57–76 (1988).

Phipps, Jr., Thomas E. and Brill, M. H., 'Bayesian entropy and interference,' *Phys. Essays* **8**, 615–625 (1995).

Piechocinska, B., 'Information erasure,' *Phys. Rev.* A **61**, (Article 62314) 1–9 (2000).

Pierce, J. R., *Symbols, Signals and Noise* (Harper and Brothers, New York, 1961), pp. 198–207.

Planck, M., *Treatise on Thermodynamics* (Dover Publications, Inc., translated from the 7th German edition, New York, 1922), pp. 105–107.

Plenio, M. B., 'The Holevo bound and Landauer's principle,' *Phys. Lett.* A **263**, 281–284 (1999).

Plenio, M. B. and Vitelli, V., 'The physics of forgetting: Landauer's erasure principle and information theory,' *Contemp. Phys.* **42**, 25–60 (2001).

Poincaré, H., 'Mechanism and experience,' *Revue de Metaphysique et de Morale* **1**, 534–537 (1893).

Poplavskii, R. P., 'Maxwell demon and the correspondence between information and entropy,' *Sov. Phys. Usp.* **22**, 371–380 (1979).

Popper, K., 'Irreversibility; or, entropy since 1905,' *Brit. J. Phil. Sci.* **8**, 151–155 (1957).

Popper, K., *The Philosophy of Karl Popper* (Open Court Publishing Co., LaSalle, Illinois, 1974), pp. 129-133; 178–179.

Popper, K. R., *Quantum Theory and the Schism in Physics* (Roman and Littlefield, New Jersey, 1982), pp. 109–117.

Porod, W., Grondin, R. O., Ferry, D. K. and Porod, G., 'Dissipation in computation,' *Phys. Rev. Lett.* **52**, 232–235 (1984).

Porod, W., Grondin, R. O., Ferry, D. K. and Porod, G., 'Porod *et al* respond,' *Phys. Rev. Lett.* **52**, 1206 (1984).

Porod, W., 'Comment on "Energy requirements in communication",' *Appl. Phys. Lett.* **52**, 2191 (1988).

Porter, T. M., *The Rise of Statistical Thinking: 1820–1900* (Princeton University Press, Princeton, 1986), pp. 11–128; 194–219.

Poundstone, W., *The Recursive Universe* (W. Morrow and Co., Inc., New York, 1985), Ch. 3, 5.

Preston, T., *The Theory of Heat* (Macmillan and Co., Ltd., London, 1929), pp. 679–680.

Prigogine, I. and Stengers, I., *Order out of Chaos* (Bantam Books, New York, 1984), pp. 175, 239.

Rapoport, A., 'Remark.' In H. P. Yockey, R. L. Platzman and H. Quastler, *Symposium on Information Theory in Biology* (Pergamon Press, New York, 1958), p. 196.

Rastogi, S. R., 'How a clever demon nearly blew up the second law of thermodynamics,' *Chem. Eng. Ed.* **26**, 78–86 (1992).

Raymond, R. C., 'Communication, entropy, and life,' *Am. Sci.* **38**, 273–278 (1950).

Raymond, R. C., 'The well-informed heat engine,' *Am. J. Phys.* **19**, 109–112 (1951).

Reese, K. M., 'Odd device may be the work of Maxwell's demon,' *Chem. Eng. News* **74**, 56 (March 1996).

Reimann, P., 'Brownian motors: Noisy transport far from equilibrium,' *Phys. Rev.* **361**, 57–265 (2002).

Rex, A. F., 'The operation of Maxwell's demon in a low entropy system,' *Am. J. Phys.* **55**, 359–362 (1987).

Rex, A. F. and Larsen, R., 'Entropy and information for an automated Maxwell's demon,' in *Proceedings of the Workshop on Physics and Computation (PhysComp '92)* (IEEE Press, Los Alamitos, 1992), pp. 93–101.

Robertson, H. S., *Statistical Thermophysics* (Prentice-Hall, Englewood Cliffs, 1993).

Robinson, A. L., 'Computing without dissipating energy,' *Science* **223**, 1164–1166 (March 1984).

Rodd, P., 'Some comments on entropy and information,' *Am. J. Phys.* **32**, 333–335 (1964).

Rosen, R., *Theoretical Biology and Complexity* (Academic Press, Inc., New York, 1985), p. 167.

Rosnay, J. de, *The Macroscope: A New World Scientific System* (Harper and Row, New York, 1975), pp. 130–167.

Rossis, G., 'La logique des expériences de pensée et l'expérience de Szilard,' *Fundamenta Scientiae* **2**, 151–162 (1981a).

Rossis, G., 'Sur la discussion par Brillouin de l'expérience de pensée de Maxwell,' *Fundamenta Scientiae* **2**, 37–44 (1981b).

Rothstein, J., 'Information, measurement, and quantum mechanics,' *Science* **114**, 171–175 (1951).

Rothstein, J., 'Information and Thermodynamics,' *Phys. Rev.* **85**, 135 (1952a).

Rothstein, J., 'Organization and entropy,' *J. Appl. Phys.* **23**, 1281–1282 (1952b).

Rothstein, J., 'A phenomenological uncertainty principle,' *Phys. Rev.* **86**, 640 (1952c).

Rothstein, J., 'Nuclear spin echo experiments and the foundations of statistical mechanics,' *Am. J. Phys.* **25**, 510–518 (1957).

Rothstein, J., 'Physical demonology,' *Methodos* **42**, 94–117 (1959).

Rothstein, J., 'Discussion: Information and organization as the language of the operational viewpoint,' *Phil. Sci.* **29**, 406–411 (1962).

Rothstein, J., 'Informational Generalization of Entropy in Physics.' In T. Bastin, *Quantum Theory and Beyond* (Cambridge University Press, London, 1971), pp. 291–305.

Rothstein, J., 'Loschmidt's and Zermelo's paradoxes do not exist,' *Found. Phys.* **4**, 83–89 (1974).

Rothstein, J., 'Generalized Entropy, Boundary Conditions, and Biology.' In R. D. L. M. Tribus, *The Maximum Entropy Formalism* (The MIT Press, Cambridge, Mass., 1979), pp. 423–468.

Rothstein, J., 'Entropy and the evolution of complexity and individuality,' Ohio State University Report OSU-CISRC-8/88-TR27 1–51 (1988a).

Rothstein, J., 'On ultimate thermodynamic limitations in communication and computation.' In F. K. Skwirzynski, *NATO ASI Series E, Vol. 142: Performance Limits in Communication Theory and Practice* (Kluwer Academic Publishers, Dordrecht, 1988b), pp. 43–58.

Saha, M. N. and Srivastava, B. N., *A Treatise on Heat* (The Indian Press, Calcutta, 1958), p. 320.

Samuelson, P. A., 'The law beats Maxwell's demon,' *Nature* **347**, 24–25 (1990).

Scarani, V., 'Quantum computing,' *Am. J. Phys.* **66**, 956 (1998).

Schack, R. and Caves, C. M., 'Hypersensitivity to perturbations in the quantum baker's map,' *Phys. Rev. Lett.* **71**, 525–528 (1993).

Schack, R., 'Algorithmic information and simplicity in statistical physics,' *Int. J. Theor. Phys.* **36**, 209–226 (1997).

Schlaffer, A. and Nossek, J. A., 'Is there a connection between adiabatic switching and reversible computing?,' in *ECCTD '97. Proceedings of the 1997 European Conference on Circuit Theory and Design* Vol. 2 (Tech. Univ. Budapest, Budapest, 1997), pp. 944–946.

Schneider, T. D., 'Sequence logos, machine channel capacity, Maxwell's demon, and molecular computers: A review of the theory of molecular machines,' *Nanotechnology* **5**, 1–18 (1994).

Schrödinger, E., *What is Life?* (Cambridge University Press, London, 1944).

Schulman, A., 'Programmer's Bookshelf,' *Dr. Dobb's J.* **17**, 133 (1992).

Schulman, L. S., *Time's Arrow and Quantum Measurement* (Cambridge University Press, Cambridge, 1997).

Schumacher, B., 'Demonic heat engines.' In J. Halliwell, J. Pérez-Mercader and W. H. Zurek, *The Physical Origins of Time Asymmetry* (Cambridge University Press, Cambridge, 1994), pp. 90–97.

Schweber, S., 'Demons, angels, and probability: Some Aspects of British Science in the Nineteenth Century.' In A. Shimony and H. Feshbach, *Physics as Natural Philosophy: Essays in Honor of Lazlo Tisza on His 75th Birthday* (MIT Press, Cambridge, MA, 1982), pp. 319–363.

Scully, M. O., 'Extracting work from a single thermal bath via quantum negentropy,' *Phys. Rev. Lett.* **87**, 220601-1–220601-4 (2001).

Scully, M. O., 'The photo-Carnot quantum engine: I. Extracting work from a single thermal reservoir via quantum coherence,' Preprint (2002a).

Scully, M. O., 'Quantum afterburner: Improving the efficiency of an ideal heat engine,' *Phys. Rev. Lett.* **88**, 050602-1–050602-4 (2002b).

Scully, M. O., 'Quantum Thermodynamics: From Quantum Heat Engines to Maxwell's Demon and Beyond.' In D. P. Sheehan, Editor, *First International Conference on Quantum Limits to the Second Law* (American Institute of Physics, 2002c).

Sekimoto, K., Takagi, F. and Hondou, T., 'Carnot's cycle for small systems: Irreversibility and cost of operations,' *Phys. Rev.* E **62**, (6) 7759–7768 (2000).

Shannon, C. E. and Weaver, W., *The Mathematical Theory of Communication* (University of Illinois Press, Urbana, 1949).

Sheehan, D. P., 'A paradox involving the second law of thermodynamics,' *Phys. Plasmas* **2**, 1893–1898 (1995).

Sheehan, D. P., 'Another paradox involving the second law of thermodynamics,' *Phys. Plasmas* **3**, 104–110 (1996a).

Sheehan, D. P., 'Response to comment on "A paradox involving the second law of thermodynamics",' *Phys. Plasmas* **3**, 706 (1996b).

Sheehan, D. P., 'Dynamically-maintained, steady-state pressure gradients,' *Phys. Rev.* E **57**, 6660–6666 (1998a).

Sheehan, D. P., 'Four paradoxes involving the second law of thermodynamics,' *J. Sci. Explor.* **12**, 303–314 (1998b).

Sheehan, D. P. and Means, J. D., 'Minimum requirement for second law violation: a paradox revisited,' *Phys. Plasmas* **5**, 2469–2471 (1998).

Sheehan, D. P., 'Reply to "Comment on Dynamically maintained steady-state pressure gradients",' *Phys. Rev.* E **61**, 4662–4665 (2000).

Sheehan, D. P., Glick, J. and Means, J. D., 'Steady-state work by an asymmetrically inelastic gravitator in a gas: A second law paradox,' *Found. Phys.* **30**, 1227–1256 (2000).

Sheehan, D. P., 'The second law and chemically-induced, steady-state pressure gradients: controversy, corroboration and caveats,' *Phys. Lett.* A **280**, 185–190 (2001).

Sheehan, D. P., Glick, J., Duncan, T., Langton, J. A., Gagliardi, M. J. and Tobe, R., 'Phase space portraits of an unresolved gravitational Maxwell demon,' *Found. Phys.* **32**, 441–462 (2002).

Sheehan, D. P., Putnam, A. R. and Wright, J. H., 'A Solid-state Maxwell demon,' *Found. Phys.* **32**, 1557–1595 (2002).

Sheehan, D. P., 'Experimentally-Testable Challenges to the Second Law.' In D. P. Sheehan, Editor, *First International Conference on Quantum Limits to the Second Law* (American Institute of Physics, New York, 2002a).

Sheehan, D. P., Editor, *First International Conference on Quantum Limits to the Second Law* (American Institute of Physics, New York, 2000b).

Shenker, O. R., 'Is $-kTr(\rho \ln \rho)$ the entropy in quantum mechanics?,' *Brit. J. Phil. Soc.* **50**, 33–48 (1999a).

Shenker, O. R., 'Maxwell's demon and Baron Munchausen: Free will as a perpetuum mobile,' *Studies Hist. Phil. Mod. Phys.* **30**, 347–372 (1999b).

Shenker, O. R., 'Logic and entropy,' PhilSci archive: PITT-PHIL-SCI00000115 (2000).

Sherwood, M., *Maxwell's Demon* (New English Library, London, 1976), pp. 93–94.

Shizume, K., 'Heat generation required by erasure,' *Phys. Rev.* E **52**, 3495–3499 (1995).

Siegfried, T., *The Bit and the Pendulum* (John Wiley and Sons, New York, 2000).

Silver, R. S., *An Introduction to Thermodynamics* (Cambridge University Press, London, 1971), pp. 42, 125.

Silverman, M. P., 'The vortex tube: A violation of the second law?,' *Eur. J. Phys.* **3**, 88–92 (1982).

Silverman, M. P., *And Yet It Moves: Strange Systems and Subtle Questions in Physics* (Cambridge University Press, Cambridge, 1993).

Simons, J. P., 'Laser-beams and molecular dreams,' *Chem. Brit.* **27**, 32–36 (1991).

Singh, J., *Great Ideas in Information Theory, Language and Cybernetics* (Dover Publications, Inc., New York, 1966), Ch. VII.

Skagarstam, B., 'On the mathematical definition of entropy,' *Z. Natur.* a **29**, 1239–1243 (1974).

Skagarstam, B., 'On the notions of entropy and information,' *J. Stat. Phys.* **12**, 449–462 (1975).

Skordos, P. A. and Zurek, W. H., 'Maxwell's demon, rectifiers, and the second law: Computer simulation of Smoluchowski's trapdoor,' *Am. J. Phys.* **60**, 876–882 (1992).

Skordos, P. A., 'Time-reversible Maxwell's Demon,' MIT Artificial Intelligence Memo (September) (1992).

Skordos, P. A., 'Compressible dynamics, time reversibility, Maxwell's demon, and the second law,' *Phys. Rev.* E **48**, 777–784 (1993).

Slater, J. C., *Introduction to Chemical Physics*, (McGraw-Hill Book Co., Inc., New York, 1939), pp. 9-12; 43–46.

Smith, C. and Wise, M. N., *Energy and Empire: A Biographical Study of Lord Kelvin*, (Cambridge University Press, Cambridge, 1989), pp. 620–628.

Smoluchowski, M. von, 'Experimentell nachweisbare der üblichen Thermodynamik widersprechende Molekularphänomene,' *Physik. Z.* **13**, 1069–1080 (1912).

Smoluchowski, M. von, 'Gültigkeitsgrenzen des zweiten Hauptsatzes der Wärmtheorie,' in *Vorträge über die Kinetische Theorie der Materie und der Elektrizität* (Teubner, Leipzig, 1914), pp. 89–121.

Smorodinsky, Y. A., *Temperature* (Mir Publishers, Moscow, 1984), pp. 241–245.

Sokirko, A. V., 'Does mass-action law breakdown occur in small thermodynamic systems?,' *J. Chem. Soc.-Faraday Trans.* **90**, 2353–2358 (1994).

Song, D. D. and Winstanley, E., 'Information erasure and the generalized second law of black hole thermodynamics,' arXiv: gr-qc/0009083 (2000).

Spielberg, N. and Anderson, B. D., *Seven Ideas that Shook the Universe* (John Wiley and Sons, Inc., New York, 1987), pp. 134–135.

Sporn, S. R., 'Air traffic controller and Maxwell's demon,' in *Proc. IEEE 1994 National Aerospace and Electronics Conference NAECON*, Vol. 2 (IEEE, New York, 1994), pp. 1309–1316.

Standish, R. K., 'Some techniques for the measurement of complexity in Tierra,' in *Proc. Advances in Artificial Life. 5th European Conference, ECAL'99 (Lecture Notes in Artificial Intelligence)*, Vol. 1674 (Springer-Verlag, Berlin, 1999), pp. 104–108.

Stasheff, C., *Her Majesty's Wizard* (Ballantine Books, New York, 1986).

Steiner, L. E., *Introduction to Chemical Thermodynamics* (McGraw-Hill Book Co., Inc., New York, 1941), p. 166.
Stix, G., 'Riding the back of electrons,' *Sci. Am.* **279**, 33–34 (1998).
Susskind, L., 'Black holes and the information paradox,' *Sci. Am.* **276**, 52–57 (1997).
Sussman, M. V., 'Seeing entropy - the incompleat thermodynamics of the Maxwell demon bottle,' *Chem. Eng. Ed.* 149–156 (1974) (Summer).
Svozil, K., 'Logic of reversible automata,' *Int. J. Theoret. Phys.* **39**, 893–899 (2000).
Szilard, L., 'On the extension of phenomenological thermodynamics to fluctuation phenomena,' *Z. f. Physik* **32**, 753–788 (1925).
Szilard, L., 'On the decrease of entropy in a thermodynamic system by the intervention of intelligent beings,' *Z. f. Physik* **53**, 840–856 (1929).
Tansjö, L., 'Comment on the discovery of the second law,' *Am. J. Phys.* **56**, 179 (1988).
Thirion, J.-P., 'Non-rigid matching using demons,' Proc. IEEE Comp. Soc. Conf. on Computer Vision and Pattern Recognition 245–251 (1996).
Thomson, W., 'The Kinetic Theory of the Dissipation of Energy,' *Nature* **9**, 441–444 (1874).
Thomson, W., 'The sorting demon of Maxwell,' *R. Soc. Proc.* **9**, 113–114 (1879).
Toffoli, T., 'Comment on "Dissipation in computation",' *Phys. Rev. Lett.* **53**, 1204 (1984).
Touchette, H. and Lloyd, S., 'Information-Theoretic Limits of Control,' *Phys. Rev. Lett.* **84**, 1156–1159 (2000).
Trebbia, P., 'Maxwell's demon and data analysis,' *Phil. Trans. Math. Phys.* **354**, 2697–2709 (1996).
Tribus, M. and McIrvine, E. C., 'Energy and information,' *Sci. Am.* **225**, 179–188 (1971).
Trincher, K. S., 'Information and Biological Thermodynamics.' In L. Kubat and J. Zeman, *Entropy and Information in Science and Philosophy* (Elsevier Scientific Pub. Co., New York, 1975), pp. 105–123.
Trupp, A., 'Second law violations in the wake of the electrocaloric effect in liquid dielectrics.' In D. P. Sheehan, Editor, *First International Conference on Quantum Limits to the Second Law* (American Institute of Physics, New York, 2002).
Tsuda, I., 'Linear demon,' Proc. 4th International Conference on Soft Computing - Methodologies for the Conception, Design, and Application of Intelligent Systems **1**, 13–14 (1996).
Uffink, J., 'Irreversibility and the Second Law of Thermodynamics.' In D. P. Sheehan, Editor, *First International Conference on Quantum Limits to the Second Law* (American Institute of Physics, New York, 2002).
van Enk, S. J. and Kimble, H. J., 'Single atom in free space as a quantum aperture,' *Phys. Rev.* A **61**, 051802 (2000).
van Kampen, N. G., 'Ten theorems about quantum mechanical measurements,' *Physica* A **153**, 97–113 (1988).
van Kampen, N. G., 'A physicist looks at quantum mechanical measurement,' in *Proc. 3rd Int. Symp. Foundations of Quantum Mechanics*, Tokyo (1989).
Vedral, V., 'Landauer's erasure, error correction and entanglement,' *Proc. R. Soc. Lond.* A **456**, 969–984 (2000).
Vedral, V., 'The role of relative entropy in quantum information theory,' *Rev. Mod. Phys.* **74**, 197–236 (2001).
Vologodskii, A., 'Maxwell demon and topology simplification by type II topoisomerases,' Proceedings of the Annual International Conference on Computational Molecular Biology, RECOMB (ACM Press, New York, 1998), pp. 266–269.
von Baeyer, H. C., *Maxwell's Demon: Why Warmth Disperses and Time Passes* (Random House, New York, 1998).
von Neumann, J., *Mathematical Foundations of Quantum Mechanics* (Princeton University Press, Princeton, 1932). Published originally in German in 1932. Ch. V.
von Neumann, J., 'Probabilistic logics from unreliable components,' Automata Studies 43–98 (1956).
Waldram, J. R., *The Theory of Thermodynamics* (Cambridge University Press, London, 1985), pp. 306–308.
Walther, T. and Trebbia, P., 'Maxwell's demon and data analysis—Discussion,' *Phil. Trans. R. Soc. London* A **354**, 2709–2711 (1996).
Wang, Y.-Z., 'Origin of the electric current in Fu's experiment - An analysis of Fu's work on realizing Maxwell's "Demon",' *Energy Conversion Management* **26**, 249–252 (1986).
Watanabe, S., *Knowing and Guessing: A Quantitative Study of Inference and Information* (John Wiley and Sons, Inc., New York, 1969), pp. 245–254.
Watanabe, S., 'Wiener on cybernetics, information theory, and entropy.' P. Masani, *Norbert Wiener: Collected Works with Commentaries* (MIT Press, Cambridge, Mass., 1985), pp. 215–218.
Weber, H. C., *Thermodynamics for Chemical Engineers* (John Wiley and Sons, Inc., New York, 1939), pp. 127–128.
Wehrl, A., 'General properties of entropy,' *Rev. Mod. Phys.* **50**, 221–260 (1978).

Weinberg, A. M., 'On the relation between information and energy systems: A family of Maxwell's demons,' Interdisciplinary *Sci. Rev.* **7**, 47–52 (1982).

Weinstein, S., 'Objectivity, information, and Maxwell's demon,' philsci-archive: PITT-PHIL-SCI00000754 (2002).

Weiss, C. and Wilkens, M., 'Particle number counting statistics in ideal Bose gases,' *Opt. Express* **1**, (1997).

Wheeler, J. A., 'Information, physics, quantum: The search for links.' In S. Kobayashi, *et al Proc. 3rd International Symposium on Foundations of Quantum Mechanics, Tokyo. In the Light of New Technology* (The Physical Society of Japan, Tokyo, 1989), pp. 354–368.

Whiting, H., 'Maxwell's demons,' *Science* **6**, 83 (1885).

Wicken, J. S., 'Entropy and information: Suggestions for a common language,' *Phil. Sci.* **54**, 176–193 (1987).

Wiener, N., *Cybernetics* (Wiley, New York, 1948).

Wiener, N., 'Entropy and information,' *Proc. Symp. Appl. Math. Amer. Math. Soc.* **2**, 89 (1950).

Wiener, N., 'Cybernetics,' *Scientia* (Italy) **87**, 233–235 (1952).

Wilczek, F., 'Maxwell's other demon,' *Nature* **402**, 22–23 (1999).

Williams, C. P., Editor, *Quantum Computing and Quantum Communications* (Springer-Verlag, New York, 1999).

Williams, C. P. and Clearwater, S. H., *Explorations in Quantum Computing* (Springer-Verlag, New York, 1998).

Wolfram, S., *A New Kind of Science* (Wolfram Media, Inc., Champaign, 2002).

Wolpert, D. H., 'Memory systems, computation, and the second law of thermodynamics,' *Int. J. Theor. Phys.* **31**, 743 (1992).

Wootton, A., 'Physics community news (obituary for Rolf Landauer),' *Phys. World* **12**, 46 (1999).

Wright, P. G., 'Entropy and disorder,' *Contemp. Phys.* **11**, 581–588 (1970).

Wright, R., 'Did the universe just happen?,' *The Atlantic Monthly* 29–44 (April 1988).

Wright, R., *Three Scientists and Their Gods* (Times Books, New York, 1988), pp. 91–93.

Yamamoto, Y. and Ueda, M., 'Quantum optics and new technology,' *J. Inst. Electronics* **73**, 63–68 (1990).

Yamamoto, Y. and Chuang, I., 'Physical limits for computing and communication,' *Information Processing* **51**, 3–14 (1994).

Yu, F. T. S., *Optics and Information Theory* (John Wiley and Sons, Inc., New York, 1976), Ch. 4, 5, 6.

Yu, F. T. S., 'Super resolution and cost of entropy,' *Optik* (Stuttgart) **54**, 1–7 (1979).

Zaslavsky, G. M., 'From Hamiltonian chaos to Maxwell's demon,' *J. Nonlin. Sci.* **5**, 653–661 (1995a).

Zaslavsky, G. M., 'From Levy flights to the fractional kinetic equation for dynamical chaos,' *Proc. International Workshop. Levy Flights and Related Topics in Physics* (Springer-Verlag, Berlin, 1995b), pp. 216–236.

Zaslavsky, G. M. and Edelman, M., 'Maxwell's demon as a dynamical model,' *Phys. Rev.* E **56**, 5310–5320 (1997).

Zaslavsky, G. M., 'Chaotic dynamics and the origin of statistical laws,' *Phys. Today* **52**, 39–45 (August 1999).

Zemansky, M. W., Abbott and van Ness, *Basic Engineering Thermodynamics, 2nd Edition* (McGraw Hill, New York, 1975), pp. 406–410.

Zemansky, M. W. and Dittman, R. H., *Heat and Thermodynamics* (McGraw-Hill Book Co., Inc., Seventh Edition, New York, 1997), pp. 330–332.

Zernike, J., *Entropy: The Devil on the Pillion* (Kluwer-Deventer, 1972), Ch. 4.

Zhang, K. and Zhang, K., 'Mechanical models of Maxwell's demon with noninvariant phase volume,' *Phys. Rev.* A **46**, 4598–4605 (1992).

Zimmerman, T. and Salamon, P., 'The demon algorithm,' *Int. J. Computer Math.* **42**, 21–31 (1992).

Zurek, W. H., 'Maxwell's demon, Szilard's engine and quantum measurements.' In G. T. Moore and M. O. Scully, *Frontiers of Nonequilibrium Statistical Physics* (Plenum Press, New York, 1984a), pp. 151–161.

Zurek, W. H., 'Reversibility and stability of information processing systems,' *Phys. Rev. Lett.* **53**, 391–394 (1984b).

Zurek, W. H., 'Algorithmic randomness and physical entropy,' *Phys. Rev.* A **40**, 4731–4751 (1989a).

Zurek, W. H., 'Algorithm randomness, physical entropy, measurements, and the second law,' *Proc. 3rd International Symposium on Foundations of Quantum Mechanics, Tokyo. In the Light of New Technology* (The Physical Society of Japan, Tokyo, 1989b), pp. 115–123.

Zurek, W. H., 'Thermodynamic cost of computation, algorithmic complexity and the information metric,' *Nature* **347**, 119–124 (1989c).

Zurek, W. H., 'Algorithmic information content, Church-Turing thesis, physical entropy, and Maxwell's demon.' In W. H. Zurek, *Complexity, Entropy, and the Physics of Information, Santa Fe Institute Studies in the Sciences of Complexity, Vol. VIII* (Addison-Wesley, Redwood, 1990), pp. 73–89.

Zurek, W. H., 'Decoherence and the transition from quantum to classical,' *Phys. Today* **44**, (October) 36–44 (1991).
Zurek, W. H., 'Algorithmic randomness, physical entropy, measurements, and the demon of choice.' In J. G. Hey, *Feynman and Computation: Exploring the Limits of Computers* (Perseus, Reading, 1999), pp. 393–410. See also arXiv: quant-ph/9807007.
Zurek, W. H., 'Einselection and decoherence from an information theory perspective,' *Ann. Phys.* **9**, 855–864 (2000a).
Zurek, W. H., 'Quantum cloning -Schrödinger's sheep,' *Nature* **404**, 130–131 (2000b).
Zurek, W. H., 'Quantum discord and Maxwell's demon,' arXiv: quant-ph/0202123 v1 (2002).

# Index

For Product Safety Concerns and Information please contact our EU
representative GPSR@taylorandfrancis.com
Taylor & Francis Verlag GmbH, Kaufingerstraße 24, 80331 München, Germany

www.ingramcontent.com/pod-product-compliance
Ingram Content Group UK Ltd.
Pitfield, Milton Keynes, MK11 3LW, UK
UKHW051834180425
457613UK00022B/1251

* 9 7 8 0 7 5 0 3 0 7 5 9 8 *